Finfish Aquaculture Diversification

We would like to honour the memory of our eminent colleague, Dr Joseph A. Brown, whose enthusiasm for aquaculture diversification provided inspiration for the book. He was greatly appreciated for his kindness, accessibility, involvement and impact on a wide range of endeavours.

FINFISH AQUACULTURE DIVERSIFICATION

EDITED BY

Nathalie R. Le François
Biodôme de Montréal, Québec, Canada

Malcolm Jobling
University of Tromsø, Norway

Chris Carter
University of Tasmania, Australia

Pierre U. Blier
Université du Québec à Rimouski, Québec, Canada

ASSISTANT EDITOR AND ILLUSTRATOR

Arianne Savoie
Université du Québec à Rimouski, Québec, Canada

www.cabi.org

CABI is a trading name of CAB International

CABI Head Office
Nosworthy Way
Wallingford
Oxfordshire OX10 8DE
UK

Tel: + 44 (0)1491 832111
Fax: + 44 (0)1491 833508
E-mail: cabi@cabi.org
Website: www.cabi.org

CABI North American Office
875 Massachusetts Avenue
7th Floor
Cambridge, MA 02139
USA

Tel: + 1 617 395 4056
Fax: + 1 617 354 6875
E-mail: cabi-nao@cabi.org

A catalogue record for this book is available from the British Library, London, UK.

Library of Congress Cataloging-in-Publication Data

Finfish aquaculture diversification / edited by Nathalie R. Le François ... [et al.].
p. cm.
Includes bibliographical references and index.
ISBN 978-1-84593-494-1 (alk. paper)
1. Fish culture. I. Le François, Nathalie R. II. Title.

SH151.F46 2010
639.3--dc22

2009015984

ISBN-13: 978 1 84593 494 1

Commissioning editor: Rachel Cutts
Production editor: Fiona Harrison

Typeset by SPi, Pondicherry, India.
Printed and bound in the UK by the MPG Books Group.

Contents

Contributors

Bruno **Archer**, Centre Aquacole de la Côte-Nord (CACN), North-Shore Aquaculture Center (NSAC), 106 rue Napoléon, Suite 200-B, Sept-Îles (Québec), Canada G4R 3L7 (cacn_dg@globetrotter.net)

Arne-Mikal **Arnesen**, Nofima A/S, Muninbakken 9–13, Postboks 6122, N-9291 Tromsø, Norway (arne-mikal.arnesen@nofima.no)

Tillman **Benfey**, Department of Biology, University of New Brunswick, PO Box 4400, Fredericton, New Brunswick, Canada E3B 6E1 (benfey@unb.ca)

Michel **Bermudes** Government of Western Australia, Department of Fisheries, Research Division, PO Box 20, North Beach, WA 6920, Australia

Björn **Björnsson**, Marine Research Institute, Skulagata 4, 121 Reykjavik, Iceland (bjornb@hafro.is)

Pierre U. **Blier**, Département de biologie, chimie et géographie, Université du Québec à Rimouski, 300 Allée des Ursulines, Rimouski, QC, Canada G5L 3A1 (pierre_blier@uqar.qc.ca)

Brian **Bosworth,** US Department of Agriculture, Agricultural Research Service, Catfish Genetics Research Unit, Stoneville, USA

Chris **Carter**, University of Tasmania, Launceston Campus, Locked bag 1399, Tasmania 7250, Australia (chris.carter@utas.edu.au)

Steven **Clarke**, Aquafin CRC, Box 120, Henley Beach, SA 5022, Australia (clarke.steven@saugov.sa.gov.au)

Michel **Couturier**, Department of Chemical Engineering, University of New Brunswick, PO Box 4400, Fredericton, NB, Canada E3B 5A3 (cout@unb.ca)

Lawrence **Dooley,** Department of Management, Centre for Wine, Food and Agribusiness, Location Room 03, Building N6, Caulfield, Monash University, Victoria 3800, Australia (Lawrence.Dooley@buseco.monash.edu.au)

Christian **Fauvel**, Ifremer, Laboratoire LALR, Station expérimentale de Palavas, Chemin de Maguelone 34250 Palavas-les-Flôts, France (christian.fauvel@ifremer.fr)

Kevin **Fitzsimmons**, Environmental Research Lab, University of Arizona, 2601 E Airport Dr, Tucson, AZ 85706-6985, USA (kevfitz@ag.arizona.edu)

Atle **Foss**, Akvaplan-Niva A/S, Bergen office, PO Box 2026, Nordnes 5, N-5817, Bergen, Norway (af@akvaplan-niva.no)

Brett **Glencross** Government of Western Australia, Department of Fisheries, Research Division, PO Box 20, North Beach, WA 6920, Australia (Brett.Glencross@fish.wa.gov.au)

Laura C. **Halfyard**, Marine Institute, Memorial University of Newfoundland, PO Box 4220, St John's, Newfoundland, Canada A1C 5R3 (laura.halfyard@mi.mun.ca)

Terrill **Hanson**, Department of Fisheries and Allied Aquacultures, Auburn University, 203 Swingle Hall, Auburn, AL 36849, USA (hansontr@auburn.edu)

Ronald **Hardy**, UI Hagerman Fish Experiment Station, 3059 F National Fish Hatchery Road, Hagerman, Idaho 83332, USA (rhardy@uidaho.edu)

Joan G. **Holt**, University of Texas at Austin, Marine Science Institute, 750 Channel View Drive, Port Aransas, TX 78373-5015, USA (joan@utmsi.utexas.edu)

Albert **Imsland**, Akvaplan-Niva A/S, Iceland Office, Akralind 4, 201 Kópavogi, Iceland and Department of Biology, University of Bergen, High Technology Centre, 5020 Bergen, Norway (Albert.Imsland@bio.uib.no)

Malcolm **Jobling**, University of Tromsø, 9037 Tromsø, Norway (malcolm.jobling@nfh.uit.no)

Robin **Katersky** University of Tasmania, Locked bag 1370, Launceston, TAS 7250, Australia (robink@utas.edu.au)

Gunnar **Knapp**, Institute of Social and Economic Research, University of Alaska Anchorage, 3211 Providence Drive, Anchorage, Alaska 99508 (gunnar.knapp@gmail.com)

Juha **Koskela**, Finnish Game and Fisheries Research Institute, Laukaa Fisheries Research and Aquaculture, 41360 Valkola, Finland (juha.koskela@rktl.fi)

Simon G. **Lamarre**, Memorial University of Newfoundland, Canada (simon.lamarre@gmail.com)

Richard **Langan**, Atlantic Marine Aquaculture Center, University of New Hampshire, Gregg Hall, 35 Colovos Road, Durham, NH 03824, USA (rlangan@unh.edu)

Juan P. **Lazo**, CICESE, Carretera Tijuana – Ensenada CP 22860, Ensenada, BC Mexico (jplazo@cicese.mx)

Nathalie R. **Le François**, Biodôme de Montréal, 4777 Avenue Pierre-De Coubertin, Montréal, Québec H1V 1B3, Canada and Université du Québec à Rimouski, 300 des Ursulines, Rimouski, QC, G5L 3A1, Canada (nle_francois@ville.montreal.qc.ca)

Cheng-Sheng **Lee**, Center for Tropical and Subtropical Aquaculture Oceanic Institute, 41–202 Kalanianaole Highway, Waimanalo, HI 96795, USA (cslee@oceanicinstitute.org)

Matthew **Litvak**, Centre for Coastal Studies and Aquaculture and Department of Biology, University of New Brunswick, P.O. Box 5050, Saint John, New Brunswick, Canada E2L 4L5 (litvak@unbsj.ca)

Chin-fa **Liu**, Consultant for Team Aqua Corporation, 14 Floor-1, No. 115, Jungshan Road, Shinying City, Tainan, Taiwan, 730, ROC (teamaqua@ms13.hinet.net)

Barbara **Nowak**, School of Aquaculture, University of Tasmania, Launceston, Australia (B.Nowak@utas.edu.au)

Robyn **O'Keefe**, Department of Biology, University of New Brunswick, PO Box 4400, Fredericton, New Brunswick, Canada E3B 5A3 (okeefe@unb.ca)

David **Otton**, 23 Thompson Drive, Tathra, New South Wales, Australia 2550 (benooka@bigpond.com)

Stefano **Peruzzi**, University of Tromso, 9037 Tromsø, Norway (stefanop@nfh.uit.no)

Loic **Quéméner,** Ifremer, Centre de Brest, BP70, 29280 Plouzané, France (Loic.Quemener@ifremer.fr)

Arianne **Savoie**, Université du Québec à Rimouski, 300 des Ursulines, Rimouski, QC, G5L 3A1, Canada (arianne_savoie@hotmail.com)

Jeffrey **Silverstein**, USDA/ARS National Center for Cool and Cold Water Aquaculture, 11861 Leetown Road, Kearneysville, WV 25430, USA (jsilvers@ncccwa.ars.usda.gov)

Marc **Suquet**, Ifremer, Laboratoire ARN, Station Expérimentale d'Argenton, Presqu'île du Vivier, 29840 Landunvez, France (marc.suquet@ifremer.fr)

Edward A. **Trippel**, St Andrews Biological Station, 531 Brandy Cove, St Andrews, New Brunswick, Canada E5B 2L9 (trippleE@mar.dfo-mpo.gc.ca)

Helge **Tveiten**, Nofima A/S, Muninbakken 9–13, Postboks 6122, N-9291 Tromsø, Norway (helge.tveiten@nofima.no)

Wade O. **Watanabe**, Centre of Marine Science, University of North Carolina at Wilmington, 7205 Wrightsville Ave, Wilmington, NC 28403, USA (watanabew@uncwil.edu)

Guillaume **Werstink**, Université du Québec à Rimouski, 300 des Ursulines, Rimouski, QC, G5L 3A1, Canada (guillaume_werstink@uqar.qc.ca)

James R. **Wilson**, Département Économie et de Gestion, Université du Québec à Rimouski, 300 Allée des Ursulines, Rimouski, QC, Canada G5L 3A1 (james_wilson@uqar.qc.ca)

Curry **Woods**, Department of Animal and Avian Sciences University of Maryland College Park, MD 20742, USA (curry@umd.edu)

James A. **Young**, Department of Marketing, University of Stirling, Stirling FK9 4LA, Scotland, UK (j.a.young@stir.ac.uk)

Acknowledgements

We would like to thank the following people:

Brian Colwill and **Dr David Otton** for their assistance with linguistic verification. David came to aquaculture as a mature aged PhD student and his single minded advocacy for the agribusiness approach and his enthusiasm for aquaculture will be greatly missed in Australia and beyond;

Marc Veillet and **Karine Bisson** of the Ministère de l'Agriculture, des Pêcheries et de l'Alimentation du Québec;

Dr Céline Audet (ISMER/UQAR, Québec, Canada) for reviewing the winter flounder section, **Dr Denis Coves** (Ifremer, France) for fruitful exchange on tuna cultivation efforts in Europe and **Dr Serge Parent** (Biodôme de Montréal, Québec, Canada) for his constructive comments;

François Larouche, Timothée Govarre and **Sarah Tremblay-Bourgeois** for punctual, highly specific tasks that provided relief when most needed.

The financial contributions of the Université du Québec à Rimouski and the Biodôme de Montréal are also acknowledged.

I Aquaculture Diversification: an Introduction

1 Fish Culture: Achievements and Challenges

Malcolm Jobling

University of Tromsø, Tromsø, Norway

1.1 Introduction

Aquaculture is the cultivation of aquatic organisms (e.g. algae, molluscs, crustaceans, fish) and implicit in this is human intervention that involves some control over the stock. The degree of human intervention may be limited in extensive culture of algae and molluscs and in extensive pond culture of fish and crustaceans, but the level of control exerted is high in intensive farming systems. Intensive culture involves intervention at all phases of the production cycle; from broodstock maintenance and egg production to the harvesting and marketing of the finished product (Fig. 1.1). The farmed animals are reared in captivity throughout their lives; they are held under relatively benign, semi-controlled conditions and rely on formulated feeds for their nutrition. Aquaculture research and development (R & D), as it relates to intensive fish farming, currently encompasses all aspects of production and postharvest processing, including culture unit design, feed delivery systems, water treatment, feed formulation and disease diagnosis and treatment. There is a growing perception that intensive aquaculture is a modern industry; this industry is becoming subject to increased scrutiny by policy makers and the general public and is becoming increasingly influenced by social attitudes that have an impact on regulations, marketing and product ranges and acceptability (Beardmore and Porter, 2003; Lee, 2003; Aerni, 2004; Kelso, 2004; Boyd *et al.*, 2005; Focardi *et al.*, 2005; Foran *et al.*, 2005; Logar and Pollock, 2005; Midtlyng, 2005; Muir, 2005; Myhr and Dalmo, 2005; Naylor *et al.*, 2005; Verbeke *et al.*, 2005; Devlin *et al.*, 2006; Huntingford *et al.*, 2006).

Intensive fish culture is benefiting from increased use of sophisticated instrumentation and management tools, and modern analytical techniques are finding increased application in aquaculture R & D. In the sections that follow, examples will be given of the ways in which some of these recently developed

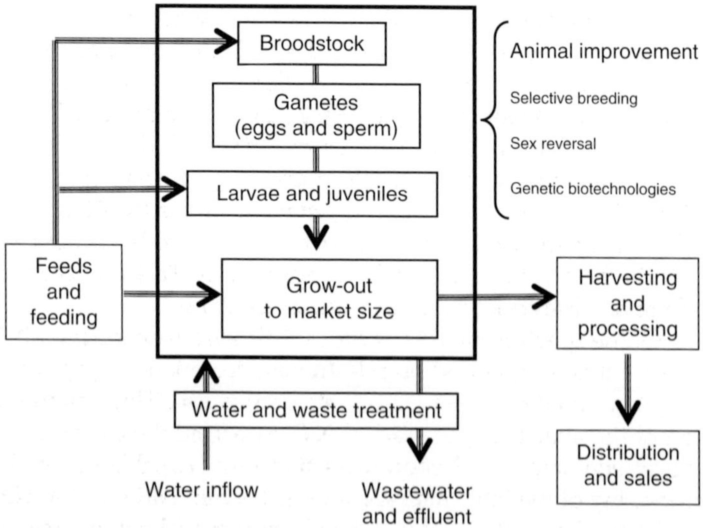

Fig. 1.1. Overview of the production cycle of farmed fish with an indication of the areas in which there is human intervention.

techniques have been used to address research problems with direct bearing on intensive aquaculture. There will also be consideration of some of the challenges and contentious issues that the industry must face.

1.2 Application of Emerging Technologies to Aquaculture Research

In recent years, there have been major advances in a variety of analytical methods and molecular technologies. For example, molecular diagnostics can play a key role in the detection and identification of food contaminants and allergens, potential pathogens and genetically modified organisms (GMOs). Within this sphere, DNA microarray technology has created the possibility for simultaneous analysis of large gene sets, making it a potent molecular diagnostics tool with a number of applications (Liu-Stratton *et al.*, 2004; Kato *et al.*, 2005; Roy and Sen, 2006). As such, the development of molecular diagnostics techniques has had major impacts on R & D in a host of applied sciences, including clinical (human) and veterinary medicine, food science, agronomy and agricultural sciences.

These disciplines interface with aquaculture across a broad range of subject areas; genetics and selective breeding, health management and disease treatment, authentication and control of feed ingredients, postharvest processing, product storage and spoilage, assurance of food safety, product development and identification and traceability of products through the food chain (Blanco and Villarroya, 2002; Cozzolino *et al.*, 2002, 2005a,b; Ali *et al.*, 2004; Carbonaro, 2004; Dunham, 2004; Huss *et al.*, 2004; Kuiper *et al.*, 2004; Pérez-Marin *et al.*, 2004; Deisingh and Badrie, 2005; Focardi *et al.*, 2005; Gorris, 2005; Guérard *et al.*, 2005; Kelly *et al.*, 2005; Lehrer and Bannon, 2005; Midtlyng, 2005; Monis *et al.*, 2005; Neira and Diaz, 2005; Pinotti *et al.*, 2005; Verbeke *et al.*, 2005; Adams and Thompson, 2006; Berrini *et al.*, 2006; Bruhn and Earl, 2006; Gasser, 2006; Reid *et al.*, 2006; Toldrá and Reig, 2006). Similar sets of methods, such as spectroscopy (e.g. near-infrared (NIR), mid-infrared (MIR), UV-, Raman-, nuclear magnetic resonance (NMR) spectroscopy), immunological assays and DNA-based technologies, are being applied increasingly to assist in the solving of research problems in medical science, food science, agriculture and livestock production and aquaculture.

Of the spectroscopic techniques, NIR spectroscopy has attained popularity because it is a rapid and non-destructive technique, equipment costs are relatively low and the equipment is quite easy to use. Thus, interest in the exploitation of the analytical potential of NIR paralleled the development of electronic optical equipment and computers that were capable of rapid extraction and processing of the information contained within NIR spectra. NIR spectroscopy can be used on both solid and liquid samples without any major pretreatment, and interpretation of the spectra can give information about both the physical and chemical properties of the sample (Blanco and Villarroya, 2002; Reid *et al.*, 2006). It currently has use as a rapid analytical tool in quality control in the petrochemical and pharmaceutical industries and as an aid in diagnostics in the clinical and biomedical fields, but NIR spectroscopy has widest application in the food science and animal feed sectors. NIR spectroscopy is, for example, used for determining chemical composition (protein, moisture and oil/fat) of feed ingredients and feeds and of crops and livestock, either prior to or at harvest/slaughter, and may also be used for identification and authentication purposes (Blanco and Villarroya, 2002; Cozzolino *et al.*, 2002, 2005a,b; Pérez-Marin *et al.*, 2004; Reid *et al.*, 2006).

As a second example, the development of molecular marker techniques created new possibilities for the identification and genetic characterization of species, populations and individuals. This has important implications for the study of interrelationships between species (taxonomy and phylogeny), population genetics, selective breeding and genetics and disease diagnosis (Melamed *et al.*, 2002; Ali *et al.*, 2004; Dunham, 2004; Deisingh and Badrie, 2005; Gjedrem, 2005; Guérard *et al.*, 2005; Monis *et al.*, 2005; Neira and Diaz, 2005; Pinotti *et al.*, 2005; Adams and Thompson, 2006; Chistiakov *et al.*, 2006; Gasser, 2006). Methods involving the polymerase chain reaction (PCR) to amplify nucleic acids have proven particularly valuable and have been applied to several types of research problem. PCR, which has been in use for almost 20 years, is a fundamental technology for genetic identification and characterization. DNA markers derived from PCR amplification are now being widely used for genetic characterization and for distinguishing between genetic variants at different hierarchical levels: species, populations, strains and individuals (Ali *et al.*, 2004; Dunham, 2004; Liu and Cordes, 2004; Gjedrem, 2005; Neira and Diaz, 2005; Chistiakov *et al.*, 2006; Reid *et al.*, 2006; Renshaw *et al.*, 2006). Major advantages of PCR over alternative techniques for nucleic acid amplification are cost and ease of use. Using PCR, enzymatic amplification of either mitochondrial or genomic DNA can be achieved *in vitro* from minute amounts of material. Highly conserved DNA regions can be used to delineate taxonomic relationships between species, examination of moderately conserved regions can be used to distinguish between populations, whereas variable DNA regions can be used for tracking individuals and 'mapping' can be used to identify genetic markers that correlate genotype to phenotype, for example, linked to important production traits. The latter finds application in the identification of molecular markers of qualitative and quantitative trait loci (QTLs) of economic interest (Melamed *et al.*, 2002; Ali *et al.*, 2004; Dunham, 2004; Liu and Cordes, 2004; Neira and Diaz, 2005; Chistiakov *et al.*, 2006). Thus, PCR encompasses amplification of a conserved region of mDNA for species identification as in PCR-RFLP, or a fingerprinting approach using PCR-RAPD for the detection of genetic variation or the definition of genetic markers.

1.3 Stock Improvement

Farmed stock can be improved by selective breeding and/or by implementing a variety of genetic biotechnologies (Maclean and Laight, 2000; Lee, 2001; Devlin and Nagahama, 2002; Melamed *et al.*, 2002; Utter and Epifanio, 2002; Beardmore and Porter, 2003; Fjalestad *et al.*, 2003; Dunham, 2004; Liu and Cordes, 2004; Gjedrem, 2005; Mignon-Grasteau *et al.*, 2005; Neira and Diaz, 2005; Chistiakov *et al.*, 2006; Goetz *et al.*, 2006). DNA marker methods, and other genetic technologies, are having increased impact on the improvement of production characteristics of aquaculture stock and they find application in broodstock management and selective breeding programmes. For example, several fish species are currently under study for genome sequencing (Dunham, 2004; Volff, 2005; Goetz *et al.*, 2006) and some commercially important species have been included in such investigations. These include channel catfish, *Ictalurus punctatus*, Atlantic salmon, *Salmo salar*, rainbow trout, *Oncorhynchus mykiss*, and Nile tilapia, *Oreochromis niloticus* (Fjalestad *et al.*, 2003; Dunham, 2004; Volff, 2005), where the focus has been on the identification of QTLs of economic interest. Such QTLs relate to those with an influence on growth, age-at-maturation and reproductive characters, body form and pigmentation, environmental tolerances, disease resistance and behaviour (Fjalestad *et al.*, 2003; Dunham, 2004; Liu and Cordes, 2004; Gjedrem, 2005; Neira and Diaz, 2005; Volff, 2005; Chistiakov *et al.*, 2006), and identified markers are being used to assist with broodstock selection for inclusion in breeding programmes.

Genetic interventions, environmental manipulations and hormonal treatments, either alone or in combination, may be used to influence sex determination, sexual differentiation and reproductive physiology and behaviour (Lee, 2001; Patiño, 2002; Melamed *et al.*, 2002; Kagawa *et al.*, 2003; Dunham, 2004; Gjedrem, 2005; Neira and Diaz, 2005; Goetz *et al.*, 2006; see Chapter 2, this volume). For example, photothermal manipulations are used to influence the onset of sexual maturation (timing of puberty) and are employed in broodstock management for production of out-of-season eggs. In addition, hormone treatments are used to induce gamete maturation and spawning in a wide range of species.

1.3.1 Sex-reversal and production of monosex populations and triploids

Genetic techniques, involving a sex-reversal step, can be used to produce single-sex (monosex) populations and gene manipulation may be employed to induce sterility through triploidization (Fig. 1.2). The production of monosex (single-sex) populations may be desirable when there are differences between the sexes in growth rate, age and size at maturity, disease resistance and other production characters, e.g. coloration, flesh characteristics, etc. The rearing of monosex populations may also be used to prevent unwanted (uncontrolled) spawning in rearing units such as ponds. The raising of sterile fish may be desirable as a measure to eliminate spawning in the farmed population and to reduce the potential negative genetic effects of farm escapees on natural populations (due to interbreeding).

Production of sterile fish may be used to reduce the negative effects of sexual maturation on growth and prevent the deterioration of flesh quality (due to the mobilization of reserves from the muscle) that accompanies maturation in

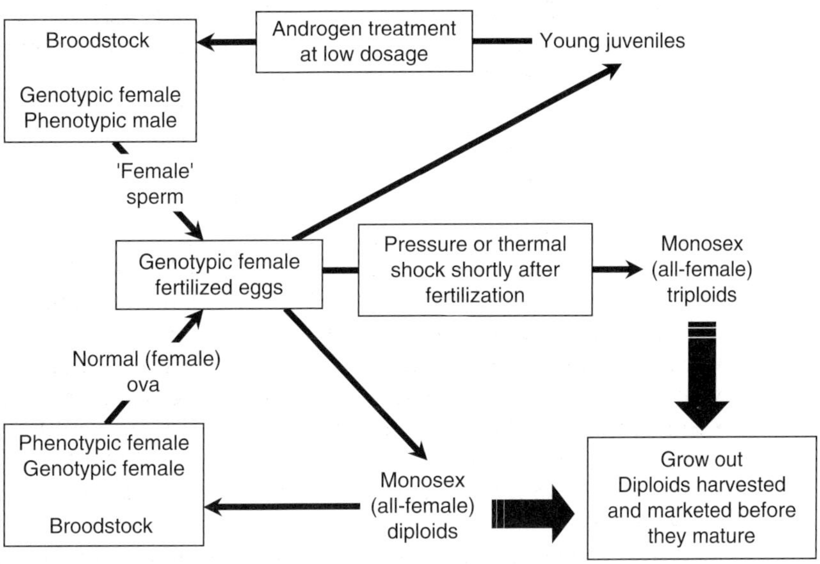

Fig. 1.2. Schematic representation of sex-reversal techniques and the methods used to produce monosex (all female) diploid and triploid fish. In this scheme, it is assumed that the female is the homogametic (XX) sex and the male is the heterogametic (XY) sex. Sex reversal is induced by treating fertilized eggs, or very young juveniles, with male sex steroid hormones (androgens). Triploidy is induced by subjecting fertilized eggs to a temperature or a pressure shock.

many fish species. Muscle protein degradation involves three major proteolytic enzyme systems: the lysosomal cathepsins, the calpain–calpastatin (calpain–CAST) system and the ubiquitin–proteasome system (Salem *et al.*, 2004, 2005a,b, 2006). Enzymes of these three systems are expressed differently in fertile diploid and sterile triploid rainbow trout, *O. mykiss*, with the cathepsins seemingly playing the major role in the proteolysis of muscle proteins seen in maturing fish. On the other hand, the calpain–CAST system appears to play an important role in the regulation of muscle protein turnover during growth and in the mobilization of reserves that occurs under a fast. This system, along with the cathepsins, is also involved in the degradation of muscle proteins that occurs post-mortem, thereby having an influence on muscle softening, textural changes and fillet 'quality' (Salem *et al.*, 2004, 2005a,b, 2006).

Although methods used for producing monosex and/or sterile populations are now considered routine for several species, they do not have universal application due to the fact that several anomalous effects have been observed. Anomalies may arise, for example, if the eggs and hatchlings are exposed to waters of particular temperatures or with certain chemical characteristics (Devlin and Nagahama, 2002; Patiño, 2002; Kagawa *et al.*, 2003; Miyamoto and Burger, 2003; Mills and Chichester, 2005; Goto-Kazeto *et al.*, 2006; see Chapter 2, this volume). This is seen in intensive commercial production of European seabass, *Dicentrarchus labrax*, in which about 75% of the farmed fish usually develop as males; male-biased populations are undesirable because males grow more slowly and mature earlier than females. A substantial proportion of the male fish mature before reaching harvest size, so production of monosex female populations would be desirable. Although sex manipulation of seabass has been achieved by exposing fish to exogenous sex steroid hormones, efforts to produce female monosex seabass populations using crosses involving sex-reversed individuals have not met with success. A major stumbling block for the application of sex-reversal techniques to this species is that the mechanisms responsible for sex determination and differentiation are not known with certainty (Piferrer *et al.*, 2005).

1.3.2 Induction of sterility

Chromosomal manipulation methods are useful for the production of triploid and tetraploid individuals, but they are not 100% successful. Consequently, the populations produced inevitably will contain a small percentage of fertile diploids. It is possible to induce sterility via the production of hybrids and by other methods. The biotechnological methods involve the production of individuals that express antagonists for a function needed for fertility, suppression of development and differentiation of germ cells, or suppression of the expression of genes required for the biosynthesis of sex steroid hormones (Maclean and Laight, 2000; Melamed *et al.*, 2002; Kagawa *et al.*, 2003; Weidinger *et al.*, 2003; Dunham, 2004; Slanchev *et al.*, 2005). For example, research into primordial germ cell (PGC) development and function has given promising results (Raz, 2003; Weidinger *et al.*, 2003; Slanchev *et al.*, 2005). PGCs are the cells that ultimately give rise to the gametes, but these cells do not have their source in the embryonic tissue that develops into the gonads. In other words, PGCs initially migrate from their source to the genital ridges of the developing embryo and there is then differentiation of the gonads. Over time, the PGCs differentiate into eggs or sperm in the ovary or testis, respectively (Raz, 2003). Work carried out on zebrafish, *Danio rerio*, provides evidence of a key role of PGCs in sex determination and gonad differentiation. Ablation of PGCs in embryonic zebrafish gave rise to sterile adults, all of which were males (as assessed using morphological and behavioural criteria, including courtship and mating with females) (Slanchev *et al.*, 2005). This research may pave the way for similar studies on commercially farmed fish species, with the aim of producing sterile populations. Such populations might have better growth and production characteristics than fertile conspecifics and sterility could also reduce the potential impact of farm escapees on wild populations.

PGCs may be transplanted from one species to another and preliminary data provide evidence that this method can be used to develop surrogate broodstock that produce xenogenic donor-derived offspring (Takeuchi *et al.*, 2004). Further, the germ cells may retain a high degree of developmental plasticity even after the commencement of differentiation (Okutsu *et al.*, 2006). For example, testicular germ cells taken from the testes of mature rainbow trout, *O. mykiss*, and transplanted into undifferentiated embryonic gonads were shown to have retained developmental plasticity. These cells were capable of displaying sexual bipotency. In other words, the testicular germ cells developed into spermatozoa in male embryonic recipients, but differentiated into eggs when the recipient embryonic rainbow trout were females (Okutsu *et al.*, 2006). This demonstrated that the differentiation and development of the testicular germ cells was not autonomous and hard-wired, but could be controlled and directed by the environment to which the cells were exposed. In addition, the gametes, either sperm or eggs, derived from the transplanted testicular germ cells were fully functional and could be used for the production of viable offspring (Okutsu *et al.*, 2006). These approaches to producing functional eggs and sperm offer a powerful tool for the study of gamete differentiation and development in fish and may also have potential for the rapid production of inbred strains that express a range of commercially desirable genetic characters.

1.3.3 Interactions between farmed and wild fish

Although the production of sterile farmed stock might eliminate problems related to the genetic impacts of farm escapees on natural populations, it would not serve to eliminate all forms of interaction. Individuals that escape from farms could be members of non-native or introduced species, highly selected domesticated stocks, hybrids or animals that have had their genetic constitutions altered using transgenic biotechnologies (genetically modified organisms – GMOs). These different categories of exotics have different characteristics and have the potential to impact wild populations in several ways. The farm escapees could act as either predators or prey, they could be vectors of parasites or disease organisms and they could compete for food or other habitat resources, resulting in displacement or exclusion of native species and natural populations. In the case of the farm escapees being fertile, there could also be interbreeding or hybridization between the exotics and native stocks, resulting in changes to genetic constitution or sterility. In all likelihood, several of these mechanisms would act simultaneously (Beardmore and Porter, 2003; Aerni, 2004; Bessey *et al.*, 2004; Devlin *et al.*, 2004, 2006; Kelso, 2004; Muir, 2004; Welcomme and Craig, 2004; Bondad-Reantaso *et al.*, 2005; Boyd *et al.*, 2005; Logar and Pollock, 2005; Murray and Peeler, 2005; Naylor *et al.*, 2005; Bekkevold *et al.*, 2006; Jonsson and Jonsson, 2006; Wessel *et al.*, 2006).

1.3.4 Transgenic fish

Hitherto, over 30 species of fish have been the subject of transgenic research and genetic engineering, and the species studied include several representatives from the major farmed groups of fishes – the carps, tilapias, catfishes and salmonids. Transgenesis may be used to obtain basic information on gene function and regulation, but also has commercial application in the improvement of animal production and in the introduction of gene constructs that result in the biosynthesis of high-value products such as recombinant proteins (Beardmore and Porter, 2003; Houdebine, 2005). The application of transgenic technology in fish has been concerned mostly with the examination of methods that can be used to enhance the growth of farmed stock, predominantly via the transfer of genes coding for growth hormone (Maclean and Laight, 2000; Beardmore and Porter, 2003; Dunham, 2004; Muir, 2004; Houdebine, 2005; Logar and Pollock, 2005; Devlin *et al.*, 2006), but additional possibilities exist for exploitation of this type of genetic manipulation. For example, gene constructs may be introduced to enhance disease resistance or modify metabolic pathways that have influences on the chemical composition of the flesh (Beardmore and Porter, 2003; Devlin *et al.*, 2006).

One such example is the modification of fatty acid metabolic pathways to enhance the biosynthesis of n-3 highly unsaturated fatty acids (n-3 HUFAs) and increase their deposition in the flesh of the farmed animal (Alimuddin *et al.*, 2005). The biosynthesis of HUFAs depends on a series of fatty acid desaturation and elongation reactions that are catalysed by a number of enzymes (Fig. 1.3). Fish species differ in their abilities to carry out the desaturation and elongation reactions and the characteristics of the enzymes differ among species (Zheng *et al.*, 2004; Agaba *et al.*, 2005; Robert, 2006). Most notably, there are differences between carnivorous marine species and herbivorous and omnivorous freshwater fish species in the ability to synthesize HUFAs from the 18C precursor fatty acids. Genes coding for key enzymes in the n-3 HUFA biosynthetic pathway may be introduced into the fish genome and fatty acid metabolic pathways modified accordingly (Alimuddin *et al.*, 2005). There is, however, the possibility that introduction of novel DNA into the fish genome could give rise to unintended side effects, such as overexpression of some genes and the silencing of others; the examination of unintended side effects of genetic engineering in fish is in its infancy, although metabolomics and other 'omics-based' methods

Enzyme	Fatty acid series			
	n-7	n-9	n-6	n-3
	16:0	18:0		
Δ9 desaturase	↓	↓		
	16:1 n-7	18:1 n-9	18:2 n-6	18:3 n-3
Δ6 desaturase	↓	↓	↓	↓
	16:2 n-7	18:2 n-9	18:3 n-6	18:4 n-3
elongase	↓	↓	↓	↓
	18:2 n-7	20:2 n-9	20:3 n-6	20:4 n-3
Δ5 desaturase	↓	↓	↓	↓
	18:3 n-7	20:3 n-9	20:4 n-6	20:5 n-3
elongase	↓	↓	↓	↓
	20:3 n-7	22:3 n-9	22:4 n-6	22:5 n-3
elongase			↓	↓
			24:4 n-6	24:5 n-3
Δ6 desaturase			↓	↓
			24:5 n-6	24:6 n-3
(β-Oxidation)			↓	↓
			22:5 n-6	22:6 n-3

Fig. 1.3. Pathways for chain elongation and desaturation of fatty acids. Fatty acids of the n-7 and n-9 series can be synthesized *de novo*, whereas fatty acids of the n-6 and n-3 series are the essential fatty acids. The Δ6 desaturase enzyme has a preference for unsaturated fatty acids as substrate. In the absence of 18:3 n-3 and 18:2 n-6, 18:1 n-9 is desaturated and elongated, leading to an accumulation of long-chain fatty acids of the n-9 series in tissues (this can be used as an indication of essential fatty acid deficiency). Carnivorous marine fish species seem to have low Δ5 desaturase activity and they have a low capacity to use the n-3 and n-6 18C fatty acids as precursors for synthesis of n-3 and n-6 HUFAs. The shorthand notations of the fatty acids give information about the number of carbon atoms, the number of double bonds and the position of the first double bond from the methyl end of the molecule (from Jobling, 2004c).

for carrying out such investigations are available (Beardmore and Porter, 2003; Kuiper *et al.*, 2004; Liu-Stratton *et al.*, 2004; Kato *et al.*, 2005; Griffin, 2006; Rischer and Oksman-Caldentey, 2006; Viant *et al.*, 2006).

1.4 Feed Ingredients and Resources

Developments in plant breeding and genetics are playing an increasingly important role in the provision of ingredients for aqua-feeds. Although a wide range of ingredients is used in the formulation of aqua-feeds (Hertrampf and Piedad-Pascual, 2000), fishmeals and fish oils traditionally have been used as major ingredients, but in recent years there has been a trend towards an increased reliance on plant-based products (see Chapter 3, this volume). Fishmeals have been used as a major ingredient because they have been a widely available source of protein of high quality. Marine fish oils have been widely used as the main lipid source in aqua-feeds because they have been readily available, relatively cheap and are an excellent source of the n-3 HUFAs deemed beneficial for human health (Graham *et al.*, 2004; Ruxton *et al.*, 2004, 2005; SACN-FSA, 2004; Bergé and Barnathan, 2005; Bourre, 2005; Cleland *et al.*, 2005; MacRae *et al.*, 2005; Napier and Sayanova, 2005; Bethune *et al.*, 2006; Kim and Mendis, 2006). The heavy reliance on fishmeals and oils as major ingredients in aqua-feeds has, however, been questioned and criticized (Naylor *et al.*, 2000; Muir, 2005). Stocks of small, pelagic fish used for fishmeal and oil production represent finite resources that are either fully exploited or are being overexploited and are at risk of collapse. Fears have also been expressed that fishmeals and oils may give rise to contamination of the flesh of farmed fish with organochlorine compounds and/or heavy metals (Easton *et al.*, 2002; Jacobs *et al.*, 2002; Hites *et al.*, 2004; Bell *et al.*, 2005; Berntssen *et al.*, 2005; Carlson and Hites, 2005; Foran *et al.*, 2005; Hamilton *et al.*, 2005; Bethune *et al.*, 2006; Montory and Barra, 2006). These contaminants may be carcinogenic and may give rise to developmental anomalies that result in the malformation of various tissues and organs in both fish and human consumers (Miyamoto and Berger, 2003; Sumpter and Johnson, 2005; Scott *et al.*, 2006; see Chapter 3, this volume).

1.4.1 Feed ingredients of plant origin

There may be major benefits to be gained by including protein sources of plant origin in fish feeds, but there are also problems associated with such feed sources (Hertrampf and Piedad-Pascual, 2000; McKevith, 2005). For example, many plants are deficient in some essential nutrients, such as some amino acids (e.g. lysine and methionine). This means that there may be a need to add amino acid supplements to feeds prepared with protein sources derived from plants. In addition, most plants contain antinutritional factors (ANFs) that interfere with digestion and absorption, reduce nutrient bioavailability and feed utilization and may have other adverse effects (Table 1.1) (Francis *et al.*, 2001; Jobling, 2004c; Acamovic and Brooker, 2005). Further, it is becoming increasingly difficult to obtain vegetable feed ingredients, for example, soya, maize and canola, guaranteed to be free from some form of biotechnological genetic modification, i.e. non-transgenic vegetable feed ingredients are becoming increasingly scarce in the global market (Kok and Kuiper, 2003; Flachowsky *et al.*, 2005; Myhr and Dalmo, 2005; Bruhn and Earl, 2006).

1.4.1.1 Transgenic plants

The genetic modification of plants may involve the introduction of genes encoding for increased disease resistance, production of insecticidal proteins or

Table 1.1. Overview of major classes of antinutritional factors (ANFs) present in plants used as protein and/or carbohydrate sources in feeds prepared for farmed fish. × indicates that the ANF is present and heat-labile ANFs (i.e. ANFs that can be destroyed or denatured by heat treatment) are shown in bold typeface (from Jobling, 2004c).

	Antinutritional factor (ANF)									
	Enzyme inhibitors									
	Protease	**Amylase**	Phytate	Phyto-oestrogens	Saponins	Tannins	**Lectins**	Alkaloids	Cyanogens	Glucosinolates
'Oilseed' meals										
Soybean	**×**		×	×	×		**×**			
Rape/canola	**×**		×	×		×				×
Cottonseed		**×**	×	×						
Sunflower	**×**				×	×				
Sesame	**×**									
Legumes										
Lupin	**×**			×	×			×		
Lucerne				×	×			×		
Faba bean	**×**		×			×	**×**			
'Grains'										
Maize	**×**		×	×						
Wheat	**×**	**×**	×	×			**×**			
Sorghum	**×**	**×**	×			×	**×**		×	
'Tubers'										
Potato	**×**	**×**		×			**×**	×		
Cassava	**×**		×						×	

increased herbicide tolerance; the agents that impart these properties are present at very low concentration and have little influence on the gross composition of the plant. Alternatively, attempts may be made to influence nutritional value by increasing the protein content or indispensable (essential) amino acids, increasing concentrations of vitamins, such as tocopherols, or by reducing concentrations of ANFs (Tucker, 2003; Flachowsky *et al.*, 2005; Lehrer and Bannon, 2005; Bruhn and Earl, 2006; Dalal *et al.*, 2006).

Plant oils are generally deficient in n-3 HUFAs, although several contain relatively high concentrations of 18C n-3 precursors (Hertrampf and Piedad-Pascual, 2000; McKevith, 2005; Robert, 2006). The fatty acid compositions of the flesh of monogastric animals frequently reflect the compositions of the feed, so aquaculture species given feeds based on plant oils may have relatively low concentrations of n-3 HUFAs in their flesh (e.g. Jobling, 2004a,b,c; Wonnacott *et al.*, 2004; Berntssen *et al.*, 2005; Torstensen *et al.*, 2005; Visentainer *et al.*, 2005; see Chapter 3, this volume). Given this, one major area of research focus has been the elucidation and modification of the metabolic pathways of fatty acid synthesis, with the aim of producing plants with higher n-3 HUFA content. The conventional metabolic pathway for the synthesis of the HUFAs is via the n-3 pathway employing the $\Delta 6$ desaturase and elongase route (Fig. 1.3), but there are alternative pathways. One of the alternatives involves the n-6 pathway and a $\Delta 17$ desaturase (C20:4 n-6 $\rightarrow$ 20:5 n-3), another involves the n-3 pathway and a $\Delta 4$ desaturase (C22:5 n-3 $\rightarrow$ 22:6 n-3) and there is also a route that relies on the n-3 pathway and $\Delta 9$ elongase and $\Delta 8$ desaturase (C18:3 n-3 $\rightarrow$ 20:3 n-3 $\rightarrow$ 20:4 n-3). The alternative pathways, found in some unicellular organisms, appear to offer most promise with regard to introduction into higher plants (Abbadi *et al.*, 2004; Graham *et al.*, 2004; Domergue *et al.*, 2005; Flachowsky *et al.*, 2005; Napier and Sayanova, 2005; Guschina and Harwood, 2006; Napier *et al.*, 2006; Robert, 2006).

Animal feeds, including aqua-feeds, may also be supplemented with n-3 HUFAs obtained by microbial fermentation (Ward and Singh, 2005; Spolaore *et al.*, 2006). The best microbial sources of these fatty acids are unicellular algae, or algal-like protists, of the order Thraustochytriales. Industrial fermentation methods have been developed for several species and/or strains of *Thraustochytrium* and *Schizochytrium* (Bergé and Barnathan, 2005; Ward and Singh, 2005). Although plants can be engineered to produce n-3 HUFAs, microorganisms have certain advantages with regard to production, extraction and purification of the fatty acids. Several fatty acid products derived from fermentation of microorganisms are currently available commercially and are being incorporated into some aqua-feeds. The microorganism fatty acids are being used particularly in feeds destined for the rearing of the youngest developmental stages of fish and crustaceans (Bergé and Barnathan, 2005; Ward and Singh, 2005; Guschina and Harwood, 2006; Spolaore *et al.*, 2006).

Genetic modification of plants and animals by transgenesis gives rise to the possibility of introducing unintentional side effects as a result of the insertion of novel DNA into the genome. Possible results could be the increased biosynthesis and accumulation of toxic metabolites, enhanced production of known or novel allergenic proteins or reductions in tissue concentrations of certain essential nutrients in the modified organism (Kok and Kuiper, 2003; Kuiper *et al.*, 2004; Liu-Stratton *et al.*, 2004; Lehrer and Bannon, 2005; Rischer and Oksman-Caldentey, 2006; Roy and Sen, 2006). Such changes could have negative effects on an animal that consumes such products. Control measures have been introduced in an attempt to reduce such risks and a number of methods, including metabolic profiling or metabolomics (the global mapping of metabolites present within cells or tissues), are being developed to address these problems (Kuiper *et al.*, 2004; Liu-Stratton *et al.*, 2004; Kato *et al.*, 2005; Lehrer and Bannon, 2005; Griffin, 2006; Rischer and Oksman-Caldentey, 2006; Müller and Steinhart, 2007). Some of these methods have been applied to examination of the metabolic profiles of fish. These novel, modern methods have, however, been used mostly in toxicological studies to identify changes in metabolism that result from the exposure of fish to toxicants or other forms of environmental stressor (e.g. Cakmak *et al.*, 2006; Viant *et al.*, 2006).

The assessment of the effects of including genetically modified feed ingredients in fish feeds is in its infancy and has mostly involved examination of growth effects and histopathology (Brown *et al.*, 2003; Glencross *et al.*, 2003; Hemre *et al.*, 2005; Sanden *et al.*, 2005, 2006); hitherto, no major negative effects have been reported. Studies are also being conducted to examine the degree to which dietary DNA is degraded in the gut and to what extent it is taken up and distributed in body tissues; such studies are needed for assessment of the potential for horizontal gene transfer from genetically modified feed ingredients to the animal that consumes them (Sanden *et al.*, 2004; Nielsen *et al.*, 2005, 2006).

1.5 Health Management and Disease Control

Infectious diseases are a major problem in all intensive animal production systems and intensive aquaculture is no exception; farm populations of aquatic animals may be decimated by a disease outbreak, leading to serious losses of production. Outbreaks of disease often occur when rearing conditions and animal welfare are poor, so a suboptimal environment is often a major contributory factor to such outbreaks. As such, health management starts with the creation of a suitable rearing environment and continues with the maintenance of that environment, including the minimizing of potential stressors (Håstein *et al.*, 2005; Keeling, 2005; Midtlyng, 2005; Murray and Peeler, 2005; Adams and Thompson, 2006; Bricknell *et al.*, 2006; Broom, 2006; Dawkins, 2006; Huntingford *et al.*, 2006; Lund, 2006; see Chapter 2, this volume).

The stock should be provided with water of good quality, sufficient space and sufficient food of high nutritional quality and should also be subjected to limited disturbance and physical handling. The difference between health and disease depends on the balance between the pathogen and the host, and that balance can be influenced greatly by environmental factors such as water temperature and water quality (Bondad-Reantaso *et al.*, 2005; Håstein *et al.*, 2005; Murray and Peeler, 2005; Huntingford *et al.*, 2006). Biosecurity has disease prevention and prophylaxis as its key points. It involves implementation of measures for prevention, control and eradication of economically important diseases. In other words, biosecurity encompasses all procedures required to protect the fish from contracting, carrying and spreading diseases and parasites (Bondad-Reantaso *et al.*, 2005; Midtlyng, 2005; Murray and Peeler, 2005).

A major source of disease-causing organisms for aquatic animals is the water flowing into the rearing units, so treatment of incoming water is often used to reduce the levels of waterborne pathogens to below infective limits (see Chapter 2, this volume). Disinfection, which must inactivate a wide range of bacteria, viruses and protozoans, may be accomplished by chemical or physical means, such as chlorination, ozonation or treatment with UV or ionizing radiation. UV radiation treatment is frequently the method of choice for the disinfection of water to be used for aquaculture purposes (Lee, 2003; Sharrer *et al.*, 2005; Hijnen *et al.*, 2006). An alternative, and not mutually exclusive alternative, to disinfection to combat disease-causing organisms involves the use of probiotics. Probiotics may be defined as microorganisms that are administered to the fish, either via the water or in the feed, and they then enter the gastrointestinal tract and serve a function in improving the health of the farmed stock (Gatesoupe, 1999; Irianto and Austin, 2002; Balcázar *et al.*, 2006; Vine *et al.*, 2006). The probiotics may act by stimulating the immune system of the host fish, or they may have direct effects on potential pathogens. The direct effects include interference with the metabolism of the pathogens via the production of inhibitory compounds by the probiotic microorganism and competition for nutrients or other resources (Gatesoupe, 1999; Irianto and Austin, 2002; Balcázar *et al.*, 2006; Vine *et al.*, 2006).

1.5.1 Diagnostics and disease treatment

Veterinary medicine has long traditions in health surveillance and disease diagnosis of terrestrial livestock, but incomplete knowledge about the environmental requirements and biology of farmed aquatic animals along with a relative dearth of information about their diseases has meant that less has been achieved to solve the medical problems faced by aquaculture stocks (Woo *et al.*, 2002; Storey, 2005; Bricknell *et al.*, 2006). Molecular markers are being used for diagnostic and epidemiological purposes in clinical and veterinary medicine (Liu-Stratton *et al.*, 2004; Monis *et al.*, 2005; Adams and Thompson, 2006; Chistiakov *et al.*, 2006; Gasser, 2006) and DNA-based technologies are finding increased use in the production of vaccines and pharmaceutical agents with human and animal health applications (Melamed *et al.*, 2002; Houdebine, 2005; Midtlyng, 2005; Myhr and Dalmo, 2005; Dalal *et al.*, 2006). As the aquaculture industry continues to grow, these areas will probably increase in priority as R & D objectives.

The development of diagnostic methods and vaccines against some of the diseases that attack aquaculture species has been instrumental in preventing disease outbreaks, thereby reducing the large-scale use of antibiotics and other drugs for the treatment of certain diseases. Nevertheless, there are relatively few commercial vaccines currently available and the pharmaceutical toolkit of the aquaculture veterinarian is also limited (Midtlyng, 2005; Sommerset *et al.*, 2005; Storey, 2005; Bricknell *et al.*, 2006; MacMillan *et al.*, 2006).

1.5.1.1 Vaccines

Most vaccines are based on inactivated bacterial pathogens and few vaccines are available for other disease organisms, such as viruses (Midtlyng, 2005; Rigos and Troisi, 2005; Sommerset *et al.*, 2005), although DNA vaccines against some viral agents are under development (Melamed *et al.*, 2002; Midtlyng, 2005; Myhr and Dalmo, 2005; Sommerset *et al.*, 2005; Adams and Thompson, 2006). DNA vaccines are gene constructs that express cloned antigens transiently in the treated individual, thereby promoting the development of protection against the disease agent; R & D into DNA vaccines includes work on viral haemorrhagic septicaemia, infectious haematopoetic necrosis, channel catfish herpes virus, infectious salmon anaemia and infectious pancreatic necrosis.

Vaccines are usually administered by injection or by bath (immersion) treatment. The former is labour-intensive and time-consuming if large numbers of animals are to be treated and bath treatments are often of reduced efficacy compared with injection, so alternative methods of vaccine delivery are being sought and developed (Vandenberg, 2004; Midtlyng, 2005; Adams and Thompson, 2006). Oral administration of vaccine via the feed offers the possibility of treating large numbers of animals with relative ease and avoids the trauma associated with handling and the injection of vaccines, but the efficacy of oral vaccination has, hitherto, proven variable (e.g. Esteve-Gassent *et al.*, 2004; Romalde *et al.*, 2004; Vandenberg, 2004; Shoemaker *et al.*, 2006). Problems with oral vaccination relate to the possible degradation of vaccine components in the digestive tract, requiring the development of microencapsulation, and other methods, for vaccine protection.

1.5.1.2 Antibiotics and other chemotherapeutic agents

At present, the vaccines available target diseases prevalent in a limited number of farmed species, so the use of other disease treatments, such as antibiotics, is widespread in the aquaculture industry (Rigos and Troisi, 2005; Bricknell *et al.*, 2006). Several antimicrobial agents, including tetracyclines, quinolones, sulphonamides and amphenicols, may be efficacious for the treatment of bacterial diseases of aquatic animals, but the range of drugs, pharmaceutical and other chemical agents approved for use in aquatic species is small compared with the number of drugs licensed for use in terrestrial livestock (Storey, 2005; MacMillan *et al.*, 2006).

One major reason for this is that the size of the aquaculture market is currently too small to offer drug companies an adequate return on the economic investment required to collect the documentation needed for approval to be granted by licensing authorities; the requirements may include documentation that the drug treatment is effective against the target disease in each specified aquatic animal, that the drug is safe for the treated animal, does not pose a hazard to humans who may consume the animal and does not give rise to environmental concerns (Midtlyng, 2005; Storey, 2005; MacMillan *et al.*, 2006). These requirements impose difficulties for the development and implementation of effective health management programmes for species that are cultured in low volume or that are under consideration for commercial farming (Storey, 2005; Bricknell *et al.*, 2006). As the aquaculture industry expands, the data needed for the licensing of pharmaceuticals are gradually being collected and are being made available to private companies and licensing authorities (Bondad-Reantaso *et al.*, 2005; Reimschuessel *et al.*, 2005; Rigos and Troisi, 2005; Samuelsen, 2006).

Prophylactic use of vaccines and probiotics, combined with good management practices, reduces the incidence of disease outbreaks (Midtlyng, 2005; Sommerset *et al.*, 2005; Balcázar *et al.*, 2006; Vine *et al.*, 2006), which, in turn, reduces the need to use antimicrobial drugs and chemotherapeutic agents (Murray and Peeler, 2005; Rigos and Troisi, 2005). Reduced reliance on such drug treatments has several benefits. For example, the treatment of water in reuse culture systems usually relies on biological action to remove metabolic wastes and the organisms responsible for the biodegradation of these wastes will be susceptible to some of the agents used to combat disease organisms. In this respect, the use of antibiotics, drugs and other chemical agents that kill water-treatment microorganisms would act as a double-edged sword.

Further, there is public concern regarding the use of drugs and other chemicals such as disinfectants, pesticides and chemotherapeutic agents in aquaculture. The general public is concerned about possible environmental contamination, and the negative effects that could result, and the possibility that development of antibiotic resistance in aquatic microorganisms could be transferred to human pathogens (Bondad-Reantaso *et al.*, 2005; Focardi *et al.*, 2005; Rigos and Troisi, 2005; Adams and Thompson, 2006; MacMillan *et al.*, 2006; Toldrá and Reig, 2006). To help allay these concerns, the use of medicated feeds should be kept to a minimum, every effort should be made to prevent the loss of such feeds to the environment and all drugs, disinfectants and other chemical agents must be used in a responsible manner (Boyd *et al.*, 2005; Focardi *et al.*, 2005; Murray and Peeler, 2005; Rigos and Troisi, 2005; Storey, 2005; MacMillan *et al.*, 2006; Toldrá and Reig, 2006).

1.6 Food Products: Nutritional Composition, Food Safety and Traceability

Aquaculture products intended as food should meet the tastes and expectations of the consumer, but there are many factors that interact to influence the composition of these products. Most of these factors exert their influence prior to harvest, but postharvest processing, transport and storage conditions can also have a major influence. For example, the nature and nutrient composition of the feed are directly influential in determining flesh characteristics, but other agents, such as chemical therapeutic agents, drug residues and waterborne pollutants and contaminants that impart off-flavours and taints to the flesh, may also have direct effects (Fig. 1.4) (e.g. Tucker, 2000; Jobling, 2004a,b,c;

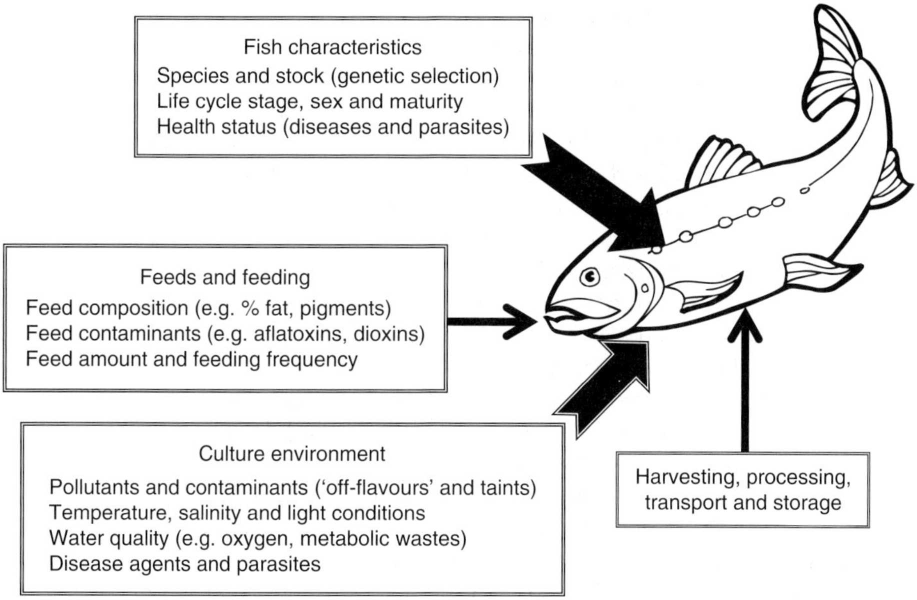

Fig. 1.4. Summary of some of the major factors influencing flesh (fillet) composition of farmed fish. The factors that play a role in determining the composition and 'quality' of the fish that reach the market are the intrinsic (relating to the fish themselves) and extrinsic factors (relating to feeds and feeding and the culture environment) that exert an influence during the growth cycle and the postharvest changes that occur during processing and fish storage.

Bell *et al.*, 2005; Berntssen *et al.*, 2005; Carlson and Hites, 2005; Torstensen *et al.*, 2005; Bethune *et al.*, 2006; Toldrá and Reig, 2006; see Chapters 2 and 3, this volume).

This means that attention must be given to food quality and safety at all links in the food supply chain from the producer to the consumer. The establishment of clear objectives and accreditation procedures is integral to food quality and safety management. Increased control over the safety and quality of food is made possible by adopting agreed practices and implementing one or more management systems, such as good manufacturing practice (GMP), good hygiene practice (GHP) and hazard analysis critical control points (HACCP) (Huss *et al.*, 2004; Sumner *et al.*, 2004; Boyd *et al.*, 2005; Gorris, 2005; da Cruz *et al.*, 2006).

1.6.1 Analysis of aquaculture products

The versatility of NIR spectroscopy, and its advantages of speed and simplicity of instrument operation, has resulted in its adoption as a rapid analytical tool in many areas (Blanco and Villarroya, 2002; Reid *et al.*, 2006), including the seafood and aquaculture industries. The main application of NIR spectroscopy within the seafood and aquaculture sectors has been for the prediction of chemical (nutritional) composition, but more recent applications include species identification and authentication, monitoring of degradation products for assessment of freshness, spoilage and 'quality' and for discrimination between fresh and frozen-thawed fish (Cozzolino *et al.*, 2002, 2005,a,b; Nilsen *et al.*, 2002; Uddin and Okazaki, 2004; Nilsen and Esaiassen, 2005; Uddin *et al.*, 2005; Lin *et al.*, 2006; Reid *et al.*, 2006).

Alternatives to NIR spectroscopy are available for the assessment of spoilage via analysis of volatile compounds (e.g. aldehydes, ketones, organic acids and their esters and ethers). The composition and concentration of volatile compounds differs between fresh, stale and spoiled fish, with spoilage volatiles developing as a result of oxidative or microbial degradation of the chemical constituents of the tissues. Thus, as fish spoilage progresses, there is increased production of basic volatile nitrogen compounds, such as trimethylamine, ammonia and dimethylamine, and it is possible to detect changes in the concentrations of these spoilage compounds remotely (Pacquit *et al.*, 2006). There are also changes in other compounds as fish become spoiled, so there is a change in the odour profile (aromagram) over time and these changes can be used in quality assessment (Olafsdottir *et al.*, 2004; Reid *et al.*, 2006; Plutowska and Wardencki, 2007).

Other methods for identifying the species and/or sources of seafood, including aquaculture, products rely on molecular diagnostic techniques that involve the detection of some form of biomarker. Several methods, such as isoelectric focusing, electrophoresis and immunoassay, rely on the identification of specific proteins or groups of proteins for discrimination between species (Berrini *et al.*, 2006), for identification of potential allergens (Lehrer *et al.*, 2003; Van Do *et al.*, 2005; Kobayashi *et al.*, 2006; Müller and Steinhart, 2007), or for detection of therapeutic agents and drug residues that may be present in the products (Toldrá and Reig, 2006). For example, the major allergens present in seafood are muscle proteins – parvalbumins in fish and tropomysins in crustaceans – and immunoassays have been developed to detect the allergens present in a range of seafood species (Lehrer *et al.*, 2003; Van Do *et al.*, 2005; Kobayashi *et al.*, 2006).

Other approaches to the rapid screening of food products for the assurance of quality and safety are based on the use of DNA microarrays (Liu-Stratton *et al.*, 2004; Kato *et al.*, 2005; Roy and Sen, 2006) and the development of electronic sensors, biosensors or 'biochips' (Toldrá and Reig, 2006; Müller and Steinhart, 2007). A biosensor consists of a chemical recognition element or biological sensor, usually an antibody or enzyme, which is connected to a data processing unit. The target chemicals contact and bind to specific biological sensors, the biomolecular interaction signals are converted into electrical signals and the electrical signals are then processed by a microprocessor. Biochip array biosensors have the advantages that the method is rapid and that food products can be screened for several chemicals simultaneously (as many as the number of specific recognition elements in the array), but initial investment costs for the development of biosensors are quite high.

A variety of other biomarker-based methods has also been developed for discrimination, authentication and quality evaluation purposes (Carbonaro,

2004; Bergé and Barnathan, 2005; Guérard *et al.*, 2005; Kelly *et al.*, 2005; Pinotti *et al.*, 2005; Reid *et al.*, 2006; Renshaw *et al.*, 2006; Toldrá and Reig, 2006; Müller and Steinhart, 2007). For example, techniques, such as PCR-RFLP and PCR-RAPD, which rely on DNA markers, have advantages over protein-based methods as they are less influenced by the tissue analysed, the age of the individual and the effects of thermal degradation. Using DNA-marker techniques, closely related species can be distinguished from each other and these techniques may enable recognition of the species compositions of complex mixtures and processed products (Chistiakov *et al.*, 2006). An additional example is the use of fatty acid profiles for the discrimination of farmed from wild fish; the fatty acids can be used as biomarkers because the fatty acid profiles of the animals generally reflect those of their food and the fatty acids present in aqua-feeds will usually differ from those present in the natural prey of the farmed species (e.g. Jobling, 2004a,b,c; Wonnacott *et al.*, 2004; Bell *et al.*, 2005; Bergé and Barnathan, 2005; Berntssen *et al.*, 2005; Torstensen *et al.*, 2005; Visentainer *et al.*, 2005; see Chapter 3, this volume).

1.7 Fish as Food: Consumer Attitudes (and the Question of GM-Fish)

Price, availability, taste, nutritional value, attitudes and moral considerations all influence consumer food choice and there is increasing interest in food-related issues among the public at large (Shanahan *et al.*, 2001; Butterworth, 2004; Huss *et al.*, 2004; Bennett *et al.*, 2005; Deisingh and Badrie, 2005; Frewer *et al.*, 2005; Gorris, 2005; Logar and Pollock, 2005; Toldrá and Reig, 2006; Müller and Steinhart, 2007; see Chapters 22 and 26, this volume). There is increasing media coverage of food-related topics but, for a variety of reasons, media reporting may be incomplete or inaccurate (Butterworth, 2004; Logar and Pollock, 2005; Bruhn and Earl, 2006). This gives cause for concern, especially as adverse messages or negative press related to food safety can have a major influence on consumers' purchasing decisions; unfavourable messages weigh much more heavily in influencing consumer decisions than do positive or favourable ones.

Recent examples of food-related topics covered in a negative light in newspapers, magazines and other media include 'mad cow disease' (BSE), use of hormones in livestock production, pesticide and other xenobiotic residues in foods, use of food additives (preservatives and colorants), the planting of GM crops and the raising of GM animals for human consumption. Thus, markets for fish and animal food products are frequently challenged with stories relating to 'health scares'. For example, although the nutritional value and potential health benefits of eating fish are regularly stated and increased consumption of fish is advised by nutritionists, many people appear to have a greater awareness of the safety risks that may be linked to fish consumption (Huss *et al.*, 2004; Verbeke *et al.*, 2005). The latter seems to be the result of a greater mass-media focus on pollution issues and contamination of fish than on communicating information about dietary recommendations and the potential health benefits of increased consumption of fish. Thus, fish and farm animal producers need to identify clear, solidly documented ‘health benefit’ messages that can be used to promote the marketing of fish and animal products and offset the negative images created by ‘health scare’ propaganda.

In the mind of the general public, transgenic technology and GMOs seem to be both food (nutritional) and animal welfare issues, although the link between the two may be rather tenuous. Although there is currently no scientific evidence for any hazards unique to GMOs, there are consumer fears about food safety, ecosystem disturbance and degradation and animal welfare (Beardmore and Porter, 2003; Bennett *et al.*, 2005):

- Food safety: is consumption of GMOs safe or are there unknown dangers?
- Environmental impacts: how might escapees influence natural populations of conspecifics and/or other species?
- Welfare and ethics: is it morally defensible to manipulate higher organisms to the extent possible using modern biotechnological techniques?

There are probably many reasons for the lack of clarity around the issue of GMOs, but one result of the confused state of affairs has been market resistance against such products (Tucker, 2003; Aerni, 2004; Kuiper *et al.*, 2004; Deisingh and Badrie, 2005; Lehrer and Bannon, 2005; Logar and Pollock, 2005; Myhr and Dalmo, 2005; Tenbült *et al.*, 2005; Bruhn and Earl, 2006; Dalal *et al.*, 2006; Saher *et al.*, 2006). In general, attitudes towards GMOs and organically grown food (OGF) are negatively related (Bennett *et al.*, 2005; Dreezens *et al.*, 2005; Saher *et al.*, 2006). Strictly speaking, GMOs and OGFs refer to different methods of production rather than to any distinct properties or unique nutritional qualities of the end product. Nevertheless, consumers tend to treat such products as being inherently different from each other and from foods produced by conventional methods. The general public usually associates OGFs with ‘naturalness and pureness’, whereas GMOs are seen as having been altered extensively and manipulated by human hands, and this often results in a negative attitude towards GM products.

Consumer acceptance of a GM product also appears to be influenced by the way the generic product type is perceived; GM is less acceptable when a product type is viewed as being ‘natural and healthy’, such as the case with fruits and vegetables, than when the product is already seen as being ‘manufactured, unnatural and unhealthy’, as with cakes, biscuits and snack-foods (Tenbült *et al.*, 2005). The resistance to GMOs appears to be particularly widespread with regard to farmed fish; this is not surprising, given that fish and fish products would normally be considered as falling within the category of ‘natural and healthy’ foods. In other words, potential consumers often view GM farmed fish as being contrary to the image of an ‘ecologically sound, natural, health food’ (Bennett *et al.*, 2005; Logar and Pollock, 2005).

Given this, there is a need for caution and increased transparency associated with the regulation, licensing and broader introduction of such products, and it is also essential that effective methods be developed for the detection of GM products that enable them to be distinguished readily from non-GM entities (Beardmore and Porter, 2003; Kok and Kuiper, 2003; Kuiper *et al.*, 2004;

Liu-Stratton *et al.*, 2004; Deisingh and Badrie, 2005; Kato *et al.*, 2005; Lehrer and Bannon, 2005; Myhr and Dalmo, 2005; Bruhn and Earl, 2006; Roy and Sen, 2006; Stewart, 2006; Müller and Steinhart, 2007). For example, regulations need to include consideration of the safeties of the inserted DNA and markers introduced to help in the identification of GMOs. There must also be evaluation of the potential toxicity and allergenicity of the gene products resulting from the introduction of novel DNA, and the possibility of horizontal gene transfer also needs to be considered.

1.8 Concluding Remarks

There has been a rapid increase in aquaculture production over the course of the past few decades (http://www.fao.org/figis) and many predict that this expansion will continue for several years to come. If the predictions are to come to fruition, several prerequisites must be fulfilled. Government agencies and public authorities must create conditions that foster advances in aquaculture R & D. They should also promote advisory services designed to serve the industry and they must issue the regulations necessary for the maintenance of the trust and confidence of the public. Research institutions currently have a range of modern techniques available that enable them to provide results in genetics and selective breeding, nutrition and aqua-feed development and management practices that can lead to the production of quality aquaculture products at affordable cost. The industry must be in a position to realize the potential of the research results in a commercial setting and farmers must be receptive to, and apply, new techniques that lead to improvements in the efficiency of their operations.

References

Abbadi, A., Domergue, F., Bauer, J., Napier, J.A., Welti, R., Zähringer, U., Cirpus, P. and Heinz, E. (2004) Biosynthesis of very-long-chain polyunsaturated fatty acids in transgenic oilseeds: constraints on their accumulation. *The Plant Cell* 16, 2734–2748.

Acamovic, T. and Brooker, J.D. (2005) Biochemistry of secondary metabolites and their effects in animals. *Proceedings of the Nutrition Society* 64, 403–412.

Adams, A. and Thompson, K.D. (2006) Biotechnology offers revolution to fish health management. *Trends in Biotechnology* 24, 201–205.

Aerni, P. (2004) Risk, regulation and innovation: the case of aquaculture and transgenic fish. *Aquatic Science* 66, 327–341.

Agaba, M.K., Tocher, D.R., Zheng, X., Dickson, C.A., Dick, J.R. and Teale, A.J. (2005) Cloning and functional characterisation of polyunsaturated fatty acid elongases of marine and freshwater teleost fish. *Comparative Biochemistry and Physiology* 142B, 342–352.

Ali, B.A., Huang, T.-H., Qin, D.-N. and Wang, X.-M. (2004) A review of random amplified polymorphic DNA (RAPD) markers in fish research. *Reviews in Fish Biology and Fisheries* 14, 443–453.

Alimuddin, Yoshizaki, G., Kiron, V., Satoh, S. and Takeuchi, T. (2005) Enhancement of EPA and DHA biosynthesis by over-expression of masu salmon Δ6-desaturase-like gene in zebrafish. *Transgenic Research* 14, 159–165.

Balcázar, J.L., de Blas, I., Ruiz-Zarzuela, I., Cunningham, D., Vendrell, D. and Múzquiz, J.L. (2006) The role of probiotics in aquaculture. *Veterinary Microbiology* 114, 173–186.

Beardmore, J.A. and Porter, J.S. (2003) *Genetically Modified Organisms and Aquaculture*. FAO Fisheries Circular 989, 35 pp.

Bekkevold, D., Hansen, M.M. and Nielsen, E.E. (2006) Genetic impact of gadoid culture on wild fish populations: predictions, lessons from salmonids and possibilities for minimizing adverse effects. *ICES Journal of Marine Science* 63, 198–208.

Bell, J.G., McGhee, F., Dick, J.R. and Tocher, D.R. (2005) Dioxin and dioxin-like polychlorinated biphenyls (PCBs) in Scottish farmed salmon (*Salmo salar*): effects of replacement of dietary marine fish oil with vegetable oil. *Aquaculture* 243, 305–314.

Bennett, B., D'Souza, G., Borisova, T. and Amarasinghe, A. (2005) Willingness to consume genetically modified foods – the case of fish and seafood. *Aquaculture Economics and Management* 9, 331–345.

Bergé, J.-P. and Barnathan, G. (2005) Fatty acids from lipids of marine organisms: molecular biodiversity, roles as biomarkers, biologically active compounds, and economical aspects. *Advances in Biochemical Engineering/Biotechnology* 96, 49–125.

Berntssen, M.H.G., Lundebye, A.-K. and Torstensen, B.E. (2005) Reducing the levels of dioxins and dioxin-like PCBs in farmed Atlantic salmon by substitution of fish oil with vegetable oil in the feed. *Aquaculture Nutrition* 11, 219–231.

Berrini, A., Tepedino, V., Borromeo, V. and Secchi, C. (2006) Identification of freshwater fish commercially labelled 'perch' by isoelectric focusing and two-dimensional electrophoresis. *Food Chemistry* 96, 163–168.

Bessey, C., Devlin, R.H., Liley, N.R. and Biagi, C.A. (2004) Reproductive performance of growth-enhanced transgenic coho salmon. *Transactions of the American Fisheries Society* 133, 1205–1220.

Bethune, C., Seierstad, S.L., Seljeflot, I., Johansen, O., Arnesen, H., Meltzer, H.M., Rosenlund, G., Frøyland, L. and Lundebye, A.-K. (2006) Dietary intake of differently fed salmon: a preliminary study on contaminants. *European Journal of Clinical Investigation* 36, 193–201.

Blanco, M. and Villarroya, I. (2002) NIR spectroscopy: a rapid-response analytical tool. *Trends in Analytical Chemistry* 21, 240–250.

Bondad-Reantaso, M.G., Subasinghe, R.P., Arthur, J.R., Ogawa, K., Chinabut, S., Adlard, R., Tan, Z. and Shariff, M. (2005) Disease and health management in Asian aquaculture. *Veterinary Parasitology* 132, 249–272.

Bourre, J.M. (2005) Where to find omega-3 fatty acids and how feeding animals with diet enriched in omega-3 fatty acids to increase nutritional value of derived products for human: what is actually useful? *Journal of Nutrition Health and Aging* 9, 232–242.

Boyd, C.E., McNevin, A.A., Clay, J. and Johnson, H.M. (2005) Certification issues for some common aquaculture species. *Reviews in Fisheries Science* 13, 231–279.

Bricknell, I.R., Bron, J.E. and Bowden, T.J. (2006) Diseases of gadoid fish in cultivation: a review. *ICES Journal of Marine Science* 63, 253–266.

Broom, D.M. (2006) Behaviour and welfare in relation to pathology. *Applied Animal Behaviour Science* 97, 73–83.

Brown, P.B., Wilson, K.A., Jonker, Y. and Nickson, T.E. (2003) Glyphosate tolerant canola meal is equivalent to the parental line in diets fed to rainbow trout. *Journal of Agricultural and Food Chemistry* 51, 4268–4272.

Bruhn, C. and Earl, R. (2006) Position of the American Dietetic Association: agricultural and food biotechnology. *Journal of the American Dietetic Association* 106, 285–293.

Butterworth, T. (2004) *PCBs in Salmon: Anatomy of a Health Scare*. STATS, Center for Media and Public Affairs, Washington, DC, 16 pp.

Cakmak, G., Togan, I. and Severcan, F. (2006) 17β-Estradiol induced compositional, structural and functional changes in rainbow trout liver, revealed by FT-IR spectroscopy: a comparative study with nonylphenol. *Aquatic Toxicology* 77, 53–63.

Carbonaro, M. (2004) Proteomics: present and future in food quality evaluation. *Trends in Food Science and Technology* 15, 209–216.

Carlson, D.L. and Hites, R.A. (2005) Polychlorinated biphenyls in salmon and salmon feed: global differences and bioaccumulation. *Environmental Science and Technology* 39, 7389–7395.

Chistiakov, D.A., Hellemans, B. and Volckaert, F.A.M. (2006) Microsatellites and their genomic distribution, evolution, function and applications: a review with special reference to fish genetics. *Aquaculture* 255, 1–29.

Cleland, L.G., James, M.J. and Proudman, S.M. (2005) Fish oil: what the prescriber needs to know. *Arthritis Research and Therapy* 8:202, 9 pp.

Cozzolino, D., Murray, I. and Scaife, J.R. (2002) Near infrared reflectance spectroscopy in the prediction of chemical characteristics of minced raw fish. *Aquaculture Nutrition* 8, 1–6.

Cozzolino, D., Chree, A., Scaife, J.R. and Murray, I. (2005a) Usefulness of near-infrared (NIR) spectroscopy and chemometrics to discriminate fishmeal batches made with different fish species. *Journal of Agricultural and Food Chemistry* 53, 4459–4463.

Cozzolino, D., Murray, I., Chree, A. and Scaife, J.R. (2005b) Multivariate determination of free fatty acids and moisture in fish oils by partial least-squares regression and near-infrared spectroscopy. *LWT Food Science and Technology* 38, 821–828.

Cruz, A.G. da, Cenci, S.A. and Maia, M.C.A. (2006) Quality assurance requirements in produce processing. *Trends in Food Science and Technology* 17, 406–411.

Dalal, M., Dani, R.G. and Kumar, P.A. (2006) Current trends in the genetic engineering of vegetable crops. *Scientia Horticulturae* 107, 215–225.

Dawkins, M.S. (2006) A user's guide to animal welfare science. *Trends in Ecology and Evolution* 21, 77–82.

Deisingh, A.K. and Badrie, N. (2005) Detection approaches for genetically modified organisms in foods. *Food Research International* 38, 639–649.

Devlin, R.H. and Nagahama, Y. (2002) Sex determination and sex differentiation in fish; an overview of genetic, physiological and environmental influences. *Aquaculture* 208, 191–364.

Devlin, R.H., D'Andrade, M., Uh, M. and Biagi, C.A. (2004) Population effects of growth hormone transgenic coho salmon depend on food availability and genotype by environment interactions. *Proceedings of the National Academy of Sciences USA* 101, 9303–9308.

Devlin, R.H., Sundström, L.F. and Muir, W.M. (2006) Interface of biotechnology and ecology for environmental risk assessments of transgenic fish. *Trends in Biotechnology* 24, 89–97.

Domergue, F., Abbadi, A. and Heinz, E. (2005) Relief for fish stocks: oceanic fatty acids in transgenic oilseeds. *Trends in Plant Science* 10, 112–116.

Dreezens, E., Martijn, C., Tenbült, P., Kok, G. and Vries, N.K. de (2005) Food and values: an examination of values underlying attitudes toward genetically modified- and organically grown food products. *Appetite* 44, 115–122.

Dunham, R.A. (2004) *Aquaculture and Fisheries Biotechnology: Genetic Approaches*. CAB International, Wallingford, UK.

Easton, M.D.L., Luszniak, D. and Von der Geest, E. (2002) Preliminary examination of contaminant loadings in farmed salmon, wild salmon and commercial salmon feed. *Chemosphere* 46, 1053–1074.

Esteve-Gassent, M.D., Fouz, B., Barrera, R. and Amaro, C. (2004) Efficacy of oral reimmunization after immersion vaccination against *Vibrio vulnificus* in farmed European eels. *Aquaculture* 231, 9–22.

Fjalestad, K.T., Moen, T. and Gomez-Raya, L. (2003) Prospects for genetic technology in salmon breeding programmes. *Aquaculture Research* 34, 397–406.

Flachowsky, G., Chesson, A. and Aulrich, K. (2005) Animal nutrition with feeds from genetically modified plants. *Archives of Animal Nutrition* 59, 1–40.

Focardi, S., Corsi, I. and Franchi, E. (2005) Safety issues and sustainable development of European aquaculture: new tools for environmentally sound aquaculture. *Aquaculture International* 13, 3–17.

Foran, J.A., Good, D.H., Carpenter, D.O., Hamilton, M.C., Knuth, B.A. and Schwager, S.J. (2005) Quantitative analysis of the benefits and risks of consuming farmed and wild salmon. *Journal of Nutrition* 135, 2639–2643.

Francis, G., Makkar, H.P.S. and Becker, K. (2001) Antinutritional factors present in plant-derived alternate fish feed ingredients and their effects in fish. *Aquaculture* 199, 197–227.

Frewer, L.J., Kole, A., van de Kroon, S.M.A. and Lauwere, C. de (2005) Consumer attitudes towards the development of animal-friendly husbandry systems. *Journal of Agricultural and Environmental Ethics* 18, 345–367.

Gasser, R.B. (2006) Molecular tools – advances, opportunities and prospects. *Veterinary Parasitology* 136, 69–89.

Gatesoupe, F.J. (1999) The use of probiotics in aquaculture. *Aquaculture* 180, 147–165.

Gjedrem, T. (ed.) (2005) *Selection and Breeding Programs in Aquaculture*. Springer-Verlag, Berlin.

Glencross, B., Curnow, J., Hawkins, W., Kissil, G.W.M. and Peterson, D. (2003) Evaluation of the feed value of a transgenic strain of the narrow-leaf lupin (*Lupinus angustifolius*) in the diet of the marine fish *Pagrus auratus*. *Aquaculture Nutrition* 9, 197–206.

Goetz, F.W., McCauley, L., Goetz, G.W. and Norberg, B. (2006) Using global genome approaches to address problems in cod mariculture. *ICES Journal of Marine Science* 63, 393–399.

Gorris, L.G.M. (2005) Food safety objective: an integral part of food chain management. *Food Control* 16, 801–809.

Goto-Kazeto, R., Abe, Y., Masia, K., Yamaha, E., Adachi, S. and Yamauchi, K. (2006) Temperature-dependent sex differentiation in goldfish: establishing the temperature-sensitive period and effect of constant and fluctuating water temperatures. *Aquaculture* 254, 617–624.

Graham, I.A., Cirpus, P., Rein, D. and Napier, J.A. (2004) The use of very long chain polyunsaturated fatty acids to ameliorate metabolic syndrome: transgenic plants as an alternative sustainable source to fish oils. *Nutrition Bulletin* 29, 228–233.

Griffin, J.L. (2006) The Cinderella story of metabolic profiling: does metabolomics get to go to the functional genomics ball? *Philosophical Transactions of the Royal Society B* 361, 147–161.

Guérard, F., Sellos, D. and Le Gal, Y. (2005) Fish and shellfish upgrading, traceability. *Advances in Biochemical Engineering/Biotechnology* 96, 127–163.

Guschina, I.A. and Harwood, J.L. (2006) Lipids and lipid metabolism in eukaryotic algae. *Progress in Lipid Research* 45, 160–186.

Hamilton, M.C., Hites, R.A., Schwager, S.J., Foran, J.A., Knuth, B.A. and Carpenter, D.O. (2005) Lipid composition and contaminants in farmed and wild salmon. *Environmental Science and Technology* 39, 8622–8629.

Håstein, T., Scarfe, A.D. and Lund, V.L. (2005) Science-based assessment of welfare: aquatic animals. *Revue Scientifique et Technique – Office International des Epizooties* 24, 529–547.

Hemre, G.-I., Sanden, M., Bakke-McKellep, A.M., Sagstad, A. and Krogdahl, Å. (2005) Growth, feed utilization and health of Atlantic salmon *Salmo salar* L. fed genetically modified compared to non-modified commercial hybrid soybeans. *Aquaculture Nutrition* 11, 157–167.

Hertrampf, J.W. and Piedad-Pascual, F. (2000) *Handbook on Ingredients for Aquaculture Feeds*. Kluwer Academic, Dordrecht, The Netherlands.

Hijnen, W.A.M., Beerendonk, E.F. and Medema, G.J. (2006) Inactivation credit of UV radiation for viruses, bacteria and protozoan (oo)cysts in water: a review. *Water Research* 40, 3–22.
Hites, R.A., Foran, J.A., Carpenter, D.O., Hamilton, M.C., Knuth, B.A. and Schwager, S.J. (2004) Global assessment of organic contaminants in farmed salmon. *Science* 303, 226–229; 305, 475–478.
Houdebine, L.-M. (2005) Use of transgenic animals to improve human health and animal production. *Reproduction of Domestic Animals* 40, 269–281.
Huntingford, F.A., Adams, C., Braithwaite, V.A., Kadri, S., Pottinger, T.G., Sandøe, P. and Turnbull, J.F. (2006) Current issues in fish welfare. *Journal of Fish Biology* 68, 332–372.
Huss, H.H., Ababouch, L. and Gram, L. (2004) *Assessment and Management of Seafood Safety and Quality*. FAO Fisheries Technical Paper 444, 230 pp.
Irianto, A. and Austin, B. (2002) Probiotics in aquaculture. *Journal of Fish Diseases* 25, 633–642.
Jacobs, M.N., Covaci, A. and Schepens, P. (2002) Investigation of selected persistent organic pollutants in farmed Atlantic salmon (*Salmo salar*), salmon aquaculture feed, and fish oil components of the feed. *Environmental Science and Technology* 36, 2797–2805.
Jobling, M. (2004a) Are modifications in tissue fatty acid profiles following a change in diet the result of dilution? Test of a simple dilution model. *Aquaculture* 232, 551–562.
Jobling, M. (2004b) 'Finishing' feeds for carnivorous fish and the fatty acid dilution model. *Aquaculture Research* 35, 706–709.
Jobling, M. (2004c) On-growing to market size. In: Moksness, E., Kjørsvik E. and Olsen, Y. (eds) *Culture of Cold-water Marine Fish*. Blackwell Science, Oxford, UK, pp. 363–432.
Jonsson, B. and Jonsson, N. (2006) Cultured Atlantic salmon in nature: a review of their ecology and interaction with wild fish. *ICES Journal of Marine Science* 63, 1162–1181.
Kagawa, H., Kobayashi, M., Okuzawa, K. and Gen, K. (eds) (2003) Reproductive physiology of fish. *Fish Physiology and Biochemistry* 28, 1–528.
Kato, H., Saito, K. and Kimura, T. (2005) A perspective on DNA microarray technology in food and nutritional science. *Current Opinion in Clinical Nutrition and Metabolic Care* 8, 516–522.
Keeling, L.J. (2005) Healthy and happy: animal welfare as an integral part of sustainable agriculture. *Ambio* 34, 316–319.
Kelly, S., Heaton, K. and Hoogewerff, J. (2005) Tracing the geographical origin of food: the application of multi-element and multi-isotope analysis. *Trends in Food Science and Technology* 16, 555–567.
Kelso, D.D.T. (2004) Genetically engineered salmon, ecological risk and environmental policy. *Bulletin of Marine Science* 74, 509–528.
Kim, S.-K. and Mendis, E. (2006) Bioactive compounds from marine processing byproducts – a review. *Food Research International* 39, 383–393.
Kobayashi, A., Tanaka, H., Hamada, Y., Ishizaki, S., Nagashima, Y. and Shiomi, K. (2006) Comparison of allergenicity and allergens between fish white and dark muscles. *Allergen* 61, 357–363.
Kok, E.J. and Kuiper, H.A. (2003) Comparative safety assessment for biotech crops. *Trends in Biotechnology* 21, 439–444.
Kuiper, H.A., Kleter, G.A., Konig, A., Hammes, W.P. and Knudsen I. (eds) (2004) Safety assessment, detection and traceabilty, and societal aspects of genetically modified foods. *Food and Chemical Toxicology* 42, 1043–1202.
Lee, C.-S. (ed.) (2001) Reproductive biotechnology in finfish aquaculture. *Aquaculture* 197, 1–321.
Lee, C.-S. (ed.) (2003) Management of aquaculture effluents. *Aquaculture* 226, 1–244.
Lehrer, S.B. and Bannon, G.A. (2005) Risks of allergic reactions to biotech proteins in foods: perception and reality. *Allergy* 60, 559–564.

Lehrer, S.B., Ayuso, R. and Reese, G. (2003) Seafood allergy and allergens: a review. *Marine Biotechnology* 5, 339–348.

Lin, M.S., Mousavi, M., Al-Holy, M., Cavinato, A.G. and Rasco, B.A. (2006) Rapid near infrared spectroscopic method for the detection of spoilage in rainbow trout (*Oncorhynchus mykiss*). *Journal of Food Science* 71, S18–S23.

Liu, Z. and Cordes, J.F. (2004) DNA marker technologies and their applications in aquaculture genetics. *Aquaculture* 238, 1–37.

Liu-Stratton, Y., Roy, S. and Sen, C.K. (2004) DNA microarray technology in nutraceutical and food safety. *Toxicology Letters* 150, 29–42.

Logar, N. and Pollock, L.K. (2005) Transgenic fish: is a new policy framework necessary for a new technology? *Environmental Science and Policy* 8, 17–27.

Lund, V. (2006) Natural living – a precondition for animal welfare in organic farming. *Livestock Science* 100, 71–83.

McKevith, B. (2005) Nutritional aspects of oilseeds. *Nutrition Bulletin* 30, 13–26.

Maclean, N. and Laight, R.J. (2000) Transgenic fish: an evaluation of benefits and risks. *Fish and Fisheries* 1, 146–172.

MacMillan, J.R., Schnick, R.A. and Fornshell, G. (2006) Stakeholder position paper: aquaculture producer. *Preventive Veterinary Medicine* 73, 197–202.

MacRae, J., O'Reilly, L. and Morgan, P. (2005) Desirable characteristics of animal products from a human health perspective. *Livestock Production Science* 94, 95–103.

Melamed, P., Gong, Z., Fletcher, G. and Hew, C.L. (2002) The potential impact of modern biotechnology on fish aquaculture. *Aquaculture* 204, 255–269.

Midtlyng, P.J. (ed.) (2005) Progress in fish vaccinology. *Developments in Biologicals* 121, 1–340.

Mignon-Grasteau, S., Boissy, A., Bouix, J., Faure, J.-M., Fisher, A.D., Hinch, G.N., Jensen, P., Le Neindre, P., Mormede, P., Prunet, P., Vandeputte, M. and Beaumont, C. (2005) Genetics of adaptation and domestication in livestock. *Livestock Production Science* 93, 3–14.

Mills, L.J. and Chichester, C. (2005) Review of evidence: are endocrine-disrupting chemicals in the aquatic environment impacting fish populations? *Science of the Total Environment* 343, 1–34.

Miyamoto, J. and Burger, J. (eds) (2003) Implications of endocrine active substances for humans and wildlife; Reports from a SCOPE/IUPAC project. *Pure and Applied Chemistry* 75, 1617–2615.

Monis, P.T., Giglio, S., Keegan, A.R. and Thompson, R.C.A. (2005) Emerging technologies for the detection and genetic characterization of protozoan parasites. *Trends in Parasitology* 21, 340–346.

Montory, M. and Barra, R. (2006) Preliminary data on polybrominated diphenyl ethers (PBDEs) in farmed fish tissues (*Salmo salar*) and fish feed in Southern Chile. *Chemosphere* 63, 1252–1260.

Muir, J. (2005) Managing to harvest? Perspectives on the potential of aquaculture. *Philosophical Transactions of the Royal Society B* 360, 191–218.

Muir, W.M. (2004) The threats and benefits of GM fish. *EMBO Reports* 5, 654–659.

Müller, A. and Steinhart, H. (2007) Recent developments in instrumental analysis for food quality. *Food Chemistry* 101, 1145–1150.

Murray, A.G. and Peeler, E.J. (2005) A framework for understanding the potential for emerging diseases in aquaculture. *Preventive Veterinary Medicine* 67, 223–235.

Myhr, A.I. and Dalmo, R.A. (2005) Introduction of genetic engineering in aquaculture: ecological and ethical implications for science and governance. *Aquaculture* 250, 542–554.

Napier, J.A. and Sayanova, O. (2005) The production of very-long-chain PUFA biosynthesis in transgenic plants: towards a sustainable source of fish oils. *Proceedings of the Nutrition Society* 64, 387–393.

Napier, J.A., Haslam, R., Caleron, M.V., Michaelson, L.V., Beaudoin, F. and Sayanova, O. (2006) Progress towards the production of very long-chain polyunsaturated fatty acid in transgenic plants: plant metabolic engineering comes of age. *Physiologia Plantarum* 126, 398–406.

Naylor, R.L., Goldburg, R.J., Primavera, J., Kautsky, N., Beveridge, M.C.M., Clay, J., Folke, C., Lubchenko, J., Mooney, H. and Troell, M. (2000) Effects of aquaculture on world fish supplies. *Nature* 405, 1017–1024.

Naylor, R., Hindar, K., Fleming, I.A., Goldburg, R., Williams, S., Volpe, J., Whoriskey, F., Eagle, J., Kelso, D. and Mangel, M. (2005) Fugitive salmon: assessing the risks of escaped fish from net-pen aquaculture. *BioScience* 55, 427–437.

Neira, R. and Diaz, N. (eds) (2005) Genetics in aquaculture VIII. *Aquaculture* 247, 1–288.

Nielsen, C.R., Berdal, K.G., Bakke-McKellep, A.M. and Holst-Jensen, A. (2005) Dietary DNA in blood and organs of Atlantic salmon (*Salmo salar* L.). *European Food Research and Technology* 221, 1–8.

Nielsen, C.R., Holst-Jensen, A., Løvseth, A. and Berdal, K.G. (2006) Persistence and distribution of intravenously injected DNA in blood and organs of Atlantic salmon (*Salmo salar* L.). *European Food Research and Technology* 222, 258–265.

Nilsen, H. and Esaiassen, M. (2005) Predicting sensory score of cod (*Gadus morhua*) from visible spectroscopy. *LWT Food Science and Technology* 38, 95–99.

Nilsen, H., Esaiassen, M., Heia, K. and Sigernes, F. (2002) Visible/near-infrared spectroscopy – a new tool for the evaluation of fish freshness? *Journal of Food Science* 67, 1821–1826.

Olafsdottir, G., Nesvadba, P., Di Natale, C., Careche, M., Oehlenschläger, J., Tryggvadóttir, S.V., Schubring, R., Kroeger, M., Heia, K., Esaissen, M., Macagnano, A. and Jørgensen, B.M. (2004) Multisensor for fish quality determination. *Trends in Food Science and Technology* 15, 86–93.

Okutsu, T., Suzuki, K., Takeuchi, Y., Takeuchi, T. and Yoshizaki, G. (2006) Testicular germ cells can colonize sexually undifferentiated embryonic gonad and produce functional eggs in fish. *Proceedings of the National Academy of Sciences of the United States of America* 103, 2725–2729.

Pacquit, A., Lau, K.T., McLaughlin, H., Frisby, J., Quilty, B. and Diamond, D. (2006) Development of a volatile amine sensor for the monitoring of fish spoilage. *Talanta* 69, 515–520.

Patiño, R. (ed.) (2002) The teleost gonad: development, function and histopathology. *Fish Physiology and Biochemistry* 26, 1–101.

Pérez-Marin, D.C., Garrido-Varo, A., Guerrero-Ginel, J.E. and Gómez-Cabrera, A. (2004) Near-infrared reflectance spectroscopy (NIRS) for the mandatory labelling of compound feedingstuffs: chemical composition and open-declaration. *Animal Feed Science and Technology* 116, 333–349.

Piferrer, F., Blázquez, M., Navarro, L. and González, A. (2005) Genetic, endocrine and environmental components of sex determination and differentiation in the European sea bass (*Dicentrarchus labrax* L.). *General and Comparative Endocrinology* 142, 102–110.

Pinotti, L., Moretti, V.M., Baldi, A., Bellagamba, F., Campagnoli, A., Savoini, G., Cantoni, C. and Dell'Orto, V. (2005) Feed authentication as an essential component of food safety and control. *Outlook on Agriculture* 34, 243–248.

Plutowska, B. and Wardencki, W. (2007) Aromagrams – aromatic profiles in the appreciation of food quality. *Food Chemistry* 101, 845–872.

Raz, E. (2003) Primordial germ-cell development: the zebrafish perspective. *Nature Reviews Genetics* 4, 690–700.

Reid, L.M., O'Donnell, C.P. and Downey, G. (2006) Recent technological advances for the determination of food authenticity. *Trends in Food Science and Technology* 17, 344–353.

Reimschuessel, R., Stewart, L., Squibb, E., Hirokawa, K., Brady, T., Brooks, D., Shaikh, B. and Hodsdon, C. (2005) Fish drug analysis – Phish-pharm: a searchable database of pharmacokinetics. *AAPS Journal* 7, E288–E327.

Renshaw, M.A., Saillant, E., Broughton, R.E. and Gold, J.R. (2006) Application of hypervariable genetic markers to forensic identification of 'wild' from hatchery-raised red drum, *Sciaenops ocellatus*. *Forensic Science International* 156, 9–15.

Rigos, G. and Troisi, G.M. (2005) Antibacterial agents in Mediterranean finfish farming: a synopsis of drug pharmacokinetics in important euryhaline fish species and possible environmental implications. *Reviews in Fish Biology and Fisheries* 15, 53–73.

Rischer, H. and Oksman-Caldentey, K.-M. (2006) Unintended effects in genetically modified crops: revealed by metabolomics? *Trends in Biotechnology* 24, 102–104.

Robert, S.S. (2006) Production of eicosapentaenoic and docosahexaenoic acid-containing oils in transgenic land plants for human and aquaculture nutrition. *Marine Biotechnology* 8, 103–109.

Romalde, J.L., Luzarodo-Alvarez, A., Ravelo, C., Toranzo, A.E. and Blanco-Mendez, J. (2004) Oral immunization using alginate microparticles as a useful strategy for booster vaccination against fish lactococcosis. *Aquaculture* 236, 119–129.

Roy, S. and Sen, C.K. (2006) cDNA microarray screening in food safety. *Toxicology* 221, 128–133.

Ruxton, C.H.S., Reed, S.C., Simpson, M.J.A. and Millington, K.J. (2004) The health benefits of omega-3 polyunsaturated fatty acids: a review of the evidence. *Journal of Human Nutrition and Dietetics* 17, 449–459.

Ruxton, C.H.S., Calder, P.C., Reed, S.C. and Simpson, M.J.A. (2005) The impact of long-chain n-3 polyunsaturated fatty acids on human health. *Nutrition Research Reviews* 18, 113–129.

SACN-FSA (Food Standards Agency) (2004) *Advice on Fish Consumption: Benefits and Risks*. The Stationary Office (HMSO), London.

Saher, M., Lindeman, M. and Hursti, U.-K.K. (2006) Attitudes towards genetically modified and organic foods. *Appetite* 46, 324–331.

Salem, M., Kenney, P.B., Killefer, J. and Nath, J. (2004) Isolation and characterization of calpains from rainbow trout muscle and their role in texture development. *Journal of Muscle Foods* 15, 245–255.

Salem, M., Nath, J., Rexroad, C.E., Killefer, J. and Yao, J. (2005a) Identification and molecular characterization of the rainbow trout calpains (Capn1 and Capn2): their expression in muscle wasting during starvation. *Comparative Biochemistry and Physiology* 140B, 63–71.

Salem, M., Yao, J., Rexroad, C.E., Kenney, P.B., Semmens, K., Killefer, J. and Nath, J. (2005b) Characterization of calpastatin gene in fish: its potential role in muscle growth and fillet quality. *Comparative Biochemistry and Physiology* 141B, 488–497.

Salem, M., Kenney, P.B., Rexroad, C.E. and Yao, J. (2006) Molecular characterization of muscle atrophy and proteolysis associated with spawning in rainbow trout. *Comparative Biochemistry and Physiology* 1D, 227–237.

Samuelsen, O.B. (2006) Pharmacokinetics of quinolones in fish: a review. *Aquaculture* 255, 55–75.

Sanden, M., Bruce, I.J., Rahman, M.A. and Hemre, G.-I. (2004) The fate of transgenic sequences present in genetically modified plant proteins in fish feed; investigating the survival of GM soybean DNA fragments during feeding trials in Atlantic salmon, *Salmo salar* L. *Aquaculture* 237, 391–405.

Sanden, M., Berntssen, M.H.G., Krogdahl, Å., Hemre, G.-I. and Bakke-McKellep, A.-M. (2005) An examination of the intestinal tract of Atlantic salmon, *Salmo salar* L., parr fed different varieties of soy and maize. *Journal of Fish Diseases* 28, 317–330.

Sanden, M., Krogdahl, Å., Bakke-McKellep, A.-M., Buddington, R.K. and Hemre, G.-I. (2006) Growth performance and organ development in Atlantic salmon, *Salmo salar* L., parr fed genetically modified (GM) soybean and maize. *Aquaculture Nutrition* 12, 1–14.

Scott, A.P., Katsiadaki, I., Witthames, P.R., Hylland, K., Davies, I.M., McIntosh, A.D. and Thain, J. (2006) Vitellogenin in the blood plasma of male cod (*Gadus morhua*): a sign of oestrogenic endocrine disruption in the open sea? *Marine Environmental Research* 61, 149–170.

Shanahan, J., Scheufele, D. and Lee, E. (2001) Attitudes about agricultural biotechnology and genetically modified organisms. *Public Opinion Quarterly* 65, 267–281.

Sharrer, M.J., Summerfelt, S.T., Bullock, G.L., Gleason, L.E. and Taeuber, J. (2005) Inactivation of bacteria using ultraviolet irradiation in a recirculating salmonid culture system. *Aquacultural Engineering* 33, 135–149.

Shoemaker, C.A., Vandenberg, G.W., Désormeaux, A., Klesius, P.H. and Evans, J.J. (2006) Efficacy of a *Streptococcus iniae* modified bacterin delivered using OralJect™ technology in Nile tilapia (*Oreochromis niloticus*). *Aquaculture* 255, 151–156.

Slanchev, K., Stebler, J., Cueva-Méndez, G. da la and Raz, E. (2005) Development without germ cells: the role of the germ line in zebrafish sex differentiation. *Proceedings of the National Academy of Sciences of the United States of America* 102, 40744079.

Sommerset, I., Krossoy, B., Biering, E. and Frost, P. (2005) Vaccines for fish in aquaculture. *Expert Review of Vaccines* 4, 89–101.

Spolaore, P., Joannis-Cassan, C., Duran, E. and Isambert, A. (2006) Commercial applications of microalgae. *Journal of Bioscience and Bioengineering* 101, 87–96.

Stewart, C.N. Jr (2006) Go with the glow: fluorescent proteins to light transgenic organisms. *Trends in Biotechnology* 24, 155–162.

Storey, S. (2005) Challenges with the development and approval of pharmaceuticals for fish. *AAPS Journal* 7, E335–E343.

Sumner, J., Ross, T. and Ababouch, L. (2004) *Application of Risk Assessment in the Fish Industry*. FAO Fisheries Technical Paper 442, 78 pp.

Sumpter, J.P. and Johnson, A.C. (2005) Lessons from endocrine disruption and their application to other issues concerning trace organics in the aquatic environment. *Environmental Science and Technology* 39, 4321–4332.

Takeuchi, Y., Yoshizaki, G. and Takeuchi, T. (2004) Surrogate broodstock produces salmonids. *Nature* 430, 629–630.

Tenbült, P., Vries, N.K. de, Dreezens, E. and Martijn, C. (2005) Perceived naturalness and acceptance of genetically modified food. *Appetite* 45, 47–50.

Toldrá, F. and Reig, M. (2006) Methods for rapid detection of chemical and veterinary drug residues in animal foods. *Trends in Food Science and Technology* 17, 482–489.

Torstensen, B.E., Bell, J.G., Rosenlund, G., Henderson, R.J., Graff, I.E., Tocher, D.R., Lie, Ø. and Sargent, J.R. (2005) Tailoring of Atlantic salmon (*Salmo salar* L.) flesh lipid composition and sensory quality by replacing fish oil with a vegetable oil blend. *Journal of Agricultural and Food Chemistry* 53, 10166–10178.

Tucker, C.S. (2000) Off-flavor problems in aquaculture. *Reviews in Fisheries Science* 8, 45–88.

Tucker, G. (2003) Nutritional enhancement of plants. *Current Opinion in Biotechnology* 14, 221–225.

Uddin, M. and Okazaki, E. (2004) Classification of fresh and frozen-thawed fish by near-infrared spectroscopy. *Journal of Food Science* 69, C665–C668.

Uddin, M., Okazaki, E., Turza, S., Yumiko, Y., Tanaka, M. and Fukuda, Y. (2005) Non-destructive visible/NIR spectroscopy for differentiation of fresh and frozen-thawed fish. *Journal of Food Science* 70, C506–C510.

Utter, F. and Epifanio, J. (2002) Marine aquaculture: genetic potentialities and pitfalls. *Reviews in Fish Biology and Fisheries* 12, 59–77.

Vandenberg, G.W. (2004) Oral vaccines for finfish: academic theory or commercial reality? *Animal Health Research Reviews* 52, 301–304.

Van Do, T., Elsayed, S., Florvaag, E., Hordvik, I. and Endresen, C. (2005) Allergy to fish parvalbumins: studies on the cross-reactivity of allergens from 9 commonly consumed fish. *Journal of Allergy and Clinical Immunology* 116, 1314–1320.

Verbeke, W., Sioen, I., Pieniak, Z., Van Camp, J. and De Henauw, S. (2005) Consumer perception versus scientific evidence about health benefits and safety risks from fish consumption. *Public Health Nutrition* 8, 422–429.

Viant, M.R., Pincetich, C.A., Hinton, D.E. and Tjeerdema, R.S. (2006) Toxic actions of dinoseb in medaka (*Oryzias latipes*) embryos as determined by in vivo ^{31}P NMR, HPLC-UV and ^{1}H NMR metabolomics. *Aquatic Toxicology* 76, 329–342.

Vine, N.G., Leukes, W.D. and Kaiser, H. (2006) Probiotics in marine larviculture. *FEMS Microbiological Reviews* 30, 404–427.

Visentainer, J.V., Souza, N.E. de, Makoto, M., Hayashi, C. and Franco, M.R.B. (2005) Influence of diets enriched with flaxseed oil on the α-linolenic, eicopentaenoic and docosahexaenoic fatty acid in Nile tilapia (*Oreochromis niloticus*). *Food Chemistry* 90, 557–560.

Volff, J.-N. (2005) Genome evolution and biodiversity in teleost fish. *Heredity* 94, 280–294.

Ward, O.P. and Singh, A. (2005) Omega-3/6 fatty acids: alternative sources of production. *Process Biochemistry* 40, 3627–3652.

Weidinger, G., Stebler, J., Slanchev, K., Dumstrei, K., Wise, C., Lovell-Badge, R., Thisse, C., Thisse, B. and Raz, E. (2003) *dead end*, A novel vertebrate germ plasma component, is required for zebrafish primordial germ cell migration and survival. *Current Biology* 13, 1429–1434.

Welcomme, R. and Craig, J.F. (eds) (2004) Nature and culture: comparative biology and interactions of wild and farmed fish. *Journal of Fish Biology* 65 (Supplement A), 1–342.

Wessel, M.L., Smoker, W.W., Fagen, R.M. and Joyce, J. (2006) Variation in agonistic behavior among juvenile Chinook salmon (*Oncorhynchus tshawytscha*) of hatchery, hybrid and wild origin. *Canadian Journal of Fisheries and Aquatic Sciences* 63, 438–447.

Wonnacott, E.J., Lane, R.L. and Kohler, C.C. (2004) Influence of dietary replacement of menhaden oil with canola oil on fatty acid composition of sunshine bass. *North American Journal of Aquaculture* 66, 243–250.

Woo, P.T.K., Bruno, D.W. and Lim, L.H.S. (eds) (2002) *Diseases and Disorders of Finfish in Cage Culture*. CAB International, Wallingford, UK.

Zheng, X., Seiliez, I., Hastings, N., Tocher, D.R., Panserat, S., Dickson, C.A., Bergot, P. and Teale, A.J. (2004) Characterization and comparison of fatty acyl Δ6 desaturase cDNAs from freshwater and marine teleost fish species. *Comparative Biochemistry and Physiology* 138B, 269–279.

2 Fish Culture: the Rearing Environment

MALCOLM JOBLING

University of Tromsø, Tromsø, Norway

2.1 Introduction

Numerous abiotic and biotic factors combine to make the rearing environment to which intensively farmed fish are exposed; all of these factors influence both the behaviour and culture performance of the fish to a greater or lesser extent (Jobling, 1994, 2004; Wedemeyer, 2001; Beveridge, 2004; Huntingford *et al.*, 2006). As the intensity of cultivation is increased, greater levels of control are exerted over the rearing environment. The factors over which most rigid control is usually exerted include feeds and feeding routines (see Chapter 3, this volume), stocking density, lighting conditions, some water quality parameters, such as dissolved oxygen and total ammonia nitrogen (TAN), and sometimes water temperature and salinity (see Chapter 24, this volume).

The characteristics of aquatic environments, and their impact on animals, differ from those seen on land and these characteristics have forced fish farmers to develop rearing practices that take these differences into account. The development of such practices was not initially stimulated by social awareness or pressures from regulatory authorities, but out of necessity; the need to ensure that the fish remained alive and healthy. The fact that fish are aquatic animals has a major influence on husbandry practices and the farming of fish is linked intimately to water characteristics in terms of quality and quantity (Stickney, 2000; Black, 2001; Wedemeyer, 2001; Beveridge, 2004). For example, the amount of oxygen that dissolves in water is much lower than the amount present in the same volume of air. Further, reduced (subsaturation) oxygen concentrations are not uncommon in water and, at times, respiration may become difficult for farmed fish. Water serves not only as the respiratory medium; its high density means that it supports and buoys up the fish and the water also acts as a diluent for the toxic metabolic waste products excreted by the fish.

(eds N.R. Le François, M. Jobling, C. Carter and P.U. Blier)

In the sections that follow, the main focus will be on the physical environment and the impacts that abiotic environmental factors may have on the survival, growth and physiological performance of intensively farmed fish.

2.2 Water Sources and Their Consequences for Rearing Systems

Fish in culture units may be exposed to various waterborne hazards that are capable of having adverse effects on health and well-being. These may be divided broadly into physical, chemical and biological hazards:

- Physical hazards are particles and solids of varying sizes.
- Chemical hazards include heavy metals, organic compounds and other chemical contaminants.
- Biological hazards encompass pathogenic microorganisms, parasites and microbial toxins.

The risk of the different types of material being present will vary depending on the water source. Waters from a range of sources are used in fish culture. These include precipitation and surface waters from rivers, lakes and canals, and ground or well water. Surface water is the source with the highest risk of contamination from heavy metals, synthetic organic chemicals and microorganisms, but ground or well water may also have hazards associated with its use. Potential hazards that may be present in well water include mineral contaminants such as iron, manganese and lead, nitrogenous compounds and pesticides that may enter the water from agricultural land.

2.2.1 Ponds

Pond culture of fish is quantitatively the most important form of production. Ponds are often extensive production systems, but vary in degree of intensification depending on stocking densities and feed input (Fig. 2.1) (see Boyd, 1995; Stickney, 2000; Muir, 2005). The water sources for ponds may be precipitation and direct surface runoff, surface water from rivers or canals, or ground or well water. Most often, several sources of water are used in combination. Ponds are open systems and this is associated with several disadvantages. For example, ponds are dynamic ecosystems that are under the influence of factors that are uncontrollable and unpredictable, e.g. weather. As such, it is difficult to exert rigorous control over the rearing environment to which the fish are exposed.

The flesh of farmed fish that are reared in open systems, such as shallow freshwater ponds and lakes, may harbour parasites that can cause disease in humans. The fish may also become tainted due to the absorption of various compounds from the water or from prey organisms (Zhou *et al.*, 1999; Tucker, 2000; Watson, 2003; Huss *et al.*, 2004; Kong *et al.*, 2005; Nie *et al.*, 2006).

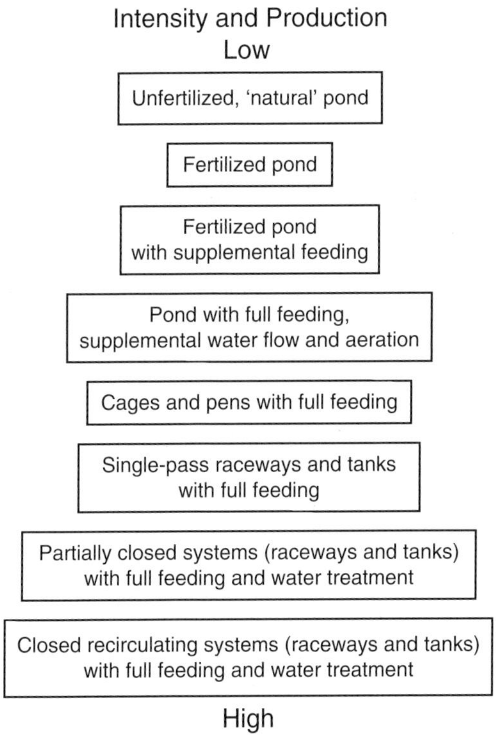

Fig. 2.1. Classification of fish culture systems along an intensity and intervention continuum. Unfertilized, natural ponds represent the culture environments with the least human input and control, whereas closed recirculating systems involve a considerable degree of human intervention and have rearing units with a highly controlled culture environment.

The taints may, for example, be organochlorine compounds, such as pesticides, and other industrial pollutants that enter the ponds in runoff or precipitation, or have accumulated in pond sediments over a period of many years. Many organic pollutants are very persistent in the environment and they also tend to bioaccumulate through the food chain. Once they enter the aquatic environment, these pollutants may accumulate in sediments and/or be taken up and concentrated by aquatic organisms. Petroleum off-flavours can develop in fish that have been exposed to polluted waters; oil spills and effluents from oil refineries can contaminate fish. Off-flavours can also develop if discharge chemicals from pulp mills, sewage works and other industrial activities enter the rearing ponds.

Several other compounds may also impart undesirable odours or off-flavours to pond-raised fish. For example, the fish can acquire earthy or musty taints from algal metabolites, derived from cyanobacteria or eukaryotic algae growing in the ponds. Terpenoids and sulphides are important metabolites produced by cyanobacteria; geosmin and methylisoborneol are two terpenoids that are implicated in the development of off-flavours in farmed fish. The compounds are produced by actinomycetes (filamentous bacteria), cyanobacteria (blue-green algae) and fungi present in pond sediment and in the water column. It is the planktonic cyanobacteria, which under some circumstances may dominate the phytoplankton, that are the major cause of the earthy and musty off-flavours in freshwater fish raised in ponds. The fish may absorb the waterborne off-flavour compounds over the gills or via the gut. Chrysophytes, cryptophytes and dinoflagellates produce several fishy-smelling fatty acid derivatives, whereas aliphatic hydrocarbon metabolites produced by green algae are associated with grassy odours (Watson, 2003). All of these types of compounds may give the farmed fish an undesirable character (Tucker, 2000).

2.2.2 Cages and pens

Cages and pens suffer from several of the same disadvantages as ponds because they are open systems in contact with the surrounding waterbody. Water exchange occurs through the net or mesh-screen walls of the cages. Rates of water exchange are dependent on local water currents or tidal flow. The water currents ensure the renewal of water within the cage and remove metabolic wastes and faecal matter.

Cages are relatively easy to construct and the cage culture of fish may represent a low-input farming practice with a high economic return. However, cages are vulnerable to natural hazards such as strong tides and storms that can result in damage, with the escape of fish as a consequence. Cage sites may also be vulnerable to deterioration in water quality resulting from oil spills, chemical pollution, etc. (see Black, 2001; Wedemeyer, 2001; Beveridge, 2004; Chapter 24, this volume).

One advantage of cages over ponds is that cage systems can make use of any open waterbody suitable for the rearing of the fish; fjords and embayments, lakes and ponds, reservoirs and quarry pits, rivers and streams. An additional advantage of cages over ponds is the ease with which most farm operations can be carried out. For example, compared to pond culture feeding, routine observation and harvesting are simpler when fish are reared in cages. Most of the disadvantages of cage culture relate to the confinement of the fish at relatively high densities. High-density rearing can create localized water quality problems and facilitates the rapid transfer of disease among the caged fish. In addition, the fact that the fish will usually not be able to obtain much nutrition from natural prey means that there is reliance on nutritionally balanced, manufactured feeds. Cages may also act as focal points for predators and poachers because access to the fish is easier than in pond systems.

2.2.3 Land-based systems

Tanks and raceways are almost always land-based systems and are connected to a water supply and drainage system (Fig. 2.2) (see Stickney, 2000; Wedemeyer, 2001). In essence, tanks and raceways are artificial ponds and streams, respec-

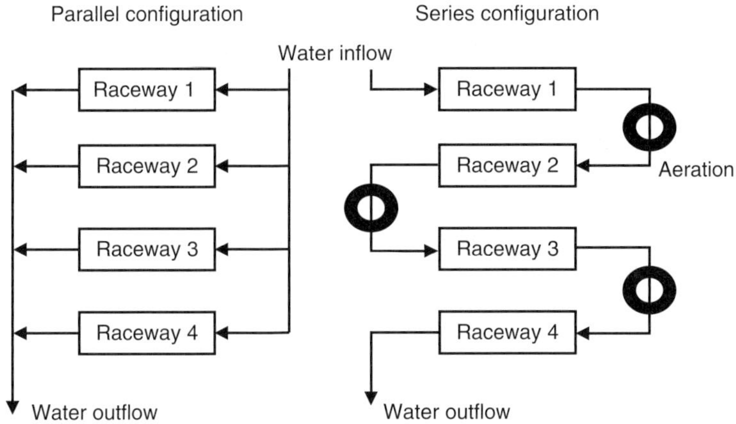

Fig. 2.2. Fish culture raceways showing series and parallel configurations. In series configuration raceway systems, there may be some treatment of the water, for example, aeration, as it flows from one raceway to the next. Water use will often be lower in a series configuration raceway system than in raceways arranged in parallel, but water that has flowed through the raceways may not run directly to waste. The effluent water may be directed to a reservoir, stored for sedimentation of large particles, reconditioned and then reused.

tively, but they lack the complex biotic environment of natural systems. Consequently, it is possible to exert a great degree of control over the environment (e.g. water temperature, flow rates, oxygen concentrations and other water quality parameters) within the rearing units. Tanks and raceways may be components of either flow-through (open) or recirculating (closed) culture systems (see Chapter 24, this volume). Flow-through systems require large quantities of water that are discharged after passing through the rearing units, whereas in a recirculating system, more than 90% of the water may be recycled.

Rearing systems incorporating tanks and raceways have a number of advantages over ponds and cages. These include simplified observation, feeding and harvest, the ability to deliver disease treatments more precisely and the possibility of collecting and treating wastes. The systems are more predictable and manageable than ponds and cages and they can be operated at steady state. They are also more secure than ponds and cages in that the risk of damage with the resulting escape of fish is reduced. Potential disadvantages include the increased risk of mechanical failure, higher energy consumption, rapid rates of disease spread among fish in rearing units and possible losses of fish due to predation and poaching. A major disadvantage with flow-through systems is the requirement for large volumes of water of good quality. Sites that have the quantities of water needed for the establishment of commercial-scale, flow-through production units are limited and are, therefore, at a premium. Both surface and groundwater sources are used to supply such production units.

In both flow-through and recirculating systems, the water is pretreated before it enters the rearing units and the effluent water from a flow-through system will usually be treated prior to discharge. The choice of water treatment will depend on the water source, but most often a combination of treatment methods will be needed to achieve the required water quality. Water pretreatment may include storage to allow the settlement of suspended solids, filtering to remove particles of various sizes, treatment with ozone or ultraviolet (UV) light to kill disease organisms, heating or cooling and aeration or degassing. The effluent leaving the rearing units will also usually be treated prior to discharge. Effluent treatment will often incorporate removal of solids, such as waste feed and faeces, treatments to remove excess dissolved nutrients and metabolic wastes and treatments to destroy potential pathogens. Water treatment usually reaches its extreme in recirculating systems, because these systems must contain water treatment units that allow a portion (often about 90%) of the water leaving the rearing units to be reconditioned and reused. Recirculating systems may be used when there is limited access to on-site water, when conservation of heat is desired or where there are regulations that impose restrictions on effluent discharge to the environment. Thus, recirculating systems usually have modules that remove solid wastes, convert ammonia to nitrite and nitrate, remove carbon dioxide, oxygenate the water, regulate pH and often include some form of disinfection unit. This means that the water in the system is reconditioned by passing it through a series of mechanical and biological filters (see Stickney, 2000; Black, 2001; Wedemeyer, 2001; Lee, 2003; Chapter 24, this volume).

Dissolved oxygen is the first limiting factor for production in such systems, so oxygen supplementation is required as part of the water treatment process. However, there is also the risk that metabolites and other wastes might accumulate to levels that would compromise fish growth and health (Lee, 2003; Colt, 2006; Eding *et al.*, 2006). Consequently, several complementary water treatment processes may be required to reduce the accumulation of wastes and maintain a satisfactory environment in the rearing units (Fig. 2.3). Most recirculating systems contain at least three basic components:

- one or more settling chambers, or sedimentation units
- a biological filter (or biofilter)
- the fish rearing units (tanks or raceways)

Recirculating systems have water treatment systems that control dissolved gases (oxygen, carbon dioxide and nitrogen), feed waste, faeces and other solids, water pH, dissolved nitrogenous compounds (ammonia and nitrite) and pathogens. As such, water reuse systems will generally require several of the following treatment processes:

- sedimentation units, or filters, to remove particulate solids
- biological filters to process dissolved wastes, such as ammonia
- gas strippers and aerators to increase dissolved oxygen, and decrease carbon dioxide and nitrogen, concentrations
- UV or ozone treatment units to disinfect (kill potential pathogens) and oxidize organic wastes and nitrite
- pH controllers that add alkaline materials for maintaining water buffering
- heaters or chilling units for controlling water temperature

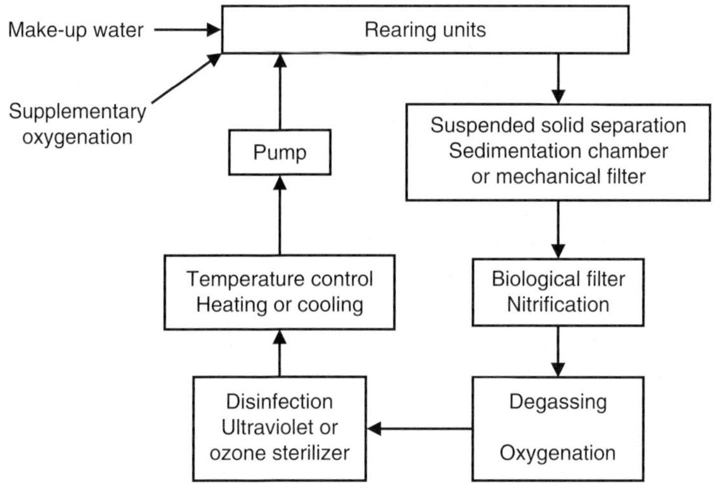

Fig. 2.3. Schematic representation of the components of recirculating and water reuse culture systems. Water conditioning in recirculating systems will usually involve physical, chemical and biological treatments. The amount of water replacement (make-up water) may be very low or can be 10–20%/day, depending on the complexity and efficiency of the water reconditioning units.

In practice, water quality control in recirculating systems is most often achieved through a combination of water treatment and water exchange. The amount of water exchange required depends on the efficiency of the water treatment components. Due to the intensity of production in recirculating systems, the ability to control the rearing environment is a prerequisite. The design of a recirculating system will, however, reflect not only the environmental requirements of the cultured species but also criteria related to subunit function and a striving to reduce operational costs, which are comparatively high with such systems (Lee, 2003; Colt, 2006; Eding *et al*., 2006; see Chapter 24, this volume).

2.3 Biosecurity: a Key for Disease Prevention

Fish that are held in captivity must be resistant to disease and they must also tolerate the rigours of handling and transport without difficulty. Maintaining the health of cultivated fish requires good management practices. The obvious goals of fish health management are to limit the frequency of disease outbreaks and reduce the severity of losses when such outbreaks occur. Health management starts with the creation of a suitable rearing environment and continues with the maintenance of that environment, including the minimizing of potential stressors. To promote health and minimize stress, the fish should be provided with water of good quality, sufficient space and sufficient food of high nutritional quality. The fish should also be subjected to limited disturbance and physical handling and be protected from predators.

Once these criteria are met, it is also important to keep the fish isolated from potential sources of infection. As such, health or disease reflects interactions between the fish (the host), potential pathogens and the environment (Wedemeyer, 2001; Woo *et al.*, 2002; Vadstein *et al.*, 2004; Bondad-Reantaso *et al.*, 2005; Murray and Peeler, 2005). In order for there to be a disease outbreak in a population of fish in culture, a number of conditions must be present. First, the disease organism, or pathogen, must be present in the environment (in the water source or infected fish) in sufficient numbers to impose a threat, the fish host must be susceptible to the pathogen and there must be a viable infection route. Thus, the difference between health and disease depends on the balance between the pathogen and the host and that balance can be influenced greatly by environmental factors such as water temperature and water quality.

Biosecurity has disease prevention and prophylaxis as its key points. Since a major source of disease-causing organisms for fish is the water flowing into the rearing units, semi-closed or recirculating systems offer significant advantages over open systems such as ponds and cages. With a closed or semi-closed system, it is possible to treat the incoming water to reduce the risk of pathogens entering the system. Further, the rearing units can be constructed of materials that can be disinfected easily should a disease outbreak occur. The units can also be constructed such that they prevent the entry of fouling organisms, potential predators and other animals and hinder the escape of the farmed fish.

Treatment of incoming water is used to reduce the levels of waterborne pathogens to below infective limits. In addition, disinfection of wastewater may be mandatory to meet regulations relating to water quality standards. Disinfection, which must inactivate a wide range of bacteria, viruses and protozoans, may be accomplished by chemical or physical means, such as chlorination, ozonation or treatment with UV or ionizing radiation. There are several disadvantages with chemical disinfectants and UV radiation treatment is frequently the method of choice for the disinfection of water entering fish culture units. UV radiation treatment is also used for disinfection of wastewater (Wedemeyer, 2001; Summerfelt, 2003; Sharrer *et al.*, 2005; Hijnen *et al.*, 2006). Microorganisms are inactivated by UV light as a result of photochemical damage to nucleic acids. UV radiation is neither transmitted through nor does it penetrate particulate matter. This means that suspended particles present in the water may shield the microorganisms from the UV radiation, thereby increasing microbial survival (Sharrer *et al.*, 2005; Hijnen *et al.*, 2006). As a consequence, UV has far greater efficacy as a disinfection treatment when applied to filtered water rather than 'raw' water (Wedemeyer, 2001). Although the UV doses used for disinfection of water in fish culture facilities are not considered to pose a risk to the fish, the potential for the production of toxic by-products should not be overlooked (Björnsson, 2004).

2.4 Water Quality

Specifying water quality characteristics that represent a good culture environment is a complicated undertaking. Water quality factors include concentrations of dissolved oxygen, concentrations of metabolic waste products (ammonia and

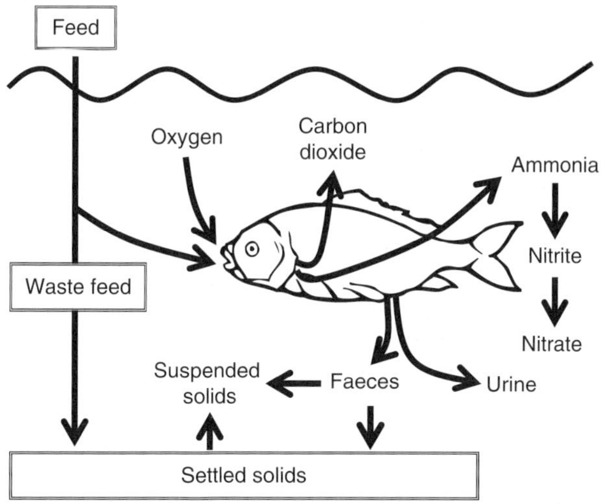

Fig. 2.4. Illustration of some of the major interactions between a farmed fish and its culture environment. Waste feed arises from overfeeding or other poor feed management practices. Faecal wastes arise from undigested, unabsorbed food; the amount and composition of the faeces will vary with feeding level (food supply) and feed formulation. Carbon dioxide, ammonia and urinary wastes result from the metabolic activities of the fish.

carbon dioxide) (Fig. 2.4), water pH and the presence of toxicants, such as heavy metals or organic pollutants. Not only do the requirements for dissolved gas concentrations, ionic concentrations, pH, etc. vary with species and life history stage, the different water quality factors may interact to influence the physiological and behavioural responses of the fish (Stickney, 2000; Wedemeyer, 2001; Jensen, 2003; Kroupova *et al.*, 2005; Colt, 2006; Eding *et al.*, 2006; Eshchar *et al.*, 2006). For example, concentrations of heavy metals, such as zinc and copper, that cause substantial damage to the gills of fish held in 'soft', acidic, fresh water may exert few toxic effects when fish are reared in 'hard', alkaline water that has a high concentration of calcium carbonate. Recommendations relating to the water quality requirements of fish are often based on the results of toxicity studies. These have the disadvantage that they are generally short-term studies that examine the survival of juvenile fish. Further, they are usually conducted as single-factor experiments with the test animals being exposed to constant concentrations of the toxicant of interest. In reality, aquatic animals will face the challenge of multifactor and/or temporally variable exposure, which raises a question about whether many of the currently used aquaculture water quality criteria are completely appropriate (Colt, 2006).

2.4.1 Salinity

Most fish can be categorized as being either freshwater or marine (Fig. 2.5), but within each category there are wide species differences in salinity tolerances. Many marine species and numerous freshwater species, such as many cyprinids,

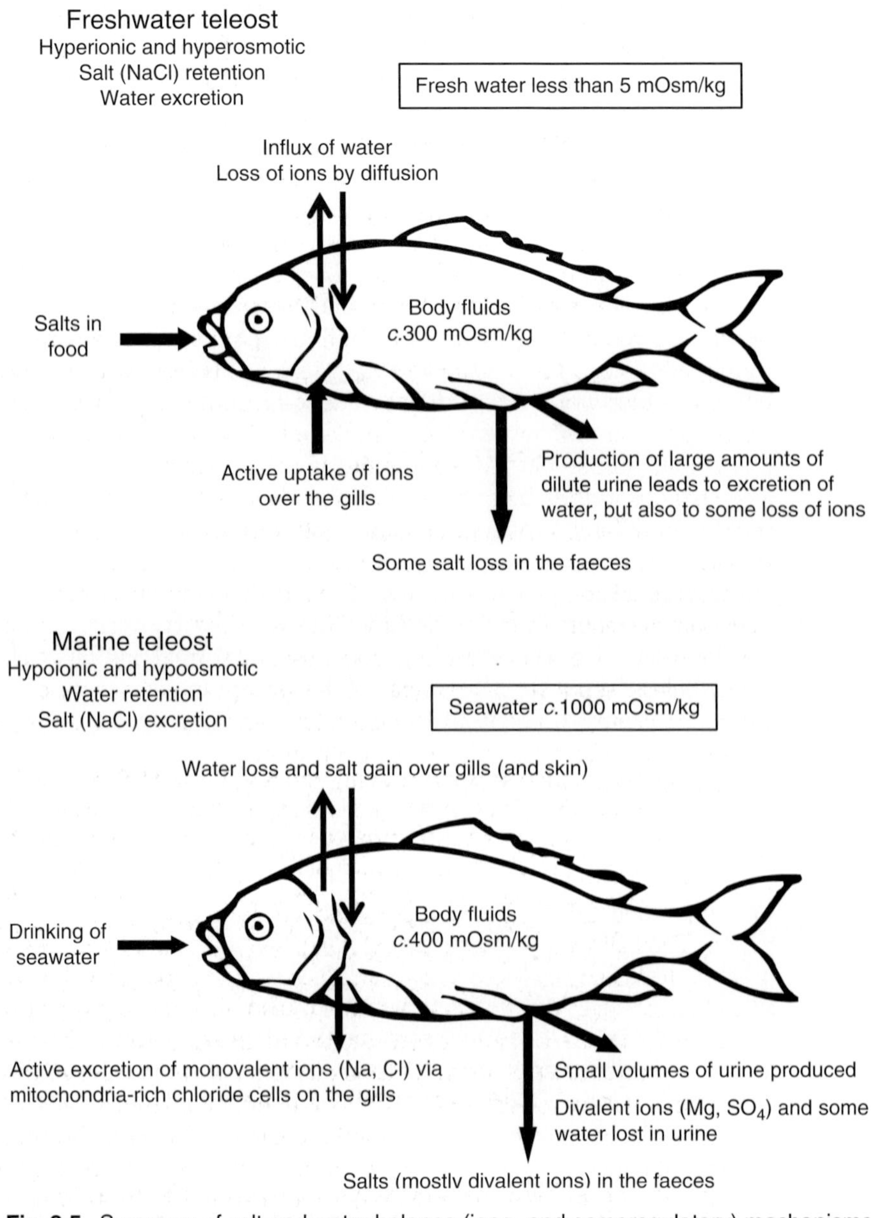

Fig. 2.5. Summary of salt and water balance (iono- and osmoregulatory) mechanisms in freshwater and marine teleost fish.

are only able to tolerate exposure to water within a limited range of salinities. These are termed stenohaline fish species. Other species often found in fresh water, some tilapias, for example, are less stringent in their salinity requirements. These freshwater species will tolerate exposure to waters that differ widely in ionic concentrations. Species with these characteristics are euryhaline species. There are also many euryhaline species that occur most often in seawater. Euryhaline species may be particularly suitable for farming, since it may be possible to raise them in fresh water, brackish water and in the sea.

Euryhalinity is defined as having the ability to tolerate and adapt to a wide range of salinities. In considering euryhalinity, it may be pertinent to draw a distinction between those species which can tolerate large short-term fluctuations in external salinity and those which migrate between freshwater and marine environments at different stages of the life cycle. Fish species showing this type of migration may be termed amphihaline.

The amphihaline fish group includes several salmonid species, many of which undergo a distinct transformation (the parr–smolt transformation) prior to migration from fresh water to the marine habitat (see Chapter 12, this volume). In other words, the fish may change from being relatively stenohaline freshwater fish (the juvenile parr) intolerant of seawater to being marine fish (the smolt, post-smolt and adult) capable of survival and growth in seawater. The changes in the ability of amphihaline species to remain in salt and water balance in different environments are both distinct and long term. An abrupt transfer from low to high salinity, or vice versa, at an inappropriate time of the life cycle could compromise survival. The fact that amphihaline species differ in their salinity requirements during the course of the life cycle complicates the development of effective rearing and husbandry methods for such species. Nevertheless, some representatives of this group, the Atlantic salmon, *Salmo salar*, for example, are farmed successfully and are among the species that have proven to be most suitable for cultivation.

The influence of salinity on growth of a number of marine and freshwater fish species with potential for farming has been examined (Boeuf and Payan, 2001; Jobling, 2004). Several of the marine species, European seabass, *Dicentrarchus labrax*, gilthead seabream, *Sparus aurata*, turbot, *Scophthalmus maximus*, and Atlantic cod, *Gadus morhua*, for example, are relatively euryhaline. They either may grow equally well at all salinities within the range from brackish to full-strength seawater or may grow best in water with salinity lower than that of full-strength seawater. The reason for the improved growth of marine fish in water of reduced salinity is a matter of debate. The osmotic and ionic concentrations of the body fluids of marine fish differ from those of seawater, so the fish are required to expend a certain amount of energy to meet the metabolic costs of ionic and osmotic regulation. There is, however, lack of agreement concerning the magnitude of these costs (Boeuf and Payan, 2001). It is, therefore, unclear whether or not the rearing of fish in an isionic, isosmotic environment would lead to significant reductions in metabolic costs. Nevertheless, an increase in growth has been observed when fish of some marine species have been raised in an isosmotic environment. As such, there may be benefits to be gained by rearing these species in water with salinity lower than that of full-strength seawater.

The relatively wide salinity tolerance of some fish species may be used to advantage in the treatment of parasite infestation. Freshwater and saltwater baths are sometimes used for treating outbreaks of ectoparasites (Woo *et al.*, 2002). The fish tolerate the abrupt salinity change, whereas the parasitic organisms do not. In addition, since some parasites and disease organisms are particularly prevalent or virulent over only a limited salinity range, the rearing of euryhaline fish in water of different salinity reduces the risk of infection and disease. For example, the parasitic dinoflagellate, *Amyloodinium ocellatus*, is found on a wide range of host species and can cause mortality in warmwater fish culture. *A. ocellatus* is most prevalent on fish held in the sea or in brackish water. Thus, by rearing euryhaline species, such as red drum, *Sciaenops ocellatus*, in water of low salinity, this disease problem can be reduced or avoided (Stickney, 2000; Wedemeyer, 2001).

2.4.2 Oxygen

The first limiting water quality parameter in fish culture will often be dissolved oxygen. Oxygen is soluble in water, but solubility declines with increasing temperature and increasing salinity. In other words, the solubility of oxygen is much lower in warm seawater than in cool fresh water. Atmospheric air contains 200–210 ml O_2/l, whereas fresh water at 15°C that is fully saturated with oxygen will contain only *c.*7 ml O_2/l (10.10 mg O_2/l). The amount of dissolved oxygen decreases by about 20% in seawater in comparison with fresh water. These differences in oxygen availability in the atmosphere and dissolved in water have a number of consequences. About 30 times as much oxygen is present in the atmosphere as in fresh water at 15°C, so a fish must ventilate its gills with a large volume of water to obtain a given amount of oxygen. Water is a dense, viscous medium with a density *c.*800 times that of air, so a fish must move about 24,000 (30 × 800) times the mass of water across its gills as the mass of air an air-breathing animal must move across its respiratory surfaces. At higher temperatures and salinity, even more water relative to air must be moved; at 30°C, the mass of seawater needed to be moved across the gills of a fish would be about 38,500 times the mass of air passed over the respiratory surface of an air-breathing animal.

Combinations of low oxygen and high carbon dioxide concentrations in the water constitute a major challenge to gas exchange in fish, especially at high temperature when metabolic rates are high. These conditions may be encountered in fishponds and also in intensive culture where both stocking densities and feeding rates are high. Supplementary aeration or oxygenation is often applied in an attempt to prevent the development of hypoxic conditions in rearing units (Stickney, 2000; Wedemeyer, 2001), but such conditions do arise occasionally and result in reduced performance. Sometimes, hypoxia is so severe that there is mass mortality of fish and other aquatic animals in the affected waterbody. Besides the mortality that is caused directly by exposure of fish to severe acute hypoxia, there may also be impacts that result from chronic exposure to sublethal levels of hypoxia. In other words, dissolved oxygen concentrations higher than those that compromise fish survival may have a major impact on performance. When fish are exposed to hypoxic conditions, they reduce feeding and this gives reduced rates of growth. Rates of feed intake and growth may be sustained close to maximal when fish are held in water that is slightly hypoxic, but the feed intake of several fish species begins to fall dramatically once oxygen saturation falls below 65–75% (Jobling, 1994, 2004). This means that an attempt should be made to ensure that dissolved oxygen concentrations remain close to saturation; under such conditions, fish feed well and display good rates of growth.

Several other effects of chronic exposure to low dissolved oxygen concentrations relate to disturbances to the reproductive development of adult fish and to effects on developing eggs and embryos. In adult fish, hypoxia may disrupt the balance of sex steroid hormones and delay egg development and maturation in females. There may also be hormonal disturbances in males that result in a reduction in sperm quality. The end result of these hypoxia-induced disturbances may be a reduction in the fertilization percentage of the eggs and a reduced percentage of fertilized eggs that survive to hatch and produce normal larvae (Wu *et al.*, 2003). Exposure of developing eggs to chronic hypoxia may give rise to abnormal embryos, for example, deformation of vertebrae and other skeletal tissues and malformations of internal organs, and can also influence sexual differentiation (Shang and Wu, 2004; Shang *et al.*, 2006).

For example, exposure of developing eggs and embryos to hypoxic water may influence the production and activities of several key enzymes required for the synthesis of sex steroid hormones. There may, for example, be downregulation of genes involved in the production of steroid-synthesizing enzymes, such as P-450 aromatase, and this can have major long-term consequences via influences on sex ratios and rates of sexual differentiation and gamete development in maturing individuals. Groups of fish exposed to hypoxia may produce higher proportions of males than those reared in normoxic water. This might be expected if the action of P-450 aromatase, with the production of oestrogens, is required for fish to develop as females. In maturing females exposed to hypoxia, the blood sex steroid profiles (ratios of androgens to oestrogens) may be disturbed, indicative of a suppression of the production of oestrogens from the androgen (testosterone) precursor. One result of this may be a delay in the rate of oocyte development. Testicular development, and the rate of progress of spermatogenesis, of males exposed to hypoxia may also be delayed in comparison with males held under normoxic conditions (Wu *et al.*, 2003; Shang *et al.*, 2006).

2.4.3 Nitrogenous products: ammonia, nitrite and nitrate

Most fish excrete the majority of their waste nitrogen over the gills as ammonia (NH_3) and the ammonia is diluted rapidly to non-toxic concentrations in the surrounding water. Ammonia is also ionized to the ammonium ion (NH_4^+). The proportions of total ammonia nitrogen (TAN) present as ammonia and the ammonium ion depend on pH, temperature and, to a lesser extent, salinity (Jobling, 1994; Wedemeyer, 2001; Colt, 2006; Eding *et al.*, 2006; Eshchar *et al.*, 2006). Proportions of ammonia, the toxic form, increase with increasing pH and increasing water temperature. A pH increase from 7 to 8 results in a tenfold increase in ammonia concentration, and a temperature increase of

5°C will give a 40–50% increase in ammonia. When ammonia concentrations in the water are high, as may occur in reuse and recirculating fish culture systems, the diffusion gradient for ammonia from the fish blood to the water is reduced. This leads to a slowing of the rate of outward diffusion of ammonia over the gills. Under such conditions, blood ammonia concentrations may start to rise.

Chronic, or long-term (for days or weeks), exposure of fish to ammonia concentrations far below those that are lethal, or highly toxic, in the short-term (over a few hours or days) may have marked effects on performance. There are, however, species differences in sensitivity. Salmonids seem to be highly sensitive to dissolved ammonia, whereas the turbot, *S. maximus*, and wolffish, *Anarhichas* spp., appear to be more resistant. Safety standards for un-ionized ammonia are not well established; a concentration of 0.0125 mg/l has sometimes been adopted as representing a safe level, although recommendations both lower and higher than this have been proposed (Stickney, 2000; Wedemeyer, 2001; Björnsson and Ólafsdóttir, 2006; Colt, 2006; Eding *et al.*, 2006). For example, in some marine species, growth suppression may not occur until concentrations of un-ionized ammonia exceed 0.06 mg/l (Atlantic cod, *G. morhua*, Dover sole, *Solea solea*), 0.1–0.15 mg/l (turbot, *S. maximus*, spotted wolffish, *A. minor*) or higher (gilthead seabream, *S. aurata*), but the growth-suppressive effects of ammonia are influenced by interactions with other water quality parameters, such as dissolved oxygen and carbon dioxide, pH, salinity and temperature (Björnsson and Ólafsdóttir, 2006; Eshchar *et al.*, 2006). Some control over concentrations of ammonia may be exerted by adjusting water flow rates through rearing units (water turnover), limiting nitrogen inputs in feeds, regulation of carbon dioxide concentration and maintaining control over fish stocking densities. Such adjustments may be employed in combination with pH regulation and ammonia removal.

In aquatic systems, the removal of ammonia from the water occurs mostly via bacterial-mediated oxidation of ammonia, first to nitrite and then to nitrate. Under most circumstances, oxidation of ammonia will proceed to nitrate but in fish culture systems, nitrite concentrations may become elevated. This occurs when rates of oxidation of ammonia to nitrite exceed the rates of oxidation of nitrite to nitrate. Increases in nitrite concentrations are of concern because nitrite is toxic to fish and negative effects have been observed following long-term exposure to concentrations as low as 0.1 mg/l (Stickney, 2000; Wedemeyer, 2001; Jensen, 2003; Kroupova *et al.*, 2005). Nitrate is also toxic to aquatic animals, but at much higher concentrations than either ammonia or nitrite. Nevertheless, when aquatic animals are exposed to high concentrations of nitrate, there may be perturbation of a number of physiological systems (Guillette and Edwards, 2005; Hamlin, 2006).

The oxidation of ammonia to nitrate (nitrification) involves two major groups of bacteria, *Nitrosomas* spp., which oxidize ammonia to nitrite, and *Nitrobacter* spp., which oxidize nitrite to nitrate. Ammonia (NH_3) inhibits the action of nitrobacters at much lower concentrations than those at which it inhibits nitrosomonads. This means that when levels of ammonia are relatively high, there may be incomplete nitrification. In other words, the ammonia is oxidized to nitrite, but the oxidation of the nitrite to nitrate is inhibited. Under these circumstances, there is the risk that nitrite could accumulate and reach toxic concentrations.

The chances of nitrite toxicity are greater for freshwater fish than for marine species. This relates both to the risk of exposure to high nitrite concentrations and to differences in the mechanisms of ion exchange between freshwater and marine fish (Fig. 2.5). Freshwater fish are hyperionic and hyperosmotic to their environment. They require an active uptake of ions over the gills to compensate for the ions lost in the urine and via passive diffusion. Some nitrite may enter over the gills via diffusion, but the major problem for freshwater fish is that nitrite has an affinity for branchial ion exchangers and can replace chloride as the ion exchanged for bicarbonate (Fig. 2.6). This means that when nitrite is present in the water, some will be taken up in exchange for bicarbonate at the expense of chloride. Freshwater fish species differ in their susceptibilities to nitrite toxicity and these differences may be related to differences in the mechanisms of ion exchange employed, particularly the rates at which chloride is normally transported across the gill epithelium (Tomasso and Grosell,

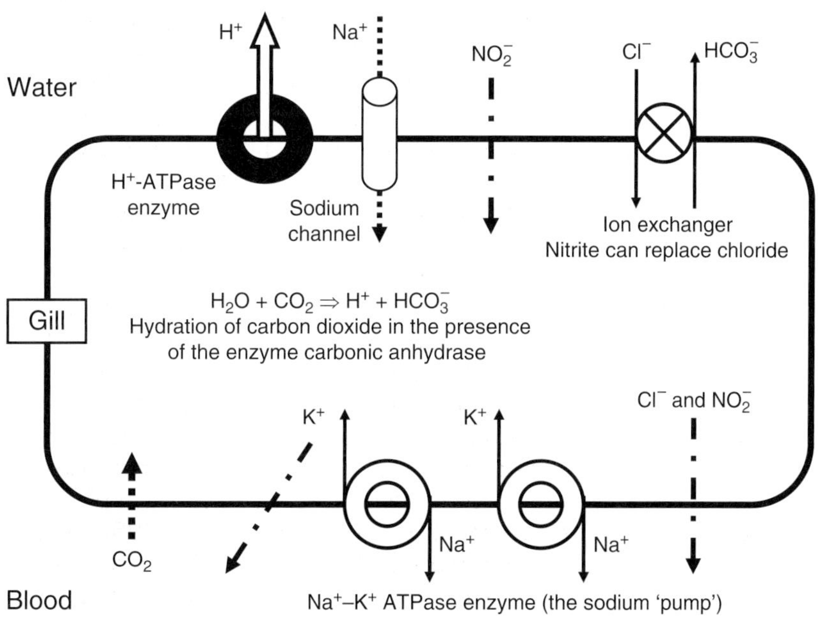

Fig. 2.6. Ion exchange over the gill of a freshwater teleost fish, illustrating the uptake of nitrite (NO_2^-). Nitrite can pass over the gill either via diffusion down its concentration gradient, or may be taken up actively by substituting for the chloride ion (Cl^-) on the chloride–bicarbonate ion exchanger. Absorption of sodium (Na^+) is aided by the excretion of hydrogen ions (H^+) from the cells of the gill membrane to the water via H^+ ATPase enzyme activity on cell apical membranes. The active transport of ions is driven by Na^+ K^+ ATPase enzyme activity (the sodium 'pump') on the basal membrane of cells.

2005). Species with high rates of chloride transport are usually less resistant to nitrite than those that have lower rates of gill chloride transport; the former tend to concentrate nitrite in the blood, whereas the latter do not (Tomasso and Grosell, 2005). The competition between chloride and nitrite for transport via the ion exchangers explains why an elevation in the concentration of chloride in the water can protect the fish against nitrite toxicity; the addition of chloride to the water is the simplest method for protecting freshwater fish against nitrite contamination.

Nitrite is generally less toxic to marine fish than to freshwater fish. This relates to the high concentration of chloride ions in seawater and to the opposite ionic gradients operating in fish that live in seawater and in fresh water (Fig. 2.5). There may be some inward diffusion of nitrite over the gills due to concentration differences, but the majority of the nitrite that enters the blood of marine fish is probably taken up from the intestine. The uptake of nitrite from the intestine of marine fish involves both passive (diffusion) and active components. The active uptake of nitrite occurs via the mechanism involved in the absorption of water from the intestine of marine fish; solute-linked water transport (Fig. 2.7).

The entry of nitrite across the gills of freshwater teleosts, and the intestine and gills of marine fish, results in an increase in plasma nitrite concentration. Nitrite can then enter the red blood cells (RBCs) from the plasma. In the RBCs,

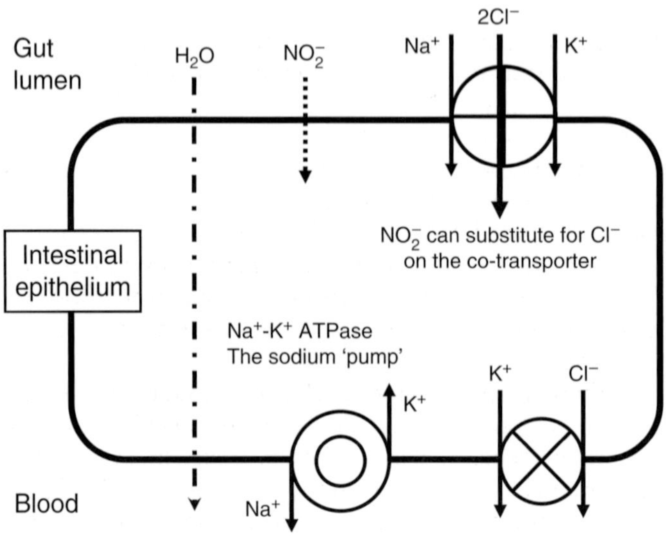

Fig. 2.7. Nitrite uptake from the intestine of marine teleosts may occur via diffusion down its concentration gradient or as part of the active mechanism involved in the absorption of water (solute-linked water transport). When there is active uptake of nitrite ions (NO_2^-), they substitute for chloride ions (Cl^-) on the sodium–potassium–chloride co-transporter. The active uptake of ions over the intestinal epithelium is driven by Na^+ K^+ ATPase enzyme activity (the sodium 'pump') on the basal membrane of the cells of the intestinal epithelium.

the nitrite oxidizes the iron in the haemoglobin (Hb with Fe^{2+}), resulting in the formation of methaemoglobin (metHb with Fe^{3+}) (Jensen, 2003; Kroupova *et al.*, 2005; Tomasso and Grosell, 2005). MetHb does not have the ability to bind oxygen reversibly, so an increase in metHb can lead to a decrease in the oxygen-carrying capacity of the blood. One sign of high metHb content is that the blood takes on a brownish, rather than red, colour. Nitrite toxicity is related to the formation of metHb and the resulting reduced oxygen-carrying capacity of the blood (Wedemeyer, 2001; Jensen, 2003; Kroupova *et al.*, 2005). This means that factors that affect either oxygen availability or the fish's oxygen demand change the levels of nitrite observed to be toxic. Thus, nitrite is more toxic when fish are held under hypoxic conditions (when oxygen concentrations are low) and nitrite toxicity is also a greater risk for well-fed, active fish that have high metabolic rates than it is for fish with lower oxygen demands.

Nitrate, the end product of the nitrification cascade, will not normally reach toxic concentrations in either natural environments or in flow-through fish rearing units that use water renewal to ensure that high water quality is maintained. On the other hand, nitrate concentrations could rise to high levels in intensive recirculating systems with limited water exchange. In this sort of system, the potential for toxic, or physiologically disruptive, effects is present (Colt, 2006; Hamlin, 2006).

2.5 Temperature

Most fish are poikilothermic ectotherms, so temperature has a marked influence on all aspects of their physiology and behaviour. Fish species differ in their thermal tolerances and temperature preferences. These tolerances and preferences are sometimes used to provide a broad classification of cultivated species. Some species, termed eurythermal, tolerate a broad range of temperatures, but others, the stenothermal species, have a much lower range of thermal tolerance (Jobling, 1994, 1997). Stenothermal species may have a preference for either warm or cold water. Warmwater species perform well in water at 25°C and above, whereas coldwater species do not thrive in waters over 20°C. Coolwater species occupy an intermediate position and are often relatively eurythermal.

Although useful, such a classification is crude. It does not take into account the fact that thermal requirements vary with life history stage and the performance criteria being considered. For example, cod, *G. morhua*, hatchlings suffer heavy mortality when temperatures exceed 16–17°C, but larger cod may tolerate temperatures up to 24–25°C. Thermal preferences also vary during ontogeny, with the preferred temperature of small, immature cod (*c.*6.5 cm) being a few degrees higher than that of large adults (Lafrance *et al.*, 2005). Such differences in thermal preferenda may be reflected in the temperatures at which feeding and growth of different life stages are at their respective maxima (Jobling, 1994, 1997, 2004).

The juveniles (under-yearlings and yearlings) of several temperate zone freshwater species (e.g. cyprinids, percids and centrarchids) appear to tolerate a wider range of temperatures than do adults of the same species. In addition, the thermal limits for feeding and growth are usually narrower than those for short-term survival (acute temperature tolerances) and the thermal requirements for reproduction are narrower still. To complicate matters further, the temperatures at which gametogenesis and spawning are possible, and at which there is successful egg and larval development, may differ from those that promote rapid rates of growth during the major part of the production cycle, i.e. thermal optima of the various physiological processes may differ (Jobling, 1994, 1997).

Fish typically respond to temperature changes by showing alterations in feed intake, growth rate and the efficiency with which they utilize the ingested feed. The rate–temperature relationships are described by asymmetric functions in which performance is maximized at an intermediate temperature, the optimum temperature. There are also lower and upper critical thermal limits that define the minimum and maximum temperatures at which given activities can be carried out (Fig. 2.8). This means that rate–temperature relationships are typically 'bell-shaped' curves. The rates at which the processes occur initially increase with increasing temperature, peak at an intermediate temperature and then decline (Fig. 2.8). The peaks, denoting the thermal optima, occur at different temperatures depending on the response being measured. When fish are fed to satiation under different thermal conditions, the temperature at which feed intake is highest is typically a few degrees above the thermal optimum for growth, and this in turn is higher than the temperature at which the fish utilize

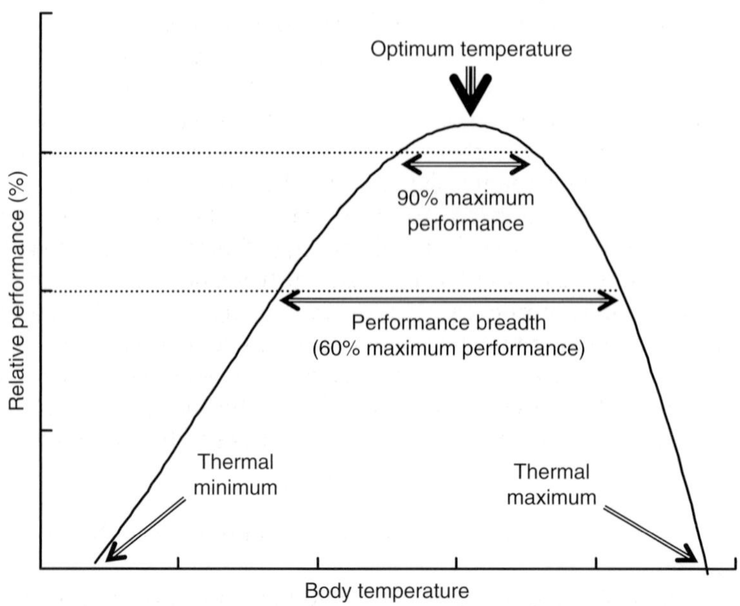

Fig. 2.8. Schematic representation of a rate–temperature curve indicating temperatures that provide key indicators of performance: optimum temperature, thermal minimum and maximum and performance breadth criteria.

feed most efficiently for growth. Within the context of fish culture operations, it will be of interest to have knowledge about thermal optima and the breadth of the performance curves; the latter define the temperature ranges over which performance is equal to or greater than a particular level, e.g. the 60% or 90% performance breadth (Fig. 2.8). Thermal optima and performance breadths for feeding and growth obviously differ from species to species, i.e. depending on whether the fish is a warmwater or a coldwater species and is stenothermal or eurythermal. Further, thermal performance criteria will usually differ within a species when fish of different sizes are considered (Jobling, 2004).

2.5.1 Temperature, reproduction and development

Temperature has a profound influence on several aspects of reproduction, from the hormonal regulation of gonad growth to the timing of spawning and the relative timing of organogenesis in developing embryos. Unfavourable temperatures experienced by broodstock may lead to disorders in gonad development, oocyte atresia and degeneration, inhibition of ovulation or a delay in the timing of spawning (Van Der Kraak and Pankhurst, 1997; Pankhurst and Porter, 2003; van der Meeren and Ivannikov, 2006). Further, an unfavourable temperature experienced during one phase of the reproductive cycle, for example during vitellogenesis, may not be revealed until a later stage, via effects on egg size, fertilization success or survival of the eggs and embryos to hatch. Thermal conditions experienced by broodstock may influence egg production and postfertilization development, but incubation temperatures experienced by developing eggs and embryos have more profound effects on the differentiation of organ systems and the timing of organogenesis.

Temperature is one of the most potent factors influencing rates of development and the overall survival of eggs and larvae (Blaxter, 1988; Kamler, 1992; Rombough, 1997). Rates of development are generally found to increase with increasing temperatures (Fig. 2.9). Incubation of eggs at high temperature may, however, result in increased incidence of malformations and abnormalities in the offspring, or the death of the eggs (Takle *et al.*, 2004, 2005, 2006). Temperature can also influence size at hatching, efficiency of yolk utilization, time to metamorphosis, behaviour, feeding rate and metabolic demands. Developmental rate is also dependent on egg size (Fig. 2.9), with the time to hatch at any given temperature being almost an order of magnitude longer for large eggs than small (Rombough, 1997). The interspecific variability in the time for eggs of marine and freshwater fish species to hatch can be summarized as:

$$\log D = 1.20 - 0.0494T + 0.203d$$

where D is incubation time in days, T is temperature (°C) and d is egg diameter in mm.

In addition to influencing the development rates of eggs, embryos and larvae, temperature may affect growth–differentiation interactions. It may have profound influences on many developmental events, such as muscle differentiation and the relative timing of organogenesis (Blaxter, 1988; Johnston *et al.*,

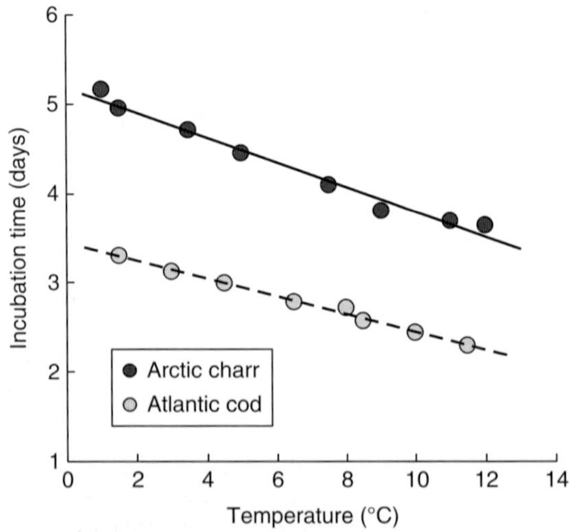

Fig. 2.9. Incubation times (days from fertilization to hatch) of Atlantic cod, *Gadus morhua*, and Arctic charr, *Salvelinus alpinus*, eggs incubated at different temperatures. Note the logarithmic scale on the *y*-axis. Both the Atlantic cod and Arctic charr are coldwater species, but the cod spawns small (1.1–1.9 mm diameter) pelagic eggs and the charr has large (4–4.5 mm) demersal eggs.

1996). In other words, developmental events can become uncoupled from each other when development occurs at different temperatures. This results in a potential for multiple phenotypes dependent on the temperature conditions experienced during early development. Unfavourable temperatures during early rearing may give rise to osteological malformations and other abnormalities in the young fish. Opercular malformations are commonly observed in juveniles that hatch from eggs incubated at unfavourable, particularly high, temperatures; such fish may be more predisposed to gill infections than normal individuals. Growth rates of malformed fish may also be reduced. Abnormal development of the head region, particularly the jaw apparatus, may interfere with feeding and have consequences for growth. Deformities of the spinal column, such as lordosis, may result in abnormal swimming and a reduced ability to capture food. The mechanisms causing these deformities remain unclear, but exposure to unfavourable temperatures can lead to the inhibition or downregulation of some genes and the upregulation of others, including those involved in the production of heat-stress proteins (Takle *et al.*, 2004, 2005, 2006).

Temperature experienced during early rearing is known to influence sex determination in several fish species (Baroiller *et al.*, 1999; Baroiller and D'Cotta, 2001; Devlin and Nagahama, 2002; Strüssmann and Nakamura, 2002; Goto-Kazeto *et al.*, 2006). This effect of temperature on sex determination has been demonstrated clearly by showing that temperature can induce a phenotypic sex change within gynogenetic monosex populations of fish (Fig. 2.10) (Baroiller *et al.*, 1999; Baroiller and D'Cotta, 2001; Devlin and Nagahama,

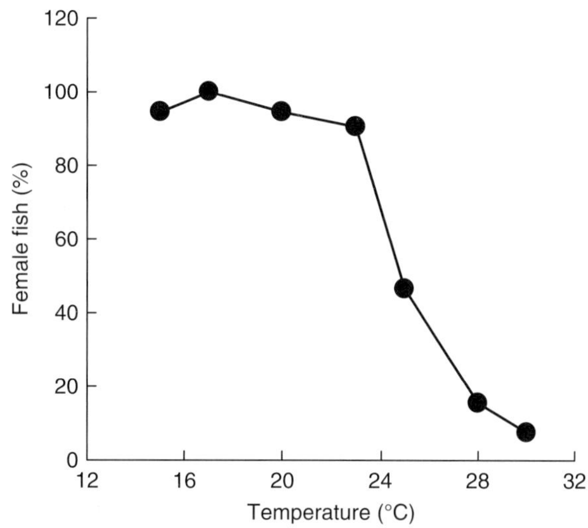

Fig. 2.10. Percentages of female goldfish, *Carassius auratus*, resulting from rearing genotypic female fish at different constant temperatures (15–30°C) during early development (from day 12 after fertilization until an age of 3 months) (data from Goto-Kazeto *et al*., 2006).

2002; Strüssmann and Nakamura, 2002; Goto-Kazeto *et al*., 2006). The physiological effects of temperature on sex determination appear to be mediated via actions on genes coding for P-450 steroidogenic enzymes, such as aromatase, and sex steroid hormone receptors (Baroiller *et al*., 1999; Baroiller and D'Cotta, 2001; Devlin and Nagahama, 2002; Strüssmann and Nakamura, 2002). Thus, the stages of development at which temperature may exert effects on the determination of phenotypic sex in fish coincide with those at which phenotypic sex may be manipulated using sex steroid hormone treatments (Piferrer, 2001; Devlin and Nagahama, 2002; Strüssmann and Nakamura, 2002). To what extent such effects may be considered beneficial or detrimental is likely to differ between species, depending on which of the two sexes is considered most desirable in culture, i.e. which of the sexes grows most rapidly and matures latest and at the largest size.

2.6 Light

There are three characteristics of light that may influence the performance of fish in culture:

- light 'quality'; the spectral characteristics of the light with respect to wavelength
- light 'quantity'; illuminance or light intensity
- light 'duration'; photoperiod, or the ratio of hours of light to hours of darkness during a 24-h period

Solar radiation is made up of photons within the visible spectrum (from violet at 400 nm to red at 700 nm), together with infrared (IR) and ultraviolet (UV). When light passes through water, its intensity decreases and the loss of intensity varies with colour, i.e. different wavelengths are absorbed to different degrees. This means that both intensity and spectral characteristics of the light change with depth. Light at wavelengths within the IR, red, violet and UV ranges is absorbed more strongly than is light of intermediate wavelengths (Bullock, 1982, 1988; Utne-Palm, 2002). Depth of penetration of light of different wavelengths is affected markedly by the presence of absorbing and scattering substances, such as particulate inorganic and organic matter and dissolved organic matter, in the water column. Both light intensity and light spectral characteristics affect the visibility of underwater objects.

The light intensity at the water surface on a bright summer day may be *c.*100,000 lux. Extensive cloud cover can reduce natural illuminance by more than 90%, so on an overcast day the light intensity at the surface may be 5000–10,000 lux. The illumination from bright sunlight may give a light intensity that is sufficiently high to induce an avoidance response in the fish, so it may be beneficial to provide fish held in shallow outdoor tanks with some shading to reduce their exposure to excessive light. Indoors, for example in a hatchery unit, artificial lighting will create light intensities that are lower than those experienced out of doors. When light is provided from cool-white fluorescent bulbs situated 50–100 cm above the water surface, intensities may exceed 1000 lux, but it is more usual that light intensities are within the range of tens to a few hundred lux. This is not problematic because most fish species are able to locate food under low light conditions. The turbidity of the water will also have an influence on the ease with which the fish are able to detect food in the water column (Utne-Palm, 2002). Turbidity is influenced by phytoplankton concentrations and the amounts of suspended particulate matter, such as silt and clay, in the water. As a generalization, it may be said that 'reaction distance' is reduced as turbidity increases, making prey detection more difficult in turbid water. Nevertheless, some planktivorous fish, including larvae and small juveniles of several species, appear to feed more effectively in slightly turbid water than they do in clear water (Utne-Palm, 2002), so raising these fish in water with a low level of turbidity may have some advantages.

The wavelengths of light to which fish are most sensitive vary with species. The visual pigments of surface-dwelling, pelagic and coastal zone marine fish often have peak sensitivities at around 490–510 nm, and several freshwater species have visual pigments with peak sensitivities within the 500–550 nm range. Some species of fish also have UV-sensitive pigments, which seem to be used in prey detection (Utne-Palm, 2002). Consequently, care should be taken to ensure that the spectral characteristics of artificial hatchery lighting are adequate. Some artificial light sources emit considerable amounts of IR and UV, whereas others are deficient in these wavelengths. Excessive IR and UV can be damaging to fish and other aquatic organisms. Excessive IR poses a risk of overheating, whereas excessive UV can induce a variety of damaging effects on aquatic organisms.

Exposure to UV can lead to alterations to the structure of proteins and lipids and can induce damage to nucleic acids, with physiological disturbance or death as the end result (Häder and Sinha, 2005). On the other hand, aquatic organisms have developed a number of mechanisms to counteract the potentially damaging effects of UV. These mechanisms include photo-protection offered by accumulation of pigments, such as melanin and carotenoids, and a range of molecular repair mechanisms. For example, DNA damage can be repaired through excision repair pathways or by photo-enzymatic repair mechanisms (Häder and Sinha, 2005). Nevertheless, 'sunburn-like' effects have been observed in fish held in shallow, unshaded outdoor tanks exposed to strong sunlight and there may also be increased mortality under such circumstances (Bullock, 1982, 1988; Zagarese and Williamson, 2001; Woo *et al.*, 2002). Small larvae are especially susceptible to damage from exposure to excessive UV (with wavelengths 280–320 nm).

Photoperiod (the ratio of the hours of light to hours of darkness in 24 h) influences several aspects of the performance of fish in culture (Boeuf and Le Bail, 1999; Boeuf and Falcón, 2001; Bromage *et al.*, 2001). Increased growth is observed when fish of some species are exposed to 'long days' (i.e. continuous or extended periods of light). This effect has been observed most frequently in salmonids, but has also been recorded for several other species, including channel catfish, *Ictalurus punctatus*, and marine flatfishes (turbot, *S. maximus*, and halibut, *Hippoglossus hippoglossus*) (Jobling, 1994, 2004; Boeuf and Le Bail, 1999). Extended photoperiods may promote feeding and growth via stimulation of the hypothalamic–pituitary axis, leading to increased production and secretion of growth hormone, thereby inducing an anabolic physiological status in the fish. The manipulation of photoperiod may also be used to influence the timing of parr–smolt transformation in salmonids. For example, certain combinations of thermal and photoperiod manipulations (photothermal manipulations) can be used to produce 0+, out-of-season Atlantic salmon, *S. salar*, smolt that may be transferred to seawater a few months after the fish hatch (Boeuf and Le Bail, 1999; see Chapter 12, this volume). Photothermal manipulation may also be used to influence the reproductive cycle, to accelerate or delay the timing of oocyte growth and development (for the production of 'out-of-season' eggs) (van der Meeren and Ivannikov, 2006). Finally, exposure of fish to continuous or extended periods of light has been used to delay the timing of sexual maturation to ensure that the fish reach market size before they mature (Bromage *et al.*, 2001; Pankhurst and Porter, 2003).

2.7 Fish Welfare Under Culture Conditions

Animal welfare is an issue that the fish farming industry cannot afford to ignore. There is increased public concern about the treatment of animals used for meat production. Consumer perceptions of animal welfare and environmental impact associated with animal production are important because they may influence product choices (Frewer *et al.*, 2005; see Chapter 26, this volume). There is a need for transparency and provision of accurate information about the rearing of animals because trust is an important component of public acceptance, or approval, of animal husbandry practices. Distrust results if information is perceived as coming from sources with vested interests, or from sources that are unreliable or unaccountable. In addition, there may be public scepticism about, and resistance to, the introduction of new technological innovations and rearing practices, even though there may not be any good reasons for taking such a position (Frewer *et al.*, 2005).

Welfare of farm animals, including fish, encompasses everything from general husbandry and disease prevention, rearing environment, feeds and feeding methods, to transport and slaughter methods (Frewer *et al.*, 2005; Håstein *et al.*, 2005; Keeling, 2005; Murray and Peeler, 2005; Broom, 2006; Dawkins, 2006; Huntingford *et al.*, 2006; Lund, 2006; Spinka, 2006). As such, promotion of good animal welfare requires that the animals do not suffer from hunger or malnutrition, injury or chronic disease, or repeated exposure to events that result in fear and distress. The most disruptive events experienced by farmed stock are linked to exposure to humans and to sudden changes in their biotic (e.g. social) and physical environments. Sudden, intense or prolonged changes in holding conditions can impact animal welfare and damage the health, productivity and product quality of the stock, and the profitability of the farming enterprise. For example, frequent exposure to conditions which induce fear reactions, emergency behaviours and pronounced escape responses not only impose metabolic costs, but can also induce undesired stress responses or give rise to injury, with increased incidence of infection and disease as a result. This means that the performance of 'natural' emergency reactions may be detrimental, rather than having a positive influence on the welfare of farmed animals (Spinka, 2006).

Discussions about what constitutes good welfare are more concerned with values than facts, so the concept of good welfare will mean different things to different people. For example, farmers usually equate animal welfare with animal health and the absence of pain and suffering, so would argue that good welfare is characterized by animals that are healthy, are growing well and show a high level of production efficiency. On the other hand, members of the general public often express concern about rearing conditions that they perceive as being unnatural in some way, without being able to provide clear distinctions between animal husbandry practices that they consider natural, unnatural, acceptable and unacceptable (Frewer *et al.*, 2005; Dawkins, 2006). Some might claim that good welfare criteria are met if the animal is able to function in a physiologically normal manner and is able to cope satisfactorily with the rearing situation. Alternative criteria defining good welfare may require that the animals are able to carry out their normal behaviour patterns without undue constraint and that they are capable of living a 'natural life' (Frewer *et al.*, 2005; Broom, 2006; Dawkins, 2006; Huntingford *et al.*, 2006; Lund, 2006; Spinka, 2006). Although there is overlap between the different standpoints, they are not completely compatible. For example, the fact that an animal has the opportunity to perform natural behaviours need not imply that welfare is optimal, and the performance of some types of natural behaviour may be detrimental to welfare. Detrimental forms of natural behaviour may include emergency or escape reactions and damaging behaviours arising from competition and aggression within social groups. As an example, extensively reared, pond-grown fish that compete for food, suffer from parasite infestation and that are subject to attack by predators might fulfil most of the criteria relating to them having a 'natural life', but these would not be considered as good welfare conditions if the criteria include a requirement for a high level of health, growth and production efficiency and avoidance of pain and suffering.

The culture environment, comprising both biotic and abiotic factors, is of critical importance in determining the health, well-being and performance of farmed fish. Interactions between environmental factors clearly have an impact on the physiology and behaviour of fish, but the results of these interactions may be difficult to predict. It is, however, clear that it is not permissible to use information about the minimum requirements that permit fish survival as criteria for the determination of the type of culture environment to which the fish should be exposed if the aim is to ensure high standards of welfare and production within a population of farmed fish.

References

Baroiller, J.-F. and D'Cotta, H. (2001) Environment and sex determination in farmed fish. *Comparative Biochemistry and Physiology* 130C, 399–409.

Baroiller, J.-F., Guigen, Y. and Fostier, A. (1999) Endocrine and environmental aspects of sex differentiation in fish. *Cellular and Molecular Life Sciences* 55, 910–931.

Beveridge, M.C.M. (2004) *Cage Aquaculture*. Blackwell Science, Oxford, UK.

Björnsson, B. (2004) Can UV-treated seawater cause cataract in juvenile cod (*Gadus morhua* L.)? *Aquaculture* 240, 187–199.

Björnsson, B. and Ólafsdóttir, S.R. (2006) Effects of water quality and stocking density on growth performance of juvenile cod (*Gadus morhua* L.). *ICES Journal of Marine Science* 63, 326–334.

Black, K.D. (ed.) (2001) *Environmental Impacts of Aquaculture*. Sheffield Academic Press, Sheffield, UK.

Blaxter, J.H.S. (1988) Pattern and variety in development. In: Hoar, W.S. and Randall, D.J. (eds) *Fish Physiology,* Volume XIA. Academic Press, London, pp. 1–58.

Boeuf, G. and Falcón, J. (2001) Photoperiod and growth in fish. *Vie Milieu* 51, 247–266.

Boeuf, G. and Le Bail, P.-Y. (1999) Does light have an influence on fish growth? *Aquaculture* 177, 129–152.

Boeuf, G. and Payan, P. (2001) How should salinity influence fish growth? *Comparative Biochemistry and Physiology* 130C, 411–423.

Bondad-Reantaso, M.G., Subasinghe, R.P., Arthur, J.R., Ogawa, K., Chinabut, S., Adlard, R., Tan, Z. and Shariff, M. (2005) Disease and health management in Asian aquaculture. *Veterinary Parasitology* 132, 249–272.

Boyd, C.E. (1995) *Bottom Soils, Sediment and Pond Aquaculture*. Chapman and Hall, New York.

Bromage, N., Porter, M. and Randall, C. (2001) The environmental regulation of maturation in farmed finfish with special reference to the role of photoperiod and melatonin. *Aquaculture* 197, 63–98.

Broom, D.M. (2006) Behaviour and welfare in relation to pathology. *Applied Animal Behaviour Science* 97, 73–83.

Bullock, A.M. (1982) The pathological effects of ultraviolet radiation on the epidermis of teleost fish with reference to the solar radiation effect in higher animals. *Proceedings of the Royal Society of Edinburgh* 81B, 199–210.
Bullock, A.M. (1988) Solar ultraviolet radiation: a potential environmental hazard in the cultivation of farmed finfish. *Recent Advances in Aquaculture* 3, 139–224.
Colt, J. (2006) Water quality requirements for reuse systems. *Aquacultural Engineering* 34, 143–156.
Dawkins, M.S. (2006) A user's guide to animal welfare science. *Trends in Ecology and Evolution* 21, 77–82.
Devlin, R.H. and Nagahama, Y. (2002) Sex determination and sex differentiation in fish; an overview of genetic, physiological and environmental influences. *Aquaculture* 208, 191–364.
Eding, E.H., Kamstra, A., Verreth, J.A.J., Huisman, E.A. and Klapwijk, A. (2006) Design and operation of nitrifying trickling filters in recirculating aquaculture: a review. *Aquacultural Engineering* 34, 234–260.
Eshchar, M., Lahav, O., Mozes, N., Peduel, A. and Ron, B. (2006) Intensive fish culture at high ammonium and low pH. *Aquaculture* 255, 301–313.
Frewer, L.J., Kole, A., Kroon, S.M.A. van de and Lauwere, C. de (2005) Consumer attitudes towards the development of animal-friendly husbandry systems. *Journal of Agricultural and Environmental Ethics* 18, 345–367.
Goto-Kazeto, R., Abe, Y., Masia, K., Yamaha, E., Adachi, S. and Yamauchi, K. (2006) Temperature-dependent sex differentiation in goldfish: establishing the temperature-sensitive period and effect of constant and fluctuating water temperatures. *Aquaculture* 254, 617–624.
Guillette, L.J. Jr and Edwards, T.M. (2005) Is nitrate an ecologically relevant endocrine disruptor in vertebrates? *Integrative and Comparative Biology* 45, 19–27.
Häder, D.-P. and Sinha, R.P. (2005) Solar ultraviolet radiation-induced DNA damage in aquatic organisms: potential environmental impact. *Mutation Research* 571, 221–233.
Hamlin, H.J. (2006) Nitrate toxicity in Siberian sturgeon (*Acipenser baeri*). *Aquaculture* 253, 688–693.
Håstein, T., Scarfe, A.D. and Lund, V.L. (2005) Science-based assessment of welfare: aquatic animals. *Revue Scientifique et Technique – Office International des Epizooties* 24, 529–547.
Hijnen, W.A.M., Beerendonk, E.F. and Medema, G.J. (2006) Inactivation credit of UV radiation for viruses, bacteria and protozoan (oo)cysts in water: a review. *Water Research* 40, 3–22.
Huntingford, F.A., Adams, C., Braithwaite, V.A., Kadri, S., Pottinger, T.G., Sandøe, P. and Turnbull, J.F. (2006) Current issues in fish welfare. *Journal of Fish Biology* 68, 332–372.
Huss, H.H., Ababouch, L. and Gram, L. (2004) Assessment and management of seafood safety and quality. *FAO Fisheries Technical Paper* 444, 230 pp.
Jensen, F.B. (2003) Nitrite disrupts multiple physiological functions in aquatic animals. *Comparative Biochemistry and Physiology* 135A, 9–24.
Jobling, M. (1994) *Fish Bioenergetics*. Chapman and Hall, London.
Jobling, M. (1997) Temperature and growth: modulation of growth rate via temperature change. In: Wood, C.M. and McDonald, D.G. (eds) *Global Warming: Implications for Freshwater and Marine Fish*. Cambridge University Press, Cambridge, UK, pp. 225–253.
Jobling, M. (2004) On-growing to market size. In: Moksness, E., Kjørsvik, E. and Olsen, Y. (eds) *Culture of Cold-water Marine Fish*. Blackwell Science, Oxford, UK, pp. 363–432.
Johnston, I.A., Vierra, V.L.A. and Hill, J. (1996) Temperature and ontogeny in ectotherms: muscle phenotype in fish. In: Johnston, I.A. and Bennett, A.F. (eds) *Animals and Temperature: Phenotypic and Evolutionary Adaptation*. Cambridge University Press, Cambridge, UK, pp. 153–181.

Kamler, E. (1992) *Early Life History of Fish: An Energetics Approach*. Chapman and Hall, London.
Keeling, L.J. (2005) Healthy and happy: animal welfare as an integral part of sustainable agriculture. *Ambio* 34, 316–319.
Kong, K.Y., Cheung, K.C., Wong, C.K.C. and Wong, M.H. (2005) The residual dynamic of polycyclic aromatic hydrocarbons and organochlorine pesticides in fishponds of the Pearl River delta, South China. *Water Research* 39, 1831–1843.
Kroupova, H., Machova, J. and Svobodova, Z. (2005) Nitrite influence on fish: a review. *Veterinarni Medicina* 50, 461–471.
Lafrance, P., Castonguay, M., Chabot, D. and Audet, C. (2005) Ontogenetic changes in temperature preference of Atlantic cod. *Journal of Fish Biology* 66, 553–567.
Lee, C.-S. (ed.) (2003) Management of aquaculture effluents. *Aquaculture* 226, 1–244.
Lund, V. (2006) Natural living – a precondition for animal welfare in organic farming. *Livestock Science* 100, 71–83.
Meeren, T. van der and Ivannikov, V.P. (2006) Seasonal shift in spawning of Atlantic cod (*Gadus morhua* L.) by photoperiod manipulation: egg quality in relation to temperature and intensive larval rearing. *Aquaculture Research* 37, 898–913.
Muir, J. (2005) Managing to harvest? Perspectives on the potential of aquaculture. *Philosophical Transactions of the Royal Society B* 360, 191–218.
Murray, A.G. and Peeler, E.J. (2005) A framework for understanding the potential for emerging diseases in aquaculture. *Preventive Veterinary Medicine* 67, 223–235.
Nie, X.P., Lan, C.Y., An, T.C., Li, K.B. and Wong, M.H. (2006) Distributions and congener patterns of PCBs in fish from major aquaculture areas in the Pearl River Delta, South China. *Human and Ecological Risk Assessment* 12, 363–373.
Pankhurst, N.W. and Porter, M.J.R. (2003) Cold and dark or warm and light: variations on the theme of environmental control of reproduction. *Fish Physiology and Biochemistry* 28, 385–389.
Piferrer, F. (2001) Endocrine sex control strategies for the feminization of teleost fish. *Aquaculture* 197, 229–281.
Rombough, P.J. (1997) The effects of temperature on embryonic and larval development. In: Wood, C.M. and McDonald, D.G. (eds) *Global Warming: Implications for Freshwater and Marine Fish*. Cambridge University Press, Cambridge, UK, pp. 177–223.
Shang, E.H.H. and Wu, R.S.S. (2004) Aquatic hypoxia is a teratogen and affects fish embryonic development. *Environmental Science and Technology* 38, 4763–4767.
Shang, E.H.H., Yu, R.M.K. and Wu, R.S.S. (2006) Hypoxia affects sex differentiation and development, leading to a male-dominated population in zebrafish (*Danio rerio*). *Environmental Science and Technology* 40, 3118–3122.
Sharrer, M.J., Summerfelt, S.T., Bullock, G.L., Gleason, L.E. and Taeuber, J. (2005) Inactivation of bacteria using ultraviolet irradiation in a recirculating salmonid culture system. *Aquacultural Engineering* 33, 135–149.
Spinka, M. (2006) How important is natural behaviour in animal farming systems? *Applied Animal Behaviour Science* 100, 117–128.
Stickney, R.R. (ed.) (2000) *Encyclopedia of Aquaculture*. Wiley-Interscience, New York.
Strüssmann, C.A. and Nakamura, M. (2002) Morphology, endocrinology, and environmental modulation of gonadal sex differentiation in teleost fishes. *Fish Physiology and Biochemistry* 26, 13–29.
Summerfelt, S.T. (2003) Ozonation and UV irradiation – an introduction and examples of current applications. *Aquacultural Engineering* 28, 21–36.
Takle, H., Baeverfjord, G. and Andersen, Ø. (2004) Identification of stress sensitive genes in hyperthermic Atlantic salmon (*Salmo salar*) embryos by RAP-PCR. *Fish Physiology and Biochemistry* 30, 275–281.

Takle, H., Baeverfjord, G., Lunde, M., Kolstad, K. and Andersen, Ø. (2005) The effect of heat and cold exposure on *HSP70* expression and development of deformities during the embryogenesis of Atlantic salmon (*Salmo salar*). *Aquaculture* 249, 515–524.

Takle, H., Baeverfjord, G., Helland, S., Kjorsvik, E. and Andersen, O. (2006) Hyperthermia induced atrial natriuretic peptide expression and deviant heart development in Atlantic salmon *Salmo salar* embryos. *General and Comparative Endocrinology* 147, 118–125.

Tomasso, J.R. Jr and Grosell, M. (2005) Physiological basis for large differences in resistance to nitrite among freshwater and freshwater-acclimated euryhaline fishes. *Environmental Science and Technology* 39, 98–102.

Tucker, C.S. (2000) Off-flavor problems in aquaculture. *Reviews in Fisheries Science* 8, 45–88.

Utne-Palm, A.C. (2002) Visual feeding of fish in a turbid environment: physical and behavioural aspects. *Marine and Freshwater Behaviour and Physiology* 35, 111–128.

Vadstein, O., Mo, T.A. and Bergh, Ø. (2004) Microbial interactions, prophylaxis and disease. In: Moksness, E., Kjørsvik, E. and Olsen, Y. (eds) *Culture of Cold-water Marine Fish*. Blackwell Science, Oxford, UK, pp. 28–72.

Van Der Kraak, G. and Pankhurst, N.W. (1997) Temperature effects on the reproductive performance of fish. In: Wood, C.M. and McDonald, D.G. (eds) *Global Warming: Implications for Freshwater and Marine Fish*. Cambridge University Press, Cambridge, UK, pp. 159–176.

Watson, S.B. (2003) Cyanobacterial and eukaryotic algal odour compounds: signals or by-products? A review of their biological activity. *Phycologia* 42, 332–350.

Wedemeyer, G. (ed.) (2001) *Fish Hatchery Management*. American Fisheries Society, Bethesda, Maryland.

Woo, P.T.K., Bruno, D.W. and Lim, L.H.S. (eds) (2002) *Diseases and Disorders of Finfish in Cage Culture*. CAB International, Wallingford, UK.

Wu, R.S.S., Zhou, B.S., Randall, D.J., Norman, W.Y.S. and Lam, P.K.S. (2003) Aquatic hypoxia is an endocrine disruptor and impairs fish reproduction. *Environmental Science and Technology* 37, 1137–1141.

Zagarese, H.E. and Williamson, C.E. (2001) The implications of solar UV radiation exposure for fish and fisheries. *Fish and Fisheries* 2, 250–260.

Zhou, H.Y., Cheung, R.Y.H. and Wong, M.H. (1999) Bioaccumulation of organochlorines in freshwater fish with different feeding modes cultured in treated wastewater. *Water Research* 33, 2747–2756.

3 Fish Culture: Feeds and Feeding

Malcolm Jobling

University of Tromsø, Tromsø, Norway

3.1 Introduction

Fish, in common with all other animals, face the challenge of acquiring sufficient nutrients to support body maintenance, growth and reproduction. Feeding involves a number of events that occur in sequence; searching for, detecting and locating food, capture, oral processing and evaluation of the food item and finally acceptance or rejection. Acceptance leads to swallowing of the food, which is followed by digestion and nutrient absorption. Although the behavioural sequence seems clear, the dynamics of feeding are influenced by many environmental variables (for reviews, see Houlihan *et al.*, 2001). For example, water temperature influences fish metabolism and also has effects on feeding activity and feeding rates (see Chapter 2, this volume). Environmental variables often exert their effects by imposing either activity or sensory limitations. For example, differences in the visual environment resulting from changes in light intensity or water turbidity can influence the ability of a fish to detect and capture prey (Utne-Palm, 2002; see Chapter 2, this volume). Biotic factors also have an important influence on feeding. For example, there may be competition among individuals, either of the same or different species, for food but, under some circumstances, feeding may be enhanced by the presence of other individuals, i.e. there is social facilitation. Social facilitation may include a learning component; copying the behaviour of successful or experienced foragers may influence the types of food or prey an individual eats and/or have an effect on the success with which an individual finds and captures food items. Moreover, animals learn about potential food items by associating them with the positive post-ingestive effects of the nutrients they contain and/or any negative toxic effects. Thus, animals consume a food item, experience positive nutritional rewards and/or negative toxic effects and may adjust their food preferences on the basis of these experiences.

Aquaculture is one of the fastest growing animal production enterprises in the world and the continued growth of the industry will depend on the development of cost-effective feeds and feeding systems (Stickney, 2000; Houlihan *et al.*, 2001; Wedemeyer, 2001; Webster and Lim, 2002). This is because feed costs are the single largest variable cost in intensively managed fish farming systems. The feeding of fish in their aquatic environment imposes some challenges not experienced by farmers of terrestrial livestock. The effective feeding of farmed fish revolves around the development of feed types, feed delivery systems and feeding routines that reduce feed losses and ensure effective digestion and utilization of the consumed feed nutrients (Houlihan *et al.*, 2001; Halver and Hardy, 2002; Webster and Lim, 2002; Jobling, 2004c). Criteria pertaining to good feeds and sound feeding strategies for farmed fish include:

- water-stable feeds that do not disintegrate, or lose nutrients due to leaching, prior to being consumed by the fish
- feeding regimes that reduce feed waste to a minimum
- highly digestible feeds that reduce faecal losses to a minimum
- nutritionally balanced feeds that maintain fish health and promote good growth
- feed formulations that give a high-quality finished product

3.2 Nutrients and Nutritional Requirements

It is beyond the scope of this chapter to provide descriptions of the nutritional requirements of farmed fish (for reviews, see Stickney, 2000; Halver and Hardy, 2002; Webster and Lim, 2002; Jobling, 2004c), but some introductory notes will be presented to enable the reader to place this aspect of nutrition into the broader perspective of feeds, feed formulation and the feeding of fish in intensive culture.

Fish are heterotrophs, meaning that they are reliant on exogenous sources of organic material for their nourishment. They are unable to synthesize organic compounds from inorganic substrates and they require organic sources of carbon, nitrogen, etc., as starting materials for the biosynthesis of the molecules that make up their body tissues. In addition, fish are required to use some of the organic materials they obtain from their food as respiratory substrates to provide the energy needed to sustain their bodily functions and drive the biosynthetic reactions. Proteins, lipids and carbohydrates are the nutrients present in the greatest proportions in both natural and formulated feeds consumed by fish, i.e. are macronutrients, whereas vitamins and minerals, present and required in lesser quantity, are the micronutrients. Nutrients have an impact on all bodily functions and metabolic and growth processes are dependent on the fish obtaining adequate supplies of certain indispensable (or essential) nutrients. An indispensable nutrient is one that cannot be synthesized by the fish *de novo* and which must be present in the diet. The indispensable nutrients are:

- some amino acids (threonine, valine, leucine, isoleucine, phenylalanine, methionine, tryptophan, arginine, histidine, lysine) (Fig. 3.1)
- fatty acids of the n-3 and n-6 series (Table 3.1)
- vitamins (Table 3.2)
- some minerals or trace elements (Table 3.3)

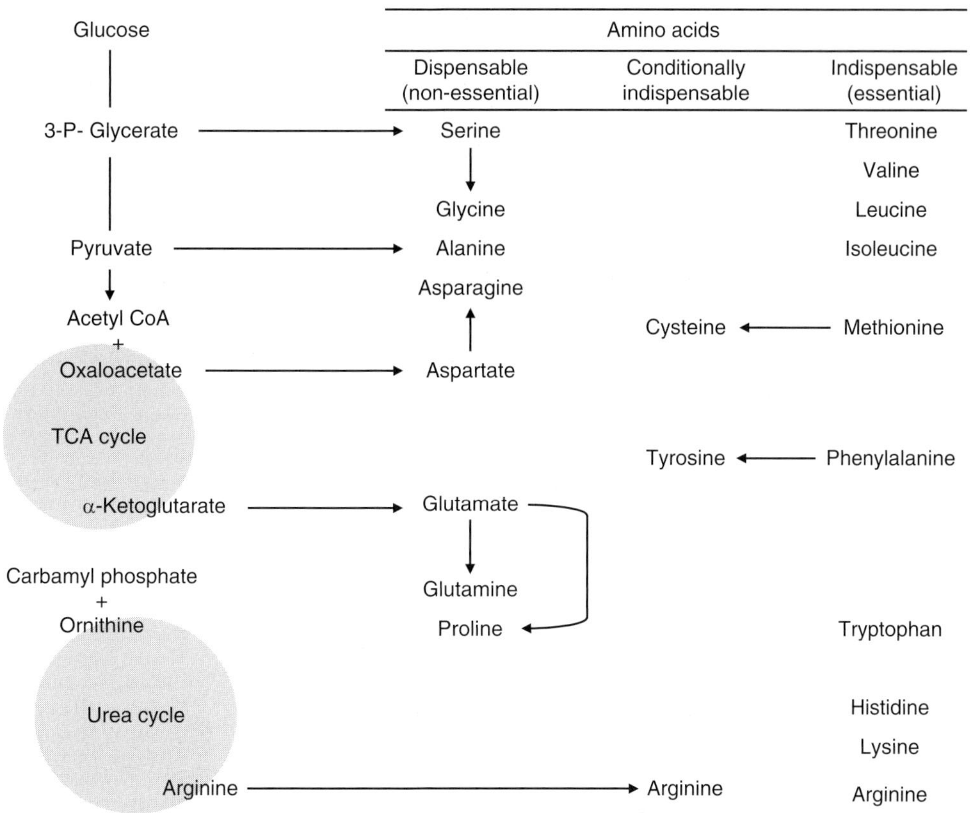

Fig. 3.1. Schematic representation of pathways for amino acid synthesis, transformation and metabolism. An overview of the indispensable (essential), conditionally indispensable and dispensable (non-essential) amino acids is given. Conditionally, indispensable amino acids can be synthesized from indispensable amino acid precursors, whereas dispensable amino acids can be synthesized from a range of organic compounds. TCA cycle = tricarboxylic acid cycle.

If the diet lacks or has insufficient amounts of one or more of the indispensable nutrients, the fish will display deficiency symptoms. Such symptoms may include reduced feed intake and growth, metabolic disturbances, abnormal development and body colour and the display of abnormal behaviour. Data about quantitative indispensable nutrient requirements are often obtained by conducting growth studies in which feeds are formulated to contain graded levels of the nutrient of interest. The growth responses of the fish are recorded and dose–response curves are constructed by equating growth with the amount of the nutrient in the feed. The growth data are then analysed to estimate the requirement for the nutrient in question.

Within the context of the preparation of feeds destined to be given to farmed fish, it should be borne in mind that individual nutrients (proteins and amino acids, lipids and fatty acids, carbohydrates, vitamins and minerals) can be viewed as being nested within different types of feed ingredients, each of which may have a surfeit of some nutrients and a dearth of others (relative to

Table 3.1. Classification and naming of fatty acids, with selected examples. In the scientific designation, anoic refers to a fatty acid without double bonds in the carbon chain, enoic to a fatty acid with one double bond, dienoic to a fatty acid with two double bonds, trienoic to three, tetraenoic to four, etc. The shorthand notation gives the number of carbon atoms in the chain, the number of double bonds and the position of the first double bond counting from the methyl end of the fatty acid molecule. Unsaturated fatty acids of the n-3 and n-6 series are the indispensable (essential) fatty acids (from Houlihan *et al.*, 2001; Jobling, 2004c).

Trivial name (scientific designation)	Number of carbon atoms	Number of double bonds	Fatty acid series	Shorthand notation
Saturated fatty acids (SFAs)				
Lauric (dodecanoic)	12	0		12:0
Palmitic (hexadecanoic)	16	0		16:0
Stearic (octadecanoic)	18	0		18:0
Monounsaturated fatty acids (MUFAs)				
Palmitoleic (hexadecenoic)	16	1	n-7	16:1 n-7
Oleic (octadecenoic)	18	1	n-9	18:1 n-9
Erucic (docosenoic)	22	1	n-9	22:1 n-9
Polyunsaturated fatty acids (PUFAs)				
Linoleic (octadecadienoic)	18	2	n-6	18:2 n-6
γ-Linolenic (octadecatrienoic)	18	3	n-6	18:3 n-6
α-Linolenic (octadecatrienoic)	18	3	n-3	18:3 n-3
Highly unsaturated fatty acids (HUFAs)				
Arachidonic (eicosatetraenoic)	20	4	n-6	20:4 n-6
EPA (eicosapentaenoic)	20	5	n-3	20:5 n-3
DHA (docosahexaenoic)	22	6	n-3	22:6 n-3

requirement). This means that intensively farmed fish are usually provided with feeds that contain a wide range of ingredients. The complete feeds given to intensively farmed fish are formulated to contain sufficient indispensable nutrients to meet all known requirements, and feeds will also contain other materials that do not play an important direct nutritional role. These compounds may be added to improve the binding, flow and storage properties of the feed, they may be feeding stimulants or attractants or they may be probiotics that are added with the aim of improving the health of the farmed stock. In contrast to intensively farmed fish, it is not uncommon for extensively held, pond-raised fish to be given incomplete feeds under the expectation that any nutritional deficits will be compensated for, and corrected, via consumption of natural prey organisms present in the pond.

3.2.1 Metabolism and body maintenance: fish versus terrestrial livestock

Conventional terrestrial livestock species, such as cattle, pigs, sheep and poultry, are all homeothermic endotherms. They maintain a constant body temperature that differs from that of the environment and they also generate their own body heat. As a result, they have a high metabolic rate and a lot of feed energy is used

Table 3.2. Overview of the lipid-soluble and water-soluble vitamins and related (vitamin-like) compounds, with an indication of their biological roles and functions (from Jobling, 2004c).

Fat (lipid)-soluble vitamins	
Retinol (vitamin A)	Normal growth, vision, reproduction
Cholecalciferol (vitamin D)	Calcium and phosphate metabolism/regulation
Tocopherols (vitamin E)	Antioxidant, muscle and RBC function
Menadione (vitamin K)	Blood clotting
Water-soluble vitamins	
Thiamin (vitamin B_1)	Energy metabolism, nerve function
Riboflavin (vitamin B_2)	Cellular energy metabolism
Niacin (nicotinic acid)	Energy metabolism, nerve function
Pantothenic acid (vitamin B_5)	Energy metabolism, nerve function
Pyridoxine (vitamin B_6)	Protein metabolism and utilization
Cyanocobalamin (vitamin B_{12})	Nerve function, RBC formation and function
Biotin (vitamin H)	Fatty acid synthesis, glucose metabolism
Folacin (folate)	Embryonic development, gut function
Ascorbic acid (vitamin C)	Antioxidant, collagen synthesis, immune responses
'Vitamin-like' substances	
Carotenoids	*Antioxidants, provitamin A*
Inositol (myo-inositol)	Cell membrane phospholipids, chemical signal transmission
Choline	*Fatty acid metabolism, cell membrane phospholipids, neurotransmission functions*
Carnitine	*Lipid/fatty acid metabolism*

Table 3.3. Overview of the biological functions of inorganic elements (minerals) in fish, with examples of the roles played by different elements (from Jobling, 2004c).

Biological or physiological role	Examples
Ionic regulation (electrolytes)	Sodium, potassium, chlorine
Acid–base balance	Calcium, sodium, chlorine
Structural functions • Bone/skeletal tissue • Cell membranes	Calcium, phosphorus, magnesium, sulphur
Nerve impulse transmission and muscle contraction	Calcium, sodium, potassium
Respiratory pigment (haemoglobin)	Iron
Component of hormones	Iodine, sulphur
Enzyme structure and function • Component of enzyme • Co-factor or component of co-factor • Activator or regulator	Zinc, selenium, cobalt, manganese, chromium, vanadium

as fuel to sustain basic life processes. This means that the maintenance requirements of homeothermic endotherms are high. Maintenance requirement is defined as the amount of food needed to meet metabolic demands, but not allowing for any increase in body mass. Given that conventional farm animals have high maintenance requirements, they are generally relatively inefficient at converting feed to body tissue (Pond *et al.*, 1995; MacRae *et al.*, 2005). Terrestrial livestock species are generally either herbivorous or omnivorous in their dietary habits (Diamond, 2002; Mignon-Grasteau *et al.*, 2005) and are capable of digesting and metabolizing feeds that contain high concentrations of the complex carbohydrates found in plants. As such, the respiratory substrates used as metabolic fuel by endothermic farmed species, such as poultry and pigs, are provided largely in the form of carbohydrates (starches and sugars). Feeds formulated for these animals usually have high carbohydrate content and relatively low protein and lipid contents. Carbohydrates are an abundant and inexpensive 'energy source', so the feeds given to conventional terrestrial farm animals are comprised largely of relatively cheap feed ingredients.

Most fish are poikilothermic ectotherms, i.e. body temperature changes as environmental temperature changes, and the main source of body heat is from the environment. Fish have low metabolic rates, often only 15–20% of mammals of similar size, and low maintenance requirements. This means that fish are relatively efficient at converting food to body tissue. When they are fed sufficient amounts of nutritionally balanced, highly digestible feeds under favourable environmental conditions, fish may deposit 50–60% of the feed nutrients as growth. On the debit side, most intensively farmed fish species are carnivorous and are unable to digest and metabolize large amounts of dietary carbohydrate. They are, therefore, reliant on feeds that contain relatively high proportions of proteins and lipids. Feed ingredients that contain these nutrients are more expensive than those that are rich in carbohydrate, so the costs of the basic feed ingredients are higher in fish feeds than in feeds formulated for terrestrial livestock.

To summarize: 'warm-blooded' (homeothermic endotherms) species, such as poultry and pigs, have high metabolic rates and are less efficient at utilizing feed for growth (weight gain) than are 'cold-blooded' (poikilothermic ectotherms) fish. This means that feed:gain ratios are higher in endothermic farmed animals than in farmed fish and the endotherms require larger quantities of digestible feed energy per unit weight gain than do fish (Pond *et al.*, 1995). The comparison between conventional farm animals and fish is, however, complicated by the fact that most terrestrial livestock are either herbivores or omnivores that can be raised on relatively cheap high-carbohydrate, low-protein feeds (protein: 12–25% of feed dry matter), whereas most intensively farmed fish are carnivores that require feeds that contain higher proportions of protein (30–55% of dry matter depending on species and life stage) and lipid.

3.3 Aqua-feeds for Farming Fish

Feed types and formulations used in fish farming can differ markedly with respect to manufacturing processes, types of raw materials and proportions of ingredients (Hertrampf and Piedad-Pascual, 2000; Stickney, 2000; Wedemeyer,

2001; Halver and Hardy, 2002; Webster and Lim, 2002; Jobling, 2004c). Much of the feed used in intensive fish culture is produced either by international companies that provide closed-formula complete feeds or by agricultural cooperatives. In some instances, aqua-feeds are produced on the farm using readily available local feed ingredients. Farm-made aqua-feeds will usually require supplementation with some feed ingredients that are not readily available in the environs of the farm. Such ingredients may include sources of additional protein or essential fatty acids. More often, the supplements will be vitamin and mineral premixes that, although nutritionally important, represent only a small proportion of the complete feed. These supplements are usually supplied to farmers by feed companies and cooperatives.

Although fish are farmed using wet, moist and dry formulated feeds, and live prey organisms, it is dry feeds that are used most widely in intensive fish culture. The widespread use of dry feeds results from the ease with which they can be transported and distributed by manufacturers, their ease of storage and handling and their relatively consistent chemical compositions and physical characteristics. The closed-formula dry feeds (crumbles, compressed pellets and extruded feeds) provided by international feed companies are formulated to meet a range of nutritional and price criteria. The inclusion levels of the various feed ingredients are usually decided on by applying computer-based, least-cost formulation techniques that employ some form of linear programming.

Dry feeds contain less than 10% moisture and this means that they are easier to transport and store than are wet and moist feeds. Dry feeds are also more water stable than either wet or moist feeds. The main methods used in the manufacture of dry pellet feeds are compressed steam pelleting and extrusion (or expansion) pelleting (Fig. 3.2). Extrusion pelleting enables the manufacture of pellets having a greater range of physical and nutritional properties than does steam pelleting. Steam pelleting produces dense pellets that sink quite rapidly, whereas extrusion pelleting allows the production of feeds that either float or sink through the water slowly. Extruded pellets tend to be harder and more stable than compressed dry pellets and can be manufactured with up to *c.*40% fat and very little carbohydrate.

3.3.1 Feeds for different life history stages

Feeds differ in their nutritional and physical characteristics depending on the species for which they are designed, and also depending on the life history stage for which they are intended (Stickney, 2000; Wedemeyer, 2001; Halver and Hardy, 2002; Webster and Lim, 2002). Broodstock feeds are formulated to ensure production of 'high-quality' eggs. Fecundity, gamete quality and larval viability are more important considerations than weight gain of the broodfish. The majority of cultivated fish species have yolk-rich (telolecithal) eggs and there is usually complete dependence on yolk nutrients (lecithotrophy) during early development. Thus, nutrients present in the egg have profound consequences for development during the embryonic and larval phases and the female broodfish is the provider of these nutrients. This indicates the importance of broodstock nutrition in determining the overall success of production.

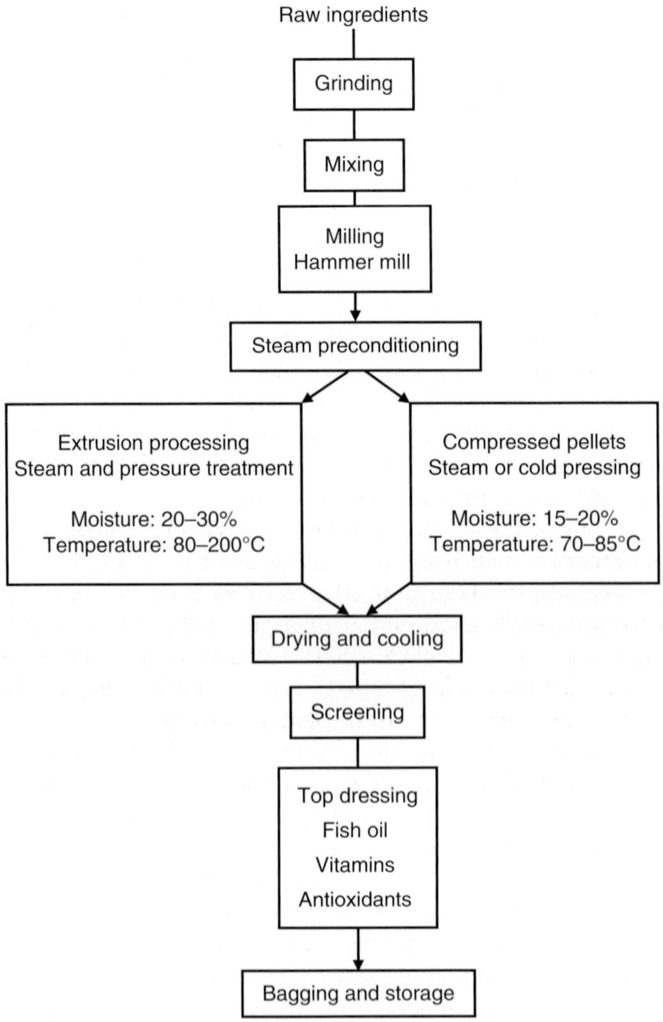

Fig. 3.2. Schematic representation of the steps involved in the manufacture of dry feed pellets – compressed pellet feeds and extruded feeds – for farmed fish. Extrusion involves the use of higher temperatures and moisture (steam) treatments than the production of pellets using compression processing methods. Extruded feeds are formulated to contain lower percentages of binders (usually starch-containing ingredients) than compressed pellets because of the improved binding of the feed ingredients that results from extrusion processing (from Houlihan *et al.*, 2001; Jobling 2004c).

Broodstock feeds may be especially fortified with vitamins and minerals, or may be formulated to contain oils that provide the broodstock with certain fatty acids in specified proportions.

The start-feeding phase is also critical in fish rearing operations and the development of nutritionally adequate feeds is a prerequisite for successful larviculture. Larval feeds are difficult to prepare because of the small size of the fish to which they are to be fed (Stickney, 2000; Langdon, 2003). Special production methods are required because of the extremely small size of the feed particles (less than 400 μm). Microbound larval feeds are held together by an internal binder, usually a complex carbohydrate or protein having adhesive and absorptive properties. Microencapsulated feeds are surrounded by material, the capsule, which usually differs from the central nutrient core; the wall may be of carbohydrate, protein or lipid, sometimes combined with other materials. The small size of the larval feed particles means that they have a high surface-to-volume ratio and this creates the risk of rapid loss of water-soluble nutrients. The feeds also need to be highly palatable and the feed ingredients must be digestible by larvae that may not have a fully functional digestive system. Given these problems, it is not unusual for fish larvae to be provided with live food organisms during the earliest phases of rearing (Stottrup and McEvoy, 2002; Olsen, 2004; Olsen *et al.*, 2004) and then be weaned on to dry feeds once they have grown to a larger size.

Starter feeds (or crumbles) may be used for the initial feeding of fish, such as salmonids, that hatch at relatively large body size (15–25 mm in length), or for weaning the young of other species from live prey to dry feeds. Starter feeds are usually produced from larger feed pellets by crushing them between rollers moving at different speeds. The pellet fragments are screened to give crumble feeds of different size ranges. Crumbles have a high surface-to-volume ratio, so disintegrate in water relatively quickly.

The economic viability of intensive fish farming depends on the ability of the farmer to deliver a product at a price that is acceptable to the consumer, so it has been traditional to develop grow-out feeds that give maximum production at minimal cost. Recently, however, there has been some change of emphasis. Increasing consideration has been given to the ways in which feeds influence certain attributes of cultured fish that may be of importance to the consumer. These attributes include nutritional quality, texture and flavour. There is also increasing concern about the eutrophication of recipient waterbodies by fish farm wastes. Consequently, in an attempt to reduce these wastes, there has been focus on developing feeds with high nutrient digestibility and improved water-stability characteristics.

Grow-out feeds are fed to fish that have passed the first-feeding stage; they represent the largest proportion of feed used during the production cycle. Grow-out feeds are most usually either steam-compressed or extruded dry pellets and are formulated to ensure rapid growth of the fish. They may be produced as high-energy (high-fat) or low-pollution (highly digestible, low phosphorus) feeds; both of these feed types may be considered relatively 'environmentally friendly' (Stickney, 2000; Wedemeyer, 2001). Finisher feeds may be given to the fish in the weeks prior to harvest to impart particular properties to the finished product, e.g. red colour in salmon fillets, a desired fatty acid profile, etc. Under some circumstances, particular microorganisms or extractives (probiotics) may be added to feeds in an attempt to stimulate the immune system of the fish or change the gastrointestinal microflora, with the aim of increasing disease resistance and improving health (Gatesoupe, 1999; Irianto and Austin, 2002; Balcázar *et al.*, 2006; Vine *et al.*, 2006). Medicated feeds are those to which drugs or antibiotics have been added. These feeds need to be highly palatable to ensure consumption by diseased fish that have a reduced appetite (Rigos and Troisi, 2005).

3.3.2 Fishmeal and fish oil: a reduced reliance on these feed ingredients

Although a wide range of ingredients is used in the formulation of dry fish feeds (Hertrampf and Piedad-Pascual, 2000; Stickney, 2000; Halver and Hardy, 2002; Webster and Lim, 2002; Jobling, 2004c), fishmeals and fish oils traditionally have been used as major ingredients in such feeds. Fishmeals have been used as a major ingredient because they represent a source of protein of high quality (Table 3.4) and marine fish oils have been widely used as the main lipid source because they have been readily available, relatively cheap and are a good source of n-3 HUFAs (Table 3.5). The heavy reliance on these feed ingredients has been questioned and criticized on grounds of sustainability (Naylor *et al.*, 1998, 2000; Muir, 2005). Fears have also been expressed that the fishmeals and fish oils used in aqua-feeds may give rise to organochlorine contamination of the flesh of farmed fish (Hites *et al.*, 2004; Berntssen *et al.*, 2005; Carlson and Hites, 2005; Foran *et al.*, 2005; Hamilton *et al.*, 2005; Bethune *et al.*, 2006). This is because the consumption of contaminated feed appears to be the main route by which the body burdens of organochlorine and other organic contaminants are increased in fish (Carlson and Hites, 2005; Cleland *et al.*, 2005; Hamilton *et al.*, 2005; Bethune *et al.*, 2006; Heiden *et al.*, 2006; Montory and Barra, 2006; Scott *et al.*, 2006). Fish may also become contaminated with other pollutants, such as the heavy metals, mercury, cadmium and lead, and these may reach concentrations that are considered to have adverse effects on human consumers (SACN-FSA, 2004). Farmed

Table 3.4. The indispensable (essential) amino acid compositions (expressed as g amino acid per kg protein) of Atlantic salmon, *Salmo salar*, muscle (fillet) and of a range of ingredients used as protein sources in fish feeds (from Jobling, 2004c).

	Atlantic salmon *Salmo salar*	Fishmeal	Soybean	Rape/ canola	Maize gluten	Wheat gluten
Arginine	71	59	73	79	37	36
Histidine	43	29	28	33	29	19
Isoleucine	60	44	47	43	35	35
Leucine	92	75	75	61	114	70
Lysine	92	81	61	66	18	15
Methionine	33	30	14	23	23	16
Phenylalanine	54	40	50	38	66	50
Threonine	49	43	40	56	38	37
Tryptophan	11	12	17	8	10	11
Valine	60	54	49	43	44	39

Table 3.5. Proportions of selected fatty acids (expressed as % total fatty acids) in an animal fat (beef tallow) and in a variety of fish and vegetable (plant) oils. SFA, saturated fatty acid; MUFA, monounsaturated fatty acid; PUFA, polyunsaturated fatty acid; HUFA, highly unsaturated fatty acid (from Jobling, 2004c).

	Fat or oil source							
Fatty acid	Beef tallow	Anchovy	Herring	Capelin	Rape/ canola	Soya	Palm	Linseed
Saturates (SFA)								
16:0	25	19.5	14.5	11	5	10	43.5	5.5
18:0	20	3.5	1	1.5	2	4	4.5	4
MUFA								
16:1	5	9	6	9	0.5	0.5	0.5	
18:1	40	15	10	14	54	23	36.5	20
20:1		2.5	15.5	13	1	0.5		
22:1		1.5	22	10.5	1			
PUFA and HUFA								
18:2 n-6	2	1	1.5	1	23	51	9	12.5
18:3 n-3	2	0.5	1.5	1	11	7	0.5	53.5
20:5 n-3		18	5	10				
22:6 n-3		11	6.5	10				

fish can accumulate heavy metals in their flesh if they are given contaminated feed. Fish that are caught in polluted waters may also have high flesh concentrations of heavy metals and other contaminants. Attempts will usually be made to avoid raising fish in polluted waters, but many contaminants are very persistent in the environment. Thus, many potential contaminant chemicals tend to accumulate in sediments over periods of years and be taken up and concentrated by aquatic organisms, even after industrial use has ceased (Nie *et al*., 2006; Shi *et al*., 2006).

3.3.2.1 Organic contaminants in fish and fish feeds

Organic contaminants, such as dioxins and PCBs, are very persistent lipophilic chemicals that are ubiquitous in the environment. These chemicals are taken up by fish and are stored in the body tissues, most often in association with fat (Cleland *et al*., 2005; Hamilton *et al*., 2005; Montory and Barra, 2006; Shi *et al*., 2006). This means that, given the lipophilic nature of dioxins and PCBs, the concentrations present in fish will be influenced by the body fat content of the fish, i.e. concentrations of these chemicals are usually higher in fatty fish than in lean ones. The fact that most fishmeal and fish oil is prepared from small pelagic fish of relatively high fat content gives, therefore, grounds for concern.

The concentrations of organic contaminants present in fish depend on geographic location and migration patterns, fish size, fish age and feeding habits (SACN-FSA, 2004; Nie *et al*., 2006; Scott *et al*., 2006). For example, levels of organic contaminants have been reported to be higher in fish from the Baltic Sea and North Sea than in those from the South Atlantic and many areas of the

Pacific Ocean (SACN-FSA, 2004). This means that levels of contaminants present in fishmeals and oils vary depending on the geographic location in which the raw fish materials are captured, for example, the North Sea and North Atlantic as opposed to the South Atlantic or off the Pacific coast of South America.

Given the lipophilic nature of the organochlorines, concentrations of these contaminants are higher in fish oils than in fishmeals (Easton *et al.*, 2002; Jacobs *et al.*, 2002). An additional factor to be considered is that the levels of contaminants present in an aqua-feed will depend on the amount of fish oil added to the feed mixture. In other words, a high-fat feed based on fish oil is likely to contain higher concentrations of organochlorine contaminants than a low-fat feed formulated with vegetable (plant) oils (Easton *et al.*, 2002; Jacobs *et al.*, 2002; Bell *et al.*, 2005; Berntssen *et al.*, 2005; Hamilton *et al.*, 2005).

The fishmeals and oils used by aqua-feed manufacturers derive from fish captured in many different locations and the source of the ingredients used in feeds varies over time, depending on quality, price and availability. This means that concentrations of organochlorine contaminants present in aqua-feeds are likely to vary widely (Carlson and Hites, 2005). They will vary with feed formulation and, for a given formulation, will vary from feed batch to feed batch, depending on the source of the ingredients used. Feeds manufactured from meals and oils prepared from fish captured in polluted waters are likely to be contaminated with heavy metals and organochlorine compounds. Further, the level of organochlorine contamination is likely to be higher in a high-fat feed than in one formulated to contain lower quantities of fish oil. This raises concern about the levels of contamination that may be present in high-fat aqua-feeds; for example, those currently used in salmon farming, prepared from fishmeals and oils of European origin (Jacobs *et al.*, 2002; Berntssen *et al.*, 2005; Carlson and Hites, 2005). It is, however, possible to reduce the concentrations of organic contaminants, such as dioxins, present in fishmeals and oils using a variety of extraction techniques. Fishmeal and fish oil production plants in a number of countries are being modernized to incorporate equipment for the extraction of contaminants, thereby enabling delivery of meals and oils low in dioxins and PCBs to the aqua-feed industry.

3.3.2.2 Physiological effects of organic contaminants

Organic contaminants in foods, and otherwise present in the environment, can exert a number of physiological effects on humans, livestock and wildlife (Miyamoto and Burger, 2003; Foran *et al.*, 2005; Mills and Chichester, 2005; Sumpter and Johnson, 2005; Waring and Harris, 2005; Basrur, 2006; Heiden *et al.*, 2006; Hiramatsu *et al.*, 2006; Milnes *et al.*, 2006; Reynaud and Deschaux, 2006; Scott *et al.*, 2006). These effects may be detrimental to the health and well-being of the affected individual. The organic contaminants may have influences on the immune system, they can be carcinogenic and the contaminants may also give rise to developmental anomalies that result in the malformation of various tissues and organs. For example, malformations and dysfunctions of the reproductive organs are frequently reported consequences of exposure to environmental contaminants.

Many contaminants can act as endocrine disrupting chemicals (EDCs). EDCs act via interference with and disruption of normal hormonal status. Several EDCs have similar structures to natural hormones and can bind to hormone receptors, thereby acting as hormone agonists or antagonists. EDCs can, however, also disrupt endocrine pathways by other mechanisms (Jalabert *et al.*, 2000; Miyamoto and Burger, 2003; Brevini *et al.*, 2005; Mandal, 2005; Mills and Chichester, 2005; Rhind, 2005; Sumpter and Johnson, 2005; Waring and Harris, 2005; Heiden *et al.*, 2006; Milnes *et al.*, 2006; Scott *et al.*, 2006). For example, some EDCs show little affinity for hormone receptors and therefore may not act directly as receptor agonists or antagonists. Such EDCs may, for example, act as enzyme inhibitors, thereby reducing the production or metabolism of endogenous hormones.

Nuclear hormone receptors bind to small, lipophilic and hydrophobic molecules, such as steroid hormones and thyroid hormones. These hormones play a role in the regulation of almost all aspects of animal physiology, including reproduction and development, homeostasis and the response to stressors. When the hormone binds to its nuclear receptor, the receptor undergoes a conformational change that leads to the transcription of genes that characterize the response to the hormone. In addition to the natural hormones, the nuclear receptors will also bind to other small lipophilic molecules. The result is a change in the affinity of the receptor for binding to DNA, co-factors and co-regulators. In this way, these molecules may interfere with a cascade of reactions and exert their effects as EDCs. Several organochlorine compounds are steroid hormone mimics that can bind to oestrogen (female sex hormones) receptors. They may, for example, either act as oestrogen receptor agonists and stimulate oestrogen-like effects, or may hinder the binding of natural oestrogens to the receptors and exert antioestrogen physiological effects. Other persistent organic pollutants act as EDCs via interaction with androgenic (male sex hormones) hormonal pathways and several of these compounds have anti-androgenic properties (Jalabert *et al.*, 2000; Miyamoto and Burger, 2003; Brevini *et al.*, 2005; Mandal, 2005; Mills and Chichester, 2005; Rhind, 2005; Waring and Harris, 2005; Basrur, 2006; Hiramatsu *et al.*, 2006; Milnes *et al.*, 2006).

On the other hand, some EDCs appear to exert their major effects more via direct influences on enzyme systems than via binding to hormonal receptors. The EDCs may inhibit enzymes involved in the steroid hormone biosynthesis cascade, or may exert their influences via effects on the metabolism of the endogenous steroids by acting, for example, as inhibitors of conjugating enzymes that result in the production of water-soluble steroid metabolites such as glucuronides and sulphates.

3.3.2.3 Fishmeals as a protein source in aqua-feeds

The tonnages of pelagic fish harvested, and amounts of fishmeal produced, have changed little since the 1980s. This has been a period over which there has been a marked expansion of aquaculture (Tidwell and Allan, 2001; Shelton and Rothbard, 2006; http://www.fao.org/figis). World landings of pelagic fish have been within the range 20–25 Mt and fishmeal production has been relatively stable at 5–6 Mt/year. In other words, there is little evidence of any direct relationship between production of farmed fish, catches of pelagic fish and fishmeal production (Tidwell and Allan, 2001). Nevertheless, fishmeal is an important ingredient in aqua-feeds, so it is paradoxical that production of farmed fish has been able to increase more than fivefold since the 1980s without seeming to have a major impact on harvests of small, pelagic fish. The amounts of fishmeal used in aqua-feed production have increased over the years, from about 0.5 Mt in 1988 to *c.*3 Mt by the early years of the 21st century. Fishmeal production has been stable, so its increased use for aqua-feeds has been the result of a reallocation of the finite amount among the various users. Increased quantities of fishmeal have been redirected from use in the production of livestock feeds or as agricultural fertilizers, to the manufacture of aqua-feeds. Even so, large quantities of fishmeal are still used for the production of feeds for livestock, particularly poultry and pigs.

The reallocation of fishmeal from livestock feeds to aqua-feeds has largely been made possible by increased reliance on 'alternative' protein sources for the manufacture of poultry and pig feeds. Fishmeal traditionally has been the preferred source of protein in livestock feeds, but gradually has been replaced by 'alternatives' of vegetable (plant) origin. In addition, many aqua-feeds are also being formulated to contain increased proportions of these 'alternative' protein sources, such as soybean meal and other meals derived from terrestrial plants (Table 3.4). Soybean meal is much more readily available on the international feedstuffs market than is fishmeal and it has a high nutritional value for many fish species. Given these properties, soybean meal has become the dominant protein source used in aqua-feeds. For example, feeds formulated for carps, tilapias and catfishes now contain very low proportions of fishmeal (often less than 5% by weight) and much higher proportions of soybean meal and meals prepared from other terrestrial plants.

3.3.2.4 'Alternative' protein sources in aqua-feeds

The search for 'alternative' protein sources will need to continue if there is to be further expansion of intensive fish culture involving reliance on dry formulated feeds. Although there may be major benefits to be gained by including vegetable protein sources in aqua-feeds, there are also a number of problems associated with these feed ingredients (Hertrampf and Piedad-Pascual, 2000; Acamovic and Brooker, 2005; McKevith, 2005). For example, many plants are deficient in some of the nutrients that are essential for fish, including some fatty acids (Table 3.5) and amino acids (e.g. lysine and methionine) (Table 3.4). This means that there may be a need to add amino acid supplements to feeds prepared with vegetable protein sources. In addition, most plants contain antinutritional factors (ANFs). ANFs are compounds that may interfere with digestion and absorption, reduce nutrient bioavailability and feed utilization or have a variety of adverse physiological and metabolic effects on consumers (Francis *et al.*, 2001; Acamovic and Brooker, 2005). This means that vegetable protein sources must be processed to remove or inactivate the ANFs they contain. The processing may involve de-hulling of seeds to remove ANFs, thermal treatment to denature ANFs or solvent extraction to remove some of these compounds (Hertrampf and Piedad-Pascual, 2000; Stickney, 2000; Halver and Hardy, 2002; Webster and Lim, 2002; Jobling, 2004c).

Phytoestrogens are one class of ANFs. These compounds are present in many vegetable protein sources, but particularly in meals prepared from legumes, such as soybeans. As their name implies, phytoestrogens, such as genistein, coumestrol, daidzein and equol, are plant-derived compounds that act as oestrogen mimics; they may exert either agonistic or antagonistic effects. Aqua-feeds that contain a high percentage of protein from vegetable sources will contain some phytoestrogens. These compounds are present in detectable quantities in several commercial aqua-feeds and, in some instances, they may be present at concentrations that can exert oestrogen-like physiological effects on fish that consume these feeds (Matsumoto *et al.*, 2004; Matsuoka *et al.*, 2005; Kelly and Green, 2006; Ng *et al.*, 2006). Under such circumstances, there may be an induction of vitellogenin synthesis in both males and females and disturbance or disruption of spermatogenesis and oogenesis. For example, there may be reductions in spermatocrit and sperm motility in male fish that consume feeds containing high concentrations of phytoestrogens, whereas in females there may be disruption of the reproductive cycle, resulting in fewer ovulating females that produce eggs with reduced fertility and that give rise to fewer viable offspring.

3.3.2.5 The need for marine fish oils in aqua-feeds

Marine fish oils are the major source of n-3 HUFAs in the human diet (Ruxton *et al.*, 2004, 2005; SACN-FSA, 2004; Bergé and Barnathan, 2005; Bourre, 2005; Cleland *et al.*, 2005; Robert, 2006). The long-chain, n-3 HUFAs are important constituents of aquatic food webs and are typically found in the flesh of oil-rich fish and in fish liver oils (Table 3.5) (Bergé and Barnathan, 2005; Cleland *et al.*, 2005). The fish do not produce significant amounts of the n-3 HUFAs themselves, but acquire them from aquatic microorganisms via the food chain.

Intensively farmed fish traditionally have been given feeds that contain oils extracted from pelagic marine fish. The oils are a rich source of n-3 HUFAs (Table 3.5), so farmed fish have acted as vehicles for delivering substantial quantities of n-3 HUFAs to human consumers. The supply of marine fish oils is finite and a marked expansion of intensive fish farming may require an alternative, sustainable source of the n-3 HUFAs. There is an additional incentive to investigate possible alternatives to fish oils in aqua-feeds, given the concerns raised about the levels of organochlorine contamination in some oils of marine origin (Easton *et al.*, 2002; Jacobs *et al.*, 2002; Berntssen *et al.*, 2005; Hamilton *et al.*, 2005; Bethune *et al.*, 2006).

Terrestrial plant oils usually contain lower concentrations of organochlorine contaminants than do marine fish oils (Jacobs *et al.*, 2002; Berntssen *et al.*, 2005). In addition, plant oils can provide a partial substitution for marine fish oils in feeds for farmed fish without any major detrimental influence on growth (Stickney, 2000; Halver and Hardy, 2002; Webster and Lim, 2002; Jobling, 2004c; Bell *et al.*, 2005). The major problem is that conventional plant oils contain very low concentrations of n-3 HUFAs (Table 3.5) (Bergé and Barnathan, 2005; Cleland *et al.*, 2005; McKevith, 2005; Robert, 2006). The result is that farmed fish given feeds containing large amounts of plant oils, and low concentrations of marine fish oils, have body fat that is rich in monounsaturates, particularly 18:1 n-9, and n-6 fatty acids, particularly 18:2 n-6, but with reduced levels of n-3 HUFAs. There are, however, efforts being made to develop transgenic plants that produce oils with increased proportions of n-3 HUFAs, with the long-term aim of using these oils as dietary supplements for humans and as ingredients in animal feeds.

The n-3 HUFAs are synthesized from 18:3 n-3 as the precursor, so, in theory, the n-3 HUFAs are not essential dietary components for either humans or farmed fish. Some plant oils, such as linseed and rape (canola), contain relatively high proportions of 18:3 n-3 (Table 3.5). Thus, it could be expected that these plant oils might be substituted for marine fish oils to meet all the requirements of farmed fish and humans for n-3 fatty acids. There are two major reasons why this may not be strictly true. First, the enzymatic desaturation and elongation pathway for conversion of 18:3 n-3 to n-3 HUFAs is relatively inefficient (Zheng *et al.*, 2004; Agaba *et al.*, 2005; Bergé and Barnathan, 2005), so a diet that is rich in 18:3 n-3 leads to an increased body burden of this fatty acid, but only a modest increase in the n-3 HUFAs (Wood *et al.*, 2003; Cleland *et al.*, 2005; Ruxton *et al.*, 2005; Visentainer *et al.*, 2005). Second, the biosynthetic pathway for n-3 HUFA production from 18:3 n-3 shares the enzyme $\Delta 6$ desaturase with the pathway for the synthesis of the n-6 HUFAs from 18:2 n-6 (Zheng *et al.*, 2004; Agaba *et al.*, 2005; Bergé and Barnathan, 2005). Although the enzyme has a preference for 18:3 n-3 over 18:2 n-6, the presence of high concentrations of 18:2 n-6 can shift enzymatic action towards the n-6 conversion pathway. The result is an inhibition of the pathway that converts 18:3 n-3 to n-3 HUFAs. Given that many plant oils contain high concentrations of 18:2 n-6 (Table 3.5), the production of n-3 HUFAs from 18:3 n-3 may be very low in animals and humans that consume diets containing plant oils (Cleland *et al.*, 2005). Consequently, manufacture of aqua-feeds with an almost complete substitution of marine fish oils by plant oils is not a viable alternative if the intention is to market farmed fish as a 'health food' with a high content of n-3 HUFAs; farmed fish given feeds containing conventional plant oils do not have high concentrations of n-3 HUFAs in their flesh.

At present, there is a dilemma. Increased use of plant oils could meet the requirements for manufacture of high-fat feeds with reduced organochlorine contamination. Use of plant oils would reduce the pressure on the limited resources of marine fish oils simultaneously. The major problem is that the fatty acid compositions of fish produced using feeds that contain high levels of conventional plant oils would not meet customer expectations. Many studies have been published on the effects of dietary fatty acids on the fatty acid compositions of fish flesh and this remains a topic of active research (e.g. Stickney, 2000; Halver and Hardy, 2002; Jobling, 2004c; Wonnacott *et al.*, 2004; Visentainer *et al.*, 2005). The earliest studies were limited to describing the influence of various dietary oils, or oil blends, on flesh fatty acids, but there has been a recent shift of focus towards examination of the ways in which flesh fatty acids can be manipulated using 'finishing feeds' that contain high concentrations of n-3 HUFAs (e.g. Jobling 2004a,b; Torstensen *et al.*, 2005).

Using a mixed feeding strategy, the farmed fish can be 'started' on feeds containing high concentrations of plant oils low in n-3 HUFAs and then given 'finishing feeds' containing high concentrations of marine fish oils that are rich in n-3 HUFAs. This boosts the n-3 HUFA concentration of the fish, relative to one fed exclusively on plant oils (Jobling, 2004a,b; Torstensen *et al.*, 2005). Such a mixed feeding strategy might have the added advantage of reducing levels of organochlorine contamination, relative to a fish fed marine fish oils throughout the entire growth cycle (Bell *et al.*, 2005). In other words, a mixed feeding strategy employing 'finishing feeds' based on marine fish oils is expected to result in:

- a final product (fish fillet) containing 'acceptable' levels of n-3 HUFAs
- a reduced risk of excessive contamination of the product with organochlorine compounds
- a more rational utilization of limited resources of marine fish oils

3.4 Feeding Behaviour, Feeding Routines and Feed Delivery Systems

Feeding behaviour and feed intake are under the influence of many biotic and abiotic factors (Fig. 3.3) (Houlihan *et al.*, 2001), and these may be categorized broadly as:

- the characteristics of the fish
- feed characteristics
- factors related to the culture environment
- factors related to feeding methods and feed availability

It is impossible to predetermine feeding routines and programme feed delivery systems to take account of all of these factors. The methods used to supply feed to farmed fish are feeding by hand and employment of automatically operated feed delivery systems (reviewed in Houlihan *et al.*, 2001; Jobling, 2004c).

3.4.1 Feeding methods and routines: assessment of satiation

When fish are fed by hand, it may sometimes be difficult to determine when they are approaching satiation and have reduced their feeding response. This may be a problem when the fish are being reared in sea cages and visibility is poor due to the water being turbid. The extent of the problem does, however, vary with the fish species that is being farmed. When salmonids are fed by hand, the water surface often appears to 'boil' as the fish engage in a feeding frenzy. Feed delivery is usually terminated once this surface activity becomes

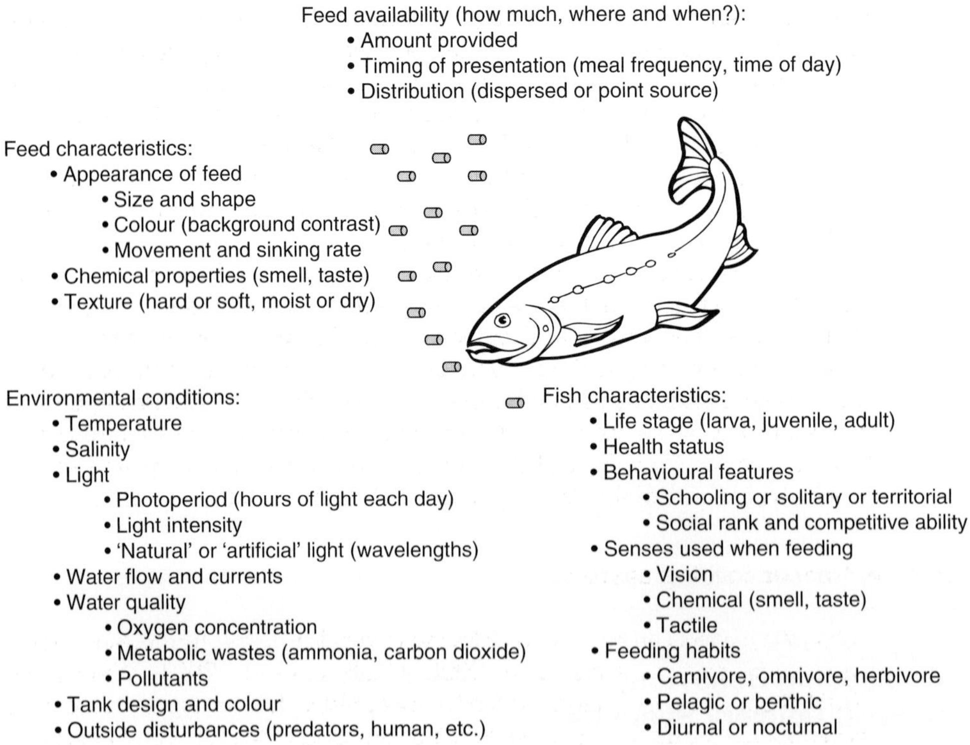

Fig. 3.3. Overview of biotic and abiotic factors that interact to influence feeding behaviour and feed intake in farmed fish. The factors may be grouped broadly into intrinsic – relating to the fish themselves – and extrinsic factors – relating to feeds, feeding and other aspects of the culture environment.

reduced, but reduced feeding activity at the surface does not necessarily mean that the fish are satiated. Even when the activity at the surface has subsided, the fish may still be willing to feed more slowly in the deeper water of the cage. Surface activity may also be influenced by conditions other than the eagerness of the fish to feed. For example, the fish may prefer to stay in deeper water when there is very bright sunlight. As such, a reduction in surface activity provides only a very crude assessment of when farmed salmon and trout are approaching satiation, so it is not a good criterion on which to base a feeding regime (Houlihan *et al.*, 2001).

Farmed fish of many other species do not show as pronounced a surface feeding response as the salmonids. Consequently, surface activity cannot be used to assess the feeding responses of these species reliably. Alternative methods are needed to monitor feeding responses and gauge whether or not the fish are responding to the feed. Several methods can be used to monitor the feeding responses and behaviour of fish in cages (Houlihan *et al.*, 2001; Jobling, 2004c). When underwater video cameras are placed deep in the cage with the lens directed towards the surface, the feeding activity of the fish can be assessed directly from the film monitor. Under some conditions, it may also be possible to see uneaten feed as it falls through the water column. Alternatively, monitoring devices placed deep in the cage can be used to detect uneaten feed pellets. These may be sensors that detect waste feed pellets using photocells or by hydroacoustic monitoring. As an alternative to sensors that detect uneaten pellets, devices for collection of feed waste can be placed directly in the cage. These usually consist of a large funnel placed in the bottom of the cage. The outlet at the base of the funnel is attached to a pipe that transports the waste to the surface using a pump or airlift system.

Additional problems encountered when feeding fish in large cages relate to rates of feed delivery and dispersal. Large quantities of feed pellets can be distributed in a short time when a feed cannon is used. The feed is usually either blown from the cannon using high gas pressures or is carried in a high-pressure jet of water. The cannon may be fitted with a 'spreading device' to ensure dispersal of the feed pellets over a wide area of water surface. These measures should ensure that the feed is delivered in sufficient quantity and over a sufficiently wide area to enable the majority of the fish to gain access to some of the feed pellets.

3.4.2 Automatic feeding systems

Automatic feeders can be used to distribute feed pellets to fish held in tanks, raceways, ponds and cages (Stickney, 2000; Houlihan *et al.*, 2001; Wedemeyer, 2001). Many designs of automatic feeder are available. Several types of automatic feeder operate by dispensing predetermined amounts of feed at preset time intervals. Automatic feeders that operate like this may be termed timed-release feeding systems. They may be conveyor-belt feeders that are either mechanically or electrically driven, feed hoppers that are vibrated at intervals to release feed, or disc feeders. Disc feeders are constructed of a rotating flat base plate and stop bar; feed is dispensed as the motor-driven flat plate rotates and the feed contacts the stop bar. Disc feeders can usually be programmed to give various rates of rotation of the plate, enabling rate of feed delivery to be varied. They can also be programmed to introduce pauses in disc rotation, enabling the feed to be dispensed in distinct meals; for example, 2 h with rotation in the 'feeding phase', followed by a 2 h pause when the plate is stationary and no feed is dispensed.

Timed-release feeding systems are those used most often for feeding small fish in the hatchery. The feeders are timed to deliver small portions of feed at short time intervals, i.e. almost continuously throughout the day (Wedemeyer, 2001). Timed-release feeding systems are also used to feed larger fish during on-growing to market size. The use of such feeders, although time-saving, may create problems for optimal feed management. There may not be optimal timing of feed dispersal and feed may be used inefficiently because:

- feed may be distributed at times of the day when the fish are not particularly eager to feed
- unless correctly adjusted, the feeder may deliver:
  - too little feed, resulting in 'underfeeding' of the fish, with poor growth as a result
  - too much feed, leading to excessive feed waste

Some of these problems can be either solved or reduced by using on-demand feeders.

On-demand feeders are feeding systems where the timing of feed release and the quantity of feed delivered by the system are controlled by the fish (Houlihan *et al.*, 2001; Jobling, 2004c). On-demand feeders may be either self-feeders or interactive feeding systems, these two types of system differing in their principles of operation (Fig. 3.4). A self-feeder is an on-demand feeding system in which feed release is controlled by a trigger mechanism operated by the fish, whereas in an interactive feeding system, feed release is controlled either via waste feed detection or by monitoring the behaviour of the fish. In an interactive feeding system, feed hoppers are programmed to release feed at regular intervals, a sensor is used to monitor behaviour or detect any uneaten feed and feedback signals from the sensor determine whether or not feed release should continue or be terminated.

When operated correctly, on-demand feeding systems, whether they be self-feeders or interactive, will give a temporal tracking of feed demand on an hourly and day-to-day basis (Fig. 3.5). They will ensure that most of the feed

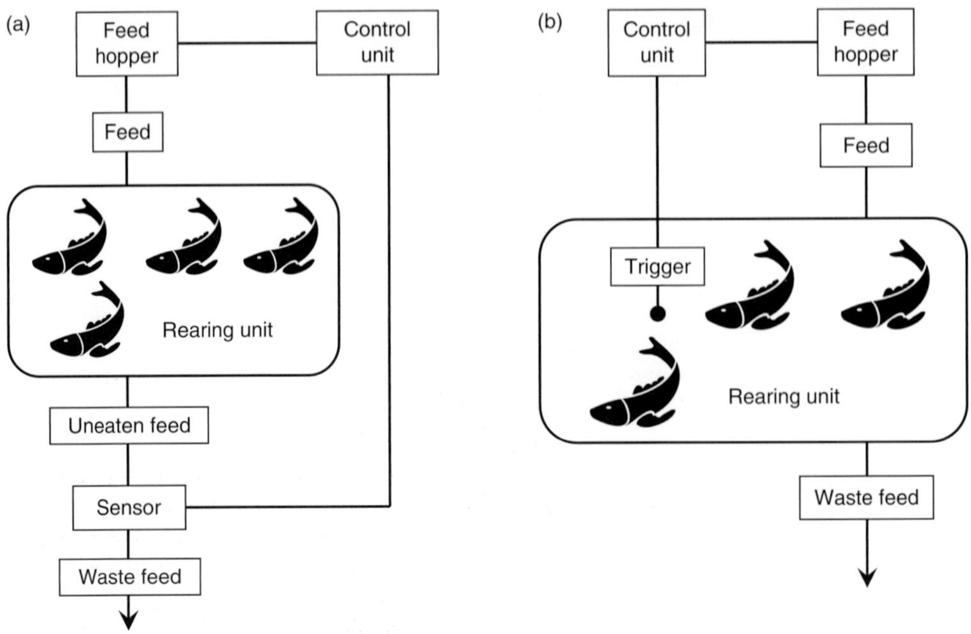

Fig. 3.4. On-demand feeders are of two types, self-feeders and interactive feeding systems. The two on-demand feeder systems differ in their principles of operation. (a) Interactive feeding system: feed release is controlled either via waste-feed detection or by monitoring the behaviour of the fish. Automatic feeders are timed to release feed at regular intervals, a sensor is used to monitor behaviour or detect any uneaten feed and feedback signals from the sensor determine whether or not feed release should be terminated. (b) Self-feeder: an on-demand feeding system in which feed release is controlled by a trigger-release mechanism operated by the fish (from Jobling, 2004c).

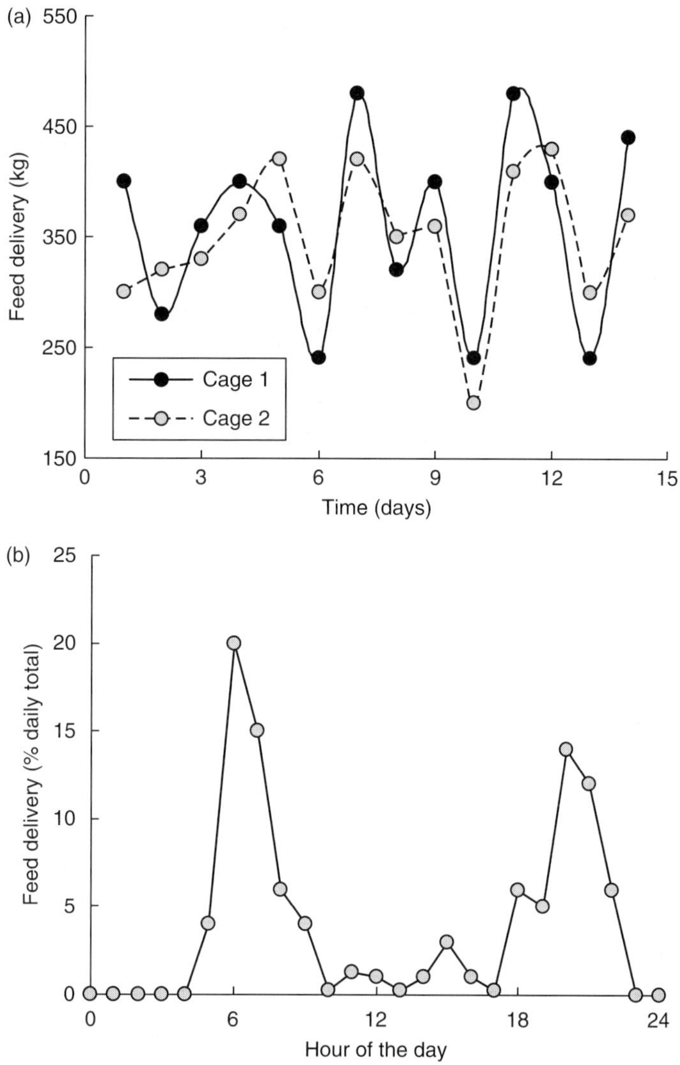

Fig. 3.5. Temporal changes in the amount of feed delivered to cages of farmed Atlantic salmon, *Salmo salar*, provisioned using interactive feeding systems. Feed delivery varied (a) between cages and from day-to-day for a given cage and (b) from hour-to-hour during the course of a given day. The recordings were made during late spring and early summer when day lengths were long and increasing. The feeders were 'idle' (inoperative) at night. Each fish cage held about 15,000 Atlantic salmon, weighing 2.5–3 kg.

pellets are dispensed at times when the fish are most eager to feed. Further, these systems may also provide adequate amounts of feed with little waste. This has dual advantages. First, the feed will be used efficiently and, second, there will be little risk of polluting the environment. Self-feeders have been used in the commercial production of channel catfish, *Ictalurus punctatus*, rainbow trout, *Oncorhynchus mykiss*, and European seabass, *Dicentrarchus labrax*, whereas interactive feeding systems are widely used in the sea-cage farming of salmonids. Interactive systems may have potential for use in the farming of many species, but they are most likely to meet with success when the fish are of a species that either displays a clear circadian rhythm in its feeding or consumes large meals at irregular intervals. This is because of the ease with which the feeding and non-feeding periods of such fish can be distinguished, enabling effective operation of an interactive feeding system.

3.5 Environmental Protection and Related Issues

Major issues that need to be addressed by the aquaculture industry relate to the rational use of limited resources, such as farm sites and water supplies, and the avoidance of environmental deterioration that could result from fish farming activities. There is a real risk that aquaculture can have negative environmental impacts and there is also some public concern about the safety of some aquaculture products. Major environmental impacts that have been attributed to aquaculture include the destruction of wetlands, excessive use of water resources and water pollution. Questions have also been raised about the possible inefficient use of natural resources. Any increase in the production of intensively farmed fish will lead automatically to an increase in the demand for aqua-feeds and the associated challenges this brings; the development of feeds that are better utilized by the fish, the search for novel feed ingredients and improvements in all aspects of feed management.

One area of public concern relates to worries about the possible eutrophication of recipient waterbodies by fish farm wastes. This has led to the introduction of water quality standards to govern the levels of suspended solids and dissolved nitrogen and phosphorus in farm effluents (Bergheim and Brinker, 2003; Lee, 2003; Boyd *et al.*, 2005; Papatryphon *et al.*, 2005). The problem of eutrophication of waterbodies that receive fish farm effluent can be addressed in three main ways. A reduction in the amount of feed waste reaching the environment can be achieved via technical improvements in feeding systems. Feeds can be formulated to reduce nutrient losses in faecal and metabolic wastes. These will be nutritionally balanced feeds incorporating highly digestible ingredients. In addition, effluent treatment can be carried out to reduce the amounts of suspended solids and dissolved phosphorus and nitrogenous compounds that are discharged to the environment (Lee, 2003; Eding *et al.*, 2006). This may be necessary to meet the mandatory requirements for wastewater discharges imposed by authorities in some countries (Bergheim and Brinker, 2003; Lee, 2003; Boyd *et al.*, 2005; Papatryphon *et al.*, 2005). Treatment of the farm effluent will also be required to reduce the risk of discharge of potentially pathogenic organisms to recipient waterbodies.

Regulation of wastewater discharge imposes a constraint on the types of rearing units that can be used. Closed and semi-closed land-based systems, based on tanks and raceways, allow relatively high levels of environmental control to be maintained. Wastewater treatment is also far easier than in open systems such as ponds and cages. In open systems, control of effluents and the release of wastes must be based on a rational choice of feeds and employment of appropriate methods for feed management.

There is doubtless considerable scope for reducing the potential negative impacts of intensive fish farming on the environment and environmentally responsible production is encapsulated within the concept of best management practices (BMPs). BMPs combine common sense, proven scientific principles, economics and good management to reduce or prevent adverse environmental impacts. The implementation and verification of BMPs should not be either difficult or too expensive to implement in practice and BMPs should also be demonstrably cost-effective. One approach to the development of BMPs is through the promotion of open dialogue between industry, research and regulatory agencies (Boyd *et al.*, 2005).

References

Acamovic, T. and Brooker, J.D. (2005) Biochemistry of secondary metabolites and their effects in animals. *Proceedings of the Nutrition Society* 64, 403–412.

Agaba, M.K., Tocher, D.R., Zheng, X., Dickson, C.A., Dick, J.R. and Teale, A.J. (2005) Cloning and functional characterisation of polyunsaturated fatty acid elongases of marine and freshwater teleost fish. *Comparative Biochemistry and Physiology* 142B, 342–352.

Balcázar, J.L., Blas, I. de, Ruiz-Zarzuela, I., Cunningham, D., Vendrell, D. and Múzquiz, J.L. (2006) The role of probiotics in aquaculture. *Veterinary Microbiology* 114, 173–186.

Basrur, P.K. (2006) Disrupted sex differentiation and feminization of man and domestic animals. *Environmental Research* 100, 18–38.

Bell, J.G., McGhee, F., Dick, J.R. and Tocher, D.R. (2005) Dioxin and dioxin-like polychlorinated biphenyls (PCBs) in Scottish farmed salmon (*Salmo salar*): effects of replacement of dietary marine fish oil with vegetable oil. *Aquaculture* 243, 305–314.

Bergé, J.-P. and Barnathan, G. (2005) Fatty acids from lipids of marine organisms: molecular biodiversity, roles as biomarkers, biologically active compounds, and economical aspects. *Advances in Biochemical Engineering/Biotechnology* 96, 49–125.

Bergheim, A. and Brinker, A. (2003) Effluent treatment for flow through systems and European Environmental Regulations. *Aquacultural Engineering* 27, 61–77.

Berntssen, M.H.G., Lundebye, A.-K. and Torstensen, B.E. (2005) Reducing the levels of dioxins and dioxin-like PCBs in farmed Atlantic salmon by substitution of fish oil with vegetable oil in the feed. *Aquaculture Nutrition* 11, 219–231.

Bethune, C., Seierstad, S.L., Seljeflot, I., Johansen, O., Arnesen, H., Meltzer, H.M., Rosenlund, G., Frøyland, L. and Lundebye, A.-K. (2006) Dietary intake of differently fed salmon: a preliminary study on contaminants. *European Journal of Clinical Investigation* 36, 193–201.

Bourre, J.M. (2005) Where to find omega-3 fatty acids and how feeding animals with diet enriched in omega-3 fatty acids to increase nutritional value of derived products for human: what is actually useful? *Journal of Nutrition Health and Aging* 9, 232–242.

Boyd, C.E., McNevin, A.A., Clay, J. and Johnson, H.M. (2005) Certification issues for some common aquaculture species. *Reviews in Fisheries Science* 13, 231–279.

Brevini, T.A.L., Cillo, F., Antonini, S. and Gandolfi, F. (2005) Effects of endocrine disrupters on the oocytes and embryos of farm animals. *Reproduction Domestic Animals* 40, 291–299.

Carlson, D.L. and Hites, R.A. (2005) Polychlorinated biphenyls in salmon and salmon feed: global differences and bioaccumulation. *Environmental Science and Technology* 39, 7389–7395.

Cleland, L.G., James, M.J. and Proudman, S.M. (2005) Fish oil: what the prescriber needs to know. *Arthritis Research and Therapy* 8:202, 9 pp.

Diamond, J. (2002) Evolution, consequences and future of plant and animal domestication. *Nature* 418, 700–707.

Easton, M.D.L., Luszniak, D. and Von der Geest, E. (2002) Preliminary examination of contaminant loadings in farmed salmon, wild salmon and commercial salmon feed. *Chemosphere* 46, 1053–1074.

Eding, E.H., Kamstra, A., Verreth, J.A.J., Huisman, E.A. and Klapwijk, A. (2006) Design and operation of nitrifying trickling filters in recirculating aquaculture: a review. *Aquacultural Engineering* 34, 234–260.

Foran, J.A., Good, D.H., Carpenter, D.O., Hamilton, M.C., Knuth, B.A. and Schwager, S.J. (2005) Quantitative analysis of the benefits and risks of consuming farmed and wild salmon. *Journal of Nutrition* 135, 2639–2643.

Francis, G., Makkar, H.P.S. and Becker, K. (2001) Antinutritional factors present in plant-derived alternate fish feed ingredients and their effects in fish. *Aquaculture* 199, 197–227.

Gatesoupe, F.J. (1999) The use of probiotics in aquaculture. *Aquaculture* 180, 147–165.

Halver, J.E. and Hardy, R.W. (eds) (2002) *Fish Nutrition*. Academic Press, San Diego, California.

Hamilton, M.C., Hites, R.A., Schwager, S.J., Foran, J.A., Knuth, B.A. and Carpenter, D.O. (2005) Lipid composition and contaminants in farmed and wild salmon. *Environmental Science and Technology* 39, 8622–8629.

Heiden, T.K., Carvan, M.J. III and Hutz, R.J. (2006) Inhibition of follicular development, vitellogenesis, and serum 17β-estradiol concentrations in zebrafish following chronic, sublethal dietary exposure to 2, 3, 7, 8-tetrachlorodibenzo-*p*-dioxin. *Toxicological Sciences* 90, 490–499.

Hertrampf, J.W. and Piedad-Pascual, F. (2000) *Handbook on Ingredients for Aquaculture Feeds*. Kluwer Academic, Dordrecht, The Netherlands.

Hiramatsu, N., Matsubara, T., Fujita, T., Sullivan, C.V. and Hara, A. (2006) Multiple piscine vitellogenins: biomarkers of fish exposure to estrogenic endocrine disruptors in aquatic environments. *Marine Biology* 149, 35–47.

Hites, R.A., Foran, J.A., Carpenter, D.O., Hamilton, M.C., Knuth, B.A. and Schwager, S.J. (2004) Global assessment of organic contaminants in farmed salmon. *Science* 303, 226–229; 305, 475–478.

Houlihan, D., Boujard, T. and Jobling, M. (eds) (2001) *Food Intake in Fish*. Blackwell Science, Oxford, UK.

Irianto, A. and Austin, B. (2002) Probiotics in aquaculture. *Journal of Fish Diseases* 25, 633–642.

Jacobs, M.N., Covaci, A. and Schepens, P. (2002) Investigation of selected persistent organic pollutants in farmed Atlantic salmon (*Salmo salar*), salmon aquaculture feed, and fish oil components of the feed. *Environmental Science and Technology* 36, 2797–2805.

Jalabert, B., Baroiller, J.-F., Breton, B., Fostier, A., Le Gac, F., Guiguen, Y. and Monod, G. (2000) Main neuro-endocrine, endocrine and paracrine regulations of fish reproduction, and vulnerability to xenobiotics. *Ecotoxicology* 9, 25–40.

Jobling, M. (2004a) Are modifications in tissue fatty acid profiles following a change in diet the result of dilution? Test of a simple dilution model. *Aquaculture* 232, 551–562.

Jobling, M. (2004b) 'Finishing' feeds for carnivorous fish and the fatty acid dilution model. *Aquaculture Research* 35, 706–709.

Jobling, M. (2004c) On-growing to market size. In: Moksness, E., Kjørsvik, E. and Olsen, Y. (eds) *Culture of Cold-water Marine Fish*. Blackwell Science, Oxford, UK, pp. 363–432.

Kelly, A.M. and Green, C.C. (2006) Binding affinities of hepatic nuclear estrogen receptors for genistein in channel catfish. *North American Journal of Aquaculture* 68, 160–167.

Langdon, C. (2003) Microparticle types for delivering nutrients to marine fish larvae. *Aquaculture* 227, 259–275.
Lee, C.-S. (ed.) (2003) Management of aquaculture effluents. *Aquaculture* 226, 1–244.
McKevith, B. (2005) Nutritional aspects of oilseeds. *Nutrition Bulletin* 30, 13–26.
MacRae, J., O'Reilly, L. and Morgan, P. (2005) Desirable characteristics of animal products from a human health perspective. *Livestock Production Science* 94, 95–103.
Mandal, P.K. (2005) Dioxin: a review of its environmental effects and its aryl hydrocarbon receptor biology. *Journal of Comparative Physiology B* 175, 221–230.
Matsumoto, T., Kobayashi, M., Moriwaki, T., Kawai, S. and Watabe, S. (2004) Survey of estrogenic activity in fish feed by yeast estrogen-screen assay. *Comparative Biochemistry and Physiology* 139C, 147–152.
Matsuoka, M., Ishibashi, H., Ushijima, M., Inudo, M., Honda, E., Iwahara, M., Cho, H.-S., Ishibashi, Y. and Arizono, K. (2005) The potential contribution of phytoestrogens and organochlorine pesticides in an experimental fish diet to estrogenic activity. *Journal of Health Science* 51, 212–219.
Mignon-Grasteau, S., Boissy, A., Bouix, J., Faure, J.-M., Fisher, A.D., Hinch, G.N., Jensen, P., Le Neindre, P., Mormede, P., Prunet, P., Vandeputte, M. and Beaumont, C. (2005) Genetics of adaptation and domestication in livestock. *Livestock Production Science* 93, 3–14.
Mills, L.J. and Chichester, C. (2005) Review of evidence: are endocrine-disrupting chemicals in the aquatic environment impacting fish populations? *Science of the Total Environment* 343, 1–34.
Milnes, M.R., Bermudez, D.S., Bryan, T.A., Edwards, T.M., Gunderson, M.P., Larkin, I.L.V., Moore, B.C. and Guillette, L.J. Jr (2006) Contaminant-induced feminization and demasculinization of non-mammalian vertebrate males in aquatic environments. *Environmental Research* 100, 3–17.
Miyamoto, J. and Burger, J. (eds) (2003) Implications of endocrine active substances for humans and wildlife; Reports from a SCOPE/IUPAC project. *Pure and Applied Chemistry* 75, 1617–2615.
Montory, M. and Barra, R. (2006) Preliminary data on polybrominated diphenyl ethers (PBDEs) in farmed fish tissues (*Salmo salar*) and fish feed in Southern Chile. *Chemosphere* 63, 1252–1260.
Muir, J. (2005) Managing to harvest? Perspectives on the potential of aquaculture. *Philosophical Transactions of the Royal Society B* 360, 191–218.
Naylor, R.L., Goldburg, R.J., Mooney, H., Beveridge, M.C.M., Clay, J., Folke, C., Kautsky, N., Lubchenko, J., Primavera, J. and Williams, M. (1998) Nature's subsidies to shrimp and salmon farming. *Science* 282, 883–884.
Naylor, R.L., Goldburg, R.J., Primavera, J., Kautsky, N., Beveridge, M.C.M., Clay, J., Folke, C., Lubchenko, J., Mooney, H. and Troell, M. (2000) Effects of aquaculture on world fish supplies. *Nature* 405, 1017–1024.
Ng, Y., Hanson, S., Malison, J.A., Wentworth, B. and Barry, T.P. (2006) Genistein and other isoflavones found in soybeans inhibit estrogen metabolism in salmonid fish. *Aquaculture* 254, 658–665.
Nie, X.P., Lan, C.Y., An, T.C., Li, K.B. and Wong, M.H. (2006) Distributions and congener patterns of PCBs in fish from major aquaculture areas in the Pearl River Delta, South China. *Human and Ecological Risk Assessment* 12, 363–373.
Olsen, Y. (2004) Live food technology of cold-water marine fish larvae. In: Moksness, E., Kjørsvik, E. and Olsen, Y. (eds) *Culture of Cold-water Marine Fish*. Blackwell Science, Oxford, UK, pp. 73–128.
Olsen, Y., Meeren, T. van der and Reitan, K.I. (2004) First feeding technology. In: Moksness, E., Kjørsvik, E. and Olsen, Y. (eds) *Culture of Cold-water Marine Fish*. Blackwell Science, Oxford, UK, pp. 279–336.

Papatryphon, E., Petit, J., Werf, H.M.G. van der, Kaushik, S.J. and Claver, K. (2005) Nutrient-balance modeling as a tool for environmental management in aquaculture: the case of trout farming in France. *Environmental Management* 35, 161–174.
Pond, W.G., Church, D.C. and Pond, K.R. (1995) *Basic Animal Nutrition and Feeding*. John Wiley and Sons, New York.
Reynaud, S. and Deschaux, P. (2006) The effect of polycyclic aromatic hydrocarbons on the immune system of fish: a review. *Aquatic Toxicology* 77, 229–238.
Rhind, S.M. (2005) Are endocrine disrupting compounds a threat to farm animal health, welfare and productivity? *Reproduction Domestic Animals* 40, 282–290.
Rigos, G. and Troisi, G.M. (2005) Antibacterial agents in Mediterranean finfish farming: a synopsis of drug pharmacokinetics in important euryhaline fish species and possible environmental implications. *Reviews in Fish Biology and Fisheries* 15, 53–73.
Robert, S.S. (2006) Production of eicosapentaenoic and docosahexaenoic acid-containing oils in transgenic land plants for human and aquaculture nutrition. *Marine Biotechnology* 8, 103–109.
Ruxton, C.H.S., Reed, S.C., Simpson, M.J.A. and Millington, K.J. (2004) The health benefits of omega-3 polyunsaturated fatty acids: a review of the evidence. *Journal of Human Nutrition and Dietetics* 17, 449–459.
Ruxton, C.H.S., Calder, P.C., Reed, S.C. and Simpson, M.J.A. (2005) The impact of long-chain n-3 polyunsaturated fatty acids on human health. *Nutrition Research Reviews* 18, 113–129.
SACN-FSA (Food Standards Agency) (2004) *Advice on Fish Consumption: Benefits and Risks*. The Stationary Office (HMSO), London.
Scott, A.P., Katsiadaki, I., Witthames, P.R., Hylland, K., Davies, I.M., McIntosh, A.D. and Thain, J. (2006) Vitellogenin in the blood plasma of male cod (*Gadus morhua*): a sign of oestrogenic endocrine disruption in the open sea? *Marine Environmental Research* 61, 149–170.
Shelton, W.L. and Rothbard, S. (2006) Exotic species in global aquaculture – a review. *Bamidgeh* 58, 3–28.
Shi, L.L., Shan, Z.J., Kong, D.Y. and Cai, D.J. (2006) The health and ecological impacts of organochlorine pesticide pollution in China: bioaccumulation of organochlorine pesticides in human and fish fats. *Human and Ecological Risk Assessment* 12, 402–407.
Stickney, R.R. (ed.) (2000) *Encyclopedia of Aquaculture*. Wiley-Interscience, New York.
Stottrup, J. and McEvoy, L. (eds) (2002) *Live Feeds in Marine Aquaculture*. Blackwell Science, Oxford, UK.
Sumpter, J.P. and Johnson, A.C. (2005) Lessons from endocrine disruption and their application to other issues concerning trace organics in the aquatic environment. *Environmental Science and Technology* 39, 4321–4332.
Tidwell, J.H. and Allan, G.L. (2001) Fish as food: aquaculture's contribution. *EMBO Reports* 2, 958–963.
Torstensen, B.E., Bell, J.G., Rosenlund, G., Henderson, R.J., Graff, I.E., Tocher, D.R., Lie, Ø. and Sargent, J.R. (2005) Tailoring of Atlantic salmon (*Salmo salar* L.) flesh lipid composition and sensory quality by replacing fish oil with a vegetable oil blend. *Journal of Agricultural and Food Chemistry* 53, 10166–10178.
Utne-Palm, A.C. (2002) Visual feeding of fish in a turbid environment: physical and behavioural aspects. *Marine and Freshwater Behaviour and Physiology* 35, 111–128.
Vine, N.G., Leukes, W.D. and Kaiser, H. (2006) Probiotics in marine larviculture. *FEMS Microbiological Reviews* 30, 404–427.
Visentainer, J.V., Souza, N.E. de, Makoto, M., Hayashi, C. and Franco, M.R.B. (2005) Influence of diets enriched with flaxseed oil on the α-linolenic, eicopentaenoic and docosahexaenoic fatty acid in Nile tilapia (*Oreochromis niloticus*). *Food Chemistry* 90, 557–560.

Waring, R.H. and Harris, R.M. (2005) Endocrine disrupters: a human risk? *Molecular and Cellular Endocrinology* 244, 2–9.

Webster, C.D. and Lim, C.E. (eds) (2002) *Nutrient Requirements and Feeding of Finfish for Aquaculture*. CAB International, Wallingford, UK.

Wedemeyer, G. (ed.) (2001) *Fish Hatchery Management*. American Fisheries Society, Bethesda, Maryland.

Wonnacott, E.J., Lane, R.L. and Kohler, C.C. (2004) Influence of dietary replacement of menhaden oil with canola oil on fatty acid composition of sunshine bass. *North American Journal of Aquaculture* 66, 243–250.

Wood, J.D., Richardson, R.J., Nute, G.R., Fisher, A.V., Campo, M.M., Kasapidou, E., Sheard, P.R. and Enser, M. (2003) Effects of fatty acids on meat quality: a review. *Meat Science* 66, 21–32.

Zheng, X., Seiliez, I., Hastings, N., Tocher, D.R., Panserat, S., Dickson, C.A., Bergot, P. and Teale, A.J. (2004) Characterization and comparison of fatty acyl Δ6 desaturase cDNAs from freshwater and marine teleost fish species. *Comparative Biochemistry and Physiology* 138B, 269–279.

4 Farmed Species and Their Characteristics

MALCOLM JOBLING

University of Tromsø, Tromsø, Norway

4.1 Introduction

Fish culture is a segment of agriculture and there are several similarities between the rearing of fish and the raising of terrestrial farm animals. There are, however, also numerous differences (see Chapter 6, this volume). These differences relate mostly to the challenges imposed by the rearing of animals in aquatic as against terrestrial environments (see Chapter 2, this volume). There are also major differences in the extent to which farmed fish and farmed terrestrial livestock have been subject to selective breeding and domestication.

Fish culture has been practised since ancient times. Some Egyptian bas-reliefs appear to show scenes of fish, possibly tilapia, *Tilapia* or *Oreochromis* sp., being reared in ponds, but South-east Asia is probably the birthplace of fish culture. For centuries, the people of the Indo-Pacific region have reared fish for consumption. Thus, throughout China, Indonesia and the Indian subcontinent, fish cultivation has a long tradition. In Europe, the Romans are known to have practised the pond rearing of fish, particularly common carp, *Cyprinus carpio* (Balon, 2004; Mignon-Grasteau *et al.*, 2005). During the Middle Ages, fish culture was carried out by monks of European monasteries; first with common carp and, later, brown trout, *Salmo trutta*. By the middle of the 19th century, fish culture had become well established in many European countries. Nevertheless, although fish farming is recorded in antiquity, its greatest expansion, worldwide, has occurred since the 1960s (Fig. 4.1). This recent upsurge in production volume and concomitant increase in numbers of farmed fish species (http://www.fao.org/figis) means that there are relatively few fish species that can be considered as domesticated animals (Balon, 2004; Mignon-Grasteau *et al.*, 2005; Shelton and Rothbard, 2006).

The recent expansion in fish farming has been driven, in part, by diminishing natural fisheries coupled to an increased consumer demand for fish products (Billard, 2003; Muir, 2005; Shelton and Rothbard, 2006). Successful

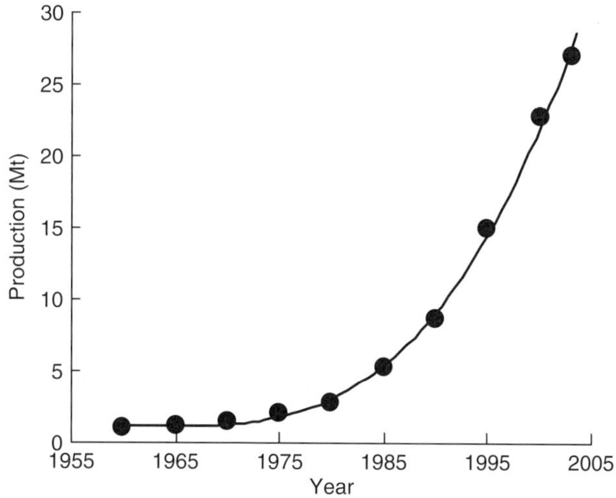

Fig. 4.1. Annual production of farmed fish in the period 1960–2003 (data source: http://www.fao.org/figis).

development of a given fish species for culture depends on a favourable mix of biological characteristics, the availability of support resources and infrastructure and market demand for the product (see Chapters 5, 7 and 8, this volume). Although these factors are common to the development of intensive fish culture in general, unique factors also contribute to the developmental success of specific sectors and species (see Chapters 6, 7, 23 and 25, this volume). The history of intensive fish culture is fraught with premature attempts to develop industrial enterprises based more on over-optimistic speculations about market demand, rather than on biological and technical knowledge and adequate information about economic feasibility. Thus, numerous factors need to be taken into consideration before embarking on an attempt to develop a species for commercial production (see Chapters 5, 7 and 8, this volume).

4.2 Characteristics Desirable in a Farmed Species

Although the cultivation of some fish species may have arisen as the result of serendipity, only a restricted number of the *c.*28,500 fish species is suitable for culture. In other words, regardless of the potential market demand for particular species, farming will be unable to develop beyond a stage of infancy if the fish is too difficult to produce in sufficiently large numbers to meet consumer demand. Some generalizations can be made regarding the characteristics that make a species suitable for farming. Fish species with potential for cultivation should:

- reproduce readily in captivity
- withstand the conditions of the rearing environment without difficulty
- tolerate the high rearing densities of intensive culture

- grow rapidly and reach market size in a short time
- accept and thrive on relatively inexpensive formulated feeds
- be resistant to disturbance and disease
- be accepted readily by consumers

Only a limited number of fish species will meet these requirements and few, if any, will fulfil all of the cultivation criteria to perfection (see Chapters 7 and 8, this volume).

A prerequisite for intensive fish farming is a ready and reliable supply of eggs that provides the juveniles needed for on-growing of the fish to a marketable size. This requires that the fish reproduce in captivity. It is obviously a major advantage if the fish mature and spawn without the need of any special conditions that are difficult to create in captivity. Having the ability to manipulate the timing of maturation, rates of reproductive development and the timing of spawning is also deemed advantageous, as this allows reproduction to be controlled for the production of 'out-of-season' eggs and juveniles. Control over reproduction can be exerted by several means, including hormonal treatments and the use of environmental manipulations, such as photothermal treatment of broodstock (Bromage *et al.*, 2001; Devlin and Nagahama, 2002; Patiño, 2002; Kagawa *et al.*, 2003).

Fish that are to be farmed successfully must have the ability to adapt to the culture environment. The culture environment is a complex of numerous abiotic factors, relating to rearing units and water characteristics, and biotic factors that may encompass conspecifics, heterospecific fish species and potential disease organisms (see Chapter 2, this volume). As such, the fish must be resistant to disease and they must also tolerate the rigours of handling and transport. In semi-intensive and intensive culture, farmers exert some control over the rearing environment by heating to increase water temperature, by manipulation of lighting conditions, by aerating the water or by applying a range of water treatments (Stickney, 2000; Wedemeyer, 2001; Lee, 2003). The high levels of production achieved in semi-intensive and intensive culture systems are based on the use of formulated feeds and much of the production cost in the intensive cultivation of fish revolves around feeds and feeding (Stickney, 2000; Houlihan *et al.*, 2001; Wedemeyer, 2001; see Chapter 3, this volume).

Fish that are intended for human consumption should reach market size rapidly; the fish should grow quickly and should be capable of being harvested before they mature sexually and growth rate slows. It should also be clear that fish intended as food should meet the tastes and expectations of the consumer, with regard to nutritional value, appearance and price (Kestin and Warriss, 2001; see Chapters 22 and 26, this volume). Fish species that achieve only small body size are usually not suitable for farming, even if they adapt readily to captivity and display high growth rates. The proviso here is that the fish are intended for consumption; exceptions exist if the fish are intended for the aquarium trade (ornamental fish), are small fish reared for the baitfish market or are fish to be used for biomedical research or environmental monitoring purposes. Production of ornamental fish and the growing of small fish for the baitfish market do, however, represent a significant part of the aquaculture industry in some countries (Shelton and Rothbard, 2006).

4.3 Fish as Food: A Comparison with Terrestrial Livestock

Fish are an important part of the human diet in many countries and are often perceived as having some nutritional advantages over the 'meat' of terrestrial livestock (Macrae *et al.*, 1993; Kestin and Warriss, 2001; Bourre, 2005). Fish are important as a protein source, are a source of fats and are also valuable sources of some vitamins and minerals (Table 4.1) (Jobling, 2004; SACN-FSA, 2004). Fish tissues contain a number of bioactive compounds, including peptides, fatty acids and enzymes, that may be beneficial within the context of human health (SACN-FSA, 2004; Bergé and Barnathan, 2005; Bourre, 2005; Cleland *et al.*, 2005). Efforts have been made to extract some of these compounds from fisheries by-products for use as food additives or for use by the pharmaceutical industry (Guérard *et al.*, 2005; Kim and Mendis, 2006; see Chapter 25, this volume).

Some of the fatty acids that humans obtain by consuming fish are considered to have a number of important roles in disease prevention and the maintenance of health (Bergé and Barnathan, 2005; Bourre, 2005; Cleland *et al.*, 2005). Human nutritionists recommend a high intake of n-3 highly unsaturated fatty acids (n-3 HUFAs) and it is recommended that n-3 HUFA consumption is increased at the expense of the intake of saturated and n-6 fatty acids (Buttriss,

Table 4.1. Chemical (nutritional) composition (as % wet weight) and fatty acid class (as % total fatty acids) compositions of the fillets of some farmed coldwater fishes. SFA, saturated fatty acid; MUFA, monounsaturated fatty acid; PUFA, polyunsaturated fatty acid; HUFA, highly unsaturated fatty acid (from Jobling, 2004).

	Atlantic halibut *Hippoglossus hippoglossus*	Atlantic salmon *Salmo salar*	Wolffish *Anarhichas lupus*	Turbot *Scophthalmus maximus*	Atlantic cod *Gadus morhua*
Moisture (%)	72	69	78	79	80
Dry matter (%)	28	31	22	21	20
Protein	16	18	19	16	18
Lipid	10	10	2.5	2.5	0.5
Fatty acid classes (as % fatty acids)					
Saturates (SFA)	16	23	22	23	22
MUFAs	72	50	39	32	18
n-3 Fatty acids (PUFAs and HUFAs)	8	20	27	36	54
n-6 Fatty acids (PUFAs and HUFAs)	2	5	7	6	3

2004; Ruxton *et al.*, 2004, 2005; SACN-FSA, 2004; Bergé and Barnathan, 2005; Cleland *et al.*, 2005; MacRae *et al.*, 2005). Fish are currently the major source of n-3 HUFAs in the human diet, whereas most plant, or vegetable, oils and the 'meat' of conventional farm animals contain only traces of these fatty acids (Macrae *et al.*, 1993; Givens and Shingfield, 2004; Bergé and Barnathan, 2005; Bourre, 2005; Cleland *et al.*, 2005; MacRae *et al.*, 2005; Ruxton *et al.*, 2005).

Fatty acid supplementation of terrestrial farm animal feeds has been attempted as a strategy to augment the n-3 HUFA content of meat, e.g. beef, lamb and pork, milk and eggs (Wood *et al.*, 2003; Givens and Shingfield, 2004; Bourre, 2005; MacRae *et al.*, 2005). The manipulation of the n-3 HUFA content of milk and meat of ruminants (cattle and sheep) by dietary means is usually less successful than in eggs and the meat of non-ruminants (pigs and poultry). This is because there is a significant metabolism of the dietary fatty acids by rumen microorganisms. Only a very small proportion of the dietary n-3 HUFAs escapes microbial metabolism and is deposited in the body fat of ruminant animals. On the other hand, the fatty acid profiles of non-ruminants are more representative of those of the diet. In these animals, there is only a limited transformation of the dietary fatty acids during digestion and absorption (Givens and Shingfield, 2004; Bergé and Barnathan, 2005; Bourre, 2005). Supplementation of pig and poultry feeds with fish oils has been the most common means used to enrich pork, poultry meat and eggs with n-3 HUFAs, but this can give rise to products with metallic taint, fishy flavour and reduced shelf life (Wood *et al.*, 2003; Givens and Shingfield, 2004; Bourre, 2005; MacRae *et al.*, 2005; Ruxton *et al.*, 2005).

To summarize: the majority of the fatty acids in the 'meat' from terrestrial animals are saturates and monounsaturates (Macrae *et al.*, 1993; Wood *et al.*, 2003; Givens and Shingfield, 2004; MacRae *et al.*, 2005) and fish fats are generally a much better source of n-3 HUFAs than are the fats present in the 'meat' of conventional farm animals. The fatty acid compositions of the storage fats of non-ruminant animals, including fish, are very much under the influence of the fatty acids contained in their feeds (Jobling, 2004; Bergé and Barnathan, 2005; Bourre, 2005). Fish fats contain high proportions of n-3 HUFAs because these fatty acids are characteristic of aquatic, particularly marine, food chains. This means that if intensively farmed fish are to deposit n-3 HUFAs in their body fats, they must be given feeds that contain fatty acids of the n-3 series.

4.4 Natural Diets of Farmed Animals

Farmed terrestrial species are either herbivorous or omnivorous (Diamond, 2002; Mignon-Grasteau *et al.*, 2005) and, therefore, naturally include a large proportion of plants in their diets (Pond *et al.*, 1995; Flachowsky *et al.*, 2005). Although a number of herbivorous and omnivorous fish species are farmed (Nash and Novotny, 1995; Stickney, 2000; Billard, 2003; Muir, 2005; Shelton and Rothbard, 2006), many of the fish that are cultivated intensively are carnivores that naturally eat other animals. In other words, many intensively farmed fish species are the aquatic equivalents of members of the dog and cat families. These differences in natural feeding habits have consequences for the formulation of feeds given to farmed fish (Pond *et al.*, 1995; Halver and Hardy, 2002; Webster and Lim, 2002).

The digestible energy (DE) that is used as metabolic fuel by terrestrial livestock, such as poultry and pigs, is provided largely in the form of carbohydrates (starches and sugars) and feeds for these animals usually have a relatively low protein content (12–25% of feed dry matter). Feeds given to farmed fish usually have a higher protein content (30–55% of dry matter, depending on species and life stage) and higher protein-to-energy ratios (protein:DE), than feeds given to terrestrial livestock. Feed ingredients that contain high concentrations of carbohydrates, for example, grains, maize and potato, are much cheaper than are ingredients that have high protein content, for example, fishmeal. This means that pig and poultry feeds are generally cheaper to produce than are fish feeds. This difference in basic feed costs does something to offset the production advantage that accrues due to the better utilization of feed for growth by farmed fish in comparison with terrestrial livestock.

4.5 Domestication and Genetic Selection

The farming of terrestrial livestock is most often based on the rearing of domesticated animals. Domestication encompasses the processes by which animals become adapted to a life in captivity by a combination of genetic changes that occur over generations and responses to environmentally induced developmental events that recur regularly over time. Domesticated animals can be considered as those that have been bred in captivity for some specific purpose, for example, as food animals, ornamentals or as pets, and where breeding and other aspects of husbandry are under human control (Diamond, 2002; Balon, 2004; Mignon-Grasteau *et al.*, 2005; Jensen, 2006). Such animals differ from the wild ancestral stock in morphology, physiology and behaviour to such an extent that they have reduced survival in the absence of human protection. Genetic mechanisms are basic to the process of domestication. Domestication involves the relaxation of certain natural selection factors, such as predation and death by starvation, there is selection for specific traits desired by humans and there is also a component of selection imposed by the captive environment. Thus, domesticated animals generally comprise strains and varieties that have been genetically selected for specific production traits over many generations and for which environmental and nutritional requirements have been well researched and documented.

Goals of breeding and selection programmes include the alteration of the characteristics of the farmed animals to make production more effective and profitable. The animals will often be selected to show fast growth, improved utilization of feed and increased resistance to disease. In addition, a domesticated phenotype possesses a range of physiological and behavioural traits that facilitate adaptation to the captive environment. These may include reduced behavioural reactivity towards humans, increased tolerance of the close presence of conspecifics and attenuated stress responses (Mignon-Grasteau *et al.*, 2005; Jensen, 2006). All of these traits are also important components with regard to considerations of animal welfare (Håstein *et al.*, 2005; Broom, 2006; Huntingford *et al.*, 2006; Jensen, 2006).

With a few exceptions – common carp, *C. carpio*, tilapias, *Oreochromis* spp., channel catfish, *I. punctatus*, rainbow trout, *Oncorhynchus mykiss*, and Atlantic salmon, *S. salar* – farmed fish have either not been subjected to concerted efforts in domestication and selective breeding or selection programmes are at a very early stage of development (Dunham, 2004; Gjedrem, 2005; Shelton and Rothbard, 2006). Only a handful of fish species can be considered as being domesticated animals and the majority of farmed fish species are best classified as being exploited captives (Balon, 2004; Mignon-Grasteau *et al.*, 2005; Muir, 2005; Shelton and Rothbard, 2006).

4.6 Culture Species

The fish, with about 28,500 living species, are represented by the largest number of species within the vertebrate lineage. Although about 300 freshwater and marine fish species are cultivated, only *c.*150 are currently produced in any quantity and the production figures tend to be dominated by representatives from a small number of fish families (Nash and Novotny, 1995; Stickney, 2000; Billard and Berni, 2004; Muir, 2005; Shelton and Rothbard, 2006; http://www.fao.org/figis). Cyprinids (carps and their close relatives) and tilapias are the most important groups and they dominate fish culture in fresh water. Catfishes are another important group in freshwater culture and the salmonids are produced in considerable quantities in both fresh and salt waters. Much of the recent increase in production of these species groups, and in aquaculture production as a whole, has resulted from transplantation of several of the major species into new geographic areas (Shelton and Rothbard, 2006).

Several species within the carp family (Cyprinidae) are farmed in South-east Asia, on the Indian subcontinent, in Europe and elsewhere (Billard, 2003; Balon, 2004; Billard and Berni, 2004; Shelton and Rothbard, 2006). There is, however, relatively little interest in cyprinid culture in North America and annual aquaculture production of cyprinids in North America amounts to no more than 5000–10,000 t (http://www.fao.org/figis). The common carp, *C. carpio*, is the species that is most widely farmed and it dominates cyprinid culture in Europe; farming of common carp represents about 145,000 t of the *c.*225,000 t of cyprinids produced in Europe (Billard, 1999; Shelton and Rothbard, 2006; http://www.fao.org/figis). In South-east Asia, various other cyprinid species, often referred to as Chinese carps, are produced in great quantity (Fig. 4.2) (Shelton and Rothbard, 2006; http://www.fao.org/figis). These include grass carp, *Ctenopharyngodon idella*, mud carp, *Cirrhina molitorella*, silver carp, *Hypophthalmichthys molitrix*, bighead carp, *Aristichthyes nobilis*, and *Carassius* spp. (crucian carp and gibel carp). Production of the *Carassius* spp. takes place in several Asian countries and in

Eastern Europe, but production is low outside China. The same can be said for several of the other cyprinids. For example, although there is production of grass carp and bighead carp in over 20 countries (mostly in South-east Asia, but also in Europe and on the American continent), fewer than 10 of these countries report an annual production of over 1000t. On the Indian subcontinent, cyprinid culture principally revolves around the rearing of catla, *Catla catla*, mrigal, *C. mrigala*, rohu, *Labeo rohita*, and calbasu, *L. calbasu*; the Indian major carps (Fig. 4.2). There is also some production of Chinese carps in this region. The vast majority of cyprinid farming occurs in ponds employing either extensive or semi-intensive methods. Pond-based, semi-intensive polyculture of cyprinids is the norm in many of the South-east Asian countries.

There is widespread cultivation of several species of tilapia (family Cichlidae). Tilapia have been nicknamed 'aquatic chicken' due to their high growth rates, feeding habits, the ease with which they can be held in captivity and their ability to adapt to a wide range of environmental conditions (Lim and Webster, 2006; see Chapter 17, this volume). As such, they are excellent candidates for aquaculture in tropical and subtropical regions and tilapia farming is currently practised in over 100 countries worldwide. Tilapia culture is carried out in Asia, on the African subcontinent, in the Middle East and Mediterranean countries and in the Americas (Beveridge and McAndrew, 2000; El-Sayed, 2006; Lim and Webster, 2006; Shelton and Rothbard, 2006). Although tilapia culture in the Americas was initiated through the introduction of tilapias to several South American countries, there is also interest in the intensive cultivation of tilapias in some parts of the USA. In other parts of the world, most tilapia farming is carried out in ponds, often using semi-intensive methods. Several species of tilapia are cultured, but the Nile tilapia, *O. niloticus*, is the most important of the farmed species; Nile tilapia represents about 80% of the total world

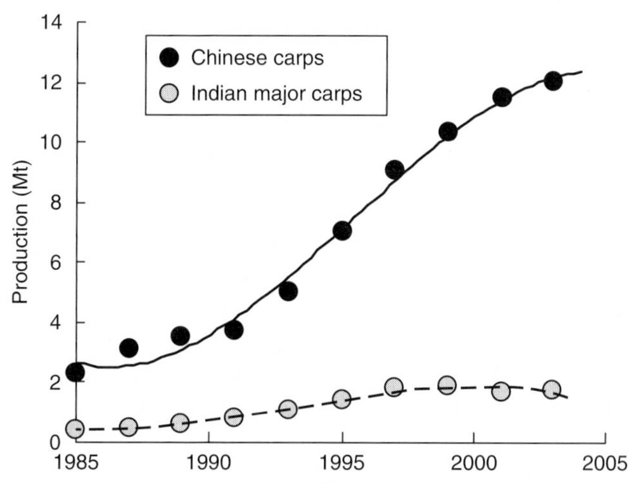

Fig. 4.2. Annual production of the main cultured cyprinids – Chinese carps and Indian major carps – in the period 1985–2003 (data source: http://www.fao.org/figis).

production of farmed tilapia. The Nile tilapia is a freshwater fish, but it has a wide salinity tolerance, so can also be cultured in brackish water; the euryhaline character of the Nile tilapia has been a contributory factor to the expansion of culture of this species in recent years. Some other species of tilapia, such as *O. mossambicus*, are cultivated in seawater in addition to fresh and brackish waters (Lim and Webster, 2006).

Freshwater fish farming in the USA is dominated by catfish culture (family Ictaluridae). Although several catfish species and their hybrids are cultured, production volume is dominated by semi-intensive and intensive pond culture of the channel catfish, *Ictalurus punctatus* (Tucker and Hargreaves, 2004; Shelton and Rothbard, 2006; see Chapter 11, this volume). North American catfishes have been introduced into Europe, Asia, Africa and South America, but cultivation of the North American catfishes occurs only on a limited scale outside the species' native range. Nevertheless, catfish culture is widespread, being practised in several Asian, African and European countries. In these regions, farming tends to be based on the rearing of native catfish species (families Siluridae, Pangasiidae, Clariidae). Pond culture is the norm, but there is some intensive cultivation of catfish in Europe. Intensive cultivation of catfish is carried out in land-based, water reuse tank and raceway systems, rather than in cages.

Traditionally, the farming of salmon and trout (family Salmonidae) has been carried out in Europe and North America, although Chile and Australia have become significant producers in recent years (Shelton and Rothbard, 2006; http://www.fao.org/figis). Salmonid farming is based principally on two species, the Atlantic salmon, *S. salar*, and the rainbow trout, *O. mykiss* (see Chapter 12, this volume). Several other salmonid species, within the genera Salmo (Atlantic salmon and trouts), Oncorhynchus (Pacific salmons and trouts) and Salvelinus (charrs), are also reared either directly for human consumption or for restocking natural waters (Pennell and Barton, 1996). Salmonid farming is carried out using intensive methods. Sea-cage culture dominates Atlantic salmon production, whereas most cultivation of rainbow trout takes place in fresh water in ponds, tanks and raceways.

Families that make important contributions to the production of farmed marine fish include the seabreams and porgies (family Sparidae), the milkfishes (family Chanidae), the drums and croakers (family Scianidae), the seabasses (family Serranidae), the temperate basses (family Moronidae) and the marine flatfishes (families Scophthalmidae, Paralichthydae, Pleuronectidae and Soleidae) (http://www.fao.org/figis; see Chapters 10, 15, 16, 18 and 21, this volume).

4.7 Concluding Comments

Expansion of intensive fish farming as a major industry will be dependent on there being a ready and reliable supply of juvenile fish for on growing, making broodstock management, controlled reproduction and genetic improvements of stock key areas for directed effort. Given that few of the species that are currently being farmed have been subjected to directed genetic selection to alter and improve production traits, there would appear to be considerable potential for significant advance in this area (Dunham, 2004; Gjedrem, 2005). In addition, production of farmed fish is dominated by relatively few species from a small number of families (http://www.fao.org/figis). Furtherance of growth of the fish farming industry cannot rely solely on increased production of these species, but must also involve diversification; to broaden the range of species cultured and provide an increase in the range of products on offer to the expanding global seafood market (see Chapters 5, 7, 8, 22, 23 and 26, this volume).

References

Balon, E.K. (2004) About the oldest domesticates among fishes. *Journal of Fish Biology* 65 (Supplement A), 1–27.

Bergé, J.-P. and Barnathan, G. (2005) Fatty acids from lipids of marine organisms: molecular biodiversity, roles as biomarkers, biologically active compounds, and economical aspects. *Advances in Biochemical Engineering/Biotechnology* 96, 49–125.

Beveridge, M.C.M. and McAndrew, B.J. (eds) (2000) *Tilapias: Biology and Exploitation*. Kluwer Academic, Dordrecht, The Netherlands.

Billard, R. (1999) *Carp: Biology and Culture*. Springer-Praxis, Chichester, UK.

Billard, R. (2003) From polyculture to co-culture in fish farming. In: Lee, C.-S. (ed.) *Aquaculture: Retrospective and Outlook*. World Aquaculture Society, Baton Rouge, Louisiana, pp. 169–200.

Billard, R. and Berni, P. (2004) Trends in cyprinid polyculture. *Cybium* 28, 255–261.

Bourre, J.M. (2005) Where to find omega-3 fatty acids and how feeding animals with diet enriched in omega-3 fatty acids to increase nutritional value of derived products for human: what is actually useful? *Journal of Nutrition Health and Aging* 9, 232–242.

Bromage, N., Porter, M. and Randall, C. (2001) The environmental regulation of maturation in farmed finfish with special reference to the role of photoperiod and melatonin. *Aquaculture* 197, 63–98.

Broom, D.M. (2006) Behaviour and welfare in relation to pathology. *Applied Animal Behaviour Science* 97, 73–83.

Buttriss, J. (2004) Oily fish: still on the menu! *Nutrition Bulletin* 29, 289–290.

Cleland, L.G., James, M.J. and Proudman, S.M. (2005) Fish oil: what the prescriber needs to know. *Arthritis Research and Therapy* 8:202, 9 pp.

Devlin, R.H. and Nagahama, Y. (2002) Sex determination and sex differentiation in fish; an overview of genetic, physiological and environmental influences. *Aquaculture* 208, 191–364.

Diamond, J. (2002) Evolution, consequences and future of plant and animal domestication. *Nature* 418, 700–707.

Dunham, R.A. (2004) *Aquaculture and Fisheries Biotechnology: Genetic Approaches*. CAB International, Wallingford, UK.

El-Sayed, A.-F.M. (2006) *Tilapia Culture*. CAB International, Wallingford, UK.

Flachowsky, G., Chesson, A. and Aulrich, K. (2005) Animal nutrition with feeds from genetically modified plants. *Archives of Animal Nutrition* 59, 1–40.

Givens, D.I. and Shingfield, K.J. (2004) Foods derived from animals: the impact of animal nutrition on their nutritive value and ability to sustain long-term health. *Nutrition Bulletin* 29, 325–332.

Gjedrem, T. (ed.) (2005) *Selection and Breeding Programs in Aquaculture*. Springer-Verlag, Berlin.

Guérard, F., Sellos, D. and Le Gal, Y. (2005) Fish and shellfish upgrading, traceability. *Advances in Biochemical Engineering/Biotechnology* 96, 127–163.

Halver, J.E. and Hardy, R.W. (eds) (2002) *Fish Nutrition*. Academic Press, San Diego, California.

Håstein, T., Scarfe, A.D. and Lund, V.L. (2005) Science-based assessment of welfare: aquatic animals. *Revue Scientifique et Technique – Office International des Epizooties* 24, 529–547.

Houlihan, D., Boujard, T. and Jobling, M. (eds) (2001) *Food Intake in Fish*. Blackwell Science, Oxford, UK.

Huntingford, F.A., Adams, C., Braithwaite, V.A., Kadri, S., Pottinger, T.G., Sandøe, P. and Turnbull, J.F. (2006) Current issues in fish welfare. *Journal of Fish Biology* 68, 332–372.

Jensen, P. (2006) Domestication – from behaviour to genes and back again. *Applied Animal Behaviour Science* 97, 3–15.

Jobling, M. (2004) On-growing to market size. In: Moksness, E., Kjørsvik, E. and Olsen, Y. (eds) *Culture of Cold-water Marine Fish*. Blackwell Science, Oxford, UK, pp. 363–432.

Kagawa, H., Kobayashi, M., Okuzawa, K. and Gen, K. (eds) (2003) Reproductive physiology of fish. *Fish Physiology and Biochemistry* 28, 1–528.

Kestin, S.C. and Warriss, P.D. (eds) (2001) *Farmed Fish Quality*. Fishing News Books (Blackwell Science), Oxford, UK.

Kim, S.-K. and Mendis, E. (2006) Bioactive compounds from marine processing by-products – a review. *Food Research International* 39, 383–393.

Lee, C.-S. (ed.) (2003) Management of aquaculture effluents. *Aquaculture* 226, 1–244.

Lim, C. and Webster, C.D. (eds) (2006) *Tilapia – Biology, Culture and Nutrition*. The Haworth Press, New York.

MacRae, J., O'Reilly, L. and Morgan, P. (2005) Desirable characteristics of animal products from a human health perspective. *Livestock Production Science* 94, 95–103.

Macrae, R., Robinson, R.K. and Sadler, M.J. (eds) (1993) *Encyclopedia of Food Science, Food Technology and Nutrition*. Academic Press, London.

Mignon-Grasteau, S., Boissy, A., Bouix, J., Faure, J.-M., Fisher, A.D., Hinch, G.N., Jensen, P., Le Neindre, P., Mormede, P., Prunet, P., Vandeputte, M. and Beaumont, C. (2005) Genetics of adaptation and domestication in livestock. *Livestock Production Science* 93, 3–14.

Muir, J. (2005) Managing to harvest? Perspectives on the potential of aquaculture. *Philosophical Transactions of the Royal Society B* 360, 191–218.

Nash, C.E. and Novotny, A.J. (eds) (1995) *Production of Aquatic Animals: Fishes*. Elsevier, Amsterdam.

Patiño, R. (ed.) (2002) The teleost gonad: development, function and histopathology. *Fish Physiology and Biochemistry* 26, 1–101.

Pennell, W. and Barton, B.A. (eds) (1996) *Principles of Salmonid Culture*. Elsevier, Amsterdam.

Pond, W.G., Church, D.C. and Pond, K.R. (1995) *Basic Animal Nutrition and Feeding*. John Wiley and Sons, New York.

Ruxton, C.H.S., Reed, S.C., Simpson, M.J.A. and Millington, K.J. (2004) The health benefits of omega-3 polyunsaturated fatty acids: a review of the evidence. *Journal of Human Nutrition and Dietetics* 17, 449–459.

Ruxton, C.H.S., Calder, P.C., Reed, S.C. and Simpson, M.J.A. (2005) The impact of long-chain n-3 polyunsaturated fatty acids on human health. *Nutrition Research Reviews* 18, 113–129.

SACN-FSA (Food Standards Agency) (2004) *Advice on Fish Consumption: Benefits and Risks*. The Stationery Office (HMSO), London.

Shelton, W.L. and Rothbard, S. (2006) Exotic species in global aquaculture – a review. *Bamidgeh* 58, 3–28.
Stickney, R.R. (ed.) (2000) *Encyclopedia of Aquaculture*. Wiley-Interscience, New York.
Tucker, C.S. and Hargreaves, J.A. (eds) (2004) *Biology and Culture of Channel Catfish*. Elsevier, Amsterdam.
Webster, C.D. and Lim, C.E. (eds) (2002) *Nutrient Requirements and Feeding of Finfish for Aquaculture*. CAB International, Wallingford, UK.
Wedemeyer, G. (ed.) (2001) *Fish Hatchery Management*. American Fisheries Society, Bethesda, Maryland.
Wood, J.D., Richardson, R.J., Nute, G.R., Fisher, A.V., Campo, M.M., Kasapidou, E., Sheard, P.R. and Enser, M. (2003) Effects of fatty acids on meat quality: a review. *Meat Science* 66, 21–32.

5 Considerations for the Selection and Commercialization of New or Alternate Species

Nathalie R. Le François,[1,3] David Otton[2] and Guillaume Werstink[3]

[1]Biodôme de Montréal, Montréal, Québec, Canada; [2]Tathra, New South Wales, Australia; [3]Université du Québec à Rimouski, Rimouski, Québec, Canada

5.1 Introduction

> Diversification in aquaculture is growing different species to produce a variety of established and new seafood products.
>
> A new aquaculture species is one not previously commercially cultivated or currently farmed, but with the potential to be successfully farmed and profitably marketed, surviving in the market place, either as a fish, or as derivatives and value-added derivatives from the fish.
>
> (David Otton)

Diversification poses two fundamental questions: is the best scope for the aquaculture industry's expansion to concentrate on existing species or develop new ones, and how should these be selected? A common view in the international aquaculture community that rightfully should be considered as an initial concern within a species selection process is described best by Naylor *et al.* (2000). These authors strongly suggested that sustainability of the aquaculture industry at the worldwide scale should be directed through: (i) expansion of the farming of low trophic-level fish; (ii) reduction of fishmeal and fish oil inputs in feed; (iii) development of integrated farming systems; and (iv) promotion of environmentally sound aquaculture practices and resource management. Any species, whether new, emerging or well established, corresponding to one or more of the above assertions is clearly well positioned in the actual trend of responsible and sustainable worldwide aquaculture development.

In regard to the first recommendation expressed by Naylor *et al.* (2000) (i.e. favour the use of low trophic species), climatic specifics acting together with the physiological limitations of herbivorous species to digest plant material in low temperature environments (see Behrens and Lafferty, 2007), as well as the very marked consumer preference for high-valued carnivorous species in the developed countries, somehow limit the capacities of the northern hemisphere to contribute significantly to the actual desired shift toward 'true' herbivorous species in the near future. It can also be reasonably assumed that given the actual environmental legislation which increasingly restrict the use of exotic species, if the aquaculture is to grow in the Nordic climates, it is essentially the culture technologies enabling the cultivation of a greater range or variety of economically valuable indigenous species that will be largely responsible (see Chapter 24, this volume). The last three recommendations of Naylor *et al.* (2000) are, however, of worldwide application and have been used as guidelines for the selection of criteria and assessment of the species covered in this book (see Chapter 8, this volume).

Basically, the two requirements for new species development are that a market exists (or can be created) and that a species can be domesticated and grown under aquaculture conditions to meet that market. Quite rightfully, Diamond (1998) observed that few domesticated terrestrial animals proved commercially successful and fewer achieved industrial levels of production. Based on Diamond (1998), six enabling criteria for domestication of terrestrial animals can be established: diet (simple and inexpensive), growth rate (rapid), reproduction in captivity (easy), farming-friendly behaviour and adequation between an animal's needs and the artificial environment that can be provided in an intensive production setting. To be domesticated, 'a candidate wild species must possess many different characteristics and lack of any single required characteristic dooms efforts at domestication' (Diamond, 1998).

Brown *et al.* (1995) gave three reasons for developing new species for aquaculture in the Canadian Atlantic Provinces. They were to broaden the base of aquaculture in the region, develop new products for a growing market and provide job opportunities for coastal regions suffering from severe declines in capture fisheries and quota restrictions. On the other hand, Pankhurst (1998) observed that new species could also be developed because there are limits to production capacity, sites and markets for existing species. These two positions are quite different, despite their common goal, i.e. growth of aquaculture production for the preservation and enhancement of the socio-economic benefits. The first vision originates from the presence of a largely unexploited potential waiting to be harnessed efficiently and profitably. The second refers to an already crowded segment of the agri-food business as a justification for the development of new species and products that will provide exclusiveness in the more secured and manageable niche markets.

5.2 Why Develop a New or an Alternate Species?

Three distinctive schools of thought on new species development exist. The first considers that research and development (R & D) resources should be invested in improving existing aquaculture species and, therefore, that new species development should be avoided. New (1999) urged caution in diversifying into new species and, drawing parallels from terrestrial livestock production, suggested concentrating developmental effort and scientific research on the major cultured species to improve growth and survival rates, feed efficiency and health through research in nutrition, genetics and health management, as 'a more efficient use of resources than dilution of effort amongst more and more species'. This school of thought, essentially declaring that aquaculture diversification is not or no longer desirable, although highly questionable and certainly controversial, is reinforced by the general concept that 'intraspecific potential', i.e. largely unknown genetic diversity resources within actual truly domesticated species, remains to be exploited to its full potential, e.g. through tailored designed and targeted genetic selection programmes (New, 1999; Diamond, 2002; Bilio, 2008).

The second school displays anxiety about existing species' position in the product life cycle and is, therefore, *keen to develop new species to accommodate a shift in consumer preferences*. In this regard, it has been suggested that new species development should proceed with caution, taking into consideration market demand, environmental issues and the level of technical knowledge required to develop the identified new species (Dr Philippe Ferlin, President, European Aquaculture Society, 4 January 1999, personal communication).

The third school takes the position that new species development should proceed deliberately but cautiously as part of continuing R & D. Sorgeloos (1999) concluded that species diversification (especially into carnivorous fish) was not a priority. 'Broad species diversification leads to an exponential growth of research requirements which are difficult to meet with limited resources' and diversification into herbivores should be examined. This third school appears to reflect a common view in the international aquaculture community (Naylor *et al*., 2000).

Both visions (second and third school of thoughts) strongly imply that careful selection is applied prior to the identification of the 'new species' among all others and that, once that choice is made, the adoption of a stepwise 'Go/NoGo' procedure is secured in order to tackle the species biological and technical bottlenecks firmly. In other words, 'scrupulous screening of candidate species and strategy are highly desirable to avoid marketwise dead-ends or pronounced biotechnological bottlenecks'. In this respect, selection methods such as the ones covered in this book (see Chapters 6, 7 and 8) and to which can be later introduced the concept of SWOT analysis (strengths, weaknesses, opportunities and threats), a tool of strategic planning for development strategies, will contribute deeply to limit, or at least accept knowingly, the risks involved in the selection of a new species for aquaculture diversification needs. In the absence of such precautions, the dangers associated with 'biotechnical doggedness', in which time and resources involved can blow out of proportion, are omnipresent. New aquaculture species development needs high capital investment and economies of scale and requires systematic evaluation. The framework for selection should be constructed around closed life cycles, available diet, fish behaviour, disease resistance, good grow-out potential, markets and the general ability to farm the chosen species. Worldwide, new species development has taken longer and required a larger resource base than is currently understood (Forster, 1999). 'The real objective, or objectives, of any new species development must be established at an early stage so that interplay between the economic and social aspects of the venture can be recognized and fully evaluated' (Ross and Beveridge, 1995).

A new species can be either new to the world or new to a region. For example, aquacultured barramundi (Asian seabass) (see Chapter 14, this volume) is old to Asia but new to Australia; wolffishes (*Anarhichas minor* and *A. lupus*) (see Chapter 19, this volume) are rather new aquaculture species in Norway (Moksness *et al.*, 1989; Falk-Petersen *et al.*, 1999) but have been introduced only recently as a candidate native species and evaluated for mariculture diversification in the east of Canada (Le François *et al.*, 2002; Desjardins *et al.*, 2007) and Iceland (Foss *et al.*, 2004). New species sometimes are introduced initially to assist the environment in which an existing species grows, thereby serving the agribusiness value chain at a point other than the end market. For example, grass carp was introduced to clean catfish ponds of rooted vegetation and milkfish, *Chanos chanos* (see Chapter 10, this volume), is currently under investigation to clean prawn settlement ponds and barramundi ponds. Similarly, in the Northern Territory of Australia, the Akul or mangrove cockle, *Polymesoda erosa*, is being trialled as a water cleaner in barramundi ponds. Also, a species can requalify as 'new' by being switched from one (farmed) environment to a different (farmed) environment. For example, one respondent observed rainbow trout, *Oncorhynchus mykiss*, was reinvented commercially by adaptation to saltwater from fresh water, remaining *Oncorhynchus mykiss* but rebranded as ocean trout or steelhead (see Chapter 12, this volume). Barramundi now farmed in (Australian) marine water could qualify as a new species by rebranding as marine barramundi to distinguish it from pond-grown freshwater barramundi, thereby differentiating *Lates calcarifer* in the market by its production environment (Otton, 2004).

5.3 Who Is Likely To Develop New Species and Products?

Not all countries display the same degree of development of their aquaculture industry, thereby explaining the numerous expressed reservations about the need to diversify aquaculture. While some countries and regions are still struggling with the identification of their non-salmonid 'winning species' for the development of a competitive marine aquaculture sector (e.g. Québec and the Atlantic regions of Canada), others are considering diversification to maintain their worldwide competitiveness (e.g. Norway, Australia). At the extreme range of this continuing development process, a country that emerges from 10–20 years of R & D of potential new species (e.g. France with seabass and seabream) will generally enter a period aimed at the consolidation of its past investments and, therefore, may restrain the intensity of activities aimed at the identification of new species.

New species and new product development in aquaculture is a slow, deliberate process and, given the investments involved, a priori there is no compulsion for a small to medium-size aquaculture venture to develop a new species, but various companies will attempt to do so. Well-established aquaculture multinationals, however, will usually have An R & D role possibly aimed at developing new products and species, whereas smaller companies understandably will favour the diversification of the range of products possibly derived from their main targeted species. Aquaculture companies may evaluate diversification as a means to maintain a market edge by developing a product mix complementary to existing business and as a hedge against species-specific diseases or market price drops (Otton, 2004). In some countries, the industry's diversification efforts are usually an initiative of governmental and university R & D programmes (see Le François *et al.*, 2002; Quéméner *et al.*, 2002; Wendling *et al.*, 2006). Aquaculture companies seeking new markets and growth of their production capacities usually will seriously consider joining these largely publicly funded efforts. The potential private partner generally will evaluate the research and efforts needs to develop a new product or species and the potential to improve current operations and products. Vigilance is recommended since, in many cases, the benefit often gained by first mover advantage in mainstream industry (Lieberman and Montgomery, 1988) could very well become a first mover disadvantage in aquaculture.

A value analysis or product redesign is a systematic investigation of a product to see how the design or materials can be changed to improve the product's performance and/or lower its cost. Value analysis offers abundant opportunities for product and process simplifications through a detailed scrutiny of the sources of non-value-added components, steps or even entire processes (Dröge *et al.*, 2000). New product development is as much about how a company operates as it is about developing something new. Unlike new industrial products, the aquaculture product usually exists already and its market environment is not necessarily characterized by rapid obsolescence and market fragmentation. In such an environment, a firm's new product development must meet two critical objectives: minimize time to market and maximize the fit between customer requirements and product characteristics (Schilling and Hill, 1998).

The following company/firm types are most likely to develop either one, or more, new aquaculture species.

Type 1. A firm, not already established, wants to start up an aquaculture venture from scratch and is looking for the ideal species or an original species.

Type 2. An established firm, possibly a large agribusiness company, not involved in aquaculture but already operating in the agribusiness value chain and seeking to exploit a market for an input they are already producing, by diversifying into aquaculture.

Type 3. An existing aquaculture firm wants to diversify its product mix by adding a new species to its product base.

Type 4. An existing aquaculture firm switching species.
Type 5. A seafood firm seeking to protect its supply chain by integrating into aquaculture.
Type 6. Farmers developing a portfolio of farming activities to offset price instability, with aquaculture as one option.

Each firm type either has or perceives an existing market for one or more species of aquacultured fish. The companies best equipped for new species development are probably those already profitable and sustainable in the aquaculture industry. How much should the private sector invest and what is the role of the government will be addressed further in Chapter 23, this volume.

5.4 When Is This Development Likely To Occur?

If the R & D segment toward successful commercialization of a new aquaculture species is led by research institutions, indications that biological, technological and economical feasibility are within reach, and that the risk levels at all steps of the production cycle have been reduced acceptably, usually and logically triggers the undertaking of pilot-scale growth trials. At this stage, best fit rearing technology identification and market positioning studies are usually run in parallel. If all goes well, the timeframe from new species identification and readiness to embark on commercialization stage activities approximates 8–10 years of concerted R & D efforts.

In the case that a company/firm type is the pro-diversification agent, the major triggering factor in new species development is market shifts, trends and fluctuations. A company is market oriented if it monitors market developments systematically, both customers and competitors, and fits products and services to these developments (Grunert *et al.*, 1997). An external market perspective throughout the company requires changes in behaviour, processes, skills, context of decision making and reward structures. It has high customer orientation and a strategic focus. According to Grunert *et al.* (1997), innovation is the detection and fulfilment of the unfulfilled needs and wants of potential customers, using the skills, resources and competencies of the company.

Essentially, tastes and habits in food are mostly part of a larger cultural heritage and change slowly, making successful innovation in the food industry difficult, but reinforcing the importance of R & D and development and market orientation skills. In Sydney, the largest Australian seafood market, Ruello and Associates Pty (2002) found fine food lovers to be drivers of new species and new products promotion, as well as providing a niche market for these products.

5.5 What Does the Development Process Entail?

Two central questions to ask about the potential of a new species are: ‘How can this new species be grown profitably?’ and ‘Is there a market for it?’

Previously, new species selection was made mainly on scientific grounds with little consideration for the commercial aspects of new species development, often until late in the development process. The 'sooner the better' strategy is now being implemented. Database and analytical tools to assist diversification of fish aquaculture are currently available (see Le François *et al.*, 2002; Quémémer *et al.*, 2002; Wendling *et al.*, 2006; Teletchea *et al.*, 2007) and a global strategy to diversify the markets through the valorization of by-products (Le François *et al.*, 2004; Desjardins *et al.*, 2007; Desrosiers *et al.*, 2008; and Chapter 25, this volume) and integrated multitrophic aquaculture to improve crops and benefit the environment is foreseeable (Troell *et al.*, 2003; Neori *et al.*, 2007).

Aquaculture is a special case for new product development where there is generally no formal structure within to decide for or against a new species for development. The cost of R & D is often a limiting factor in new species development, suggesting a compromise solution of focusing on assessing costs, benefits and impacts of domestication and subsequent commercialization. Hence, using a combination of agribusiness and new product development attempts to generate a more structured approach to new species vetting. The method of assessment can be refined continually as more knowledge emerges and the industry, market, science, environmental and social contexts change (Williams, 1999). The commercialization success of a new species will reside essentially in its capacity to find its 'niche market' and to achieve recognition and adequate market demand successfully. Generally, aquaculture products are positioned initially in comparison to similar or related fishery products (Fauconneau, 2004).

5.6 How Should Commercialization Be Conducted?

The industrial new product development (NPD) issue assists in the assessment of new aquaculture products by applying mainstream business techniques rigorously to the aquaculture NPD process, which integrates science, marketing and industrial technology. Also, the technological, logistical and infrastructure solutions for caging, feeding, handling and processing aquacultured fish come from regular industry. This application of customer-driven, contemporary business and agribusiness techniques is different from the (current) largely scientific approach to new species development in aquaculture. NPD is defined as: '*the overall process of strategy, organisation, concept generation, product and marketing plan creation and evaluation, and commercialisation of a new product*' (Otton, 2004). Applied to aquaculture, Table 5.1 shows the process of new species development synthesized from the results of a survey juxtaposed with the process of NPD and combined with the science of aquaculture and the concept of agribusiness.

The theory is that a new species or species derivatives should be subject to exactly the same rigorous selection criteria as any product because, at the stage of product launch, the aquacultured fish is just another item that consumers are not obliged to buy. Two NPD processes are necessary for developing and commercializing a new species or species derivative: first, the screening

Table 5.1. Overview of the process of new species development (adapted from Otton, 2004).

Stage	Knowledge	Decision	Action	
Idea source and generation	Species identity? Business environment? Supply and demand?	Yes	Screen the idea.	Check success of same or similar species elsewhere and position in the product life cycle. Assess attributes of uniqueness, attractiveness and flexibility.
		No	Either abandon or put on hold for technological developments.	
Idea screen	Market?	Yes	Size and scope.	
		No	Can a market be built?	
	Ease of farming?	Yes	Assess cost of production and profit.	Assess how technical problems can be solved, for example closed life cycle. Assess species robustness, environmental tolerance and potential mortality rates.
		No	Does a potential market justify further technical evaluation?	
Scoping	Cost of production?	Yes	What is the financial and business risk?	Identify which market is to be served high value/high volume. Is the technology available? Identify and consult stakeholders. Establish a cross-functional development team.
		No	Does the species have the merit to justify large research and development expenditure?	
	Government support?	Yes	Investigate a contract or arrangement to source juveniles.	
		No	Source non-government funds or reconsider project.	
Second screen	Does a preliminary economic model suggest species profitability?	Yes	Establish value chain function.	Construct a virtual agribusiness value chain considering: closed life cycle and ease of producing juveniles, disease resistance, feed development, FCR (Feed Conversion Ratio), growth cycle time, quality and shelf life, consumer knowledge of species, versatility of carcass use, recovery rate and potential to value add.
		No	Remodel the species on a revised value chain.	
	Does production fit environmental models?	Yes	Proceed	
		No	Rework	

continued

Table 5.1. Continued

Stage	Knowledge	Decision	Action	
Build business case	Synergies with current operations?	Yes	Fit new species to current value chain and site. Establish alternate site availability.	Define and justify the market offering.
		No	Rework value chain to fit species and rework existing infrastructure.	
	Can the value chain deliver the species to the target market?	Yes	Assess competition. Can quality and availability be maintained?	Make investment decisions.
		No	Production or marketing problem?	
	Does the species have a sustainable competitive advantage?	Yes	Go to development.	Identify personnel to implement development.
		No	Consider abandoning the project.	
Go to development	Do the reviews, consultations and modelling thus far indicate the species is worthy of development?	Yes	Proceed.	
		No	Reconsider project or seek another species.	
Development	Implement development plan?	Yes	Place species in grow-out cages for assessment.	Check production performance and monitor environmental compliance.
		No	Put species technology on the shelf for later uptake.	
Go to testing	Assess species in marketplace?	Yes	Subtly put species to the market.	Where and how will the species be tested, if at all?
		No	Species already available as wild caught.	

Testing and validation	Commercialize product, especially value-added offering?	Yes	Begin positioning the product and establishing outlets.	Ensure sufficient supply to back early promotion.
		No	Save for formal launch or ease on to market.	
Go to launch	Venue, food and beverage support, co-launching, media coverage?	Yes	Advertise and go ahead with launch.	Contact all stakeholders and invite to launch
		No	Ease product quietly on to the market.	
Launch	How to assess launch impact?			Ensure the new species message gets taken up by stakeholders.
Commercialization	Sustained agribusiness value chain performance.	Observe supply and cold chain function. Monitor domestic market performance and assess export potential.	Assess how agribusiness value chain is working.	

process, which chooses the species/product and, second, the development process, which commercializes the species/product. These combine aquaculture success criteria with success criteria from mainstream business and agribusiness. This method is derived from the case studies using qualitative data (Yin, 1994; Sterns *et al.*, 1998; Westgren and Zering, 1998; Murphy *et al.*, 1998) to establish a set of minimum criteria to answer the question: 'What are the enabling criteria for successful finfish production (and marketing) under aquaculture conditions?' Table 5.2 shows what industry players in

Table 5.2. Multiple selection criteria ranked according to industry representatives (Otton, 2004).

Criterion/attribute/success factor	Rank	Comment
Marketability	1	Essential criteria
Adaptability to aquaculture/ease of farming	2	
Well priced/profitable	3	
Short growth cycle time	4	
Market and consumer knowledge of species	5	Highly important criteria
Versatility of carcase use, available technology	6	
Robust/environmentally tolerant, availability	7	
Economically produced	8	
Good FCR – easy to produce juveniles – disease and parasite resistant – quality	9	
Environmentally acceptable production – able to develop diet	10	
Attractive – shelf life – potential to value add	11	Important criteria
Serve an established or potential market – closed life cycle – herbivorous euryhaline – uniqueness – flexibility – serve a high-value or high-volume market	12	
Customer safe perception – competitive advantage – well marketed – fit the agribusiness value chain – suitable to environment	13	
Not easily duplicated – innovative/marketed in a new form – government support – improvement on existing product – recovery rate/fillet yield – globally competitive/serve a global market – knowledge of species biology/hatchery cycle	14	Relevant criteria
Value – fashionable – similar species not grown overseas – shortage of supply – adaptable to environment – site availability – chemical free production – able to achieve first mover advantage – potential to become a mainstream fish – synergies with current operations – possessing scales – live market appeal – low mortality rate – advantage of being produced in Australia – achieve a regular price	15	Noted criteria

the successful channel catfish, Atlantic salmon and barramundi industries (surveyed between 2000–2002) deduce from their own observations are appropriate selection criteria for new aquaculture species.

Ideally, four selection assessment activities could run parallel in the candidate species screening process: a desktop review; a taste test by selected food panels; a trial run that cages wild-caught species to ascertain their performance in captivity; and a scan of the international scene to establish a database on the candidate or similar species to test the selection criteria and project profits. The selection and development of channel catfish in the USA is transparent and well documented. Catfish was very well known as a sporting and eating fish, but not developed for aquaculture like a new species. It evolved as a generic with no strategy for development but with an element of serendipity. Shell (1993) observed 'in the late 1950s no one appreciated or could imagine the potential of aquaculture'. Efforts leading to domestication were initiated to develop it further as a sport fish and replenish its number in natural waterways. After developing an interest in fish-for-food, Dr Homer Swingle conducted a series of experiments which led to three publications on the aquaculture of channel catfish (Swingle, 1954, 1956, 1958; reviewed in Shell, 1993), forming the information base for catfish development. A powerful stimulant to catfish aquaculture was expanding the agribusiness dimension of the industry by construction of the first catfish processing plant in Greensboro, Alabama (USA), which opened in 1964. Already aware of the concept of agribusiness, Swingle concluded that to be profitable, artificial feed had to be developed for catfish and that stocking rates would need to be high (Swingle, 1954; Shell, 1993).

New species development is also assisted by the environment in which the species is developed and the environment the new species product enters during and after development. As a new species, catfish entered a less competitive and less cluttered marketplace than possibly any new species developed would enter. Catfish aquaculture in the USA lives in what Chapter 11, this volume, describes as a business ecosystem with ample sites (Mississippi Delta) and intellectual support (Auburn University, Mississippi State University). The states in which the catfish industry operates, particularly Mississippi, fit Pouder and St John's definition of a 'hot spot' which is 'fast growing geographic clusters of competing firms' and 'fast growing geographically clustered firms within industries' (Pouder and St John, 1996). All of the above concepts tend to protect the competitive position of species under cultivation and sustain an enabling environment for new species development.

Some outstanding questions (for which complete answers may only be available as a species proceeds along the pathway of development) arise from the process model.

1. Do the technical aspects appear solvable?
2. Can the nexus between science (technology) and business be overcome?
3. Is enough information available to make an investment decision?
4. What can be done to minimize risk?

5.7 Conclusion

Actual established aquaculture species were chosen based on few oversimplified criteria, such as high sale price and availability of juveniles and broodstock in the wild. Nowadays, the wide range of environmental conditions, the emergence of new production technologies, the quest for increased productivity, the market trends, sustainability and environmental concerns and regulations and disease prevention reinforce the tendency for a sustained diversification of the industry. Likewise, in the abundance of choice at the market level and given the investments required by a new species to achieve the commercialization stage, the introduction and future success of a new product on the market deserves and commands the orchestration of a global, multifaceted and innovative strategy, which is covered briefly in this chapter. For further exploration of the challenges of a new or alternate species development and introduction on the consumer market, the reader is also strongly encouraged to see Chapters 6, 7, 8 and 22, all this volume.

References

Behrens, M.D. and Lafferty, K.D. (2007) Temperature and diet effects on omnivorous fish performance: implications for the latitudinal diversity gradient in herbivorous fishes. *Canadian Journal of Fisheries and Aquatic Sciences* 64, 867–873.

Bilio, M. (2008) Controlled reproduction and domestication in aquaculture. The current state of the art. Part IV (Final Discussion and Conclusions). *Aquaculture Europe* 33, 12–24 (http://www.easonline.org/files/various/domestication_in_aquaculture_bilio_web.pdf. Accessed 24 July 2009).

Brown, J., Helm, M. and Moir, J. (1995) New candidate species for aquaculture. In: Boghen, A.D. (ed.) *Cold Water Aquaculture in Canada*. The Canadian Institute for Research on Regional Development, The Tribune Press Ltd., Sackville, New Brunswick, Canada, pp. 3–22.

Desjardins, M., Le François, N.R., Fletcher, G.L. and Blier, P.U. (2007) High antifreeze protein levels in Atlantic wolffish (*Anarhichas lupus*) make them an ideal candidate for potentially ice-laden waters. *Aquaculture* 272, 667–674.

Desrosiers, V., Le François, N.R. and Blier, P.U. (2008) Trypsin-like enzyme from Atlantic wolffish (*Anarhichas lupus*) viscera: purification and characterization. *Journal of Aquatic Food Product Technology* 17, 11–26.

Diamond, J. (1998) *Guns, Germs and Steel. A Short History of Everybody for the Last 13,000 Years*. Random House, Sydney, Australia.

Diamond, J. (2002) Evolution, consequences and future of plant and animal domestication. *Nature* 418, 700–707.

Dröge, C., Jayaram, J. and Vickery, S.K. (2000) The ability to minimise the timing of new product development and introduction: an examination of antecedent factors in the North American automobile supplier industry. *Journal of Product Innovation Management* 17, 17–40.

Falk-Petersen, I.-B., Hansen, T.K., Fieler, R. and Sunde, L.M. (1999) Cultivation of the spotted wolffish (*Anarhichas minor* Olafsen) – a new candidate for cold-water fish farming. *Aquaculture Research* 30, 711–718.

Fauconneau, B. (2004) Diversification, domestication et qualité des produits aquacoles. *Productions Animales* 17, 227–236.

Forster, J. (1999) Aquaculture chickens, salmon – a case study. *World Aquaculture* 30, 33, 35–38, 40, 69–70.

Foss, A., Imsland, A.K. Falk-Petersen, I.-B. and Oiestad, V. (2004) A review of the culture potential of spotted wolffish *Anarhichas minor* Olafsen. *Reviews in Fish Biology and Fisheries* 14(2), 277–294.

Grunert, K.G., Harmsen, H., Meulenberg, M., Kuiper, E., Ottowitz, T., Declerck, F., Traill, B. and Göransson, G. (1997) A framework for analysing innovation in the food sector. In: Traill, B. and Grunert, K.G. (eds) *Product and Process Innovation in the Food Industry*. Blackie Academic, London, pp. 1–37.

Le François, N.R., Lemieux, H. and Blier, P.U. (2002) Biological and technical evaluation of the potential of marine and anadromous fish species for cold-water mariculture. *Aquaculture Research* 33, 95–108.

Le François, N.R., Desjardins, M. and Blier, P.U. (2004) Enhancement of profitability perspectives of an aquaculture production by the extraction of high commercial value biomolecules: evaluation of the potential of wolffish cultivation. In: Shahidi, F. and Simpson, B.K. (eds) *Seafood Quality and Safety: Advances in the New Millennium*. ScienceTech Publishing Company, St John's, Newfoundland, Canada, pp. 61–69.

Lieberman, M.V. and Montgomery, D.B. (1988) First mover advantages. *Strategic Management Journal* 9, 41–58.

Moksness, E., Gjosaeter, J., Reinert, A. and Ingvard, S. (1989) Start feeding and on-growing of wolffish (*Anarhichas lupus*) in the laboratory. *Aquaculture* 77, 221–228.

Murphy, E., Dingwall, R., Greatbatch, D., Parker, S. and Watson, P. (1998) Qualitative research methods in health technology assessment: a review of the literature. *Health Technology Assessment* 2(16) (http://www.hta.ac.uk/fullmono/mon216.pdf. Accessed 24 July 2009).

Naylor, R.L., Goldburg, R.J., Primavera, J.H., Kautsky, N., Beveridge, M.C.M., Clay, J., Folke, C., Lubchenco, J., Mooney, H. and Troell, M. (2000) Effect of aquaculture on world fish supplies. *Nature* 405, 1017–1024.

Neori, A., Troell, M., Chopin, T., Yarish, C., Critchley, A. and Buschmann, A.H. (2007) The need for a balanced ecosystem approach to blue revolution aquaculture. *Environment* 49, 36–43.

New, M. (1999) Global aquaculture: current trends and challenges for the 21st century. *World Aquaculture* 30, 18–13, 63–79.

Otton, D.D. (2004) Establishment of an agribusiness model for assessing the commercial viability of new species for aquaculture. PhD thesis, University of Tasmania, Launceston, Australia.

Pankhurst, N.W. (1998) Aquaculture: an international perspective. *Proceedings of Maori and the Business of Aquaculture Conference*. Kia Ora Promotions, Auckland, pp. 1–10.

Pouder, R. and St John, C.H. (1996) Hot spots and blind spots: geographical clusters of firms and innovation. *Academy of Management Review* 21, 1192–1225.

Quéméner, L., Suquet, M., Mero, D. and Gaignon, J.-L. (2002) Selection method of new candidates for finfish aquaculture: the case of the French Atlantic, the Channel and the North Sea coasts. *Aquatic Living Resources* 15, 293–302.

Ross, L.G. and Beveridge, M.C.M. (1995) Is a better strategy necessary for development of native species for aquaculture? A Mexican case study. *Aquaculture Research* 26, 539–547.

Ruello and Associates Pty Ltd (2002) Retail sale and consumption of seafood. Revised edition. Fisheries Research and Development Corporation, pp. 19. http://www.wafic.com.au/images/5-Retail_Sales_and_Consumption.pdf (accessed 24 July 2009).

Schilling, M.A. and Hill, W.L.C. (1998) Managing the new product development process: strategic imperatives. *The Academy of Management Executive* 12, 67–81.

Shell, E.W. (1993) *The Development of Aquaculture: An Ecosystems Perspective*. Craftmasters Printers, Opelika, Alabama.
Sorgeloos, P. (1999) Challenges and opportunities for aquaculture research and development in the next century. *World Aquaculture* 39, 13–15.
Sterns, J.A., Schweikhardt, D.B. and Peterson, H.C. (1998) Using case studies as an approach for conducting agribusiness research. *The International Food and Agribusiness Management Review* 1(3), 311–327.
Teletchea, F., Fostier, A., Le Bail, P.Y., Jalabert, B., Gardeur, J.N. and Fontaine, P. (2007) STOREFISH: a new database dedicated to the reproduction of temperature freshwater teleost fishes. *Cybium* 31(2), 237–245.
Troell, M., Halling, C., Neori, A., Chopin, T., Buschmann, A.H., Kautsky, N. and Yarish, C. (2003) Integrated mariculture: asking the right questions. *Aquaculture* 226, 69–90.
Wendling, D., Werstink, G., Dubosc, J., Goimier, Y. and Abes., M. (2006) *Sélection de produits aquacoles présentant un potentiel d'élevage et un potentiel économique pour Mayotte : une analyse multicritère appliquée à la sélection d'espèces pour l'aquaculture*. Creufop-Station méditerranéenne de l'environnement du littoral, Sète, France.
Westgren, R. and Zering, K. (1998) Case study research methods for firm and market research. Paper presented to the 1998 Food and Agricultural Marketing Consortium Research Conference, Park City, Utah.
Williams, M.J. (1999) Sustainable production: selecting species to domesticate. Paper presented at the World Aquaculture Conference, Sydney, Australia, Book of Abstracts, pp. 812.
Yin, R.K. (1994) *Case Study Research: Design and Methods*, 2nd edn. Sage Publications, Thousand Oaks, California.

6 A Systematic Market Approach to Species Diversification: a French Case Study

Marc Suquet

Ifremer, Laboratoire Adaptation Reproduction Nutrition, Station Expérimentale d'Argenton, Landunvez, France

6.1 Introduction

Marine fish farming in the Mediterranean regions has focused intensively on seabass and seabream, *Dicentrarchus labrax* and *Sparus aurata* (see Chapters 15 and 16, this volume, for more information). Since the first attempts at artificially reproducing seabass, reported by Barnabé and Tournamille in 1972, the production cycles of both species have been finely mastered. Aquaculture production of seabass and seabream reached 49,000 and 91,000 t, respectively, in 2004 (FAO, 2006). Both species were selected for their high selling price and their good reputation among consumers.

After a 30-year period, and in spite of the large expansion in the commercial aquaculture of these selected species, this choice is questionable from the following points of view:

- From the point of view of the farming industry, the growth performances of seabass and seabream are low (0.1–0.2 kg after 1 year) compared to fast-growing fish such as bluefin tuna, *Thunnus thynnus*, or cobia, *Rachycentron canadum* (from 5 to 10 kg after 1 year).
- From the point of view of the processing industry, aquaculture development must take into account new trends in the European market for fish, and especially the soaring role of fish flesh at low cost, ready to cook and the increasing role of supermarkets in the distribution chain (Paquotte, 1998). Aquaculture has to follow this industrial pattern by focusing on cuts of fish (steaks or fillets) rather than on whole fish. Because of their small size and their intermediate fillet yield ranging from 40 to 45%, seabass and seabream are not well adapted to the requirements of the transformation process.
- From the point of view of the consumer, the main concerns are a reduction in time spent in meal planning and preparation, low prices and fish quality

(eds N.R. Le François, M. Jobling, C. Carter and P.U. Blier)

issues (Mariojouls and Paquotte, 2001). The two first arguments explain the increasing large consumption of cuts of fish in most European countries, for which seabass and seabream are not well adapted, and the last sustains consumer suspicion about aquaculture production.
- From the point of view of the economist, aquaculture follows a classic economic model of demand and supply, i.e. dramatic price falls when production volumes increase. For example, in Greece, the selling price of seabass and seabream decreased from €12/kg in 1990 to €4/kg in 2003, while production increased from 20,000 t to 200,000 t (Dallimore, 2005).

Regarding these four arguments, seabass and seabream can no longer be considered as 'star candidates' for future aquaculture growth.

The need to diversify and to control the rearing techniques of 'new species' was claimed by several authors (Jones, 1972; Avault, 1993; Paquotte, 1998; Suquet *et al.*, 2002). The expression 'new species' is conflicting. For example, Atlantic cod, *Gadus morhua*, is considered as a promising 'new' species for aquaculture, but has been an important economic commodity in the international market since the Viking period (around 800 AD) and is still the object of intense targeted fisheries (see Chapter 13, this volume). Seabass and seabream were also considered as 'new species' at the very beginning of their farming industry in the 1980s, so this expression is clearly correlated with the aquaculture status or production volumes of the considered species at a given period and not to their relative novelty per se.

The origin of fish species diversification may be related to the prohibition of common carp culture by the Emperor Li of the Tang Dynasty, causing fish farmers to culture other carp species (Liao, 2000). A selection procedure for new candidates will avoid a cul-de-sac in the expansion of new species farming, as observed for common sole (*Solea solea*). Despite extensive studies carried out from the 1960s and 694 publications listed in ASFA (Aquatic Sciences and Fisheries Abstracts) providing a valuable database to stimulate interest in sole fish farming (Howell, 1997), its aquaculture production was only 62 t in 2004 (FAO, 2006). Problems observed were a low growth rate of juveniles using formulated feeds, the species adapted poorly to high stocking densities and had a high susceptibility to disease, thereby diverting research to other fish species.

Diversification faces the complex problem of diversity expressed in terms of fish species (more than 20,000 teleostean species), environmental conditions and farming technologies (ranging from traditional ponds to sophisticated processes such as recirculating systems). The choice of a new species has to avoid selection based on a limited number of criteria, as in the case of seabass or seabream, but must also preclude the manifestation of an emotional 'fish appeal', sometimes observed in farmers electing their own pet species using subjective and limited arguments, as the best solution for aquaculture development.

A few selection methods of new candidates have been reported previously (Table 6.1). From the first one suggested by Fan Lee (500 BC) to the last one developed by Le François *et al.* (2002) (see Chapter 8, this volume), significant improvements have been suggested considering complementary

Table 6.1. Previously published selection methods of new fish species candidate to aquaculture development.

Authors	Date of publication	Contributing	Lacking
Fan Lee	500 BC*	First method	Few criteria suggested
Jones	1972	Considered market price, growth rate, ease of rearing	Limited to 10 species
Webber and Riordan	1975	Considered biological and zootechnical criteria	No species ranking
Menu	1987	Included biological and economical aspects	No species ranking
Wu	1989	Bioeconomic computer model	Limited to 6 species
Cook and Walmsley	1990	Computer criteria decision method, rating coefficients	Limited to 16 species
Avault	1993	Suggested a panel of criteria	No criteria and species ranking
Kentouri *et al.*	1995	Considered biological, economical and zootechnical aspects	No species ranking
Lensi	1995	Included enquiries and biological, economical criteria	Limited to 44 species
Brummet	1996	Used economic criteria and management decision method	Intended only for smallholder
Parfouru *et al.*	1997	Included biological and economical aspects	No species ranking
Benetti *et al.*	1999	Included biological and economical aspects	Limited to 15 species
Le François *et al.*	2002	Included three production strategies	Limited to 45 species

Note: *In Mann, 1984.

topics, increasing the number of selection criteria, integrating market demand, using ranking systems, taking account of the level of uncertainty and R & D efforts needed until commercialization, etc. Selection methods have also been published for molluscs (Mann, 1984), marine invertebrates of the east coast of Canada (Lemieux *et al.*, 2002) and for species adapted to environmental conditions (mainly water temperature, since rearing performances of poikilotherm species depend on this parameter), such as in the tropical Pacific area (Bell and Gervis, 1999).

This chapter describes a three-phase selection method of fish species as candidates for aquaculture development, initially proposed by Quemener *et al.*

(2002). This selection method of new candidates for finfish aquaculture takes into account a panel of 32 economical, biological and zootechnical criteria suggested by enquiries and surveys conducted with the French professionals of the production, transformation, distribution and consumption sectors. Applied to 20,000 species, it indicated strong attributes of gadoids for aquaculture development on the French Atlantic, the Channel and the North Sea coasts. Atlantic cod, *G. morhua*, was identified as the best candidate and its selection by Quemener *et al.* (2002) coincided with the rising interest in the aquaculture of this species in northern European countries. The expansion of its cultivation is sustained and stimulated by a fall in wild catches, an increase in its market value, good characteristics for transformation and a worldwide market. Fish flesh at low cost and ready to cook was the selected market share of this large-scale selection exercise. Therefore, species presenting a high growth rate and low production costs have been logically targeted.

6.2 Methodology

The selection method was based on both enquiries conducted with the main professionals of the French production–consumption line and on a three-step selection procedure.

6.2.1 Enquiries

Enquiries were conducted in 1999 with farmers (representing 28% of the French aquaculture companies), with 13 companies (selected from the most important ones) of the French transformation channel and with 11 companies of the French distribution channel (representing 80% of the catering activity and 80% of the national turnover of this sector). They aimed to define the requirements of these four professionals of the production–consumption line concerning the aquaculture of 'new fish species'. Selection criteria and their associated weighting coefficients used during the selection procedure were based on information collected during these enquiries.

6.2.2 Selection procedure

The three successive phases of the selection procedure were as follows (Table 6.2):

- Phase 1: a first elimination step among 20,000 teleost fish species, allowing the settlement of a mother population that could be used for different subsequent selection cases.
- Phase 2: a geographical elimination phase applied to the case of the French Atlantic, the Channel and the North Sea coasts.
- Phase 3: a final classification phase for the same geographical area.

Table 6.2. General scheme of the three-phase selection procedure.

Data origin	Phases of the selection procedure	Fish species number at the end of the phase
Cites (1999) Fishbase (1998)	Phase 1: Settlement of a mother population	8063
Hureau (1996) Quéro and Vayne (1997) Stat P (2000)	Phase 2: Geographical elimination	375
ASFA base Fishbase (1998) Hureau (1996) Ofimer (1999) Quéro and Vayne (1997) Stat P (2000)	Phase 3: Final classification phase	32

Data sources are indicated in Table 6.2. Additional information was recorded from French wholesale fishmongers (personal communication, 1999: G. Dubois, Charlie Guennec Company; B. Jolivet, Furic Marée and V. Mulak, Cevpm). A set of 32 selection criteria was developed according to the requirements of the four professionals of the production–consumption line (Table 6.3). Weighting coefficients were attributed (see Quemener *et al.*, 2002) to each criteria and for each professional according to the results of the enquiries previously conducted. Data were managed using Access software (1997) for Phases 1 and 2 and Electre 3 (1994), a multiple-criteria, decision-making software, for Phase 3. Species for which aquaculture production was higher than 100 t were not considered as 'new candidates'. However, five cultured species (Atlantic salmon, seabass, seabream, sea trout and turbot) were maintained as control. They aimed at validating the selection procedure developed in this study. Four intermediate classifications were established for each of the four professionals of the production–consumption chain and a final one presenting ranking obtained by adding the results recorded for each actor.

6.3 Constraints and Limitations of the Model

The advantages of the selection procedure described in this chapter were developed previously (Quemener *et al.*, 2002): a fine integration of the transformation–consumption line requirements sustained by enquiries conducted with these sectors and weighting coefficients attributed to selection criteria for each of these sectors; a selection procedure based on a wide panel of 32 criteria taking into account biological, fisheries, sociological and economical aspects; an exhaustive initial population of 20,000 fish species which supports the emergence of true 'new species' without any previous experience in aquaculture; and, finally, an objective data management using databases and a decision-making methods software.

Table 6.3. 32 criteria panel used for the three phases of the selection procedure (technical aspects are reported in Quemener *et al.*, 2002).

Phase	Criteria	Criteria description
1	Systematic	Archaic species and species whose morphology is distant from commercial species
	Dangerous and non-edible	Poisonous and traumatogenic species
	Electrogenic activity	Species generating electric fields
	Environment	Species living in polar and boreal environments and below 200 m
	Salinity	Species never occurring in seawater
	Minimum weight	Species never reaching 50 g
	Minimum length	Species never reaching 16 cm
	Threatened species	Species threatened with extinction
2	Geographic distribution	Species living in the targeted FAO zone
	Wholesale price	Species whose wholesale price is lower than €1.5/kg
3	Biological knowledge	Number of publications in ASFA base
	Catch potential	Capacity to catch juveniles in the wild
	Geographic recovery	Recovery percentage between the geographic distribution and the rearing area
	Temperature adaptation	Difference between the north and south latitudes of the species' geographic distribution area
	Weight at 1, 2 and 3 years	Calculated from von Bertalanffy function
	Rearing potential	Combine the wholesale price and the age reached at 3 kg
	Size	Body length adaptation to processing requirements
	Profile	Body shape adaptation to processing requirements
	Section	Body section adaptation to processing requirements
	Fillet yield	Fillet yield value
	Bones	Presence of bones in the flesh
	Fish aspect	External species aspect taking into account morphology and colour
	Reputation	Difference of the species' selling price with the mean price of group species with similar landing
	Species knowledge	Capacity of consumers to cite and identify species
	Presentation	Presentation methods variety
	Consumption price	Low selling price as required by the transformation and distribution channels
	Taste	Fish flesh taste
	Colour	Frank aspect of flesh colour
	Lipids	Favours intermediate lipid content
	Proteins	Favours a high protein content

However, the present method has several constraints and limitations. It is based on a combination of market studies and biological data which may progress with time. The first evidence is the change of fish selling price depending on landing and demand: the wholesale price of fresh cod on the French market, estimated in constant euros, resulted in a twofold increase from 1995 to 2004 (Fig. 6.1). That is why the mean value of the wholesale prices recorded during 3 suc-

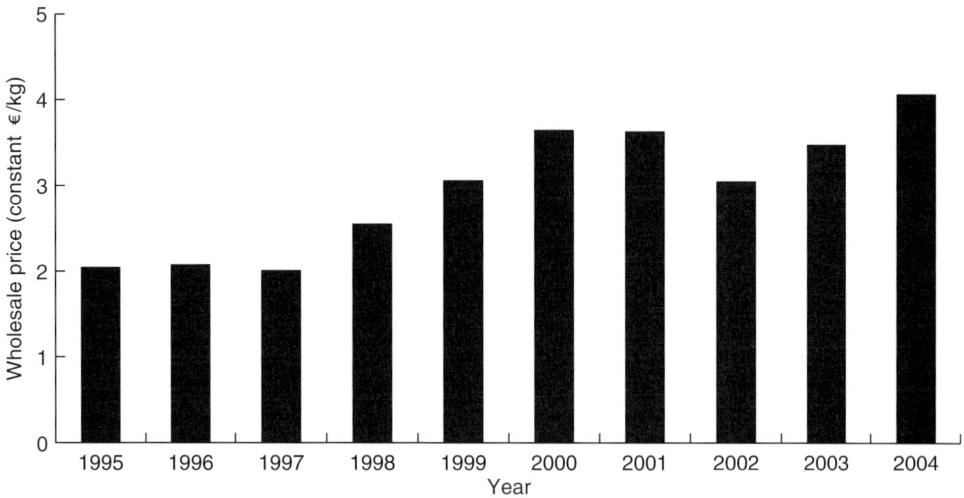

Fig. 6.1. Changes over time in cod wholesale price on the French market (constant €, base 2004; after Ofimer from 1996 to 2004).

cessive years (1996–1998) was used in this study. Biological knowledge of species increases with time: between 1975 and 2005, the number of published papers related to cod in the ASFA (Aquatic Sciences and Fisheries Abstracts) database was increased by a 6.8 factor (Fig. 6.2). Biological knowledge may also be strongly improved with time: the age of the European hake, *Merluccius merluccius*, was overestimated because of difficulties in reading otoliths. As a consequence, the growth rate estimated in a tagging–recapture experiment was twice greater than that predicted by the model (De Pontual *et al*., 2006).

The first phase of the selection procedure is common to all geographical cases. This is not the case for the second and third phases, which are related strongly to a geographical zone. Three criteria, geographic distribution (Phase 2), geographic recovery and temperature adaptation (Phase 3), refer to

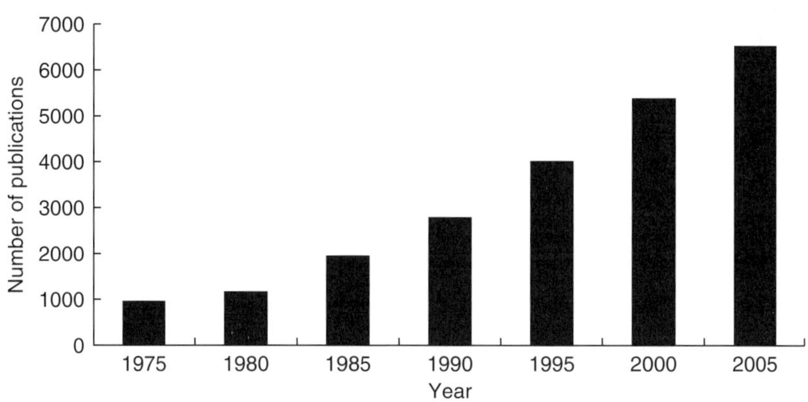

Fig. 6.2. Changes over time in the number of publications related to cod in ASFA base.

the latitude dependence of fish species and their adaptation capacity to environmental conditions. However, local changes in some environmental parameters, and mainly in temperature but also salinity, may modulate the rearing performances of the selected species.

A lot of criteria could be added to the selection process described in this chapter, and especially parameters related to the aquaculture ability of candidate species. Some criteria related to aquaculture capacity were suggested by Cook and Walmsley (1990): growth rate, nutritional requirements, resistance to stress, availability of technology… but others may be added: egg size, disease susceptibility, adaptation to captivity… Since aquaculture performances are poorly reported in databases and for a limited number of species, they could not be included in the present study. This is also the case for fecundity: a very low number of eggs or larvae is reported in sharks or rays (from 10 to 200; Quéro and Vayne, 1997). This could prevent aquaculture development of these species.

In order partially to prevent the lack of parameters in the selection procedure, some complex criteria were calculated from information reported in databases: temperature adaptation of candidate species was estimated by the difference in the northern latitude and southern latitude of the geographic distribution area of each species. Rearing potential was assessed by a combination of the wholesale price of a given species and its age at 3 kg, since this weight was well adapted to the requirements of the transformation channel (formula in Quemener *et al.*, 2002). Farming some new species was restrained by bad 'surprises' observed when they were maintained in captivity. Wild red porgy, *Pagrus pagrus*, exhibits a red skin colour. In captivity conditions, the skin turns dark grey, which is accepted badly by consumers and is associated with a lower market value (Kalinowski *et al.*, 2005). In Senegal sole, *Solea senegalensis*, 44% of the individuals sampled from hatching up to 75 days posthatching showed skeletal malformations (Gavaia and Cancela, 2002). Bluefin tuna, *Thunnus thynnus*, farming expanded rapidly along the Mediterranean coasts. However, handling of tuna results in very high mortality rates, lowering ranching expansion (Ticina *et al.*, 2004). When reared in captivity, reproductive dysfunctions are observed in some fish species. This was observed in European eel, *Anguilla anguilla*, and amberjack, *Seriola dumerili*, which failed to undergo vitellogenesis in captivity (Zohar and Mylonas, 2001). Because of limited information available in databases for these new species, these bad 'surprises' cannot be integrated in a selection process developed before their farming expansion. As a consequence, a post-selection phase must be included in the aquaculture development of previously selected candidates, studying some complementary elements of their adaptation to captivity which cannot be taken into account during the selection process.

In order to avoid a subsequent waste of time, this phase must be shortened by updating the selection procedure regularly: this will be carried out by incorporating recent information in databases and integrating new criteria, and especially those assessing the rearing performance of candidate species. Supplementary criteria not related to aquaculture may also be added to the selection process: seafood allergy is potentially severe since its prevalence was estimated at 5.9% of households by a random survey performed in 2002 in the

USA (Sicherer *et al.*, 2004). Fish allergy was reported in 0.4% of households. Including these criteria in the present selection procedure was not possible because studies were limited, recent and focused on a low number of fish to which allergy was reported.

Then, some criteria could be considered as subjective. This is especially the case with regards to flesh taste, which is driven by consumer group features and mainly their demographic and socio-economic characteristics. For the present study, data concerning these criteria were collected from objective sources: databases (Frimodt, 1995), scientific publications (Bykov, 2000) and wholesale fishmongers (cited in Section 6.2.2).

6.4 Case Study: the Case of the French Atlantic, the Channel and the North Sea Coasts

6.4.1 Enquiries

The French fish farmers line required mainly high-rearing potential species, but did not take new market trends into consideration by neglecting to ensure a high fillet yield of new candidates and targeting high selling price species (Fig. 6.3).

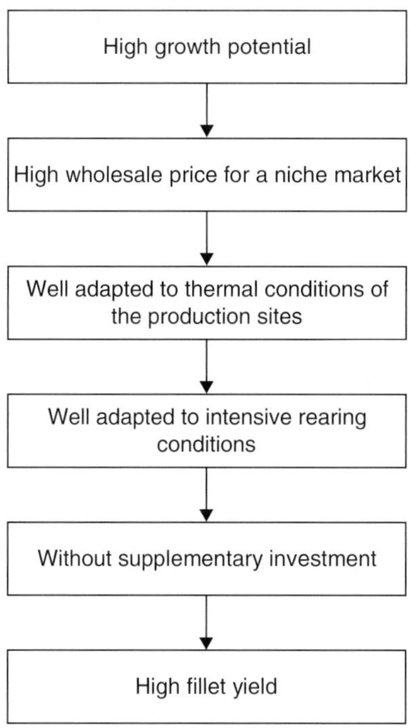

Fig. 6.3. Main requirements of French fish farmers (from Gaignon and Quemener, unpublished results, 1999; from highly required species characteristics at the upper part of the figure to the lowly required ones at the lower part).

However, farmers with a high annual production (> 150 t) paid more attention to these new issues than did small producers.

The main results from a list of 21 and 28 questions asked of the transformation and distribution channels, respectively, are reported in this chapter. The mean weight required by transformation and distribution channels is, respectively, 2.6 kg and comprises between 1 and 5 kg. Both channels require a standard quality product ready to cook and corresponding to new market trends (Figs 6.4 and 6.5). Cod was elected as a 'star species', by both the transformation and distribution lines (Fig. 6.6). Then, the first three to five species elected respectively by the distribution and transformation channels belonged to Gadoids.

Consumer preferences were established after the Institut d'Observation et de Décision (IOD) study (Ofimer, 1999). The study showed an extensive

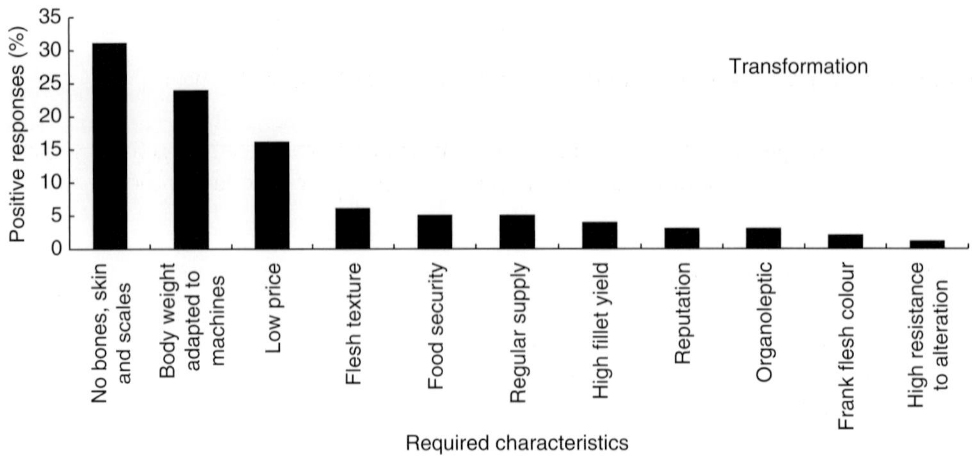

Fig. 6.4. Characteristics of new species required by the transformation lines.

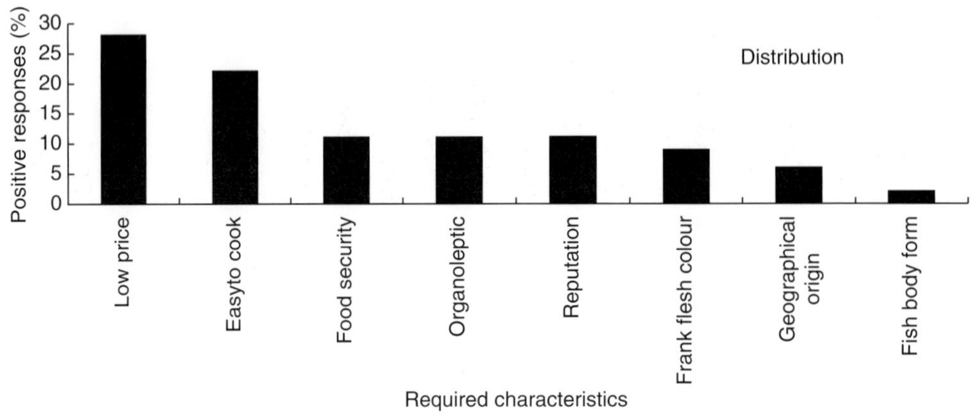

Fig. 6.5. Characteristics of new species required by the distribution lines.

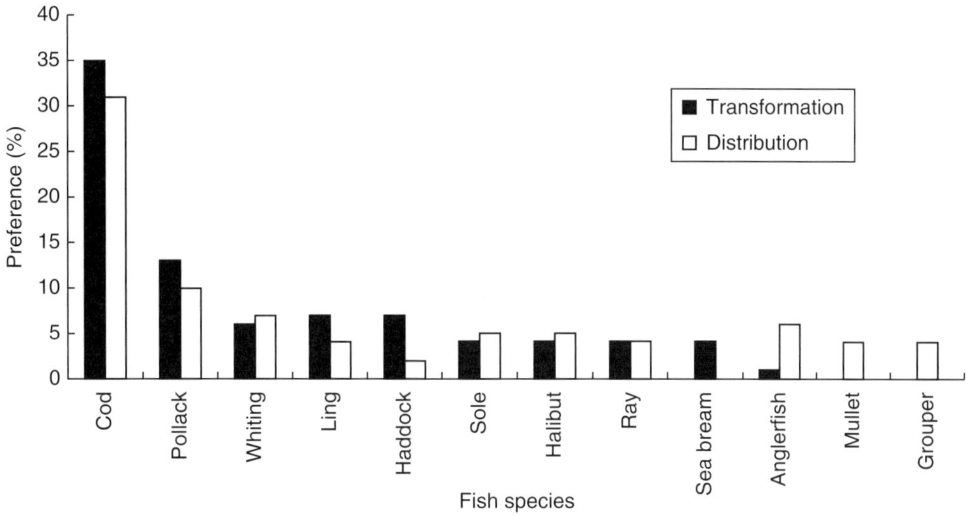

Fig. 6.6. Fish species selection of the transformation and distribution channels.

knowledge of some of the 'new' candidates and especially cod, which is the third species after salmon and trout, in terms of consumer knowledge (Fig. 6.7).

6.4.2 Selection procedure

From 20,000 fish species considered at the beginning of the selection procedure, successive elimination of species is reported in Table 6.2. Only 32 species were considered during the third phase of final classification. Cod was

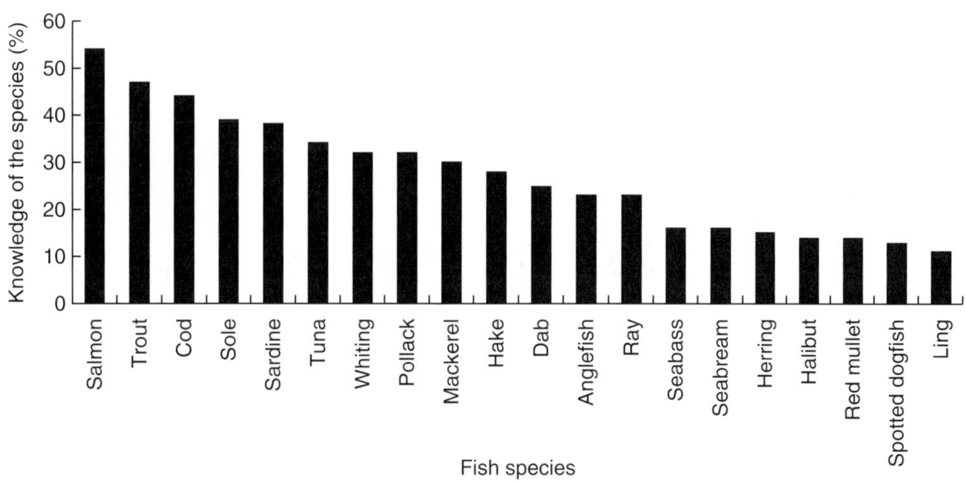

Fig. 6.7. Knowledge of fish species by consumers (percentage of consumers knowing the species; after Ofimer, 1999).

selected by the procedure as the first candidate for aquaculture development on the French Atlantic, the Channel and the North Sea coasts (Fig. 6.8). This rank was sustained by the very high position of cod in the intermediate classifications established for each of the four professionals of the production–consumption line (production: 3rd, distribution: 1st, transformation: 1st and consumption: 1st). Cod aquaculture production in 2004 represented only 4.2% and 7.8% of the seabass and seabream production, respectively, as estimated in tonnes. However, from 2001 its expansion was very rapid (Fig. 6.9). This confirms the candidature of this species, which is strongly sustained by a good knowledge of its biology, a high growth potential, a morphology well adapted to the transformation process, good consumer knowledge, a dramatic decrease of its worldwide landing and a good adaptation to the requirements of the transformation, distribution and consumption sectors. In contrast to flatfish aquaculture, cod farming may use the same facilities as those used by the salmon industry (Roselund and Skretting, 2006). Cod aquaculture will benefit from seabass and seabream farming experience, but also from its own experience: investigations into cod aquaculture were first carried out by Georg Ossian Sars in 1864 and then by Dannevig in 1883 (Svasand *et al.*, 2000). Recent mastering of larval-rearing techniques gave rise to the traditional bottleneck of fish species aquaculture (Brown *et al.*, 2003). Furthermore, a high growth capacity was recorded in captivity conditions: 2 kg in 20 months (Huse, 1991). Regarding the local area targeted by this study, cod is well adapted to the thermal conditions of the northern French coasts of the Channel and the Atlantic, in which water temperature ranges from 5.5 to 18.5°C (Ifremer, 2006): thermal preference is 16–17°C for 2 g cod and 6°C for 5000 g fish (Björnsson and Steinarsson, 2002). However, the context of the French market is different for cod compared to farmed salmon and seabass: when aquaculture was launched, salmon and seabass were considered as luxury products and their price was €10–20/kg. Cod does not have the same image: it is considered as a popular fish and its price should be lower than these two species (Girard and Paquotte, 2003). Other Gadoids were well ranked in the present work: European hake (6th) and ling (8th).

According to Le François *et al.* (2002), the persistence of control species and the very high ranking of Atlantic salmon (2nd) validated the proposed selection procedure. Some original candidates were identified, strengthening the capacity of the method to select 'true new species'. The aquaculture potential of these species could be confirmed by a subsequent phase, focusing on their adaptation to captivity.

The development of fish farming in France is lowered by a negative perception of this activity, illustrated by a large consumer preference for wild products (Fig. 6.10): farmed fish is perceived mainly as being less safe than wild fish. This negative image of farmed fish is particularly important in France (Girard and Paquotte, 2003). Furthermore, aquaculture expansion competes strongly with tourism: France is a top destination, with more than 70 million visitors each year (Kerourio, 2004). French aquaculture also faces strong environmentalist pressures, most often opposed to the expansion of fish farming. Furthermore, administrative procedures are long and complex. As a consequence, French fish farming has remained stagnant since 1998 at

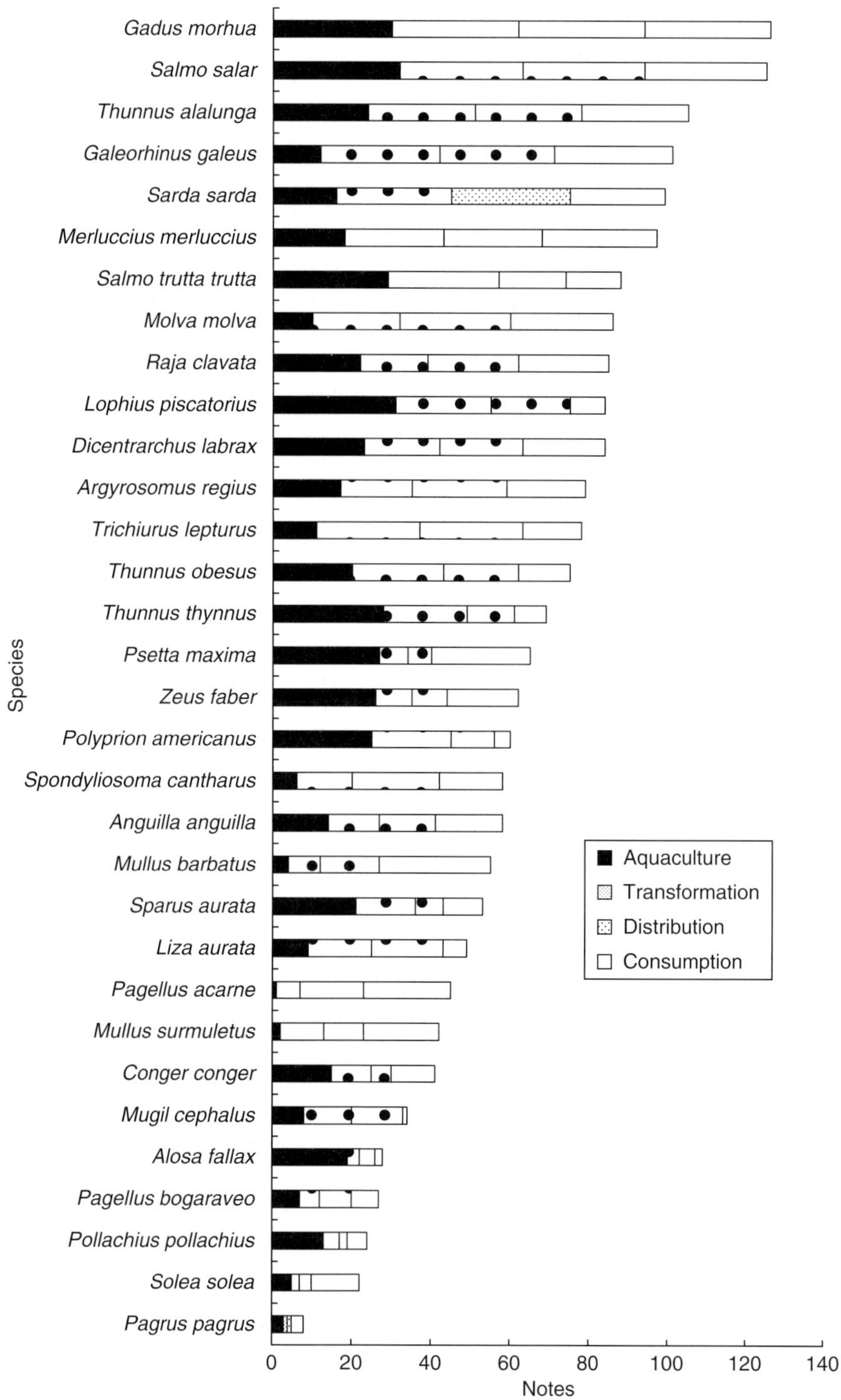

Fig. 6.8. Final classification of 32 candidate species taking into account the four profiles (aquaculture, transformation, distribution and consumption) requirements (longest bars corresponds to best positions).

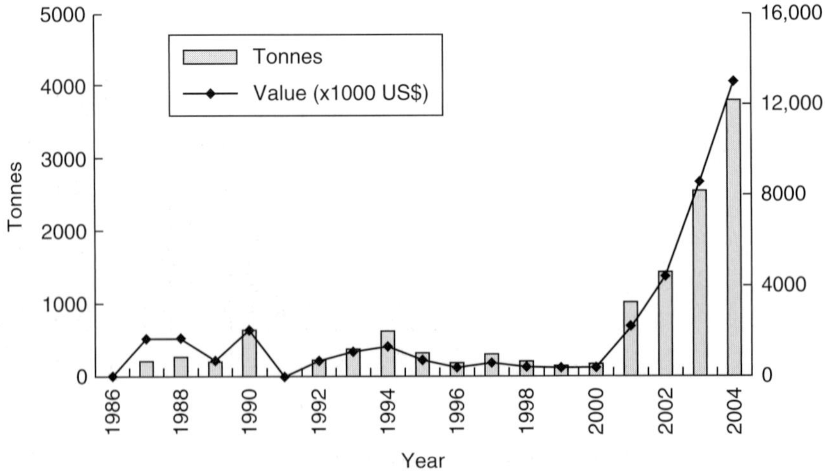

Fig. 6.9. Expansion of cod aquaculture from 1986 to 2004 (after FAO, 2006).

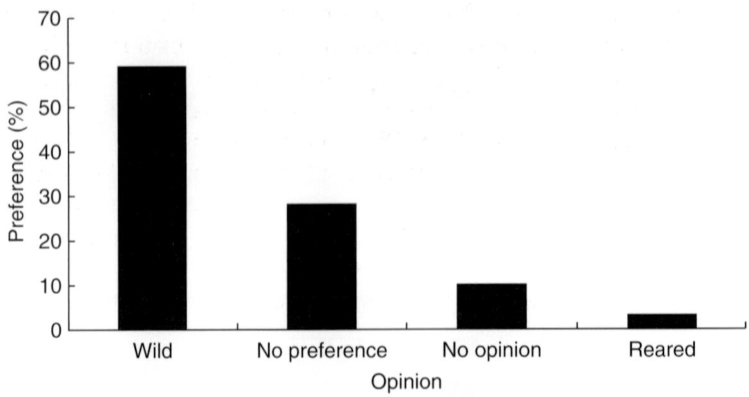

Fig. 6.10. Consumers' preferences in terms of product origin (after Ofimer, 1999).

a production rate ranging between 5000 and 6000t (Fig. 6.11). French aquaculture of 'new fish species' is limited to meagre, *Argyrosomus regius*, (450t in 2004; FAO, 2006) and red drum, *Sciaenops ocellatus* (300t in 2005; Aquamay, 2006).

In conclusion, new candidates for aquaculture development cannot be selected by their selling price only, as was the case at the beginning of the marine farming industry. Because aquaculture expansion results in a rapid decrease of this parameter, the reason for their initial choice for farming expansion is cancelled. Selecting a 'new' fish species has a large subsequent impact in terms of money and time invested. As a consequence, a careful selection process including complementary criteria must be agreed. It avoids a subjective estimation of species' characteristics which chooses a new candidate using only an emotional 'fish appeal', far removed from aquaculture reality.

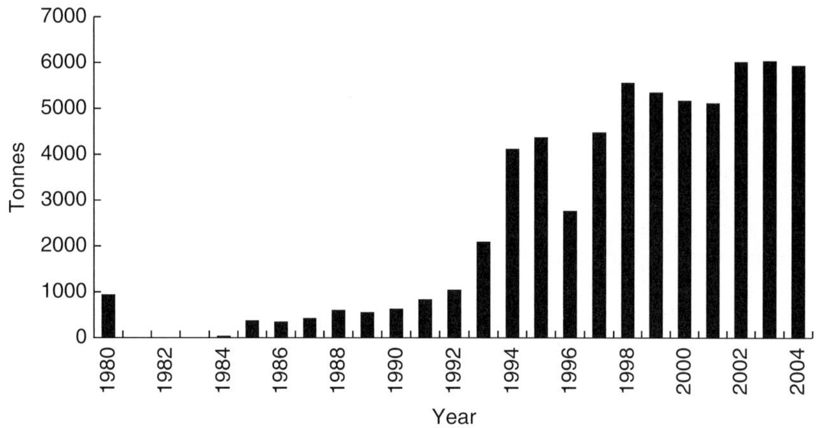

Fig. 6.11. Changes over time of marine fish French aquaculture production (after FAO, 2006).

The selection process presented in this chapter is based on the requirements of the production–consumption lines. It is a complex procedure comprising three successive phases, an initial population of 20,000 fish species and a panel of 32 criteria. It was validated by considering the case of the northern coasts of France. Cod has been selected by this work as the top candidate for aquaculture development in this area. This selection is highly sustained by the rapid development of cod farming in Northern Europe, the production of which could reach similar levels to that of farmed salmon within 15–20 years (Roselund and Skretting, 2006). Cod farming could be the third 'wave' of marine aquaculture, after salmon in Northern Europe and seabass or seabream around the Mediterranean coasts.

References

Aquamay (2006) Situation des filières aquacoles en outre-mer, Corse et régions ultra périphériques. Mayotte, France.

Avault, J.W. (1993) Ten requirements for culturing a 'new' species: a checklist. *Aquaculture Magazine* 19, 68–73.

Barnabé, C. and Tournamille, J.C. (1972) Expériences de reproduction artificielle de loup, *Dicentrarchus labrax* Linné, 1758. *Revue des Travaux de l'Institut des Pêches Maritimes* 36, 185–189.

Bell, J.D. and Gervis, M. (1999) New species for coastal aquaculture in the tropical Pacific – constraints, prospects and considerations. *Aquaculture International* 7, 207–223.

Benetti, D.D., Clark, A.M. and Feeley, M.W. (1999) Feasibility of selected candidates species of marine fish for cage aquaculture development in the Gulf of Mexico with novel remote sensing techniques for improved offshore systems monitoring. In: Stickney, R.R. (ed.) *Proceedings of the 3rd International Conference.* Sea Grant College Program Publishers, Bryan, Texas, pp. 103–119.

Björnsson, B. and Steinarsson, A. (2002) The food-unlimited growth rate of Atlantic cod (*Gadus morhua*). *Canadian Journal of Fisheries and Aquatic Sciences* 59, 494–502.

Brown, J.A., Minkoff, G. and Puvanendran, V. (2003) Larviculture of Atlantic cod (*Gadus morhua*): progress, protocols and problems. *Aquaculture* 227, 357–372.

Brummet, R.E. (1996) Species selection for smallholder aquaculture. *Naga* 19, 19–22.

Bykov, V.P. (2000) *Chemical Composition and Processing Properties of Marine and Oceanfishes*. Handbook. VNIRO, Moscow.

Cites – Convention on International Trade in Endangered Species of wild fauna and flora (http://www.cites.org/, accessed 15 December 1999).

Cook, P.A. and Walmsley, R.D. (1990) Evaluation of a computer model for the selection of candidate species for aquaculture. *Southern African Journal of Aquatic Sciences* 16, 90–100.

Dallimore, J. (2005) The commercialisation of the European aquaculture sector and future commercial development. In: Howell, B. and Flos, R. (compilers) *Proceedings Aquaculture Europe 2005*. EAS Special Publication 35, pp. 19–27.

De Pontual, H., Groison, A.L., Pineiro, C. and Bertignac, M. (2006) Evidence of underestimation of European hake growth in the Bay of Biscay, and its relationship with bias in the agreed method of age estimation. *ICES Journal of Marine Science* 63, 1674–1681.

FAO (2006) Figis data collection (http://www.fao.org/figis/servlet/static?dom=collection&xml=global-production.xml, accessed 7 November 2006).

Fishbase (1998) Fishbase 98, CD Rom. Iclarm, Manila.

Frimodt, C. (1995) *Illustrated Multilingual Guide to the World's Commercial Coldwater Fish*. Fishing News Book, Oxford, UK.

Gavaia, P.J. and Cancela, M.L. (2002) Osteological development and abnormalities of the vertebral column and caudal skeleteon in larval and juvenile stages of hatchery-reared Senegal sole (*Solea senegalensis*). *Aquaculture* 211, 305–323.

Girard, S. and Paquotte, P. (2003) The French market for fresh fish: an opportunity for farmed cod? In: *Proceedings of EAFE 2003, The Fifteenth Annual Conference of European Association of Fisheries Economists*. Ifremer, Brest, France, 7 pp.

Howell, B.R. (1997) A re-appraisal of the potential of the sole, *Solea solea* (L.), for commercial cultivation. *Aquaculture* 155, 355–365.

Hureau, J.C. (1996) Expert center for taxonomy identification, fishes of the North-Eastern Atlantic and the Mediterranean. *The World Biodiversity Database*, CD Rom. Unesco, Paris.

Huse, I. (1991) Culturing of cod *Gadus morhua*. *CRC Handbook of Mariculture, Volume II Finfish Aquaculture*. CRC Press, Boca Raton, Florida, pp. 43–51.

Ifremer (2006) http://www.ifremer.fr/nausicaa/roses/private/browser.htm (accessed 20 November 2006).

Jones, A. (1972) Marine fish farming: an examination of the factors to be considered in the choice of the species. *Laboratory leaflet (New series)* 24. Ministery of Agriculture, Fisheries and Food, Lowestoft, Suffolk, UK, 16 pp.

Kalinowski, C.T., Robaina, L.E., Fernandez-Palacios, H., Schuchardt, D. and Izquierdo, M.S. (2005) Effect of different carotenoid sources and their dietary levels on red porgy (*Pagrus pagrus*) growth and skin colour. *Aquaculture* 244, 223–231.

Kentouri, M., Papandroulakis, N. and Divanach, P. (1995) Specific diversification in Greek finfish mariculture. *Cahiers options méditerranéennes* 14, 129–136.

Kerourio, P. (2004) http://geotourweb.com/nouvellepage4.htm (accessed 20 November 2006).

Le François, N.R., Lemieux, H. and Blier, P.U. (2002) Biological and technical evaluation of the potential of marine and anadromous fish species for cold-water mariculture. *Aquaculture Research* 33, 95–108.

Lemieux, H., Blier, P.U., Parent, B. and Le François, N.R. (2002) *Potentiel d'élevage de différentes espèces d'invertébrés marins à des fins de diversification de l'aquaculture dans l'Est du Canada*. Bureau du Commissaire au développement de l'aquaculture, MPO-DFO Moncton, New Brunswick, Canada, 425 pp.

Lensi, P. (1995) La politique de diversification de l'aquaculture en Corse. La démarche 'nouvelles espèces'. *Cahiers Options Méditerranéennes* 14, 137–148.

Liao, I.C. (2000) The state of finfish diversification in Asian aquaculture. *Cahiers Options Méditerranéennes* 47, 109–125.

Mann, R. (1984) On the selection of aquaculture species: a case study of marine molluscs. *Aquaculture* 39, 345–353.

Mariojouls, C. and Paquotte, P. (2001) Consumer trends: what opportunities for aquaculture products? In: Coimbra, J. (ed.) *Modern Aquaculture in the Coastal Zone*. IOS Press, Amsterdam, pp. 61–69.

Menu, B. (1987) *Aquaculture sur les côtes méditerranéennes françaises: proposition d'étude de nouvelles espèces de poissons marins téléostéens*. Ifremer, Palavas, France.

Ofimer (1996–2004) *Bilan annuel: données de commercialisation sous criées*. Paris, France.

Ofimer (1999) *Images, usages et attitudes en matières de poissons frais d'aquaculture*. 2nd Volume: Approche quantitative. Paris, France.

Paquotte, P. (1998) New species in the Mediterranean aquaculture: is it an answer to the market demand for differentiated products? In: Enne, G. and Greppi, G.F. (eds) *Proceedings of the 33th International Symposium on New Species for Mediterranean Aquaculture*. Elsevier, Paris, pp. 73–83.

Parfouru, D., Paquotte, P. and Gaignon, J.L. (1997) Methodology for new species selection. In: *Proceedings of Martinique 97, Island Aquaculture and Tropical Aquaculture*. EAS, Oostende, Belgium, pp. 227–228.

Quemener, L., Suquet, M., Mero, D. and Gaignon, J.L. (2002) Selection method of new candidates for finfish aquaculture: the case of the French Atlantic, the Channel and the North Sea coasts. *Aquatic Living Resources* 15, 293–302.

Quéro, J.C. and Vayne, J.J. (1997) *Les poissons de mer des pêches françaises*. Delachaux et Niestlé, Lausanne, Switzerland.

Roselund, G. and Skretting, M. (2006) Worlwide status and perspectives on gadoid culture. *ICES Journal of Marine Science* 63, 194–197.

Sicherer, S.H., Munoz-Furlong, A. and Sampson, H.A. (2004) Prevalence of seafood allergy in the United States determined by a random telephone survey. *Journal of Allergy and Clinical Immunology* 114, 159–165.

Stat, P. (2000) Statistiques de pêche (http://w3.ifremer.fr/isih/login.do).

Suquet, M., Divanach, P., Quemener, L. and Gaignon, J.L. (2002) Criteria for cost-effective diversification for European finfish mariculture. *Cahiers Options Méditerranéennes* 59, 119–128.

Svasand, T., Kristiansen, T.S., Pedersen, T., Salvanes, A.G.V., Engelsen, R., Noevdal, G. and Nodtvedt, M. (2000) The enhancement of cod stocks. *Fish Fisheries* 1, 173–205.

Ticina, V., Grubisi, L. and Katavi, I. (2004) Sampling and tagging of live bluefin tuna in growth-out floating cages. *Aquaculture Research* 35, 307–310.

Webber, H.H. and Riordan, P.F. (1975) Criteria for candidate species for aquaculture. *Aquaculture* 7, 107–123.

Wu, R.S.S. (1989) Biological and economic factors in the selection of cultured fish species and the development of a bio-economic model. *Advances in Tropical Aquaculture* 9, 437–444.

Zohar, Y. and Mylonas, C.C. (2001) Endocrine manipulations of spawning in cultured fish: from hormones to genes. *Aquaculture* 197, 99–136.

7 The Agribusiness Approach

David Otton[1] and Lawrence Dooley[2]

[1]Tathra, New South Wales, Australia; [2]Department of Management, Centre for Wine, Food and Agribusiness, Monash University, Victoria, Australia

7.1 The Concept of Agribusiness

This chapter explains the workings of the concept of agribusiness introduced in Chapter 5, this volume. The term agribusiness was first used publicly by Professor John H. Davis in a paper presented at the Boston Distribution Conference in 1955. In 1957, he and fellow Harvard academic, Professor Ray Goldberg, published *A Concept of Agribusiness*, in which they defined agribusiness as 'the sum total of all operations involved in the manufacture and distribution of farm supplies; production operations on the farm; and the storage, processing and distribution of farm commodities made from them'. Thus, according to Davis and Goldberg, 'agribusiness essentially encompasses today the functions which the term agriculture denoted 150 years ago' and 'modern agribusiness is the product of a complex of evolutionary forces spontaneously at work without central guidance or direction' (Davis and Goldberg, 1957; Otton, 2004).

7.2 Method

The agribusiness system is explained and then related to information gleaned from the studies in Chapter 5, this volume, to show how aquaculture industries and their component companies understand and use the concept of agribusiness in their operations (Otton, 2004).

7.3 The Agribusiness System

Based on Davis and Goldberg's (1957) work, the Schroder–Pollard model (Fig. 7.1) explains the basic operation of the agribusiness system.

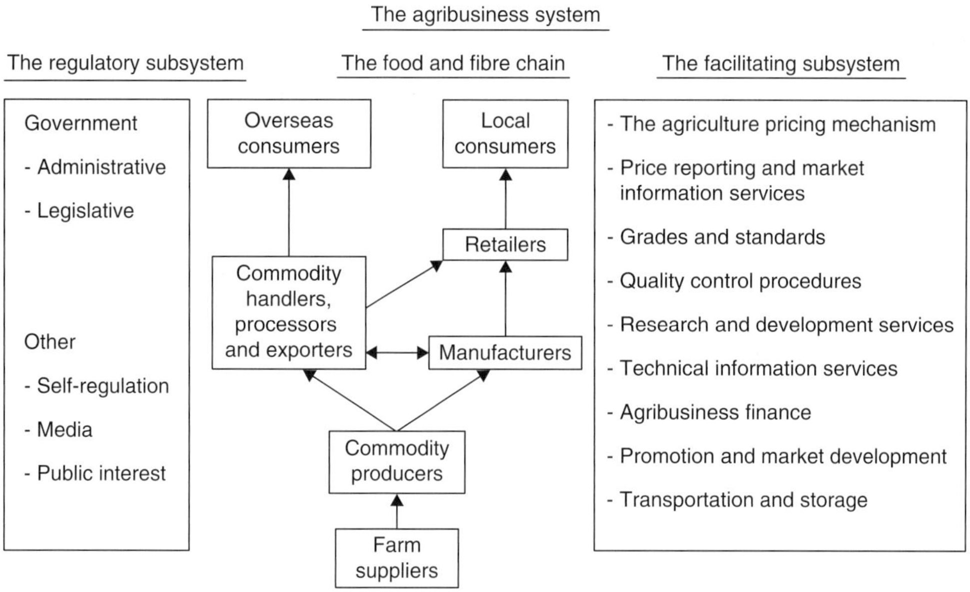

Fig. 7.1. The Schroder–Pollard agribusiness model (Schroder and Pollard, 1989).

In addition to this model, the term 'agribusiness value chain' describes the means of delivering the value offering to the end-user customer. A simple chain follows: *Input supply > Production > Output handling > Processing > Marketing > Customer end user*. This chain is paralleled by public and private sector infrastructure, which includes advisory and financial services plus government regulation and policy (Schroder and Pollard, 1989; Schroder, 2003).

The concept of agribusiness describes the value creation system in food and fibre production and, according to Walters and Lancaster, 'the only source of competitive advantage is the ability to conceive the entire value creation system and make it work by coordinating the activities among actors so that actor and activity are better matched' (see Walters and Lancaster, 1999a,b). A competitive advantage can be created at any stage in the aquaculture chain; for example, designing feed formulation in response to consumer demand to enhance the nutritional and/or sensory characteristics of aquaculture products by mimicking the fatty acid composition of wild fish in farmed fish, or producing low-fat salmon (New, 2001; Otton *et al.*, 1997; Otton, 2004). In fresh food production, there are 11 identifiable factors or basics for an efficient fresh supply chain between growers and customers. They are scale of operation, strategic alliances, production flexibility, continuity of supply, quality control, communications, supermarket strategies, food safety legislation, supply chain integrity, rationalization of the supply base and innovation. Key elements of the food system are transparency and traceability, suggesting it is prescription driven as well as consumer driven, resulting in triple bottom line accounting of economic success, environmental sustainability and social benefit (Grimsdell, 1996; Urban, 1998; Fearne and Hughes, 1999; Otton, 2004).

All players in the chain take a percentage of the final consumer price; therefore, operating costs must be monitored and reduced where possible. The modern concept of triple bottom line accounting means that savings made on traditional costs may be replaced by social and environmental costs (Gifford *et al.*, 1998; Hayes *et al.*, 2000).

Any proposed agribusiness operation should begin with the construction of a diagrammatic virtual agribusiness value chain and model based on the above frameworks to consider the operation of key production and marketing components (Otton *et al.*, 1997). The value chain not only serves the end consumer but also several other players and customers in the chain, for example, suppliers, manufacturers and distributors. Therefore, virtual modelling enables an analysis of how the chain should work in reality and how agribusiness trends and issues may affect the overall agribusiness operation. The model, whether virtual or real, is constrained and limited only by what is possible and when operating is regularly reconfigured to meet changing circumstances (Davis and Goldberg, 1957; Schroder and Pollard, 1989; Frederiksen and Bremner, 2001; Otton, 2004). Good chain management captures efficiencies, controls costs, reduces risks, responds to consumer demands for product attributes and removes as many links in the chain as possible (Boehlje *et al.*, 1999a).

Aquaculture is different from agriculture in that farmed fish compete in the marketplace with wild harvested fish, whereas terrestrially farmed flora and fauna rarely competes with that which is wild harvested. Unlike farmed food, many aquacultured products can be produced all year round with consistent size and quality. The availability of wild-caught fish fluctuates, which may cause price volatility (Engle and Quagrainie, 2006).

In aquaculture, as in agriculture, a species' commercial survival depends not only on its adaptability to farming, but also on its ability to remain within the agribusiness system. Therefore, when diversifying aquaculture by growing new species, this virtual chain must consider the biology and physiology of the species under cultivation and the differing operating environments in which the species is produced and marketed. Overall, chain operation can make or break a species' profitability, so virtual chain construction not only conceptualizes a species' progress through the chain, but enables analysis and reconfiguration of the chain once the species passes the novelty stage of market entry (Otton, 2004).

The concept of agribusiness enables an over-arching view of how food and fibre are produced without viewing any one segment in isolation and regarding each as part of a process within a marketing channel (Schroder and Pollard, 1989). Marketing channels are 'sets of interdependent organizations involved in the process of making a product or service available for use or consumption' (Stern *et al.*, 1989), where the expectations of consumers and all other players are satisfied through continual improvement of processes and relationships in the channel (Schroder and Pollard, 1989; Stern *et al.*, 1989). If these are under the control of one firm, the channel is described as vertically integrated and channel activities are coordinated by administrative control. Vertical integration (VI) facilitates communication and risk sharing and usually requires common ownership of firms at successive stages of the channel. VI improves operational efficiency and reduces the costs of marketing, but may result in a system that is less flexible and less responsive to the needs of consumers (Schroder and Pollard, 1989; Schroder, 2003).

A firm may respond to the selling power of its suppliers or the buying power of its buyers by integrating backward or forward in the chain (Schroder and Pollard, 1989). For example, a producer may purchase a market outlet, or a retail business may purchase a farm (Otton, 2004). In most agricultural industries, other forms of alliance such as supply contracts between producers, processors and retailers are common, as are joint ventures and business-to-business (B2B) (Schroder and Pollard, 1989; Cooper, 2000; Otton, 2004). B2B trading is a process of disintermediation, which means reconfiguring the value chain to eliminate redundant chain segments which previously acted as middlemen. Cooper described this as the 'scissors economy', where the middlemen would disappear unless they could add value (Cooper, 2000; Otton, 2004). Transaction costs are developing as an underlying factor in modern agribusiness where industries become vertically integrated when the transaction costs associated with market exchange outweigh the benefits of such an exchange (Hayes *et al*., 2000).

In mainstream business, industrialization provides a management system that allows the integration of each step in the economic process to achieve increasing efficiencies in the use of capital, labour and technology. Industrialization of agriculture means continued consolidation of farms, growing use of production and marketing contracts and vertical integration among input suppliers, lenders, agricultural producers, processors and distributors of food and fibre products, both domestically and globally.

It involves large-scale production units that use standardized technology and management. The application of modern industrial manufacturing, production, procurement, distribution and coordination concepts to the food and industrial product chain means negotiated coordination replaces market coordination in the system. Size and standardization lower production costs, produce goods to fit processor specifications, consumers' needs (for specific product attributes) and allay food safety concerns. Industrialization is a last stage in the development of agribusiness in which smaller producers not associated with an industrialized system have difficulty gaining economies of scale and access to the technology required to be competitive, with the possible exception of niche markets (Urban, 1991; Drabenstott, 1994; Barry, 1995; Boehlje, 1996; Boehlje and Sonka, 1998; Boehlje *et al*., 1999a,b).

'The inevitable development for agribusiness is the prescription food system, a concept beyond industrialization and characterized by more tightly aligned value chains from genetics through producers and processors to end use customers' (Urban, 1998).

In 2000, Professor Goldberg asserted the genetic revolution was leading to an industrial convergence of the food, health, fibre and energy business and that the implication of this life science revolution was to redefine the boundaries of the agribusiness system to those of an agriceutical system (Goldberg, 2000). This structure should follow pharmaceutical standards for research, production and pricing, with transparency and traceability enabling the customer to trace each food item to its earliest production step. Global traceability is becoming a feature of the system where each food source will require an ISO similar to ISO9000 of global, policed manufacturing standards of process and quality, which qualify goods for world trade. Therefore, all chain participants have a responsibility for food safety that not only covers handling but ingredient inputs (Urban, 1998; Goldberg, 2000). For example, the purchaser of a fish product will know which animal it came from, what the animal ate, its growing conditions and how it was killed and packaged (Frederiksen and Bremner, 2001; Otton, 2004). Farmers may need to adopt the same liability for food as does the hotel, restaurant and institution trade. The world's food system will further industrialize and integrate because 'consumers will define food as an input or a prescription for their physical condition, mental health and safety, as well as a template for beneficial environmental practices in food production' (Urban, 1998).

A mature agricultural product is one consumed by most of the population (Schroder and Pollard, 1989). Many are mature, but not fish, the market for which is expanding, giving aquaculture a double benefit because as general consumption rises, so does acceptability and demand for aquacultured fish. A second feature of many aquaculture products is that one supplier's production is not differentiated from that of another supplier. This applies at the level of individual producer and at higher levels of aggregation; for example, an individual country supplying a world market. The solution is branding including country of origin labelling (COOL) by the supplier to differentiate commodities. This may be easier for aquacultured products than wild harvested products; however, the postharvest or cold chain presents the same problems of handling and traceability for fish from either source (Ruello, 1999, 2000; Frederiksen and Bremner, 2001; Otton, 2004).

7.4 Case Studies: Catfish, Atlantic Salmon and Barramundi

These case studies, conducted on the US catfish industry and the Australian salmon and barramundi industries between 2000 and 2003, showed similarities and individual traits in the way each industry and some companies within the industries applied the concept of agribusiness. From an industry development view, the two most important criteria across the industries were species' market acceptance and ease of farming. A species with high market appeal may be unsuitable for farming and, conversely, a species well suited to farming may lack market appeal. Any new species must either have an established market or have the physiological characteristics and supporting agribusiness operation to create a market for it. The most popular species are not always ideal for aquaculture, but rather species compatible with the agribusiness system for which technology can be developed. This sometimes may make selection and development of new species a compromise. The top ranking agribusiness criteria were common across all three industries, diverging as the criteria lessened in ranking (Otton, 2004).

Atlantic salmon, the globally established marine-farmed fish, and its industry is a model for closed life cycle finfish aquaculture. Channel catfish, a freshwater species regionally established in the USA, and its industry is the agribusiness model. Barramundi serves as a link between the two. Whereas salmon usually grow in a caged marine environment and catfish grow mainly in freshwater ponds, the euryhaline barramundi, farmed in most Australian states and Asia, grows in marine and fresh water, enabling production in both environments using cages, ponds or tanks, or combinations of all three. These options enable widely dispersed and less geographically confined production in mainland Australia than either catfish in America or salmon in Tasmania. The case studies covered multiple layers of responsibility in aquaculture company structures and identified critical success factors for the industry species, value chain operation and what to look for in new species, particularly screening data and predicted agribusiness value chain performance. All study industries applied the concept of agribusiness to their operations, sometimes not conscious of how it was working for them. All three species and their supporting industries were suitable models for new species development (Otton, 2004).

7.4.1 Channel catfish (*Ictalurus punctatus* Rafinesque 1818)

The case study showed that catfish emerged as an unlikely contender from a suite of potential new freshwater aquaculture species. This assessment did not fit a preconceived model but satisfied some obvious criteria; well known, fast growth and, though omnivorous, is mainly herbivorous with a food conversion rate of 2:1. Catfish transparent industry data and evidence of a new product development process conducted by Professors Swingle and Shell at Auburn University, Alabama, plus other industry players and stakeholders (Shell, 1993; Otton, 2004). A major difference between catfish and the other study species is that the model slotted into an existing agribusiness system, whereas salmon and barramundi (in Australia) had an agribusiness model built around them (Otton, 2004).

The American south-eastern states have the optimum environment, climate, water supply and water quality for growing catfish, a species native but not endemic to the region. Catfish had huge green field assets from which to build its industry production segment. These, particularly in the Mississippi Delta, include cheap and available resources of groundwater, flat land and soil suitable for the construction of catfish ponds. An important competitive advantage in the farm supplies segment is the availability of cheap coarse grains from the Midwest that are transported down the Mississippi River and milled in the catfish producing states. These input supplies integrate grain farmers and millers to the catfish value chain, facilitating supply of high-quality, well-priced feed.

The catfish industry is an example of industrialized agribusiness with high public exposure and acceptance. Industry support comes from the Catfish Institute which, financed by a miller's levy on feed sold to farmers, promotes the industry and supervises industry research. Catfish Farmers of America represents the industry to the government and related stakeholders and the Catfish Bargaining Association negotiates the prices that farmers and processors receive from their customers. The Thad Cochran Warmwater Aquaculture Center (sic) at Mississippi State University provides scientific support together with other institutions, including Auburn University and Louisiana State University.

Though catfish has the characteristics Americans like of easy-to-portion white, flaky meat, little or no bones and mild flavour, its image was that of a bottom-feeding, scavenger fish. Growing catfish by aquaculture and market development by the Catfish Institute changed consumers' perception of the species from a bottom dweller to a unique, nationally acceptable fish enhanced by aquaculture to improve and control its growth, size, quality, freshness and availability at an affordable stable price. Retailers then saw the opportunity to promote catfish, thereby creating demand and new markets, giving the industry a multiplier effect of 7:1 (Otton, 2004).

The current agribusiness issues of industry stability and profitability should be solved by consolidation of production and processing units, improving strategic alliances and developing the ability to compete in the contemporary global economy by broadening markets domestically and overseas. During its development as an industry, catfish was marketed successfully beyond its regional boundaries; however per annum consumption is only around half a kilogram per capita in the USA. A new product development process to market new catfish products in new areas or develop a new species complementary to catfish are both options for the catfish industry (Jepsen, 2000; Otton, 2004; Roger Barlow, President of the Catfish Institute, personal communication, May 2004).

7.4.2 Atlantic salmon (*Salmo salar* Linnaeus 1758)

Salmon farming began in Norway in the 1960s and increased steadily in the 1970s because of technological breakthroughs, high profits and support from government agencies promoting industry development. The industry in Norway and then Scotland faced problems of all-year-round supply, quality, industry unity and agribusiness supply chain management. An important feature of the northern hemisphere model was the research link between public authorities and industry. The parties invested in both research centres and programme definition for genetics, pathology and nutrition.

Similar government support was critical to the establishment and maintenance of Atlantic salmon aquaculture in Tasmania, with the benefit of northern hemisphere scientific and agribusiness models adaptable to the southern hemisphere. As with catfish aquaculture, government funding provided secure finance for the provision of stock and advisory services during the infant years of a hitherto non-existent industry (Harache and Paquotte, 1998; Morgan, 1999; Otton, 2004).

The agribusiness model for Tasmanian salmon aquaculture has sheltered production sites remote from urban development and tourism, containing pristine water, with a favourable, warmer temperature profile that assists high growth rates, enabling faster growth rates than in the northern hemisphere. Feed, labour and smolts represent the major cost of salmon production in Tasmania. In the wild, salmon are carnivorous, whereas under culture, they eat pelletized diets that account for about 40% of costs.

Though not as cheap as catfish feed, relatively inexpensive protein and ingredient sources deliver a high-quality diet. Food conversion ratios vary between farms, depending on site quality and managerial capability, but industry average appears to be 1.8:1. Tasmanian producers have higher average costs, lower disease control costs and lower stocking rates than overseas growers. Scientific understanding of genetics, nutrition and fish metabolism is more advanced for Atlantic salmon than any other species under cultivation in Australia (Jungalwalla, 1991; Otton, 2004).

Recent mergers and acquisitions of salmon companies have achieved economies of scale in production, harvesting, processing, distribution and marketing. The basis of yield per production unit across all three study industries is fish survival and growth, which is affected by environmental issues, husbandry and disease control. This in turn affects short- and long-term returns. As Tasmanian salmon aquaculture serves a domestic market in Tasmania and continental Australia, plus an export market (mainly in Asia), financial returns are vulnerable to fluctuations in exchange rates (Otton, 2004).

7.4.3 Barramundi (*Lates calcarifer* Bloch 1790)

In Australia, the Queensland Department of Primary Industries started research into barramundi aquaculture during 1984, with the aim of adapting to Australian conditions culture techniques developed in Thailand. The euryhaline barramundi thrives in fresh, brackish or marine water, a capability well suited to aquaculture, enabling a wide range of production alternatives. Barramundi aquaculture in Australia is a cross between an industrial marine model (salmon) and an industrial freshwater model (catfish). Like catfish in the USA, freshwater barramundi aquaculture in Australia slotted into an existing (albeit underdeveloped) agribusiness system. The species is now farmed in fresh, estuarine and marine water, mainly in Queensland, the Northern Territory and South Australia, using four farming systems; purpose-built ponds, cage culture in ponds, cage culture in estuarine and marine water and recirculation (Griffin *et al.*, 1993; Wingfield, 2001; Otton, 2004).

Though now an established aquaculture industry, barramundi compared to salmon and catfish is still in the early stages of development. Research needs are in the areas of stock improvement, nutrition, health, production system technology and the quality of water available for and used in barramundi aquaculture. The current and future production system alternatives enable the sequential multiple use of water by co-locating barramundi aquaculture with water-intensive agricultural operations, which may lead to steady industrialization (Rogers, 1999a,b; McVeigh, 2000; Rimmer *et al.*, 2001; Otton, 2004).

7.5 Conclusion

All case study industries have established agribusiness value chains. The catfish chain existed in a form for modification from agriculture to aquaculture; the salmon and barramundi chains were already constructed overseas, imported with the technology, then modified and adapted for Australian conditions. The salmon industry reconfigured the northern hemisphere model for Tasmania and the marine component of the barramundi industry adapted the coldwater salmon model for tropical Australian conditions, thereby creating a submodel. The role of government was critical to industry development in all three cases, particularly at start-up, where the industries benefited from government assisted hatchery establishment. In general, governments provided encouragement, money and physical and intellectual resources (Jungalwalla, 1991; Rogers, 1999a,b; Otton, 2004).

The case studies' respondents opined that a source of competitive advantage in aquaculture was doing, at each stage, everything better than their competitors, which equated to continual improvement in the value chain. The barramundi industry identified a gain in competitive advantage by assigning different methods of ownership to the production units, either by company ownership, leasing, privately owned farms or companies contracted to supply fish. In production, the critical links are maximizing survival and growth and optimizing food conversion rates (FCR). Feed is around half the cost of production and a trend is for either industry integration with a feed company or a feed company developing an aquaculture company to secure an additional market for its product. An emerging trend and future necessity is minimizing or eliminating reliance on fishmeal and fish oil in aquaculture diets (Otton, 2004). A competitive advantage can be created at any stage in the chain, for example, designing feed formulation in response to consumer demand to enhance the nutritional and/or sensory characteristics of aquaculture products by mimicking the fatty acid composition of wild fish in farmed fish, or producing low-fat salmon (Otton *et al.*, 1997; New, 2001; Otton, 2004).

Atlantic salmon and catfish have enjoyed sustained competitive advantage, as described by Porter (1985), but now face competition from overproduction and cheaper imports, respectively. Therefore, these industries must now look at their competitive situation and reconfigure their value chains to fit contemporary demands and rebalance other areas of the chain, for example, feed and diet, to reduce costs and improve output.

7.6 Future Trends and Issues

A recent investigation into the future trends in the food component of Australian agribusiness indicated a shift away from family farms to strategically located corporate production units. This trend continues towards strategically located processing plants and alliances with global supermarkets and brands with an emphasis on innovation (McKinna, 2006). Aquaculture is agribusiness and it follows that trends apparent in Australia are trends evident in the global economy, bearing in mind that the channel catfish case study was conducted in the USA and Atlantic salmon and barramundi, though case studied in Australia, are linked to similar industries overseas.

Other trends and issues for future investigation identified by the case studies are:

1. What is the next cycle of value growth in aquaculture and are there strategic moves to be made by aquaculture companies within aquaculture regions to capture this growth?
2. How are changing consumer tastes catered for by exploring the commercial possibilities of new species and products? This is a widely acknowledged problem in the food industry.
3. How is a new species for aquaculture assessed and should it be developed for export, domestic or both?
4. How does the aquaculture industry rationalize its operations to cope with real or perceived changes and develop more efficient inputs, particularly diet development.
5. How does the aquaculture industry increase per capita/per annum consumption and promote the high health benefits of aquaculture products?

An additional future trend in aquaculture not identified by the case study but evident in the industry is biomarine farming, where new species are grown for medicine or medicines are extracted from existing aquaculture species for use by humans. This is described as the next wave of super medicines from the ocean (Benkendorff, 2006).

References

Barry, P.J. (1995) Industrialisation of US agriculture: policy, research and education needs. *Agriculture and Resource Economic Review* 24, 128–135.

Benkendorff, K. (2006) Overview of the marine bio-resource industry. Paper presented at the Australasian Aquaculture (Innovation in Aquaculture) Conference, Adelaide, Australia (28 August).

Boehlje, M. (1996) The industrialization of agriculture: what are the implications? *Choices* First Quarter, 30–33.

Boehlje, M. and Sonka, S. (1998) Structural realignment in agriculture: how do we analyse and understand it? University of Illinois, Champaign-Urbana, Illinois (http://www.ag.uiuc.edu/famc/program98/sonka.htm, accessed 30 December 1998).

Boehlje, M.D., Hofing, S.J. and Schroeder, R.C. (1999a) Value chains in agricultural industries. Staff paper No. 99-10. Department of Agricultural Economics, Purdue University, West Lafayette, Indiana (http://www.centrec.com/, accessed 30 December 1999).

Boehlje, M.D., Hofing, S.J. and Schroeder, R.C. (1999b) Financing and supplying inputs to the 21st century producer. Staff paper No. 99-11. Department of Agricultural Economics, Purdue University, West Lafayette, Indiana (http://www.centrec.com/, accessed 30 December 1999).

Cooper, S. (2000) Global risk management. Paper presented at the 10th World Food and Agribusiness Congress, Chicago, Illinois (28 June).

Davis, J.H. and Goldberg, R.A. (1957) *A Concept of Agribusiness*. Division of Research, Graduate School of Business Administration, Harvard University, Boston, Massachusetts. The Alpine Press, Harvard University, Boston, 136 pp.

Drabenstott, M. (1994) Industrialisation: steady current or tidal wave. *Choices* Fourth Quarter, 4–8.

Engle, C.R. and Quagrainie, K. (2006) *Aquaculture Marketing Handbook*, First Edition. Blackwell Publishing, Ames, Iowa, 271 pp.

Fearne, A. and Hughes, D. (1999) Success factors in the fresh produce supply chain: insights from the UK. *Supply Chain Management* 4, 120–128.

Frederiksen, M. and Bremner, A. (2001) Fresh fish distribution chains: an analysis of three Danish and three Australian chains. *Food Australia* 54, 117–123.

Gifford, D., Hall, L. and Ryan, W. (eds) (1998) *Chains of Success: Case Studies on International and Australian Food Business Co-operating to Compete in the Global Market*. Agribusiness and Community Branch, Department of Primary Industries and Energy, Canberra, 144 pp.

Goldberg, R.A. (2000) The Challenge for IAMA (International Food and Agribusiness Management Association). The genetic revolution: transforming our industry, its institutions and its functions. *Proceedings of the World Food and Agribusiness Congress* (26 June), Chicago, Illinois (http://www.ifama.org/conferences/2000Congress/Congress/GoldbergSpeech.PDF, accessed 31 December 2000).

Griffin, R., Morrissy, N., Rimmer, M. and Russell, J. (1993). Barramundi. In: Kailola, P.J., Williams, M.J., Stewart, P.C., Reichelt, R.E., McNee, A. and Grieve, C. (eds) *Australian Fisheries Resources*. Bureau of Resource Sciences, Department of Primary Industries and Energy and the Fisheries Research and Development Corporation, Canberra, pp. 252–255.

Grimsdell, K. (1996) The supply chain for fresh vegetables: what it takes to make it work. *Supply Chain Management* 1, 11–14.

Harache, Y. and Paquotte, P. (1998) European marine fish farming: an emerging industrial activity. *World Aquaculture* 29, 42–48.

Hayes, D., Hayenga, M. and Thompson, S.J. (2000). Transaction costs, economics and the evolving structure of agricultural production. Paper presented at the International Food and Agribusiness Congress. Chicago, Illinois (25 June).

Jepsen, L.-A. (2000) Sector review: the catfish industry in Vietnam. Prepared on behalf of the Catfish Institute, Belzoni, Mississippi, 26 pp.

Jungalwalla, P.J. (1991) The development of an integrated saltwater salmonid farming industry in Tasmania, Australia. In: Cook, R.H. and Penell, W. (eds) *Proceedings of the Special Session on Salmonid Aquaculture, World Aquaculture Society* Los Angeles, California. Canadian Technical Report Fisheries and Aquatic Sciences, November 1991, pp. 65–73.

McKinna, D. (2006) The six mega-trends driving the Australian agrifood industry. *Proceedings of the 3rd Innovative Food Centre Conference. Food Innovation: Emerging Science, Technologies and Applications – FFIESTA* (www.mckinna.com.au, accessed 31 December 2006)

McVeigh, P. (2000). Multiple water use: getting more from a 'bucket' of water. *Proceedings of the Australian Bureau of Agricultural and Resource Economics (ABARE) Outlook 2000 Conference. Natural Resources*. Canberra, pp. 123–126.

Morgan, A. (1999) Aquaculture: the commercial reality. Paper presented at the World Aquaculture Conference. Sydney, Australia (27 April).

New, M. (2001) GMOs, single cells and hazards in Europe's aquafeed future. *Fish Farmer* 24, 41–42.

Otton, D.D. (2004) Establishment of an agribusiness model for assessing the commercial viability of new species for aquaculture. PhD thesis, University of Tasmania, Australia, 660 pp.

Otton, D.D., Drysdale, G., Leopold, I., Greenwood, C. and McCann, C. (1997) Unpublished Master of Business (Agribusiness) retorted packaging value chain exercise. Monash University, Frankston, Victoria, Australia (9 April).

Porter, M.E. (1985) *Competitive Advantage: Creating and Sustaining Superior Performance.* The Free Press, New York, 557 pp.

Rimmer, M., Wingfield, M. and Young, C. (2001). Farmed barramundi research and development plan. Paper presented at the Australian Barramundi Farming Workshop, Cairns, Queensland, Australia (6 August).

Rogers, H. (1999a) The Pivot Northern Territory barramundi sea cage project and market assessment. *Proceedings of the Australian Barramundi Farming Workshop*, Bribie Island, Queensland, Australia, pp. 12–19.

Rogers, H. (1999b) Anticipating future demand; the impact of demographics. *Proceedings of the Aquaculture Beyond 2000 Changing Direction Conference and Workshop*, 23–24 August 1999, Canberra.

Ruello, N. (1999). A study of the retail sale and consumption of seafood in Sydney. A report to the Fisheries Research and Development Corporation (FRDC Project 98/345), Volumes 1 and 2. Sydney. New South Wales, 173 and 177 pp.

Ruello, N. (2000) Retail sale and consumption of seafood. A report prepared for the Fisheries Research and Development Corporation, Canberra, 20 pp.

Schroder, B. and Pollard, V. (1989) The concept of agribusiness. *Agricultural Science* 2, 13–17.

Schroder, W. (2003) Recent developments in agribusiness: their implications for business and government. Presentation to the Faculty of Huazhong Agricultural University, Peoples Republic of China, October 2003.

Shell, E.W. (1993) *The Development of Aquaculture: An Ecosystems Perspective.* Craftmasters Printers, Opelika, Alabama, 265 pp.

Stern, L.W., El-Ansary, A.I. and Brown, J.R. (1989) *Management in Marketing Channels.* Prentice Hall, New Jersey, 514 pp.

Urban, T.N. (1991) Agricultural industrialisation: it's inevitable. *Choices* Fourth Quarter, 4–6.

Urban, T.N. (1998) Beyond industrialisation: the prescription food system. *Choices* Fourth Quarter, 43–44.

Walters, D. and Lancaster, F. (1999a) Value and information – concepts and issues for management. *Management Decision* 37, 643–656.

Walters, D. and Lancaster, F. (1999b) Value-based marketing and its usefulness to customers. *Management Decision* 37, 697–708.

Wingfield, M. (2001) The Australia barramundi farming industry. *2000 Australian Aquaculture Forum Yearbook*. National Aquaculture Council, Executive Media, Melbourne, Australia, pp. 39–40.

II Finfish Species Description and Biotechnical Analysis

8 Quantitative Approaches for Identifying Finfish Species Suited for Sustainable and Productive Aquaculture

Nathalie R. Le François,[1,2] Arianne Savoie[1] and Pierre U. Blier[1]

[1]*Université du Québec à Rimouski, Rimouski, Québec, Canada;*
[2]*Biodôme de Montréal, Montréal, Québec, Canada*

8.1 Introduction

The worldwide development and growth of the aquaculture industry has, in many ways, relied on a continual quest for new or alternate species that can be cultured in order to exploit new market opportunities, environments or production technologies fully. Undeniably, the risks involved in commercializing new species are considerable and, to some extent, unappealing for the private aquaculture sector. Such impediments can be reduced significantly in the presence of clear orientations, concerted actions and adequate financial support. Within companies or agencies, species selection often poses an enormous challenge for the people who are in charge of and accountable for determining which species to prioritize and where to allocate long-term research and investment efforts. Decision-making tools that take into account biological, technical, economic, marketing, environmental, political and regulatory considerations can consequently be designed and applied so as to evaluate candidate species objectively for future R & D activities and/or investments from the private sector.

Section 8.2 presents a species selection model previously developed by Le François *et al.* (2002) that was designed to identify the most suitable marine or anadromous fish species among **a wide range of exclusively native species (cultivated or not)** from the St Lawrence estuary system (Eastern Canada) (n = 47). This analytical model relied heavily on site-specific criteria (e.g. temperature, salinity, presence of native populations) and could be used to identify the best production strategy (complete land-based life cycle, on-growing of cage or land-based juveniles or stock enhancement activities). This type of approach provides a fair degree of independence and freedom. It was based essentially on the concept that biological and technical bottlenecks are often underestimated when species are selected solely on market assumptions,

(eds N.R. Le François, M. Jobling, C. Carter and P.U. Blier)

whereas a market can reasonably be created and developed for a relatively unknown but more climatically adapted species, provided they conform to the strong market trends, i.e. fresh, mild-tasting, boneless, white-fleshed fish fillets that are available in a variety of shapes and sizes, can be blended with sauces and can hold various coatings (Johnson and Halfyard, 2002; Muir, 2005).

Site-specific farming, whether in the agriculture or aquaculture sectors, can be defined as doing the right thing, the right way, at the right location and time. In turn, the full expression of the outputs from a given animal or plant production can be harvested efficiently and the profits can be maximized. Site-specific farming has reached its highest level of complexity in the agriculture business (see Edwards *et al.*, 1993). In aquaculture, this concept is less developed, but varying degrees of sophistication and applications can be found: from integrated fish farming at the highest level – which requires the use of multiple species and is highly dependent on existing agricultural activities and/or prevailing climatic conditions (see Bao-tong and Hua-Zhu, 1984; Chopin *et al.*, 2006) – to recirculation technology at the lowest level which, through the effective management of production variables (temperature, oxygen, salinity, etc.), offers a fairly high degree of independence from the external environment (see Chapter 24, this volume) and provides a significant range of possibilities.

Section 8.3 presents another selection model derived from the study by Le François *et al.* (2002). It differs from the previous model as it **includes only species already cultivated, either experimentally or commercially**, and does not rely heavily on site-specific criteria but rather includes productivity- and sustainability-related criteria in its modus operandi. Indeed, as world demand for food from aquatic environments continues to rise, the importance of environmentally responsible aquaculture also increases. There are numerous programmes aimed at seafood retailers, buyers and consumers to guide them in selecting products that meet sustainability criteria (e.g. Blue Ocean Institute [http://www.blueocean.org/seafood/seafood-guide], Sea Choice [www.seachoice.org] and Seafood Watch [www.montereybayaquarium.org/cr/seafoodwatch.aspx]). Sustainability criteria, such as a given production's degree of reliance on marine resources, the risks of escape, disease and parasite transfer to wild stocks and the risks of pollution and habitat effects, are proposed. In the end, an overall seafood recommendation for a type of aquaculture, product or fishery is given to the consumer/buyer. Ultimately, a clearer evaluation of the environmental implications of a given aquaculture activity, technology or process is possible in the form of life cycle assessment (LCA), as is actually the case for agriculture and capture fisheries (see Ayer and Tyedmers, in press).

Accordingly, species selection methods that take into account ecological and sustainability considerations balanced with productivity concerns should be developed (Muir, 2005; Leung *et al.*, 2007). Issues and constraints related to aquaculture are diverse, ranging from environmental concerns to market, political and regulatory inconsistencies, productivity and health concerns, to name just a few. In this complex, dynamic and politically charged context, Anderson (2007) was justified in asking: 'How will economically sustainable aquaculture emerge and flourish worldwide?' One way of contributing directly to this rising concern within the industry is first and foremost to select or promote the selection of species that strongly display the fundamental characteristics needed to improve sustainability in a given environment when warranted (e.g. high density, disease tolerance and level of herbivory). Beyond that first step, the application of genetic-based technologies to improve the available strains should necessarily be considered (see Muir, 2005; Sevilleja, 2007).

In this analysis, all 27 species covered in this book (see Chapters 9–21, this volume) were evaluated and compared in terms of their aquaculture potential based on a complete life cycle production strategy, i.e. regardless of their optimal temperature or salinity, landing or market price information. Productivity indices and sustainability criteria were, for the first time, introduced into a selection model, e.g. kg of flesh produced per year, eggs per kg of female, stock status and percentage of animal protein in the diet. Up-to-date information on all species covered is available in technical sheets provided as appendices to this book (Appendices 1–24). In addition, 6 of the 27 species covered are examined extensively in regard to their biology and rearing technologies in the recent work of Moksness *et al.* (2004) (e.g. Atlantic cod, haddock, spotted wolffish, Atlantic halibut, turbot and sole).

In this chapter, we describe briefly two models for selecting finfish species (Sections 8.2 and 8.3) and specify the limits and constraints of each model. The two models strongly advocate that new or alternate species selection procedures be designed to identify the species that display the strongest attributes for domestication and that match productivity, marketability and sustainability targets. Readers will be able to consult the results of these two different analytical procedures throughout this chapter (see Sections 8.2.3 and 8.3.3 for the results and discussion of the site-specific and the sustainability/productivity selection procedure, respectively).

Among the points to be considered are the following:

- All species selection models are perfectible and should offer flexibility.
- Species selection models can be considered more or less as the initial steps that facilitate the decision-making process leading to the choice of one species over another.
- A well-designed species selection model for which the criteria have been studied extensively on the basis of their relevance and power of differentiation (i.e. discerning qualities and recognizing differences) ensures a more reliable outcome, or at least provides incentives for focused dialogue among the different parties involved in R & D or implementation activities.

8.2 Site-specific Selection of the Most Suitable Species and Production Strategy: a Québec (Canada) Case Study

The model for selecting candidate species for coldwater mariculture activities presented here was developed by Le François *et al.* (2002). It was developed originally at the request of the government of Québec (Canada) in order to identify the marine or anadromous fish species best suited for diversifying coldwater aquaculture that could be economically sustainable in the maritime environment of eastern Québec. That water system is characterized by local and large-scale seasonal variations along its axis (temperature, salinity, tidal currents, etc.) and by extensive ice coverage during the winter months (El-Sabh and Silverberg, 1991; Koutitonsky and Bugden, 1991). It includes the St Lawrence River, estuary and gulf that together form a semi-enclosed sea separated from the Atlantic by the island of Newfoundland, but which are connected with the Atlantic by the Cabot and Belle-Isle Straits. Potential mariculture sites in Québec (Canada) are distributed over more than 12,000 km of coastline with largely unexploited potential, where only a few small and mid-sized commercial facilities are in operation (mainly invertebrate cultivation: blue mussels, *Mytilus edulis*, and giant scallops, *Placopectens magellanicus*).

8.2.1 Collection of species-specific biotechnical data

The first step in this selection process was to generate a list of a wide range of marine and anadromous fish species present on the east coast of North America and based on Scott and Scott (1988) and Mahon *et al.* (1998). The list of species was then validated by mariculture and fisheries experts. Species without any apparent potential for cultivation (e.g. Cottidae, Rajidae) and those that are cultivated commercially (e.g. Salmonidae) were included in the list to test the outcome of the selection procedure (Table 8.1).

The relevance of the screening procedure was confirmed by whether those families were rejected or retained. Information regarding critical or highly relevant aquaculture characteristics for each species was collected from scientific literature and/or by questioning recognized experts directly and was then presented in technical sheets. The sheets were essential to feed the model and apply the analytical procedure to a large number of species.

8.2.2 Description of the model

Three successive steps – each of which gradually used more specific criteria – were applied in order to select species (Steps 1, 2 and 3). Three production strategies were examined: stock enhancement, grow-out and/or complete life cycle (reliance at any stage of the production cycle on recirculation technology was not considered). The first selection step either rejected a given species or directed it towards a plausible scenario for a second selection stage. The species oriented toward a specific production strategy were submitted to the second selection stage, during which their respective potential was evaluated more closely using discriminating criteria adapted to the given strategy. Based on the results of the first and second selection stages, species were ranked according to other criteria specific to each scenario in a third selection stage (Fig. 8.1).

8.2.2.1 Step 1

The first step used simple, highly discerning criteria that were very important in the particular environmental, technical and economic context of eastern Québec. The initial four criteria (1. Marketability: yes or no; 2. Food requirements: filter feeding or not filter; 3. Reproduction: ovoviviparous or oviparous;

Table 8.1. Criteria for the final classification of finfish species within the complete cycle, marine on-growing and stock enhancement production strategies (from Le François *et al.*, 2002).

	40 points	30 points	25 points	20 points	15 points	10 points	5 points	0 point
Complete cycle production								
Growth rate[a]	≤ 2			2–3				> 3
Optimal growth temp. (°C)	≤ 8			8–12				> 12
Larval size at hatching (mm)	> 15		10–15			5–10		< 5
Incubation survival (%)		> 75		60–75		45–60		< 45
First-feeding survival (%)		> 75		50–75		25–50		< 25
Weaning period (days)		< 2[b]		2–4		4–6		> 6
Reproduction control[c]		Controlled		< 5				> 5
Density (kg/m^3 or /m^2)[d]		> 80		40–80				< 40
Incubation temp. (°C)				< 7	7–11	11–15		> 15
Live feed needed				None		Size ≥ *Artemia*		Size < *Artemia*
Min. lethal temp. (°C)				≤ –1				> –1
Min. salinity tolerance				< 15		15–25		> 25
Flesh yield (%)				> 50	40–50		30–40	< 30
Egg yield (%)				> 25		15–25		< 15
Landing price (CAN$/kg)				> 6		3–6		< 3
On-growing								
Growth in 4 years (cm/year)	> 10			5–10				< 5
Optimal growth temp. (°C)	≤ 8			8–12				> 12
Capture survival (%)		> 75		50–75		25–50		< 25
Max. size reached (cm)				> 60		40–60		< 40
Min. salinity tolerance				< 15		15–25		> 15
Flesh yield (%)				> 50	40–50		30–40	< 30
Egg yield (%)				> 25		15–25		< 15
Landing price (CAN$/kg)				> 5		3–5		< 3
Stock enhancement								
Population status	Decreasing		Stable and weak			Abundant		Veryabundant
Studies-stock enhancement	Effective			Doubtful results				Ineffective
Landing price CAN($/kg)				> 5		2–5		1–2

Note: [a]Years to reach commercial size; [b]Or absence of weaning. [c]Or length of reproduction in nature; [d]Flatfish species = kg/m^2, round species = kg/m^3.

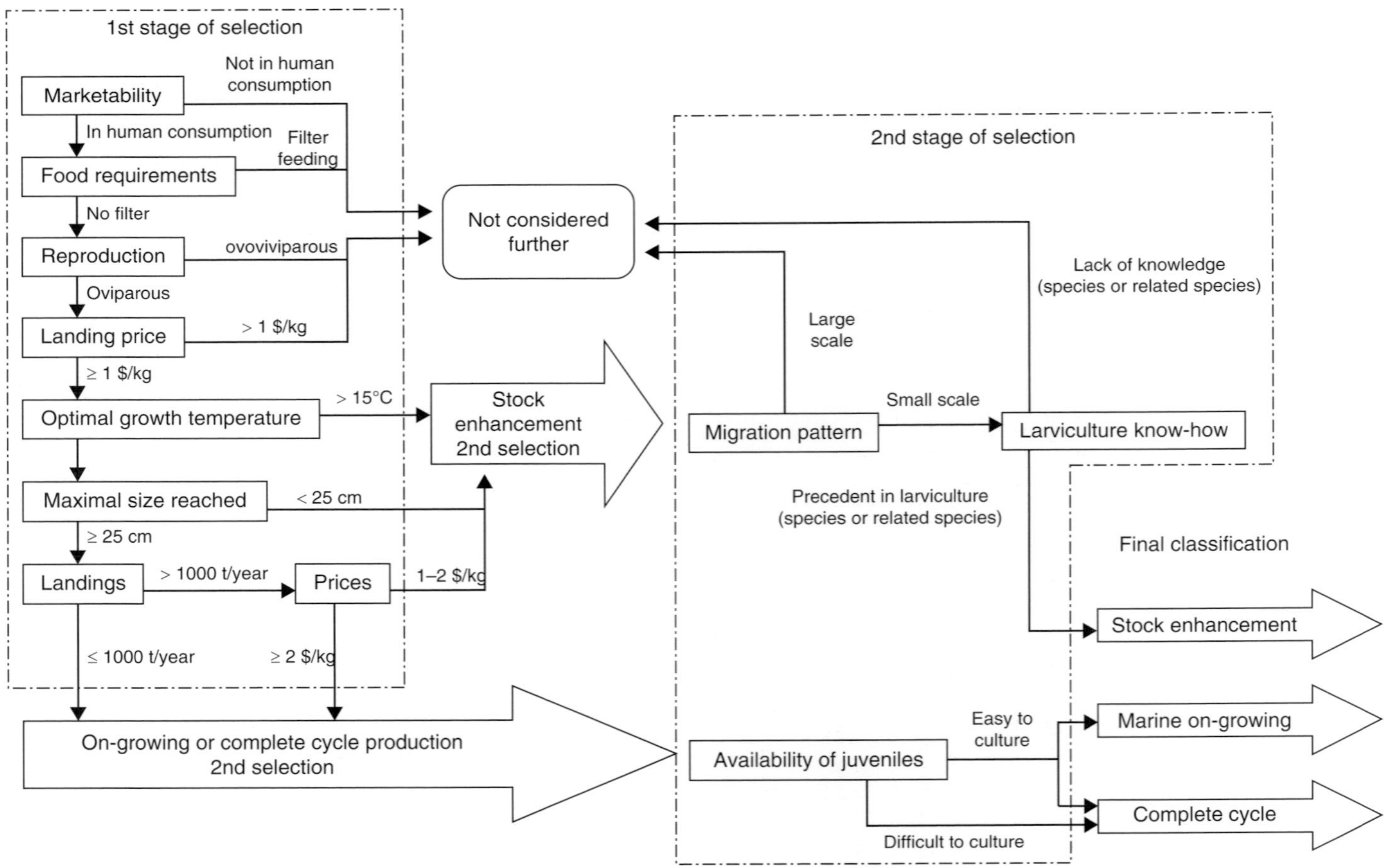

Fig. 8.1. Schematic diagram of the selection process covering three production strategies (complete cycle, on-growing and stock enhancement) (from Le François *et al.*, 2002).

4. Landing price: over CAN$1/kg or under CAN$1/kg) could lead to rapid rejection of a given species. The remaining criteria (optimal growth temperature, maximum size reached, volume of landings and landing price) were subject to a wide range of values, depending on local market trends and environmental conditions. The species that progressed to that level of analysis were oriented toward the most plausible scenario (stock enhancement, on-growing or complete production cycle) so that they could be evaluated more closely (Step 2) and ultimately ranked (Step 3).

8.2.2.2 Step 2

Species chosen for complete life cycle marine on-growing and stock enhancement production were submitted to a second selection stage according to associated criteria. The stock enhancement strategy was evaluated according to the migration pattern (small scale or large scale) and larviculture know-how (lack of knowledge or previous larviculture), while the on-growing and complete cycle strategies were evaluated according to the availability of juveniles (easy or difficult). All species that remained after the second selection stage were submitted to a final classification procedure based on the production strategy considered (Step 3).

8.2.2.3 Step 3

The criteria for a given production strategy and their respective ranges of values are presented in Table 8.1. Each criterion was given a maximum score that reflected its relative importance in a given situation and its prospect for improvement. The total score was then expressed as a percentage. Gaps in knowledge (i.e. lack of information for a given criterion) did not influence the final score but, preferably, allowed for the rapid identification of areas that could benefit from applied research efforts. For further details on the methodology, refer to Le François *et al.* (2002).

8.2.3 Results and discussion

Our species selection approach identified two species of wolffish (spotted and Atlantic wolffish) as having the highest potential for commercialization using a complete life cycle production strategy. Both species have large, hardy, 'live prey free' larvae, i.e. they do not require the production of live feed such as *Artemia* and rotifers to initiate and complete the critical exogenous feeding stage successfully. Those species also display very high tolerance to cold water (Desjardins *et al.*, 2006, 2007), have good growth and survival performance (Imsland *et al.*, 2006; Lamarre *et al.*, 2009) at very high rearing densities according to Foss *et al.* (2004) and Tremblay-Bourgeois *et al.* (in press) and do not show high susceptibility to diseases (Espelid, 2002).

Since 1998, both wolffish species have been the subject of an extensive research and development programme that covers the principal biotechnical constraints. The spotted wolffish clearly outperforms the Atlantic wolffish, mainly by expressing higher growth potential at lower temperatures and displaying more farming-friendly behaviour. Since the domestication attributes of the spotted wolffish are unequivocal, plans have been made to conduct pilot-scale on-growing of juveniles and search for potential investors. As with all the species of marine fish considered, a fair level of production cycle refinement is needed in the long run in order to achieve economical sustainability. The most important challenges for spotted wolffish cultivation are market-related because this species is largely unknown to potential buyers and consumers (see Chapter 19, this volume).

Arctic and brook charr (*Salvelinus alpinus* and *S. fontinalis*, respectively) also had interesting potential. Both species are already cultivated commercially and are used essentially as control species in our analysis. Our results provide evidence that wolffishes – which obtained a higher score than charrs – offer good potential as new or alternate marine fish aquaculture species. Wolffishes have many of the salmonids' desirable traits for domestication (early-stage hardiness and size; ease of first-feeding operations; growth performance in cold temperatures; straightforward fertilization and incubation operations; simple rearing practices, etc.) but are better suited to the environmental conditions of the St Lawrence water system. Both species of charr are cultivated mainly in freshwater conditions in Québec (Canada). Further investigations of the bioeconomic feasibility of raising them at varying salinities have been conducted in Québec (see Chapter 12, this volume).

The on-growing production strategy positioned Arctic charr and Atlantic cod in first and second place, followed by Atlantic halibut (see Chapters 12, 13 and 21, respectively, all this volume). Optimal growth temperature was a decisive criterion and, in the absence of the aquaculture production of juveniles, survival following the capture of broodstock fish was an important consideration. The three best-suited species for the stock enhancement production strategy were striped bass, haddock and Atlantic sturgeon, based mainly on previous impact studies and the extent of their migration patterns (see Chapters 15, 13 and 9, respectively, all this volume).

This species selection procedure was extremely valuable in terms of the mariculture diversification needs of the aquaculture sector in Québec coastal communities. Essentially, by reducing risk to an acceptable level and providing decision makers and development agencies with legitimacy for concerted actions, this model strongly supported the selection of species for which future research and development activities should be undertaken or prioritized, given that government investment capacity is limited.

8.2.4 Particularities, constraints and limitations of the model

This selection process proposes a choice of a limited number of fish species based on a list of species selected on objective grounds. Therefore, unknown species that were not used in commercial fisheries or for which no rearing experience was available were not considered extensively. A drawback to species selection methods in general is the unavoidable subjective nature of the total point allocation process for the different criteria. Careful evaluation of their relative importance in an aquaculture context is crucial and is bound to present interregional variability. However, the relative importance of the criteria and the ranges of values can be readily modified or evaluated in line with potential users' technological, environmental or economic profiles.

Major restrictions were imposed by the authorities who launched the study: non-native species had to be excluded totally, due to regulatory concerns, and recirculation technology could not be considered as a rearing technology for any of the production strategies covered because of the level of investment required. If the data were re-evaluated by ignoring those restrictions, the outcomes and the relative importance of several key criteria would probably have changed. For example, the optimal temperature (40 points maximum allocation) and the salinity tolerance range (20 points) would be relegated to the second or even third level of importance, whereas density (30 points) would gain in relevance and disease susceptibility would be introduced as a decisive criterion in the final classification step. Doing so most likely would have placed the sturgeons (both Atlantic and shortnose) in final ranking positions equivalent to or even higher than the wolffishes and charrs, instead of redirecting them toward stock enhancement activities at the first selection stage due to warmer temperature requirements (Chapter 9, this volume; and Section 8.3.3.3).

The methodology for the information-gathering phase of the study by Le François *et al.* (2002) may appear tedious compared to the approach used by Quéméner *et al.* (2002), who relied on the Fishbase database (www.fishbase.org) as an information source for a computer-based analytical procedure (see also Chapter 6, A Systematic Market Approach to Species Diversification: A French Case Study). However, that methodology provides a significant amount of production-related information that is easy to consult and can be updated readily according to advances in the fields of fish biology, fisheries or rearing technologies. Furthermore, the model proposed by Le François *et al.* (2002), unlike that of Quéméner *et al.* (2002), did not use detailed information on the organoleptic characteristics and suitability of the seafood transformation channels in their evaluation but rather used simple, basic, straightforward but effective criteria such as marketability (not for human consumption = rejection), maximum size reached (≤ 25 cm = rejection) or flesh yield (20 points maximum allocation: from > 50% = 20 points to < 30% = 0 points) that either could lead to the rejection of a species or be used to refine a given species' ranking further.

8.3 Evaluation of Commercial Species' Attributes in Terms of Productivity/Sustainability

Identifying aquaculture species and adapted cultivation systems that are expected to be profitable is an essential step in developing economically sustainable aquaculture. Profitability takes precedence over environmental or socio-political considerations since the whole idea of aquaculture diversification is irrelevant if profits are absent or insufficient, according to Anderson (2007). The proposed model should be viewed as a tentative classification tool that introduces productivity and sustainability criteria in balance with the intrinsic domestication attributes of a wide range of species that have reached the commercial stage. This selection procedure was constructed in order to classify already cultivated species in terms of their compliance with sustainable aquaculture principles. A range of production intensities was considered, i.e. from the use of intensive-rearing technologies (such as recirculation aquaculture systems) to extensive pond-rearing systems. Freshwater, brackish and seawater species were considered all together and were clustered according to temperature requirements: cold–temperate water species and warmwater species, which predictably clustered to some extent; the different species according to their feeding preferences (carnivorous–omnivorous or herbivorous); and the level of complexity of their cultivation technology (open, semi-closed or closed systems). Economic criteria, such as landing prices, market prices and production costs, per se were not considered directly in our classification procedure, given the wide variety of species and regional particularities associated with market trends.

A continuum of species that present a range of production levels was used,[1] i.e. high-volume established species such as catfishes and tilapia in freshwater ponds or Atlantic salmon from freshwater land-based juvenile production facilities transferred for on-growing in serial large-volume sea cages; lower volume established species such as European seabass, seabream, turbot; and low-volume novel species such as tuna, codfishes, meagre, wolffishes and several flatfishes. The first group includes species for which the biotechnical and profitability characteristics needed for successful aquaculture are known. The second group likely includes species for which the market outlooks or production costs have been misjudged or underrated, thereby limiting growth and development of their production for the time being. The third group includes species for which future outlooks are highly dependent on R & D initiatives and which present a moderate to high level of risk and a fair degree of uncertainty. All groups of fish (i.e. varying degrees of production levels) are represented in the proposed temperature categories.

8.3.1 Collection of species-specific biotechnical data

A technical sheet was produced for all species covered in the book (see Appendices 1–24). Closely related species shared the same technical sheet (e.g. sturgeons: Atlantic and shortnose sturgeon; wolffishes: spotted and Atlantic wolffish; sole: common and Senegalese sole; and tuna: Atlantic

[1] The order of appearance of the species covered in this book is based on their evolutionary relationships according to Nelson (2006).

bluefin, southern bluefin and Pacific bluefin tuna). Relevant information on the following was compiled and referenced by the expert/specialist assigned to the relevant species chapter in this book: (i) gametes and developing eggs (e.g. egg, milt and sperm characteristics, egg yield, incubation time and optimal temperature, fertilization success and overall incubation success); (ii) larval and juvenile stages (larval size at hatching, optimal temperature, rearing units, time until first feeding, live feed needed, survival from hatching to weaning); (iii) on-growing stage (commercial size, years to reach commercial size, rearing units, rearing density, susceptibility to disease, optimal growth temperature, diet); (iv) reproduction (period, fertilization mode and protocol, time to reach sexual maturity, optimal temperature and control of reproduction); (v) commercialization (product characteristics, flesh and/or egg yield, price/kg of whole fish, gutted, other commercialization options); and (vi) research avenues.

8.3.2 Criteria selection and description

In this section, we present and discuss the criteria and proposed composite indices used to assess the respective potential of the species targeted in our analytical exercise. The criteria and indices are classified under the three main activities of a typical finfish aquaculture production cycle and are ranked on a scale of 100%: (i) reproduction and incubation (20 points); (ii) larval and juvenile production (35 points); and (ii) on-growing and commercialization (45 points) (Table 8.2).

In contrast to the analytical procedure of Le François *et al.* (2002), the total point allocation was based on the fact that all the species under analysis had a closed life cycle (with the exception of some species of tuna), thereby shifting the relative importance of classification criteria to the on-growing and commercialization stage rather that the reproduction and juvenile production stages. The various criteria were selected by taking into account their accessibility in bibliographic databases, peer-reviewed literature or various books that summarized large amounts of information about the individual or closely related species. This precaution enabled us to produce a comprehensive score sheet. To understand fully the selection procedure outcomes, readers should refer to the species technical sheets for the values of the different criteria associated with a given species (Appendices 1–24 at the end of this book).

8.3.2.1 Reproduction and incubation (total of 20 points)

Traits associated with reproductive physiology are critical in order to domesticate teleost fish species successfully. Reproduction control is a widely acknowledged biotechnical constraint that severely limits the growth and development of commercial cultivation for numerous fish species. This area of zootechnology generally involves extensive and costly operations and thus is a determining factor as regards a particular species' relative suitability for reaching the commercialization stage. In this respect, we suggest that readers consult: (i) STOREFISH, a new database for fishery and aquaculture scientists which currently focuses on European temperate freshwater teleost fishes but should

Table 8.2. Criteria (n = 11), composite indices (in italic) and points allocation (0–15 points) of the productivity/sustainability selection.

		Points allocated													
Phase	Criteria/Indices	0	1	2	2.5	3	4	5	6	7	7.5	9	10	12	15
Reproduction	Age at sexual maturity	≥ 7	5.0–6.9	4.0–4.9		3.0–3.9	2.0–2.9	< 2.0							
	Reproduction control	Natural spawning: 10 pts; if not, 2 points for each of these factors: external fertilization, photoperiodic and temperature shifts of spawning, hormonal induction of spawning and existence of cryopreservation protocol.													
	Incubation time (days)	≥ 100		60–99		20–59	6–19	< 6							
Larval production	Larval size (mm)	≤ 2.5			2.6–3.7			3.8–5.5			5.6–9.9		≥ 10		
	Live feed needed	Yes		*'Eggs/kg female' × 'Fertilization success' × 'Incubation success' × 'Larval survival'*									No		
	Juveniles/kg female	< 999				1000–4999			5000–9999			10,000–59,999		60,000–599,999	> 600,000
On-growing and commercialization	Grow-out length (years)	> 3.0	2.6–3.0	2.1–2.5		1.6–2.0	1.1–1.5	≤ 1.0							
	Productivity	*['Commercial size'/'Grow-out length'] × 'Flesh yield' × 6.6 (maximal value: 15 points)*													
	Density (kg/m^3)	< 9		10–30				31–50		> 50					
	Density (kg/m^2)	< 9		10–49				50–64		> 65					
	Density (kg/ha)					5600–6100									
	Disease susceptibility	High	Medium	Low–medium		Low	Very low								
	Sexual maturation before commercial size	Yes					No								
	Stock status	Good	Near threatened	Vulnerable		Endangered	Critically endangered								
	% Animal protein part	≥ 50		40–49			30–39		0–29						

be expanded to northern hemisphere freshwater and marine fish species (Teletchea *et al.*, 2007); and (ii) a recent review entitled 'Controlled Reproduction and Domestication in Aquaculture' proposed by Bilio (2007a,b; 2008a,b) (http://www.easonline.org/files/various/domestication_in_aquaculture_bilio_web.pdf). The criteria selected for the evaluation/comparison of reproductive features among our pool of species are: age at sexual maturity, reproduction control and incubation time, for a total of 20 points.

AGE AT SEXUAL MATURITY (5 POINTS). Broodstock management practices and productivity are facilitated for species that mature early. In addition to a reduced fish cost in the end, generation time is faster and targets for genetic selection results are therefore achieved more rapidly. A long generation time impedes the application of genetic improvement techniques such as selection, monosex culture development and strain development (Davis *et al.*, 2005). A long-term vision for productivity improvement implies that a given species be allocated a higher score for this criterion when sexual maturation occurs sooner. Early maturation, however, can be considered a problem when it occurs before the commercial size is reached as it frequently involves growth impairment, increased disease susceptibility and flesh-quality reduction. Consequently, this factor is taken into consideration in the on-growing and commercialization phase (Section 8.3.2.3).

Score:

Age at sexual maturity	< 2.0	2–2.9	3–3.9	4–4.9	5–6.9	≥ 7
Points	5	4	3	2	1	0

REPRODUCTION CONTROL (10 POINTS). Artificial fertilization requires that a person extracts the eggs and semen from the broodstock, ensures a thorough mixing of the sexual products, avoids contamination and then proceeds with the incubation procedures. It is a multi-step, time-consuming operation. Therefore, when a species was reported to produce fertilized eggs naturally in tanks, it was given the highest score. If not, factors indicating our level of control over a species' reproduction were evaluated.

First, if a species displays an external mode of fertilization, it is given more points. Internal fertilization (*Anarhichas* sp.) implies that contact of the eggs with water before fertilization must be absolutely avoided. It is also imperative that skilled personnel be in charge of the broodstock because the exact timing of ovulation must be predicted correctly prior to the release of the egg masses.

Second, extending or shifting the spawning period can improve the productivity and profitability of an aquaculture operation greatly. Consequently, when protocols for photoperiodic or temperature shift of spawning, hormonal induction of spawning and cryopreservation techniques were available for a given species, they were deemed to confer a greater degree of control over the reproduction events and were considered indisputable advantages for accelerating domestication activities such as the development of domestic certified broodstock and strains. Unquestionably, as in terrestrial husbandry (agriculture), only when reproduction independence is achieved through the established availability of domestically produced broodstock and juveniles can a species' future cultivation be sustainable.

Score:
– If natural reproduction occurred naturally in tanks: 10 points.
– If not, 2 points for each of these factors: external fertilization, photoperiodic and temperature shifts of spawning, hormonal induction of spawning, cryoconservation protocol existing.

INCUBATION TIME (5 POINTS). Taking care of eggs involves the removal of unfertilized, abnormal or dead eggs and periodic disinfections. In species with long incubation times, premature hatching can occur as a result of damage to the egg membrane caused by protozoan activity and bacterial degradation (Pavlov, 1986; Moksness and Pavlov, 1996, Falk-Petersen *et al.*, 1999). Intensive incubation often results in bacterial overgrowth on the eggs, leading to high larval mortality. The necessary operations are frequently tedious and time-consuming. Since the eggs and developing embryos are very fragile, operational failures that affect water distribution and cause unwanted sudden variations in temperature or oxygen concentration can have great impacts on the quantity and quality of the juvenile production. A longer incubation period increases the risks that such events will occur. Species with shorter incubation times were therefore attributed more points for that criterion.

Score:

Incubation time (days)	< 6	6–19	20–59	60–99	≥ 100
Points	5	4	3	2	0

8.3.2.2 Larval and juvenile production (total of 35 points)

LARVAL SIZE AT HATCHING (10 POINTS). Larval size gives a good indication of the complexity of larval rearing and the robustness of the larvae. Small or poorly developed larvae at hatching will undergo many changes as they develop their swimming, visual and digestive capacities and will be more sensitive to environmental variations. Mortality in the larval period is most commonly attributable to a mismatch between digestive capacity development and the feed offered, as well as a certain degree of weakness in their immune systems. Larger larvae are generally more robust and require a shorter weaning period. As a result, well-developed larvae should require less complex rearing techniques and be more successful. This criterion, however, must be evaluated in relation to the potential number of offspring produced per female in order to indicate clearly its significance for a given species.

Score:

Larval size (mm)	≤ 2.5	2.6–3.7	3.8–5.5	5.6–9.9	≥ 10
Points	0	2.5	5	7.5	10

LIVE FEED NEEDED (10 POINTS). Production of *Artemia*, rotifers and copepods is labour- and equipment-intensive, space consuming and cost prohibitive. To ensure a reliable and continuous supply of high quality, nutritious zooplankton as live prey year round, a 'food chain' that relies on the production of algae is often required. The supply can be variable and nutritional value can be inconsistent. Person-Le Ruyet *et al.* (1993) evaluated the cost of feeding live prey to *Dicentrarchus labrax* larvae and found that in the first 3 months posthatching, live feed represented 50% of the feed cost, even though it represented only 1.6% of the dry weight of the food administered. Therefore, when a species can be weaned without live feed, it presents a great advantage.

Score:
If a species does not require live feed: 10 points, if it does: 0 points.

NUMBER OF JUVENILE PRODUCED PER KG FEMALE (15 POINTS). The main constraint for fish aquaculture development, successful diversification and commercial sustainability is the controlled production of juveniles (both in quantity and quality) for commercial-scale on-growing operations. Consequently, species with high fecundity and high early-stage survival rates should be preferred. Additionally, maintaining and growing broodstock fish is expensive. If less broodstock fish are required to obtain the desired number of juveniles per year, considerable savings can be achieved. However, many highly fecund species also have a low survival rate for the eggs and/or larvae due to their fragility and our limited understanding of their essential basic requirements in terms of nutrition and rearing environment. More often than not, species with the greatest potential for aquaculture are those that can be weaned to commercially available feed particulate diets early or omnivorous and filter-feeding species that can feed partly or totally from plants and zooplankton produced naturally in pond systems, for example. We calculated the number of *live* larvae or juvenile per kg of female that survived past the critical weaning stage as follows:

Number of live larvae/kg female = Number of eggs/kg female
× Fertilization success (%)
× Incubation success (%)
× Larval survival (%)

Score:

Juveniles/kg female	< 999	1000–4999	5000–9999	10,000–59,999	60,000–599,999	> 600,000
Points	0	3	6	9	12	15

8.3.2.3 On-growing and commercialization (total of 45 points)

GROW-OUT TIME (5 POINTS). The 'grow-out' phase of a fish aquaculture production cycle is defined as the time period needed by the organism under cultivation, from transfer from the nursery until harvest. This period typically is associated with a stabilized mortality rate, increased feed rations, frequent size grading operations, management of stocking density and monitoring of the feed conversion ratio (FCR) and water quality. Increasingly, welfare considerations generally are associated with that particular phase of the production cycle and usually include transportation, slaughtering and transformation concerns. Feed is the largest variable cost (more than 50% for some carnivorous species) and tends to be considered a critical and decisive expenditure affecting commercial success (see Chapter 3, this volume). The size of the fish at harvest is a business and marketing consideration that is based principally on growth rate, market demand and production costs. The longer the time (expressed in years) a fish spends in an aquaculture facility, the higher the production costs will be and the more likely that a system failure or disease breakout could occur. Species with shorter production cycles (expressed in years) were attributed more points.

Score:

Grow-out time	≤ 1	1.1–1.5	1.6–2	2.1–2.5	2.6–3	> 3
Points	5	4	3	2	1	0

PRODUCTIVITY (15 POINTS). The main engine for growth in aquaculture production is the implementation of semi-intensive and intensive farming practices where producers actively influence the growing condition of their fish stocks (Asche *et al.*, 2008). Increased productivity in terms of improved growth of a particular fish species – in particular the edible portion (flesh [whole fish, fillets] and/or egg yield [caviar products]) – reduces production costs (shortens the production cycle) but, most importantly, makes the production more profitable at a given price. Hence, an especially relevant component of this multi-factor equation is the quantity of flesh or eggs that can be produced per year by a particular fish species, strain or hybrid. This information is a productivity indicator and, when multiplied by the price per kg of flesh or egg, reflects the profitability outlook. However, as stated in the introduction section, market and landing prices were not considered due to those values' pronounced fluctuations according to season, location and trends.

Score = Productivity = Commercial size/Grow-out length × Flesh yield (%)[2]

DENSITY (7 POINTS). In aquaculture, density is considered to be one of the important factors affecting fish growth, feed utilization and fish yield (Liu and Chang, 1992). It is highly species-specific (Ashley, 2007) and varies with age and life stage (Huang and Chiu, 1997). The optimal utilization of space for maximum fish production can improve a fish farm's profitability greatly (Ridha, 2006) because it reduces the costs related to infrastructures, even if the production level is the same. Density is a particularly important production parameter when evaluating the use of recirculation technologies for commercial cultivation purposes (see Chapter 24, this volume) and can be considered a compromise between rapid growth and maximum occupation of available tank space, while taking into account growing consumer concern for animal welfare (Ellis *et al.*, 2002).

Rearing density is expressed differently, depending on fish swimming patterns and behaviours. Accordingly, density can be measured in terms of biomass per surface area (kg/m^2 for tanks and cages or number of fish/ha for cultivation ponds), layers of fish on the bottom (Bjornsson, 1994) and per cent coverage (Kristiansen *et al.*, 2004) for demersal species such as flatfishes and wolffishes. In the case of pelagic species (codfishes, salmonids, etc.), density is

[2] Multiplied by the value of 6.6 for a maximum of 15 points allocated to this criterion for our selected species.

usually expressed in terms of biomass per volume area (kg/m^3). Three density groups were therefore created (see Table 8.2).

Score:
Density

Kg/m^3	< 9	10–30		31–50	> 50
Kg/m^2	< 9	10–49		50–64	> 65
Kg/ha			5600–6100		
Points	0	2	3	5	7

DISEASE SUSCEPTIBILITY (4 POINTS). Fish diseases frequently cause economic losses in aquaculture (Toranzo *et al.*, 1993) and increasingly are being recognized as a major constraint in aquaculture production and trade, affecting the sector's economic development in many countries (Aly *et al.*, 2008). All organisms are susceptible to various pathogens and parasites during their lives. Stressful conditions must be avoided and the species that are more tolerant and adaptable to captive conditions (i.e. less vulnerable) are less likely to develop diseases. Disease control and management can have profound impacts on the profit margin of an operation and need to be considered when selecting a species for aquaculture. Some strains and inter- or intraspecific hybrids display improved resistance to otherwise devastating diseases in comparison with other strains or purebred populations. In addition, reduced disease susceptibility improves a species' appeal for the label 'organic' or 'green', as it refers to reduced usage of antibiotics or other therapeutants.

Score:

Disease susceptibility	Very low	Low	Low–medium	Medium	High
Points	4	3	2	1	0

SEXUAL MATURATION BEFORE COMMERCIAL SIZE (4 POINTS). Precocious sexual development is an undesirable trait in many commercially farmed species because early sexual maturation is often correlated with negative aspects such as reduced growth, reduced feed conversion, low resistance to infectious diseases (Bromage *et al.*, 2001; Weltzien *et al.*, 2004; Felip *et al.*, 2006) and lower flesh quality (especially for Atlantic salmon). Several strategies to prevent or control sexual maturation of cultured stocks have been developed for commercial operations: production of monosex populations (males usually mature at a younger age), triploidization and selection for late maturing strains. Score: for a given species, when sexual maturation does not occur before commercial size is reached, 4 points are allocated; otherwise, no points are given.

STOCK STATUS (4 POINTS). This criterion refers to the presence or absence of competition between cultivated products and wild fisheries in the marketplace. In the case of a threatened or vulnerable species,[3] the absence of competition between an aquaculture species and a tightly regulated, overfished or closed wild fishery could influence the market price and public perception positively, given proper product introduction and marketing efforts. Furthermore, if specific fish stocks

[3] The data were taken from the International Union for Conservation of Nature (IUCN, 2008).

are depleted or overfished, aquaculture operations involving massive juvenile production could be called on to participate in species-specific recovery efforts (stock enhancement and reintroduction programmes) or could serve as alternatives or replacements in the consumer market (cultured rather than wild fish), as is currently the case with the tuna fisheries (see Chapter 20, this volume). Last, but not least, research and development activities usually associated with new species development offer increased opportunities for a better understanding of the physiological impacts of habitat degradation and/or global warming on threatened fish populations (e.g. wolffishes, sturgeons). Furthermore, from a global perspective, the cultivation of species that rely intensively or solely on collecting wild broodstock or juveniles should not score very high on this criterion.

Score:

Stock status	Good	Near threatened	Vulnerable	Endangered	Critically endangered
Points	0	1	2	3	4

ANIMAL PROTEIN PORTION OF THE FEED (6 POINTS). Naylor *et al.* (2000) claim that aquaculture is environmentally degrading because increased demand for feed leads to increased fishing efforts to produce fish oil and fishmeal and thereby threatens the viability of wild fish stocks. It follows from this argument that the availability of marine feed and its high inclusion level in fish diets will limit how much the aquaculture sector can produce, since most marine resources are currently exploited to the maximum (FAO, 2002). Producers of species that require high quantities of fishmeal will find that feed costs will become gradually more volatile and their operations may reach a threshold where profitability is problematic (Asche and Bjorndal, forthcoming). However, the rather one-dimensional 'fishmeal trap hypothesis' described above is examined from all angles by Asche and Bjorndal (forthcoming) and Asche *et al.* (2008). Those authors argue that aquaculture is not the sole cause of fishmeal's reduced availability and price fluctuations. The role of captive fisheries regulations and management, as well as the market for protein meals, must also be examined. Nevertheless, fishmeal scarcity and prices probably will hold back the development of carnivorous species such as cod and limit them to the more specialized and lucrative niche market. However, semi-vegetarianism as practised in salmoniculture and the development of alternative protein sources (plants, krill) could partly alleviate the problem (for more information, see Chapter 3, this volume). In the current context, cultivation of a species that is herbivorous or has low animal protein requirements would be perceived as more sustainable and more 'socially acceptable', according to marketplace trends and initiatives such as Sea Watch and Sea Choice (see Section 8.1). The less a species relies on animal protein, the higher the score granted for this criterion.

Score:

% Animal protein part	0–29	30–39	40–49	≥ 50
Points	6	4	2	0

8.3.3 Results and discussion of cold, temperate and warmwater fish species analysis

Readers are strongly advised to refer to Tables 8.2 and 8.3 for the classification and scores results, to the species chapters and to the technical sheets in the Appendices.

8.3.3.1 Coldwater fish species

Three representatives of the family Salmonidae, viewed among all as the successful coldwater aquaculture species, obtained total scores of over 48% (1st to 3rd positions with 52.5, 49.2 and 48.5 for Arctic charr, Atlantic salmon and rainbow trout, respectively). They were followed closely by the Atlantic cod and the spotted wolffish (4th and 5th position with 45.0 and 44.2%, respectively). The relative position achieved by the spotted wolffish, the colder adapted species that shares many of the salmonid's desirable domestication attributes, is comprehensible, despite a relatively poor background in applied research studies and commercial activities (See Chapter 19, this volume). These results are similar to the ranking resulting from the previous study of Le François *et al.* (2002). In the latter study, salmonid species were included in their analytical procedure as reference or control species, based on their worldwide success as aquaculture species. The position of Atlantic cod, a metamorphic species, is interesting and in all probabilities reflects the outcomes of the faithful application of intense and well-structured research efforts in north-belt countries that, in light of the prolonged uncertainties associated with the larval stages bottleneck and astronomic costs of research initiatives that covered that particular life stage for numerous years, were justified by cod's historical fishery importance. The same reasoning could also be applied to the Atlantic halibut (43.7%), but this time justified by the perceived high-priced market value of this product. The case of winter flounder is interesting. This metamorphic flatfish species is characterized by the smallest of eggs and larvae at hatching and slow growth and yet achieves a score of 40.1%. Close examination of the analytical process and experimental data bring into light certain limitations of the model, which will be dealt with in further detail in the next section (Section 8.3.4).

Species that scored lower than 41% (from 36.3 to 40.7%) include metamorphic species such as the winter flounder, haddock and pollack (in decreasing order). They remain alternative species that are the object of limited aquaculture research and development initiatives worldwide (winter flounder, brook charr, haddock in North America [eastern Canada and north-east USA] and pollack [France and Spain]). For the outlier species, brook charr (38.8%) and Atlantic wolffish (38.2%) (both non-metamorphic species), scores achieved at the on-growing and commercialization stage largely explain their relative position in comparison with the other representatives of their family. In general, the top-scoring species (salmonids and wolffishes) all share a marked simplicity of their early-life stages (from fertilized egg to free-feeding, well-developed larvae and juveniles). Salmonidae are regarded as the controlled species (mainly Atlantic salmon and rainbow trout) for the coldwater fish category.

REPRODUCTION AND INCUBATION PHASE. The species that achieved the highest scores under that production phase are natural spawners with a reduced egg incubation period and somewhat precocious sexual maturation (Atlantic cod, haddock, pollack and winter flounder) (70–90% total score). Salmonid species on average obtained scores of 60–65%, negative attributes include mandatory artificial fertilization procedures and a relatively long incubation period in comparison to most marine fish species of our selection. In comparison to the salmonids and all of the metamorphic species present in our selection of species (cod, halibut, haddock, winter flounder and pollack), the wolffishes scored the least points (35%).

Age at sexual maturity was variable among our selection of coldwater species, ranging from 1.5 years for the brook charr to 4 years for the Atlantic halibut. The criteria control of reproduction allowed a clear distinction between codfishes and flatfishes in captivity conditions. The former acquired the maximal points (10 points) for that reproductive characteristic by being natural batch spawners, whereas flatfishes' (Atlantic halibut and winter flounder) dependence on artificial reproduction for effective fertilization in captivity was very much like salmonids or wolffishes and therefore did not score as well. Contrary to the former, flatfishes present very complex and '*technology demanding*' early-life stages that actual reproduction control techniques (photoperiodic control, hormonal induction, etc.) do not necessarily compensate for fully. Very long incubation time (over 150–180 days for wolffishes, compared to 71–86 days for salmonids and 12–18 days for the other fish families under scrutiny) and currently poorly advanced reproduction control protocols are at the origin of the low score of the wolffishes.

LARVAL AND JUVENILE PRODUCTION PHASE. The metamorphic species in general did not rank particularly well in the larval and juvenile production phase, with a mean score of 28.9% compared to 58.6% for the species characterized by well-developed digestive, visual and locomotive capacities during early life. Larval size values ranged from 2.4–6.5 mm for the winter flounder and Atlantic halibut, respectively, to 20–22 mm for the salmonid/wolffish subgroup and lower and intermediate values of 3.8 mm for the representatives of the codfishes featured in our analysis (Atlantic cod, haddock and pollack).

Easiness of passage from endogenous to exogenous feeding not surprisingly confers a superior advantage to a species. Accordingly, representatives of the subgroup salmonids/wolffishes that do not need live feed obtained a maximal score for that criterion, whereas all the other species considered in our exercise obtained no points. Duration of that critical phase varied slightly among species that required live feed and therefore was considered having a lesser impact on the overall costs within the production cycle than having to possess and operate the facilities and equipment to produce live feed. Accordingly, no intermediate values were included in our evaluation for that criterion. The last criterion considered in the larval and juvenile production phase is a composite index that translates fecundity, fertilization, incubation and larval survival values in the evaluation of the percentage of juveniles that will survive per kg of female at the end of the critical early stages (juvenile/kg of female). Species with fairly high fecundity (mostly metamorphic species) and for which rearing conditions, reproduction and feeding protocols are fully optimized represent, for most finfish commercial productions, the ultimate species from an early-stage perspective. Such species in cold environments do not currently exist, as is the case for warmer water species that, on average, earn 93% for that criterion. All codfishes were allocated 40% for that index. However, once again, of all the species encountered, the winter flounder, a questionable aquaculture species (see Le François *et al*., 2002; Chapter 21, this volume), obtained the highest score (60%) and will be discussed later in Section 8.3.4. Salmonids and wolffishes obtained 0% and Atlantic halibut 20%, principally due to their lower fecundity.

ON-GROWING AND COMMERCIALIZATION PHASE. The decisive criterion in that particular phase of the production cycle is the calculation of the productivity composite index and involves commercial size, grow-out time and flesh yield considerations (15 points on a total of 45). The highest ranked, top five coldwater fish species were: the Atlantic halibut (47%), followed by the spotted wolffish and Atlantic salmon, which both obtained 38.2%, then Arctic charr (36.7%) and the Atlantic cod (35.5%). The lowest ranked species are, in decreasing order, Atlantic wolffish (24.9%), haddock (24.2%) and brook charr (21.1%). The top five species share in common a productivity index value superior to 21%, with an average value of 35.3% (from 61.5 to 21%), in comparison to all the remaining species, for which an average value of 9.1% can be calculated (from 2.1% to 16.8%).

Other criteria used for the evaluation of the production efficiency and sustainability considerations of the on-growing and commercialization phase (grow-out time [5 points], density [7 points], disease susceptibility [4 points], sexual maturation before commercial size [4 points], stock status [4 points] and per cent animal protein in the feed [6 points]). Both concepts were not automatically antagonists as they sometimes presented strong justification overlaps. In general, the values associated to these multiple but low-weight criteria are somewhat influenced by the degree of commercial rearing experience of a given species. These criteria therefore conceal an unavoidable relativization exercise in the presence of both established and novel species for evaluation and comparison.

Density considerations allowed discrimination of species that truly tolerate crowded conditions from those that require larger surface area or volume to express their best growth performances. The benefits of choosing aquaculture species adapted to elevated rearing density are multiple and include: an increased yield per production unit; a reduction of production costs; the financial feasibility of using capital-expensive technologies that abide with environmental regulations and sustainability principles, such as recirculation technologies. Atlantic halibut and Arctic charr obtained the highest score (100%), followed by the wolffishes and Atlantic cod with 5 points each (71.4%). All the remaining species were allocated 2 points (28.6%).

Disease susceptibility allowed the identification of the wolffishes as the most tolerant species under evaluation. Novelty of their commercial cultivation, coupled with their marked robustness early in life (big eggs, developed larvae, exogenous feeding on hatching), might partly explain their perfect score (4/4 points or 100%). In opposition, the salmonids, a family of fish widely distributed and cultivated worldwide, did not gain any points under that criterion (0 points). All metamorphic species (i.e. displaying vulnerable early-life stages) considered in our study scored 1.2/4 points (30%). The more a species is exposed to stressful conditions, such as the conditions encountered in intensive aquaculture facilities, and involves common or related sources of eggs and/or juveniles, the more disease occurrence will be documented and the consequences reported and quantified.

The criteria that evaluate the size achieved by a given species at sexual maturation (4 points) give an indication of the predisposition of a given species for growth and flesh quality reduction and low resistance to opportunistic parasites or disease prior to harvest. On closer evaluation of this criteria, we acknowledge that more consideration be given to the status of the commercial production of a given species. For example, spotted wolffish is not actually considered as presenting early sexual maturation in the scientific literature. However poorly adapted on-growing conditions and feeding practices have led to the appearance of undeniable signs of precocious maturation in the unique wolffish cultivation facility currently in operation. Novel species such as the wolffishes, winter flounder and pollack, which are currently categorized as late-maturing species, could well turn out to be precocious, given the effects of not yet well-defined levels of environmental or physiological cues that they will experience in captive environments. Numerous factors can act on the onset of sexual maturation and it is particularly well documented for salmonids (see King *et al*., 2003: photoperiod; Bushman and Burns, 2005: social status; McLay *et al*., 2008: rearing density; and Shearer and Swanson, 2000: nutritional status). Sexual maturation, however, can be controlled somewhat effectively by the use of triploid fish populations and/or the production of monosex fish stocks, genetic selection, etc.

In relation to stock status of a given species, the final ranking allocates some points exclusively to the Atlantic halibut and the haddock. We used an evaluation strategy that avoided regional specificities affecting particular populations. The 2008 IUCN Red List of threatened species was used (http://www.iucnredlist.org) to assign the value associated to the status of a given species worldwide. However, it appears that the fact that a species is not the object of significant commercial fisheries plays a significant role in its status designation. The case of wolffishes is particularly interesting in this respect since their biology makes them particularly vulnerable to habitat degradation and global warming. In Canada, wolffishes have being designated recently as species at risk (Kulka *et al*., 2007) and are actually covered by a national recovery plan (http://dsp-psd.pwgsc.gc.ca/collection_2008/ec/En3-4-52-2008E.pdf, accessed January 2008). In the USA, the Atlantic wolffish is currently under evaluation and, in the absence on any fishery management plan, conservation authorities are taking action (www.clf.org/programs/cases.asp?id=1058, accessed January 2008). Furthermore, there is an obvious lack of information on the stock status of wolffishes in Norwegian, Icelandic and European waters. Hence, the geographical area of the distribution of wolffishes is entirely covered and could be rightfully considered a subject of some concern. Yet, the IUCN Red List does not give any indication that this species might be threatened to any degree. We should investigate the use of more sensitive indicators of species stock status, especially for 'data-limited' fisheries. Scandol (2005) presented some options that were related directly to fisheries that could signal trends in stock status indicators such as catch, catch per unit effort, mean age, mean length, recruitment data and fishery-independent surveys. Fishery-independent surveys had the best overall performance, followed by age or length of the commercial data. The proportion of the catch that was less than a specific age or size was also linked strongly to stock status. Fishery-independent surveys are increasingly important in monitoring the aquatic resources exposed to changes in management and to obtain improved scientific assessments that are consistent with the principles of ecologically sustainable development (Rotherham *et al.*, 2007).

The evaluation of the actual degree of reliance of a species on protein originating from animal sources (6 points) did not give any points to true carnivorous species such as the wolffishes, codfishes and Atlantic halibut. In general, only coldwater fish species for which research efforts have yielded equilibrated and adapted diet partially composed of vegetable sources of proteins qualified for some points: Atlantic salmon (4 points), brook charr (2 points), rainbow trout (2 points) with, respectively, an average of 38, 40 and 44% animal protein in their diets. Winter flounder also gained 2 points, with 45% of animal protein required in the diet. The experience of the established salmoniculture industry tells us that nutrition research aimed at lowering the animal protein fraction in the diet of coldwater fish species becomes conceivable when volumes of production are significant, that public perception is concerned and that the introduction of the cultured product on the more lucrative niche market is considered.

8.3.3.2 Temperate fish species

The final ranking in regard to the suitability of our selection of temperate species as candidates for aquaculture in respect to productivity and sustainability targets positioned the Atlantic sturgeon in 1st position (54.6%), followed by the European whitefish (a salmonid species) in 2nd position (with a total score of 53.6%). The shortnose sturgeon, in 3rd position, obtained a similar score with 52.6%. The red drum, the meagre, the turbot and, finally, the sole completed our list with an average score of 43.8% (from 47.2 to 38.4%) (Table 8.3). Sturgeons and whitefish obtained the lowest scores for the reproduction and incubation phase with, respectively, 9.0 and 8.0%. The high potential of sturgeons for sustainable and productive cultivation is obvious, in particular when evaluating their characteristics during the on-growing and commercialization phase (an average of 29.6% on a total of 45% calculated for both species). The relatively good ranking of the European whitefish, a salmonid species with the smallest eggs and hatchlings, is interesting. In comparison to true marine fish species, newly-hatched European whitefish can be weaned advantageously after not more than 1 week posthatch on microparticulates or dry feed, with very high survival rates in comparison to meagre, turbot and sole, the other

Table 8.3. Score obtained by the fish species at three distinct phases of a production cycle according to temperature: (i) Reproduction and incubation (20 points); (ii) Larval and juvenile production (35 points); and (iii) On-growing and Commercialization (45 points).

Fish species	T (°C)	Reproduction and incubation	Larval and juvenile production	On-growing and commercialization	Score (%)
Coldwater species					
Arctic charr, *Salvelinus alpinus*	12	13	23	16.5	**52.5**
Atlantic salmon, *Salmo salar*	15	12	20	17.2	**49.2**
Rainbow trout, *Oncorhynchus mykiss*	15	13	20	15.5	**48.5**
Atlantic cod, *Gadus morhua*	15	18	11	16	**45**
Spotted wolffish, *Anarhichas minor*	8	7	20	17.2	**44.2**
Atlantic halibut, *Hippoglossus hippoglossus*	12	12	10.5	21.2	**43.7**
Brook charr, *Salvelinus fontinalis*	15	13	20	9.3	**42.3**
Winter flounder, *Pseudopleuronectes americanus*	15	14	12	14.1	**40.1**
Atlantic wolffish, *Anarhichas lupus*	10	7	20	11.2	**38.2**
Haddock, *Melanogrammus aeglefinus*	15	18	8.5	10.9	**37.4**
Pollack, *Pollachius pollachius*	13	16	8.5	11.8	**36.3**
Temperate water species					
Atlantic sturgeon, *Acipenser oxyrhynchus*	20	9	16.5	29.1	**54.6**
European whitefish, *Coregonus lavaretus*	18	8	31	14.6	**53.6**
Shortnose sturgeon, *Acipenser brevirostrum*	20	9	13.5	30.1	**52.6**
Red drum, *Sciaenops ocellatus*	17	15	15	17.5	**47.5**
Meagre, *Argyrosomus regius*	19	17	12	18.2	**47.2**
Sole, *Solea solea*	19	16	9	13.4	**38.4**
Warmwater species					
Tilapia sp., *Oreochromis* sp.	27	20	18	23.5	**61.5**
Milkfish, *Chanos chanos*	27	16	17.5	27.5	**61**
Channel catfish, *Ictalurus punctatus*	27	17	20.5	16.2	**53.7**
Gilthead seabream, *Sparus aurata*	23	19	14.5	16.8	**50.3**
European seabass, *Dicentrarchus labrax*	23	19	17	8.6	**44.6**
Barramundi, *Lates calcarifer*	31	17	11.5	14.2	**42.7**
Striped bass, *Morone saxatilis*	27	15	6	18.8	**39.8**

species of temperate climates included in our analysis. Accordingly, whitefish achieved a high score during the larval and juvenile production phase.

REPRODUCTION AND INCUBATION PHASE. The species that achieved the highest scores for that particular phase of the production cycle were the meagre, sole and red drum, with values that ranged from 15% to 17% on a possible total of 20%. Natural spawning of meagre and sole in comparison to the other truly marine species considered (turbot and red drum), sturgeon and whitefish, provided a significant advantage. Sturgeons and whitefish obtained the lowest scores for the reproduction and incubation phase with, respectively, 9.0 and 8.0%. The venerable ages at sexual maturity of the Acipenseridae and the less advanced level of reproduction control of the whitefish are the main factors explaining their low scores.

LARVAL AND JUVENILE PRODUCTION PHASE. The species that achieved the highest scores under that item was the European whitefish (31%), followed by the Atlantic sturgeon (16.5%), the red drum (15%) the shortnose sturgeon (13.5%), the meagre (12%), the turbot (11.0%) and the sole (9.0%). A larval size superior to 3 mm could allocate a maximum of 10 points. The European whitefish reached 10 points (13 mm), sturgeons 7.5 points (7–9 mm) and turbot 2.5 points (3 mm). Red drum, meagre, turbot and sole earned no points (larval size < 3 mm). With the exception of the European whitefish (10 points), all temperate species under evaluation required live feed (0 points). In regard to the number of juvenile surviving past the critical stages, the red drum obtained the highest score with 15 points, followed by the meagre with 12 points on a total of 15. High fecundity (by an order of 100 compared to the featured sturgeon and salmonid species), coupled with a good fertilization rate, incubation and first-feeding survival of these true marine species likely explain the outcomes of the calculation of this composite index.

ON-GROWING AND COMMERCIALIZATION PHASE. Careful examination of the results at the on-growing and commercialization phase positioned the sturgeons at the top, with 30.1 and 29.1% for the Atlantic and shortnose sturgeons, respectively. Turbot, meagre, red drum, European whitefish and sole follow, with 19.7, 18.2, 17.5, 14.6 and 13.4%, respectively, on a maximum of 45%. The sturgeons really distinguish themselves due to their impressive productivity (6.93/15 points), coupled with elevated rearing densities (7/7 points), the actual status of their populations within their range of distribution (vulnerable to endangered) and their low animal protein requirements (4/6 points). In comparison, the temperate water marine fish species included in our analysis did not score very high, due mainly to their relatively much lower productivity, lesser tolerance to high rearing densities in general (with the exception of the turbot) and higher reliance on animal protein in their feed (0 points, with the exception of meagre: 2/6 points). The European whitefish displayed a predisposition to a reduced animal protein content in its feed and gained 4 points.

The criteria disease susceptibility and sexual maturation prior to commercial size were not very efficient at discriminating species in the warmwater category. With the exception of the European whitefish (1 point), all species gained 3 points in regard to their tolerance to crowding, whereas all the other species matured past the commercial size and gained 4 points.

8.3.3.3 Warmwater fish species

In the warmwater finfish species category, we find two reference or control species: *Tilapia* sp. and channel catfish, and both can be considered commodity products on the wholesale market. The aquaculture production of both species is straightforward and encounters few biological or technical constraints. Milkfish, also a robust and low trophic-level species, can be considered a 'regional species' of low value, produced mainly in Indonesia, the Philippines and Taiwan, but which abides strongly with the concepts of sustainability; whereas gilthead seabream, European seabass and striped bass could be considered as high-value, low-volume established species and barramundi and tunas should be considered high-value new or novel species currently being developed through focused research initiatives.

Final ranking of the warmwater finfish species included in our selection analysis positioned the tilapia and the milkfish in 1st and 2nd position, respectively, followed by the channel catfish in 3rd position, with overall scores of 61.5, 61.0 and 53.7%, respectively, then the gilthead seabream, the European seabass, the barramundi, the striped bass and both species of tuna (*ex aequo* in last position (see Table 8.3)).

Tilapia and channel catfish have in common their tolerance of low water quality and crowding. Catfishes and *Tilapia* sp. can be considered, to some extent, as herbivorous species and to necessitate low-tech rearing equipment at all life stages, which makes them ideal candidates for sustainable aquaculture.

Barramundi scored relatively low overall (42.7%), despite indications that this species is on the verge of a dramatic increase in global aquaculture production based on its good marketability as an established quality product for the restaurant market. It is a robust species that performs well over a wide range of environmental conditions, including temperature, salinity and water quality, and which displays elevated growth rates (see Chapter 14, this volume).

REPRODUCTION AND INCUBATION PHASE. The top position for the reproduction and incubation phase goes to tilapia, with a perfect score (20/20 points). *Tilapia* sp. reproduces naturally in tanks or ponds, matures very early and has a very short egg incubation time. In equal second position, we find gilthead seabream and European seabass (19/20 points), because they mature later than tilapia.

Both species of tuna obtained a very low score (7 and 8/20 points for *Thynnus maccoyii* and *T. thynnus*, respectively). They both mature fairly late (6 years old or more), no natural spawning is observed in tanks and no protocol of practical application currently exists for the hormonal induction of spawning, cryopreservation of semen and photoperiodic and/or temperature shift of spawning. The main focus of actual research on tuna cultivation is aimed at reproduction and larval rearing, so these figures hopefully should change in the near future. Comparatively, barramundi scored well, with 17/20 points.

LARVAL AND JUVENILE PRODUCTION PHASE. The best-performing species for this phase of the production cycle include the two control species occurring in the warmwater group: channel catfish (20.5/35 points), followed by tilapia (18/35 points) and milkfish (17.5/35 points). European seabass also performed well with 17 points. Tilapia and channel catfish do not require that live feed be presented to their newly hatched larvae, while milkfish and European seabass display a very high fecundity (high number of eggs produced per female with good survival rates from egg to weaned larvae), which allows them to achieve high scores.

Both species of tuna have a very low score for this phase also. The only criterion for which they actually gained points was 'larval size' (2.5/35 points). They necessitate live feed at the larval stage and produce a very low number of viable juveniles per female, due to very high mortality rates at young stages. Increasing survival at the early stages is the object of intensive research worldwide. Striped bass also had a relatively low score for this phase (6/35 points). This is due to the fact that larvae of this species are very small and also that live feed should be distributed to the first stages.

ON-GROWING AND COMMERCIALIZATION PHASE. Best score for this stage goes to milkfish, which display a relatively high productivity index and can tolerate high densities (27.5/45 points). It is a species very resistant to diseases, which does not require animal protein in its diet, making it an 'environmentally friendly' and sustainable species for aquaculture. Milkfish, however, suffer from a limited market (essentially, Indonesia) and a poor consumer perception, with the associated low price (see Chapter 10, this volume).

Both species of tuna obtained very good scores for this phase: 26.1 and 25.1/45 points for *T. maccoyii* and *T. thynnus*, respectively. They possess a very high growth rate, produce a very high fillet yield, fetch high prices on the market and are both designated as endangered species. However, they require a very high proportion of animal protein part in their feed. Tuna aquaculture presently is based mainly on on-growing of caught wild juveniles. As seen in the analysis of the score, this species displays very good attributes for on-growing, especially in terms of growth rates. Due to stock status and increased consumer awareness, there is a definite need for a cultivated source of juveniles for commercial on-growing operations. Closure of the life cycle of this species is a huge challenge, but the great demand for this product, coupled with the reduction of the fishery quotas, commands very high prices, which consequently justifies the intensive and well-funded R & D that is actually occurring worldwide.

European seabass obtained a particularly low score for this section, due to its low productivity, early maturation and the need for a high percentage of animal protein in the feed.

8.3.4 Particularities, constraints and limitations of the model

The contexts in which a species will be produced vary considerably according to the country of reference. Some countries (often developing countries) may benefit more from a low-input production of a low-priced species. On the other hand, richer countries may prefer the option of 'niche' species, rare and expensive on the market but requiring more sophisticated aquaculture techniques (Bilio, 2007).

In the productivity/sustainability species analysis, we chose not to include site-specific criteria such as optimal temperature or salinity or too specific market indicators. When choosing a species for culture in a specific site, one has to take these factors into account to comply best with fish needs and consequently minimize production costs (see the model in Section 8.2). We chose to consider primarily a biologically based species-grouping criteria: temperature range. In our analytical process, we examine species that are cultivated and distributed all around the world and therefore discriminating them according to market price value was considered impossible, given seafood price fluctuations due to seasonal, supply and market demand fluctuations, consumer preference shifts, varying landing and energy costs, to name but a few.

Production costs were not evaluated directly because they were dependent on the rearing technologies used (e.g. cages, ponds, land-based tanks), the prevailing temperature in a particular site, etc. However, the reader will notice that some of the criteria reflect some economical parameters. For example, a criterion such as 'live feed needed' is linked directly to production costs. A species that can be weaned on a commercial powdered feed will require less labour and less equipment, and consequently be less costly to produce.

Despite important gaps in biological and technical knowledge, the inclusion of tunas in this book was attributable largely to significant recent biotechnical breakthroughs and the worldwide intensity of R & D efforts (Japan, Australia, European Union) aimed at resolving the major bottlenecks governing the production of juvenile tuna. The perspectives of achieving economical and environmental sustainability through: (i) the achievement of controlled life cycle; and (ii) reduced pressure on the targeted wild fisheries (mainly governed by the growing demand for juvenile wild tuna for commercial on-growing operations) guided our decision to include these emerging species. Consequently, for the species of tunas covered in our analysis, and for a few other species as well, some information was considered unavailable. Mean scores or equivalent values from closely related species were attributed in order to be able to complete our analysis. Furthermore, when no written information was available, the values used originated from a consultation with a tuna specialist, Dr Denis Coves, (IFREMER, Palavas-les-flots, France).

References

Aly, S.M., Mohamed, M.F. and John, G. (2008) Effect of probiotics on the survival, growth and challenge infection in *Tilapia nilotica* (*Oreochromis niloticus*). *Aquaculture Research* 39, 647–656.

Anderson, J.L. (2007) Sustainable aquaculture: what does it mean and how do we get there? In: Leung, P., Lee, C.S. and O'Bryen, P.J. (eds) *Species and System Selection for Sustainable Aquaculture*. Blackwell Publishing Ltd, Oxford, UK, pp. 9–19.

Asche, F. and Bjorndal, T. (in press) Aquaculture: production and markets. In: Grafton, R.Q., Hilborn, R., Squires, D., Tait, M. and Williams, M. (eds) *Handbook of Marine Fisheries Conservation and Management*. Oxford University Press, Oxford, UK.

Asche, F., Roll, K.H. and Tveteras, S. (2008) Future trends in aquaculture: productivity growth and increased production. In: Holmer, M., Black, K., Duarte, C.M., Marbà, N. and Karakassis, I. (eds) *Aquaculture in the Ecosystem*. Springer, Dordrecht, The Netherlands, pp. 271–292.

Ashley, P.J. (2007) Fish welfare: current issues in aquaculture. *Applied Animal Behaviour Science* 104, 199–235.

Ayer, N.W. and Tyedmers, P.H. (in press) Assessing alternative aquaculture technologies: life cycle assessment of salmonid culture systems in Canada. *Journal of Cleaner Production*.

Bao-tong, H. and Hua-Zhu, Y. (1984) Characteristics of integrated fish farming in China. Sixth Session of the IPFC Working Party, 19–25 January 1984, New Delhi.

Bilio, M. (2007a,b) Controlled reproduction and domestication in aquaculture. The current state of the art Part I and II. *Aquaculture Europe* 32(1), 5–14; 32(3), 5–23.

Bilio, M. (2008a,b) Controlled reproduction and domestication in aquaculture. The current state of the art. Part III and IV. *Aquaculture Europe* 33(1), 5–19; 33(2), 12–34.

Bjornsson, B. (1994) Effects of stocking density on growth rate of halibut (*Hippoglossus hippoglossus* L.) reared in large circular tanks for three years. *Aquaculture* 123, 259–270.

Bromage, N., Porter, M. and Randall, C. (2001) The environmental regulation of maturation in farmed finfish with special reference to the role of photoperiod and melatonin. *Aquaculture* 197, 63–98.

Bushman, P.J. and Burns, J.R. (2005) Social control of male sexual maturation in the swordtail characin, *Corynopoma riisei*. *Journal of Fish Biology* 44, 263–272.

Chopin, T., Robinson, S., MacDonald, B., Haya, K., Page, F., Ridler, N., Szemerda, M., Sewuster, J. and Boyne-Travis, S. (2006) Integrated multi-trophic aquaculture: seaweeds and beyond . . . the need of an interdisciplinary approach to develop sustainable aquaculture. *Journal of Phycology* 42, 11.

Davis, K.B., Simco, B.A. and Silverstein, J.T. (2005) Relationship of gonadal development to body size and plasma sex hormone concentrations in female channel catfish. *North American Journal of Aquaculture* 67(3), 259–264.

Desjardins, M., Le François, N.R., Fletcher, G.L. and Blier, P.U. (2006) Seasonal modulation of plasma antifreeze protein levels in Atlantic (*Anarhichas lupus*) and spotted wolffish (*A. minor*). *Journal of Experimental Marine Biology and Ecology* 335, 142–150.

Desjardins, M., Le François, N.R., Fletcher, G.L. and Blier, P.U. (2007) High antifreeze protein levels in wolffish (*Anarhichas lupus*) make then an ideal candidate for culture in cold, potentially ice laden waters. *Aquaculture* 272, 667–674.

Edwards, C.A., Grove, T.L., Harwood, R.R. and Pierce Colfer, C.J. (1993) The role of agrobiology and integrated farming systems in agricultural sustainability. *Agriculture, Ecosystems and Environment* 46, 99–121.

Ellis, T., North, B., Scott, A.P., Bromage, N.R., Porter, M. and Gadd, D. (2002) The relationships between stocking density and welfare in farmed rainbow trout. *Journal of Fish Biology* 61, 493–531.

El-Sabh, M.I. and Silverberg, N. (1991) Oceanography of a large-scale estuarine system: the St Lawrence. *Coastal and Estuarine Studies* 39, 1–9.

Espelid, S. (2002) Susceptibility of spotted wolffish to infectious diseases and use of immunoprophylaxis. *Bulletin of the Aquaculture Association of Canada* 102(2), 37–38.

Falk-Petersen, I.B., Hansen, T.K., Fieler, R. and Sunde, L.M. (1999) Cultivation of the spotted wolffish *Anarhichas minor* (Olafsen) – a new candidate for cold-water fish farming. *Aquaculture Research* 30(9), 711–718.

FAO (Food and Agriculture Organization) (2002) *The State of Fisheries and Aquaculture*. FAO, Rome.

Felip, A., Zanuy, S. and Carrillo, M. (2006) Comparative analysis of growth performance and sperm motility between precocious and non-precocious males in the European sea bass (*Dicentrarchus labrax*, L.). *Aquaculture* 256, 570–578.

Foss, A., Imsland, A., Falk-Petersen, I.-B. and Oiestad, V. (2004) A review of the culture potential of spotted wolfish, *Anarhichas minor* Olafsen. *Reviews in Fish Biology and Fisheries* 2, 277–294.

Huang, W.B. and Chiu, T.S. (1997) Effects of stocking density on survival, growth, size variation and production of *Tilapia* fry. *Aquaculture Research* 28, 165–173.

Imsland, A.K., Foss, A., Sparboe, L.O. and Sigurdsson, S. (2006) The effect of temperature and fish size on growth and feed efficiency ratio of juvenile spotted wolffish *Anarhichas minor. Journal of Fish Biology* 68, 1107–1122.

IUCN (International Union for Conservation of Nature) (2008) IUCN red list of threatened species (http://www.iucnredlist.org, accessed 10 June 2008).

Johnson, H. and Halfyard, L.C. (2002) Marketing issues and opportunities for farmed wolffish in the United States and Europe. *Bulletin of the Aquaculture Association of Canada* 102-2, 11–12.

King, H.P., Lee, P.S. and Pankhurs, N.W. (2003) Photoperiod-induced precocious male sexual maturation in Atlantic salmon (*Salmo salar*). *Fish Physiology and Biochemistry* 28, 427–428.

Koutikonsky, V.G. and Bugden, G.L. (1991) The physical oceanography of the Gulf of St Lawrence: a review with emphasis on the synoptic variability in motion. In: The Gulf of St Lawrence: small ocean or big estuary? *Canadian Special Publication of the Fisheries and Aquatic Sciences* 113, 57–90.

Kristiansen, T.S., Ferno, A., Holm, J.C., Privitera, L., Bakke, S. and Fosseidengen, J.E. (2004) Swimming behaviour as an indicator of low growth rate and impaired welfare in Atlantic halibut (*Hipppoglossus hippoglossus* L.) reared at three stocking densities. *Aquaculture* 230, 137–151.

Kulka, D., Hood, C. and Huntington, J. (2007) Recovery strategy for northern wolffish (*Anarhichas den*ticulatus) and spotted wolffish (*Anarhichas minor*) and management plan for Atlantic wolffish (*Anarhichas lupus*) in Canada. Fisheries and Oceans Canada: Newfoundland and Labrador region, St John's, Newfoundland (http://saregistry.gc.ca).

Lamarre, S.G., Le François, N.R., Dreidzic, W. and Blier, P.U. (2009) Low acclimation temperature induces compensation of proteasome 20S but not protein synthesis in juvenile spotted wolfish (*Anarhichas minor*, Olafsen). *Journal of Experimental Biology* 212, 1294–1301.

Le François, N.R., Lemieux, H. and Blier, P.U. (2002) Biological and technical evaluation of the potential of marine and anadromous fish species for cold-water mariculture. *Aquaculture Research* 33, 95–108.

Leung, P., Lee, C.-S. and O'Bryen, P.J. (2007) *Species and System Selection for Sustainable Aquaculture*. Blackwell Publishing Ltd., Oxford, UK.

Liu, K.M. and Chang, W.B. (1992) Bioenergetic modeling of effect of fertilization, stocking density and spawning on growth of the Nile tilapia, *Oreochromis niloticus* (L.) *Aquaculture and Fisheries Management* 23, 291–301.

McLay, H.A., Youngson, A.F., Wright, R.S. and Johnstone, R. (2008) Effects of rearing density on sexual maturation and growth in sea-cage reared Atlantic salmon, *Salmo salar* L. *Aquaculture Research* 23(3), 353–365.

Mahon, R., Brown, S.K., Zwanenburg, K.C.T., Atkinson, D.B., Buja, K.R., Claflin, L., Howell, G.D., Monaco, M.E., O'Boyle, R.N. and Sinclair, N. (1998) Assemblages and biogeography of demersal fishes of the east coast of North America. *Canadian Journal of Fisheries and Aquatic Sciences* 55, 1704–1738.

Moksness, E. and Pavlov, D.A. (1996) Management by life cycle of wolffish, *Anarhichas lupus* L., a new species for cold-water aquaculture: a technical paper. *Aquaculture Research* 27, 865–883.

Moksness, E., Kjorsvik, E. and Olsen, Y. (2004) *Culture of Cold-water Marine Fish*. Fishing News Books, Blackwell Publishing Ltd., Oxford, UK.

Muir, J. (2005) Managing to harvest? Perspectives on the potential of aquaculture. *Philosophical Transactions of the Royal Society B* 360, 191–218.

Naylor, R.L., Goldburg, R.J., Primavera, J.H., Kautsky, N., Beveridge, M.C., Clay, J., Folke, C., Lubchenco, J., Mooney, H. and Troell, M. (2000) Effect of aquaculture on world fish supplies. *Nature* 405, 1017–1024.

Nelson, J.S. (2006) *Fishes of the World* 4th Edition. John Wiley and Sons Inc., New York.

Pavlov, D.A. (1986) Developing the biotechnology of culturing White Sea wolffish; *Anarhichas lupus marisalbi*. II. Ecomorphological pecularities of early ontogeny. *Journal of Ichthyology* 26, 156–169.

Person-Le Ruyet, J., Alexandre, J.C., Thébaud, L. and Mugnier, C. (1993) Marine fish larvae feeding: formulated diets of live prey? *Journal of World Aquaculture Society* 24, 211–224.

Quéméner, L., Suquet, M., Mero, D. and Gaignon, J.L. (2002) Selection method of new candidate for finfish aquaculture: the case of the French Atlantic, the Channel and the North Sea Coasts. *Aquatic Living Resources* 15, 293–302.

Ridha, M.T. (2006) Comparative study of growth performance of three strains of Nile tilapia, *Oreochromis niloticus*, L., at two stocking densities. *Aquaculture Research* 37, 172–179.

Rotherham, D., Underwood, A.J., Chapman, M.G. and Gray, C.A. (2007) A strategy for developing scientific sampling tools for fishery-independent surveys of estuarine fish in New South Wales, Australia. *ICES Journal of Marine Science* 64, 1512–1516.

Scandol, J. (2005) Use of quality control methods to monitor status of fish stocks. In: Kruse, G.H., Gallucci, V.F., Hay, D.E., Perry, R.I., Peterman, R.M., Shirley, T.C., Spencer, P.D., Wilson, B. and Woodby, D. (eds) *Fisheries Assessment and Management in Data-limited Situations*. Alaska Sea Grant, University of Alaska, Fairbanks, Alaska, pp. 213–233.

Scott, W.B. and Scott, M.G. (1988) *Atlantic Fishes of Canada*. Canadian Bulletin of Fisheries and Aquatic Sciences 219, University of Toronto Press, Canada.

Sevilleja, R.C. (2007) Genetics-based technologies for sustainable development in Philippine aquaculture: the case of tilapia. In: Leung, P., Lee, C.-S. and O'Bryen, P.J. (eds) *Species and System Selection for Sustainable Aquaculture*. Blackwell Publishing Ltd, Oxford, UK, pp. 357–372.

Shearer, K.D. and Swanson, P. (2000) The effect of whole body lipid on early sexual maturation of 1+ age male Chinook salmon (*Oncorhynchus tshawytscha*). *Aquaculture* 190, 343–367.

Teletchea, F., Fostier, A., Le Bail, P.Y., Jalabert, B., Gardeur, J.N. and Fontaine, P. (2007) STOREFISH: a new database dedicated to the reproduction of temperate freshwater teleost fishes. *Cybium* 31(2), 237–245.

Toranzo, A.E., Novoa, B., Romalde, J.L., Nunez, S., Devesa, S., Marino, E., Silva, R., Martinez, E., Figueras, A. and Barja, J.L. (1993) Microflora associated with healthy and diseased turbot (*Scophthalmus maximus*) from three farms in northwest Spain. *Aquaculture* 114, 189–202.

Weltzien, F.A., Andersson, E., Andersen, Ø., Shalchian-Tabrizi, K. and Norberg, B. (2004) The brain–pituitary–gonad axis in male teleosts, with special emphasis on flatfish (Pleuronectiformes). *Comparative Biochemistry and Physiology A* 137, 447–477.

9 The Sturgeons (Family: Acipenseridae)

MATTHEW LITVAK

Centre for Coastal Studies and Aquaculture and Department of Biology, University of New Brunswick, Saint John, New Brunswick, Canada

9.1 General Introduction

Order Acipenseriformes consists of two extant families: Acipenseridae, comprised of five genera with 25 species, and Polydontidae, with one genus and two species. Figure 9.1 shows the general shape of the family. The order has a circumpolar distribution in the northern hemisphere. Many exhibit some form of diadromy; however, all spawn in fresh water. This wide distribution within the species makes them difficult to manage as they often cross state, provincial and international boundaries. They are among the most endangered fishes in the world, with all 27 species listed on the IUCN's (The International Union for Conservation of Nature and Natural Resources or The World Conservation Union) Red List; 6 species are critically endangered, 11 species are endangered, 9 species are vulnerable and 1 species is at low risk (IUCN, 2006). All the species are listed under the Convention on International Trade in Endangered Species (CITES), either on Appendix I (2 species) or II (25 species). This is a result of human activities including damming, water regulation, pollution and fisheries persecution, either through directed fisheries and/or by-catch.

Sturgeons are sought for both their caviar and their meat. Caviar is one of the most valuable fish products in the world; often associated with royalty and fine foods. Unfortunately for sturgeon, their rarity makes them even more valuable, a very bad 'positive feedback' cycle pushing them to the brink of extinction. The history of caviar is curious; dependent on the region and time, sturgeons have been viewed as either trash fish or fish to capture for royalty only. Caviar, viewed largely as a Russian national dish, was most likely first eaten by Neotlithic peoples around the Black and Caspian Seas (Saffron, 2002). Records of Atlantic sturgeon (*Acipenser oxyrinchus oxyrinchus*) scutes in high abundance have been in found in the Augustine and Oxbow mounds (middens) found along the Miramichi in New Brunswick, Canada, suggesting the importance of sturgeon to Mi'Kmaq

Fig. 9.1. Acipenseridae.

First Nations over 3000 years ago. Denys (1968) described Mi'Kmaq using torches and harpoons to capture large sturgeon (presumably Atlantic sturgeon) during the 1600s along the east coast of Canada.

9.2 Farming of Sturgeons

Current world production from wild capture is extremely low and is often not possible because of low numbers in the wild and/or it is prohibited; there has been a catastrophic decrease in wild production since the 1980s (Fig. 9.2). However, recent developments in culture techniques have allowed aquaculture production to increase at a tremendous rate (Fig. 9.2), but the demand for their products is not even close to being met. In North America, caviar from white sturgeon, *Acipenser transmontanus*, is being produced in California and is also under development in British Columbia by Target Marine. Lake, *A. fluvescens*, shortnose, *A. brevirostrum* and Atlantic sturgeons, *A. oxyrinchus oxyrinchus,* are under development in various regions of North America. Caviar from beluga, *Huso huso*, starry, *A. stellatus*, Danube sturgeon, *A. gueldenstaedii*, Siberian, *A. baerii*, and the hybrid *guldenastaedii x baerii*, are all used to produce caviar around the world.

The development of sturgeon for aquaculture has not only the unique promise of great economic value and value-added opportunities, but also the

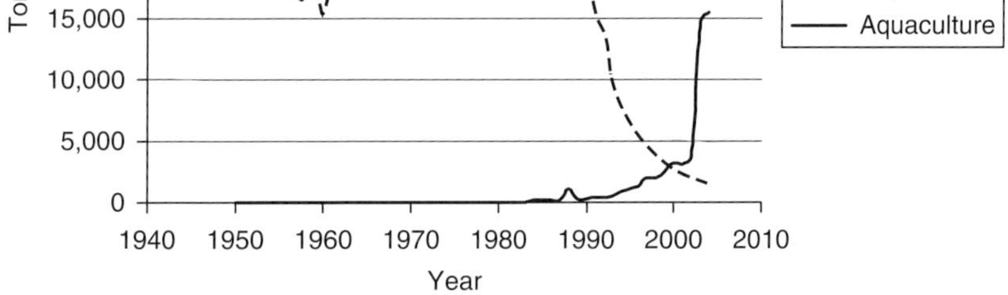

Fig. 9.2. World sturgeon and paddlefish production. *Source* : FISHSTAT Plus, Fisheries Data Analysis Software for Windows, FAO, Rome; accessed 2007.

potential to protect the remaining wild stocks before they become extinct. The IUCN and their sturgeon specialist group, in addition to a number of environmental NGOs (World Wildlife Fund, TRAFFIC [Wildlife Trade Monitoring Network], Caviar Emptor), indicate the important role that aquaculture will play in saving the remaining wild stocks of sturgeon by providing 'non-wild' sources of caviar. In addition, because of the commercial interest and development of broodstocks, there is greater opportunity for the rehabilitation of stocks. Aquaculture development is not a simple task; sturgeons are long-lived species that often take many years to produce caviar and the current importation and export restrictions are strict and often complicated. This chapter provides a brief introduction to sturgeon biology and their potential for further aquaculture development, focusing on two species which occur along the east coast of North America: the shortnose and Atlantic sturgeons.

Currently there are two companies working on shortnose sturgeon aquaculture in Canada, Supreme Sturgeon and Caviar and Acadian Sturgeon and Caviar (http://www.acadian-sturgeon.com/). Supreme Sturgeon and Caviar (Pennfield, New Brunswick) have been working on shortnose sturgeon development since 1998 (B. Tucker, New Brunswick, 2007, personal communication). They are now starting to produce caviar at their facility and are currently obtaining CAN$70/ounce for it in Québec and at local hotels in New Brunswick. Their major concern is not the market or the price, as demand is high and the market is large; their main concern is the constant supply of caviar from their facility. It is very difficult to time the harvest and it is becoming apparent that shortnose sturgeon grown under current culture regimes exhibit a high degree of variance in the timing of caviar production. Currently, shortnose sturgeon will provide up to 15% yield by body weight for caviar (B. Tucker, personal communication). This is similar to other species (Saffron, 2002).

The second company, Acadian Sturgeon and Caviar (C. Ceapa, New Brunswick, 2007, personal communication), started in 2005 and is currently trying to expand their site as their stock of shortnose sturgeon grows. They too have identified shortnose sturgeon as an excellent candidate for aquaculture production. Currently, they have fingerlings for sale for both shortnose and Atlantic sturgeon, market wild sturgeon caviar and value-added meat products and have sold Atlantic sturgeon fingerlings to groups from the USA and Europe.

9.3 Shortnose Sturgeon, *Acipenser brevirostrum*

9.3.1 Biology and culture

Scientific name: *Acipenser brevirostrum* (LeSeur 1818).
Common names: shortnose sturgeon, shortnosed sturgeon and little sturgeon (Saint John River, New Brunswick), pinkster and roundnoser (Hudson River, New York), bottlenose or mammose (Deleware River), salmon sturgeon (Carolinas) and soft-shell or lake sturgeon (Altamaha River) (Dadswell *et al*., 1984).

Taxonomy: Class: Osteichthyes; Subclass: Actinopterygii; Infraclass: Chondrostei; Order: Acipenseriformes; Family: Acipenseridae; Genus: Acipenser; Species: brevirostrum. Maliseet name: Buzgus (C. Atwin, personal communication, Oromocto First Nation).

9.3.1.1 Gross anatomy

The largest recorded specimen was caught on the Saint John River. It was a female fish weighing 23.6 kg with a fork length of 122 cm (Dadswell, 1979). Although not reported to be sexually dimorphic, females tend to be heavier than males at similar lengths (COSEWIC, 2005).

Their bodies are elongated and cylindrical in the abdomen and tail, but somewhat depressed anterior of the pectoral girdle (Fig. 9.3). They, like other sturgeons, are a heavily armoured fish with five rows of bony scutes (also called plates or bucklers), an inferior elongated protrusible wide mouth, four barbels anterior to the mouth on the ventral surface of the rostrum, small eyes and spiracles. Scutes are sharper on juveniles and become blunted with age. The caudal peduncle is narrow, with the dorsal fin posterior of the paired pelvic fins. Their caudal fin is heterocercal with a prominent long upper lobe with a short and broad lower lobe. Pectoral fins are large and fixed. First pectoral fin ray is thick and ossified. The body has no scales but has minute denticles. Their colour is variable; they have a darker mottled chain pattern on the dorsal surface of the head region over an olive brown or green background. Their lateral surfaces are lighter moving ventrally, with the ventral surface white in colour (Fig. 9.3). Dorsal scutes are lighter brown. Lateral, ventral and pre-anal scutes have a yellow tinge. Leading edges of fins are lighter and sometimes white.

Their skeleton is cartilaginous, with the exception of the bones of the skull, jaws and pectoral girdle. Notochord is unconstricted (Schmitz, 1998). They

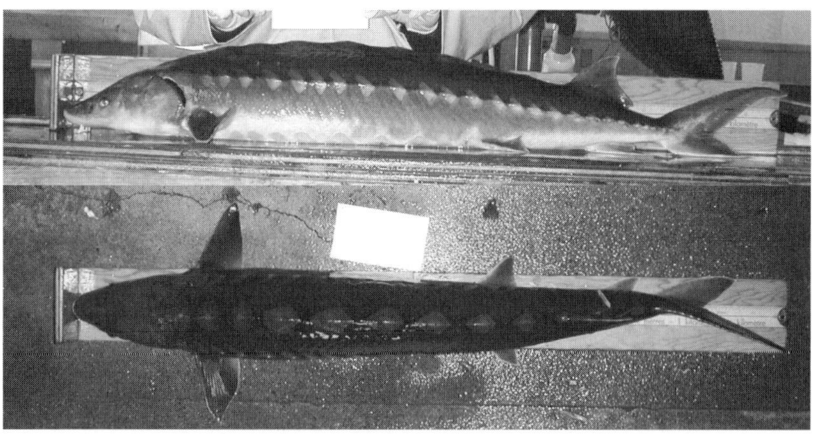

Fig. 9.3. Lateral and dorsal views of wild shortnose sturgeon broodstock caught in the Saint John River, New Brunswick, Canada (images taken by M. Litvak; measuring board is 1 m).

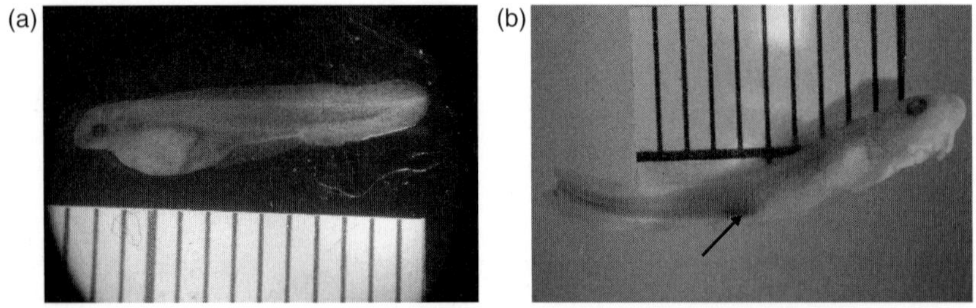

Fig. 9.4. (a) Young, recently hatched shortnose larva. (b) Older yolk-sac larva clearly showing yolk plug in spiral valve. Divisions on the rulers are mm.

possess a swim bladder that is joined to the oesophagus (physostomous). The oesophagus is muscular and acts like a crop or gizzard to crush invertebrates. They have a spiral valve for an intestine, similar to that of Class Chondrichthyes (Vladykov and Greeley, 1963).

Shortnose sturgeon larvae are approximately 7–11 mm total length (TL) at hatch (Buckley and Kynard, 1981). Larvae possess a large yolk sac (Fig. 9.4a, b) with a yolk plug in the spiral valve, which is released when feeding begins at 9–12 days posthatch (dph) (NMFS, 1998a). They are viewed as direct developers and do not experience a metamorphic stage. They resemble the juvenile/adult form at approximately 20–25 mm in TL (Bain, 1997).

9.3.1.2 Distribution (see Fig. 9.5)

Found only in North America; 19 distinct shortnose sturgeon population segments have been identified along the eastern coast of North America (NMFS, 1998a) from Florida to New Brunswick. Populations range from 100 in the Merrimack River, Massachusetts, to ~ 38,000 in the Hudson River, New York. Their only known occurrence in Canada is in the Saint John River, New Brunswick.

9.3.1.3 General life history

Like other sturgeons, the shortnose sturgeon is long-lived. The oldest fish caught from the Saint John River population was a 67-year-old female (Dadswell, 1979). The oldest male caught on the Saint John River was 32 years old. Von Bertalanffy growth curves based on adult fork length and age were determined for females and males of the Saint John population to be L_t(female) = $127.0(1\text{-}e^{-0.047(t+1.107)})$ and L_t(males) = $108.7(1\text{-}e^{-0.063(t-0.79)})$, respectively (Dadswell, 1979). Sexes are gonochoristic and intersexes or hermpahrodites are rare (Atz and Smith, 1976). Although they have not been viewed as sexually dimorphic, females tend to be heavier than males of the same length (Dadswell, 1979; COSEWIC, 2005) and recently, Vescei *et al.* (2003) suggested that the shape of the urogenital opening could be used to sex this species.

Shortnose sturgeon has been described as anadromous or amphidromous (Kynard, 1997). In their southern range, they are estuarine and

Fig. 9.5. World distribution of *Acipenser brevirostrum*.

anadromous, but in the more northern regions, they are freshwater amphidromous (Kieffer and Kynard, 1993). In the Saint John River, adults are found in both fresh water and areas under tidal influence (Dadswell, 1979, 1984; Dadswell *et al.*, 1984). They have been caught along the coast, but this is a rare occurrence (Dadswell *et al.*, 1984). They are generally restricted to brackish and freshwater sections of their natal rivers (Buckley and Kynard, 1985; Hall *et al.*, 1991; Kieffer and Kynard, 1993; O'Herron *et al.*, 1993; Kynard, 1997).

Spawning is thought to occur in areas with high flow rates. Flow rates for spawning in the US populations range between 0.4 and 0.7 m/s and are generally over gravel and/or boulder substrate and at a temperature of 8–13°C (Buckley and Kynard, 1985; Kieffer and Kynard, 1993; Kynard, 1997; NMFS, 1998a; COSEWIC, 2005).

Males and females from the Saint John River become reproductive on average at 11 and 13 years of age, respectively; later than southern populations (Dadswell, 1984). Little information exists on spawning periodicity. However, it has been suggested that females spawn less frequently, once every 3–5 years and males every other year (Dadswell, 1979). In the more southern populations, they tend to spawn more often (Dadswell *et al.*, 1984). However, males from the Saint John River held in captivity do produce semen annually (personal observation).

Shortnose sturgeon is a highly fecund, iteroparous fish, producing many demersal and adhesive eggs (27,000–208,000; Dadswell, 1979). Egg production is related directly to female size; each female on average produces 11,600 eggs per kg of fish (Dadswell, 1979). Eggs are black to brown and are approximately 3.5 mm in diameter at release and expand to 4 mm after fertilization and adhesion to the substrate (Kynard, 1997). There are only anecdotal accounts of spawning behaviour in captivity and none in the wild. There have been two studies on the fine ultrastructure of shortnose sturgeon sperm (Dilauro *et al.*, 1999, 2000) using transmission electron microscopy to provide a description of shortnose sturgeon sperm. The cell is comprised of an acrosome, head region, midpiece and single falgellum. The sperm body is 9.71 μm long (acrosome + nucleus + midpiece) and the length of the flagellum is 37 μm (Dilauro *et al.*, 1999, 2000).

Sturgeons are bottom suctorial feeders. The name sturgeon has its origin in the German verb 'störer', which means 'to root' (Saffron, 2002). Shortnose sturgeon juveniles feed mainly on crustaceans and insects and the adults eat mainly molluscs, particularly *Mya arenaria* (Dadswell, 1979; Pottle and Dadswell, 1979; Dadswell, *et al.*, 1984).

Hardy (2000) examined growth and starvation resistance of larval shortnose sturgeon in response to delayed feeding. He found starvation affected growth and survival, yet, despite the degree of starvation, larvae were able to resume growth and experience high survivorship following feeding. Specific growth rates (SGRs) (dry weight) were highest in the longest food-deprived groups directly following feeding, suggesting that a possible compensatory mechanism may exist in response to starvation. A point-of-no-return (PNR) of 56% was reached 41 days post-fertilization, which is long compared to many other *r*-selected fish species.

9.3.1.4 Behaviour/adaptability

Although sturgeons are one of the most primitive teleost fishes (Bemis *et al.*, 1997), they may have a more complex system of social behaviours and interactions than previously thought. Shortnose sturgeon appears to be a social fish. I have seen them exhibit shoaling behaviour a few days after hatch. This shoaling behaviour exists only when there is a flow; larvae form tight, well-spaced schools to swim against the current. This schooling behaviour breaks down when there is no flow. Scuba diving observations of adults in the wild suggest that they also exhibit shoaling behaviour (L. Sabatis, Oromocto First Nation, 2003, personal communication).

There is other strong but unusual evidence to suggest that adult shortnose are extremely social and may even exhibit pairing. Dadswell (1979) made the interesting observation that when groups of tagged fish were recaptured using a gill net, they were often recaptured with fish in the same order as the initial capture. The probability that pairs of fish would be recaptured together and removed from the gill net in the same order by chance is infinitesimally low. Members of my lab have also made the same observation.

9.3.1.5 Fishery status

No directed fishery is allowed, as they have been listed as endangered in the USA since 1967 and a species of special concern in Canada since 1980. Historically, shortnose were most likely a by-catch during the early Atlantic sturgeon fishery in the USA and Canada (Smith and Clugston, 1997; Stein *et al.*, 2004).

9.3.2 Broodstock management and hatchery operations

The main objective in the culture of shortnose sturgeon is caviar. Compared with other sturgeon species, shortnose is rather small and has the potential to reach maturity earlier than many of its larger cousins. The intent is to reduce greatly the natural time to maturity and reproductive condition to 5 years of age or earlier. This is a realistic goal because the time to reproduction has been reduced dramatically with other sturgeon currently grown in culture conditions. Shortnose caviar is relatively large (approximately 3.5 mm in diameter), which would make them a good size for the caviar market. They adapt well to the culture environment and tend to grow fast when reared in warm water.

In general, rearing of sturgeon is not difficult compared to many of the *r*-selected marine species which require the production of live feed. Shortnose sturgeons adapt well to culture conditions, are social and therefore are tolerant of high stocking densities and can also tolerate lower oxygen levels.

Shortnose sturgeons become reproductive during the spring. The ideal production cycle is to grow females only (see the section on future R & D below); however, at this time, this is impractical. Therefore, a potential production cycle will involve growing the males for meat and harvesting them at year 2, when sexes are separated more easily. These fish will be headed for the meat and value-added market. The remaining females from that year class then will be on-grown until maturity, when their caviar will be harvested.

Females do not necessarily breed every year, so their status must be determined by biopsying fish as they reach maturity. A number of attempts have been made to use vitellogenin (VTG) assays in combination with E1 and E2 and calcium to indicate the reproductive state of other sturgeon, and this technique shows great promise in identifying shortnose female sturgeon that have reached maturity. Eggs are staged to see if the fish are ready for natural (not often) or hormone injections. A biopsy (Conte *et al.*, 1988) or using a hollow tube inserted through a circular puncture hole (as in Williot *et al.*, 2005; not unlike a trocar) is an easy and efficient way to sample eggs from the ovaries. I have used the trocar technique and found it to be much more efficient than the surgical method described by Conte *et al.* (1988) and it can be done in just a few seconds. The eggs are boiled, bisected and then examined under a dissecting scope to see the position of the germinal vesicle (GV) relative to the length between the animal pole (AP) and vegetal pole (VP) (Conte *et al.*, 1988), or using an improved technique of incubation in progesterone prior to boiling (Mohler, 2003). Females that are generally receptive to hormone manipulation have a ratio distance of AP to GV divided by the distance between AP and VP of less than 0.07 (stage IVc) (Conte *et al.*, 1988). Hormones are used to stimulate ovulation; either carp pituitary extract in a fish Ringer's solution or a synthetic analogue of LhRh, either in pellet (Flynn *et al.*, 2006) or solution, is inserted or injected into the musculature. Females usually produce eggs within 24 h of the injection and must be monitored carefully. The eggs are then removed from the ovaries by gentle massaging, remembering that the opening of the oviduct is in the anterior section of the ceolomic cavity – it is like a scoop. Eggs are expressed into a bowl. Milt is procured by gentle squeezing of the testes and collected with a tygon tube attached to a 20 ml syringe, which is inserted into the urogenital opening.

Sperm is diluted in fresh water and added to the eggs. The eggs become adhesive after exposure to water and are 'de-glued' using either silt from the river that the fish are from (sieving and autoclaving is strongly recommended before using this technique) or Fuller's earth. The earth is added in a slurry and the eggs are stirred gently (some people recommend a feather) and then rinsed after they lose their adhesiveness.

The fertilized eggs (~ 500 ml) are then placed in an up-welling MacDonald type jar (7 l) equipped with an overflow screen. Water is added at a high enough rate to keep the embryos in suspension (3–4 l/min). The overflows from the jars are directed into the larvae-rearing tank on hatching. After approximately 120 degree days, the larvae begin to hatch. Survival to hatch is variable as they are prone to fungal infections; however 70–100% hatch is not uncommon. Stocking density ranges from 0.15 to 0.3 larvae/cm^2.

Larvae can be reared in almost any shaped tank, but circular tanks are preferred because of the 'self-cleaning' and potential for continuous swimming. Larvae are negatively phototactic. They swim in polarized shoals against the tank current flow. After 8–12 days at a temperature of 17°C, they can be fed live *Artemia*. The timing of this was thought initially to be critical; however, Hardy (2000) found that they were rather resistant to starvation and their point-of-no-return was up to 33 days posthatch at 17°C. Frozen *Artemia* can be used as a substitute (personal observation) when live *Artemia* supplies are interrupted.

They are fed at least every 3 h per 24 h period. They can be co-fed within 20 days of the initiation of feeding and can be weaned over a 1 week period or less. Microdiet replacement of '*Artemia*' size is fine and juveniles can be moved on to a trout crumble within the first few months. Larval and weaning to juvenile survival is high; it is not uncommon for over 70% of the eggs fertilized to become juveniles. After they are weaned, mortality rates are extremely low.

9.3.3 On-growing to market size

Juveniles have been grown in our facility at a density of 180 kg/m^3 (Fig. 9.6); however, without oxygen injection, this high a density will stunt their growth. Shortnose sturgeon juveniles fed once per day grew less than those fed either 4 or 8 times per day at a feed ration of 3% body weight per day (Giberson and Litvak, 2003). Their SGR was 1.6%/day and feed conversion efficiency was 93%. This was in a recirculation system at a temperature of 14°C and no oxygen injection; this is therefore an extremely conservative estimate of their growth potential. The fact that they grew well at a higher frequency bodes well for intensive recirculation culture systems when the desire is to smooth out the production of nitrogenous waste products in order not to overwhelm the biofilters. With culture in its infancy, to date there have been no epizootics that have been identified.

Fig. 9.6. Shortnose sturgeon can be grown at high densities (180 kg/m^3).

9.4 Atlantic Sturgeon, *Acipenser oxyrinchus oxyrinchus*

9.4.1 Biology and culture

Scientific name: *Acipenser oxyrinchus oxyrinchus* (Mitchill 1815).
Common names: Atlantic sturgeon, American Atlantic sturgeon, Atlantischer Stör (German), common sturgeon, Esturgeon noir (French), Esturión del Atlántico (Mexico: Spanish), Jeseter ostrorypý (Czech), Jesiotr ostronosy (Polish), Sinisampi (Finnish), Stör (Swedish), Vestatlantisk stør (Danish).
Taxonomy: Class: Osteichthyes; Subclass: Actinopterygii; Infraclass: Chondrostei; Order: Acipenseriformes; Family: Acipenseridae; Genus: Acipenser; Species: oxyrinchus oxyrinchus.

9.4.1.1 Gross anatomy

The largest recorded specimen was caught in the Saint John River, New Brunswick, Canada. It was a female fish weighing 368 kg and 430 cm in TL (Vladykov and Greeley, 1963). The body of the Atlantic sturgeon is elongated and cylindrical in the abdomen and tail, but somewhat depressed anterior of the pectoral girdle (Figs 9.7 and 9.8). They are armoured like other sturgeons,

Fig. 9.7. Underwater injection of LhRh to induce ovulation of wild-caught Atlantic sturgeon from the Saint John River, New Brunswick, Canada.

Fig. 9.8. Atlantic sturgeon juvenile grown in culture at the University of New Brunswick, Canada.

with five rows of bony scutes, an inferior elongated mouth (not as a wide as the shortnose sturgeon) relative to the inter-orbital width (a key character for discrimination between the species when under 1 m in TL) on the ventral surface of the rostrum – younger representatives have long rostrums/snouts that are V-shaped, but this changes as they age, with the rostrum becoming more blunt through time – four barbels anterior to the mouth on the ventral surface, small eyes and spiracles. Scutes are very sharp on the juveniles but become blunted with age, although those on the caudal peduncle maintain a certain degree of sharpness that should be respected when handled. The caudal fin is heterocercal with a prominent upper lobe and a shorter and broad lower lobe. Pectoral fins are large and fixed. The first pectoral fin ray is thick and ossified. The body has no scales but has minute denticles. Their colour is variable; the dorsal surface is usually light to dark brown with mottled pattern when occurring in rivers (Figs 9.7 and 9.8) and with a more silver sheen when found in the ocean. The colour is darkest dorsally, becoming lighter laterally and almost white on the ventral surface.

9.4.1.2 Distribution (see Fig. 9.9)

Until recently, Atlantic sturgeon was thought to have occurred only in North America. However, a genetic analysis of museum collections in Europe suggests that, until recently, they also occurred in Europe (Ludwig *et al.*, 2002). There are 30 or more systems along the coast of North America where there have been, or could have been, spawning populations of Atlantic sturgeon

Fig. 9.9. World distribution of *Acipenser oxyrinchus oxyrinchus*.

(NMFS, 1998b; Dadswell, 2006). They occur from the Hamilton Inlet on the coast of Labrador to the St Johns River in eastern Florida (Vladykov and Greeley, 1963). They often occur sympatrically with shortnose sturgeon; however, they differ in that they are truly anadromous and migrate to sea as juveniles and return to spawn as mature adults.

9.4.1.3 General life history

Atlantic sturgeons, like other sturgeons, are difficult to age; however, they are reported to live over 60 years (NMFS, 1998b; Dadswell, 2006). Von Bertalanffy growth curves based on adult fork length are $L_F = 259.4(1\text{-}e^{-0.0639(t+1.0172)})$ for females and $L_F = 201.3\ (1\text{-}e^{-0.01131(t+0.8696)})$ for males (Van Eenennaam and Doroshov, 1998). Sexes are separate and although intesexes do occur, it is rare (Van Eenennaam and Doroshov, 1998). Although females tend to be larger and mature later than males (Van Eenennaam and Doroshov, 1998), gender identification is difficult to determine out of the reproductive season. Depending on latitude, males mature between 7 and 24 years and females between 8 and 24 years of age (Dadswell, 2006). Recently, Vescei *et al.* (2003) suggested that the shape of the urogenital opening could be used to sex this species.

Atlantic sturgeon is an anadromous species returning to its natal rivers to spawn (Dadswell, 2006). Spawning locations are not known; however, it is believed that they will be flowing water over hard substrates (rocks, rubble, shale and sand) (Smith and Clugston, 1997). Atlantic sturgeons are fecund; ripe ovaries may make up as much as 25% of the total fish weight. The relationship between egg number (E_N) and fork length (L_F) is: $E_N = 4{,}678{,}387 + 29{,}182\ L_F$. Spawning begins in summer when water temperatures reach 18–20°C (Smith and Clugston, 1997). Eggs hatch between 94 h and 140 h at 20 and 18°C, respectively (Smith *et al.*, 1980). The yolk sac is absorbed in about 10 days. Atlantic sturgeons grow much faster and have higher conversion efficiency right from the yolk-sac stage. Hardy and Litvak (2004) found that Atlantic sturgeon used yolk more efficiently to gain size than did shortnose sturgeon. Juveniles are found in fresh and brackish waters and move out to coastal waters at around 80–120 cm (Bain, 1997).

9.4.1.4 Feeding

Atlantic sturgeon, like shortnose sturgeon, is a suctorial feeder. Freshwater prey of the juveniles includes aquatic insect larvae, amphipods, isopods and oligchaetes and in brackish water includes shrimp, amphipods, isopods, gastropods and polychaete worms. Adults feed on polychaetes, shrimp, amphipods, isopods and small fish (Vladykov and Greeley, 1963). Adults do not feed during spawning migration (personal observation).

9.4.1.5 Behaviour/adaptability

Very similar to shortnose sturgeon; however, they do not appear to be as social. Despite their enormous size, these fish are usually very calm and docile in captivity. The only exception is when broodstock initially are removed from the tanks; they can become quite agitated and this can be very dangerous for the handlers. Fish are calmed eventually by restricting their movement gently and placing them on their backs on a padded V-shaped trough or sling.

9.4.1.6 Fishery status

They have been an important species to first nations on the North American coast (Denys, 1968), thousands of years before the arrival of Europeans. Major exploitation of Atlantic sturgeon began in the last quarter of the 19th century (Smith and Clugston, 1997). Peak harvest occurred in the 1880s when 3350 Mt were landed, precipitating a collapse in 1901 (Smith and Clugston, 1997). During the late 1800s, Atlantic sturgeon was one of the major world sources of caviar (Saffron, 2002). By-catch of this species is a serious problem for their recovery (Smith and Clugston, 1997; Stein *et al.*, 2004). US Fishery closure is to last 41 years in order to protect 20-year classes of females in the remaining spawning stocks (NMFS, 1998b). There are still two small regulated fisheries for Atlantic sturgeon in Canada; in the Gulf of Saint Lawrence (no mature adults taken) and the Saint John River (no immature fish taken), New Brunswick, together accounting for approximately 50 Mt in 2002 (Dadswell, 2006).

9.4.1.7 Aquaculture attributes and challenges

Atlantic sturgeon aquaculture is directed towards the development of caviar, meat and value-added products. These fish are long-lived, fecund and have been marketed already for their caviar; however, not from aquaculture production. The major problem with this species is the potential length of time it will take to reach maturation and produce caviar; however, white sturgeon is already in production and has a very similar life history, so they provide an excellent example for the development of Atlantic sturgeon for aquaculture. Sanders *et al.* (2003) developed a series of economic models that examined the effect of altering a number of parameters, including: the effect of small fish market (early male and underperforming female harvested for sale), increase in the number of females in the system and holding females for a second maturation cycle (in order to produce greater caviar yield). Their model was based on a very conservative estimate of the price of the meat produced from the smaller fish (US$4.80/kg) and did not include any potential benefits of value-added products such as smoked sturgeon, which will in all likelihood have an impact on cash flow and present and future values. It was interesting that the potential increase of 10% in caviar production for holding the females for an extra 2 years did not appear to be a good strategy.

For Atlantic sturgeon, the intent is to have them mature at age 9 or earlier, as is the case with white sturgeon. Their eggs are 2.5 mm in diameter and their yield in the wild has been as high as 25% (Smith and Clugston, 1997) and would be of interest to caviar connoisseurs that appreciate the smaller caviar of Sevruga, *A. stellatus*. Atlantic sturgeon also adapt well to culture conditions and grow very fast when reared in warm water. We have seen growth rates of up to 6%/day and have a food conversion efficiency (FCE) close to 100% – these are very fast-growing fish when grown in culture.

9.4.2 Broodstock management and hatchery operations

In general, rearing of sturgeon is not difficult compared to many of the *r*-selected marine species, which require the production of live feed. Atlantic sturgeons adapt well to culture conditions; they are social and therefore are tolerant of high stocking densities and can also tolerate lower oxygen levels.

Atlantic sturgeons become reproductive during the summer. The ideal production cycle is to grow only females (see the section on future R & D below); however, that is impractical at this time. Therefore, a potential production cycle will involve growing the males for meat and harvesting them at year 2, when sexes are separated more easily. These fish will be headed for the meat and value-added market. The remaining females from that year class then will be on-grown until maturity, when their caviar will be harvested.

Females do not necessarily breed every year, so their status must be determined by biopsying the fish when reaching maturity. Eggs are staged to see if the fish are ready for natural (not often) or hormone injections, as with the shortnose sturgeon. The major difference in approach for the two species is because of the large size and difficulty in handling the Atlantic sturgeon. For sturgeon this large, it is much easier to do the injection while the fish is in water (see Fig. 9.7). Females usually produce eggs within 24 h of the injection and must be monitored carefully. Mature Atlantic sturgeons are very large and the removal of eggs is not an easy task. In the past, females were sacrificed and the eggs removed at death (Conte *et al.*, 1988). However, eggs can be stripped from these fish with a lot of help. The fish are placed ventral side up into a large V-shaped padded trough, equipped with a hood over the head, and the fish is incubated and secured gently at the head and caudal fin. The eggs are manipulated gently through massaging first anteriorly (to load the oviduct with ovulated eggs) and then posteriorly to push the eggs out into a large collection vessel. Copious amounts of milt can be collected in fashion similar to the shortnose sturgeon, except a very large syringe or 1000 ml beaker is needed.

Like the shortnose sturgeon, sperm is diluted in fresh water and added to the eggs. The eggs become adhesive after exposure to water and are 'de-glued' using either silt from the river that the eggs are from (sieving and autoclaving before using this technique is strongly recommended) or Fuller's earth. The earth is added in a slurry and the eggs are stirred gently (some people recommend a feather) and then rinsed after they loose their adhesiveness.

The fertilized eggs (~ 500 ml) are then placed in an up-welling MacDonald type jar (7 l) equipped with an overflow screen. Water is added at a high enough rate to keep the embryos in suspension (3–4 l/min). The overflow from the jars is directed into the larvae-rearing tank. After the embryos begin to hatch, the overflow screen is removed, which allows the larvae to flow into the larvae-rearing tank. After 121–140 h at 16–19°C, the larvae begin to hatch (Smith *et al.*, 1980). Survival to hatch is variable as they are prone to fungal infections; however, 70–100% hatch is not uncommon.

Larvae are reared in circular tanks and are negatively phototactic. They swim in polarized shoals against the tank current flow. After 7–10 days posthatch at a temperature of 17–19°C, they can be fed live *Artemia* at a density between

1–2 /ml. The timing of this was thought initially to be critical. However, Hardy (2000) found that they were rather resistant to starvation and their point-of-no-return was up to 42 days at 1°C. Frozen *Artemia* can be used as a substitute (personal observation) when live *Artemia* supplies are interrupted. They can be co-fed within 20 days of the initiation of feeding and can be weaned over a 1 week period or less. Microdiet replacement of *Artemia* size is fine and they can be moved on to a trout crumble within the first few months. Larval and weaning to juvenile survival is high; it is not uncommon to have over 70% of the eggs fertilized to become juveniles. After they are weaned, mortality rates are extremely low.

Giberson and Litvak (2003) examined the importance of feeding frequency on the growth, conversion efficiency and meal size of juvenile Atlantic and shortnose sturgeon. Overall, Atlantic sturgeon grew better, ate more and exhibited greater feeding efficiencies than shortnose sturgeon, regardless of feeding frequency. Atlantic sturgeon exhibited SGRs of 2.3%/day and conversion efficiencies of 100%. They had similar growth and feeding efficiencies in all feeding frequency regimes and this growth rate was achieved in a recirculation system that was not equipped with oxygen injection – we have seen much higher growth rates from this species.

9.4.2.1 Disease

Overinflated swim bladders in 0+ progeny cultured from eggs acquired from wild Atlantic sturgeon broodstock have been observed. This was also reported to occur in the US hatcheries for 0+, 2- and 3-year-old cultured Atlantic sturgeon (NMFS, 1998b). This causes a great deal of distress for the fish, as they float upside down at the surface. This can be relieved through the use of a syringe equipped with a 16–18 gauge needle and inserted through the body wall laterally, withdrawing the gas to relieve the pressure. The cause of this over-inflation has not yet been identified; however, the disease has been treated using injectable oxytetracycline (Liquimycin LA-200; Pfizer, Incorporated, New York) with good results (NMFS, 1998b), suggesting that a bacterial infection may be the root cause.

Cipriano (1996) conducted experimental challenges on Atlantic sturgeon with the bacterium, *Aeromonas salmonicida* (furunculosis), which did infect and kill sturgeon. Experimental challenges have been conducted with herpes virus type-2 (WSHV-2) (Mohler, 2003). Hedrick and McDowell (UC Davis) have used waterborne exposure to produce mortality as well as clinical signs of infection, including haemorrhagic lesions and ulcers on both dorsal and ventral surfaces and particularly around the mouth (cited in Mohler, 2003). It is unclear at this time whether white sturgeon iridovirus (Drennan *et al.*, 2005) will be a problem for Atlantic sturgeon.

9.5 Product Description

Sturgeons have not only been known for caviar; historically, the body was used for oil, and isinglass from the swim bladders and notochord was used to clarify wine and beer. However, today they are known mainly for the most valuable of fish products – caviar. There is an art and history to caviar production. In Russia, the 'ikryanchik', an artisan who dresses the sturgeon and procures and processes the caviar, removes the ovaries and presses them through a mesh screen to separate the eggs from the ovarian folds (Saffron, 2002). Loose salt is then applied at a ratio of 4% per weight of eggs and thus caviar is born (Saffron, 2002). Caviar is classified into three categories: Beluga, Osetra and Sevruga. The designation on caviar packages, *Malossol*, which translates from Russian as 'little salt', indicates the low salt content in high-quality caviar. Caviar is then packed in cans, glass or porcelain and sometimes is pasteurized for longer-term storage. The price on 18 March 2007, reported at Marky's, for these three classes was US$150, US$82 and US$85 per ounce, respectively (Marky's, 2007). In addition to caviar, sturgeon meat is sold as bullets (head-off, gutted) and fillets; value-added products include smoked, sausages, mousse, roulade and jerky. Like other sturgeons reared for their meat, Atlantic and shortnose sturgeon yield is approximately 70%.

9.6 Markets

The market for caviar is huge, as current demand exceeds supply by a large margin. Meat and its value-added products are highly valued and well worth the investment. However, the major problem with sturgeon aquaculture, and for shortnose sturgeon in particular, is the ability to sell it outside of the state in which it is grown. Shortnose sturgeon is listed as endangered in the USA and as a vulnerable species by the IUCN. This is covered by CITES, which is an international agreement between governments. Its aim is to ensure that international trade in specimens of wild animals and plants does not threaten their survival. Shortnose sturgeon is listed in Appendix I, which does not permit trade, and Atlantic sturgeon is listed in Appendix II, which allows regulated trade of these species (CITES Convention, Article VII, Paragraph 4). Shortnose has the potential to be moved to Appendix II if the nation state in which it is grown in an aquaculture setting certifies the facility. Although CITES clearly indicates the path for certification and international trade, not all of the 160+ countries follow the agreement. The Endangered Species Act (ESA) in the USA prohibits the cultivation within the USA and importation of any species its lists as endangered. Fernandes (2005) suggests a re-evaluation of the ESA goal protecting endangered species by prohibiting their cultivation. He suggests that this approach is counterproductive for the conservation of this species. If either the legislation is not changed or shortnose sturgeon is provided an exemption, as was the case with alligators, export into the USA will not be possible, which will restrict the market for shortnose sturgeon caviar.

Although aquaculture production has been identified by the IUCN and the Sturgeon Specialist group as a potential way to reduce the demand on wild-caught caviar, from both legal and illegal sources, the ability of companies to move their product is severely impaired by these regulations.

Exacerbating this issue is the possibility that nation states use these CITES regulations as a trade barrier to 'protect' their own economic interests in sturgeon production.

9.7 Future R & D for Shortnose and Atlantic Sturgeons

The major objective of sturgeon aquaculture is the production of caviar and it is imperative to maximize profit per production unit, suggesting the importance of early gender identification and/or development of all-female populations. The first attempts at control of gender through gynogenetic manipulation indicated that the females of this species were in all likelihood the heterogametic sex (Flynn *et al.*, 2006). If this were true, then a 'super-female' genotype would be a potential result of gynogentic manipulations (Flynn *et al.*, 2006). Future effort in this area is important and may lead to very significant results for caviar production. Until that time, we still need to have a technique that will allow for early gender identification.

There are a number of protocols that need to be assessed and then developed for shortnose and Atlantic sturgeon aquaculture, including plasma hormones (T, KT, E2) and VTG, boroscopy (Kynard and Kieffer, 2002), ultrasonography (Colombo *et al.*, 2004) and laparoscopic surgery (Hernandez *et al.*, 2004). Development of a predictive relationship between plasma hormone production and egg developmental stage will permit aquaculturists to track the status of each individual fish in culture in order to know when to harvest the caviar (see Cuisset *et al.*, 1991;Van Eenennaam *et al.*, 1996; Linares-Casenave *et al.*, 2003). Currently, this is done by biopsies and egg staging, a very labour-intensive and potentially riskier approach for the fish than a simple blood sample. Feist *et al.* (2004) were able to identify gender in white sturgeon through plasma steroids (T, KT and E2) at 21 months of age, which is 1–2 years earlier than the surgical techniques used in Calfornia farms.

Presently, commercial operations use feeds designed for salmonids (Hung *et al.*, 1997). These salmonid diets may not be optimal for sturgeon (Hung and Deng, 2002), neither for promoting fast growth, nor proper flesh and caviar properties at harvest. In addition, excess or limited nutrients from poorly designed diets can lead to excess fish wastes entering the environment due to preferential requirement for energy over growth by the fish. Steve Leadbeater (MSc student with Santosh Lall NRC/IMB and Litvak UNB) is trying to obtain an accurate representation of protein, lipid, FA composition, AA profiles, mineral composition, vitamin and trace metal requirements for shortnose sturgeon by examining their egg and body composition in order to determine the nutritional needs of the shortnose sturgeon and use this information to build cost-effective diets. He is currently testing six diet formulations for feed development.

Broodstock for both species are long-lived, large and not always available in large numbers; therefore, it is important to develop an understanding of their genetics, specifically at the facility level, as F1s are generated and crossed. The ability to avoid inbreeding will become more important as the industry ages because of the limited space to hold such large fish, and this information should be developed now to ensure the industry's long-term health.

References

Atz, J.W. and Smith, C.L. (1976) Hermaphroditism and gonadal teratoma-like growths in sturgeon (*Acipenser*). *Bulletin of Southern California Academy of Sciences* 75, 119–126.

Bain, M.B. (1997) Atlantic and shortnose sturgeons of the Hudson River: common and divergent life history attributes. *Environmental Biology of Fishes* 48, 347–358.

Bemis, W.E., Findeis, E.K. and Grande, L. (1997) An overview of Acipenseriformes. *Environmental Biology of Fishes* 48, 25–71.

Buckley, J. and Kynard, B. (1981) Spawning and rearing of shortnose sturgeon from the Connecticut River. *Progressive Fish Culturist* 43, 74–76.

Buckley J. and Kynard, B. (1985) Yearly movements of shortnose sturgeon in the Connecticut River. *Transactions of the American Fisheries Society* 114, 813–820.

Cipriano, R.C. (1996) Infection of Atlantic sturgeon (*Acipenser oxyrinchus*) with the bacterial fish pathogen *Aeromonas salmonicida*, cause of furunculosis. Abstracts of 52nd Annual Northeast Fish and Wildlife Conference, Farmington, Connecticut. Call Number: 96-095// PY96//FH.

Colombo, R.E., Wills, P.S. and Garvey, J.E. (2004) Use of ultrasonography to determine sex of shovelnose sturgeon. *North American Journal of Fisheries Management* 24, 322–326.

Conte, F.S., Doroshov, S.I., Lutes, P.B. and Strange, E.M. (1988) *Hatchery Manual for the White Sturgeon* (Acipenser transmontanus *Richardson) with Application to Other North American Acipenseridae.* US Fish and Wildlife Service 3322, Publications Division of Agriculture and Natural Resources, University of California, Oakland, California, 104 pp.

COSEWIC (2005) *COSEWIC Assessment and Update Status Report on the Shortnose Sturgeon* Acipenser brevirostrum *in Canada*. Committee on the Status of Endangered Wildlife in Canada, Ottawa (www.sararegistry.gc.ca/status/status_e.cfm, accessed March 2007).

Cuisset, B., Pelissero, C., Le Menn, F. and Nunez-Rodriguez, J. (1991) ELISA for Siberian sturgeon (*Acipenser baeri* Brandt) vitellogenin. In: Williot, P. (ed.) *Acipenser. Proceedings of the First International Sturgeon Symposium*, 3–6 October 1989. CEMAGREF, Bordeaux, France, pp. 107–111.

Dadswell, M.J. (1979) Biology and population characteristics of the shortnose sturgeon, *Acipenser brevirostrum*, LeSeur 1818 (Osteichthyes: *Acipenseridae*), in the Saint John river estuary, New Brunswick, Canada. *Canadian Journal of Zoology* 57, 2186–2210.

Dadswell, M.J. (1984) Status of the shortnose sturgeon, *Acipenser brevirostrum*, in Canada. *Canadian Field Naturalist* 98, 75–79.

Dadswell, M.J. (2006) A review of the status of Atlantic sturgeon in Canada, with comparisons to populations in the United States of America. *Fisheries* 31(5), 218–229.

Dadswell, M.J., Taubert, B.D., Squiers, T.S., Marchette, D. and Buckley, J. (1984) *Synopsis of Biological Data on Shortnose Sturgeon,* Acipenser brevirostrum *LeSueur 1818*. National Oceanographic and Atmospheric Administration Technical Report NMFS 14, Washington, DC, 45 pp.

Denys, N. (1968) *The Description and Natural History of the Coasts of North America (Acadia)*. Champlain Society Publication 2. Greenwood Press, New York, 625 pp.

Dilauro, M.N., Kaboord, W.S. and Walsh, R.A. (1999) Sperm-cell ultrastructure of North American sturgeons. II. The shortnose sturgeon (*Acipenser brevirostrum* LeSueur, 1818). *Canadian Journal of Zoology* 77, 321–330.

Dilauro, M.N., Kaboord, W.S. and Walsh, R.A. (2000) Sperm-cell ultrastructure of North American sturgeons. II. The lake sturgeon (*Acipenser fulvescens* Rafinesque, 1817). *Canadian Journal of Zoology* 78, 438–447.

Drennan, J.D., Ireland, S., LaPatra, S.E., Grabowski, L., Carrothers, T.K. and Cain, K.D. (2005) High-density rearing of white sturgeon *Acipenser transmontanus* (Richardson) induces white sturgeon iridovirus disease among asymptomatic carriers. *Aquaculture Research* 36, 824–827.

FAO Fishery Information, Data and Statistics Unit (2006) Aquaculture production: values 1984–2004. FISHSTAT Plus – universal software for fishery statistical time series [online or CD-ROM]. Food and Agriculture Organization of the United Nations (http://www.fao.org/fi/statist/FISOFT/FISHPLUS.asp, accessed March 2007).

Feist, G., Van Eenennaam, J.P., Doroshoc, S.I., Schreck, C.B., Schneider, R.P. and Fitzpatrick, M.S. (2004) Early identification of sex in cultured white sturgeon *Acipenser transmontanus*, using plasma steroid levels. *Aquaculture* 232, 581–590.

Fernandes, S.M. (2005) A market approach to sturgeon conservation under the US endangered species act. *Fisheries* 30(12), 20–27.

Flynn, S.R., Matsuoka, M., Reith, M., Martin-Robichaud, D.J. and Benfey, T.J. (2006) Gynogenesis and sex determination in shortnose sturgeon, *Acipenser brevirostrum* Lesuere. *Aquaculture* 253, 721–727.

Giberson, A. and Litvak, M.K. (2003) Effects of feeding frequency on growth, conversion efficiency and meal size in Atlantic (*Acipenser oxyrinchus*) and shortnose sturgeon (*A. brevirostrum*) juveniles. *North American Journal of Aquaculture* 65(2), 99–105.

Hall, J.W., Smith, T.I.J. and Lamprecht, S.D. (1991) Movements and habitats of shortnose sturgeon, *Acipenser Brevirostrum,* in the Savannah River. *Copeia* 1991(3), 695–702.

Hardy, R. (2000) Factors influencing survival, development, and behaviour of shortnose (*Acipenser brevirostrum*) and Atlantic (*A. oxyrinchus*) larvae. MSc thesis, University of New Brunswick, Saint John, New Brunswick, Canada.

Hardy, R. and Litvak, M.K. (2004) The effect of temperature on the early development of larval shortnose and Atlantic sturgeon. *Environmental Biology of Fishes* 70, 145–154.

Hernandez, S.J., Bakal, R.S., Hickson, B.H., Rawlings, C.A., Wilson, H.G., Radlinsky, M., Hernandez-Divers, S.M. and Over, S.R. (2004) Endoscopic sex determination and gonadal manipulation of Mexico gulf sturgeon (*Acipenser oxyrinchus desotoi*). *Journal of Zoo and Wildlife Medicine* 35(14), 459–470.

Hung, S.S.O. and Deng, D.F. (2002) Sturgeon, *Acipenser* spp. In: Webster, C.D. and Lim, C. (eds) *Nutrient Requirements and Feeding of Finfish for Aquaculture.* CAB International, Wallingford, UK, pp. 344–357.

Hung, S.S.O., Storebakken, T., Cui, Y., Tain, L. and Einen, O. (1997) High-energy diets for white sturgeon, *Acipenser transmontanus* Richardson. *Aquaculture Nutrition* 3, 281–286.

IUCN (2006) IUCN Red List of Threatened Species (http://www.iucnredlist.org/, accessed 21 December 2006).

Kieffer, M. and Kynard, B. (1993) Annual movements of shortnose and Atlantic sturgeon in the Merrimack River, Massachussets. *Transactions of the American Fisheries Society* 122, 1088–1103.

Kynard, B. (1997) Life history, latitudinal patterns, and status of the shortnose sturgeon, *Acipenser Brevirostrum. Environmental Biology of Fishes* 48, 319–334.

Kynard, B. and Kieffer, M. (2002) Use of a borescope to determine sex and egg maturity of sturgeon and the effect of borescope use on reproductive structures. *Journal of Applied Ichthyology* 18, 505–508.

Linares-Casenave, J., Kroll, K.J., Van Eenennaam, J.P. and Doroshov, S.I. (2003) Effect of ovarian stage on plasma vitellogenin and calcium in cultured white sturgeon. *Aquaculture* 221, 645–656.

Ludwig, A., Debus, L., Lieckfeldt, D., Wirgin, I., Benecke, N., Jenneckens, I., Williot, P., Waldman, J.R. and Pitra, C. (2002) When the American sea sturgeon swam east. *Nature* 419, 447–448.
Marky's (2007) http://www.markys.com/caviar/customer/home.php (accessed March 2007).
Mohler, J.W. (2003) *Culture Manual for the Atlantic Sturgeon*, Acipenser oxyrinchus oxyrinchus. Region 5 US Fish and Wildlife Service Publication. US Fish and Wildlife Service, Northeast Region, Hadley, Massachusetts.
NMFS (1998a) Recovery plan for the shortnose sturgeon (*Acipenser brevirostrum*). Prepared by the Shortnose Sturgeon Recovery Team for the National Marine Fisheries Service, Silver Spring, Maryland, 104 pp.
NMFS (1998b) *Status Review of Atlantic Sturgeon* (Acipenser oxyrinchus). National Marine Fisheries Service, North-east Regional Office, Gloucester, Maryland, 174 pp.
O'Herron, J.C., Able, K.W. and Hastings, R.W. (1993) Movements of shortnose sturgeon (*Acipenser Brevirostrum*) in the Delaware River. *Estuaries* 16, 235–240.
Pottle, R. and Dadswell, M. (1979) Studies on larval and juvenile shortnose (*Acipenser brevirostrum*). Report to the North-east Utilities Company, Hartford, Connecticut, 87 pp.
Sanders, B.J., Fadel, J.G. and Wade, E.M. (2003) Economic optimization modeling of white sturgeon (*Acipenser transmontanus*) caviar and meat production under different management conditions. *Aquaculture* 217, 409–430.
Saffron, I. (2002) *Caviar: The Strange History and Uncertain Future of the World's Most Coveted Delicacy*. Broadway Books, New York, 270 pp.
Schmitz, R.J. (1998) Comparative ultrastructure of the cellular components of the unconstricted notochord in the sturgeon and the lungfish. *Journal of Morphology* 236, 75–104.
Smith, T.I.J. and Clugston, J.P. (1997) Status and management of Atlantic sturgeon, *Acipenser oxyrinchus*, in North America. *Environmental Biology of Fishes* 48, 335–346.
Smith, T.I.J., Dingley, E.K. and Marchette, D.E. (1980) Induced spawning and culture of Atlantic sturgeon. *Progressive Fish Culturist* 42, 147–151.
Stein, A.B., Friedland, K.D. and Sutherland, M. (2004) Atlantic sturgeon marine by-catch and mortality on the continental shelf of the northeast United States. *North American Journal of Fisheries Management* 24, 171–183.
Van Eenennaam, J.P. and Doroshov, S.I. (1998) Effects of age and body size on gonadal development of Atlantic sturgeon. *Journal of Fish Biology* 53, 624–637.
Van Eenennaam, J.P., Doroshov, S.I., Moberg, G.P., Watson, G.J., Moore, D.S. and Linares-Casenave, J. (1996) Reproductive conditions of the Atlantic sturgeon (*Acipenser oxyrinchus*) in the Hudson River. *Estuaries* 19, 769–777.
Vescei, P., Litvak, M.K., Noakes, D.L.G., Rien, T. and Hochleithner, M. (2003) A non-invasive technique for determining sex of live adult North American sturgeon. *Environmental Biology of Fishes* 68, 333–338.
Vladykov, V.D. and Greeley, J.R. (1963) Family *Acipenseridae*. In: Olsen Y.H. (ed.) *Fishes of the Western North Atlantic*. Memoirs of the Sears Foundation for Marine Research 1(3). Yale University, New Haven, Connecticut, pp. 24–60.
Williot, P., Brun, R., Rouault, T., Pelard, M., Mercier, D. and Ludwig, A. (2005) Artificial spawning in cultured starlet sturgeon, *Acipenser ruthenus* L., with special emphasis on hermaphrodites. *Aquaculture* 246, 263–273.

10 Milkfish (Family: Chanidae)

Cheng-Sheng Lee[1] and Chin-Fa Liu[2]

[1]*Oceanic Institute, Waimanalo, Hawaii, USA;* [2]*Team Aqua Corporation, Shinying City, Tainan, Taiwan, ROC*

10.1 General Introduction

Milkfish, *Chanos chanos*, is the only species in the family Chanidae. It is characterized by a compressed and oblong body and a small, toothless and terminal mouth (Fig. 10.1). The caudal fin is deeply forked; the dorsal and pelvic fins opposite are small to moderate. It has cycloid scales and a distinct lateral line. The overall colour is silvery with bluish or olive tints on the back. Milkfish have a broad geographic distribution virtually throughout the entire tropical Indo-Pacific Ocean (Fig. 10.2). Milkfish are found as far west as the Red Sea and along the coast of Africa, far east to the Pacific waters off Central America, far north to southern Japan, Hawaii and Mexico, and far south to southern Australia, New Zealand and the Norfolk Islands (Fowler, 1959; Schuster, 1960; Rabanal and Ronquillo, 1975). There are many different scientific and local common names given to milkfish (Lee, 1995). The genetic population structure is distinctly diverse in the Pacific, as indicated by biochemical (Winans, 1980), molecular (Ravago-Gotanco and Juinio-Menez, 2004) and morphological studies (Senta and Kumagai, 1977; Winans, 1985). The natural stocks of milkfish are isolated and do not interchange between different locations, even when as close as the islands of Oahu and Hawaii in the Hawaiian Islands chain.

The milkfish is not a targeted species in capture fisheries. The annual yield of milkfish from capture was only 1479 Mt in 2004 (FAO, 2006). Milkfish are euryhaline fish and tolerate a wide range of salinity from fresh water to hypersaline water (125‰). Their tolerance to extreme environmental conditions, as well as their reliance on lower trophic level in food chains make milkfish a good aquaculture species.

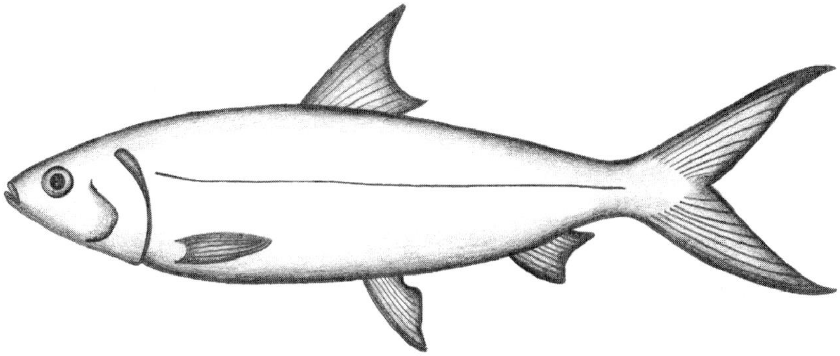

Fig. 10.1. Chanidae.

10.2 Farming of Milkfish

Milkfish culture began over 700 years ago in Indonesia (Schuster, 1952; Ronquillo, 1975), at least 400 years ago in Taiwan and the Philippines and approximately 300 years ago in Hawaii (Ling, 1977). Commercial milkfish farming has been practised continually in Indonesia, the Philippines and Taiwan, with a record total production of 573,732 Mt in 2004 (Fig. 10.3). The total milkfish production from aquaculture fluctuated between 300,000 and 400,000 Mt from 1981 to 1998, but has increased continually since then (Fig. 10.3, FAO 2006). Small-scale farming at several other locations is also practised for subsistence purposes. Milkfish are important traditional food fish in three major milkfish-farming countries and several Pacific Islands, but has a limited market. Milkfish farming is the second most important culture species after shrimp in Indonesia (Sugama, 2007). It is a bony fish and is not desirable to new consumers. The industry finds it difficult to expand the market. Most milkfish are consumed locally in Indonesia and the Philippines and less than 30% is exported to other countries from Taiwan. Any excess supplies immediately drive down the market price (Su *et al.*, 2002). With the increasing opportunity of culturing other high-valued aquatic species, many milkfish farms were converted to other fish farms. Milkfish farming is facing another challenge of decreasing farming areas (Su *et al.*, 2002). According to our survey in 2005, the farming area in Taiwan was 50,000 ha and 280,000 ha in the Philippines.

To resolve various challenges, farming technology has improved greatly from the traditional extensive practice to intensive farming to improve unit production by a 5- to 8-fold increase. Following the establishment of milkfish hatchery technology in Indonesia in the 1990s and the recovery of production to the 1980s level in the Philippines, total milkfish production showed an annual increase of approximately 7.1% between 1998 and 2004. However, the farm gate unit value decreased as production increased (FAO, 2006). The average unit value from 1993 to 1998 was US$1799.60 and was US$1269.10

Fig. 10.2. World distribution of *Chanos chanos*.

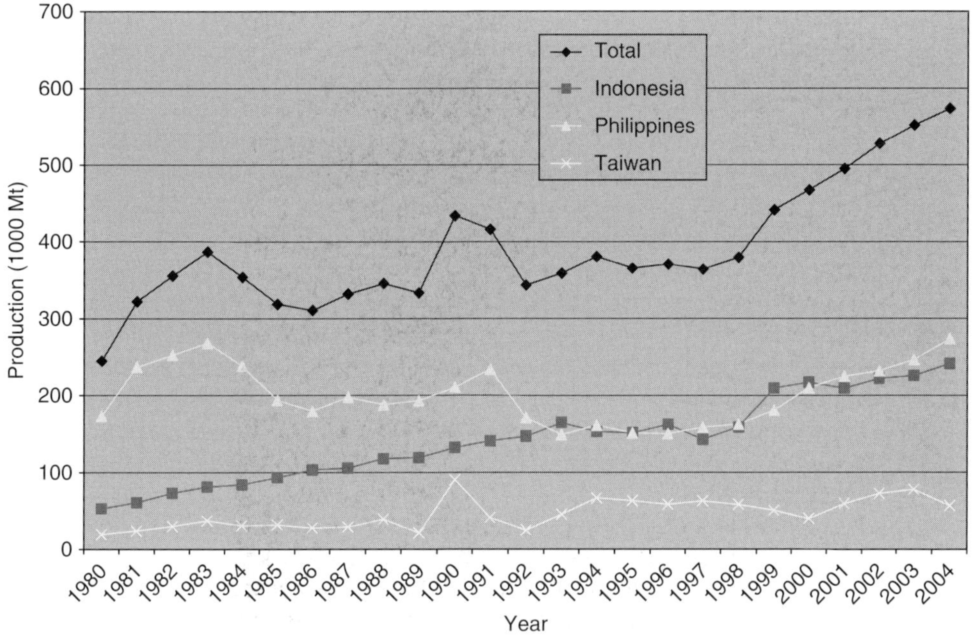

Fig. 10.3. Total milkfish production (in Mt) from aquaculture from 1980 to 2004 (FAO, 2006).

from 1999 to 2004, based on production and value data from the FAO (2006). Apparently, expansion of milkfish markets is essential to stabilize milkfish prices and to encourage more milkfish production.

10.3 Broodstock Management and Hatchery Operations

Analyses of gut contents to determine the food preferences and nutritional requirements of milkfish in the wild have been conducted by several studies (see review by Lee, 1995). Liao (1971) reported that crustaceans were found in the gut of wild-caught adult milkfish. Other feeding and gut analyses studies (Chacko, 1949; Tampi, 1958; Schuster, 1960; Poernomo, 1976; Villaluz *et al.*, 1976; Vicencio, 1977; Crear, 1980) also revealed planktonic organisms and fish larva in the gut of adult milkfish in the wild. Small-sized milkfish ingested diatom, blue-green algae and filamentous green algae, and different types of feed were eaten as fish got larger (see review by Lee, 1995). Therefore, milkfish can be grouped as omnivorous fish, although they are treated as herbivorous fish in farming practices.

A diet with a crude protein content of 32% could meet the minimal nutritional requirements of broodstock, although a balanced diet with 40% protein content is recommended (Lee, 1995). Samsi (1979) reported that an animal protein-based diet gave better growth, but Shiau *et al.* (1988)

indicated replacement of up to 67% of fishmeal with plant protein did not cause an adverse effect on growth. Seneriches and Chiu (1988) also recommended that, for optimal growth, a formulated diet for milkfish fry should contain not less than 15% of the protein from fishmeal. Lysine, leucine and arginine are most likely to be the limiting amino acids (Coloso *et al.*, 1983). The dietary lipid requirement was estimated to be about 7–10% (Alava and de la Cruz, 1983).

The growth of milkfish has been documented from the time of hatch to adult fish in the wild based on information from culture and samples from the wild (see review by Lee, 1995). The growth of milkfish less than 2 years old was based on actual measurement during the farming period. The growth of adult wild-caught milkfish was estimated through age determination by vertebral ring (Kumagai, 1990). Kumagai (1990) documented sexual difference in growth and showed that female fish tended to be larger among the same age group in the wild. However, there was no report on sexual difference in size among farm fish.

Milkfish appear to be a gonochoristic species, a conclusion based on gonad morphology and the absence of hermaphroditism. Until now, there were no easily identifiable morphological differences between males and females. Histologically, the milkfish testis is a lobular type and the ovary is a 'cystovarian' type, as in most teleost fishes. Fecundity increases as fish size increases and varies from less than 1 million eggs to as many as 6 million (Bagarinao, 1991). Gonad development has been described on the basis of macroscopic appearance, histological examination and gonadosomatic index (Liao and Chen, 1984; Tan, 1985). The gonadosomatic index (GSI) for males is less than for females. GSI range varies according to the stage of maturity, genetic variation or environmental conditions. The highest GSI for female milkfish was reported on Christmas Island as 25 (Crear, 1980) and 5.4 for male milkfish in Hawaii (Kuo and Nash, 1979). The minimal mature age in captivity was suggested to be 5 years for males and 6 years for females (Lee, 1995). Lee *et al.* (1986b) stated the minimal size for maturation should be 3 kg or more. As biological and non-biological parameters change, the required age and size can be changed.

It has been confirmed that milkfish are capable of spawning more than once during the annual spawning season (Lee *et al.*, 1986a). The average daily increase of oocyte diameter was between 21.4 and 27.4 μm (Lee *et al.*, 1986c). Therefore, a fish with an oocyte diameter of 350 μm could reach the mature stage (750 μm in diameter) for hormonal induction of maturation in 30 days. The spawning season falls in the months of warmwater temperatures and long photoperiods. The season lasts almost throughout the whole year in Indonesia and the southern Philippines because of year-round warm weather conditions (Lee, 1995). Stimulation of photoperiod change might be necessary for maturation, based on preliminary studies showing no maturation under a constant long photoperiod regime (Lee *et al.*, 1987). A longer spawning season allows the production of milkfish fry throughout the year and has positive impacts on production. Mature milkfish were found in natural environments with salinity ranging from 8 to 125‰. However, spawning was not reported or documented at extreme salinity conditions.

Milkfish farming has been reviewed recently by several authors (Liao, 1991; Lee, 1995; Bagarinao, 1998; Stickney, 2000). Milkfish can be farmed in fresh water, brackish water and seawater. Bautista *et al.* (1991) found that fish reared in brackish water had a higher content of highly unsaturated fatty acid than those reared in freshwater conditions. Farming technology has been modified as other culture-related technology has advanced. This report highlights the farming technology from hatchery production to final harvest and product-processing technology. Available detailed information will not be repeated in this document. Information on current farming practices provided by one of the authors of this chapter (C.-F. Liu) is included to give state-of-the-art commercial operation.

A reliable source of healthy and mature broodstock is essential for the successful operation of artificial propagation. An initial attempt to collect mature broodstock in the wild for propagation was proved to be unreliable. Following the initial success of induced spawning of pond-reared broodstock (Liao and Chen, 1979; Lin, 1982) and the report on natural spawning (Lin, 1984, 1985), maturation of milkfish in captivity was confirmed. The maturation facility, ranging from an 80 m^2 cage to a 1500 m^2 dirt pond (Lee, 1995), could be stocked with milkfish of 3 kg in body weight or more at one fish per 2 m^2 and with a sex ratio of 2–3 females to 1 male. Adult milkfish can be collected from the culture pond or ocean. Diets for broodstock vary among different farms and are kept confidential and unavailable to the public. A variety of feeds including rice grain, wheatmeal, soybean meal, formulated eel feeds and trash feed were used to obtain the first case of natural spawning (Lin, 1985). Emata *et al.* (2000) found broodstock given a 36% protein-formulated diet supplemented with 0.1% vitamin C and 0.05% vitamin E or vitamin C alone for 3 years had a higher percentage of spawns with higher percentage (> 90%) egg viability, hatching and cumulative survival than a no vitamin supplement group. Based on the information available, Lee (1995) concluded a diet containing 41% crude protein with more than 20% from animal sources should be sufficient for maturation under natural environmental conditions. After fish reached mature size (3 kg or more), age (preferably 5 years or older) and acclimated to culture conditions, they would mature or respond to induction of maturation. There were no defined ranges of temperature and photoperiod for maturation. Based on the studies and our observations, the water temperature should be 27°C or higher and long photoperiod.

Hormonal induction of maturation of female milkfish via intramuscular (IM) implantation has been reported by Lee *et al.* (1986b). An LHRH-a cholesterol pellet (containing 20 μg LHRH-a) either alone or in combination with a 17α-methyltestosterone (17-MT) silastic capsule (containg 250 μg 17-MT) once a month resulted in oocyte growth from 250 μm to 755 μm or more in diameter in 1–2 months. The same fish can mature more than once in a season, based on Lee's report (1995), showing the same fish matured up to seven times in one season.

Both induced and natural spawning technologies were studied further and established by many researchers in the 1980s (see review by Lee, 1995). Hormone therapy includes fish pituitary, human chorionic gonadotropin (HCG) or LHRH-a. Lee *et al.* (1986c) demonstrated that spontaneous spawning of milkfish occurred following a single LHRH-a cholesterol pellet implantation at a dosage of 41.7 ± 3.3 μg/kg body weight or a single injection of 58.7 ± 9.3 μg/kg body weight LHRH-a. Further studies indicated that the stage of maturity most receptive to an acute LHRH-a therapy was that of an average oocyte diameter of 750 μm or more, with a single oocyte diameter distribution mode or two modes, but not exceeding 400 μm for the second mode (Tamaru *et al.*, 1988). However, spontaneous spawnings did not result in fertilized spawns at all times due to the failure of gametes release by males at the right time. Spontaneous spawning did not always trigger the spawning behaviour of males. Lee *et al.* (1986c) concluded that a higher chance of fertilized spawn and high fertilization rate could be obtained after males received a 200 μg/kg LHRH-a injection.

Compared to induced spawning, natural spawning is a more reliable source of obtaining fertilized eggs and is currently practised in all milkfish hatcheries. Natural spawning usually takes place after sunset from 7 p.m. to 4 a.m. and does not relate to the lunar cycle (Lin, 1984; Lee *et al.*, 1986b). Fertilized eggs are collected with a fixed V-shaped plankton net. Water current is created by several paddle wheels in the maturation pond to lead floating fertilized eggs into the pocket at the centre of the V-shaped plankton net. The spawning season in Taiwan usually lasts from May to August, but almost year-round in the Philippines and Indonesia. Although spawnings were not found every day, it was common to have at least 25 spawnings each month, based on our survey. To use all the collected fertilized eggs effectively, a group of hatcheries works together and shares resources to reduce the production cost further, as described by Lee *et al.* (1997). As the total production of fertilized eggs from a broodstock holding facility exceeds the need of a single hatchery, having one broodstock-holding facility provides for the needs of several hatcheries and can reduce the overall hatchery production cost. This model has been adapted by backyard hatcheries in Indonesia (Sugama, 2007) and makes Indonesia a major milkfish fry provider in South-east Asia.

Milkfish fertilized eggs (about 1.10–1.25 mm) can be incubated in various sizes of containers, depending on the individual hatchery and different practices. The outdoor larval rearing system uses tanks or a net cage (5 × 5 × 1 m each) to incubate fertilized eggs. The size of incubation tanks can range from 1 t to 5 t. Each net cage can stock 3–5 kg of eggs in weight and each kg of egg mass has 750,000–800,000 eggs. Fertilized eggs sink to the bottom of the tank and lose buoyancy when the salinity of the incubation water is 30‰ or below. The optimal salinity range for incubation is 30–40‰. Abnormal larvae were found when salinity was lower than 20‰. The recommended aeration was 30–40 ml/min and the best stocking density was 400 eggs/l or less during incubation. For the best hatching success, unfertilized eggs should be discarded from the incubation tank and all fertilized eggs suspended in the water column. Incubation temperature should be kept between 23 and 30°C. At 28°C and 35‰, hatching time was about 28 h and the average total length of newly hatched larvae was 3.7 mm (Walsh *et al.*, 1991). Newly hatched larvae have large yolk sacs of approximately 2.20 mm in length and 0.28 mm in width.

Milkfish larval rearing follows standard marine finfish hatchery operation procedure that has been studied and reviewed (Liao *et al.*, 1979; Eda *et al.*, 1990; Lee, 1995; Liu and Kelley, 1995). Rotifers, brine shrimp nauplii and formulated feed are the major feed items. Gapasin and Duray (2001) reported that enrichment of live feed with docosahexaenoic acid (DHA) reduced the percentage of operculum deformity significantly. Microbound diet and flake using a drum drier were proved to have better buoyancy, water stability and particular size (Borlongan *et al.*, 2000). They further confirmed that milkfish larvae be co-fed artificial diet with rotifers from the beginning or day 8 and fed alone from day 15 onwards.

Based on indoor hatchery technology, an outdoor hatchery technology was developed in Taiwan (Chang *et al.*, 1993). Outdoor hatchery technology has proved to be the most cost-effective way of producing milkfish fry (Lee *et al.*, 1997) and is currently practised in major milkfish-farming countries. Milkfish fry production from hatcheries has become the main source of fry for farming and has replaced wild-caught fry almost completely. With the establishment of hatchery technology, the milkfish industry in Taiwan has been able to change from a fry importer to an exporter in the 1990s (Su *et al.*, 2002). However, hatchery-bred fry usually have more incidences of morphological abnormality than wild-caught fry. The predominant abnormalities are a cleft on the branchiostegal membrane and a deformed operculum (Hilomen-Garcia, 1997). Besides slow growth and development, abnormal fry suffered higher mortality during handling and transfer of stock.

Outdoor hatchery technology uses natural productivity in outdoor ponds (range from 200–300 m^2 to 1000–2000 m^2 in size or 1–1.2 m depth) to rear newly hatched larvae. Additional rotifers are added if the rotifer density in ponds is too low. Fertilized eggs can be stocked directly in larval-rearing ponds and hatch in 45–50 h at a temperature of 28–30°C. The common initial larval stocking density is 2–3 larvae/l, or 10–15% of the stocking density at indoor hatchery. Rotifers enriched with highly unsaturated fatty acids are introduced with phytoplankton to the rearing pond on the second day after hatching. Rotifer density is kept between 5–8 individuals/ml. The transparency of the pond water is kept at 40–50 cm. From day 10 onwards, ground formulated feed is given four times daily in addition to rotifers and is increased gradually according to fish development. Depending on water temperature and feeding status, fish of a total length of 1.1 cm or larger could be harvested at day 20 after hatch, or even earlier. Compared to indoor intensive hatchery operations (Eda *et al.*, 1990), outdoor hatchery could shorten hatchery time by up to 10 days. The survival rate varies according to the operator's skill and can be expected to reach 30% or more. Before fry are transported to a distant farm, they are stocked in clean water for 15–20 h and grouped into classes of 2–3 sizes. After fry reach the farm site, they are cared for intensively in a nursery pond until they reach 3 g or the desired size. The inclusion of a nursery phase after hatchery will also lower the incidence of deformities in grow-out ponds (Sumagaysay *et al.*, 1999).

10.4 On-growing to Market Size

Milkfish grow-out systems reviewed in many reports include shallow water, deep-water pen culture and cage culture systems (Lee and Banno, 1990; Liao, 1991; Lee, 1995; Cruz, 1998; De la Vega, 1998; Marte *et al.*, 2000; Stickney, 2000). Milkfish also can be monocultured or polycultured with other aquatic species to use the space and food in ponds fully. Selection of the type of culture system for the operation should consider various factors carefully, from the technical competency of the workers, infrastructure to support the operation, the ambient environmental conditions and the market status. The advantages and disadvantages of different milkfish culture systems were compared and reported by Su *et al.* (2002).

Traditional milkfish farming is a shallow-water system or extensive culture system and relies on productivity from the pond. Benthic algae or 'lab-lab' produced in the pond are the natural food for milkfish. 'Lab-lab' is the term used in the Philippines for algal mats and all associated with microorganisms. Farming procedures including pond preparation, seedstock management and harvest schedule are well established in Taiwan according to local environmental conditions (Liao, 1991). Basically, the milkfish pond is sun-dried until the soil cracks, then filled with 5–20 cm of water and sun-dried again. This procedure has to be repeated several times to stimulate the growth of benthic algae. Finally, seawater of 25–34 ‰ salinity is added to the pond at 20–40 cm depth and fertilizers are added to maintain the growth of benthic algae. The continuous growth of benthic algae is critical for the success of shallow-water farming. In April, or when the ambient temperature rises above 22°C, overwinter fish with a body weight of 25–30 g/fish and/or 3–5 g/fish are stocked at 0.15–0.2 fish/m^2. New fry (1.3–1.5 cm in body length) are stocked in May and thereafter at 0.25–0.3 fish/m^2 monthly, depending on management strategy. Generally, the total stocking density is about 7000 fish/ha. Fish are harvested at size 250–300 g from June and thereafter (Fig. 10.4). New fry are stocked after each harvest throughout the growing season. The total stocking number depends on the availability of new fry and the primary productivity of the pond. Typically, 1800–2200 kg of fish can be harvested from each ha. This type of operation is environment friendly but has been replaced gradually by the intensive culture system or other type of operation for better economic return.

Once, milkfish pen culture in the Philippines took advantage of high primary productivity in Laguna de Bay and reached a productivity as high as 6–7 t/ha/year, but reduced to 1 t/ha/year after the farm area exceeded the carrying capacity of the Bay (Bagarinao, 1998). The water depth at Laguna de Bay is less than 5 m. Fish pens can be constructed using posts cut from bamboo, coconut palm or wood to enclose an area from 1–5 ha with a circular, square or rectangular shape. Inside the posts, nylon and kuralon nets or bamboo mats are used to surround the area to prevent the fish from escaping. Because of high profit return and no operation regulations, farming areas increased exponentially and production reached 82,000 Mt in 1983–1984. Consequently, mass fish kills occurred due to overcrowding and deterioration

Fig. 10.4. Milkfish harvest from a pond production in Taiwan.

of water quality. Thereafter, the farming area reduced and unit production decreased to accommodate environmental change.

To utilize open-water areas, milkfish were cultured in sea cages at high density and used only 1/30th the area of an extensive pond (Cruz, 1998). Production from a 1000 m^2 cage with a 6-m depth reached 5.7 t after 138 days (De la Vega, 1998). Marte *et al.* (2000) reported that a Norwegian-made circular-type cage was used for farming milkfish at stocking density typically of 30–60 pieces/m^3. It yielded 15–25 kg/m^3, with a 70–80% survival rate. Total milkfish cage production in the Philippines accounted for 90% of the total production from marine cages, or overall 10,000 Mt/year from approximately 1000 cages.

In the mid-1970s, a milkfish deep-water culture system was developed in response to technical and socio-economic changes such as availability of formulated feed and aeration devices, limited farm areas and manpower, as well as the necessity of increasing unit area production. The basic difference of a deep-water system from a shallow-water system includes 2–3 m water depth, aeration equipment and automatic feeding of formulated feed. Because of the advanced support infrastructure, more fish (2–5 times that of a shallow-water system) can be stocked in deep-water systems. The stocking density ranges from 1.2 to 2 fish/m^2. The deep water not only allows a high stocking density, but also provides a shelter area during the Taiwan cold season. The salinity of pond water ranges from fresh water to normal strength seawater. However, it is better to increase salinity to 15‰ or higher before harvesting for quality consideration if the salinity of rearing water is less than 5‰. Water transparency should be kept around 20–30 cm to increase dissolved oxygen during daytime. Paddle wheels should be used if dissolved oxygen (DO) is lower than 5 ppm. For every 800–1000 kg target amount of harvesting, one 1 hp paddle wheel is recommended. Pellet feed (with 23–27% crude protein) is provided daily at 12% and gradually reduced to 0.5% as fish grow larger. Many factors, such as weather and pond conditions, affect the appetite of fish. It is important to keep a close watch on feeding behaviour. Feeding time is from 8 a.m. to 4 p.m. daily. Feeding should be stopped if fish do not come to feeding sites in 15 min or less. The number of automatic feeders can be calculated based on one feeder for every 1000 fish. The feed conversion rate was 1.5–1.6. From stocking (average body weight 45 g/fish) to harvest (average body weight 600g/ fish) it takes 4–5 months, with an average survival rate of 95%.

Polyculture of milkfish has also been practised to increase the total return from fish ponds. The selection of species to be polycultured with milkfish would not compete for food and space. Red seaweed, agarophyte, grey mullet, tiger prawn, *Penaeus mondon* and *P. indicus*, and all male Nile tilapia were reported to have positive returns when they were polycultured with milkfish (Eldani and Primavera, 1981; Guanzon *et al.*, 2004). Polyculture, however, requires more skill and working experience from operators.

Disease outbreaks reported in farming milkfish were reviewed by Lio-Po (1984) and Lee (1995), but has not been a major issue up to now. There have been no virus-related diseases reported. However, operators should not overlook the importance of health issues associated with farming practice because disease issues have threatened the survival of many farming practices in other species. Milkfish fry have been inter-transferred among countries. Unwanted pathogens could be passed to other countries unintentionally, such as happened in the marine shrimp industry.

Many economic studies on milkfish farming were conducted between the mid-1970s and the mid-1980s and concluded that the price of natural-caught fry was the most important cost item (Shang, 1986, reviewed by Lee, 1995). The fluctuation of milkfish fry supplies from the wild affects not only availability but also price and final total production. Many of the conditions on which those studies were based have changed over the years. Milkfish hatchery technology has been established and provides a reliable source of fry supply. To encourage more milkfish farming, the fluctuation of product prices must be minimized through market expansion and reduction of the production cost.

Like other farming species, the major cost items for milkfish production include fry, feed, labour and utilities. As anticipated, production cost varies from country to country and also varies with culture system, farm size and management strategy. Since most milkfish farming relies on hatchery fry, the production cost of fry will affect the production cost of the final product. Lee *et al.* (1997) reported that the outdoor hatchery operation system produced more fry at less cost than the indoor hatchery system. The outdoor hatchery operation system takes advantage of sharing essential resources, such as broodstock, fertilized eggs, labour and knowledge, to reduce the production cost.

Comparing the overall production cost in Taiwan in 1979 and 1990, the percentage of the fry cost in the total production cost decreased, as more fry came from outdoor hatcheries (Lee, 1993). On the other hand, labour cost and fertilizer or feed cost contributed to a higher percentage of the total production cost. With the value increase in land and the cost increase of labour, one way to reduce the production cost was to increase the yield from the unit area or to intensify production. The development of the deep-water culture system is in response to this situation. Proper planning is important to avoid any serious environmental problems resulting from high-intensity farms. A preliminary benefit–cost ratio analysis of available farming data showed traditional shallow milkfish farming and monoculture were not profitable operations (see review by Lee, 1995). Polyculture and deep-water culture had better benefit–cost ratios. Polyculture of milkfish with white shrimp, *Litopenaeus vannamei*, provided additional income from the harvest of the white shrimp while no additional feed was needed for the shrimp during the grow-out period.

The price of milkfish is not determined by the production cost but by supply and demand. It was found that the relatively slow response of the wholesale price to retail price changes in Taiwan caused low price efficiency (Ling, 2003). The horizontally integrated relation of wholesale markets in the south-western region was also found. The fluctuation of product price was suspected to be from factors on the supply side rather than those on the demand side. Salayo (2006) also found prices in Manila were related with Lucena, Dagupan, Iloilo and Zamboanga prices, but not with Cebu prices. There was a strong seasonal price change. Therefore, it would be beneficial to producers to understand the pattern of price movements.

Milkfish is an important food item in the producing countries and is consumed locally, with a small proportion (about 4500 Mt) exported to other countries. Milkfish can be also harvested and used as live or frozen bait for tuna and other fisheries. About 65% of the total milkfish bait were demanded during the peak seasons between February and March and August and September. Frozen milkfish bait was packed in a 10 kg box with three different size classes; i.e. 45–50 pieces/10 kg, 50–55 pieces/10 kg or 55–60 pieces/10 kg, to be shipped to major fishing ports. Milkfish of 13–15 cm in body length were used as live bait. The FOB (free on board) price ranged from US$0.8 to US$1/kg. The annual demand for milkfish bait was estimated at 4500 Mt.

Milkfish for human consumption are sold as fresh products in traditional markets, but in both fresh and frozen form in supermarkets. In response to consumer preference, fish are processed into back, tail, head and belly portions for different needs. Usually, the belly portion commands the highest price. In the Philippines, markets provide a de-boning service to customers, for a fee. Fish are cut open from the dorsal side and split to the butterfly style. After the fish are cleaned of internal organs and the vertebrae removed, intermuscular small bones are removed with forceps. In total, there are about 196–208 tiny bones to be removed. This procedure requires time, skill and labour and is not easily adapted by other countries, although it is welcomed by consumers.

To expand the local markets further or extend to other markets, different milkfish products have been developed as food-processing technology has advanced. Other than in fresh or frozen form, milkfish are processed to different food products to extend shelf life and expand to other markets. Processed milkfish products include fish ball, smoked fish, barbecue fish, canned fish, marinated fish, de-boned fish fillets and other forms. More value-added products must be developed and marketed in order to stabilize the sale price.

10.5 Future Perspectives

Milkfish farming technology has been well developed during its farming history to produce as many fish as demanded at a very affordable cost. However, the production efficiency can still be improved through the application of biotechnology. The government should continually invest money in research and development in the areas such as biosecurity, and genetic selection efforts for fast growth and good flesh quality. Milkfish is a good affordable animal protein source for many consumers in South-east Asia. Therefore, milkfish will continue to be an important farmed fish in Taiwan, the Philippines and Indonesia. Because milkfish possesses many good characteristics as a farmed fish species and can be an environmentally friendly culture species, the farm area should extend to other locations where milkfish is an indigenous species. The main bottlenecks for the extension are that milkfish have many intramuscular tiny bones that can only be consumed by a special eating custom. Consequently, it has only limited market demand. To further expand milkfish market demand, food processing technology must be developed to overcome the issues associated with intramuscular tiny bones. As discussed in this chapter, several new processed products have been developed. With further market promotion, milkfish can truly be a commodity fish not only in the Philippines but in many other locations as well.

Reference

Alava, V.R. and Cruz, M.C. de la (1983) Quantative dietary fat requirement of *Chanos chanos* fingerlings in a controlled environment. Paper presented during the Second International Milkfish Aquaculture Conference, 4–8 October 1983, Iloilo City, Philippines.

Bagarinao, T.U. (1991) *Biology of Milkfish* Chanos chanos *(Forsskal)*. Aquaculture Department, South-east Asian Fisheries Development Center, Philippines.

Bagarinao, T. (1998) Historical and current trends in milkfish farming in the Philippines. In: DeSilva, S.S. (ed.) *Tropical Mariculture*. Academic Press, London, pp. 381–422.

Bautista, M.N., Del Valle, M.J. and Orejana, F.M. (1991) Lipid and fatty acid composition of brackishwater- and freshwater-reared milkfish (*Chanos chanos* Forskal). *Aquaculture* 96, 241–248.

Borlongan, I.G., Marte, C.L. and Norcillado, J.N. (2000) Development of larval diets for milkfish (*Chanos chanos*). *Journal of Applied Ichthyology* 16(2), 68–72.

Chacko, P.I. (1949) Food and feeding habits of the fishes of the Gulf of Mannar. *Proceedings of the Indian Academy of Science* 29B, 83–97.

Chang, S.L., Su, M.S. and Liao, I.C. (1993) Milkfish fry production in Taiwan. In: Lee, C.-S. Su, M.S. and Liao, I.C. (eds) *Finfish Hatchery in Asia: Proceedings of Finfish Hatchery in Asia 1991*. Tungkang Marine Laboratory, Tungkang, Taiwan, pp. 157–171.

Coloso, R.M., Tiro, L.B. and Benitez, L.V. (1983) A reference amino acid pattern for milkfish, *Chanos chanos* juveniles. Paper presented during the Second International Milkfish Aquaculture Conference, 4–8 October 1983, Iloilo City, Philippines.

Crear, D. (1980) Observations on the reproductive state of milkfish populations (*Chanos chanos*) from hypersaline ponds on Christmas Island (Pacific Ocean). *Proceedings of the World Mariculture Society* 11, 548–556.

Cruz, P.S. (1998) Prospects of milkfish sea cage farming. SEAFDE *Asian Aquaculture* 20(5), 14–16.

De la Vega, A. (1998) Milkfish farming in marine pens and cages. SEAFDEC *Asian Aquaculture* 20(5), 26–28.

Eda, H., Murashige, R., Eastham, B., Wallace, L., Bass, P., Tamaru, C.S. and Lee, C.S. (1990) Survival and growth of milkfish (*Chanos chanos*) in the hatchery. I. Feeding. *Aquaculture* 89, 233–244.

Eldani, A. and Primavera, J.H. (1981) Effect of different stocking combinations on growth, production and survival of milkfish (*Chanos chanos* Forsskal) and prawn (*Penaeus mondon* Fabricius) in polyculture in brackishwater ponds. *Aquaculture* 23, 59–72.

Emata, A., Borlongan, I. and Damaso, J. (2000) Dietary vitamin C and E supplementation and reproduction of milkfish (*Chanos chanos* Forsskal). *Aquaculture Research* 31, 557–564.

FAO (Food and Agriculture Organization of the United Nations) (2006) Aquaculture Production Database. Food and Agriculture Organization of the United Nations, Rome, Italy.

Fowler, H.W. (1959) *Fishes of Fiji*. Government of Fiji, Suva, 670 pp.

Gapasin, R.S.J. and Duray, M.N. (2001) Effects of DHA-enriched live food on growth, survival and incidence of opercular deformities in milkfish (*Chanos chanos*). *Aquaculture* 193, 49–63.

Guanzon, N.G., De Castro-Mallare, T.R. and Lorque, F.M. (2004) Polyculture of milkfish *Chanos chanos* (Forsskal) and the red seaweed *Gracilariopsis bailinae* (Zang et Xia) in brackish water earthen ponds. *Aquaculture Research* 35, 423–431.

Hilomen-Garcia, G.V. (1997) Morphological abnormalities in hatchery-bred milkfish (*Chanos chanos* Forsskal) fry and juveniles. *Aquaculture* 52, 155–166.

Kumagai, S. (1990) Reproduction and early life history of milkfish *Chanos chanos* in the waters around Panay Island, Philippines. PhD dissertation, Kyushu University, Fukuoka, Japan.

Kuo, C.-M. and Nash, C.E. (1979) Annual reproductive cycle of milkfish, *Chanos chanos* (Forsskal), in Hawaii waters. *Aquaculture* 16, 247–251.

Lee, C.S. (1993) Socio-economic impact of transferring the milkfish hatchery technology in Taiwan. International Symposium on Socio-economics of Aquaculture, 14–17 December 1993, Tungkang, Taiwan.

Lee, C.-S. (1995) Aquaculture of milkfish (*Chanos chanos*). TML *Aquaculture Series No 1*. Tungkang Marine Laboratory, Taiwan Fisheries Research Institute. Taiwan.

Lee, C.-S. and Banno, J.E. (1990) Milkfish culture and production in South-east Asia – present and future. In: Tanaka, H., Uwate, K., RJuario, J.V., Lee, C.-S. and Foscarini, R. (eds) *Proceedings of the Regional Workshop on Milkfish Culture Development in the South Pacific*, 21–25 November 1988, Tarawa, Kiribati, pp. 43–65.

Lee, C.-S., Tamaru, C.S., Banno, J.E., Kelley, C.D., Bocek, A. and Wyban, J.A. (1986a) Induced maturation and spawning of milkfish, *Chanos chanos*, Forsskal, by hormone implantation. *Aquaculture* 52, 199–205.

Lee, C.-S., Tamaru, C.S., Banno, J.E. and Kelley, C.D. (1986b) Influence of chronic administration of LHRH-analogue and/or 17α-methyltestosterone on maturation in milkfish, *Chanos chanos*. *Aquaculture* 59, 147–159.

Lee, C.-S., Tamaru, C.S., Kelley, C.D. and Banno, J.E. (1986c) Induced spawning of milkfish, *Chanos chanos*, by a single application of LHRH-analogue. *Aquaculture* 58, 87–98.

Lee, C.-S., Tamaru, C.S. and Weber, G.M. (1987) Studies on the maturation and spawning of milkfish, *Chanos chanos*, in a photoperiod-controlled room. *Journal of the World Aquaculture Society* 18(4), 253–259.

Lee, C.-S., Leung, P.-S. and Su, M.-S. (1997) Bioeconomic evaluation of different fry production systems for milkfish (*Chanos chanos*). *Aquaculture* 155, 367–376.

Liao, I.C. (1971) Note on some adult milkfish from the coast of southern Taiwan. *Aquaculture* 1(3), 1–10 (in Chinese).

Liao, I.C. (1991) Milkfish culture in Taiwan. In: McVey, J.P. (ed.) *CRC Handbook of Mariculture. II Finfish Aquaculture*. CRC Press, Boca Raton, Florida, pp. 91–115.

Liao, I.C. and Chen, T.I. (1979) Report on the induced maturation and ovulation of milkfish (*Chanos chanos*) reared in tanks. *Proceedings of the World Mariculture Society* 10, 317–331.

Liao, I.C. and Chen, T.I. (1984) Gonadal development and induced breeding of captive milkfish in Taiwan. In: Juario, J.V., Ferrari, R.P. and Benitez, L.V. (eds) *Advances in Milkfish Biology and Culture.* Island Publishing House Inc, Manila, Philippines, pp. 41–52.

Liao, I.C., Juario, J.V., Kumagai, S., Nakajima, H., Natividad, M. and Buri, P. (1979) On the induced spawning and larval rearing of milkfish, *Chanos chanos* (Forsskal). *Aquaculture* 18, 75–93.

Lin, L.T. (1982) Further success in induced spawning of pond-reared milkfish. *China Fisheries Monthly* 357, 17–19 (in Chinese).

Lin, L.-T. (1984) Studies on the induced breeding of milkfish *Chanos chanos* (Forsskal) reared in ponds. *China Fisheries* 378, 3–29.

Lin, L.-T. (1985) My experience in artificial propogation of milkfish – studies on natural spawning of pond-reared broodstock. In: Lee, C.-S. and Liao, I.C. (eds) *Reproduction and Culture of Milkfish*. Oceanic Institute, Hawaii, USA and Tungkang Marine Laboratory, Tungkang, Pingtung, Taiwan, pp. 185–203.

Ling, B.-H. (2003) Price cointegration in spatial markets: an application to milkfish markets in Taiwan. *Aquaculture Economics and Management* 7, 85–94.

Ling, S.-W. (1977) *Aquaculture in Southeast Asia: A Historical Overview*. University of Washington Press, Seattle, Washington.

Lio-Po, G.D. (1984) Diseases of milkfish. In: Juario, J.V., Ferraris, R.P. and Benitez, L.V. (eds) *Advances in Milkfish Biology and Culture*. Island Publishing House Inc, Manila, Philippines, pp. 145–155.

Liu, K.K.M. and Kelley, C.D. (1995) *The Oceanic Institute Hatchery Manual Series. Milkfish (Chanos chanos)*. The Oceanic Institute, Honolulu, Hawaii.

Marte, C.L., Cruz, P. and Flores, E.E.C. (2000) Recent developments in freshwater and marine cage aquaculture in the Philippines. In: Liao, I.C. and Lin, C.K. (eds) *Cage Aquaculture in Asia*. Asian Fisheries Society, Manila, Philippines and World Aquaculture Society, Southeast Asian Chapter, Bangkok, Thailand, pp. 83–96.

Poernomo, A. (1976) Notes on food and feeding habits of milkfish (*Chanos chanos*) from the sea. *Proceedings of the International Milkfish Workshop Conference*, 19–22 May 1976, Tigbauan, Iloilo, SEAFDEC-IDRC, pp. 162–166.

Rabanal, H.R. and Ronquillo, I.A. (1975) Distribution and occurence of milkfish, *Chanos chanos* (Forsskal). *Proceedings of the National Bangus Symposium* SEAFDEC, Iloilo, Philippines, pp. 20–33.

Ravago-Gotanco, R.G. and Juinio-Menez, M.A. (2004) Population genetic structure of the milkfish, *Chanos chanos*, based on PCR-RFLP analysis of the mitochondrial control region. *Marine Biology* 145, 789–801.

Ronquillo, I.A. (1975) Biological studies on banos (*Chanos chanos*). *Philippine Journal of Fisheries* 9, 18–37.

Salayo, N.D. (2006) Price relationships in Philippine milkfish markets: univariate and causality analysis. *Aquaculture Economics and Management* 10(1), 59–80.

Samsi, S. (1979) Effects of various protein sources on the growth and survival rates of milkfish (*Chanos chanos* Forsskal) fingerlings in a controlled environment. MSc thesis, University of the Philippines in the Visayas.

Schuster, W.H. (1952) Fish culture in brackishwater ponds of Java. IPFC Special Publication (1), XII. Indo-Pacific Fisheries Council, FAO, Rome.

Schuster, W.H. (1960) Synopsis of biological data on milkfish *Chanos chanos* (Forsskal), 1775. *FAO Fisheries Biology Synopsis No. 4*. Fisheries Division, Biology Branch. FAO, Rome.

Seneriches, M.L.M. and Chiu, Y.N. (1988) Effect of fishmeal on the growth, survival, and feed efficiency of milkfish (*Chanos chanos*) fry. *Aquaculture* 71, 61–69.

Senta, T. and Kumagai, S. (1977) Variation in the vertebral number of milkfish *Chanos chanos*, collected from various localities. *Bulletin of the Faculty of Fisheries of the Nagasaki University* 43, 35–40.

Shang, Y.C. (1986) Economic aspects of milkfish farming in Asia. In: Lee, C.-S., Gordon, M.S. and Watanabe, W.O. (eds) *Aquaculture of Milkfish (Chanos chanos): State of the Art*. The Oceanic Institute, Hawaii, pp. 263–275.

Shiau, S.-Y., Pan, B.S., Chen, S., Yu, H.L. and Lin, S.L. (1988) Successful use of soybean meal with methionine supplement to replace fish meal in diets fed to milkfish *Chanos chanos* (Forsskal). *Journal of the World Aquaculture Society* 19, 14–19.

Stickney, R.R. (2000) Milkfish culture. In Stickney, R.R. (ed.) *Encyclopedia of Aquaculture*. Texas Sea Grant, Bryan, Texas, pp. 530–532.

Su, M.-S., Lee, C.-S. and Liao, I.C. (2002) Technical responses to challenges in milkfish aquaculture. *Reviews in Fisheries Science* 10(3–4), 451–464.

Sugama, K. (2007) Public policy for sustainable development of milkfish (*Chanos chanos*) aquaculture in Indonesia. In: Leung, P.S., Lee, C.-S. and O'Bryen, P.J. (eds) *Species and System Selection for Sustainable Aquaculture*. Blackwell Publishing, Ames, Iowa, pp. 347–356.

Sumagaysay, N.S., Hilomen-Garcia, G.V. and Garcia, L.M.B. (1999) Growth and production of deformed and non-deformed hatchery-bred milkfish (*Chanos chanos*) in brackishwater ponds. *Israeli Journal of Aquaculture* 51(3), 106–113.

Tamaru, C.S., Lee, C.-S., Kelley, C.D., Aida, K. and Hanyu, I. (1988) Characterizing the stage of maturity most receptive to an acute LHRH-analogue therapy for inducing milkfish (*Chanos chanos*). *Aquaculture* 74, 147–163.

Tampi, P.R.S. (1958) On the food of *Chanos chanos* (Forsskal). *Indian Journal of Fisheries* 5, 107–117.

Tan, J.D. (1985) A histological study of the hypophysial–gonadal system during sexual maturation and spawning in the milkfish *Chanos chanos* (Forsskal). *Journal of Fish Biology* 26, 657–688.

Vicencio, Z.T. (1977) Studies on the food habits of milkfish, *Chanos chanos* (Forsskal). *Fisheries Research Journal of Philippine* 2, 3–18.

Villaluz, A.C., Tiro, L.B., Ver, L.M. and Vanstone, W.E. (1976) Qualitative analysis of the contents of the anterior portion of the oesphagus from adult milkfish, *Chanos chanos*, captured in Pandan Bay from May 10–June 16, 1975. *Proceedings of the International Milkfish Workshop Conference*, 19–22 May 1976, Tigbauan, Iloilo, Philippines, pp. 228–231.

Walsh, W.A., Swanson, C. and Lee, C.-S. (1991) Effects of development, temperature, and salinity on metabolism in eggs and yolk-sac larvae of milkfish, *Chanos chanos* (Forsskal). *Journal of Fish Biology* 39, 115–125.

Winans, G.A. (1980) Geographic variation in the milkfish, *Chanos chanos*. I. Biochemical evidence. *Evolution* 34, 558–574.

Winans, G.A. (1985) Geographic variation in the milkfish *Chanos chanos*. II. Multivariate morphological evidence. *Copeia* 4, 890–898.

11 The Catfish (Family: Ictaluridae)

Jeffrey Silverstein,[1] Terrill Hanson[2] and Brian Bosworth[3]

[1]*USDA/ARS National Center for Cool and Cold Water Aquaculture, Kearneysville, West Virginia, USA;* [2]*Department of Fisheries and Allied Aquacultures, Auburn University, Auburn, USA;* [3]*US Department of Agriculture, Agricultural Research Service, Catfish Genetics Research Unit, Stoneville, USA*

11.1 General Introduction

The channel catfish has a cylindrical body shape with sharp spines in the pectoral and dorsal fins (Fig. 11.1). There is an adipose fin on the back between the dorsal and caudal fins. Channel catfish, like other catfishes, have conspicuous sensory appendages located around the mouth, called barbels; there are four under the jaw and one on each end of the maxilla (Wellborn, 1988). The channel catfish's geographical distribution ranges from the Hudson Bay region, south to Florida and northern Mexico, north through New Mexico, Colorado and Montana to southern Manitoba (Fig 11.2). It has been widely introduced in the Atlantic Ocean and the Pacific Ocean drainages, and beyond.

In nature, channel catfish generally live in moving streams, but they are abundant in reservoirs, lakes and slow-moving streams as well. Channel catfish are crepuscular feeders, preferring the times just after sunset and just before sunrise. The natural diet generally includes both plant and animal material. Insects, algae, crawfish, seeds and small fish have all been found in the stomach contents of channel catfish. Main feed differences are related to whether the fish are living in riverine or lake systems.

Channel catfish are migratory, with males generally moving greater distances than females (Wendel and Kelsch, 1999). They seem to have the capability of homing. In river systems, channel catfish will move great distances for access to food and return to home areas, though in reservoirs their movements appear to be more random.

Spawning occurs in late spring and early summer, when the water temperatures are about 21–28°C. The male selects a spawning site and cleans a cavity to attract a female. A complex courtship follows and mating can continue for 6h. The male generally will then drive the female away and guard the nest until hatching between 6–10 days later, depending on temperature (Jackson, 2004).

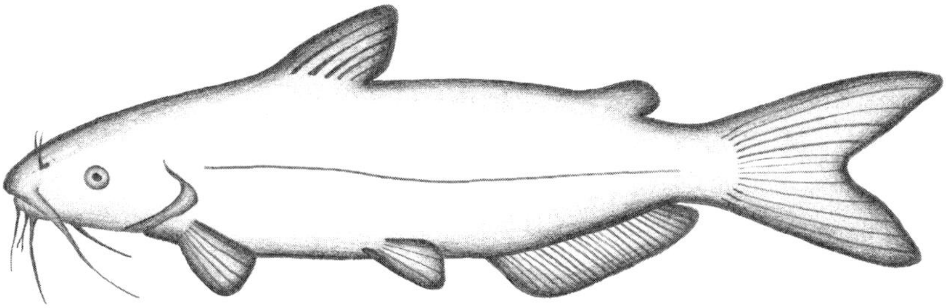

Fig. 11.1. Ictaluridae.

11.2 Farming of Catfish

Catfish farming in the USA is dominated by production of the channel catfish, *Ictalurus punctatus*, but also includes minor production of the blue catfish, *I. furcatus*, and F1 hybrids among channel and blue catfish. Channel and blue catfish both belong to the family Ictaluridae and are native to the central drainages of North America. Channel catfish populations now exist throughout the USA, and in many other countries through introductions.

Although it is difficult to define specifically when 'commercial farming' of catfish began, the first significant commercial production began in Mississippi in the mid-1960s. Catfish farming experienced a period of rapid increase in areas devoted to ponds and production during the next three decades in south-eastern USA, especially Alabama, Arkansas, Louisiana and Mississippi. Currently, approximately 270,000 t of catfish are processed annually in the USA. Catfish have a firm, white flesh with a mild flavour, preferred by many consumers in the USA. Favourable biological characteristics of the fish, physical and climatic characteristics of the region and socio-economic conditions all contributed to the rapid expansion of catfish production.

Aquaculture production of channel catfish is straightforward. They readily accept manufactured diets at all life stages and demonstrate good growth and feed conversion on relatively low-cost diets. They tolerate a wide range of water quality conditions, can be raised at high densities and are not considered to be cannibalistic at any life stage. Spawning and hatching of catfish is simple and efficient and does not require the use of expensive hatchery facilities or technologies.

South-eastern USA has large areas of flat, rural land with high clay content that make construction of large earthen ponds used for growing catfish economically feasible. The region also has either sufficient groundwater or rainfall to meet the water quality and quantity requirements needed for catfish farming. The familiarity of farmers and lending institutions in the region with other production agriculture enterprises (cotton, soybeans, etc.), the availability of a large pool of unskilled labourers and early cooperative efforts to develop processing plants and feed mills have been instrumental in the development and growth of the catfish industry.

Fig. 11.2. World distribution of *Ictalurus punctatus.*

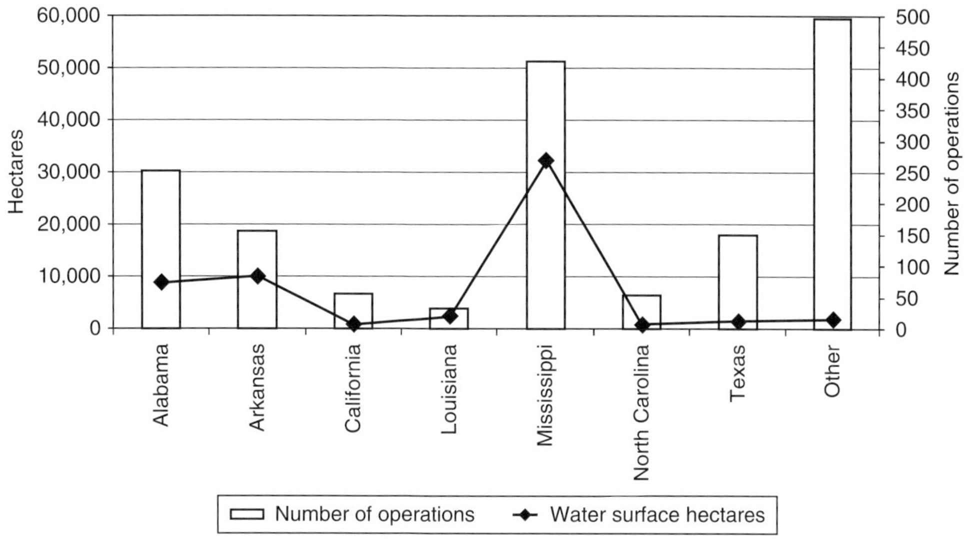

Fig. 11.3. Water surface area (hectares) under catfish production and number of catfish producers in 2008. *Source*: USDA/NASS, Catfish Processing Report, various months.

Presently, the catfish industry is the largest aquaculture industry in the USA. In 2008, 231 million kg of farm-raised catfish were processed at a sales value of US$389 million (NASS, various years). This production came from 59,450 water ha, located primarily in Mississippi, Alabama, Arkansas and Louisiana, and an additional production area distributed from California to North Carolina (Fig. 11.3).

11.3 Broodstock Management and Hatchery Operations

Producers have developed an efficient method for open-pond spawning of catfish that takes advantage of the catfish's natural inclination to spawn in some type of cavity. Mature male and female catfish (typically fish ≥ 3 years old) are placed in brood ponds prior to the spawning season at densities of 800–1200 kg/ha, with a male to female ratio between 1:1 and 1:2. Rising water temperature in the spring is the cue for catfish to begin spawning and the majority spawn in May and June, when water temperatures are between 25 and 30°C. As the water temperatures approach the levels needed for spawning, farmers place 'spawning containers' in ponds and male–female pairs of catfish will spawn in these containers voluntarily. Metal milk cans commonly have been used as spawning containers, but a wide variety of containers have been used, including ammunition cans, wooden boxes and plastic barrels. Many producers now use some type of plastic spawning container because they are lightweight and durable.

The male fish cleans out the container prior to spawning, then the female and male both enter the container and engage in courtship behaviour, followed by the eggs being laid and fertilized. The fertilized eggs are adhesive and form a large, gelatinous mass that sticks to the bottom side of the container. After spawning, the male chases the female out of the container and protects and cares for the eggs. In nature, the male would guard the eggs until hatch but, for commercial production, the containers are checked every 2–4 days and the egg masses are removed. Because the egg mass is adhesive, a plastic spatula, old credit card or some similar item may be needed to loosen the mass from the container. Several egg masses are placed in a tub or cooler filled with aerated water and then the egg masses are transported to the hatchery for incubation and hatching (Fig. 11.4).

Broodfish typically are fed the same commercial diets used for food-fish production (28–32% protein) at a rate of 0.5–1% of body weight daily or every other day when water temperatures are warm enough for feeding activity. Some producers supplement broodfish diets with forage fish by adding fathead minnows, shiners or tilapia to brood ponds (Torrans and Lowell, 2001).

Catfish hatcheries typically consist of a series of elevated hatching troughs supplied with flow-through water and aeration. Sufficient water quality and quantity are critical factors for the operation of a successful hatchery. Avery and Steeby (2004) suggest that a water flow of 380–570 l/min is required for a catfish hatchery producing 10 million fry/year. Water sources used for catfish hatcheries include groundwater, surface water or a mixture of the two. Groundwater is preferred since it is generally free of contaminants, pathogens and other fish and has consistent temperature and quality. However, the capacity of the aquifer, costs of drilling a well and pumping water and needs to aerate, add calcium and

Fig. 11.4. Terry Bates with spawn just retrieved from spawning can (on right). The credit card (on ground) is used to scrape the spawn off the can bottom; the spawn is then put in the iodine solution in the bucket to disinfect it while en route to the hatchery. (Photo courtesy Les Torrans.)

remove iron can be disadvantages of using groundwater. Surface water has some advantages over groundwater, including lower pumping costs, lower iron content and higher dissolved oxygen; nevertheless, most hatchery operators prefer groundwater, primarily due to the potential presence of pathogens in surface water and the associated increased risk of disease outbreaks (Tucker, 1991).

Water flows through the trough and eggs are agitated by paddles connected to a rotating shaft or by diffuse air supplied through air stones. The agitation is thought to mimic the action of the male catfish, which protects the egg mass in nature and uses its fins to move water over the eggs until they hatch. During incubation, fungal and bacterial infections of egg masses can reduce hatch, and therefore various chemical treatments are used to control these infections. Commonly used chemical treatments include povidone iodine, formalin or hydrogen peroxide. Egg treatments typically are discontinued as the eggs near hatch.

Time required for hatching is temperature dependent, but hatching typically occurs 6–8 days postfertilization at common hatching temperatures of 25–28°C. As fry hatch, they fall through the mesh of the hatching basket and collect on the bottom of the hatching trough. Fry are pinkish in colour after hatch and are commonly referred to as 'sac fry' because of their large, attached yolk sac, which will be the fry's source of nutrients and energy during the next 4–5 days. As the fry develop, they become more darkly pigmented and, when the yolk sac has been absorbed, they begin to swim to the surface and initiate feeding activity (Fig. 11.5). At this point, they are called 'swim-up fry' and it is

Fig. 11.5. Channel catfish swim-up fry in the fry-rearing trough in the hatchery. Most farmers stock them out to the ponds as soon as they swim up. (Photo courtesy Les Torrans.)

critical to feed them an appropriate diet frequently. Swim-up fry typically are fed finely ground feed containing 45–50% protein with a high fishmeal content 24 h a day (Robinson *et al.*, 1989). Careful feeding is required to ensure that fry receive adequate nutrition and that they are not overfed, which can lead to water quality problems. Due to the importance of proper feeding, sac fry typically are fed by hand, although some producers supplement hand feeding with the use of automated feeders.

When fry have been on feed for several days, they are transported and stocked into fertilized earthen ponds. The amount of time fry are fed in the hatchery is dictated primarily by the availability of hatching and rearing trough space. If the quantity of eggs and fry being produced at a given time is low, fry can be held and fed for a few more days, but typically fry are stocked into ponds 4–5 days after initiating feeding. Recently, some hatchery managers have begun to transfer fry to large net pens suspended in raceways and feed them for 1–2 weeks longer than normal prior to pond stocking. Feeding fry longer in the hatchery allows stocking of larger fry, with the rationale that larger fry will have improved survival in ponds.

Catfish swim-up fry are enumerated either by weight or volume and then transferred to nursery ponds in tanks supplied with compressed oxygen aeration. Fry are acclimated to the receiving pond water and then drained into the pond. Prior to stocking, ponds are fertilized with inorganic and/or organic fertilizers to promote phytoplankton and zooplankton blooms. These blooms serve to provide a food source for the fry and also reduce the growth of aquatic vegetation, which can have a negative effect on production and complicate fish harvest. Various fertilization strategies are used, but a good bloom is required for adequate fry survival and growth (Mischke and Zimba, 2004).

Fry are commonly stocked into nursery ponds at 250,000–500,000 fish/ha. After stocking, fry are fed finely ground feed 2–3 times daily at 25–50% of their biomass. Much of the feed given during the period immediately following fry stocking acts primarily as a fertilizer to continue promoting production of the plankton blooms supplying the fry with nutrition. Usually, about 2–3 weeks after stocking, the fish will be observed feeding at the surface. At this point, the fish are referred to as 'on-feed' and are commonly called fingerlings (Tucker *et al.*, 2004) (Fig. 11.6).

11.4 On-growing to Market Size

When the small fingerlings begin to feed, they are given a small floating pellet (typically 35% protein) one to two times daily. The use of a floating pellet allows estimation of catfish feeding activity, which is important because feeding activity can vary widely daily due to various factors (temperature, dissolved oxygen, etc.). Overfeeding results in wasted feed, which can impact profit negatively and result in degraded water quality, while underfeeding can result in reduced fish growth and suboptimal production. As the fingerlings grow, feed intake increases and most producers develop feeding strategies in which they will feed to satiation up to a set maximum level, at which point feeding is stopped (e.g. feed to

Fig. 11.6. 2–3 inch (5–7.5 cm) catfish fingerlings. Approximately 3 months of age, though if stocked at high densities, this may be the size they reach by 6 months. (Photo courtesy Les Torrans.)

satiation up to a maximum of 150 kg/ha/day). Producers may change feeding rates based on the size of fingerling they are trying to produce at harvest and their knowledge based on past experience in feeding fish.

Fingerling catfish are susceptible to several pathogens, including bacterial pathogens such as *Edwardsiella ictaluri*, the causative agent of enteric septicaemia of catfish (ESC), and viral pathogens like the herpes virus known to cause the disease, channel catfish virus disease (CCVD). Restricting feeding during outbreaks of ESC, the most problematic disease of catfish, has been demonstrated to reduce mortalities (Wise and Johnson, 1998). ESC is commonly more of a problem in fingerlings in autumn, when water temperatures are conducive to outbreaks and many producers employ restricted feeding strategies during this time to reduce ESC-related mortalities (i.e. feeding fingerlings every other day or every third day). Obviously, restricting feeding has a negative impact on fish growth and producers must consider the needs to maximize growth and reduce disease losses in developing feeding strategies.

At the end of the first growing season (typically October in south-eastern USA), feeding activity declines due to decreased water temperatures and fingerlings are commonly harvested by seining the ponds. Fingerlings are weighed, enumerated by sample counts and transported to new ponds for grow-out to food fish. Survivals of 70% and harvest sizes of 20–30 g (20–30 kg/thousand) would be considered good for fingerling production (Figs 11.7 and 11.8).

Fig. 11.7. 4–5 inch (10–12.5 cm) catfish fingerlings, approximately 4 months of age. (Photo courtesy Les Torrans.)

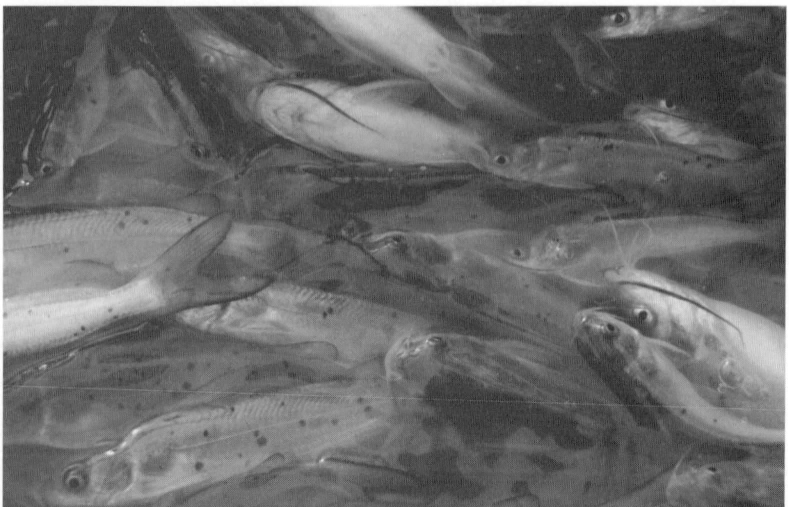

Fig. 11.8. 6–8 inch (15–20 cm) channel catfish fingerlings, approximately 6–7 months of age. (Photo courtesy Les Torrans.)

The majority of the area in catfish production is devoted to growing fingerlings to marketable size food fish (~ 0.5–2 kg). Although farmers have developed a variety of food-fish production strategies, they can be classified generally as either 'multiple-batch' or 'single-batch' scenarios. In multiple-batch production, fish of different ages and sizes are mixed and grown in the same pond. Fish are harvested using a seine with a size-selective mesh that retains fish of marketable size, but allows smaller fish to escape back into the pond. Harvested fish are replaced by stocking new fingerlings, and the cycle of partial harvest and restocking continues. The main advantage of the multiple-batch system is the wide range of fish sizes, including harvestable size fish, available in most ponds throughout the year. This was important when the industry was experiencing rapid growth and market expansion in the 1980s and 1990s because it allowed consistent harvest and processing of fish throughout the year, an important requirement for developing and expanding large, stable markets. Although the production capacity of the current industry has reduced concerns about having sufficient quantities of marketable fish available year round, the advantage of the multiple-batch production method, having marketable size fish in many ponds, remains important due to problems with off-flavour fish (described below) (Engle and Gayle, 1993; Tucker, 2000). Off-flavoured catfish are a consumer taste issue and caused primarily by naturally occurring algae populations in commercial catfish ponds.

Off-flavour, undesirable tastes/odours associated with the fish, is a critical issue for catfish producers because it frequently causes harvest delays of marketable size fish and increases production costs. Processing plants require catfish producers to subject catfish to a series of 'flavour checks' prior to accepting fish for processing. If fish are not acceptable, the producer is not allowed to sell the fish and they must go through the flavour check process again at a later date. The majority of off-flavours in farm-raised catfish are due to waterborne organic compounds, produced by cyanobacteria, that accumulate in the fish flesh and are commonly described as muddy, earthy, mouldy or musty; although other sources of objectionable flavours exist (Tucker, 2000). Options available to manage off-flavour resulting from cyanobacteria/algae generally involve allowing the fish to purge the off-flavour after moving them to 'clean' ponds, treating the pond with chemicals that will reduce or eliminate the cyanobacteria/algae source of the off-flavour (Shrader *et al*., 2005) or simply waiting until natural cycles in algal blooms reduce the cyanobacteria/algae populations responsible for the off-flavour.

Moving fish to clean ponds to reduce off-flavour is effective but requires labour to seine and move the fish, requires that the new pond remains 'clean' while the fish purge and the time needed to purge the off-flavour depends on both water temperature (purging is faster in warmer water) and type of compound causing the off-flavour. Copper sulphate and diuron are commonly used algicides for managing off-flavours in catfish ponds. Diuron appears to be more selective for the algal species that produce off-flavours, whereas copper sulphate is a broad-spectrum algicide, and treatment with copper can result in oxygen depletions due to elimination of the entire algal bloom. Copper sulphate is also toxic to fish at levels not greatly higher than those required for use as an algicide.

Waiting for the off-flavour incidents to pass naturally is effective eventually but is not predictable and, if a large number of ponds are off-flavour, cash flow can be seriously interrupted while waiting for fish to become 'on-flavour'.

There is currently a trend in the industry towards increased use of single-batch production (Southworth *et al.*, 2006). In single-batch production, fish of approximately the same size and age are stocked into a pond, grown and harvested as a group. Ponds are harvested entirely and then restocked with a new batch of fingerlings. Single-batch production typically involves a three-tiered production scheme, with ponds devoted to fingerling, stocker (large fingerling) and food-fish production. Single-batch production requires more grading and moving of fish and may be more prone to harvest delays associated with off-flavour fish, but generally results in better net production and feed conversion and more accurate inventory of size and number of fish than multi-batch production (Engle and Gayle, 1993; D'Abramo *et al.*, 2006, 2008).

It is difficult to define a typical food-fish production scenario because producers use a variety of fingerling sizes, stocking rates and feeding strategies in both multiple- and single-batch production. Stocking larger fingerlings reduces the time to harvest, but requires that the producer devote pond space to the production of larger fingerlings or buy more expensive, larger fingerlings. Stocking of smaller fingerlings reduces stocking costs, but lengthens the time to harvest. Tucker and Robinson (1990) report that stocking rates used in commercial production range from 1000 fish/ha to over 20,000 fish/ha. Producers base stocking rates on experience and production records and the effects of stocking rate on production and profits may vary due to differences in feeding rates, production goals and incidence of disease.

Food fish are typically fed a floating pelleted feed containing 28–32% protein, although some producers have begun to feed 26% protein diets. Feed typically is placed in a large hopper, either mounted on a truck or on a trailer pulled by a tractor, and the feed is blown on to the surface of the pond as the feeder drives along the upwind side of the pond. Food fish are usually fed once a day, 7 days a week during periods when water temperatures are warm enough to induce good feeding activity (> 20°C). Feeding rates typically are based on feeding to satiation, feeding a per cent of the estimated biomass in the pond or feeding to satiation up to a pre-set daily maximum (i.e. up to 120 kg/ha/day). A variety of factors influence feeding activity; therefore, feeding rates usually are adjusted based on feeding activity, water quality and experience of the feeder. An experienced feeder who adjusts feeding rates based on feeding behaviour and maximizes feed inputs without overfeeding is a valuable asset to a catfish farmer.

Harvest of food-size catfish typically involves seining the pond with a seine net made of heavy-duty polyethylene twine with a mesh size of 4.13–4.45 cm ($1^5/_8$ to $1^3/_4$ inches). Typical food-fish seines are about 3 m deep, have a weighted 'mud-line' to keep the bottom of the net near the bottom of the pond, floats attached to the top of the net to keep the top of the net on the surface and are pulled through the pond by tractors attached to each end. A nylon mesh net with an enclosed bottom, referred to as a 'live car' or 'sock', is attached to an opening in the centre of the seine. While the principle of seining is fairly straightforward, efficient seining requires an experienced crew of 5–7 people and although some farms have their own seining crew, many farmers hire seine operators to harvest fish.

As the pond is seined, the fish move through the opening in the seine and are concentrated in the sock. When the seine has been pulled through the pond, the sock is detached from the seine and is staked out away from the bank. The fish are frequently allowed to self-grade in the sock overnight. Small, submarketable size fish will swim through the mesh and marketable size fish will be retained. Catfish are active in warm water and small fish grade out of the sock fairly efficiently at warm temperatures; however, fish do not grade well at colder temperatures. Fish are held at high densities in the sock and it is very important to ensure dissolved oxygen levels are adequate in the sock, especially in warmwater conditions during the night, when oxygen levels are low. A high incidence of mortalities immediately prior to selling the fish to the processor will have a devastating effect on profits.

An in-pond horizontal bar grader, which was developed at the University of Arkansas at Pine Bluff, USA (Heikes, 1999; Trimpey *et al.*, 2004), is widely used to grade fingerlings, but is also gaining popularity for grading food-size fish immediately after seining. Advantages of the in-pond grader to traditional sock grading are the ability to grade immediately after seining, more accurate grading, inventory of submarketable fish and the ability to grade effectively at coldwater temperatures.

After food-size fish have been graded, they are loaded on to transport trucks and delivered live to processing plants. On arrival at the plant, fish are subjected to a final flavour check and, if they pass this check, they are unloaded, weighed, transported by conveyor into the plant and stunned by electrical shock. Electrical stunning makes it easier to handle and slaughter the fish humanely.

Catfish products include fillets, whole dressed fish, steaks, strips (fillets cut into strips) and nuggets in both fresh and frozen forms. In addition, various types of breading and marinades are used to increase the variety of products available to the consumer. Product breakdown for typical large processors are 60–70% fillets, 10–15% whole dressed fish, 10–15% nuggets and 10–15% value-added products (Engle and Hanson, 2004; Hanson and Sites, 2008).

Catfish processors adhere to strict regulations to ensure delivery of safe, high-quality products to consumers. These regulations include Good Manufacturing Processes, Standard Sanitary Operating Procedures and Hazard Analysis Critical Control Points, as well as a voluntary seafood inspection programme administered by the US Department of Commerce (Marshall, 2004). Adherence to these regulations and adoption of strict quality controls has played an important role in expanding catfish markets by assuring delivery of a variety of high-quality, nutritious, safe products to the consumer.

Catfish has become one of the top ten fish and seafood products preferred by US consumers, behind shrimp, canned tuna, salmon, pollock and tilapia (National Fisheries Institute, 2006) (Table 11.1). Catfish consumption has increased by 25% over 1990–2007 (while other US favourite seafoods such as shrimp and salmon have doubled in consumption over the same period). Harvested catfish are 0.5–2.0 kg and are processed into fresh and frozen product forms, including whole, fillet, nugget, strip, and steak products (NASS, various years). In the mid-1980s,

Table 11.1. Top 10 US consumed fish and seafood species (per capita consumption, in kg).

2004	2000	1995	1990
Shrimp (1.90)	Canned tuna (1.59)	Canned tuna (1.54)	Canned tuna (1.68)
Canned tuna (1.50)	Shrimp (1.45)	Shrimp (1.13)	Shrimp (1.00)
Salmon (1.00)	Pollock (0.72)	Pollock (0.69)	Cod (0.63)
Pollock (0.59)	Salmon (0.72)	Salmon (0.54)	Pollock (0.58)
Catfish* (0.50)	Catfish (0.49)	Cod (0.44)	Salmon (0.33)
Tilapia (0.32)	Cod (0.34)	Catfish (0.39)	Catfish (0.32)
Crab (0.27)	Clam (0.21)	Clam (0.0.26)	Clam (0.28)
Cod (0.27)	Crab (0.17)	Crab (0.15)	Flatfish (0.26)
Clam (0.23)	Flatfish (0.19)	Flatfish (0.14)	Scallop (0.14)
Flatfish (0.14)	Scallop (0.12)	Scallop (0.11)	Crab (0.13)

Note: *Catfish consumption has been recalculated to reflect the change in US law that prohibits imported 'catfish' – basa, tra, etc. – from being called catfish.
Source: National Fisheries Institute (http://www.aboutseafood.com/media/top_10.cfm).

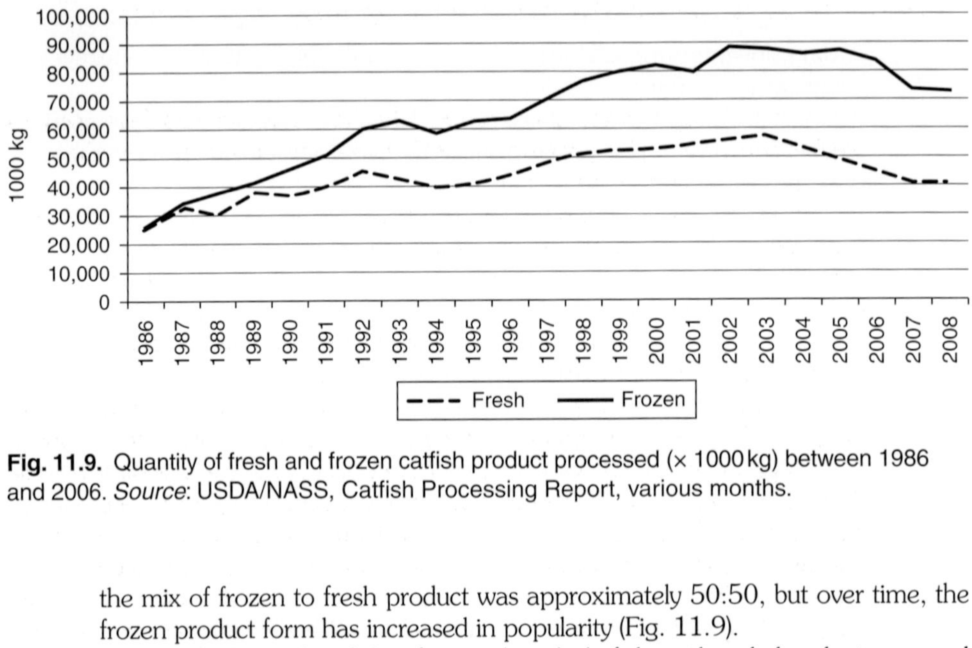

Fig. 11.9. Quantity of fresh and frozen catfish product processed (× 1000 kg) between 1986 and 2006. *Source*: USDA/NASS, Catfish Processing Report, various months.

the mix of frozen to fresh product was approximately 50:50, but over time, the frozen product form has increased in popularity (Fig. 11.9).

Whole dressed catfish refers to the whole fish, with only head, viscera and skin removed, and is about 60% of the live fish weight. Whole fish was once the largest segment of product forms processed, but has been surpassed by fillet and other product forms (Fig. 11.10). Other catfish product forms include steak cuts, nuggets, strips and all other product forms, including breaded and added-ingredient products. The catfish steak is the cross-section cut from larger dressed fish; the catfish nugget product is derived from small fillet cuts from below the rib section of the fish; and the strip is the finger-sized pieces of fish cut from fillets and usually includes breading or added ingredients. The fillet

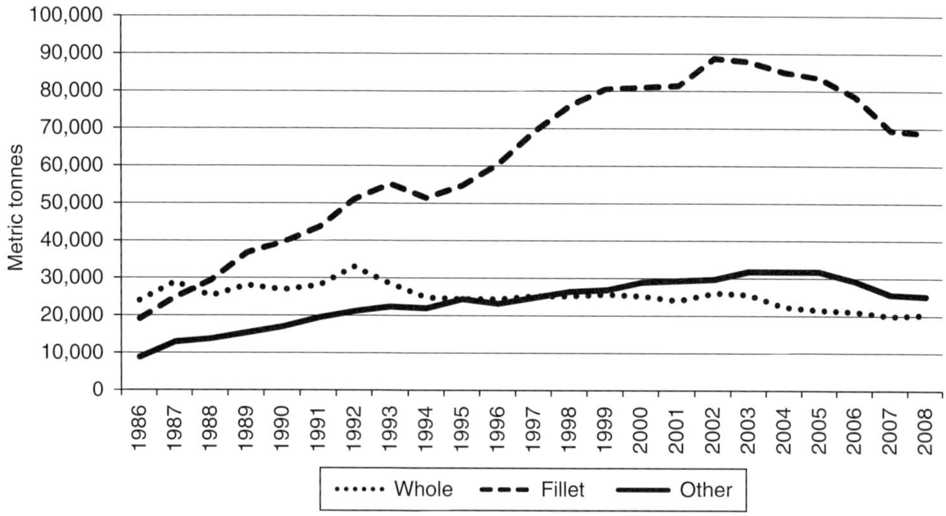

Fig. 11.10. Quantity of catfish product processed by product form (metric tonnes) between 1986 and 2008. *Source*: USDA/NASS, Catfish Processing Report, various months.

forms include regular, shank and strip fillets. As the market has developed, the market for individually quick-frozen (IQF) fillets and added-value products has grown as a per cent of total catfish sales.

Estimated variable costs for the multiple-batch production system range from US$7008 to US$8268/ha (Table 11.2). Fixed costs for the multiple-batch production system range from US$581 to US$993/ha. Costs of food-size catfish production range from US$1.39/kg to US$2.03/kg. When variable and fixed costs are included, the majority of cost references place the cost of production in the US$1.50/kg to US$1.65/kg range.

In 2003, the average nominal price paid to catfish producers was US$1.29/kg, which is marginal at best for returning a profit to producers. Catfish production methods and efficiency on farms vary, so it is difficult to say there is only one break-even sales price for the industry. In general, a production cost in the US$1.28–1.32/kg range is considered the break-even level for profitability, though this depends on the unit cost of feed, other production inputs and level of fixed costs. The farm bank price paid to catfish producers averaged US$1.69/kg in 2007 and increased to US$1.72/kg in 2008, but this improved price differential did not show the monthly volatility between the first and second halves of 2007 and 2008 (Fig. 11.11) (Hanson and Sites, 2008). The average January–June 2007 price paid to producers was US$1.84/kg and between July and December the average price fell to US$1.54/kg, a US$0.30/kg decline, equating to a 16.5% price decline within a 6-month period. The average January–June 2008 price paid to producers was US$1.62/kg and between July and December the average price increased to US$1.81/kg, a US$0.19/kg increase, or a 12.0% price increase within a 6-month period. Such price volatility makes it difficult for producers to plan effectively for their future.

Table 11.2. Food-size catfish cost of production, net return and break-even price information obtained from published literature.

Cost type	Cost item	Reference
Farm-level cost of production	Fixed expenses: US$993/ha	Heikes *et al.* (2006)
	Total expenses: US$6328–11,876/ha	
	Annual operating costs: US$7008–7383/ha/year	
	Total annual costs: US$1.52–1.61/kg	Heikes *et al.* (2006)
	Total variable cost: US$8268/ha (607 ha farm size)	Hanson *et al.* (2004b)
	Total fixed cost: US$608/ha	
	Total cost: US$8876/ha	
Unit cost of production	US$1.39–1.65/kg	Heikes *et al.* (2006)
	US$1.50/kg over total costs	
	Small farm (4.9 water ha) total annual production costs were US$2.03/kg	Tucker *et al.* (2004)
	Cost per kg: US$1.61/kg (65 ha farm size) US$1.59/kg (130 ha farm size)	
	US$1.52/kg (259 ha farm size)	
	Fingerling cost per ha:	
	– 7.6 cm fingerling, US$8468	Engle (2001)
	– 12.7 cm fingerling, US$9605	
	– 17.8 cm fingerling, US$9946	
	Stocker cost of production from:	
	Low (50,000/ha) fingerling stocking rate was: US$0.29/stocker	Pomerleau and Engle (2003)
	Medium (100,000/ha) stocking rate: US$0.21/stocker	
	High (150,000/ha) was US$0.18/stocker	
Net return	US$27–1166/ha when market price was US$01.65/kg	Heikes *et al.* (2006)
	For a 607 ha farm: net return above variable costs was US$2031/ha and net return above total costs was US$1421/ha	Hanson *et al.* (2004b)

11.5 Future Perspectives

Future research directions are focused in genetic, nutrition, production and health fields. Measurement of genetic variation for production traits in catfish lines and characterization of correlations between traits is a major effort of breeders (Bosworth *et al.*, 2004). Developing genomic resources for integrating functional genomics into the catfish applied breeding programme is an additional focus (Li and Waldbieser, 2006). One area of focus for nutrition research is on improving the economic performance of catfish feeds (Li *et al.*, 2006). Additionally, interactions among fish size and feeding schedule on economic performance of catfish farming will be measured. Managing ponds to reduce

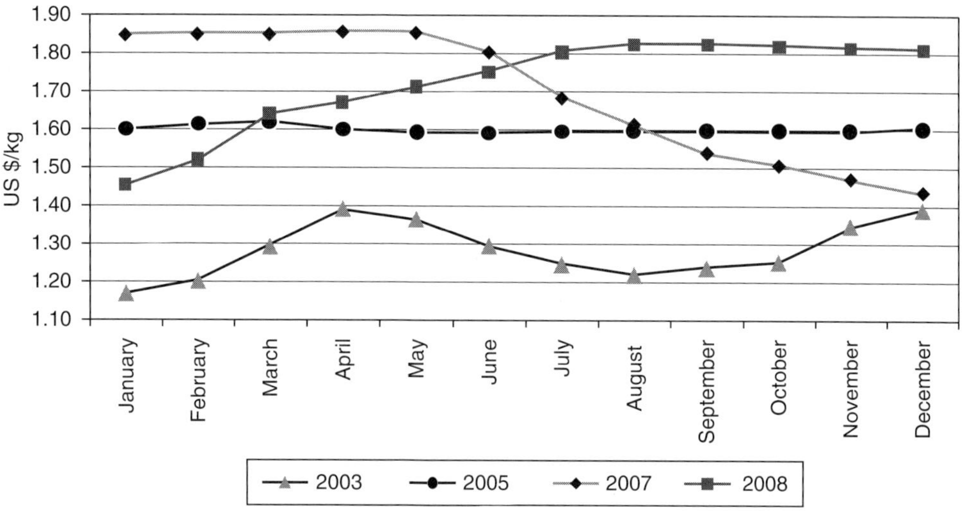

Fig. 11.11. Prices paid (US$/kg) at pond bank for aquacultured catfish between 2003 and 2008. *Source*: USDA/NASS, Catfish Processing Report, various months.

off-flavour episodes is another active area of research (Zimba and Gitelson, 2006). For fish health, candidate genes such as the toll-like receptors are being tested for association with disease resistance (Baoprasertkul *et al*., 2006) and *in vitro* immune function assays are being developed.

References

Avery, J.L. and Steeby, J.A. (2004) Hatchery management. In: Tucker, C.S. and Hargreaves, J.A. (eds) *Biology and Culture of Channel Catfish*. Elsevier, Amsterdam, pp. 145–165.

Baoprasertkul, P., Peatman, E., Somridhivej, B. and Liu, Z. (2006) Toll-like receptor 3 and TICAM genes in catfish: species–specific expression profiles following infection with *Edwardsiella ictaluri*. *Immunogenetics* 58, 817–830.

Bosworth, B.G., Wolters, W.R., Wise D.J. and Klesius, P.H. (2004) Genetic effects for growth and resistance to *Edwardsiella ictaluri* estimated from diallel cross among USDA-102 strain channel catfish *Ictalurus punctatus*, blue catfish *Ictalurus furcatus*, and their F1 hybrid. *Journal of the World Aquaculture Society* 35, 420–426.

D'Abramo, L.R., Steeby, J.A., Hanson, T.R. and Gerard, P.D. (2006) Evaluation of a fingerling to stocker phase as part of a new production strategy for the commercial farming of channel catfish *Ictalurus punctatus*. *North American Journal of Aquaculture* 69(1) January, 22–35.

D'Abramo, L.R., Hanson, T.R. and Steeby, J.A. (2008) Production and associated economics of channel catfish *Ictalurus punctatus* fed different practical diets in a fingerling-stocker phase of culture. *North American Journal of Aquaculture* 70(2) April, 154–161.

Engle, C.R. (2001) The economics of producing and understocking different sizes of catfish fingerlings on growout farms. University of Arkansas at Pine Bluff, Pine Bluff, Arkansas, p. 18.

Engle, C.R and Gayle, L.P. (1993) Trade-offs between single and multiple-batch production of channel catfish, *Ictalurus punctatus*: an economics perspective. *Journal of Applied Aquaculture* 3(3/4), 311–332.

Engle, C.R. and Hanson, T.R. (2004) Marketing and economics. In: Tucker C.S. and Hargreaves J.A. (eds) *Biology and Culture of Channel Catfish*. Elsevier, Amsterdam, pp. 601–633.

Engle, C.R. and Quagrainie, K. (2006) *Aquaculture Marketing Handbook*. Blackwell Publishing, Ames, Iowa.

Engle C.R., Capps, J.O., Dellenbarger, L.E., Dillard, J.G., Hatch, L.U., Kinnucan, H.W. and Pomeroy, R.S. (1990) In: Station, A.A.E. (ed.) *The US Market for Farm-raised Catfish: An Overview of Consumer, Supermarket and Restaurant Surveys*. University of Arkansas at Pine Bluff, Arkansas.

Hanson, T.R. (2001) Catfish marketing strategies. In: *World Aquaculture Society Meetings Book of Abstracts*. World Aquaculture Society, Orlando, Florida.

Hanson, T.R. (2006a) Catfish farming in Mississippi, Mississippi history now. In: Online publication of the Mississippi Historical Society.

Hanson, T.R. (2006b) Economic impact of off-flavor to the US catfish industry. In: Schrader, K. and Rimando, A. (eds) *Off-flavors in Aquaculture*. American Chemical Society, Washington, DC, pp.13–29.

Hanson, T.R. and Sites, D. (2008) 2007 Catfish Database. Information Report, 2008-01, March. Department of Agricultural Economics, Mississippi State University, Mississippi (http://www.agecon.msstate.edu/research/catfishdb/2007/whole%20document.pdf).

Hanson, T.R., Dean, S. and Spurlock, S.R. (2004a) Economic impact of the farm-raised catfish industry on the Mississippi state economy. *Journal of Applied Aquaculture* 3, 311–332 1993 FR 39(3) 15, 11–28.

Hanson, T.R., Steeby, J. and Delta, A.J. (2004b) *Catfish Production – Assumptions for 250-, 750- and 1,500-Acre Farms Using Multiple-batch Stockings*. Department of Agricultural Economics, Mississippi State University, Mississippi.

Hatch, L.U., Engle, C.R., Zidack, W.E. and Olowolayemo, S.O. (1991) Retail grocery markets for catfish. *Experiment Station Bulletin 611,* Auburn University, Alabama.

Heikes, D.L. (1999) Development of in-pond grading technology for commercial aquaculture. In: *World Aquaculture Society, Book of Abstracts 2001*. World Aquaculture Society, Baton Rouge, Louisiana, p. 287.

Heikes, D.L., Engle, C.R. and Kouka, P.J. (2006) Catfish yield verification trials, final report, May 1993–December 1996. University of Arkansas Cooperative Extension Program, University of Arkansas at Pine Bluff, Arkansas.

House, L., Hanson, T.R., Sureshwaran, S. and Selassie, H. (2003) Opinions of US consumers about farm-raised catfish: results of a 2000–2001 survey. Mississippi State University MAaFES, Mississippi.

Jackson, D.C. (2004) Natural history and fisheries. In: Tucker C.S. and Hargreaves J.A. (eds) *Biology and Culture of Channel Catfish*. Elsevier, Amsterdam, pp. 15–35.

Li, M.H., Peterson, B.C., Janes, C.L. and Robinson, E.H. (2006) Comparison of diets containing various fish meal levels on growth performance, body composition, and insulin-like growth factor-I of juvenile channel catfish *Ictalurus punctatus* of different strains. *Aquaculture* 253, 628–635.

Li, R.W. and Waldbieser, G.C. (2006) Production and utilization of a high-density oligonucleotide microarray in channel catfish, *Ictalurus punctatus. BMC Genomics* 7, 134.

Marshall, D.L. (2004) Processing. In: Tucker C.S. and Hargreaves J.A. (eds) *Biology and Culture of Channel Catfish*. Elsevier, Amsterdam, pp. 585–600.

Mischke, C.C. and Zimba, P.V. (2004) Plankton community responses in earthen channel catfish nursery ponds under various fertilization regimes. *Aquaculture* 233, 219–235.

NASS (various years) Catfish processing. National Agriculture Statistics Service, US Department of Agriculture, Washington, DC.

National Fisheries Institute (2006) Top 10 US Consumption by Species Chart. National Marine Fisheries Service, USA.

Perez, K.R. (2006) *Fishing for Gold – The Story of Alabama's Catfish Industry*. The University of Alabama Press, Tuscaloosa, Alabama.

Pomerleau, S. and Engle, C.R. (2003) Production of stocker size channel catfish: effect of stocking density on production characteristics, costs, and economic risk. *North American Journal of Aquaculture* 65, 112–119.

Robinson, E.H., Steeby, J.A. and Brent, J.R. (1989) Evaluation of three feeds for hatchery rearing channel catfish fry. *Journal of the World Aquaculture Society* 6, 256–260.

Schrader, K.K., Tucker, C.S., Hanson, T.R., Gerard, P.D., Kingsbury, S.K. and Rimando, A.M. (2005) Management of musty off-flavor in channel catfish from commercial ponds with weekly applications of copper sulfate. *North American Journal of Aquaculture* 67, 138–147.

Southworth, B.E., Stone, N. and Engle, C.R. (2006) Production characteristics, water quality, and costs of producing channel catfish *Ictalurus punctatus* at different stocking densities in single-batch production. *Journal of the World Aquaculture Society* 37, 21–31.

Torrans, L. and Lowell, F. (2001) Use of tilapia as supplemental forage for channel catfish broodstock. *North American Journal of Aquaculture* 63, 215–221.

Trimpey, J., Engle, C.R., Heikes, D., Goodwin, A. and Davis, K. (2004) A comparison of new in-pond grading technology to live car grading for food-sized channel catfish (*Ictalurus punctatus*). *Aquaculture Engineering* 31, 263–276.

Tucker, C.S. (1991) Water quantity and quality requirements for channel catfish hatcheries. Southern Regional Aquaculture Center, Stoneville, Mississippi.

Tucker, C.S. (2000) Off-flavor problems in aquaculture. *Reviews in Fisheries Science* 8, 45–88.

Tucker, C.S. and Hargreaves, J.A. (2004) *Biology and Culture of Channel Catfish*. Elsevier, Amsterdam.

Tucker, C.S. and Robinson, E.H. (1990) *Channel Catfish Farming Handbook*. Van Nostrand Reinhold, New York.

Tucker, C., Avery, J., Engle, C. and Goodwin, A. (2004) *Industry Profile: Pond-Raised Channel Catfish*. A review developed for the National Risk Management Feasibility Program for Aquaculture. Mississippi State University, Agreement No RMA 01-IE-0831-127.

Tucker, C.S., Avery, J.L. and Heikes, D. (2004) Culture methods. In: Tucker C.S. and Hargreaves, J.A. (eds) *Biology and Culture of Channel Catfish*. Elsevier, Amsterdam, pp. 166–195.

Wellborn, T.L. (1988) *Channel Catfish Life History and Biology*. Southern Regional Aquaculture Association, Texas Agricultural Extension Service, University of Florida, Publication No 180.

Wendel, J. and Kelsch, S. (1999) Summer range and movement of channel catfish in the Red River of the North. In: Irwin, E., Hubert, W., Rabeni, C., Schramm, J. and Coon, T. (eds) *Catfish 2000: Proceedings of the International Ictalurid Symposium*. American Fisheries Society, Bethesda, Maryland, pp. 203–214.

Wise, D.J. and Johnson, M.R. (1998) Effect of feeding frequency and Romet-medicated feed on survival, antibody response, and weight gain of fingerling channel catfish *Ictalurus punctatus* after natural exposure to *Edwardsiella ictaluri*. *Journal of the World Aquaculture Society* 29, 169–175.

Zimba, P.V. and Gitelson, A. (2006) Remote estimation of chlorophyll concentration in hyper eutrophic aquatic systems: model tuning and accuracy optimization. *Aquaculture* 256, 272–286.

12 The Salmonids (Family: Salmonidae)

MALCOLM JOBLING,[1] ARNE-MIKAL ARNESEN,[2] TILLMAN BENFEY,[3] CHRIS CARTER,[4] RONALD HARDY,[5] NATHALIE R. LE FRANÇOIS,[6] ROBYN O'KEEFE,[3] JUHA KOSKELA[7] AND SIMON G. LAMARRE[8]

[1]*University of Tromsø, Tromsø, Norway;* [2]*Nofima A/S, Tromsø, Norway;* [3]*Department of Biology, University of New Brunswick, Fredericton, New Brunswick, Canada;* [4]*University of Tasmania, Launceston, Tasmania, Australia;* [5]*UI Hagerman Fish Culture Experiment Station, Hagerman, Idaho, USA;* [6]*Biodôme de Montréal, Montréal, Québec, Canada;* [7]*Finnish Game and Fisheries Research Institute, Laukaa Fisheries Research and Aquaculture, Valkola, Finland;* [8]*Memorial University of Newfoundland, Canada*

12.1 General Introduction

As a taxon, the salmonids (Fig. 12.1) comprise 11 genera, with 65–70 species. All species occur naturally in the northern hemisphere but several salmonid species have been introduced to the southern hemisphere, where they often form the basis of sport fisheries or aquaculture enterprises (Nelson, 1994; Lever, 1996). There are both freshwater and anadromous species, but some of the anadromous species also have populations that are strictly confined to fresh waters.

The most important species for sport and commercial fisheries, and for aquaculture, are representatives of the genera *Salmo* (Atlantic salmon and trouts) and *Oncorhynchus* (Pacific salmons and trouts), although some species within other genera, e.g. *Salvelinus* (charrs) and *Coregonus* (whitefishes and ciscoes), are also exploited commercially (Pennell and Barton, 1996). Aquaculture production of salmonids increased rapidly during the latter years of the 20th century, had reached 1.5 million t (Mt) by the turn of the century and continued to expand during the early years of the 21st century (Fig. 12.2). Just over half of the production is Atlantic salmon, *Salmo salar*, and much of the remainder is cultivation of rainbow trout, *Oncorhynchus mykiss*. Annual production of farmed rainbow trout is around 0.5 Mt (FAO, 2007).

Several of the salmonid species have attributes and biological characteristics that make them highly suitable for intensive farming, although the number

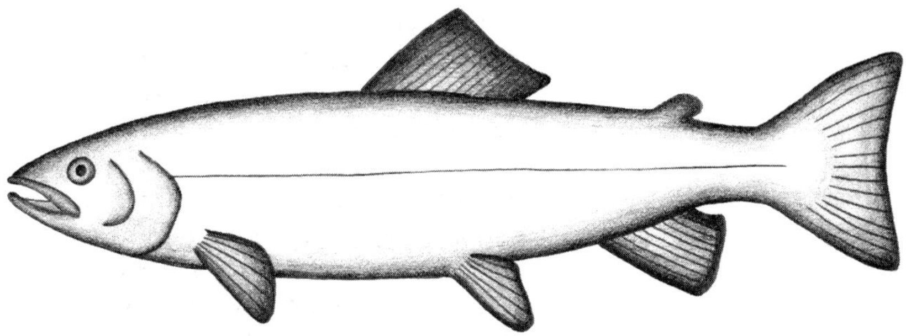

Fig. 12.1. Salmonidae.

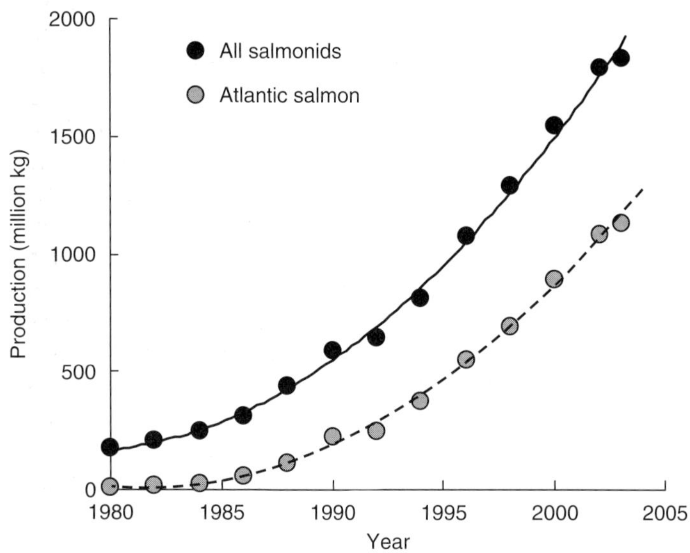

Fig. 12.2. Aquaculture production of salmonid fish, 1980–2003. *Source*: http://www.fao.org/figis.

of positive attributes varies from species to species. The characteristics possessed by salmonids that are deemed desirable in a cultured species can be summarized as follows:

- The incubation of eggs, and the hatchery rearing of juveniles, is relatively straightforward. Eggs can usually be obtained quite readily from captive broodstock without the need to induce spawning. Further, the timing of spawning can be controlled using thermal and photoperiod manipulations to give 'out-of-season' eggs and enable a continuous cycle of production, for example, in freshwater rearing of rainbow trout. The eggs of salmonid species are large (some *c*.5 mm diameter), are easy to incubate and survival to hatching is generally high. Large egg size results in large offspring (many *c.*15–25 mm at hatch); the offspring will usually feed directly on dry, particulate feeds that are produced commercially and are readily available.

- Several species adapt well to farm conditions and can be fed dry, pellet feeds throughout the grow-out phase. The fish tolerate a moderate degree of crowding and handling and are moderately resistant to disease, even though several parasite and disease problems have been recognized (Roberts and Shepherd, 1997; Woo *et al.*, 2002).
- Some species grow to a relatively large body size quite quickly. For example, Atlantic salmon may reach a weight of 4–5 kg in 18 months from the time a 50–100 g smolt is put into the sea. When held at an ideal growth temperature (*c.*15°C), rainbow trout reach a market size of 400–650 g in 10–13 months from the time of hatching.
- Large body size is an advantage when it comes to processing the harvested fish; it is easier to process large fish than small fish and the range of products that can be produced from large fish is also greater. The meat yield of salmonids is high, with the fillet representing 50–60% of the body mass.
- The meat quality of salmonids is good; it has characteristics that are generally appealing to a broad range of potential consumers. Salmonids are well known as food fish and they have good acceptability in the market.

This chapter combines sections on whitefishes, charrs, Atlantic salmon and Pacific salmons and trouts, with a view to detailing key aspects of their biology and culture.

12.2 Whitefishes: Biology and Culture

Whitefishes, representatives of the genus *Coregonus*, are all freshwater (occasionally anadromous) fish species with a natural occurrence in the northern regions of the northern hemisphere. Several of the species are extremely polymorphic and display considerable diversification in morphology, habitat, feeding habits and life history. For example, isolated lake populations have often evolved their own individual and distinctive characters and several morphotypes may even occur within a single water body (Lu and Bernatchez, 1999; Todd and Fleischer, 2002; Heikinheimo *et al.*, 2004; Østbye *et al.*, 2005). Interpretation of this diversity traditionally has represented a great challenge to taxonomists and there have been suggestions that a great many whitefish species inhabit both northern European and North American waters.

Several of the species designations remain open to debate, as do the phylogenetic relationships between the species (Nelson, 1994; Eckmann *et al.*, 1998; Todd and Fleischer, 2002; Heikinheimo *et al.*, 2004). Insights into the evolution of morphological types, and the relationships between closely related taxa, are being revealed by genetic analysis involving examination of mitochondrial DNA (mtDNA) and microsatellites (Eckmann *et al.*, 1998; Lu and Bernatchez, 1999; Todd and Fleischer, 2002; Heikinheimo *et al.*, 2004; Østbye *et al.*, 2005). As a generalization, it can be said that the dominant whitefishes are the *Coregonus lavaretus* complex (common or European whitefish) in Eurasia and the *C. clupeaformis* complex (lake whitefish) in North America.

There have been numerous attempts to transplant whitefish into lakes and watercourses both within and outside the species' native range (Järvinen, 1988; Todd and Luczynski, 1992; Luczynski *et al.*, 1995; Lever, 1996; Eckmann *et al.*, 1998; Todd and Fleischer, 2002; Heikinheimo *et al.*, 2004). Often, transplantations have been carried out in an attempt to establish exploitable populations of fish within oligotrophic lakes, but the success of such transplantations has been variable. Both the common and lake whitefish are of commercial importance and often form the basis of fisheries throughout their native ranges. Whitefish of some other species, e.g. vendace, *C. albula*, are also fished commercially. Annual catches of coregonid fishes are about 40,000 t, whereof around 5000 t are vendace and 14,000 t lake whitefish (FAO, 2005). Common whitefish make up most of the remainder of the commercial catches.

There is cultivation of a number of whitefish species in several countries including Finland, Russia, Czech Republic, Italy and Japan (Flüchter, 1980; Dabrowski *et al.*, 2002). Cultivation is primarily for stocking and conservation purposes, but there is also a limited aquaculture industry that produces fish directly for the table. Annual production of whitefish was 2000–5000 t in the period 2000–2004 (FAO, 2007). When reared for stock enhancement purposes, whitefish are usually held semi-intensively in ponds and are released as one-summer fish. Stocking common whitefish in natural ponds has a long history in Finland, although the number produced in this way is slowly decreasing. In contrast, intensive production is currently increasing and currently around 800 t. There has also been some work carried out to examine the efficacy of more intensive production methods, particularly in Japan and Italy (Dabrowski *et al.*, 2002).

Diversification of salmonid production is the main driving force behind commercial whitefish aquaculture initiatives in northern Europe. Exclusivity for this potential aquaculture production based on *Coregonus* sp. is presently limited to Scandinavia (*C. lavaretus*) and Canada (*C. clupeaformis*) (COSEPAC/COSEWIC, 2005), where recurrent restocking activities have provided the knowledge required to sustain commercial production through the control of reproduction and larval quality and quantity. The cultivation of *Coregonus* species can be qualified as a research and development activity that is appealing, particularly through the perspective of cultivating a white-fleshed salmonid species that corresponds to the trends currently observed in consumer markets, i.e. white fleshed products. However, in contrast to the high-volume cultured salmonids (salmon, trout), whitefishes produce small eggs and have a longer larval stage (egg diameter and length at hatch of 2–3.5 mm and 12–14 mm, respectively, compared to 4–6 mm and 15–20 mm) that required, until recently, the use of live prey in combination with adapted larval feeds (Harris and Hulsman, 1991; Pangle *et al.*, 2003).

12.2.1 Common whitefish, *Coregonus lavaretus*

The common whitefish is an exploited species in several European countries, where it forms the basis of commercial freshwater lake fisheries. World distribution of *C. lavaretus* is shown in Fig. 12.3. Fisheries are often most concentrated during the breeding season, when the mature fish form large spawning

Fig. 12.3. World distribution of *Coregonus lavaretus*.

congregations. At that time, large numbers of fish can be taken by gill netting. Catches are often sent to fresh-fish markets, but some common whitefish are smoked prior to being sold and the roe (eggs) may be used to produce caviar.

As a species, the common whitefish displays great diversity and may occur as several ecotypes, or morphs, within a single lake. Each ecotype usually exhibits distinct differences in morphology and feeding habits and there is often some reproductive isolation among the various morphs that inhabit a lake.

12.2.1.1 Farming of common whitefish

Few publications are available on the cultivation of this species and most originate from Finland. Rearing protocols appear to be successful, at least for the production of early stages for restocking purposes. Commercial production of commercial-size common whitefish appears promising and is possibly a viable alternative to rainbow trout farming in Scandinavia.

12.2.1.2 Broodstock management and hatchery operations

The timing of the breeding season is variable, depending on specific stock and population characteristics and on regional climatic conditions. Most populations of common whitefish seem to spawn during the autumn and early winter, but some are spring spawning. In the cultivation of common whitefish, it has been traditional to take gametes (eggs and sperm) from wild fish, but captive broodstocks are being established. For example, stock comparisons have been undertaken in Finland, there has been some mass selection and a family-based selective breeding programme was initiated in 2004.

Sexual maturity is reached after 3–4 years under commercial culture. Common whitefish eggs are 3–3.5 mm in diameter and females generally have an egg yield of 10–15% of their body mass, but the egg yield from large females may be higher. When common whitefish are being cultured, ripe eggs are usually stripped from the females. The eggs are then fertilized artificially with milt from running-ripe males by the dry method. The milt from several males is mixed and then a portion is added to the unfertilized eggs. Immediately after the milt is added, the eggs and milt are mixed thoroughly to ensure maximum fertilization. Following mixing, the eggs are left to stand for a short time, a small amount of fresh water is then added and the eggs are left to stand again. Finally, the eggs are washed in fresh water to remove excess milt. The newly fertilized eggs are usually hardened for 3–4 h in water and are then transferred to the incubation units. Incubation is usually carried out in hatchery jars; water flowing through the jars is regulated in such a way as to ensure that the eggs remain suspended and that there is a slow and steady mixing of the mass of eggs. Eggs may also be incubated in hatchery trays similar to those used for the incubation of the larger eggs of other salmonids, such as Atlantic salmon, *S. salar*, and rainbow trout, *O. mykiss*.

Egg incubation should be carried out in cool water (4–7°C); at temperatures exceeding 12°C, there may be increased incidence of deformity of the developing embryos and reduced survival. Temperatures as low as 1°C can be used initially, but the temperature should be increased to at least 5°C prior to hatching of the eggs (Rösch, 1995). The rate of development depends on incubation temperature, with development occurring more slowly at a low temperature. There are also differences in developmental rates between stocks and ecotypes (Eckmann, 1987), so the time required to a given developmental stage may vary widely at a given incubation temperature. This means that the time from fertilization to hatch is very variable. This is the result of a combination of stock differences in developmental rates and the fact that fewer degree days are needed when incubation temperatures are low. Degree days is an expression that refers to the incubation temperature in °C multiplied by time in days. For a range of common whitefish stocks and incubation temperatures, the eyed-egg stage is often reached after about 100–160 degree days and the eggs may hatch after 300–360 degree days. After they have reached the eyed stage, the eggs are tolerant of handling and can be moved from one incubator to another and/or transported between hatcheries.

The hatchlings (*c.*11–12 mm; 5–6 mg) have small yolk sacs that are absorbed within 3–5 days of hatching. Consequently, the fish should be provided with food shortly after hatching. The food may be living prey organisms or dry microfeeds manufactured for the early rearing of fish and crustaceans (Dabrowski *et al.*, 2002). Live prey may be small zooplanktonic organisms, e.g. crustacean nauplii and copepodites, collected from nearby lakes, or the nauplii of the brine shrimp, *Artemia salina* (Flüchter, 1980). Descriptions of rearing methods for live-feed organisms and collection methods for zooplankton are well known (Stickney, 2000; Stottrup and McEvoy, 2002; Olsen, 2004) and the development of microfeeds has also been described (Stickney, 2000; Langdon, 2003). Larval whitefish will accept dry feeds soon after hatching and can be weaned satisfactorily using microfeeds developed for marine fish. Once the fish reach a body mass of *c.*50 mg, they may be weaned on to starter feeds used for the production of trout and salmon and, over time, these can be replaced by dry feeds of progressively larger size. Thus, from a body mass of about 50 mg onwards, common whitefish may be reared exclusively on dry feeds. The larvae and small juveniles can be reared at temperatures of 12–18°C and it is usual to hold these small fish under conditions of continuous illumination and very frequent feeding (Koskela and Eskelinen, 1995; Rösch, 1995; Koskela *et al.*, 2002). Commercial producers estimate 60–80% survival from hatch to weaning (Lankinen, personal communication).

12.2.1.3 On-growing to market size

Feeds for the on-growing of common whitefish should be formulated to contain a minimum of 40% protein, as growth cannot be sustained at high rates when feeds of lower protein concentration are provided (Dabrowski *et al.*, 2002; Ruohonen *et al.*, 2003). In practice, good rates of growth may be achieved when feeds contain 45–55% protein, about 25% fat and a low concentration of carbohydrate; common whitefish do not appear to tolerate high percentages of dietary carbohydrate (Ruohonen *et al.*, 2003; Vielma *et al.*, 2003). In common with other fish species, the body composition of farmed common whitefish may be influenced by changing the composition of the feed. For example, increasing the concentration of fat in the feed leads to increased fat deposition in the body (Koskela, 1995; Koskela *et al.*, 1998, Ruohonen *et al.*, 2003).

This restricts the amount of fat that can be included in a feed if negative effects on the final product are to be avoided. When dietary fat concentration is increased, there is increased deposition of fat within the visceral cavity, leading to a high dress-out loss when the fish are harvested and eviscerated. In addition, the use of high-fat feeds results in the production of a fish with a fattier carcass and higher fillet fat. At the same time, there is a reduction in the proportion of moisture; in common with other fish species, the body composition of common whitefish displays an inverse relationship between percentage fat and percentage moisture (Koskela *et al.*, 1998).

In nature, common whitefish are found in cold and cool waterbodies; during on-growing in culture, good rates of growth can be achieved if the fish are held in well-oxygenated water at 12–18°C. Rates of feeding and growth are depressed in warmer water and common whitefish do not tolerate exposure to high temperatures for long periods. Common whitefish can be raised in both fresh water and brackish water. Several types of rearing unit are used for culturing common whitefish; tanks, Danish-style earth ponds and net cages. Danish-style ponds are frequently the rearing unit of choice for the freshwater cultivation of common whitefish. Most often, common whitefish are held in fresh water throughout their life, but in Finland both freshwater and brackish-water culture is practised. In brackish-water culture, the fish are raised in fresh water for about 1 year and are then transferred to net cages at brackish-water farm sites during the spring. Stocking densities are generally 20–30 kg/m^3. The fish are then reared in the net cages for 1–2 years before being harvested at 0.8–1.0 kg. The fish can be fed using a range of automatic feeding systems, but feeding by hand is also used. The whitefish is a visual feeder, so is fed during daylight hours. When feeding by hand, the daily ration is regulated by assessment of satiation, but when automatic feeders are used, the amounts to be fed can be calculated from growth and feeding charts (Koskela *et al.*, 2002).

Common whitefish are susceptible to bacterial diseases such as vibriosis (*Vibrio anguillarum*) and furunculosis (*Aeromonas salmonicida*). The fish are usually vaccinated against these diseases before being moved from the freshwater nursery sites to the cage farms. Vaccines against both vibriosis and furunculosis are produced commercially and these appear to stimulate the immune system and give good protection against these diseases (Lönnström *et al.*, 2001; Koskela *et al.*, 2004).

Whitefish is well known and has a good reputation in Scandinavia; in Finland, production is increasing and predicted to reach 4000 t in the near future. Production and market prospects in other areas may be limited by lack of recognition in other regions of the world.

12.3 Charrs: Biology and Culture

Charrs, genus *Salvelinus*, are fascinating salmonid species with a wide variety of developmental patterns and a diversity of life histories. As a group, charrs have a natural distribution restricted to the northern hemisphere. They occur in the Holarctic and include the most northerly distributed fish found in fresh water. The charrs are primarily fish of fresh waters, but they also occur as anadromous forms. Anadromous charr typically occur towards the northern limits of their species range. In these northern regions, resident, non-migratory charr often occur in the same watercourses as the anadromous forms. The morphological and ecological diversity of the charrs has created problems for taxonomic definition and has made a distinct delimitation of species difficult (Balon, 1980; Nelson, 1994; Groot, 1996). Nelson (1994) recognizes three subgenera of charrs:

- *Salvethymus* is represented by a single species, *Salvelinus svetovidovi* (the long-finned charr), found in north-eastern Russia. The species has a very restricted distribution, inhabiting Lake El'gygytgyn, an oligotrophic lake formed in a meteorite crater.
- *Baione* comprises two charr species, both of which have their natural distributions in North America. The brook charr, *S. fontinalis*, occurs as both anadromous and freshwater forms along the eastern seaboard of North America, whereas the lake charr, *S. namaycush*, is a freshwater species of northern North America.
- *Salvelinus* is the most speciose of the subgenera, with about eight species that include the Arctic charr, *S. alpinus*, the Dolly Varden, *S. malma*, and the bull trout, *S. confluentus*.

Charrs of several species, most notably brook charr (also called brook trout and speckled trout), have been introduced into areas outside of their native range. The North American brook charr has been introduced into Europe, South and Central America, Australasia and Oceania and a few African countries. Lake charr (also called lake trout) have been introduced from North America to Europe, South America and New Zealand. Introduction of lake trout has been made mostly to deep, cold lakes at high latitude and/or high altitude. Arctic charr have been introduced into several European countries and there has also been extensive translocation between waterbodies within the species' native range (Lever, 1996).

There is artificial propagation of several charr species. Most are raised for stock enhancement in connection with lake rehabilitation programmes or for sport-fishing purposes. Hybrids between some *Salvelinus* species, such as splake (♀ lake charr × ♂ brook charr), are also produced for stocking purposes, the fish often being released into lakes as a sport-fishing resource. In addition, some of the charr species, particularly the brook charr and Arctic charr, are farmed as table fish for direct human consumption.

12.3.1 Arctic charr, *Salvelinus alpinus*

The Arctic charr has a circumpolar distribution, where it is found in cool and cold waters within the Arctic, boreal and temperate regions of the Holarctic (Fig. 12.4). It is the charr with the most northern occurrence of the *Salvelinus* species. At the northernmost edge of the species distribution, Arctic charr may be found inhabiting lakes that remain ice-covered throughout the summer in

Fig. 12.4. World distribution of *Salvelinus alpinus*.

some years. Although populations of Arctic charr are often landlocked, typically within deep lakes, many northern populations occur in watercourses with access to the sea and some of these have an anadromous component (Balon, 1980; Groot, 1996; Jonsson and Jonsson, 2001; Klemetsen *et al.*, 2003; Rikardsen *et al.*, 2004).

The anadromous Arctic charr generally migrate to the sea during the summer months and spend up to 8 weeks in seawater before returning to fresh water. Both immature and sexually mature fish take part in the seasonal migrations between fresh water and the sea. The age at which immature fish first participate in the seasonal migration is variable and is largely dependent on local growth conditions. Some immature fish may undertake their first migration to the sea at an age of 2–3 years, whereas others do not start to migrate until they are 4–5 years old. The immature fish return to fresh water to overwinter each year and they may participate in several annual migrations before they mature. The mature charr that migrate to the sea in spring return to fresh water to breed and then overwinter, but the mature individuals may not spawn each year. Anadromous and resident components of charr populations can interbreed and each can give rise to the other (Klemetsen *et al.*, 2003; Rikardsen *et al.*, 2004). In addition to existing as migratory and non-migratory fish, Arctic charr also display considerable variations in body form and meristic characters. These differences are often related to the food sources on which the fish feed, resulting in trophic polymorphism (Jonsson and Jonsson, 2001; Klemetsen *et al.*, 2003).

Spawning generally takes place in freshwater lakes and streams during autumn and early winter, although there are some spring-spawning populations. The numbers of eggs produced by the mature females is variable, but relative fecundity is usually within the range of 1500–3000 eggs for each kg body weight. The eggs (4–4.5 mm diameter) are demersal, remain buried in the gravel substrate throughout the winter, and hatch during spring. When the nutrients present in their yolk sac have been used, which takes several weeks, the young fish emerge from the gravel. The fish are 20–25 mm long at emergence and they remain close to the bottom, hiding among stones and gravel. To feed, they swim quickly to intercept small food items, such as crustaceans and insect larvae, and then return to their hiding place. As the fish grow larger, they move to different habitats; some move into deeper water, some move to rocky areas in the littoral zone and some adopt a pelagic way of life.

Some lakes may contain four sympatric morphs – planktivorous, small and large benthivorous and piscivorous – which have some degree of reproductive isolation from each other. Preferential mating within morphs and adherence to different spawning sites may lead to behavioural isolation, which results in the maintenance of the trophic morphs. Further, as intermediate offspring, resulting from interbreeding between two different morphs, may be inferior competitors to the trophic specialists, they are expected to be eliminated from the population. The inherent developmental plasticity of the Arctic charr (Jonsson and Jonsson, 2001; Klemetsen *et al.*, 2003) has both advantages and disadvantages for the establishment of the species in culture. The fish may, for example, be able to adapt to a range of feeding conditions, but variability in developmental rates, growth and size and age at maturity within a cohort may be deemed disadvantageous from a farming viewpoint.

The Arctic charr is a valuable food fish in certain areas within its geographic range and it is also prized as a sports fish. Commercial fisheries are developed in North America and parts of Europe, and there is also cultivation for restocking and some farming for the table. Arctic charr have been farmed in some quantity in Scandinavia and Canada for several years and there is also farming to supply local markets in some alpine regions of Europe.

12.3.1.1 Farming of Arctic charr

Charr of anadromous populations tend to grow larger (frequently to several kg) and mature later than their landlocked conspecifics and several studies have been carried out on anadromous charr to assess their potential for commercial cultivation (Jobling *et al.*, 1993, 1998; Delabbio, 1995; Johnston, 2002). Given the wide variation in growth rates and sizes and ages of maturity among wild charr, and among representatives of different populations in culture, it is important to exercise some care when making the choice of a charr population that will form the basis of a farmed stock. Domestication, and selective breeding, programmes are currently under way in some countries, e.g. Sweden, Iceland and Canada, and commercial producers may be provided with the offspring from selected stocks for on-growing. Aquaculture is small and annual production is less than 2000t (FAO, 2007).

At present, Arctic charr from several sources, and from both anadromous and landlocked populations, are being used in pilot-scale and commercial farming operations. Small-scale producers may rely on farming fish taken from local waters and there is also some capture-based culture involving the on-growing of wild-caught fish to market size. Although farming of charr has proven to be commercially viable in several countries, there are several problems that may complicate successful production in areas outside its native range (Jobling *et al.*, 1993, 1998; Johnston, 2002). One major problem that hinders the establishment of charr farming in some regions relates to the requirement of the species for water of low temperature. Low water temperatures are required, particularly during the final stages of the reproductive cycle, during egg incubation and for the early development of the hatchlings. In addition, high water temperatures may increase the risk of disease outbreaks in larger fish, either broodstock or those that are close to market size. This means that the successful holding of broodstock, and hatchery operations, are only possible in locations where river and lake temperatures are low at the appropriate times of the year, or where suitable well bore-water is available.

12.3.1.2 Broodstock management and hatchery operations

If cultured broodstock females are held at high temperature during the late summer and autumn, gamete quality may be compromised. There may be disorders in gonad development, oocyte atresia and degeneration, inhibition of ovulation or a delay in the timing of spawning. For example, ovulation may be delayed in females held at temperatures over 8°C for long periods and may be inhibited at 11°C and over. In addition, female broodstock held in water with a temperature over 5°C during the final stages of the reproductive cycle may produce eggs of poor quality due to the rapidity with which overripening and egg deterioration occurs at elevated temperature (Jobling *et al.*, 1993, 1998).

Eggs stripped from ovulated females are fertilized with milt stripped from running-ripe males. Most often, the dry method, common for the fertilization of the eggs of most salmonid species held in culture, is used, although the wet method of fertilization can also be employed (Pennell and Barton, 1996; Johnston, 2002; Stead and Laird, 2002). Milt stripped from the males is first checked for sperm motility, and milt showing poor activity is discarded. The milt from several males is mixed and a portion of the milt (at least 1 ml of fresh milt per litre of eggs, and more if the milt has been stored for any length of time) is then added to the unfertilized eggs. Immediately after the milt has been added, the eggs and milt are mixed thoroughly to ensure maximum fertilization. Following mixing, the eggs are left to stand for about 1 min. An equal volume of fresh water is then added and the eggs are left to stand for about 1.5 min. Finally, the eggs are washed in fresh water to remove excess milt. Fertilization success using this method should exceed 90%. Eggs may be transferred to incubators immediately or, as is more usual, they may be hardened prior to being placed in incubators. Egg hardening, which takes about 1 h, is carried out in fresh water. The eggs are sensitive to disturbance during hardening so should not be handled or subjected to movement for about 3–4 h. During hardening, the eggs absorb water and increase in size. Once hardening is complete, the eggs will usually be disinfected with a buffered iodophor prior to being placed in the incubators. Incubators can be of several types; silo or cylinder, vertical-flow stack incubator (Heath incubator) or horizontal flow troughs with baskets, boxes or trays.

Exposure of the eggs to temperatures over 10°C is deleterious to the development and survival of the embryos. Within the hatchery, an incubation temperature of 4–6°C will usually give satisfactory results with respect both to rates of development and survival. The rate of development depends on incubation temperature and embryos reach the eyed stage after 200–230 degree days. Once they have reached the eyed stage, the eggs are more resistant to disturbance than at earlier stages of incubation and can be handled, disinfected, moved and transported within the hatchery. The time from egg fertilization to hatch is 350–500 degree days, depending on the source of the eggs and incubation conditions. Immediately following hatching, the young fish depend on their yolk sac for their nutrient supply. Exogenous feeding occurs around the time of swim-up, when the fish have little of the yolk sac remaining; swim-up occurs at about 250–300 degree days after hatching. Suitable temperatures for the start-feeding and early rearing of charr lie within the range 6–8°C, although temperatures may be increased to 12–18°C as the juveniles grow larger (Jobling *et al.*, 1993, 1998; Johnston, 2002).

The young fish will accept dry feeds from the time they start to feed (Jobling *et al.*, 1993; Dick and Yang, 2002; Johnston, 2002), initially of small particle size but of increasing size as the fish grow. Given the low production volumes of Arctic charr, most feed companies do not manufacture dry feeds specifically for farming charr. There has, however, been some production of feeds fortified with astaxanthin, a carotenoid pigment, to enhance the development of red coloration in the fillet of farmed charr in the weeks prior to harvest. Nevertheless, it is salmon and trout feeds that are used most commonly for farming charr. Feed particle sizes used for charr generally follow those recommended for rainbow trout in the feeding guides published by the feed companies. As a general rule of thumb, feed pellets given to juveniles, from about 2 g and upwards, should have a diameter that is equivalent to about 2% of the fork length of the fish and increasing to about 2.5% when the fish reach 100 g. Feeding is generally carried out using automatic feeders timed to release pellets at frequent intervals, but feeding by hand is also quite commonly practised.

12.3.1.3 On-growing to market size

On-growing of Arctic charr to market size is carried out in several types of rearing unit; in Danish-style ponds, in cages placed in lakes and brackish waters and in tanks and raceways. Although rearing in fresh water is the most common form of farming, some charr are raised in brackish waters and there is some use of full-strength seawater for the production of charr during the summer months. Young charr of anadromous origin appear to undergo a parr–smolt transformation, similar to that seen in Atlantic salmon during the course of the spring, and this enables them to survive and grow in seawater during the summer. Year-round production of charr has, however, been restricted due to the low tolerance of the fish to seawater during the winter months, and particularly when temperatures are low. In addition, sexual maturation inhibits the ability of the fish to hypo-osmoregulate.

During on-growing, good rates of growth may be achieved at temperatures within the range 8–15°C and feed:gain ratios should be comparable to those of other farmed salmonids, e.g. Atlantic salmon and rainbow trout. The Arctic charr is a relatively robust species and it thrives at, and may even prefer, high stocking densities. Nevertheless, low-density culture (5–25 kg/m^3) is the norm when charr are held in ponds and cages, whereas stocking densities can be much higher (40–150 kg/m^3) when the charr are raised in tanks and raceways (Jobling *et al.*, 1993, 1998). The fact that the Arctic charr is tolerant of such high stocking densities makes it a candidate for rearing in water reuse and recirculation systems, where effective use of rearing unit volume is at a premium. When charr are held in such systems, it is essential that there is an adequate reserve, in the event of a failure of components in the rearing system. In the absence of an adequate back-up, a failure in the system will lead almost invariably to substantial losses of stock within a very short period of time, even though charr can withstand short-term exposure to low dissolved oxygen concentrations.

Farmed Arctic charr may be subject to attack by a variety of disease organisms, but the charr does not appear to be more susceptible to disease outbreaks than other commonly-cultured salmonids. Bacterial diseases encountered in charr farming include enteric redmouth (ERM), *Yersinia ruckeri*, furunculosis, *A. salmonicida*, and bacterial kidney disease (BKD), *Renibacterium salmoninarum*; all of these are relatively common salmonid bacterial pathogens (Roberts and Shepherd, 1997). When held in brackish water or seawater, the fish may be subject to vibriosis (*V. anguillarum*), but vaccines that provide protection against this pathogen are available. Farmed charr may also be attacked by several skin and gill micro and macroparasites.

Charr are often grown to *c*.300 g and are then marketed as portion- or pan-size fish, but larger fish are also produced. Arctic charr weighing 1–2 kg may be sold fresh as whole round fish (eviscerated, but with head on) or they can be processed into fillets, steaks and cutlets. Some of the large Arctic charr are smoked prior to sale. One problem encountered in farming charr is that some of the fish may become sexually mature before they reach market size. The problem of early maturation is more prevalent among males than females, as, in general, the males mature at a younger age and smaller body size than do the females. Not only does sexual maturation lead to a reduction in the rate of growth, but there is also a mobilization of material from the fillet, resulting in poorer quality fish with respect to both nutritional and textural properties.

An additional problem faced by Arctic charr farmers is that markets are already well supplied with farmed salmonids of other species, most notably Atlantic salmon and rainbow trout. These latter species are farmed in large quantities, are marketed internationally and are well known as food fish of good quality. On the other hand, the eating qualities of charr are not particularly well known outside the area of its natural distribution, even though the charr has much to commend it as a culinary commodity. For example, the flesh of charr is more delicate in both texture and flavour than that of most other farmed salmonids and fillet fat also tends to be lower, particularly than that of farmed Atlantic salmon. In addition, charr have smaller heads relative to their body size than many other salmonids and this means that the fillet yield from Arctic charr is high (56–65%) (Jobling *et al.*, 1993, 1998; Delabbio, 1995; Johnston, 2002). Despite its favourable characteristics, the Arctic charr is unlikely to be farmed in volumes that approach those of its relatively low-cost salmonid competitors, the rainbow trout and Atlantic salmon. The requirement for fresh water of good quality will pose one limitation on the potential for increasing the scale of charr farming in the future, and this will preclude the possibility of developing charr to the same extent as that of Atlantic salmon and Pacific salmons, *Oncorhynchus* spp., that are on-grown in the sea. As a consequence, production of farmed Arctic charr is likely to remain low to moderate, with the charr being farmed to meet the needs of niche markets that include the restaurant trade. Under such conditions, it is expected that the Arctic charr could achieve relatively high wholesale prices, due to limited supply and the perception of the fish as being of high-value and representing a product that possesses unique attributes.

12.3.2 Brook charr, *Salvelinus fontinalis*

The brook charr (*S. fontinalis*) is also commonly referred to as brook trout or speckled trout, and in French as truite mouchetée or omble de fontaine. Its native range covers temperate waters throughout north-eastern North America (Scott and Crossman, 1973) (Fig. 12.5). As with other salmonids, it has been widely

Fig. 12.5. World distribution of *Salvelinus fontinalis*.

introduced throughout the world (Groot, 1996). It is closely related to other charrs (*Salvelinus* spp.) and hybridizes readily with Arctic charr (*S. alpinus*), both in nature (Hammar *et al.*, 1991; Bernatchez *et al.*, 1995; Glémet *et al.*, 1998) and under artificial conditions (Dumas *et al.*, 1992, 1995, 1996). Other common artificial hybrids include the 'splake' (male brook charr × female lake charr, *S. namaycush*) and 'tiger trout' (male brook charr × female brown trout, *S. trutta*).

Compared to most other salmonids, the relatively stubby shape and square tail of brook charr make it less efficient at fast swimming but is ideal for life in shallow, slow-flowing waters (Power, 1980). It is carnivorous, feeding on aquatic and terrestrial invertebrates, fish and, occasionally, small mammals (Scott and Crossman, 1973). Although most populations remain resident in fresh water, there are many coastal anadromous populations that migrate to the sea for part of the summer but then overwinter in fresh water (Power, 1980; Groot, 1996). Seawater-adapted brook charr take on the silvery appearance characteristic of anadromous salmonids during their marine phase (Power, 1980). Brook charr spawn naturally in the autumn (Scott and Crossman, 1973; Power, 1980), although this can range from late summer through to early winter at their northern and southern range limits, respectively (Groot, 1996). Eggs are smaller than in salmon or trout, ranging from 3.5 to 5.0 mm in diameter (Scott and Crossman, 1973).

12.3.2.1 Farming of brook charr

Brook charr was the first fish species to be cultured in North America (Stickney, 2001). In their classic overview of aquaculture, Bardach *et al.* (1972) called it one of the 'big three' species in commercial trout culture (together with rainbow trout and brown trout). From an historical perspective, it is interesting to note that Arctic charr was not mentioned by Bardach *et al.* (1972) and Atlantic salmon was mentioned only briefly. Power (1980) began his comprehensive review of the biology of brook charr by stating that it 'is unquestionably the best known and most studied of the charrs'. This statement is perhaps no longer valid, given the more recent interest in the biology and culture of Arctic charr, but the fact remains that there is a large body of scientific and practical information available on the biology and culture of brook charr. Annual aquaculture production has risen slightly during 2003–2004 to around 1000 t (FAO, 2007).

12.3.2.2 Broodstock management and hatchery operations

Photoperiod manipulation can be used to advance or delay brook charr spawning (Henderson, 1963; Carlson and Hale, 1973; Holcombe *et al.*, 2000), making it relatively easy to obtain an 'off-season' supply of eggs (Fig. 12.6). Broodstock are easy to sex, with the males developing much deeper colour, more prominent white leading edges on their ventral fins and a hooked lower jaw. Gametes are collected from broodstock by stripping in the typical salmonid fashion. Milt can be cryopreserved for use at a later date by dilution in buffered physiological saline containing cryoprotectant, followed by rapid freezing in liquid nitrogen vapour (Lahnsteiner, 2000). Eggs are typically 3.3–5.0 mm

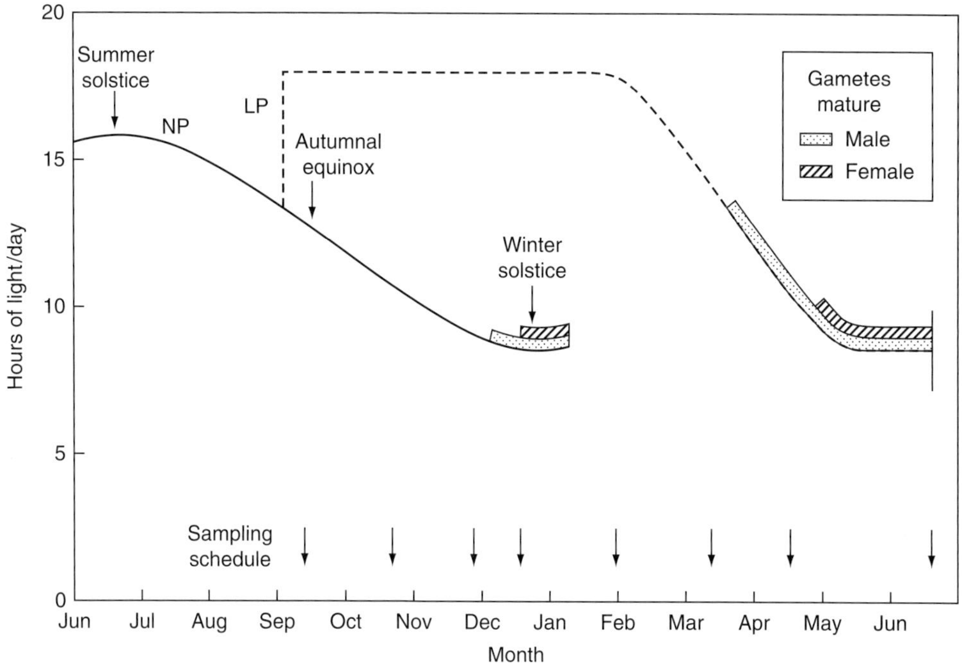

Fig. 12.6. Example of off-season spawning in brook charr through photoperiod manipulation. (NP and LP = natural and artificial (long) photoperiods, respectively; reproduced with permission from Holcombe *et al.*, 2000.)

diameter and usually fertilized using the dry method, whereby milt is added to the eggs and they are mixed gently in the initial absence of water. The spermatozoa become activated once they are diluted in ovarian fluid. Within 1–2 min, excess milt is rinsed off the eggs and they are left in clean water to harden prior to surface disinfection and transfer to standard salmonid incubators. Fertilization can be close to 100% but, as with other salmonids, fertilization success of eggs from first-time spawners is often lower than for eggs from second-time spawners (Dumas *et al.*, 1996).

The optimum temperature for brook charr egg incubation is 6°C; temperatures up to 12°C are tolerable, but with reduced survival and higher abnormality rates (Hokanson *et al.*, 1973). Within this temperature range, the effect on survival is most pronounced during early incubation, prior to eggs reaching the eyed stage; after this, temperatures can be raised above 8°C (Marten, 1992). However, higher temperatures during the interval from eyed egg to hatch result in shorter alevins at hatch (Marten, 1992). The duration of the incubation period, when standardized through degree days, is longest at the optimum egg incubation temperature, falling off rapidly at lower temperatures and more slowly at higher temperatures. Marten (1992) derived a formula with high predictive power for estimating time to eyed egg and hatch for a domesticated strain of brook charr. Although this demonstrates the clear role of temperature as the primary determinant of developmental rate prior to hatch, the absolute duration of the incubation period in brook charr varies among strains (Baird *et al.*, 2002).

Alevins can be reared at higher temperatures than eggs, but mean and maximum temperatures should not exceed 16 and 20°C, respectively (Hokanson *et al.*, 1973). When incubated at 8–13°C, brook charr and Arctic charr have similar yolk weights at hatch (78 and 76% of total weight, respectively) and do not differ in yolk conversion efficiency (Dumas *et al.*, 1995). When reared at 8°C, brook charr fry weigh approximately 160 mg when they start feeding, increasing to 3.3 g in 16 weeks when fed commercial salmonid starter diets (Gunther *et al.*, 2005). There is no change in body composition as the fish become larger during this initial 16 weeks of growth (Gunther *et al.*, 2005). Survival from hatching to weaning was reported to be 95% (Dumas *et al.*, 1992).

12.3.2.3 On-growing to market size

Standard salmonid rearing units and protocols are used for brook charr culture. These can range from simple raceway and pond systems to more sophisticated and costly cage systems. Densities are typically 25–30 kg/m^3 (Vijayan and Leatherland, 1988; Vijayan *et al.*, 1990). Most brook charr culture is carried out in fresh water, but the marine culture potential of this species has been considered in Québec since the early 1980s. The first experimental trials occurred in the Baie de Gaspé (Lafleur *et al.*, 1986). The main interest for sea-cage production is linked to the gradual introduction of more stringent environmental regulation for the preservation of the freshwater biota. Brook charr reared in sea cages also take on a silvery colour (Power, 1980), which may add to market value.

Brook charr grow well when fed diets formulated for rainbow trout, attaining higher tissue protein and fat levels than rainbow trout (Rasmussen and Ostenfeld, 2000). Although juvenile and adult brook charr grow at temperatures as low as 5°C (Power, 1980), the best growth rates are attained at 13–19°C, with 13°C being the optimum for fish in the 0.6–30 g range (Dwyer *et al.*, 1983) and 16°C the optimum for larger juveniles and adults (Hokanson *et al.*, 1973). Sexual maturation is inhibited at temperatures above 19°C and broodstock females should be kept below 12°C (ideally at around 9°C) to ensure good egg quality (Hokanson *et al.*, 1973). The upper incipient lethal temperature of brook charr is 25°C (Power, 1980), although brief exposure to temperatures as high as 30°C can be tolerated (Benfey *et al.*, 1997; Galbreath *et al.*, 2004, 2006).

At transfer to seawater, brook charr should be at least 19 cm long (McCormick and Naiman, 1984), sexually immature (Sutterlin *et al.*, 1976) and moved into water at a salinity below 35‰ (Boeuf and Harrache, 1984). Saltwater adaptability is also dependent on season (Besner and Pelletier, 1991) and temperature (Saunders *et al.*, 1975), and is limited by sexual maturation (Le François *et al.*, 1997; Le François and Blier, 2000). Survival is maximized with spring transfer and decreases gradually to a minimum with autumn transfer (Besner and Pelletier, 1991). Gonadal development during the summer is linked tightly with a reduction in branchial osmoregulatory function and seawater survival. The use of sterile fish resulted in 100% survival and the conservation of adequate seawater adaptability during the critical period of sexual maturation in the autumn (Le François *et al.*, 1997; Le François and Blier, 2000, 2003).

In 2000, a comprehensive biological, technical and socio-economic feasibility study was launched by the government authorities of Québec (Canada) in collaboration with the private sector to evaluate seasonal on-growing of 0^+ triploid (sterile) brook charr in sea cages (Le François *et al.*, 2002). Triploidy is viewed generally as an adequate genetic containment measure for the mass production of fish in coastal waters. However, it can also cause a significant reduction in seawater adaptability in several salmonid species (Galbreath and Thorgaard, 1995; Withler *et al.*, 1995). The compulsory use of suboptimal-sized triploids meant that two important negative attributes for successful seawater adaptability constrained the trial. Although the use of salty diets reduced post-transfer mortality of 35–40 g triploids by 50%, chronic mortality was observed (Lamarre and Le François, 2003). The economical viability and social acceptability of seasonal on-growing activities was seriously challenged and the project was abandoned in 2005.

Compared even to other salmonids, brook charr is a very hardy species that adapts easily to culture conditions. It exhibits the typical salmonid physiological responses to acute handling stress (Biron and Benfey, 1994; Benfey and Biron, 2000) and exhaustive exercise (Hyndman *et al.*, 2003a,b), but can be reared at much higher densities than Atlantic salmon (> 200 kg/m^3; Benfey and O'Keefe, unpublished data). However, an evaluation of performance at 30, 60 and 120 kg/m^3 showed that the best growth and food conversion efficiency was obtained at 30 kg/m^3 (Vijayan and Leatherland, 1988; Vijayan *et al.*, 1990).

Numerous parasites affect wild brook charr (Scott and Crossman, 1973; Power, 1980) and these can be transferred to cultured stock through the incoming water supply. Of notable concern is the freshwater copepod, *Salmincola edwardsii*, an ectoparasite that can cause extensive gill damage, leading to mortality in farmed brook charr. Ememectin benzoate, commonly used to treat farmed Atlantic salmon for sea lice infestations, is also effective for ridding brook charr of parasitic *Salmincola* (Duston and Cusack, 2002). Brook charr are also susceptible to a number of viral and bacterial salmonid diseases, including bacterial kidney disease (Mitchum and Sherman, 1981) and furunculosis (Cipriano *et al.*, 2002; Perry *et al.*, 2004). As with other salmonids, it is possible to protect brook charr against furunculosis through either immunization (Marquis and Lallier, 1989) or immunostimulation (Anderson and Siwicki, 1994).

The current resource status of brook charr is very much more for sport fishing (stocking and U-fish operations) than for the retail and restaurant markets. Many Canadian provinces and US states maintain significant brook charr stocking programmes to support angling. For instance, an average of 1.36 million brook charr were stocked per year in Ontario (Canada) between 1990 and 1999 (Kerr, 2000). As a commercial aquaculture species, 1.2 t of farmed brook charr were produced in Canada in 2003, with a value of CAN$12.9 million

(DFO, 2006). Québec dominates the commercial production of this species in Canada, with 54% of total freshwater production in 2005. The fact that most brook charr culture is for stocking purposes has limited the effort put into breeding programmes. A selection programme based on the Rupert strain and aimed at the commercialization of brook charr for the consumer market was launched in Québec recently in order to reduce early maturation problems in males and, to a lesser extent, improve growth rate and disease resistance.

Product yield and appearance can be a problem for farmed brook charr. Gutted and fillet yields are lower than for farmed rainbow trout (Rasmussen and Ostenfeld, 2000), but not different from Arctic charr (Dumas *et al.*, 1996). The attractive coloration of live brook charr disappears rapidly after slaughter, making them less attractive than other salmonids when sold whole or as unskinned fillets.

12.3.2.4 Future perspectives

Probably the biggest hurdle to successful brook charr aquaculture is the tendency for fish of both sexes, but especially males, to mature before reaching market size (Carlson and Hale, 1973; McCormick and Naiman, 1984; Boulanger, 1991). Early maturation is recognized as a significant constraint to the commercial viability of Arctic charr aquaculture and comparative studies have shown that brook charr mature even earlier than Arctic charr when cultured under the same conditions (Dumas *et al.*, 1996). Early maturation of males can be addressed by the production of all-female populations (Boulanger, 1991; Galbreath *et al.*, 2003; Sacobie and Benfey, 2005). When combined with triploidy induction, early maturation of females can also be eliminated (Boulanger, 1991; Dubé *et al.*, 1991; Galbreath and Samples, 2000; Schafhauser-Smith and Benfey, 2001). Triploid brook charr grow well in fresh water (Boulanger, 1991; O'Keefe and Benfey, 1997, 1999) and can be purchased commercially in Québec.

A clear research priority for the development of brook charr as a significant commercial aquaculture species should be to determine the extent of natural variability and heritability of commercially important production traits and to take advantage of gains possible through selection for traits that vary among strains, such as development rate, growth rate and disease resistance (e.g. Baird *et al.*, 2002; Perry *et al.*, 2004; Volkman *et al.*, 2004). Related to this, very limited work has been done to date on developing species-specific microsatellites and quantitative trait loci for brook charr (Perry *et al.*, 2005); the application of such molecular tools has proven very beneficial for the development of commercial breeding programmes in Atlantic salmon and rainbow trout.

Another research priority for brook charr is the development of formulated feeds specific to this species. There is little information available on diet utilization and nutritional requirements of farmed brook charr and they are generally fed diets formulated for rainbow trout. This is in spite of differences between these two species in thermal optima, physiology, life-history characteristics and behaviour. Although brook charr clearly grow well on rainbow trout diets, their flesh quality and culture performance would likely be enhanced if fed diets formulated to meet their species-specific requirements.

The brook charr is a hardy salmonid species that adapts readily to artificial culture. It has been reared in captivity in North America for over 200 years. However, as was the case two centuries ago, the primary focus today remains on hatchery production of fish for stocking purposes rather than for the retail market. Although the biology of brook charr is well studied, there is little information on optimal conditions for commercial culture compared to species such as Atlantic salmon and rainbow trout. Advances in breeding programmes, diet formulation and marine culture strategies will help to increase the potential of brook charr as an aquaculture species.

12.4 Atlantic Salmon and Trouts: Biology and Culture

The Atlantic salmon and trouts are freshwater and anadromous salmonids that occur naturally in the North Atlantic basin. They are found on both sides of the Atlantic Ocean; in north-eastern North America and in Europe. Atlantic salmon, *S. salar*, and brown trout, *S. trutta*, are the best-known of the species; both are autumn spawning and both species have anadromous and freshwater populations (Pennell and Barton, 1996; Klemetsen *et al.*, 2003). The migratory, anadromous brown trout is commonly called seatrout.

Both Atlantic salmon and brown trout are important for recreational fisheries and both have been introduced to waters outside the native range. For example, there have been successful introductions of brown trout to at least 25 countries and self-sustaining populations of brown trout are found on all continents, with the exception of Antarctica (Lever, 1996). The success or failure of an attempted introduction of brown trout to a given waterbody is often governed by temperature conditions; in common with other salmonids, the brown trout is a relatively stenothermal coldwater species. There is some cultivation of brown trout, but most is for stocking purposes rather than directly for the table. Most brown trout cultivated for the table are of the seatrout form and farming amounted to 5000–10,000 t annually in the period 1995–2003 (FAO, 2007).

12.4.1 Atlantic salmon, *Salmo salar*

The Atlantic salmon is usually anadromous, although there are some non-anadromous freshwater resident populations. The geographic distribution of the species encompasses the northern regions of the Atlantic basin. Populations occur naturally in countries that lie along both the eastern and western edges of the North Atlantic Ocean (Lever, 1996; Klemetsen *et al.*, 2003) (Fig. 12.7). Both within and among populations, Atlantic salmon may display differences in freshwater habitat use, in the length of time they remain in fresh water before undergoing the parr–smolt transformation and migrating to the sea and

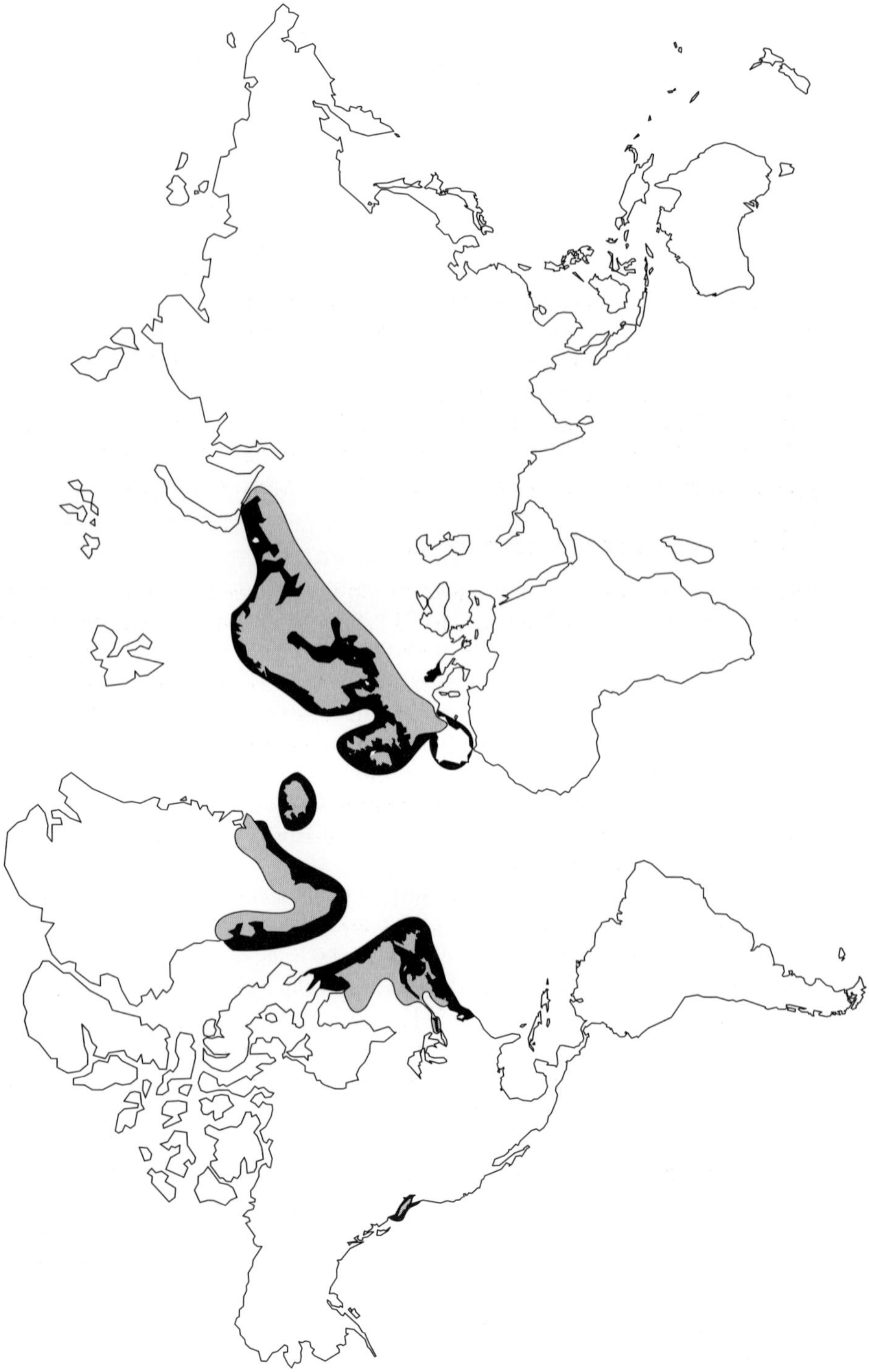

Fig. 12.7. World distribution of *Salmo salar*.

age at maturity. These differences result in the revelation of a considerable diversity in life history when individuals and populations are compared (Jones, 1959; Gibson, 1993; Groot, 1996; Klemetsen *et al.*, 2003).

Attempts have been made to introduce the Atlantic salmon to areas outside its natural range. Most attempts have been unsuccessful, with the fish being unable to establish self-sustaining populations (Lever, 1996). Introductions of the species to Australia (Tasmania) and Chile have given rise to the development of salmon aquaculture in these two countries. The Atlantic salmon is a fish of considerable economic importance. It is important for sport fisheries, it supports limited commercial fisheries and it forms the basis of aquaculture industries in northern Europe, Tasmania, Chile and North America, particularly Canada. The production and commercial importance of farmed salmon far exceeds that of the exploitation of wild stocks. Annual aquaculture production exceeded 1 Mt in 2001 and reached 1.24 Mt in 2004 (FAO, 2007). A number of books covering the biology, conservation and management of Atlantic salmon have been published (e.g. Jones, 1959; Mills, 2000, 2003; Prévost and Chaput, 2001) and details of salmon cultivation are described in several volumes (e.g. Pennell and Barton, 1996; Stead and Laird, 2002).

12.4.1.1 Life cycle of Atlantic salmon

Atlantic salmon spawn in fresh water during the autumn and early winter. The maturing adult fish, which are usually 3–6 kg in body weight, may enter fresh water from the sea several months prior to the time at which they will spawn. The Atlantic salmon is iteroparous, meaning that some individuals survive to spawn several times, i.e. over several breeding seasons. Thus, Atlantic salmon differ from most species of Pacific salmon, *Oncorhynchus* spp., which are semelparous, and die after a single spawning season. Nevertheless, relatively few (probably less than 10%) Atlantic salmon survive to spawn for more than one breeding season; the survival rate of females is higher than that of males. The Atlantic salmon are usually sexually mature as 1–3-year-old sea-winter fish, and sometimes at older sea ages. In addition, some male Atlantic salmon become mature as parr without having undertaken a migration to the sea. This means that Atlantic salmon show a wide range of age and size at maturity and individuals with markedly different ages at maturity may occur within the same population (Fig. 12.8) (Groot, 1996; Klemetsen *et al.*, 2003; Esteve, 2005).

The number of eggs produced (fecundity) and egg size tend to increase with the increasing size of the female. Although absolute fecundity varies greatly among females, owing to the highly variable adult body size, relative fecundity varies much less. On average, female salmon produce around 1500 eggs/kg body weight, although relative fecundity can vary from *c.*1200–3000 eggs/kg female body weight. Atlantic salmon lay their eggs in gravel nests called redds. The process of redd-cutting creates pockets of eggs overlain by loose gravel, from which fine sediment (sand, silt and clay) has been removed. Sediment removal from the gravel, usually by being transported away in flowing water, ensures that the interstitial pore spaces are sufficiently large to guarantee passage of oxygenated water through the redd. Successful egg incubation requires that the oxygen concentration within the redd is sufficient to support diffusive

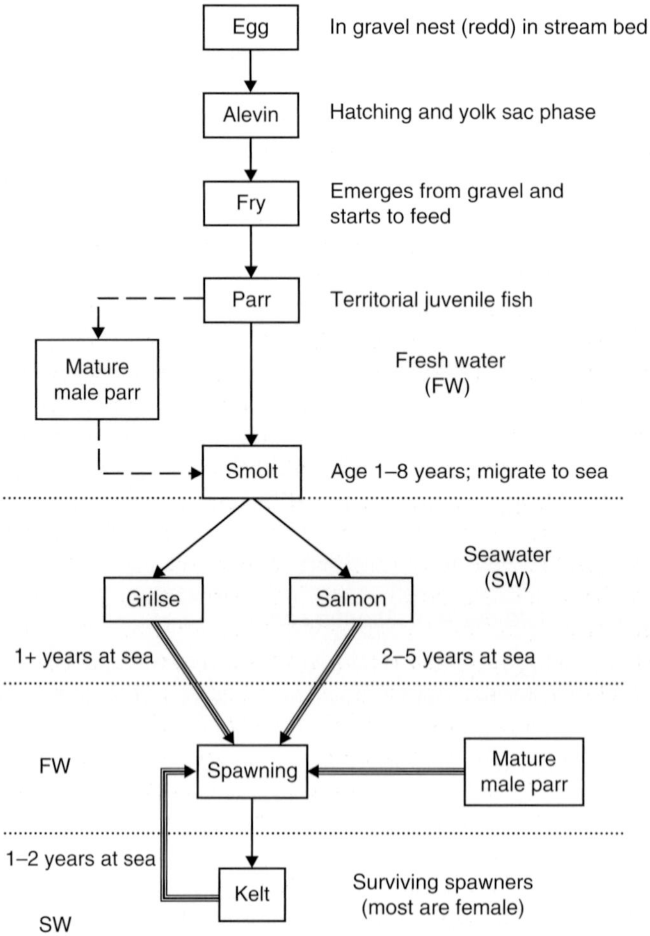

Fig. 12.8. Life cycle of the Atlantic salmon, *Salmo salar,* illustrating different developmental (life history) pathways.

oxygen exchange across the egg membrane at the different stages of embryonic development. Water flow through the redd must also be adequate to flush potentially harmful metabolites from the vicinity of the developing eggs (Groot, 1996; Esteve, 2005).

Atlantic salmon eggs are large (5–6 mm diameter) and incubation time is long; approximately 230 degree days are required from the time of fertilization to the eyed-egg stage and it takes 500–530 degree days from fertilization to hatch. The fish hatch at a relatively advanced stage of development. They are quite large (15–25 mm) at the time of hatching, although they still have a large yolk sac. Following hatch, the young fish, or alevins, remain buried in the gravel of the redd and use the yolk as their source of nutrition. When most of the yolk sac nutrients have been used, the fish emerge from the gravel (swim-up phase) and, during this brief period, the fish respond positively, instead of negatively, to light. Emergence from the gravel redd precedes first-feeding; once the fish emerge, they begin to respond to potential food items and commence feeding. The young fish are exclusively carnivorous; their diet usually contains a high proportion of insect larvae, although other organisms that make up the aquatic drift are also taken (Jones, 1959; Gibson, 1993).

The freshwater parr stage can last from 1 to 8 years, after which the majority of the survivors (5–10% of the initial number of fertilized eggs) undergo the physiological and behavioural changes (parr–smolt transformation) that are required for survival in the marine environment (Fig. 12.8). Some of the male parr do not undergo the parr–smolt transformation but become sexually mature and produce sperm that can fertilize the eggs produced by adult female salmon. Whether or not a male parr matures will depend on both environmental and genetic factors. A good food supply and water temperatures that allow rapid growth will tend to lead to an increase in the proportion of male fish that mature at the parr stage. In addition, male offspring of early-maturing males are more likely to mature as parr than are the offspring of large late-maturing male salmon. During the parr stage, the young fish are territorial and tend to stay close to the bottom, remaining concealed from potential predators in areas away from very strong currents. The parr emerge from their hiding places to feed on the drift fauna, small animals that are carried in the water current. The young salmon have characteristic camouflage coloration, comprising a series of dark blue-grey parr markings superimposed on a lighter brownish and yellow-green background (Jones, 1959; Gibson, 1993; Groot, 1996).

During parr–smolt transformation, the cryptically coloured, stream-dwelling juvenile (parr) changes to become a silvery, streamlined, pelagic fish (smolt) that is adapted for a life in the marine environment. In the typical parr–smolt transformation, there are a series of changes in morphology, physiology and behaviour (Hoar, 1988; Pennell and Barton, 1996; Stickney, 2000; Stead and Laird, 2002). These changes develop over the course of a few weeks during spring and culminate in downstream migration, followed by a period of residence in the marine environment. While the changes are interrelated, they may not be linked particularly tightly to each other. In other words, the parr–smolt transformation is not a single process in which all the changes are controlled strictly by the same limited set of effectors; different characteristic features of the parr–smolt transformation develop at different rates and can become uncoupled from each other (Hoar, 1988).

If the smolt is to survive in the marine environment, it must be capable of effective hypo-osmoregulation. The parr–smolt transformation involves changes in the osmoregulatory capabilities of the fish, from hyper-osmoregulation in fresh water to hypo-osmoregulation in the sea. The transition from a freshwater to a marine existence requires a reversal in the ionic regulatory mechanisms to deal with a change from a diffusive ion loss in the freshwater environment to an ion influx when the fish is in seawater. The most marked changes occur in the gills, particularly in the numbers and structure of the mitochondria-rich cells, which are the primary ionocytes. At the same time, the organs involved in water balance, especially the kidney and the gastrointestinal tract, must adapt to cope with a reversal of osmotic water flow from a water gain to osmotic water loss. The structural and functional changes in the tissues involved in salt and water balance are initiated while the fish are still resident in fresh water. In other words, changes occur prior to, and in anticipation of, exposure to the hyperosmotic marine environment. This means that the fish is, at least in part, preadapted to life in seawater before it commences downstream migration (Hoar, 1988; Pennell and Barton, 1996).

The major morphological and physiological changes that occur during parr–smolt transformation are:

- Loss of parr marks and a silvering of the body due to increased deposition of purines (guanine and hypoxanthine) in the skin. The formation of purines relates to changes in nitrogen and protein metabolism.
- Changes in the visual pigments, including a loss of ultraviolet (UV) visual sensitivity.
- Increased streamlining of the body, an elongation of the caudal peduncle region and a reduction in condition factor ($K = [W/L^3] \times 100$). The body becomes more slender because the rate of length increase is more rapid than the rate of weight gain.
- Changes in red muscle myosin heavy chains (MHC), muscle contraction kinetics and swimming behaviour. Two MHCs are found in the parr, but only a single MHC is present in the smolt. Red muscle contraction cycles are longer in the smolt than in the parr. At a given swimming speed, the smolt have lower tail-beat frequencies and longer 'stride lengths' than the parr.
- Proliferation of mitochondria-rich cells (MRCs) in the gills. The MRCs have high 'sodium pump' (Na^+-K^+ ATPase) activity and a high succinic dehydrogenase enzyme activity (indicative of increased numbers of mitochondria and potential for increased metabolic activity).
- Increased salinity tolerance and improved ability to regulate salt and water balance in seawater.
- Metabolic changes leading to an increased proportion of body water and a reduction in percentage body fat. Liver and muscle glycogen tends to decrease, there is an increase in the activity of glycogenolytic enzymes and blood glucose concentration changes.
- Changes in tissue fatty acid composition, especially increased deposition of n-3 highly-unsaturated fatty acids (n-3 HUFAs) in the membrane phospholipids of the gills and gastrointestinal tract.

The behavioural changes that are observed at this time include a marked reduction in territorial defence, movement from the bottom to higher in the water column, formation of schools, a decreased ability to hold station against water currents, a negative rheotaxis and the commencement of downstream migration.

The smolts migrate downstream and leave fresh water and they start the marine phase of their life during the spring. The smolts usually enter seawater during May or June. Given that some of the male fish mature in fresh water, there are more females than males among the migratory smolts. On entering the sea, the fish migrate to the feeding grounds, where the majority of growth occurs. During the marine phase of their life, the Atlantic salmon feed, predominantly, on crustaceans and small, pelagic fish. Some Atlantic salmon return to their native rivers after spending one winter at sea; these fish are known as grilse and are usually 1.5–4 kg in body weight. The majority of salmon return to fresh water, as maturing individuals, after two or three winters at sea, although some may remain in the sea for up to 5 years (Groot, 1996; Klemetsen *et al.*, 2003; Esteve, 2005).

12.4.1.2 Farming of Atlantic salmon

Farming of Atlantic salmon expanded during the latter years of the 20th century and it is now an established industry in several regions of the world. All of the major production areas lie within the latitudes 40–70° in the northern hemisphere and 40–50° in the southern hemisphere. Salmon farming is established in several countries in northern Europe (e.g. Norway, the UK, Faeroe Islands and Iceland), North America (mostly on the north-eastern seaboard of the USA and Canada, but also on the Pacific coast), Australia (Tasmania) and Chile in South America. The cultivation techniques are similar wherever salmon farming is practised, being based on egg incubation, early rearing and smolt production in freshwater hatcheries and smolt production units followed by on-growing to market size (3–7 kg) in seawater. The on-growing phase is dominated by sea-cage culture, although small quantities of salmon are raised onshore in tanks and raceways that receive pumped seawater. There are many recent summaries concerning broodstock and hatcheries (Pennell and Mclean, 1996; Fitzgerald *et al.*, 2002; Kindness, 2002), smolt production (Clarke *et al.*, 1996; Fitzgerald *et al.*, 2002) and on-growing (Fitzgerald *et al.*, 2002).

12.4.1.3 Broodstock management and hatchery management

After fertilization, the eggs are incubated in the hatchery either before or after water hardening. To remove potential pathogens present at the egg surface, the eggs are disinfected in buffered iodophor solution, either as newly fertilized and water hardened or at the 'eyed-egg' stage. Modern salmon hatcheries are usually equipped with incubators for eggs and alevins and with rearing tanks in various sizes for on-growing fry up to the smolt stage. During 150 years of hatchery practice, technology has evolved from small units used to produce fry for enhancement purposes, to larger facilities designed especially to meet the demands of a large and growing aquaculture industry.

Atlantic salmon eggs are *c.*6 mm in diameter and there are typically 5000 eggs/l (egg volume measured without water). In principle, four types of incubators are used in modern hatcheries: (i) the hatching silo; (ii) the trough and basket systems; (iii) trough and flow systems; and (iv) the hexhatch (Kindness, 2002). To achieve a good result during the incubation process, several environmental requirements should be fulfilled. The most important are: low light intensity (eggs are usually incubated in total darkness); favourable water chemistry with pH close to 7 (not below 6); low levels of suspended solids in the water; minimum 7 mg/l oxygen level in the incubator outlet; evenly distributed water flow through the egg mass and sufficient flow to ensure removal of waste products from the eggs; no mechanical stress on the eggs caused by high water flow (especially during the 'green stage').

The developmental rate of the eggs is controlled by temperature, although the number of degree days to a certain developmental stage will decrease slightly when water temperature increases, and vice versa. The development of eggs and larvae can be described as follows (degree days indicated at 6°C incubation temperature): the 'green stage' (eggs referred to as green eggs) starts 24h after fertilization and water hardening and lasts to the start of the 'eyed-egg' stage, when the black eyes of the embryo can be seen within the egg (245 degree days). The eyed-egg stage lasts for 265 degree days and when it terminates, the eggs will start to hatch with gradually increasing frequency. The duration of the hatching period is inversely correlated with water temperature but, in most cases, the main bulk of eggs will hatch within 2–3 days. After hatching, the larvae are called alevins, or yolk-sac fry, due to their large mass of yolk gathered in a sac attached to the abdomen. The alevins will normally stay relatively inactive on the bottom and somatic growth will occur at the expense of the yolk mass. About 290 degree days following hatch, the yolk sac will be absorbed almost completely and the alevins are ready for start-feeding. In sum, 800 degree days will elapse from fertilization to first-feeding.

Six to 36h after fertilization, the eggs are relatively robust and can (after water hardening) be transported to the incubation facilities. However, during the green stage, fertilized eggs are very vulnerable and must therefore be treated carefully. The most critical period is between 20 and 120 degree days. Therefore, extreme conditions, such as mechanical stress due to handling or high water flow, or exposure to air or sunlight, should be avoided. To avoid infections by fungus (*Saprolegnia* sp.) during the green stage, dead eggs either must be removed carefully each day manually (if feasible) or the eggs must be treated with a suitable fungicide, the only option when using hatching silos. During the eyed-egg stage, the eggs are again robust and can be handled and transported. Midway through this stage, the eggs are usually 'shocked', whereby mechanical stress from pouring or siphoning and stirring causes the yolk membrane of infertile or poor eggs to rupture. When the yolk membrane has ruptured, the yolk proteins will coagulate and the eggs will turn white within 1 or 2h so that dead eggs can easily be removed manually or by using an egg-sorting machine.

Salmon eggs have been incubated at water temperatures from 4 to 12°C and, since the rate of development depends on temperature, the duration of the incubation period and time of hatching may be controlled by manipulating water temperature. This, combined with manipulation of time of spawning and smoltification, allows a more even production and supply of market-sized fish throughout the year. There is, however, a positive relationship between increasing incubation temperatures and egg mortality and frequency of individuals with malformations occurring later in production. Today, incubation of Atlantic salmon eggs at temperatures above 8°C is not recommended due to the risk of increasing frequencies of fish with deformities in the skeleton and organs (Baeverfjord *et al*., 1998, 1999).

When the eggs hatch, the alevins fall through slots in the egg tray and settle on the bottom. Usually, this is covered with a substrate that gives the fish support so that they can stay in an upright position easily. This will reduce both bunching and the overall activity level. The environmental requirements at this stage are quite similar to those during egg incubation, but increased water flow as the alevins grow is recommended; though not so much that the juveniles are forced to swim too actively and cause yolk-sac damage. Water quality is managed to avoid some complications of blue sac and gas bladder disease due to elevated levels of ammonia and gas super saturation in the water, respectively.

When most of the yolk mass is absorbed, the activity level of the alevins increases and some individuals tend to swim against the surface in search for food (swim-up). This swim-up behaviour is not as marked in Atlantic salmon as in rainbow trout and may be difficult to assess. Therefore, feeding with a formulated diet is usually initiated when about 10% of the yolk sac remains. The correct timing is important to avoid aggregations of the fish and gill problems if feeding starts too early, or high mortality and a high frequency of pinheads if feeding starts too late.

First-feeding of alevins is carried out in tanks of varying sizes (1–4 m diameter) at low water levels (10–30 cm depending on tank size). Successful start-feeding requires that the environmental conditions are controlled and optimized. Good water quality (pH between 6 and 7) with low levels of heavy metals and suspended solids is important to avoid problems with gills. Gill damage may lead to bacterial infections, hyperplacia in the gill filaments and subsequent mortality. In this context, a high standard of hygiene in tanks is crucial. Dead fish and excess feed and faeces should be removed at least once per day and the tank walls brushed when necessary (depending on water temperature). Water flow should be adjusted to maintain the required levels of oxygen and to ensure efficient transport of suspended solids (but not live fish) to the tank outlet. The flow pattern should ensure an even distribution of juveniles over the entire area of the tank bottom. In this phase, growth is stimulated by light and therefore the fish are exposed to continuous light (*c.*1000 lux) during first-feeding. In nature, juvenile salmon start feeding at temperatures around 8–10°C, but usually higher temperatures are employed in salmon farming to ensure high rates of growth. However, the use of high temperatures during start-feeding and on-growing up to 60 g may increase the frequency of fish with spinal deformities later in production (Baeverfjord and Wibe, 2003).

During first-feeding, automatic feeders are used to deliver the formulated feed either continuously or in short and frequent intervals. At this stage, most fish do not seek feed particles actively, but depend on having them brought to them by the water current. After a couple of days, an increasing number of actively feeding fish can be observed. As this number increases, the water level in the tank is raised. At completion of first-feeding, juveniles (fry) have a body weight of *c.*0.2 g. In principle, if large tanks are used, the fish may be held in the same tank from first-feeding to smoltification.

12.4.1.4 Smolt production

When fry reach a body weight of 2–3 g, they develop the typical dark 'finger marks' on their sides and are thereafter referred to as 'parr'. Since salmon are visual feeders, parr are fed with dry feed during the day and not during darkness. However, in the hatcheries, juvenile salmon are usually held under continuous light (LD24:0) from start-feeding until a shorter day length is introduced to induce smoltification (at body weight > 8 g). In tanks, social interactions may occur and access to sufficient amounts of feed combined with an even flow-pattern and a current speed of 1–2 body lengths/s is a prerequisite to avoid aggressive behaviour as, for example, fin- and eye-nipping. This will also assure optimal growth and food conversion. Pellet size is adjusted according to the size of the fish and the approximate amounts of feed to be delivered to each tank are estimated on the basis of total fish biomass, average individual body size and water temperature. As the fish grow, individual variability in body size, and total biomass in the tanks, will increase. Therefore, on several occasions during smolt production, the fish population is size-graded and different groups are transferred to new tanks to keep fish density at acceptable levels and to reduce individual variability in each tank. This will improve overall growth rate, reduce cannibalism and other social interactions, enable a better management and may increase the overall percentage of smolts achieved. Parr and smolts may show good survival and growth at densities as high as 100 kg/m^3, providing that a high water quality is maintained, but usually densities are held at lower levels (20–60 kg/m^3).

After on-growing of parr in the hatchery, the fish eventually will go through the parr–smolt transformation and be ready for seawater transfer, providing that the environmental conditions have been right. Atlantic salmon differs from Pacific salmon in having a relatively slow growth in fresh water and a relatively large body size at smoltification. This will increase the cost of production (Clarke *et al.*, 1996). The global aquaculture production of Atlantic salmon was 1.24 Mt in 2004 (FAO, 2007). This corresponds to a yearly transfer to seawater of at least 400–500 million smolts. Today, the largest hatcheries may have an annual production of up to 8 million smolts.

Production of salmon smolts has, during the past four decades, evolved from relatively small-scale, low intensity production, mainly for stocking purposes, to large-scale, highly intensive production of sea-ready fish for sea-cage on-growing. Smolts are produced mainly in land-based hatcheries using tank- or raceway-based flow-through systems. Where ambient water temperatures are too low to support sufficient growth, the production water is heated using heat pumps and heat exchangers. To a limited extent, smolts may be produced in cages in freshwater lakes from a body size of approximately 2 g (e.g. Chile, Norway and Scotland), but this production is now decreasing in magnitude because of the negative effects this has on the environment and because of a higher risk of disease. Due to restrictions on freshwater use, and a growing demand for smolts, the industry is looking for new production strategies and technology, which may contribute to increased production per volume water used. Until now, development has gone from the use of flow-through systems with excess water use, through systems with low turnover and reuse of water in varying degrees, to today's increasing interest in smolt production in classic recirculated aquaculture systems.

The onset of the parr–smolt transformation is regulated by changes in photoperiod, whereas water temperature mainly has an effect on the rate of physiological changes that occur and also defines the range within which a normal transformation can take place. Up to the late 1980s, the natural changes in photoperiod and water temperature (fresh water and seawater) were the main factors controlling the annual production pattern in the industry. For example, most of the smolts produced in Norway were transferred to seawater in May–June, 16–18 months after hatching. This resulted in a strong seasonality in harvesting, and thereby large amounts of fish going into the market at certain times of the year. To achieve better utilization of infrastructure and equipment, and a more even supply of fish to the market throughout the year, different strategies for smolt production have been developed. By manipulation with photoperiod and freshwater temperature, it is now possible to produce smolts ready for seawater transfer from 8 months after hatching, and by combining two year-classes, farmers may have access to sea-ready smolts throughout the whole year. For example, to produce fish that will smolt within the same year as the eggs hatched ('underyearlings'; S0+), fry/parr are reared at LD24:0 until they reach a body length of 8 cm in the summer. Then a period of short day length lasting for a minimum of 6 weeks is introduced ('winter signal'; LD12:12 or shorter), followed by a new period of LD24:0 until smoltification.

In many ways, the seawater-ready smolts can be regarded as the fish farmer's seed and, in line with this, the quality of the smolts will be decisive for the following survival and growth in seawater. The notion 'smolt quality' refers in its simplest meaning to the ability of the fish to survive and grow after transfer to seawater. This trait is influenced by several factors that characterize the fish at the time of seawater transfer; for example, health and vaccination status, genetic origin, body size, growth history, body shape and appearance and seawater tolerance. However, smolt quality is often used in the more specific meaning of physiological smolt status, especially referring to the ability of the fish to acclimate rapidly to seawater following transfer from fresh water (seawater tolerance). Good smolt quality implies that the fish have gone through a normal parr–smolt transformation and a high seawater tolerance is a prerequisite for a successful transfer of fish to the sea cages. In the industry, seawater challenge tests are used to assess seawater tolerance and to predict the optimal time for seawater transfer. These are performed by exposing representative samples of fish to seawater at standardized salinity and temperature for 24 h (e.g. 32–35 g/l; salinity, 6–10°C) and by monitoring blood plasma electrolyte levels (chloride or sodium) and mortality (if any) at the termination of the test. This method may be referred to as the 'ion regulation test'. Alternatively, the fish may be exposed to 40 g/l; salinity and mortality recorded after 96 h (survival test). In recent years, measurements of enzyme activity (Na^+-K^+ ATPase) in gill tissue have been used as an alternative to seawater challenge tests, with a varying degree of success.

Today, knowledge of the nutritional needs of Atlantic salmon fry and parr, and fish in the grow-out phase in seawater, is quite good, but detailed knowledge about the needs of fish during the parr–smolt transformation is insufficient. So far, experience from commercial farming and research indicates that fish held on quite different diets with respect to total lipid and protein content, fatty acid composition, vitamins and additives such as salt and betain are capable of completing a normal parr–smolt transformation.

To ensure that the fish are capable of withstanding the pathogenic agents they encounter in the sea phase, all smolts are vaccinated. Usually, the fish are immunized by an injection of polyvalent vaccine, containing both viral and bacterial antigens, into the abdominal cavity in good time before the onset of the parr–smolt transformation. If the vaccination is carried out too close in time to the transition, this may result in reduced immunization and also reduced handling tolerance of the fish. Vaccination by injection may, in some cases, result in adverse effects such as limited mortalities, appetite reduction, inflammation on the injection site, intra-abdominal adhesions and pigmentation and granulomata. However, these effects do not in any way counterbalance the advantages of immunization.

Transportation of the smolts from the hatcheries to the sea sites is usually carried out by road (in tanks) or by well-boats. Sometimes, a helicopter may be used for shorter transportation distances. The duration of transport may vary from 1 to 24 h. This is a critical event since handling and transport impose severe stress to the fish. During transport, it is important to keep water quality parameters such as temperature, pH, ammonia, pO_2 and pCO_2 within acceptable limits. It is generally believed that a large proportion of the disease outbreaks that occur during the first months after transfer to sea cages can be related to the stress imposed on the fish during transport.

The first 24 h following transfer to the sea cages, especially when transported in fresh water, are critical with respect to survival and successful seawater acclimation. The fish encounter a new environment with respect to chemical composition and soluble gases and will also often be exposed to a change in water temperature. The osmotic gradient between the body fluids and the surrounding water is completely reversed and, consequently, a series of physiological adaptations must be completed within a short period of time in order to ensure survival. In addition, the immune system is impaired due to the hormonal changes associated with smoltification. It is crucial to get the fish to eat as soon as possible and the fish therefore require especially high attention from the farmers during this period.

12.4.1.5 On-growing to market size

On-growing of Atlantic salmon takes place in many regions across both hemispheres and, while practice is influenced strongly by factors specific to both site and geographic region, the approaches used are broadly similar. This is partly explained by high investment from traditional salmon-growing countries and companies and the widespread availability of technology and knowledge, coupled with effective technology transfer.

Atlantic salmon on-growing is carried out typically in floating sea cages, although there is some use of land-based tanks, ponds and raceways and development of submersible and submerged cages aimed at using more exposed and offshore sites. Anchored cage systems are either a circular or rectangular (square) type from which bag-nets, 5–20 m deep, are suspended from floating frames. Cages are usually anchored together and are, depending on the design, either joined by walkways or only accessible individually by boat. Site characteristics obviously influence the choice of system. For example, in regions of Norway with steep, deep fjords and small tidal ranges, sea cages are often grouped around a pier that is attached to land. In Scotland, where sites are shallower and tidal ranges greater, sea cages are not attached to land and are only accessible by boat. Rectangular galvanized steel sea cages are available in 5–30 m square sections, a 24 m section might consist of a 15 × 15 m net surrounded by walkways. Some flexibility is afforded; for example, in situations where increased water flow is required, four such sections might be joined to make one larger section with four times the volume and fewer nets for water to pass through. The other main type of sea cage is the circular ring cage, in which a buoyant circular ring structure is used to hold the bag-net, predator net, mooring cables and other ancillary structures such as bird netting or feeders. These structures can range in circumference from less than 60 m to over 160 m and there is a trend to increase the size of ring cages further, with 200 m cages commercially available. Compared to square cages, they are usually placed further apart, allow greater water flow through a group of cages and are easier to handle independently of each other. Plastic ring cages tend to be more resistant to harsher environmental conditions and survive high wave intensity, strong currents and icing better. There are a number of commercial variations of square and ring cages holding from a few tonnes to over 200 t of fish. Stocking densities are quite variable and reflect a range of factors, many relate to maintaining high water quality but other influences include the time in the production cycle, site-specific knowledge and the type of farming, organic, for example. In the late 1980s, typical stocking densities for sea cages were *c.* 15–20 kg/m^3, although some may have exceeded 50 kg/m^3, whereas higher densities of 35–40 kg/m^3 were used in tanks on Icelandic land-based farms (Isaksson, 1991). Currently, stocking densities are legislated to be below 25 kg/m^3 in Norway and are usually below this; for example, final densities of 15 and 5–10 kg/m^3 were recommended for inshore and offshore cages, respectively (Fitzgerald *et al.*, 2002). In other regions, stocking densities are around 8 and 10 kg/m^3 for Australia (Tasmania) and Chile, respectively.

On-growing takes 50–100 g smolt to harvest weight of between 3 and 5 kg in 12–18 months. As detailed above, managing smolt quality, particularly timing smolt production in relation to on-growing harvest strategies and transfer to seawater, are critical to successful on-growing. Smolt can be transferred into brackish water in estuary locations and moved, when reaching a larger size, to 'downstream' sites with increased salinity. However, transfer directly into seawater is more usual. Mature salmon are not desirable in a production run and it is particularly important to reduce the number of grilse, salmon that mature in the first year at sea, since they have the same problems as older fish that mature but are also smaller and therefore lost from final harvest. Disadvantages of mature salmon include higher mortality in seawater, 'unattractive' external secondary sexual characteristics, feed resources are partitioned into unwanted reproductive growth rather than somatic growth and pigmentation is lost from flesh. Strategies to control maturation include the use of lights to manipulate photoperiod and night-time illumination over winter reduces grilsing significantly; the effect of lights is thought to be mediated by plasma melatonin remaining below a threshold level during night-time when under illumination (Bromage *et al.*, 2001). Other strategies, which are not used extensively but are useful in some regions, include the production of sterile triploid fish to reduce gonadal growth, or all-female stock, since females show lower rates of maturity than males. Although grading to achieve more even size distributions is accomplished relatively easily with current technology, it nevertheless has a financial cost and is stressful to fish, so its frequency is reduced as far as is practical. A major function of grading is to manage the supply of harvest-size fish in an attempt to produce a more constant rate of supply at market weights. While the number of individuals in a sea cage would prevent stable linear hierarchies and 'pecking orders', it can also be argued that 50,000 salmon in a sea cage inhabit a complex social environment where the ability and motivation of individuals to feed is a result of many factors, including relative size, individual history and genome. In relation to overall production costs, the feed costs are significant, quoted to range between 40 and 60% (Sinnott, 2002). Even under conditions of high growth and feed efficiency, feed costs can be around 50%. Careful attention is paid to effective feeding and a variety of feeding strategies are used. Monitoring feeding response and hand feeding can be done, or adaptive feeding systems are available whereby the amount of feed supplied depends on the response of the fish (and how much is wasted) (Purser and Forteath, 2003). Pellet sizes are matched to fish size and strategies such as partially introducing new feeds through mixing new with old are advised (Sinnott, 2002).

The annual production of Atlantic salmon feeds is over 1 Mt and more than for any other intensively farmed fish. Commercial feeds are very effective despite published information on the nutrient requirements for Atlantic salmon being far from complete. Successful feeds are explained by the use of a considerable amount of commercially protected information, selective use of information from other animal and fish species (Guillaume *et al.*, 1999; Halver and Hardy, 2002) and a recent focus on Atlantic salmon nutrition research (Storebakken, 2002). Atlantic salmon are fed extruded pelleted feeds that are constantly being redefined and refined. Principal drivers for change are greater knowledge of salmon nutrition, the need to reduce feed costs and to find suitable alternatives for fish oil and fishmeal and the aim to increase feed utilization efficiency and to reduce environmental impacts. There have been large improvements in feed formulation and feeding practices so that commercial feed utilization efficiency has improved markedly over the past 20 years. Typical farm feed conversion ratios (FCR) during on-growing show these improvements and are in the region of 1.2–1.4, compared to 1.5–2.0 achieved in the 1990s (Sinnott, 2002). Protein retention efficiency provides an excellent indication of diet quality, particularly the protein to energy ratio and amino acid balance. Values of over 50% protein retention are achieved experimentally and demonstrate how well nutritional requirements and diet formulation are understood for Atlantic salmon on a practical level. A value of 25% protein retention was given for Norwegian salmon farming in the 1980s (Håkanson *et al.*, 1988) and can be compared to 35% calculated assuming a current farm FCR of 1.2 and use of feeds with an average of 38% protein in growing a 4 kg fish. Dietary protein decreases with salmon size so that feeds might contain 40, 38 and 36% protein for 1.5, 3 and

5 kg fish, respectively. Corresponding increases in oil levels in salmon feeds have been made possible by extrusion and vacuum coating technology and feeds with 40% oil are commercially available. Dietary protein is decreased and dietary lipid increased with increasing salmon size, this results in the dietary protein to energy ratio decreasing from 19 to 16–17 g protein/MJ for 1–2.5 and 2.5–5.0 kg fish, respectively (Einen and Roem, 1997; Sinnott, 2002).

Global aquaculture and Atlantic salmon production continue to increase, whereas fishmeal and fish oil production are, at best, stable. This means there is a strong commercial imperative to reduce the proportion of marine fish products in salmon feeds. Alternatives to fishmeal depend partly on nutritional qualities such as digestible protein content, amino acid balance and presence of antinutritional factors (Carter and Hauler, 2000; Carter, 2007). However, the selection of ingredients also depends on local legislation, industry guidelines and marketplace agreements. For example, the use of terrestrial animal by-products is more constrained in the European Union and Scandinavian countries than in some other regions. Poultry meals and porcine blood products provide manufacturers with high-protein ingredients and make formulation more straightforward: currently, they are used in Canada but not in the EU. A variety of vegetable products ranging from low-protein dehulled ingredients to high-protein isolates, including those made from soybean, maize, field pea, rapeseed, lupin and cereals, are at various stages of development and commercial evaluation. The replacement of 50% fish oil with vegetable oils such as soy, canola and palm is widely accepted as a viable strategy for reducing fish oil usage while maintaining the fatty acid profile of salmon. Feeding strategies, such as high fish oil 'finisher' feeds, may be used to manage the use of vegetable oils better. In the future, oils with higher amounts of long chain omega 3 fatty acids and based on microorganisms, genetically modified oil seeds and even weeds will be used increasingly (Carter, 2007). It should be noted that there may be a consumer preference for salmon flesh that is less 'marine'; salmon fed a blended plant oil for the entire production cycle were preferred by a trained taste panel over salmon fed entirely fish oil (Torstensen *et al.*, 2005). Encouragingly, all the fish oil could be replaced without compromising flesh quality and 75% could be replaced without compromising the long-chain polyunsaturated fatty acids composition in relation to human health (Torstensen *et al.*, 2005). Synthetic astaxanthin is the predominant pigment used in feeds: effective and economical pigmentation regimes are of critical importance because flesh colour is a major marketing characteristic, pigments are very expensive and salmon are not efficient at retaining them in muscle. A variety of factors affect pigmentation, including genetic variation, size, age, life stage, dietary fat, dietary oil source, as well as the dietary pigment concentration, and the time over which it is fed. Sinnott (2002), for example, recommends including relatively high concentrations of 60–80 mg astaxanthin/kg feed at the start of pigmentation with 50–80 g smolt and then reducing this when muscle concentrations reach 7–8 mg/kg muscle.

Nutritional diseases caused by nutrient deficiency are rare but appear to be due to unpredicted events, such as when interactions between components in novel ingredients reduce nutrient bioavailability, or when growth rates exceed the supply of essential nutrients, or when environmental and nutritional factors interact negatively. The incidence of screamers disease in Chile is a good example of a negative interaction between dietary and environmental factors reducing nutrient (phosphorus) bioavailability to induce a disease (Roberts *et al.*, 2001). A large number of infectious diseases have been recorded for Atlantic salmon and, although only a few have caused major commercial or welfare issues, they can be a major cause of loss on fish farms (Ellis, 2002). However, these pathogens are opportunistic and, where farms and fish are well managed, do not normally cause disease (Ellis, 2002). In addition, our understanding of fish health, immunology, vaccines and other strategies to reduce infection are advancing rapidly (Ellis, 2002). Commercially important diseases include furunculosis (vaccine available), rikettsia (vaccine under development), various vibriosis (vaccines available for *V. anguillarum* and *V. salmonicida*), bacterial kidney disease, enteric redmouth disease (vaccine available), infectious pancreatic necrosis (vaccine under development), winter ulcers (vaccine under development) and infectious salmon anaemia, as well as some larger ectoparasites, particularly sea lice species and amoeba. As indicated, the use of vaccines is widespread while others are being developed, including ones aimed at sea lice and amoeba. Other strategies range from managing farms within the local area to selective breeding.

There is considerable detailed information available on pre- and postharvest influences on flesh quality in Atlantic salmon (Kestin and Warriss, 2001; Howgate, 2002). Of particular importance are the influences on flesh quality of the relationships between feed composition and feeding regimes (e.g. Johnston, 2001; Sargent *et al.*, 2001; Torrissen *et al.*, 2001), the impact of handling and killing methods (e.g. Sorensen *et al.*, 2004) and postharvest processing (Kestin and Warriss, 2001). Harvest follows a period, usually between 10–12 days, without feeding, mainly to clear the digestive tract of contents. At harvest, salmon from an entire cage are usually transferred to the processing plant; this can be via well boat or by towing the cage there directly. Attention is paid to reducing stress prior to slaughter because this addresses industry codes of practice, animal welfare issues and ensures a higher quality product. Salmon are pumped out of the boat well, or holding race, and along the harvest line. Several methods have been used to render the fish unconscious, the most frequently used being carbon dioxide anaesthesia and percussive stunning. Other procedures include electro-stunning, brain spiking and anaesthesia using drugs or chilling in ice, and may be combined. Following a cut to the gill arch, the fish are bled and immersed in an ice-water slurry for transport and processing. Adoption of ethical slaughter practices are of importance to the industry and more rapid killing methods, such as percussive stunning, are favoured over traditional methods using carbon dioxide and ice emersion followed by bleeding (Sorensen *et al.*, 2004). Anaesthesia using drugs can affect taste, depending on the drug, and electro-stunning can cause bruising and blood spots (Howgate, 2002; Sorensen *et al.*, 2004). Rested harvest methods that use sedation prior to slaughter have advantages, for both animal welfare and product quality reasons, and are widespread in the industry.

Selective breeding has been used as a strategy to improve Atlantic salmon and rainbow trout in Norway since 1971 (Gjedrem, 2000, 2005). Initially, selection was based on growth, then the frequency of grilse was incorporated to increase the age at maturation. More recent selection strategies have used disease challenge to increase disease resistance and measurements of flesh colour, fat content and fat distribution to improve meat quality (Gjedrem, 2000, 2005). There has also been development and testing of genetically modified or transgenic fish. However, it is unlikely that such fish will gain wide market acceptance given the niche that Atlantic salmon occupies. Climate change and the resultant exposure of Atlantic salmon to increased water temperature at various times in the production cycle arguably present a major challenge for the industry and a need to develop a range of strategies to cope with its various effects.

12.5 Pacific Salmons and Trouts: Biology and Culture

The Pacific salmons and trouts, genus *Oncorhynchus*, are representatives of a taxon of anadromous, occasionally freshwater fish species that have their natural distribution in the North Pacific region. They occur along the western seaboard of North America, from California in the south to the Arctic Ocean, and along the eastern seaboard of Asia from the coast of Japan northwards (Groot and Margolis, 1991; Nelson, 1994; Pennell and Barton, 1996; Quinn, 2005).

Seven species of Pacific salmons are generally recognized; *O. masou* (masu or cherry salmon), *O. rhodurus* (amago salmon or biwamasu), *O. kisutch* (coho or silver salmon), *O. tshawytscha* (Chinook), *O. keta* (chum salmon), *O. gorbuscha* (pink salmon) and *O. nerka* (sockeye). There are also species designated Pacific trouts, most notably the rainbow trout, *O. mykiss*, but also the cutthroat trout, *O. clarki*, and a few other species of limited distribution.

Several of the Pacific salmon and trout species form the basis of major commercial fisheries and the biology and life cycles of the commercially important species have been much studied (Groot and Margolis, 1991; Groot *et al.*, 1995; Pennell and Barton, 1996; Esteve, 2005; Quinn, 2005). There is large-scale cultivation of several of the Pacific salmon species. The vast majority of cultivation is for the enhancement of existing populations, for artificial stocking of 'barren' watercourses and 'sea ranching', but there is also farming of some species directly for human consumption as table fish.

The Pacific salmon breed during the autumn and they are typically semelparous (i.e. they die after spawning). The amago and masu salmon are restricted to the Asian side of the Pacific basin, both species occurring in Japan. The amago salmon has a complete freshwater life cycle and, uncharacteristically for Pacific salmon, not all fish die after spawning. The masu salmon occurs as both anadromous and freshwater forms; the anadromous fish die after spawning, whereas those that spend their entire life in fresh water may survive to spawn several times.

The other five species of Pacific salmon are found on both the Asian and North American sides of the Pacific Ocean. The Chinook is the largest (up to 45 kg) of the species. Two major forms of Chinook salmon are recognized.

Stream-type Chinook spend one or more years in fresh water before migrating to the sea. These fish make extensive migrations before returning to fresh water in spring or summer, several months prior to spawning. On the other hand, ocean-type Chinook migrate to the sea within 3 months of emerging from the gravel redd. Chinook salmon of this type spend most of the marine phase of their life cycle in coastal waters and return to fresh water in autumn, shortly before spawning.

Coho salmon typically spend 1–2 years in fresh water before migrating to the sea and then spend about 18 months in the marine environment before returning to fresh water to spawn. There are two main forms of coho salmon that differ in migratory pattern. During their migrations, ocean-type coho move further offshore than do fish of the inshore-type. The latter tend to remain in coastal and near-shore waters throughout the marine phase of the life cycle (Groot and Margolis, 1991; Groot, 1996; Quinn, 2005).

The chum salmon undertakes the most extensive oceanic migrations of all Pacific salmon species. It is also the most widespread of the species, although it is more numerous on the Asian than on the North American side of the Pacific Ocean. The life cycle of the chum salmon is relatively simple; the fish migrate to the sea shortly after emerging from the redd and the adults return to spawn after 2–4 years at sea. The maturing fish most usually return to fresh water during summer or early autumn. Although spawning generally occurs in rivers and streams, some chum salmon appear to spawn in tidal areas and complete the life cycle in seawater.

The pink salmon, which is probably the most abundant of the species, migrates to the sea at a very young age, shortly after emergence from the gravel redd. This species is distinct in having a fixed 2-year life cycle. This can give rise to separate even- and odd-year populations within a given watercourse. There are pink salmon runs in alternate years only in rivers where a single ecotype occurs.

Sockeye salmon traditionally have been the most economically important of the Pacific salmon species. They also have the most complex life-cycle characters of all the species. This results from the numerous possible combinations of residence times in fresh and salt waters. Although most sockeye occur as anadromous populations, some spend their entire lives in fresh water, this ecotype being known as the kokanee. In anadromous populations of sockeye, the young fish may spend up to 5 years in fresh water before migrating to the sea. Sockeye may reside in seawater for from 1 to 6 years before returning to fresh water to spawn. Some males mature after 1 year in the sea and return to spawn as 'jacks', whereas other males and the females mature later. Spawning may occur in streams or along the shorelines of lakes, leading to a range of migratory patterns when the juveniles leave for the sea; and the entire life cycle can be completed in fresh water, as in the case of the kokanee (Groot and Margolis, 1991; Groot, 1996; Quinn, 2005).

The Pacific trouts differ from the Pacific salmons in that they tend to spawn during the spring. The rainbow trout is the best known of the Pacific trout species. It is endemic to the western states of North America and eastern Asia, but it has been introduced to so many countries outside of its native range that it is now virtually cosmopolitan (Lever, 1996; Pennell and Barton, 1996). The rainbow trout is now one of the most widely distributed and highly prized freshwater game and aquaculture species.

Some of the Pacific salmon species have also been introduced to areas outside their natural distribution range. For example, Chinook and coho salmon have been introduced into the Great Lakes region of North America and coho salmon have also been transplanted to Chile in South America. There is farming of Chinook salmon as table fish in Canada, with an annual production of 20,000–25,000 t. Coho salmon are farmed in some quantity in Chile, with production of the coho exceeding 0.1 Mt by the early years of the 21st century (FAO, 2007).

12.5.1 Rainbow trout, *Oncorhynchus mykiss*

The natural distribution of rainbow trout extended across the temperate regions of the north Pacific rim, from Baja California through Alaska, the Aleutians and the western Pacific areas of the Kamchatka Peninsula and Okhotska Sea drainages (Fig. 12.9). Rainbow trout are generally classified as a freshwater fish, but over their natural range, most stocks possess an anadromous component. Anadromous fish spend 1–2 years as fry and juveniles in fresh water, 18–30 months in the ocean and return as adults to spawn in their natal river systems, although considerable plasticity exists in the time spent in freshwater or the ocean. Anadromous forms are called steelhead due to their silvery colour and originally were thought to be a separate species from freshwater trout, which have been divided into coastal and inland (redband) forms. However, both coastal and redband trout exhibit freshwater and anadromous forms. Subspecies that are morphologically distinct from coastal and redband forms are found in isolated inland drainages and in parts of Arizona and Mexico (Behnke, 1992). Compared to Pacific salmon, rainbow trout exhibit great variety in appearance and life history.

Today, rainbow trout are distributed on all continents except Antarctica as a result of over 100 years of transplantation. Rainbow trout egg collection stations were first established in California in 1870 and, over the next two decades, over 2.5 million eggs were shipped to state and federal hatcheries in the USA. Subsequently, broodstock stations were established in California, Oregon, Virginia and Michigan, from which numerous shipments were made throughout the USA and to European, Asian and South American countries, plus Japan and New Zealand (Benhke, 2002). Virtually all farmed rainbow trout can be traced back to these early shipments, which were a blend of redband and steelhead trout (Behnke, 2002).

The current extensive distribution of the species is the result of a combination of biological and economic factors. First, the species is relatively hardy, tolerating a wide range of water qualities and temperatures. Second, rainbow trout are highly esteemed as a sport fish for recreational angling; this applies both to smaller fish in streams and lakes and the larger anadromous steelhead form. Third, rainbow trout are easy to culture compared to many other species

Fig. 12.9. World distribution of *Oncorhynchus mykiss*.

of fish and can be produced economically by farming. Finally, rainbow trout are considered a good food fish. As such, cultivation of the species both for direct human consumption and for stocking purposes has made it a valuable economic commodity.

12.5.1.1 Life cycle of rainbow trout

Wild rainbow trout spawn in spring when winter water temperatures begin to exceed 6–7°C (Behnke, 1992). In warmer coastal areas, spawning can occur in late winter (January–February), but in colder climates, spawning is delayed until April or May. As is the case with Pacific salmon, eggs are deposited by females in excavated gravel (redds) in streams, fertilized by males and covered by the female. Females spawn 300–3000 relatively large eggs (50–100 mg); the number of eggs is related to body size and age of the female (25,000 eggs were once spawned by a very large farmed trout at the University of Washington). Time to hatching varies with water temperature: 80 days at 4.5°C, 31 days at 10°C and 19 days at 15°C, or 360, 310 and 285 degree days, respectively (Leitritz and Lewis, 1980). Eggs are sensitive to handling and shock from 2 days postfertilization until blastophore closure, approximately 10 days at 10°C. Once the embryo's eyes become pigmented at about 17 days (10°C), the period of sensitivity is over. At hatching, the embryo is attached to a large yolk sac and remains burrowed in the gravel. Development of the embryo continues until the yolk is greater than 95% absorbed, at which time the alevins emerge from the gravel and become oriented to the surface, where they seek food. This is called the 'swim-up' stage and the time required to reach the swim-up fry stage from fertilization depends on water temperature: *c.*50 days at 15°C to 90–110 days at 9–10°C. Young trout fry subsist on zooplankton and small aquatic insects, switching to larger prey as they grow. Growth rate is highly variable in nature, depending on water temperature and food supplies. Trout residing in small streams may reach only 10–20 g in weight after 1 year, whereas fish living in rivers and lakes may be ten times larger at 1 year of age. Trout are opportunistic feeders, preying on aquatic and terrestrial insects, crustaceans, small fish and salmon eggs. Anadromous forms living in the ocean consume squid, crustaceans (shrimp) and pelagic, schooling fish such as herring. Freshwater forms typically spawn at 3 years and may mature at 50–100 g if living in small streams, or 500–700 g if living in rivers or lakes. Lifespans of coastal strains vary from 3–4 years for stream-dwelling fish to 6–8 years for fish in rivers and lakes. Inland strains have lifespans of 3–10 years. Anadromous fish attain larger sizes at maturity, averaging 1.1–5.4 kg, with lifespans of 4–7 years, with a recorded maximum age of 9 years (Behnke, 1992).

12.5.1.2 Farming of rainbow trout

Rainbow trout eggs were first collected at the McCloud River trout station in California, established in 1879 by Livingston Stone (Stickney, 1996). Prior to this, brook trout had been spawned and cultured in the eastern states of the USA, following procedures in publications from Europe on the culture of brown trout. In 1881, 179,000 eggs were collected and shipped to ten locations in the eastern USA, and from these locations, rainbow trout broodstock were established and subsequently eggs were shipped all over the world. Records show, for example, that in 1901, rainbow trout eggs were shipped from the USA to Argentina (Stickney, 1996).

Rainbow trout can be farmed anywhere suitable water is found in sufficient quantities and quality to make farming feasible. Rainbow trout grow best at 14–15°C, but can be farmed in water that is between 10°C and 18°C. Trout can tolerate colder water, but grow slowly. Warmer water is problematic for trout, although they can tolerate temperatures above 20°C for brief periods. Rainbow trout are farmed in flow-through systems, with rearing densities depending on water flow rather than raceway or tank surface area. In the USA, raceways are operated in sequence, with between three and seven raceways in a series. Typical rearing densities are 1.8 kg/l/min water flow in raceways receiving first-use water and up to 9.6 kg/l/min when all raceways in a series are combined (Brannon and Klontz, 1989). In Europe, long earthen raceways, sometimes more than 1000 m long and resembling wide, shallow streams, are sometimes used to raise trout. Major trout-producing countries are France, Italy, Spain, Denmark, Norway, Chile, Japan, the UK and the USA and annual aquaculture production in the early years of the 21st century was around 0.5 Mt (FAO, 2007).

12.5.1.3 Broodstock management and hatchery operations

Rainbow trout reared in broodstock farms first spawn as 2- or 3-year-old fish, depending on water temperature and stock, and fish are generally held after spawning, reconditioned and then spawned once or twice again in subsequent years, if possible. Rainbow trout normally spawn in late winter or early spring, but spawning time can be shifted by adjusting photoperiod. Some stocks have been developed that spawn in late autumn or early winter. Egg development is temperature-dependent, e.g. slowed by incubating in cold water. Through the use of selected stocks, altered photoperiod and coldwater incubation, rainbow trout eggs are available year round.

Maturation in rainbow trout occurs over many months and, during maturation, somatic growth slows and gonadal growth accelerates. The number of mature eggs per female depends on fish size and nutrition status before and during maturation. The final phase of maturation involves hydration of eggs and their release from ovaries into the body cavity. Readiness to spawn can be judged manually by feeling the degree of thinness of the abdominal musculature and by feeling if eggs are loose in the abdominal cavity, or by trying to express eggs manually from the vent with the fish in a tail-down position. Maximum egg fertility is short-lived; trout at 15°C can become overripe in a few days if they are not spawned, resulting in very low fertility or complete infertility (blank egg lots). At cooler water temperatures, egg fertility is extended over a longer period. For decades, poor fertility of eggs from trout reared at water temperatures above 12°C was thought to be due to problems in egg maturation at higher water temperatures, but recent work has shown that excellent egg fertility can be obtained from trout spawning at 15°C, providing fish are checked daily and spawned when eggs are ripe.

Artificial spawning of trout is a relatively simple affair. Eggs are expelled from anaesthetized females using manual pressure, flushing with physiological saline, or air injected into the abdominal cavity. Eggs are collected in a dry bowl or pail, blood, faeces or other material is removed, milt is added and mixed with the eggs and water is added. Fertilized eggs take up water between the layers of the egg shell, a process called water hardening. Milt is collected easily from ripe males by manual pressure on the abdomen. Milt can be diluted in extender solution or cryopreserved for later use. Refrigerated milt in extender solution, kept under oxygen at 3–5°C, can remain active for 10–14 days.

After fertilization and water hardening, eggs are incubated in jars or trays supplied with upwelling water in darkness. As mentioned, eggs are susceptible to shock from about 48h after water hardening until blastophore closure after 95 degree days, so water flows must be controlled such that eggs roll slightly and gently in jars or trays (Leitritz and Lewis, 1980). Excess water flow that causes eggs to be agitated will result in egg loss. When the embryos reach the eyed stage (eye pigment visible, 170 degree days), they are removed from trays or jars by pouring into another container, then returned. Dead eggs turn white and are removed manually or mechanically to prevent fungal growth that might smother adjacent live eggs. At hatching (308 degree days), yolk-sac fry are separated from egg shells and dead fry and kept in jars or trays with substrate while they absorb yolk material. When yolk material is *c*.95% absorbed, the fry can be introduced to exogenous feed, usually in the form of dry, powdered feed. At this stage of production, the emphasis is on ensuring a successful transition to exogenous feeding and thereby ensuring high survival. Fish are fed small amounts nearly continuously by hand or mechanical feeders, but care must be taken to avoid overfeeding as uneaten feed particles decompose and irritate gills, making the fish susceptible to bacterial gill disease. After 10–14 days, feeding frequency can be reduced to eight times per day and reduced gradually to three times per day when fish reach approximately 5g. Slight overfeeding of fish during fry and fingerling rearing results in maximum growth, as long as fish are fed to apparent satiation at each feeding. Frequent feeding of small amounts tends to increase size variability within a group of trout because aggressive fish eat a disproportionate amount of feed. Once trout are consuming pelleted feed, they should be fed about 1% of tank or raceway biomass at each feeding to ensure that all fish receive feed. Of the total amount of feed used during a production cycle, over 90% is fed during the on-growing phase, so it is unwise to restrict feed during early rearing or to use inexpensive feed as these factors do not increase the cost of production markedly but they can lower growth rate and increase mortality, to the detriment of production costs.

Fry feeds are finely ground, pelleted mixtures of fishmeal, fish oil, vitamin and mineral premixes and wheat flour or other gelatinized starch (Hardy, 2002). Trout fry grow best when fed high-protein (50%) diets; dietary protein content is lowered as fish grow and dietary energy levels are increased by increasing the amount of fish and plant oil by top-dressing, i.e. adding oil to pellets.

12.5.1.4 On-growing to market size

In freshwater trout rearing, fish are stocked in outdoor raceways or ponds when they reach 15–30g. Fish remain in raceways or tanks until reaching harvest size (600–750g in the USA), a period of 10–13 months from spawning. In Norway,

France and Chile, rainbow trout are transferred to marine net pens when they reach at least 100–150 g. Introduction to marine net pens can be abrupt if fish are transferred in early summer, but usually fish are introduced to seawater gradually over a 1–2 week period to facilitate seawater acclimation. Trout are then grown to harvest size (2.5–3.5 kg) over a period of 15–24 months.

The emphasis during the on-growing phase of rainbow trout production is on economical growth. Rainbow trout feeds are similar in composition to feeds for salmon; however, total dietary lipid levels are lower. The upper limit for total lipid in trout feeds is about 25%; higher levels do not increase fish growth rate or protein retention, but result in high amounts of visceral and fillet fat. High visceral fat levels lower yields of edible product and increase the cost of production. Protein levels in trout feeds vary between 40 and 45%, depending on dietary lipid level. FCR in commercial farms average 1.1–1.2, protein retention averages slightly over 40% when expressed on the basis of total dietary protein intake and values are higher when expressed as total digestible protein intake. Commercial feed formulations for rainbow trout vary around the world, but in the USA, fishmeal generally constitutes 20–30% of the diet. Other protein ingredients used in trout feeds include by-products of animal and poultry processing and plant protein ingredients such as soybean meal, soy protein concentrate, maize gluten meal and protein concentrates. Starch, supplied by ground whole wheat or barley, or by grain milling products, is relatively indigestible to trout. However, starch digestibility is increased by heat treatment during feed pelleting. Cooking extrusion is the principle method of pelleting trout feeds and this process gelatinizes starch, thereby increasing starch digestibility and increasing water stability and hardness of pellets. Freshwater trout farms face restrictions on levels of soluble and insoluble phosphorus in farm effluent water. Phosphorus is an essential dietary nutrient for trout, but excessive dietary levels are excreted either in faeces if the phosphorus is indigestible or as soluble phosphorus in urine if levels of available phosphorus in the diet exceed the needs of the fish. Lowering total phosphorus levels in diets and increasing available phosphorus to levels slightly above the dietary requirement is the strategy used to comply with regulations on phosphorus levels in farm effluent water. Efforts are under way around the world to reduce levels of marine protein (fishmeal) in trout feeds and increase levels of plant protein concentrates. Such concentrates tend to have high concentrations of phytate, the storage form of phosphorus in seeds. Phytate in indigestible to all monogastric animals, including fish, and passes through the gastrointestinal tract intact, thus adding phosphorus to farm effluent water. Adding microbial phytase to trout feeds by top-dressing after pelleting releases phosphorus from phytate and lowers phosphorus levels in faeces (Hardy, 2002). The effectiveness of phytase to release phosphorus from plant products varies with the source of the product (Cheng and Hardy, 2002). Trout fed feed supplemented with astaxanthin, a carotenoid pigment found in the natural foods of wild salmonids, develop salmon-coloured fillets, whereas in fish fed diets without carotenoid pigments, fillets are white. In the USA, trout production is split evenly between red and white fillets, whereas elsewhere in the world, mainly red fillets are produced.

On-growing of rainbow trout in sea cages is similar to on-growing of salmon. Fish are held in floating sea cages located in areas with sufficient tidal flow and current to maintain oxygen levels in pens at levels suitable for trout (> 5 mg/l). Fish growth rates depend on water temperature and photoperiod. Feeding rates are calculated such that feed is not limiting to growth rate. Pelleted, slowly sinking feed is delivered mechanically using pneumatic blower systems or, on some farms in Chile, by hand feeding. Systems that detect pellets near the bottom of pens and, in some systems, collect and return them to the surface of the pen are used in Norway and other countries to minimize feed loss. Trout are harvested when they reach 2.0–4.0 kg, depending on the market. Larger fish are filleted, whereas smaller fish are sold in some markets with gills and viscera removed.

Rainbow trout are affected by parasites and by bacterial and viral diseases, both in nature and in farms (Roberts, 2001). All trout diseases were present in wild fish before farming ever began, but farming practices have moved trout diseases among continents, resulting in outbreaks of some non-native diseases in native stocks. The most problematic parasites of rainbow trout are *Ichthyophthirius multifiliis*, commonly known as 'Ich', or 'white-spot', *Ichthyododo necatrix* (also known as *Costia necator*), causing the disease known as 'costiasis' in trout reared in fresh water, and copepods, which attach to gills. The myxozoan, *Myxobolus cerebralis*, causes the well-known disease of wild trout, whirling disease. The life cycle of this myxozoan includes a phase in bottom sediments, so whirling disease is found only in trout reared in earthen raceways. A final important protazoan disease is proliferative kidney disease (PKD), caused by the myxozoan, *Tetracapsula bryosalmonae*, which results in anaemia, gross swelling of kidney tissue and substantial losses when the water temperature increases to the point where anaemia lowers the oxygen-carrying capacity of the blood.

There are many bacterial and viral diseases of rainbow trout and losses to several of these diseases are economically significant to the trout farming industry. Major bacterial pathogens of rainbow trout cause bacterial gill disease and fin rot, *Flavobacterium branchiophilum*, columnaris, *F. columnare*, coldwater disease, *F. psychrophilum*, enteric red mouth (ERM), *Y. ruckeri*, furunculosis, *A. salmonicida*, bacterial kidney disease, *R. salmoninarum*, and salmonid rickettsiosis, *Piscirickettsia salmonis* (Roberts, 2001). Effective vaccines are available for ERM and furunculosis. The most serious viral diseases of rainbow trout are infectious pancreatic necrosis (IPN), infectious haematopoietic necrosis (IHN) and viral haemorrhagic septicaemia (VHS) (Roberts, 2001).

Fish health management in rainbow trout farming is based on prevention; once a disease outbreak occurs, it is very difficult to treat or control. Elements of prevention include sanitation (clean raceways), high-quality feed, prevention of overcrowding, elimination of disease vectors (biosecurity) and vaccination. Many fish pathogens are present in water or can be isolated from fish, but disease outbreaks generally occur in connection with poor rearing conditions or handling stress. Crowding associated with size-grading, vaccination, moving fish between raceways or pens, low dissolved oxygen content and high particulate levels in water are examples of conditions that can result in a disease outbreak in rainbow trout. Health management by prevention is the most economical and effective approach in rainbow trout farming. Birds are major disease vectors in trout farms because they move from farm to farm, catching and consuming diseased fish, both from farms and rivers. Disease organisms pass through the gut of birds and remain pathogenic. Most freshwater farms in the USA use netting to restrict bird access to trout raceways. Cars and trucks, equipment and boots should be disinfected when moved between farms. Build-up of uneaten feed and faeces in raceways should be avoided and dead fish should be removed promptly. Treating rearing water with antimicrobial compounds is conceivably a very effective means of preventing fish-to-fish transmission of pathogens; one such compound is approved for this use in the UK.

As mentioned, vaccination is extremely effective in preventing several important bacterial diseases in rainbow trout. Single or multivalent vaccines are widely used before fish are moved outdoors but, to date, vaccine delivery in trout is by batch immersion. Injecting individual rainbow trout with vaccines is too expensive to be used in commercial production. Treatment of disease in rainbow trout with antibiotics delivered via the feed is relatively uncommon. Only two antibiotics are approved for use in the USA to treat fish and they are no longer effective against many trout disease strains. There are strict regulations concening antibiotic use in trout farming that, in combination with the cost of adding antibiotics to feeds, limit their use.

Rainbow trout have been bred in captivity for 140 years and numerous strains have been identified and characterized (Kincaid *et al.*, 2002). Performance evaluation studies demonstrate that selection of some rainbow trout stocks for increased growth compared to maintenance of other stocks for fisheries enhancement purposes has resulted in differences in performance when fish are reared under identical, controlled conditions (Overturf *et al.*, 2004). Further, there are many values for genotypic and phenotypic correlations for various traits in rainbow trout (see Table 4.5 in Tave, 1993) but, to date, there are few published studies describing the results of systematic breeding programmes for rainbow trout stocks to increase growth, efficiency or disease resistance, or to select for some other desirable trait for trout farming. One reason for this is that commercial companies traditionally have supplied eggs to trout growers and most companies are loath to publish the details of their breeding programme.

Activity in trout breeding is increasing with the relatively recent establishment of several government research programmes to pursue improvement in trout performance through selection and breeding. Trout breeding systems are based on several selection schemes, including: (i) individual selection; (ii) between-family selection; (iii) within-family selection; (iv) a combination of between- and within-family selection; (v) pedigree selection (selection of offspring based on parental performance); and (vi) progeny testing (selection based on performance of offspring and keeping high-performing offspring as broodstock) (Gjedrem, 2005). Individual selection, also known as mass selection, involves ranking fish on a trait of interest, such as weight at a certain size, and only keeping fish in the selection programme that are in the top proportion. Selection pressure depends on the proportion of the fish that are allowed to become breeders; keeping the top 5%, for example, results in higher selection pressure than does keeping the top 50%. Between-family and within-family selection requires the establishment of family lines and the adoption of some means of maintaining and confirming family identity. This can be done using genetic testing of individual fish, or by using passive integrated transponder (PIT) tags implanted in each fish.

Molecular biology and genomics are expanding the potential for gains in trout selective breeding programmes because they offer techniques to measure the response of the genome to selection rather than limiting response variables to phenotypic traits, such as weight gain. For example, researchers are investigating the use of quantitative trait loci (QTLs) as a basis for trout selection. However, QTLs have been found to be unstable in crosses between different stocks, thereby limiting their utility in aquaculture. Increasingly, expression levels of individual genes can be correlated to specific traits used to measure the effects of within- and between-family crosses. Expression levels that are correlated to conventional performance indices can be measured quickly compared to the length of time and extensive fish-rearing facilities required to conduct medium- or long-term growth trials in offspring of a large number of crosses. Expression levels are also important to identify negative traits that may be correlated to traits being selected. For example, selecting for fast-growing salmon is often a goal of breeding programmes. However, selecting for weight gain inadvertently may select for increased susceptibility to disease. Such inadvertent selection can be tested by conducting a series of challenge tests for diseases of concern, or by measuring survival and cause of death in selected fish over a production cycle in production systems where exposure to diseases may occur. However, it is much more efficient to measure expression of genes associated with various components of the immune system known to provide resistance to bacterial and/or viral diseases (Overturf and LaPatra, 2006; Purcell *et al.*, 2006). Moreover, measuring expression of a gene associated with muscle growth, for example myosin (Overturf and Hardy, 2001), provides a quantitative means of assessing the effects of selection for muscle accretion. Measuring expression of genes associated with immune function provides a quantitative means of assessing the effects both of direct selection for disease resistance and of inadvertent selection for increased susceptibility to disease, obviously a trait unwanted in the selection programme.

12.6 Acknowledgements

The help from Matt Bransden, Rhys Hauler, David Mitchell, Atle Mortensen and Tage Rochman in providing technical information is gratefully acknowledged.

References

Anderson, D.P. and Siwicki, A.K. (1994) Duration of protection against *Aeromonas salmonicida* in brook trout immunostimulated with glucan or chitosan by injection or immersion. *The Progressive Fish-Culturist* 56, 258–261.

Baeverfjord, G. and Wibe, Å. (2003) Shorttail deformities in Atlantic salmon – effect of freshwater production temperature. In: *Beyond Monoculture*. Abstracts at Aquaculture Europe 2003, Trondheim. EAS Special Publication No 33. European Aquaculture Society, Oostende, Belgium, pp. 121–122.

Baeverfjord, G., Lein, I., Åsgård, T., Rye, M. and Storset, A. (1998) High temperatures during egg incubation may induce malformations in Atlantic salmon (*Salmo salar* L.). In: *Aquaculture and Water*. Abstracts at Aquaculture Europe 98, Bordeaux. EAS Special Publication No 26. European Aquaculture Society, Oostende, Belgium, pp. 24–25.

Baeverfjord, G., Lein, I., Åsgård, T., Rye, M., Storset, A. and Siikavuopio, S.I. (1999) Vertebral deformations induced by high temperatures during embryogenesis in Atlantic salmon (*Salmo salar* L.). In: *Towards Predictable Quality*. Abstracts at Aquaculture Europe 99, Trondheim. EAS Special Publication No 27. European Aquaculture Society, Oostende, Belgium (ISBN 90-71625), pp. 6–7.

Baird, H.B., Kreuger, C.C. and Josephson, D.C. (2002) Differences in incubation period and survival of embryos among brook trout strains. *North American Journal of Aquaculture* 64, 233–241.

Balon, E.K. (1980) *Charrs, Salmonid Fishes of the Genus Salvelinus*. Dr. W. Junk Publishers, The Hague, The Netherlands.

Bardach, J.E., Ryther, J.H. and McLarney, W.O. (1972) *Aquaculture: The Farming and Husbandry of Freshwater and Marine Organisms*. Wiley-Interscience, New York.

Behnke, R.J. (1992) *Native Trout of Western North America*. American Fisheries Society Monograph 6, Bethesda, Maryland.

Behnke, R.J. (2002) *Trout and Salmon of North America*. The Free Press, New York.

Benfey, T.J. and Biron, M. (2000) Acute stress response in triploid rainbow trout (*Oncorhynchus mykiss*) and brook trout (*Salvelinus fontinalis*). *Aquaculture* 184, 167–176.

Benfey, T.J., McCabe, L.E. and Pepin, P. (1997) Critical thermal maxima of diploid and triploid brook charr, *Salvelinus fontinalis. Environmental Biology of Fishes* 49, 259–264.

Bernatchez, L., Glémet, H., Wilson, C.C. and Danzmann, R.G. (1995) Introgression and fixation of Arctic char (*Salvelinus alpinus*) mitochondrial genome in an allopatric population of brook trout (*Salvelinus fontinalis*). *Canadian Journal of Fisheries and Aquatic Sciences* 52, 179–185.

Besner, M. and Pelletier, D. (1991) Adaptation of the brook trout, *Salvelinus fontinalis*, to direct transfer to sea water in spring and summer. *Aquaculture* 97, 217–230.

Biron, M. and Benfey, T.J. (1994) Cortisol, glucose and hematocrit changes during acute stress, cohort sampling, and the diel cycle in diploid and triploid brook trout (*Salvelinus fontinalis* Mitchill). *Fish Physiology and Biochemistry* 13, 153–160.

Boeuf, G. and Harrache, Y. (1984) Adaptation osmotique à l'eau de mer de différentes espèces espèces (*Salmo trutta, Salmo gairdneri, Salvelinus fontinalis*) et hybrides (*Salmo trutta* × *Salvelinus fontinalis*) de salmonidés. *Aquaculture* 40, 353–358.

Boulanger, Y. (1991) Performance comparison of all-female, diploid and triploid brook trout. *Canadian Technical Report of Fisheries and Aquatic Sciences* 1789, 111–119.

Brannon, E.L. and Klontz, G. (1989) The Idaho aquaculture industry. *The Northwest Environmental Journal* 5, 23–35.

Bromage, N., Porter, M. and Randall, C. (2001) The environmental regulation of maturation in farmed finfish with special reference to the role of photoperiod and melatonin. *Aquaculture* 197, 63–98.

Carlson, A.R. and Hale, J.G. (1973) Early maturation of brook trout in the laboratory. *The Progressive Fish-Culturist* 35, 150–153.

Carter, C.G. (2007) Sustainable intensive aquaculture. In: Hui, Y.H. (ed.) *Handbook of Food Processing Manufacturing*. John Wiley and Sons, Inc., New York, pp. 967–979.

Carter, C.G. and Hauler, R.C. (2000) Fish meal replacement by plant meals in extruded feeds for Atlantic salmon, *Salmo salar* L. *Aquaculture* 185, 299–311.

Cheng, Z. and Hardy, R. (2002) Effect of microbial phytase on apparent nutrient digestibility of barley, canola meal, wheat and wheat middlings, measured *in vivo* using rainbow trout (*Oncorhynchus mykiss*). *Aquaculture Nutrition* 8, 271–277.

Cipriano, R.C., Marchant, D., Jones, T.E. and Schachte, J.H. (2002) Practical application of disease resistance: a brook trout fishery selected for resistance to furunculosis. *Aquaculture* 206, 1–17.
Clarke, W.C., Saunders, R.L. and McCormick, S.D. (1996) Smolt production. In: Pennell, W. and Barton, B.A. (eds) *Principles of Salmonid Culture*. Elsevier, Amsterdam, pp. 517–567.
COSEPAC/COSEWIC (2005) Évaluation et rapport de situation du COSEPAC sur le grand corégone (*Coregonus clupeaformis*) au Canada-Mise à jour. Comité sur la situation des espèces en péril au Canada, Ottawa.
Dabrowski, K., Czensy, S. and Matusiewicz, M. (2002) Coregonids. In: Webster, C.D. and Lim, C.E. (eds) *Nutrient Requirements and Feeding of Finfish for Aquaculture*. CAB International, Wallingford, UK, pp. 230–244.
Delabbio, J. (1995) Arctic charr culture in Canada. In: Boghen, A.D. (ed.) *Cold-water Aquaculture in Atlantic Canada*. Canadian Institute for Research on Regional Development, Moncton, New Brunswick, Canada, pp. 83–106.
DFO (2006) Aquaculture freshwater finfish species: brook trout (www.dfo-mpo.gc.ca/Aquaculture/finfish/brook_trout_e.htm, accessed 29 September 2006).
Dick, T.A. and Yang, X. (2002) Arctic char, *Salvelinus alpinus*. In: Webster, C.D. and Lim, C.E. (eds) *Nutrient Requirements and Feeding of Finfish for Aquaculture*. CAB International, Wallingford, UK, pp. 203–218.
Dubé, P., Blanc, J.-M., Chouinard, M. and Noüe, J. de la (1991) Triploidy induced by heat shock in brook trout (*Salvelinus fontinalis*). *Aquaculture* 92, 305–311.
Dumas, S., Blanc, J.M., Audet, C. and Noüe, J. de la (1992) The early development of hybrids between brook charr (*Salvelinus fontinalis*) and Arctic charr (*Salvelinus alpinus*). *Aquaculture* 108, 21–28.
Dumas, S., Blanc, J.M., Audet, C. and Noüe, J. de la (1995) Variation in yolk absorption and early growth of brook charr, *Salvelinus fontinalis* (Mitchell), Arctic charr, *Salvelinus alpinus* L., and their hybrids. *Aquaculture Research* 26, 759–764.
Dumas, S., Blanc, J.M., Vallée, F., Audet, C. and Noüe, J. de la (1996) Survival, growth, sexual maturation and reproduction of brook charr, *Salvelinus fontinalis* (Mitchell), Arctic charr, *Salvelinus alpinus* L., and their hybrids. *Aquaculture Research* 27, 245–253.
Duston, J. and Cusack, R.R. (2002) Emamectin benzoate: an effective in-feed treatment against the gill parasite *Salmincola edwardsii* on brook trout. *Aquaculture* 207, 1–9.
Dwyer, W.P., Piper, R.G. and Smith, C.E. (1983) Brook trout growth efficiency as affected by temperature. *Progressive Fish-Culturist* 45, 161–163.
Eckmann, R. (1987) A comparative study on the temperature dependence of embryogenesis in three coregonids (*Coregonus* spp.) from Lake Constance. *Swiss Journal of Hydrology* 49, 353–362.
Eckmann, R., Appenzeller, A. and Rösch, R. (1998) Biology and management of coregonid fishes – 1996. *Archiv für Hydrobiologie, Advances in Limnology* 50, 1–533.
Einen, O. and Roem, A.J. (1997) Dietary protein/energy ratios for Atlantic salmon in relation to fish size: growth, feed utilization and slaughter quality. *Aquaculture Nutrition* 3, 115–126.
Ellis, T. (2002) Health and disease in Atlantic salmon farming. In: Stead, S.M. and Laird, L. (eds) *Handbook of Salmon Farming*. Springer Praxis Publishing, Chichester, UK, pp. 373–401.
Esteve, M. (2005) Observations of spawning behaviour in Salmonidae: *Salmo*, *Oncorhynchus* and *Salvelinus*. *Reviews in Fish Biology and Fisheries* 15, 1–21.
FAO (2005) *Capture Production* (http://www.fao.org/fi/statist/FISHOFT/FISHPLUS.asp).
FAO (2007) Fishery Information, Data and Statistics Unit (FIDI). Fishery Statistical Collections. *FIGIS Data Collection*. FAO, Rome http://www.fao.org/figis/servlet/static?dom=collection&xml=global-aquaculture-production.xml).

Fitzgerald, R., Stefansson, S.O., Garforth, D. and Irwin, S. (2002) From egg to market size: onrearing in freshwater and marine environments. In: Stead, S.M. and Laird, L. (eds) *Handbook of Salmon Farming*. Springer, London, pp. 65–104.

Flüchter, J. (1980) Review of the present knowledge of rearing whitefish (Coregonidae) larvae. *Aquaculture* 19, 191–208.

Galbreath, P.F. and Samples, B.L. (2000) Optimization of thermal shock protocols for induction of triploidy in brook trout. *North American Journal of Aquaculture* 62, 249–259.

Galbreath, P.F. and Thorgaard, G.H. (1995) Saltwater performance of all-female triploid Atlantic salmon. *Aquaculture* 138, 77–85.

Galbreath, P.F., Adams, N.D. and Sherrill, L.W. III (2003) Successful sex reversal of brook trout with 17α-methyldihydrotestosterone treatments. *North American Journal of Aquaculture* 65, 235–239.

Galbreath, P.F., Adams, N.D. and Martin, T.H. (2004) Influence of heating rate in measurement of time to thermal maximum in trout. *Aquaculture* 241, 587–599.

Galbreath, P.F., Adams, N.D., Sherrill, L.W. III and Martin, T.H. (2006) Thermal tolerance of diploid versus triploid rainbow trout and brook trout assessed by time to chronic lethal maximum. *Environmental Biology of Fishes* 75, 183–193.

Gibson, R.J. (1993) The Atlantic salmon in freshwater: spawning, rearing and production. *Reviews in Fish Biology and Fisheries* 3, 39–73.

Gjedrem, T. (2000) Genetic improvement of cold-water fish species. *Aquaculture Research* 31, 25–33.

Gjedrem, T. (2005) *Selection and Breeding Programs in Aquaculture*. Springer, New York.

Glémet, H., Blier, P. and Bernatchez, L. (1998) Geographical extent of Arctic char (*Salvelinus alpinus*) mtDNA introgression in brook char populations (*S. fontinalis*) from eastern Québec, Canada. *Molecular Ecology* 7, 1655–1662.

Groot, C. (1996) Salmonid life histories. In: Pennell, W. and Barton, B.A. (eds) *Principles of Salmonid Culture*. Elsevier, Amsterdam, pp. 97–230.

Groot, C. and Margolis, L. (1991) *Pacific Salmon Life Histories*. UBC Press, Vancouver, Canada.

Groot, C., Margolis, L. and Clarke, W.C. (1995) *Physiological Ecology of Pacific Salmon*. UBC Press, Vancouver, Canada.

Guillaume, J., Kaushik, S., Bergot, P. and Metailler, R. (1999) *Nutrition and Feeding of Fish and Crustaceans*. Springer Praxis Publishing, Chichester, UK.

Gunther, S.J., Moccia, R.D. and Bureau, D.P. (2005) Growth and whole body composition of lake trout (*Salvelinus namaycush*), brook trout (*Salvelinus fontinalis*) and their hybrid F_1 splake (*Salvelinus namaycush* × *Salvelinus fontinalis*), from first-feeding to 16 weeks post first-feeding. *Aquaculture* 249, 195–204.

Halver, J.E. and Hardy, R.W. (2002) *Fish Nutrition*, Third Edition. Academic Press, San Diego, California.

Håkanson, L., Ervik, A., Mäkinen, T. and Möller, B. (1988) *Basic Concepts Concerning Assessments of Environmental Effects of Marine Fish Farms*. Nordic Council of Ministers, Copenhagen.

Hammar, J., Dempson, J.B. and Verspoor, E. (1991) Natural hybridization between Arctic char (*Salvelinus alpinus*) and brook trout (*S. fontinalis*): evidence from northern Labrador. *Canadian Journal of Fisheries and Aquatic Sciences* 48, 1437–1445.

Hardy, R.W. (2002) Rainbow trout, *Oncorhynchus mykiss*. In: Webster, C.D. and Lim, C.E. (eds) *Nutrient Requirements and Feeding of Finfish for Aquaculture*. CAB International, Wallingford, UK, pp. 184–202.

Harris, K.C. and Hulsman, P.F. (1991) Intensive culture of lake whitefish (*Coregonus clupeaformis*) from larvae to yearling size using dry feeds. *Aquaculture* 96, 255–268.

Heikinheimo, O., Amundsen, P.-A., Auvinen, H., Bodaly, D., Eshenroder, R., Huusko, A., Mills, K., Müller, R., Todd, T. and Winfield, I. (2004) Biology and management of coregonid fishes. *Annales Zoologici Fennici* 41, 1–390.

Henderson, N.E. (1963) Influence of light and temperature on the reproductive cycle of the eastern brook trout, *Salvelinus fontinalis* (Mitchill). *Journal of the Fisheries Research Board of Canada* 20, 859–897.

Hoar, W.S. (1988) The physiology of smolting salmonids. In: Hoar, W.S. and Randall, D.J. (eds) *Fish Physiology XIB*. Academic Press, San Diego, California, pp. 275–343.

Hokanson, K.E.F., McCormick, J.H., Jones, B.R. and Tucker, J.H. (1973) Thermal requirements for maturation, spawning, and embryo survival of the brook trout, *Salvelinus fontinalis*. *Journal of the Fisheries Research Board of Canada* 30, 975–984.

Holcombe, G.W., Pasha, M.S., Jensen, K.M., Tietge, J.E. and Ankley, G.T. (2000) Effects of photoperiod manipulation on brook trout reproductive development, fecundity, and circulating sex steroid concentrations. *North American Journal of Aquaculture* 62, 1–11.

Howgate, P. (2002) Post-harvest handling and processing. In: Stead, S.M. and Laird, L. (eds) *Handbook of Salmon Farming*. Springer Praxis Publishing, Chichester, UK, pp. 187–202.

Hyndman, C.A., Kieffer, J.D. and Benfey, T.J. (2003a) The physiological response of diploid and triploid brook trout to exhaustive exercise. *Comparative Biochemistry and Physiology* 134A, 167–179.

Hyndman, C.A., Kieffer, J.D. and Benfey, T.J. (2003b) Physiology and survival of triploid brook trout following exhaustive exercise in warm water. *Aquaculture* 221, 629–643.

Isaksson, A. (1991) Culture of Atlantic salmon. In: Stickney, R.R. (ed.) *Culture of Salmonid Fishes*. CRC Press, Boca Raton, Florida, pp. 85–129.

Järvinen, A. (1988) Proceedings of the International Symposium on Biology and Management of Coregonids. *Finnish Fisheries Research* 9, 1–527.

Jobling, M., Jørgensen, E.H., Arnesen, A.M. and Ringø, E. (1993) Feeding, growth, and environmental requirements of Arctic charr: a review of aquaculture potential. *Aquaculture International* 1, 20–46.

Jobling, M., Tveiten, H. and Hatlen, B. (1998) Cultivation of Arctic charr: an update. *Aquaculture International* 6, 181–196.

Johnston, G. (2002) *Arctic Charr Aquaculture*. Fishing News Books, Oxford, UK.

Johnston, I.A. (2001) Implications of muscle growth patterns for the colour and texture of fish flesh. In: Kestin, S.C. and Warris, P.D. (eds) *Farmed Fish Quality*. Blackwell Science, Oxford, UK, pp. 13–30.

Jones, J.W. (1959) *The Salmon*. Collins, London.

Jonsson, B. and Jonsson, N. (2001) Polymorphism and speciation in Arctic charr. *Journal of Fish Biology* 58, 605–638.

Kerr, S.J. (2000) Brook trout stocking: an annotated bibliography and literature review with an emphasis on Ontario waters. Fish and Wildlife Branch, Ontario Ministry of Natural Resources: Peterborough, Ontario, Canada.

Kestin, S.C. and Warriss, P.D. (2001) *Farmed Fish Quality*. Blackwell Science, Oxford, UK.

Kincaid, H.L., Mengel, L.J. and Brimm, S. (2002) National Fish Strain Registry – Trout. US Geological Survey.

Kindness, R.B. (2002) Broodstock management and early freshwater stages. In: Stead, S.M. and Laird, L. (eds) *Handbook of Salmon Farming*. Springer, London, pp. 37–64.

Klemetsen, A., Amundsen, P.-A., Dempson, J.B., Jonsson, B., Jonsson, N., O'Connell, M.F. and Mortensen, E. (2003) Atlantic salmon *Salmo salar* L., brown trout *Salmo trutta* L. and Arctic charr *Salvelinus alpinus* (L.): a review of aspects of their life histories. *Ecology of Freshwater Fish* 12, 1–59.

Koskela, J. (1995) Influence of dietary protein levels on growth and body composition of whitefish (*Coregonus lavaretus*). *Archiv für Hydrobiologie, Advances in Limnology* 46, 331–338.

Koskela, J. and Eskelinen, U. (1995) Growth of European whitefish (*Coregonus lavaretus* L. s.l.) at different temperatures. *Polskie Archiwum Hydrobiologii* 39, 677–682.

Koskela, J., Jobling, M. and Savolainen, R. (1998) Influence of dietary fat level on feed intake, growth and fat deposition in the whitefish *Coregonus lavaretus. Aquaculture International* 6, 95–102.

Koskela, J., Määttä, V., Vielma, J., Rahkonen, R., Forsman, L., Setälä, J. and Honkanen, A. (2002) *Siian kasvatus ruokakalaksi (Table fish farming of whitefish).* Riista- ja kalatalouden tutkimuslaitos, Helsinki (in Finnish).

Koskela, J., Rahkonen, R., Pasternack, M. and Knuutinen, H. (2004) Effect of immunization with two commercial vaccines on feed intake, growth and lysozyme activity in European whitefish (*Coregonus lavaretus* L.). *Aquaculture* 234, 41–50.

Lafleur, P.E., Larrivée, M. and St-Gelais, S. (1986) La production de salmonidés en eau salée. In: Colloque sur l'aquiculture. 14, 15 March 1986. Perspectives d'avenir en aquaculture, Saint-Hyacinthe, QC, Canada, MAPAQ, pp. 143–156.

Lahnsteiner, F. (2000) Semen cryopreservation in the Salmonidae and in the northern pike. *Aquaculture Research* 31, 245–258.

Lamarre, S.G. and Le François, N.R. (2003) Seawater adaptability optimization of triploid 0+ brook charr (*Salvelinus fontinalis*) using experimental diets prior to direct transfer in estuarine conditions. Poster presented at the 133th Annual Meeting of the American Fisheries Society, 10–14 August 2003, Québec City.

Langdon, C. (2003) Microparticle types for delivering nutrients to marine fish larvae. *Aquaculture* 227, 259–275.

Le François, N.R. and Blier, P.U. (2000) Branchial $Na^{+}K^{+}$ATPase activity in brook charr (*Salvelinus fontinalis*): effect of gonadal development in hypo- and hyperosmotic environments. *Journal of Experimental Zoology* 286, 647–655.

Le François, N.R. and Blier, P.U. (2003) Reproductive events and associated reduction in the seawater adaptability of brook charr (*Salvelinus fontinalis*): evaluation of gill metabolic adjustments. *Aquatic Living Resources* 16, 69–76.

Le François, N.R., Blier, P.U., Adambounou, L.T. and Lacroix, M. (1997) Alteration of gonadal development of brook charr (*Salvelinus fontinalis*): impact on salinity tolerance following transfer in estuarine conditions. *Journal of Experimental Zoology* 279, 273–283.

Le François, N.R., Lemieux, H. and Blier, P.U. (2002) Biological and technical evaluation of the potential of marine and anadromous fish species for cold-water mariculture. *Aquaculture Research* 33, 95–108.

Leitritz, E. and Lewis, R.C. (1980) *Trout and Salmon Culture*. California Fish Bulletin Number 164. Agricultural Sciences Publications, University of California, Berkeley, California.

Lever, C. (1996) *Naturalized Fishes of the World*. Academic Press, London.

Lönnström, L.-G., Rahkonen, R., Lundén, T., Pasternack, M., Koskela, J. and Gröndahl, A. (2001) Protection, immune response and side-effects in European whitefish (*Coregonus lavaretus* L.) vaccinated against vibriosis and furunculosis. *Aquaculture* 200, 271–284.

Lu, G. and Bernatchez, L. (1999) Correlated trophic specialization and genetic divergence in sympatric lake whitefish ecotypes (*Coregonus clupeaformis*): support for the ecological speciation hypothesis. *Evolution* 53, 1491–1505.

Luczynski, M., Bodaly, R.A., Bond, W.A., Eckman, R., Kamler, E., Mills, K.H., Reist, J.D., Rösch, R., Segner, H. and Todd, T.N. (1995) Biology and management of coregonid fishes – 1993. *Archiv für Hydrobiologie, Advances in Limnology* 46, 1–485.

McCormick, S.D. and Naiman, R.J. (1984) Some determinants of maturation in brook trout, *Salvelinus fontinalis. Aquaculture* 43, 269–278.

MAPAQ (2006) Présentation des pêches et de l'aquaculture commerciales: l'aquaculture en eau douce (www.mapaq.gouv.qc.ca/Fr/Peche/Profil/pecheaquaculture/aqucutlure, accessed 29 September 2006).

Marquis, H. and Lallier, R. (1989) Efficacy studies of passive immunization against *Aeromonas salmonicida* infection in brook trout, *Salvelinus fontinalis* (Mitchill). *Journal of Fish Diseases* 12, 233–240.

Marten, P.S. (1992) Effect of temperature variation on the incubation and development of brook trout eggs. *The Progressive Fish-Culturist* 54, 1–6.

Mills, D. (2000) *The Ocean Life of Atlantic Salmon – Environmental and Biological Factors Influencing Survival*. Blackwell Science, Oxford, UK.

Mills, D. (2003) *Salmon At The Edge*. Blackwell Science, Oxford, UK.

Mitchum, D.L. and Sherman, L.E. (1981) Transmission of bacterial kidney disease from wild to stocked hatchery trout. *Canadian Journal of Fisheries and Aquatic Sciences* 38, 547–551.

Nelson, J.S. (1994) *Fishes of the World*. John Wiley and Sons, New York.

O'Keefe, R.A. and Benfey, T.J. (1997) The feeding response of diploid and triploid Atlantic salmon and brook trout. *Journal of Fish Biology* 51, 989–997.

O'Keefe, R.A. and Benfey, T.J. (1999) Comparative growth and food consumption of diploid and triploid brook trout (*Salvelinus fontinalis*) monitored by radiography. *Aquaculture* 175, 111–120.

Olsen, Y. (2004) Live food technology of cold-water marine fish larvae. In: Moksness, E., Kjørsvik, E. and Olsen, Y. (eds) *Culture of Cold-water Marine Fish*. Blackwell Science, Oxford, UK, pp. 73–128.

Østbye, K., Næsje, T.F., Bernatchez, L., Sandlund, O.T. and Hindar, K. (2005) Morphological divergence and origin of sympatric populations of European whitefish (*Coregonus lavaretus* L.) in Lake Femund, Norway. *Journal of Evolutionary Biology* 18, 683–702.

Overturf, K. and Hardy, R. (2001) Myosin expression levels in trout muscle: a new method for monitoring specific growth rates for rainbow trout *Oncorhynchus mykiss* (Walbaum) on varied planes of nutrition. *Aquaculture Research* 32, 315–322.

Overturf, K. and LaPatra, S. (2006) Quantitative expression (Walbaum) of immunological factors in rainbow trout, *Oncorhynchus mykiss* (Walbaum), after infection with either *Flavobacterium psychrophilum*, *Aeromonas salmonicida*, or infectious haematopoietic necrosis. *Journal of Fish Diseases* 29, 215–224.

Overturf, K., Bullock, D., LaPatra, S. and Hardy, R. (2004) Genetic selection and molecular analysis of domesticated rainbow trout for enhanced growth on alternative diet sources. *Environmental Biology of Fishes* 69, 409–418.

Pangle, K.L., Sutton, T.M. and Brown, P.B. (2003) Evaluation of practical and natural diets for juvenile lake herring. *North American Journal of Aquaculture* 65, 91–98.

Pennell, W. and Barton, B.A. (1996) *Principles of Salmonid Culture*. Elsevier, Amsterdam.

Pennell, W. and Mclean, W.E. (1996) Early rearing. In: Pennell, W. and Barton, B.A. (eds) *Principles of Salmonid Culture*. Elsevier, Amsterdam, pp. 365–465.

Perry, G.M.L., Tarte, P., Croisetière, S., Belhumuer, P. and Bernatchez, L. (2004) Genetic variance and covariance for 0+ brook charr (*Salvelinus fontinalis*) weight and survival time of furunculosis (*Aeromonas salmonicida*) exposure. *Aquaculture* 235, 263–271.

Perry, G.M.L., King, T.L., St-Cyr, J., Valcourt, M. and Bernatchez, L. (2005) Isolation and cross-familial amplification of 41 microsatellites for the brook charr (*Salvelinus fontinalis*). *Molecular Ecology Notes* 5, 346–351.

Power, G. (1980) The brook charr, *Salvelinus fontinalis*. In: Balon, E.K. (ed.) *Charrs: Salmonid Fishes of the Genus Salvelinus*. Dr. W. Junk bv Publishers, The Hague, The Netherlands, pp. 141–203.

Prévost, E. and Chaput, G. (2001) *Stock, Recruitment and Reference Points: Assessment and Management of Atlantic Salmon*. INRA, Paris.

Purcell, M.K., Nichols, K.M., Winton, J.R., Kurath, G., Thorgaard, G.H., Wheeler, P., Hansen, J.D., Herwig, R.P. and Park, L.K. (2006) Comprehensive gene expression profiling following DNA vaccination of rainbow trout against infectious hematopoietic necrosis virus. *Molecular Immunology* 43, 2089–2106.

Purser, J. and Forteath, N. (2003) Salmonids. In: Lucas, J. and Southgate, P. (eds) *Aquaculture: Farming of Aquatic Animals and Plants*. Blackwell Science, Oxford, UK, pp. 295–320.

Quinn, T.P. (2005) *The Behaviour and Ecology of Pacific Salmon and Trout*. American Fisheries Society (University of Washington Press), Bethesda, Maryland.

Rasmussen, R.S. and Ostenfeld, T.H. (2000) Effect of growth rate on quality traits and feed utilization of rainbow trout (*Oncorhynchus mykiss*) and brook trout (*Salvelinus fontinalis*). *Aquaculture* 184, 327–337.

Rikardsen, A.H., Thorpe, J.E. and Dempson, J.B. (2004) Modelling the life-history variation of Arctic charr. *Ecology of Freshwater Fish* 13, 305–311.

Roberts, R.J. (2001) *Fish Pathology*, 3rd edn. W.B. Saunders, London.

Roberts, R.J. and Shepherd, C.J. (1997) *Handbook of Trout and Salmon Diseases*. Fishing News Books, Oxford, UK.

Roberts, R.J., Hardy, R.W. and Sugiura, S.H. (2001) Screamer disease in Atlantic salmon, *Salmo salar* L., in Chile. *Journal of Fish Diseases* 24, 543–549.

Rösch, R. (1995) Rearing of coregonid (*Coregonus* sp.) larvae in tanks: a review. *Archiv für Hydrobiologie, Advances in Limnology* 46, 293–300.

Ruohonen, K., Koskela, J., Vielma, J. and Kettunen, J. (2003) Optimal diet composition for European whitefish (*Coregonus lavaretus*): analysis of growth and nutrient utilisation in mixture model trials. *Aquaculture* 225, 27–39.

Sacobie, C.F.D. and Benfey, T.J. (2005) Sex differentiation and early gonadal development in brook trout. *North American Journal of Aquaculture* 67, 181–186.

Sargent, J.R., Bell, J.G., McGhee, F., McEvoy, J. and Webster, J.L. (2001). The nutritional value of fish. In: Kestin, S.C. and Warris, P.D. (eds) *Farmed Fish Quality*. Blackwell Science, Oxford, UK, pp. 3–12.

Saunders, R.L., Muise, B.C. and Henderson, E.B. (1975) Mortality of salmonids cultured at low temperature in sea water. *Aquaculture* 5, 243–252.

Schafhauser-Smith, D. and Benfey, T.J. (2001) The reproductive physiology of three age classes of female diploid and triploid brook trout (*Salvelinus fontinalis*). *Fish Physiology and Biochemistry* 25, 319–333.

Scott, W.B. and Crossman, E.J. (1973) *Freshwater Fishes of Canada*. Supply and Services Canada, Ottawa.

Sinnott, R. (2002) Fish farming and feed companies. In: Stead, S.M. and Laird, L. (eds) *Handbook of Salmon Farming*. Springer Praxis Publishing, Chichester, UK, pp. 105–185.

Sorensen, N.K., Tobiassen, T. and Carlehoeg, M. (2004) How do handling and killing methods affect ethical and sensory quality of farmed Atlantic salmon? In: Sakaguchi, M. (ed.) *More Efficient Utilization of Fish and Fisheries Products*. Developments in Food Science 42 Elsevier, Amsterdam, pp. 301–316.

Stead, S.M. and Laird, L. (2002) *Handbook of Salmon Farming*. Springer-Praxis, Chichester, UK.

Stickney, R.R. (1996) *Aquaculture in the United States, A Historical Survey*. John Wiley and Sons, Inc., New York.

Stickney, R.R. (2000) *Encyclopedia of Aquaculture*. Wiley-Interscience, New York.

Stickney, R.R. (2001) History and purpose of fish culture. In: Wedemeyer, G.A. (ed.) *Fish Hatchery Management*, 2nd edn. American Fisheries Society, Bethesda, Maryland, pp. 1–30.

Storebakken, T. (2002) Atlantic salmon, *Salmo salar*. In: Webster, C.D. and Lim, C.E. (eds) *Nutrient Requirements and Feeding of Finfish for Aquaculture*. CAB International, Wallingford, UK, pp. 79–102.

Stottrup, J. and McEvoy, L. (2002) *Live Feeds in Marine Aquaculture*. Blackwell Science, Oxford, UK.

Sutterlin, A.M., Harman, P. and Barchard, M. (1976) The culture of brook trout in salt water. Ministere de l'Environment. Service des Peches at Science de la mer. Rapport Technique 636: 1–21.

Tave, D. (1993) *Genetics for Fish Hatchery Managers*, 2nd edn. An Avi Book, Van Nostrand Reinhold, New York.

Todd, T.N. and Fleischer, G. (2002) Biology and management of coregonid fishes – 1999. *Archiv für Hydrobiologie, Advances in Limnology* 57, 1–689.

Todd, T.N. and Luczynski, M. (1992) Biology and management of coregonid fishes – 1990. *Polskie Archiwum Hydrobiologii* 39, 247–894.

Torrissen, O.J., Sigurgisladottir, S. and Slinde, E. (2001) Texture and technological properties of fish. In: Kestin, S.C. and Warris, P.D. (eds) *Farmed Fish Quality*. Blackwell Science, Oxford, UK, pp. 42–57.

Torstensen, B.E., Bell, J.G., Rosenlund, G., Henderson, R.J., Graff, I.E., Tocher, D.R., Lie, O. and Sargent, J.R. (2005) Tailoring of Atlantic salmon (*Salmo salar* L.) flesh lipid composition and sensory quality by replacing fish oil with a vegetable oil blend. *Journal of Agricultural and Food Chemistry* 53, 10166–10178.

Vielma, J., Koskela, J., Ruohonen, K., Jokinen, I. and Kettunen, J. (2003) Optimal diet composition for European whitefish (*Coregonus lavaretus*): carbohydrate stress and immune parameter responses. *Aquaculture* 225, 3–16.

Vijayan, M.M. and Leatherland, J.F. (1988) Effect of stocking density on the growth and stress-response in brook charr, *Salvelinus fontinalis*. *Aquaculture* 75, 159–170.

Vijayan, M.M., Ballantyne, J.S. and Leatherland, J.F. (1990) High stocking density alters the energy metabolism of brook charr, *Salvelinus fontinalis*. *Aquaculture* 88, 371–381.

Volkman, E.T., Pangle, K.L., Rajchel, D.A., Sutton, T.M. and Bast, D.R. (2004) Comparison of hatchery performance attributes among three strains of age-0 coaster brook trout. *North American Journal of Aquaculture* 66, 278–284.

Withler, R.E., Beacham, T.D., Solar, I.I. and Donaldson, E.M. (1995) Freshwater growth, smolting, and marine survival and growth of diploid and triploid coho salmon (*Oncorhynchus kisutch*). *Aquaculture* 136, 91–107.

Woo, P.T.K., Bruno, D.W. and Lim, L.H.S. (2002) *Diseases and Disorders of Finfish in Cage Culture*. CAB International, Wallingford, UK.

13 The Codfishes (Family: Gadidae)

Björn Björnsson,[1] Matthew Litvak,[2] Edward A. Trippel[3] and Marc Suquet[4]

[1]*Marine Research Institute, Reykjavik, Iceland;* [2]*University of New Brunswick, New Brunswick, Canada;* [3]*St Andrews Biological Station, St Andrews, Canada;* [4]*Ifremer, Landunvez, France*

13.1 General Introduction

The family Gadidae (Fig. 13.1), also called gadoids, are most abundant in the shallow waters of the continental shelf, especially in the northern seas. These are soft-rayed fishes with small and cycloid scales, mouth usually large and gill openings wide. They feed mainly on various invertebrates and fishes. Many of the species are of great commercial value, e.g. Atlantic cod, *Gadus morhua* L., haddock, *Melanogrammus aeglefinus* (L.), saithe, *Pollachius virens* (L.), pollack, *Pollachius pollachius* (L.), whiting, *Merlangius merlangus* L., cusk, *Brosme brosme* (Müller), ling, *Molva molva* (L.), blue ling, *M. dypterygia* (Pennant), blue whiting, *Micromesistius poutassou* (Risso), Norway pout, *Trisopterus esmarki* (Nilsson), and silver hake, *Merluccius bilinearis* (Mitchill) in the North Atlantic Ocean. In the North Pacific Ocean, the most valuable gadoids are Alaska pollock, *Theragra chalcogramma* (Pallas), and Pacific cod, *G. macrocephalus* Tilesius (Leim and Scott, 1966; Hart, 1973).

Atlantic cod, haddock, and to a lesser extent pollack, are the main candidates of this family, which have been considered for aquaculture purposes.

13.2 Atlantic Cod, *Gadus morhua*

Cod has an elongated, slightly compressed shape, tapering from the vent to the caudal fin with a pale lateral line (Fig. 13.2). Normally, the back is brown, sides yellow with numerous brown spots and the belly is white. The head and mouth are relatively large. The caudal, three dorsal, two anal and two pectoral fins are large (Leim and Scott, 1966). Cod is a moderate swimmer, capable of fast bursts to capture swimming prey (Björnsson, 1993) but cannot sustain high swimming speeds. There are sensitive barbels at the tip of the pelvic fins and

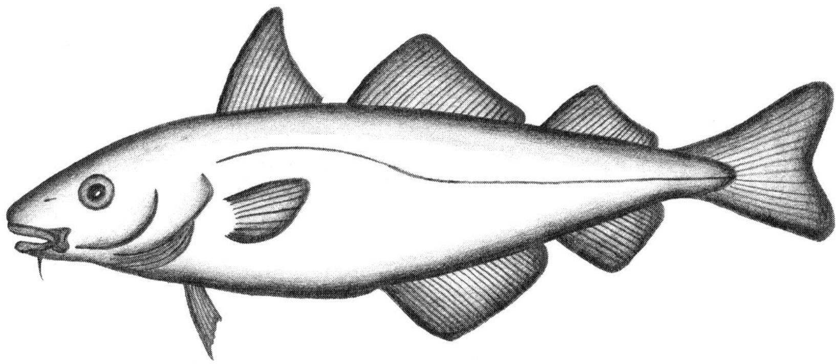

Fig. 13.1. Gadidae.

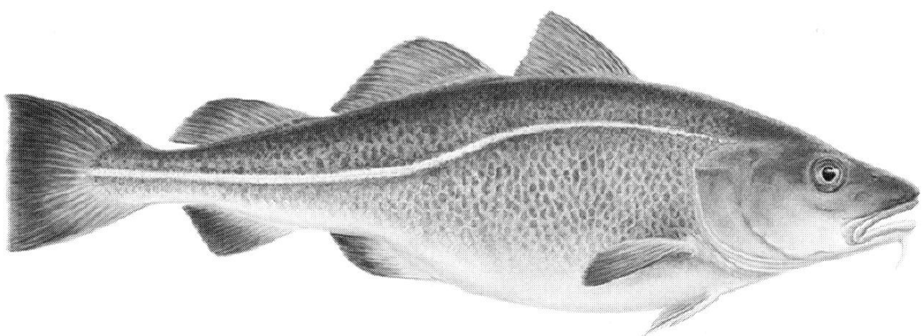

Fig. 13.2. Atlantic cod, *Gadus morhua* L. A drawing by Jón B. Hlídberg, Reykjavík.

on the lower jaw to assist in locating stationary prey. The flesh is lean, but fat is deposited in the liver.

There are several cod stocks on both sides of the North Atlantic Ocean, ranging in distribution from 40–80°N (Fig. 13.3). The main stocks are (or have been) found in the Gulf of Maine, Nova Scotia, Gulf of St Lawrence, Newfoundland, Labrador, Greenland, Iceland, the Faroe Islands, North Sea, Baltic Sea, Norway and the Barents Sea. Cod, which usually spawn in spring from March to May, are batch spawners, each female spawning every 2–3 days over a period of 4–6 weeks. They are highly fecund and the number of eggs increases with body size. A 100 cm-long female will produce about 4 million eggs (Kjesbu *et al.*, 1998). The eggs are small, 1.2–1.8 mm in diameter, buoyant and hatch near the surface of the sea, usually in the low salinity coastal current. The pelagic larvae are transported from the spawning grounds to the nursery area over a period of 3–4 months, during which they metamorphose into juveniles (ICES, 2005). Demersal life starts in late summer, when the juveniles seek the bottom at a size of 3–7 cm.

During the pelagic phase, the larvae feed on zooplankton, mainly copepod eggs and nauplii of *Calanus finmarchicus*. During the first 2 years of demersal life, the diet consists of various invertebrates such as euphausids, shrimps,

Fig. 13.3. World distribution of *Gadus morhua*.

crabs, amphipods, molluscs and small fish (Pálsson, 1994). The proportion of fish in the diet increases with the size of the cod.

Atlantic cod has been exploited for centuries (Kurlansky, 1998) but it was only with the technological development of the trawler in the 20th century that the stocks were overexploited. Prior to the decades of heavy exploitation, cod were generally long-lived (> 10 years) and matured late (> 7 years), but currently, most of the catch consists of 4–6-year-old cod (2–4 kg) (ICES, 2005). The total catch of Atlantic cod increased from 2.1 to 3.9 million tonnes (Mt) from 1950–1968 but, since then, there has been a gradual decrease to 0.9 Mt in 2004 (Fig. 13.4). The decline is thought to be due both to overexpoitation (Myers *et al.*, 1996) and environmental changes (Planque and Frédou, 1999; Drinkwater, 2005). The main cod fishing nations have been: Norway, Iceland, Russia (USSR), Canada and the UK (Fig. 13.4).

From economic and biological perspectives, cod has both positive and negative attributes as an aquaculture species.

Postitive attributes:

- Large markets. World catches of wild cod have declined from 3.9 to 0.9 Mt/year in the period from 1968 to 2004, forming a void for farmed cod.
- Low fat diet. The fillets of cod with only 1% fat content may fulfil the growing demand from the consumer market for a low fat diet.

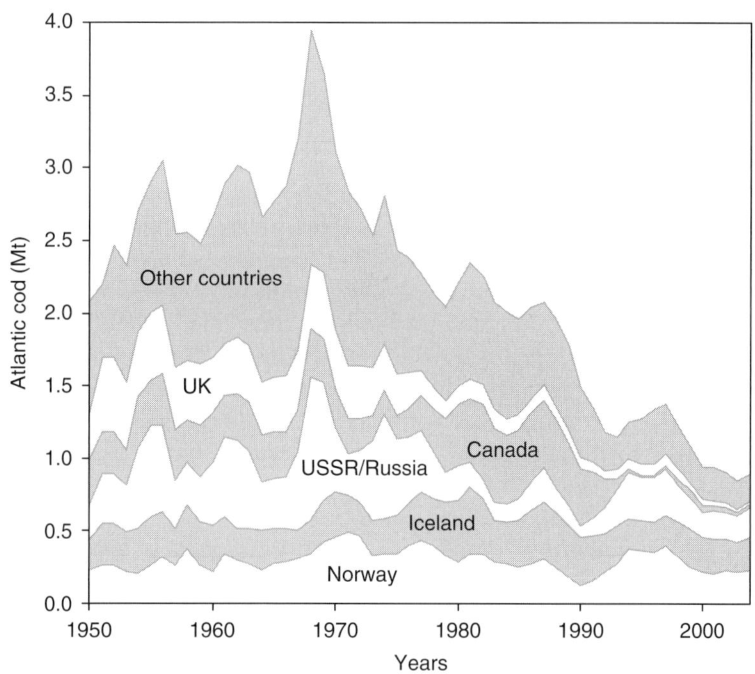

Fig. 13.4. Total catch of cod in the North Atlantic Ocean by nations in the period 1950–2004 (www.fao.org).

- Hardy species. Cod adapts easily to rearing conditions, can withstand size grading and tolerates a wide range of temperatures and salinities.
- Good growth. For a coldwater species it has a good growth potential, reaching market size (2–4 kg) in 2–3 years at high stocking densities.
- Excellent feed conversion. At optimal temperature, the feed conversion ratio is as low as 0.6–0.8 dry weight feed/wet weight fish.
- Low feeding frequency. Cod has a large stomach and can eat large amounts of feed in a single meal, which decreases the feeding cost.
- High fecundity. Small broodstock required, large breeding potential.

Negative attributes:

- Low to moderate price.
- Low to moderate fillet yield. The fillet yield as a percentage of whole fish is about 45% for immature farmed cod (*c.*40% for mature cod) compared to about 60% for farmed Atlantic salmon.
- Gaping in fillets. Farmed cod is more sensitive to handling and gaping in fillets is a larger problem than in wild cod.
- Early sexual maturation.
- Lack of vaccines.
- An escape artist. Cod has the ability to find and make small holes in the net to escape from sea cages.
- Cannibalism.
- Small larvae. The larvae must be start-fed on live feed, which increases the production cost of the juveniles.

13.2.1 Farming of Atlantic cod

In the late 19th century, large cod hatcheries were built in the USA and Norway to produce yolk sac larvae for stock enhancement purposes. In the period 1924–1950, the American hatcheries produced between 1.5–2.5 billion yolk sac larvae/year and the Norwegian hatchery between 20–400 million larvae/year (Solemdal *et al.*, 1984). These releases of cod larvae were stopped in 1971 when no improvements in the year-class strength could be confirmed (Tveite, 1971). Subsequently, methods were developed to produce cod juveniles both extensively (Kvenseth and Øiestad, 1984) and intensively (Howell, 1984). In Norway, large numbers of juvenile cod were produced extensively in lagoons from 1983 to 1997 for large-scale stock enhancement studies. The conclusion was that stock enhancement was not economically feasible (Otterå *et al.*, 1999; Kristiansen, 1999; Svåsand *et al.*, 2004).

Currently, cod farming is under development in seven countries: Norway, Canada, Iceland, the UK (Scotland), the USA, Ireland and the Faroe Islands. From 1987 to 2001, the total production of cod was only about 300 t/year, based almost entirely on on-growing of wild cod (Fig. 13.5). In recent years, there has been a rapid increase in cod production in Norway, reaching 12,000 t in 2007, which is more than 80% of the total production of farmed cod. Full-cycle farming with juveniles produced in hatcheries started only recently, first in the UK and Canada (1999), then Norway (2001) and finally Iceland and the USA

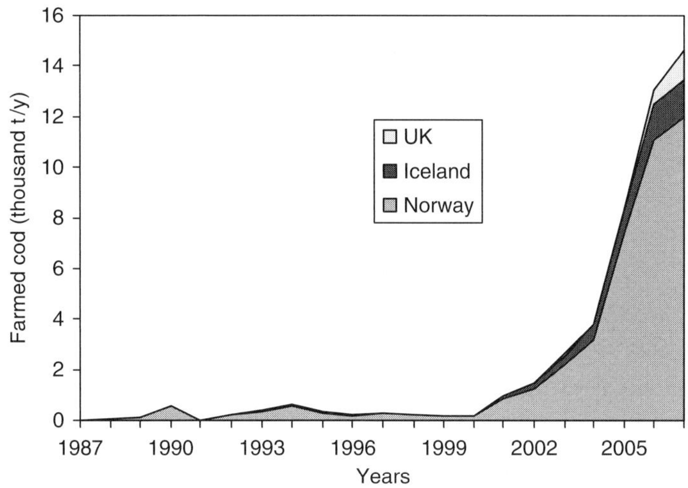

Fig. 13.5. Production of farmed cod in Norway, Canada, Iceland and the UK, both on-growing of wild cod and full-cycle cod farming (www.fao.org; www.fiskeridirektoratet.no; www.fishaq.gov.nl.ca; Christopher I. Hendry, St Johns, 2006, personal communication; www.fisheries.is; www.frs-scotland.gov.uk).

(2003). The largest juvenile production has been in Norway, reaching 14 million juveniles in 2007, followed by the UK with a production of about 1 million juveniles (Fig. 13.6). In the next few years, it is expected that the largest increase in cod production will come from full-cycle cod farming.

The largest increases in cod farming in the foreseeable future will be in Norway when considering the ideal environmental conditions for cod farming,

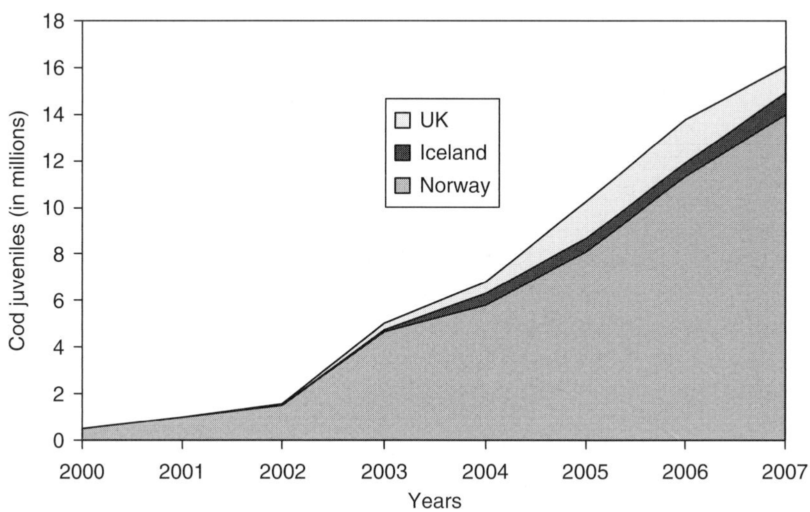

Fig. 13.6. Intensive production of juvenile cod in hatcheries in Norway, Canada, Iceland and the UK (www.torsk.net/index.php/site/content/download/627/2455/file/Steien.pdf; www.fishaq.gov.nl.ca; Christopher I. Hendry, St Johns, 2006, personal communication; www.fisheries.is; www.frs-scotland.gov.uk).

the active presence of the salmon farming industry, the major government-funded research and development programmes aimed at cod farming and based on the fact that Norwegian fish farming companies are willing to invest in large-scale industrial cod farming. However, the prognosis by Rosenlund and Skreeting (2006) that farmed cod would reach levels similar to those of farmed salmon within the next 15–20 years was probably too optimistic.

13.2.2 Rearing technologies and practices: full-cycle versus on-growing

The production methods used in farming cod must be highly cost-effective, considering the relatively low market price of farmed cod. Therefore, it is unlikely that on-growing of cod in land-based facilities can compete with on-growing in sea cages (Knútsson, 1997; Engelsen *et al.*, 2005). Two basic methods of cod farming have been developed, one based on capturing and on-growing wild fish (Fig. 13.7a) and the other one using hatchery produced juveniles (Fig. 13.7b).

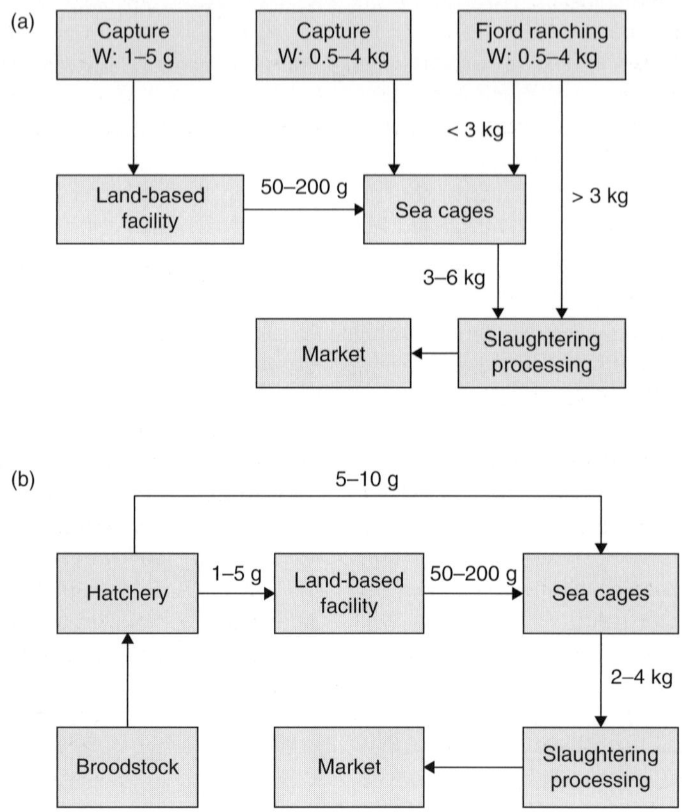

Fig. 13.7. Schematic presentation of the production cycle. (a) Capture-based cod farming. (b) Full-cycle cod farming.

13.2.2.1 Capture-based cod farming

There are some important advantages in capturing inshore cod for on-growing instead of practising the traditional fisheries. In the commercial inshore fishery, it is inevitable that a large number of young cod and haddock will be caught and discarded (Pálsson, 2003). Instead of killing and bringing to shore relatively small cod, they could be captured and reared to a larger and more valuable size. Both juvenile (1–5 g) and older cod (0.5–4 kg) have been captured for on-growing in sea cages (Fig. 13.7a). The largest mortalities have taken place during the weaning process and normally about 50% of the juveniles have survived until transfer to sea cages. The estimated cost of each weaned juvenile is about US$0.15. The greatest advantage of using wild juveniles for on-growing is their low cost, and the absence of visual deformities of the fish but diseases can be a problem. By using larger cod, on-growing costs associated with the rearing of juveniles are avoided and instead of killing the fish at 1–2 kg (US$1.6/kg gutted weight), it can be reared to 3–5 kg (US$2.1/kg gutted weight). The economy of this method depends to a large extent on the local conditions, e.g. access to young cod, availability of cheap feed (such as capelin), shelter and favourable temperatures during the on-growing period.

A special case of capture-based cod farming is the so-called fjord ranching, where herds of wild cod are formed by regular feeding in an area closed for commercial fishing. In one study, it was established that the annual weight gain of fish in the herds was three times that of the wild fish outside the herds (Björnsson, 2002). It has also been proposed that large-scale anthropogenic feeding of a cod stock can increase substantially the fisheries yield from the stock (Björnsson, 2001).

13.2.2.2 Full-cycle cod farming

The greatest advantage of full-cycle farming over capture-based farming is the possibility of selective breeding. Although both extensive and intensive production of cod juveniles started in the early 1980s, there has been limited market for juveniles for on-growing until in recent years. In the period 1999–2005, methods of large-scale intensive production of cod juveniles were developed in Norway, Iceland, the UK, Canada and the USA and now the supply of juveniles for on-growing is not a limiting factor in these countries.

Eggs from the broodstock are fertilized and hatched in indoor facilities where larval and juvenile production takes place. In locations with good shelter and warm temperatures, for example, during the summer months, the small juveniles can be moved directly to special sea cages for small fish (Watson *et al.*, 2006). More commonly, the small juveniles are moved to a land-based facility where they are reared for several months before transportation to sea cages, where they will be grown to market size (Fig. 13.7b). Most of the full-cycle cod farming in sea cages will be limited to countries which have access to sheltered inshore areas with a good flow of cool seawater. A submerged sea cage has been developed and tested for cod at an exposed location 14 km off the coast of New Hampshire, USA (Chambers and Howell, 2006). The economics of these cages have not been demonstrated but, if cod can be farmed profitably in offshore cages, the potential farming area would be virtually unlimited.

13.2.3 Broodstock management and hatchery operations

In full-cycle cod farming, it is desirable to rear the broodstock in indoor tanks under quarantine conditions to prevent diseases from the wild. It is relatively easy to maintain the broodstock in land-based tanks. Cod adapts well to life in captivity and will develop viable eggs and sperm. The eggs and sperm are easy to obtain by stripping, but fertilized eggs floating in the broodstock tanks can also be collected. However, cod females sometimes have difficulty in releasing the ripening eggs and may swell enormously and die (up to 30%) (van der Meeren and Ivannikov, 2001).

It is desirable to keep the broodstock at low density (10 kg/m^3) in indoor tanks (> 25 m^3). Good water quality must be maintained, preferably using a flow-through system with clean and cool seawater, 1 l/kg/min, > 90% oxygen saturation, temperature 6–8°C and salinity 30–34‰ (Steinarsson, 2004; Kjesbu and Norberg, 2005).

The broodstock must be reared in several tanks with different photoperiods to have access to fertilized eggs throughout the year to optimize the production capacity of the hatchery and minimize the production cost of juveniles. The spawning time of cod is determined by decreasing day length in late summer, which triggers the formation of several hormones required for the development of gonads (Norberg *et al.*, 2004; Pavlov *et al.*, 2004; Skjæraasen *et al.*, 2004).

Experiments with cod have shown that restricted rations may result in reduced fecundity and low egg quality and, therefore, the broodstock should be well fed prior to and during gonadal development (Pavlov *et al.*, 2004). It has also been found for cod that the gonadosomatic index increases with increased fat content in the feed (Karlsen *et al.*, 2006a). The broodstock diet must be rich in protein, unsaturated fatty acids, vitamins and minerals (Hemre *et al.*, 2002).

The broodstock must be part of a selective breeding programme, preferably based on family selection to avoid inbreeding, and allow selection for traits such as fillet quality, which can be estimated only after slaughtering the fish. When starting selective breeding, it is necessary to bring in large number of males (> 100) and females (> 200) originating from different areas to ensure sufficient genetic variability in the broodstock (Grundy *et al.*, 2000). Since breeding programmes are expensive and the fecundity of cod high, it is sufficient to have one breeding programme in each country from which eggs or larvae are distributed to commercial farms.

Usually, the hatched eggs are produced by stripping, normally one male used to fertilize two females to produce half-siblings. It is easy to identify mature males by running milt, but males and females can also be distinguished by ultrasound inspection of the developing gonads and it is convenient to use tags with different colours for males and females. Each female usually produces 10–20 batches of developed eggs (ovulation) in each spawning season, with 2–3 days between batches, normally 1.0–1.5 l of eggs/kg female (Pavlov *et al.*, 2004; Kjesbu and Norberg, 2005). The sexual development of each fish can be assessed visually and it is important to check them regularly to ensure that they are stripped and the eggs fertilized shortly after ovulation (Brooks *et al.*, 1997).

To fertilize the eggs, it is a common procedure to mix about 1 ml of sperm with 100 ml of cool seawater and then add it to about 200 ml of freshly stripped eggs in a pitcher. After initial stirring and subsequent storing for 10 min, the fertilized eggs are poured into a strainer, washed with a few litres of cool seawater and disinfected (Pavlov *et al.*, 2004). After washing in cool seawater, the eggs are incubated in the dark at 5–8°C and 34‰ salinity, usually in silos with a constant flow of seawater and sufficient aeration. Good results have been obtained by placing 3–6 eggs/ml in each silo (Støttrup, 2002; Steinarsson, 2004). The egg quality can be estimated from the percentage of normal cell division after 4–6 h from fertilization.

At optimal temperature (7–8°C), the eggs hatch after 12–13 days from fertilization (Steinarsson, 2004; Kjesbu and Norberg, 2005). Every day, dead eggs are decanted from the bottom of the silo. Then the aeration is stopped and the eggs collected at the surface of the silo, disinfected and put in a clean silo. After hatching is completed, larvae are transported to the larval rearing unit, usually a circular tank with a flat bottom and a slow-rotating bottom scraper, adjustable light levels and subsurface feeders. An initial stocking density of 100–150 larvae/l is commonly used in commercial hatcheries (Steinarsson, 2004).

In the larval tanks, the temperature is increased gradually from 8°C, a few days after hatching, to 12–15°C during metamorphosis, which occurs after 40–60 days. Three days after hatching, the larvae are fed for the first time on rotifers, 4 animals/ml. Cod larvae are visual feeders and long daylength during the first few weeks is required to promote good growth and survival (Puvanendran and Brown, 2002; Brown *et al.*, 2003). After 3 weeks, small quantities of *Artemia* are introduced to the tanks and, for the next 4–6 weeks, the amounts are increased gradually, whereas the amounts of rotifers are decreased and finally stopped about 4–5 weeks posthatch. Formulated feed is usually introduced to the larval tanks at the same time as *Artemia* (Steinarsson, 2004; Stoss *et al.*, 2004).

Currently, most of the larvae for cod farming are produced intensively in indoor hatcheries. However, cod juveniles produced extensively or semi-extensively on copepods are found to survive better and grow faster than cod juveniles produced intensively and without the deformities often seen in the intensive production. This difference may be due to the low nutritional value of the rotifers and *Artemia* used in the hatcheries (Imsland *et al.*, 2006).

There is much interest in developing formulated start-feed for cod larvae, but the main problem is the small size of the larvae (50 μg dry weight). In comparison, salmon alevins, which are routinely start-fed on dry feed, are 400 times heavier than cod larvae. The main challenge is to reduce the excessive nutrient leakage caused by the high surface-to-volume ratio of the small food particles required for cod larvae (see Chapter 3, this volume). Production of cod juveniles in commercial hatcheries still depends on the supply of live prey such as rotifers and *Artemia*, but many producers have replaced *Artemia* with formulated feeds.

13.2.4 On-growing to market size

Once the metamorphosed juveniles have been weaned onto dry feed, they can be size graded and moved to land-based tanks. Size grading is important to reduce cannibalism (Folkvord, 1991; Otterå and Folkvord, 1993). The optimal size of juveniles to be put in sea cages varies with environmental conditions, such as shelter and temperature. As the tolerance to low temperature and high turbulence in sea cages increases with the size of the juveniles, it is clear that smaller juveniles can be put in sea cages in Norway than in Iceland. The same basic cage technology which has been developed for salmon is suitable for cod (Engelsen *et al.*, 2004; Jobling, 2004), although lower feeding frequencies are required for cod than for salmon (Lambert and Dutil, 2001; Solberg *et al.*, 2006). Cod have been found to tear nylon netting with their teeth. Therefore, stronger and more expensive nets may be required for cod than for salmon (Moe *et al.*, 2005).

Juveniles and older cod grow equally well on commercial dry feed with different proportions of protein, fat and carbohydrate (Lie *et al.*, 1988; Morais *et al.*, 2001; Rosenlund *et al.*, 2004; Hamre and Mangor-Jensen, 2006; Karlsen *et al.*, 2006a). The natural diet of wild cod consists mainly of protein and fat, but only small amounts of carbohydrates. The commercial feed usually contains 10–15% carbohydrates such as wheat flour to ensure-sufficient binding qualities of the pellet. Cod are able to digest starches efficiently in diets with less than 20% carbohydrates (Hemre *et al.*, 2003). For young cod, the protein requirement may be less than 50% of dry weight (Morais *et al.*, 2001; Hamre and Mangor-Jensen, 2006) or between 50 and 60% according to Rosenlund *et al.* (2004). The recommended fat content in the diet of young cod ranges from 10 to 20% of dry weight (Morais *et al.*, 2001; Rosenlund *et al.*, 2004; Hamre and Mangor-Jensen, 2006). It has been shown that inclusion of fish bone and crab by-products in the feed can have a growth-promoting effect on young cod (Toppe *et al.*, 2006).

Grower feeds must be produced as economically as possible as they constitute the single largest expenditure in cod farming. Feeds should contain only enough protein to fulfil the minimum protein requirement for maximum growth, because protein is more expensive than fat and because protein gives less energy (4.3 kcal/g) than fat (9.4 kcal/g) for metabolism (Schmidt-Nielsen, 1979). Commercial dry feed of cod is mainly produced from fishmeal and fish oils. Experimental studies have shown that inexpensive proteins and oils from plants can be used, to some extent, to replace proteins and oils of marine origin (Hansen *et al.*, 2006; Mørkøre, 2006; Refstie *et al.*, 2006).

Apparently, genetic differences in growth rate between cod stocks are relatively minor. The few studies where offspring from different cod stocks have been reared under identical condititions do not suggest a large difference in growth rate between stocks (Svåsand *et al.*, 1996; Purchase and Brown, 2001). Optimal temperature for growth ($T_{opt.G}$) varies with body weight (W) of cod. It increases with weight during the larval period from about 8°C at hatching to about 16°C 50–60 days after hatching (Otterlei *et al.*, 1999; Steinarsson and Björnsson, 1999). However, from small juveniles to adulthood, $T_{opt.G}$ decreases from about 15°C for 2g cod to about 9°C for 2kg cod (Björnsson *et al.*, 2001, 2007). The optimal temperature for feed conversion ($T_{opt.FC}$) of juvenile to adult fish appears to be about 2°C lower than $T_{opt.G}$. Above a certain temperature threshold, mortality in a group of cod increases exponentially with increased temperature (Björnsson *et al.*, 2001). Therefore, the recommended rearing temperatures are 2–3°C below $T_{opt.G}$.

Growth rate at optimum temperature (G_{max}) is 5–10%/day for the newly hatched larvae, increasing to 15–20%/day after 30–40 days from hatch (Steinarsson, 2004) and then decreasing to 7%/day for 2g cod and to 0.4%/day for 2kg cod (Björnsson *et al.*, 2007). For juvenile to adult cod, G_{max} can be estimated by the following function of body weight: $G_{max} = 9.36W^{-0.409}$. For juvenile to adult cod, the minimum feed conversion ratio is very low, 0.6–0.8kg dry weight of feed required to produce each kg of cod (wet weight) (Björnsson *et al.*, 2001; Toppe *et al.*, 2006).

Sea temperatures lower in Iceland than in Norway mean that juveniles generally must be reared longer and to a larger size in land-based facilities in Iceland ($W \approx 50$–200g) than in Norway ($W \approx 10$–100g) before entering sea cages. The cost per rearing volume is much higher in land-based facilities than in sea cages and it is therefore of vital importance to maximize production in the land-based facilities. One way to achieve that goal is to optimize the environmental conditions in the rearing tanks, such as temperature (Jobling, 1988; Björnsson *et al.*, 2001), salinity (Lambert *et al.*, 1994; Dutil *et al.*, 1997), photoperiod (Hansen *et al.*, 2001; Karlsen *et al.*, 2006b), water quality (Foss *et al.*, 2004; Björnsson and Ólafsdóttir, 2006) and fish density (Lambert and Dutil, 2001). Juvenile cod can be reared at high stocking densities (> 40kg/m^3) without reduction in growth rate, provided that water quality is acceptable; in recirculation systems, total ammonium nitrogen (TAN) should be kept below 1mg/l (Björnsson and Ólafsdóttir, 2006; Foss *et al.*, 2006; Le François *et al.*, 2008).

As cod farming is in its initial stages of development, there are several disease problems to be solved. More research is required to understand the diseases of cod and how to minimize their detrimental effects, but it is clear that cod farming is not going to be a viable industry until specific vaccines have been developed to control the most serious diseases. Cod is a poor antibody responder compared to the salmonids, for example, and its specific immune system develops relatively late. However, many of its non-specific or innate defence parameters are active from the time of hatching (Magnadóttir *et al.*, 2004).

There are many diseases which are likely to affect cod under farming conditions; either bacterial, viral, fungal or parasitic diseases (Bricknell *et al.*, 2006). Currently, the two pathogens, vibriosis, *Listonella anguillarum* (formerly called *Vibrio anguillarum*), and the atypical furunculosis, *Aeromonas salmonicida* subsp. *achromogenes*, pose the greatest risk to cod farming (Kristmundsson *et al.*, 2005b; Bricknell *et al.*, 2006). They are commonly associated with stessful conditions, such as transportation, sorting, extreme temperatures, poor water quality and high stocking density. The standard vaccine developed for vibriosis in salmon farming does not provide sufficient protection against *L. anguillarum* serotype 02β and a vaccine against the atypical furunculosis has not yet been developed. Other bacteria have also caused problems in cod farming: *Mycobacterium* spp., winter ulcers, *Moritella viscosa* (Bricknell *et al.*, 2006), red mouth disease, *Yersinia ruckeri* (Bjarnheidur K. Gudmundsdóttir, Reykjavík, 2006, personal communication), *Tenacibaculum* spp. (formerly called Flexibacter) (Bergh *et al.*, 2005) and *Francisella* sp. (Olsen *et al.*, 2006).

There are four viral diseases of concern in cod farming: infectious pancreatic necrosis virus (IPNV), nodavirus, viral haemorrhagic septicaemia virus (VHSV) and cod ulcus syndrome (CUS). The risk of these diseases to the cod farming industry has not been established (Bergh *et al.*, 2005; Bricknell *et al.*, 2006).

A total of 107 parasites have been found in wild cod, 7 of those suggested to be specific to Atlantic cod (Hemmingsen and MacKenzie, 2001). Some of the parasites of wild cod, such as the seal worm, *Pseudoterranova decipiens*, are transmitted through the food web. The cost for the fish industry to remove the seal worm from cod fillets is very high, but this parasite will not be transmitted to farmed cod. Several parasites can be transmitted to larval and juvenile cod by consumption of wild zooplankton, but in modern hatcheries, these parasites can be eliminated. The so-called black spot disease which is caused by the trematode, *Cryptocotyle lingua*, has been found in farmed cod. The microsporidia *Loma* spp., usually found in the gills, can cause mortalities in farmed cod (Kristmundsson *et al.*, 2005a). The four following ectoparasites, which all have a direct life cycle, are likely to multiply in cod farms: the fish louse, *Caligus* spp., the ciliate, *Trichodina* spp., the flagellate, *Ichthyobodo* sp. (formerly called *Costia*), and *Gyrodactylus* spp. The fish louse can be controlled to some extent by oral medication, whereas the three remaining ectoparasites by formalin treatment (Bergh *et al.*, 2005).

There are some other factors, apart from infectious diseases, which can affect the health of farmed cod, such as algal blooms, jellyfish, swim bladder syndrome, gasbubble disease, various eye injuries, body deformities, stress and malnutrition. It is likely that many of the diseases of farmed cod can be controlled by prophylaxis and proper husbandry routines.

13.2.5 Commercialization

The optimal slaughtering size of farmed cod may be 3–5 kg. Loins and fillets are the most valuable part of the fish. Cod is a lean fish with low fat content (1%) and high water (79–88%) and protein (10–19%) contents in the muscle (Holdway and Beamish, 1984; Lambert and Dutil, 1997). Besides the fillets of cod, there are several less valuable by-products such as head, liver, gonads, stomach and enzymes.

Farmed cod differs from wild cod in many ways. It has a higher condition factor, and thus relatively larger and thicker fillets, but also larger livers and gonads. The weight loss by gutting immature farmed cod is about 17%, compared to 9% for Atlantic salmon (the weight loss by gutting mature cod can be as high as 35%). The head of immature farmed cod is about 16% of total weight, compared to 8% for rainbow trout (Mørkøre, 2005). Therefore, fillet yield is much less for immature farmed cod (*c.*45%) than salmon (*c.*60% of whole weight), which is of economic consequence as head, bones, liver, gonads, stomach and enzymes amount to only about 10% of the total revenue of farmed cod.

Compared with wild cod, farmed cod has a higher protein and glycogen content in muscle and lower pH. After cooking, it becomes whiter, with a milder taste and firmer texture than wild cod. However, farmed cod has been found to be more sensitive to rough handling and gaping in fillets is a greater problem than in wild cod. Gaping can be reduced by pre-rigor filleting (Gunnarsson and Jóakimsson, 2004; Mørkøre, 2005).

In recent years, wild Atlantic cod resource disposition has been: frozen fillets/blocks 33%, frozen whole (headed) 18%, salted/dried 30% and fresh 19% (Alda Möller, Reykjavík, 2006, personal communication). The highest value has been obtained by selling cod as fresh fillets.

It is likely that farmed cod is going to be marketed as fresh cod, mostly as fillets. Farmed cod can be slaughtered at any time of the year according to demand. Generally, the quality will be higher for farmed than wild cod; for example, because farmed cod can be starved and cooled and stress minimized before slaughtering to ensure maximum freshness and because the fillets are thicker and whiter than for wild cod. However, the flesh of farmed cod is more sensitive to handling than the flesh of wild cod. Wild cod fillets have to be inspected carefully for parasites, especially the seal worm. Farming of cod opens a new market for whole cod, which can be guaranteed as being free of nematodes. This may encourage chefs to develop new ways of preparing farmed cod, e.g. to make sushi or for broiling and baking whole. Farmed cod is going to be a somewhat different product than wild cod and must be marketed separately due to increasing demand for traceability.

Mean prices of fresh fillets of wild cod exported from Iceland have been increasing from US$6.38/kg in 2001 to US$9.57/kg (free on board [FOB]) in 2005 (Statistical Series, 2007). In 2005, the mean price in Icelandic fish markets was US$2.00/kg and US$2.45/kg of ungutted and gutted cod, respectively. At the same time, the mean price of ungutted farmed cod obtained by Icelandic fish farmers was US$2.50/kg. In a recent economic analysis, it was assumed that the prices obtained by Norwegian fish farmers were about US$3.00/kg ungutted cod (Engelsen *et al.*, 2005). Although the prices have been somewhat higher for farmed than wild cod, it is likely that this difference will diminish with larger volumes of farmed cod. The price of farmed cod will also depend on the supply of wild cod. The current low catches of cod are giving historically high prices (Engelsen *et al.*, 2004).

13.2.6 Future perspectives

For full-cycle cod farming, breeding programmes are critical to improve growth rate, fillet yield, feed conversion, flesh quality and other important traits. Since the breeding programme for Atlantic salmon started in Norway, the growth rate has improved by about 10% per generation (Gjøen and Bentsen, 1997).

A family selection programme for cod has started recently in Norway (first generation 2005) and Iceland (first generation 2006). Family groups are reared in separate tanks until the fish are large enough for individual tagging (Fjalestad and Mortensen, 2005).

Grower feeds constitute the largest expenditure in cod farming. They must be nutritious to ensure fast growth and high growth efficiency, but they also have to be inexpensive. The most economic proportions of protein, fat and carbohydrates must be determined and studies carried out to find out how much of the fat and protein from marine origin can be replaced by raw material from terrestrial plants to reduce feed cost. It is also important to develop good start-feeds for cod larvae to improve their quality and reduce the production cost of juveniles. Feeding systems developed for salmon farming must be tested and adapted to cod farming to reduce labour cost and minimize waste feeding (Jobling, 2004).

Premature sexual maturation must be avoided in farmed cod. Currently, farmed cod usually develop gonads twice, in their second and third year, before reaching market size (2–4 kg). In particular, the second sexual development is of serious economic consequences due to the detrimental effects on growth rate, feed conversion, fillet yield and quality. By selecting mainly for fast growth, breeding programmes can, with time, produce cod which will reach market size before becoming sexually mature for the second time. However, since the first sexual maturation of farmed cod occurs at a small size (0.5–1.0 kg) and high frequency (> 70%), it may not be practical to select against premature sexual maturation in cod. On the other hand, trials with light manipulation to prevent or delay sexual maturation have given promising results (Karlsen *et al.*, 2006b; Taranger *et al.*, 2006). There are other more controversial ways of preventing sexual maturation, such as administration of hormones (Shelbourn *et al.*, 1992), triploidy (Diaz *et al.*, 1993; Tiwary *et al.*, 2004) and use of transgenetic fish (Devlin *et al.*, 1994; Sin, 1997). Finally, it is urgent to develop vaccines against the most common and virulent diseases of farmed cod.

Currently, the market price is similar or lower than the production cost of farmed cod. Therefore, it seems clear that the production costs of farmed cod must go down if cod farming is going to be a viable industry in the future. Perhaps the greatest threat to the cod farming industry comes from the restoration of wild stocks.

13.3 Haddock, *Melanogrammus aeglefinus*

Haddock differs externally from cod, *G. morhua*, in that its lateral line is darker, it has a more silvery body and possesses a dark blotch on each side over the middle of the pectoral fin just below the lateral line (Bigelow and Schroeder, 1953). The body of haddock is heavy, elongate and laterally compressed. The head is large, possessing a rounded snout with a slightly inferior and small mouth with a single small barbel near the tip of the lower jaw. Their eyes are moderately large. The caudal is slightly forked, the second ray of the

Fig. 13.8. Adult haddock at the St Andrews Biological Station, Department of Fisheries and Oceans in St Andrews, New Brunswick, Canada. Photo by E.A. Trippel, St Andrews Biological Station, Department of Fisheries and Oceans.

pelvic fins are elongated and the position of the pelvic fins are dorsal and anterior to the pectoral fins. Their colour is dark purplish grey, turning silver laterally and white ventrally. It is silver in colour with a dark patch on the shoulder (Fig. 13.8).

Haddock occurs in the Atlantic Ocean (Fig. 13.9). In the eastern North Atlantic, it ranges from the Bay of Biscay to Spitzbergen; in the Barents Sea to Novaya Zemlya; occurs around Iceland; is rare in southern Greenland; and in the western North Atlantic from Cape May, New Jersey to the Strait of Belle Isle (Bigelow and Schroeder, 1953; Scott and Scott, 1988).

Haddock is a demersal species associated with hard, smooth sand or gravel bottoms. On the Scotian shelf, they are found within a temperature range of 1–13°C (Scott and Scott, 1988). Haddock populations undertake extensive migrations in the Barents Sea and Iceland and more restricted movements in the north-western Atlantic, mostly to and from the spawning grounds (Cohen *et al.*, 1990). Generally, age at maturity is reached at 4 years for males and 5 years for females; however, males and females from the North Sea stock reach maturity at 2 and 3 years, respectively. Haddock are fecund; a 50 cm fish produced 1,773,000 eggs (Scott and Scott, 1988). Spawning occurs in typically marine waters (35‰ salinity) between 50 and 150 m depth, in the north-western Atlantic from January to July (depending on the areas) and in the north-eastern Atlantic from February to June (mostly in March–April) (Cohen *et al.*, 1990). The eggs are 1.3–1.6 mm transparent, spherical, pelagic and lacking an oil globule (Fahay, 1983). The eggs are buoyant and float to the surface (Scott and Scott, 1988). Their development rate is controlled by temperature; they hatch in 20 days at

Fig. 13.9. World distribution of haddock.

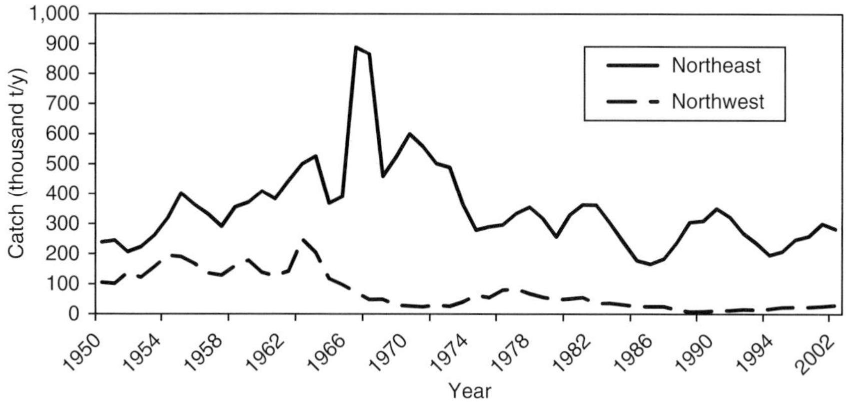

Fig. 13.10. Haddock landings in the northeast and northwest Atlantic Ocean. Data from FAO Fisheries Department, Fishing Information, Data and Statistics Unit collected with FISHSTAT Plus.

2°C and in 9 days at 10°C (Martell *et al.*, 2005). The growth rate varies considerably between regions, in the North-west Atlantic those of Georges Bank typically grow the fastest, with 0.9 kg and 2.0 kg attained on average at age 3 and 6 years, respectively; though considerable variability in annual growth exists (Van Eeckhaute and Brodziak, 2006). Haddock is an omnivorous bottom-feeding fish, feeding mainly on relatively small bottom-living organisms including crustaceans, molluscs, echinoderms, polychaetes and fishes.

Haddock stocks have been classified by the IUCN Red List as vulnerable, facing a high risk of extinction in the wild in the medium-term future (Sobel, 1996). Current landings of haddock are still low (Fig. 13.10) and their value has increased in North America during the past 50 years. The total catch reported to FAO for this species for 1999 was 249,317 t; the countries with the largest catches were the UK (72,001 t) and Norway (53,232 t) (FISHSTAT, 2000, accessed March 2007).

13.3.1 Farming of haddock

Haddock, *M. aeglefinus* (L.), is one of the most highly prized fish caught for the north-eastern North American market (Scott and Scott, 1988). Interest in ha dock culture had its origins in New Brunswick, Canada, in the late 1980s when the salmon industry started looking for alternatives to Atlantic salmon, *Salmo salar*, that could be grown with the existing cage systems (Waiwood, 1994; Litvak, 1998). Since then, there has been a lot of research and development on haddock aquaculture by industry, government and university scientists, but commercial productions are not actually in operation.

13.3.2 Broodstock management and hatchery operations

Wild fish are caught with hook and line. There continues to be a problem with exopthalmia and other symptoms of gas saturation related to bringing adult haddock up from depth. Unfortunately, haddock are not easily sexed. Fish are followed through a breeding cycle to allow for identification of gender and then tagged. However, ultrasonography may be used to help in the identification of gender prior to the spawning season (Martin-Robichaud and Rommens, 2001).

Haddock, like many marine finfish, is a highly fecund batch spawner, producing many buoyant small eggs with limited yolk reserves. Although haddock can be hand-stripped, they are much more prone to handling damage than other gadids. Haddock is a serial volitional spawner like cod (see Section 13.2 of this chapter). Spawning in tanks can be communal or in pairs, which allows for better broodstock management (Trippel, 2003; Trippel *et al.*, 2009). Monitoring of the seasonal egg production of captive females (paired mating) indicated that, on average, a 53 cm female produced 535,000 eggs (Trippel and Neil, 2004). Haddock produce more batches with lower batch-specific fecundity than cod, such that rhythmic spawning of up to 16–20 batches with one every 2–3 days is possible (Rideout *et al.*, 2004a; Trippel and Neil, 2004). Embryos are collected from tanks with a collection system similar to that developed for cod. Natural spawning of Atlantic Canadian populations occurs from April to June (Scott and Scott, 1988). The breeding season can be advanced successfully by photoperiod manipulation (Martin-Robichaud and Berlinsky, 2004).

A major problem with haddock broodstock has been the loss of fish at or around spawning. Most current broodstock are wild and may have been damaged already during capture. Unfortunately, the proximate cause of death is unknown, although in almost all cases hyperinflation of the swim bladder was observed. Haddock also exhibit courtship swimming behaviour and vocalization during mating and have the most interesting sound production of gadids (Hawkins and Amorim, 2000; Bremner *et al.*, 2002). Alternatively, spawn could be collected directly from cages, where a large number of individuals would undergo communal spawning. After collection, embryos are disinfected routinely with glutaraldehyde (400 ppm 5–10 min) following the protocol for cod; other disinfectants, such as 0.1% sodium hypochlorite (Peck *et al.*, 2004), have also been used.

Embryos have been incubated in a variety of set-ups, e.g. static, recirculation and flow-through (upwelling). For production purposes, an upwelling cylindroconical tank equipped with a banjo filter is preferred. Depending on temperature, the 3–4 mm larvae will hatch in 2–3 weeks (5–8°C).

A major problem during the egg incubation stage is complete batch failure (Rideout *et al.*, 2004b). This has been related to handling damage and/or egg and possibly sperm quality (Trippel *et al.*, 2000). Recent work suggests that we can use an early diagnostic of blastomere cleavage pattern (Rideout *et al.*, 2004b) to determine hatching potential of a batch at a very early stage in development. There has been very little work on the prediction of egg quality from broodstock quality. Seasonal changes in semen characteristics occur in haddock, particularly in spermatocrit (Rideout *et al.*, 2004a), though haddock semen tends to contain fewer sperm per unit volume than cod semen (Trippel *et al*., 1998). Sperm cryopreservation protocols have been developed for haddock (Degraaf and Berlinsky, 2004; Rideout *et al*., 2004c).

Haddock larvae are reared in intensive land-based systems using either flow-through or recirculation. Initially, stocking density was set at the standard cod model of 20 larvae/l. However, Hamlin and Kling (2001) demonstrated that gadids experienced higher survival and growth at very high densities of 150–300 larvae/l. Light spectra, intensity and photoperiod all affect growth and survival of haddock larvae (Downing and Litvak, 1999a,b, 2000, 2001, 2002; Downing, 2002).

At this time, no artificial food can be used to feed haddock from hatch. Thus, haddock, like many marine species being developed for aquaculture, depend on the production of live rotifers. Rotifer enrichment protocols vary and recent results suggest that many satisfy the haddock minimum requirements for fatty acids (Castell *et al*., 2003). Generally, larvae are fed at least twice per day on rotifers enriched with algae and/or a variety of enrichment media until 20–25 days post hatch (dph) (grown at 8–10°C). They are then switched to *Artemia* and, at approximately 40 dph, weaned to a microdiet (Hamlin and Kling, 2001).

Although weaning success of haddock from live to artificial diets is achieved, an early weaning process has not yet been refined; many larvae die during this stage (Blair *et al*., 2003). Juvenile growth is enhanced and activity reduced by exposure to continuous dim light (Tripple and Neil, 2003).

13.3.3 On-growing to market size

Currently, juveniles are transferred to cages at 3–5 g in either May or June (Frantsi *et al*., 2002). A cage within a cage model has been used. Ten thousand juveniles are stocked into a small 6 × 6 × 3 m deep cage of 1.27 cm stretch treated mesh placed inside a larger salmon-type cage (Frantsi *et al*., 2002). The salmon cage acts as a predator net and also as a breakwater to minimize the smaller cage's movements due to wave action (Dr Chris Frantsi, personal communication). Fish are graded as they grow and then are placed into larger 'salmon' cages. Haddock does not have the same level of protection conferred from the antifreeze or glycerol that is found in the blood plasma of cod (Ewart *et al*., 2000). Therefore, locations for haddock culture will be restricted to areas that do not experience temperatures below −0.8°C.

13.3.4 Commercialization

The existing market for haddock is fresh dressed, fresh and frozen fillets; fillets are sold as combinations of skin-on, skinless, pinbone in and boneless. It is also smoked (finnan haddie) and sometimes canned. Prices vary with season and year, but generally products are sold in the following ranges: fresh dressed US$1.20–1.70/lb; fresh fillets US$4.00–7.00/lb; frozen fillets US$3.50–4.25/lb.

13.3.5 Future perspectives

Haddock grown with our current capability attain a size of 2–2.5kg in 3 years after hatch, which is almost twice as heavy as a wild fish of the same age. However, it is hoped that growth rates will become much higher with further domestication through traditional breeding selection practices and improved husbandry and diets. Presently, there is a hiatus in the attention given to haddock culture though, as new developments are made in cod.

13.4 Pollack, *Pollachius pollachius*

Pollack (Fig. 13.11) looks like saithe, *P. virens*, but has a darker lateral line and a less forked caudal fin.

The body colour of pollack is variable: from yellow on their side to dark dorsally. The lateral line is greenish, continuous and bended close to the pectoral fins. Pollack has a lower jaw projecting beyond the upper one. In opposition to cod, it has no chin barbel. Its maximal size is 1.3m (Quéro and Vayne, 1997). Pollack is distributed in the North-east Atlantic, from north of Norway to Portugal (Fig. 13.12). The species occurs close to the coast at depths ranging from 0 to 200m. It is often associated with rocks or wrecks.

The biology of pollack is poorly known. Basic data such as age, size at first maturity or fecundity of wild fish are not yet reported. Spawning time varies from February in Spain to June in Norway, at a water temperature of 10°C. Eggs are pelagic, have no lipid droplets and their diameter ranges from 1.1 to 1.2mm (Russell, 1976).

Fig. 13.11. Pollack in aquaculture tank (photo O. Dugornay, Ifremer).

Fig. 13.12. World distribution of pollack.

Newly hatched larvae are found near the surface. From 0.5 to 3.9 cm, the diet is composed mainly of *Acartia* and *Calanus* species (Tully and Ceidigh, 1989). Between 5 and 7 cm, pollack feed on crustaceans. From 10 cm, juveniles move to deeper waters and feed on small fish and from 60 cm, pollack migrate to waters deeper than 100 m (Potts, 1986). For a temperate fish species, the growth of pollack is rapid (Fig. 13.13). Compared to cod and whiting which are more abundant in sheltered waters, pollack are found at more exposed locations (Fromentin *et al.*, 1997).

Pollack are commonly caught with bottom or pelagic trawl, long lines and gill nets. After a sharp increase in 1980 up to 18,000 t, the world landing decreased to 10,000 t in 2004 (Fig. 13.14), when the main fishing nations were France (3066 t), Norway (2869 t), the UK (2291 t) and Ireland (1119 t) (FAO, 2006).

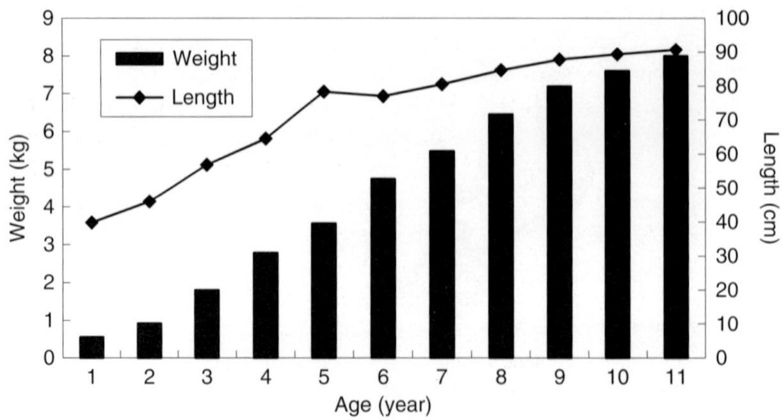

Fig 13.13. Growth rate of pollack from the Celtic Sea (from Dupouy *et al.*, 1990).

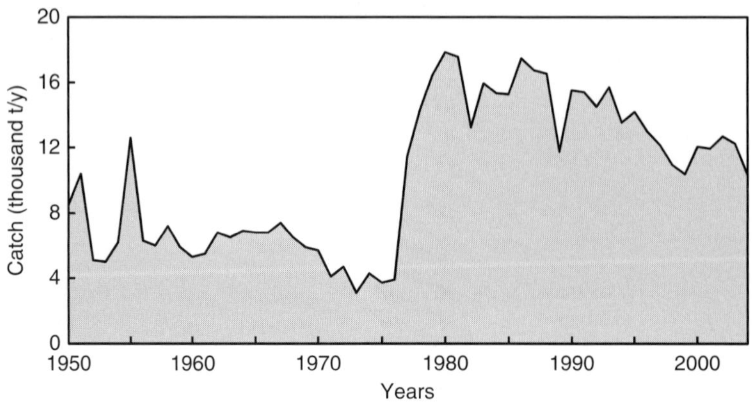

Fig.13.14. World catch of pollack (from FAO, 2006).

13.4.1 Farming of pollack

Its aquaculture potential is based on rapid growth and high flesh quality. Taking into account several biological, economical and technical criteria, pollack was considered of intermediate interest for aquaculture development on the Atlantic coast of France (Quéméner *et al.*, 2002). However, aquaculture production of pollack remains limited, reaching about 200,000 juveniles and 200 t of market sized fish in 2003 (Rosenlund and Skretting, 2006). First rearing attempts indicated that reproduction in captivity and larval rearing was easy, in spite of the lack of established culture methods (Suquet *et al.*, 1996; Gatesoupe, 2002). Interest for the development of its commercial rearing is limited and presently only conducted in Spain.

13.4.2 Broodstock management and hatchery operations

Rearing studies on pollack have been conducted mainly by Ifremer (Brest, France). Reproductive cycles of pollack were described and changes in oestradiol and 11-ketotestosterone were recorded respectively in females and males (Omnes *et al.*, 2002). Pollack spawned spontaneously in 15 m^3 tanks at daylengths of 9–14 h and temperatures of 9.5–12°C. Pollack, which is a batch spawner, gave 600,000 eggs/kg of female in captivity (Suquet *et al.*, 1996). Compared to other marine fish species reared in aquaculture, sperm concentration is low (Mutelet, 2003). Total lipid content of eggs is 26 ± 9% of dry matter, including 1/4 neutral lipids and 3/4 phospholipids (Mutelet, 2003). At hatching, reported larval length is 3 mm. The reproductive activity was highly affected by water temperature: increasing this parameter from 8 to 12°C decreased the total number of eggs collected by 96% and the number of viable eggs by 98% (Suquet *et al.*, 2005).

A better survival of pollack larvae was observed at 12 and 15°C, compared to 18°C. Between 100 and 1000 lux, light intensity had no effects on larval survival (Buchet *et al.*, 2002). Opening of the mouth and metamorphosis were observed 3–4 days and 25–30 days, respectively, posthatching for water temperature ranging from 12 to 15°C. The growth of newly hatched larvae was increased significantly using a probiotic treatment of *Artemia* nauplii (Gatesoupe, 2002). A high mortality was recorded during weaning, 28% survival being observed 90 days posthatching.

13.4.3 On-growing to market size

The growth of pollack juveniles (mean weight 143 ± 2 g) was maximal at temperatures ranging from 12–15°C (Fig.13.15). Daily feed intake was higher at 15–18°C compared to 21°C. Feed conversion ratio was significantly higher at 18°C compared to 12 and 15°C. Pollack showed a high capacity to recover from a prolonged period of low growth induced by high temperature (Person-le Ruyet *et al.*, 2006).

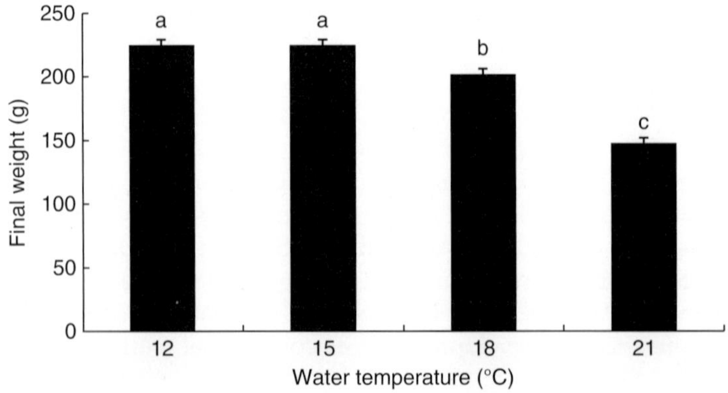

Fig.13.15. Effect of water temperature on the final weight of pollack juveniles after an 84-day experiment (from Person-Le Ruyet *et al.*, 2006).

13.4.4 Future perspectives

There are several characteristics which make pollack a good candidate for aquaculture: rapid growth rate, high flesh quality, good adaptation to the temperate waters of the Atlantic coast, availability of juveniles in the wild for broodstock and high fillet yield close to 50% (Suquet *et al.*, 2000). Furthermore, spontaneous spawn can be collected in captivity. First rearing trials are encouraging and the onset of sexual maturity is observed in 3-year-old animals at a weight close to 1.5 kg higher than the required market size (0.3 kg). On the other hand, the wholesale price of pollack is low (€3.89/kg for the French market in 2005; Ofimer, 2006) compared to other reared species (seabass €9.02/kg; seabream €7.95/kg; turbot €14.79/kg).

Pollack aquaculture is hampered by the rapid success of cod farming. Compared to pollack, cod has higher growth rates, higher selling prices due to decreased landings and more is known about its biology and rearing technique. In conclusion, despite its interesting potential, pollack aquaculture currently cannot be considered as promising. This fact is confirmed by the slow development of pollack farming, with only one company (in Spain) being involved in research and development (Rosenlund and Skretting, 2006).

References

Bergh, Ø., Samuelsen, O.B. and Korsnes, K. (2005) Sykdommer hos torsk, forebygging og behandling (Diseases of cod, prevention and treatment). In: Otterå, H., Taranger, G.L. and Borthen, J. (eds) *Oppdrett av torsk (Farming of cod)*. Norsk Fiskeoppdrett AS, Bergen, Norway, pp. 177–182.

Bigelow, H.B. and Shroeder, W.C. (1953) Fishes of the Gulf of Maine. *US Fish and Wildlife Service Fish Bulletin* 53, 1–577.

Björnsson, B. (1993) Swimming speed and swimming metabolism of Atlantic cod (*Gadus morhua*) in relation to available food: a laboratory study. *Canadian Journal of Fisheries and Aquatic Sciences* 50, 2542–2551.

Björnsson, B. (2001) Can fisheries yield be enhanced by large-scale feeding of a predatory fish stock? A case study of the Icelandic cod stock. *Canadian Journal of Fisheries and Aquatic Sciences* 58, 2091–2104.

Björnsson, B. (2002) Effects of anthropogenic feeding on the growth rate, nutritional status and migratory behaviour of free-ranging cod in an Icelandic fjord. *ICES Journal of Marine Science* 59, 1248–1255.

Björnsson, B. and Ólafsdóttir, S.R. (2006) Effects of water quality and stocking density on growth performance of juvenile cod (*Gadus morhua* L.). *ICES Journal of Marine Science* 63, 326–334.

Björnsson, B., Steinarsson, A. and Oddgeirsson, M. (2001) Optimal temperature for growth and feed conversion of immature cod (*Gadus morhua* L.). *ICES Journal of Marine Science* 58, 29–38.

Björnsson, B., Steinarsson, A. and Árnason, T. (2007) Growth model for Atlantic cod (*Gadus morhua*): effects of temperature and body weight on growth rate. *Aquaculture* 271, 216–226.

Blair, T., Castell, J., Neil, S., D'Abramo, L., Cahu, C., Harmon, P. and Ogunmoye, K. (2003) Evaluation of microdiets versus live feeds on growth, survival and fatty acid composition of larval haddock (*Melanogrammus aeglefinus*). *Aquaculture* 225, 451–461.

Bremner, A.A., Trippel, E.A. and Terhune, J.M. (2002) Sound production by adult haddock, *Melanogrammus aeglefinus*, in isolation, pairs and trios. *Environmental Biology of Fishes* 65, 359–362.

Bricknell, I.R., Bron, J.E. and Bowden, T.J. (2006) Diseases of gadoid fish in cultivation: a review. *ICES Journal of Marine Science* 63, 253–266.

Brooks, S., Tyler, C.R. and Sumpter, J.P. (1997) Egg quality in fish: what makes a good egg? *Reviews in Fish Biology and Fisheries* 7, 387–416.

Brown, J.A., Minkoff, G. and Puvanendran, V. (2003) Larviculture of Atlantic cod (*Gadus morhua*): progress, protocols and problems. *Aquaculture* 227, 357–372.

Buchet, V., Suquet, M., Omnes, M.H., Petton, B., Normant, Y., Severe, A., Quéméner, L., Barone, H., Fauvel, C. and Gaignon, J.L. (2002) Pollack (*Pollachius pollachius*): acquisition of biological data in a breeding purpose. In: Lane, A. (ed.) *Seafarming Today and Tomorrow* European Aquaculture Society, Special Publication No 32, Oostende, Belgium, pp. 165–166.

Cahu, C. and Zambonino-Infante, J. (2001) Substitution of live food by formulated diets in marine fish larvae. *Aquaculture* 200, 161–180.

Castell, J., Blair, T., Neil, S., Howes, K., Mercer, S., Reid, J., Young-Lai, W., Gullison, B., Dhert, P. and Sorgeloos, P. (2003) The effect of different hufa enrichment emulsions on the nutritional value of rotifers (*Brachionus plicatilis*) fed to larval haddock (*Melanogrammus aeglefinus*). *Aquaculture International* 11, 109–117.

Chambers, M.D. and Howell, W.H. (2006) Preliminary information on cod and haddock production in submerged cages off the coast of New Hamshire, USA. *ICES Journal of Marine Science* 63, 385–392.

Cohen, D.M., Inada, T., Iwamoto, T. and Scialabba, N. (1990) Gadiform fishes of the world (Order Gadiformes). An annotated and illustrated catalogue of cods, hakes, grenadiers and other gadiform fishes known to date. *FAO Fisheries Synopsis* 125 (10), FAO, Rome.

DeGraaf, J.D. and Berlinsky, D.L. (2004) Cryogenic and refrigerated storage of Atlantic cod (*Gadus morhua*) and haddock (*Melanogrammus aeglefinus*) spermatozoa. *Aquaculture* 234, 527–540.

Devlin, R.H., Yesaki, T.Y., Biagi, C.A., Donaldson, E.M., Swanson, P. and Chan, W.-K. (1994) Extraordinary salmon growth. *Nature* 371, 209–210.

Diaz, N.F., Iturra, P., Velso, A., Estay, F. and Colihueque, N. (1993) Physiological factors affecting triploid production in rainbow trout, *Oncorhynchus mykiss*. *Aquaculture* 114, 33–40.

Downing G. (2002) Impact of spectral composition on larval haddock, *Melanogrammus aeglefinus* L., growth and survival. *Aquaculture Research* 33, 251–259.

Downing, G. and Litvak, M.K. (1999a) The effect of photoperiod, tank colour and light intensity on growth of larval haddock (*Melanogrammus aeglefinus*). *Aquaculture International* 7, 135–140.

Downing, G. and Litvak, M.K. (1999b) The influence of light intensity on growth of larval haddock (*Melanogrammus aeglefinus*). *North American Journal of Aquaculture* 61, 135–140.

Downing, G. and Litvak, M.K. (2000) Influence of spectral composition on first feeding, growth, and survival of larval haddock (*Melanogrammus aeglefinus*). *Aquaculture Association of Canada Special Bulletin* 4, 37–40.

Downing, G. and Litvak, M.K. (2001) The effect of light intensity and spectrum on the incidence of first feeding by larval haddock. *Journal of Fish Biology* 59, 1566–1578.

Downing, G. and Litvak, M.K. (2002) Effects of light intensity, spectral composition and photoperiod on development and hatching of haddock (*Melanogrammus aeglefinus*) embryos. *Aquaculture* 213, 265–278.

Drinkwater, K.F. (2005) The response of Atlantic cod (*Gadus morhua*) to future climate change. *ICES Journal of Marine Science* 62, 1327–1337.

Dutil, J.-D., Lambert, Y. and Boucher, E. (1997) Does higher growth rate in Atlantic cod (*Gadus morhua*) at low salinity result from lower standard metabolic rate or increased protein digestibility? *Canadian Journal of Fisheries and Aquatic Sciences* 54 (Suppl. 1), 99–103.

Engelsen, R., Asche, F., Skjennum, F. and Adoff, G. (2004) New species in aquaculture: some basic economic aspects. In: Moksness, E., Kjørsvik, E. and Olsen,Y. (eds) *Culture of Coldwater Marine Fish*. Blackwell Publishing, Oxford, UK, pp. 487–515.

Engelsen, R., Skjennum, F.C., Ryland, P.I., Jenssen, E. and Adoff, G. (2005) Økonomi i torskeoppdrett-basert på driftserfaringer (Economy in cod farming-based on farming experience). In: Otterå, H., Taranger, G.L. and Borthen, J. (eds) *Oppdrett av torsk (Farming of cod)*. Norsk Fiskeoppdrett AS, Bergen, Norway, pp. 213–233.

Ewart, K.V., Blanchard, B., Johnson, S.C., Bailey, W.L., Martin-Robichaud, D.J. and Buzeta, M.I. (2000) Freeze susceptibility in haddock (*Melanogrammus aeglefinus*). *Aquaculture* 188, 91–101.

Fahay, M.P. (1983) Guide to the early stages of marine fishes occurring in the western North Atlantic Ocean, Cape Hatteras to the southern Scotian Shelf. *Journal of the Northwest Atlantic Fisheries Science* 4, 1–423.

FAO (2006) Figis data collection (www.fao.org/figis/servlet/static?dom=collection&xml=globalproduction.xml, accessed 4 October 2006).

FISHSTAT Plus Version 2.3 (2000) Universal software for fishing statistical and time series. Food and Agriculture Organization of the United Nations.

Fjalestad, K.T. and Mortensen, A. (2005) Avlsarbeid (Breeding). In: Otterå, H., Taranger, G.L. and Borthen, J. (eds) *Oppdrett av torsk (Farming of Cod)*. Norsk Fiskeoppdrett AS, Bergen, Norway, pp. 49–61.

Folkvord, A. (1991) Growth, survival and cannibalism of cod juveniles (*Gadus morhua*): effects of feed type, starvation and fish size. *Aquaculture* 97, 41–59.

Foss, A., Siikavuopio, S.I., Sæther, B.-S. and Evensen, T.H. (2004) Effect of chronic ammonia exposure on growth in juvenile Atlantic cod. *Aquaculture* 237, 179–189.

Foss, A., Kristensen, T., Åtland, Å., Hustveit, H., Hovland, H., Øfsti, A. and Imsland, A.K. (2006) Effects of water reuse and stocking density on water quality, blood physiology and growth rate of juvenile cod (*Gadus morhua*). *Aquaculture* 256, 255–263.

Frantsi, C., Lanteigne, C., Blanchard, B., Alderson, R., Lall, S., Johnson, S., Leadbeater, S., Martin-Robichaud, D. and Rose, P. (2002) Haddock culture in Atlantic Canada. *Bulletin of the Aquaculture Association of Canada*, 102(1), 31–34.

Fromentin, J.M., Stenseth, N.C., Gjosaeter, J., Bjornstad, O.N., Falck, W. and Johannessen, T. (1997) Spatial patterns of the temporal dynamics of three gadoid species along the Norwegian Skagerrak coast. *Marine Ecology Progress Series* 155, 209–222.

Gatesoupe, F.J. (2002) Probiotic and formaldehyde treatments of *Artemia* nauplii as food for larval pollack (*Pollachius pollachius*). *Aquaculture* 212, 347–360.

Gjøen, H.M. and Bentsen, H.B. (1997) Past, present, and future of genetic improvement in salmon aquaculture. *ICES Journal of Marine Science* 54, 1009–1014.

Gunnarsson, V.I. and Jóakimsson, K.G. (2004) Gæðastjórnun, slátrun og vinnsla á eldisþorski (Quality Management, Slaughtering and Processing of Farmed Cod). In: Björnsson, B. and Gunnarsson, V.I. (eds) *Þorskeldi á Íslandi (Cod Farming in Iceland)*. Marine Research Institute, Report 111, pp. 127–144.

Grundy, B., Villanueva, B. and Woolliams, J.A. (2000) Dynamic selection for maximizing response with over-lapping generations. *Animal Science* 70, 373–382.

Hamlin, H.J. and Kling, L.J. (2001) The culture and early weaning of larval haddock (*Melanogrammus aeglefinus*) using a microparticulate diet. *Aquaculture* 201, 61–72.

Hamre, K. and Mangor-Jensen, A. (2006) A multivariate approach to optimization of macronutrient composition in weaning diets for cod (*Gadus morhua*). *Aquaculture Nutrition* 12, 15–24.

Hansen, A.-C., Rosenlund, G., Karlsen, Ø., Olsvik, P.A. and Hemre, G.-I. (2006) The inclusion of plant protein in cod diets, its effects on macronutrient digestibility, gut and liver histology and heat shock protein transcription. *Aquaculture Research* 37, 773–784.

Hansen, T., Karlsen, Ø., Taranger, G.L., Hemre, G.-I., Holm, J.C. and Kjesbu, O.S. (2001) Growth, gonadal development and spawning time of Atlantic cod (*Gadus morhua*) reared under different photoperiods. *Aquaculture* 203, 51–67.

Hart, J.L. (1973) *Pacific Fishes of Canada*. Fisheries Research Board of Canada, Ottawa, Bulletin 180.

Hawkins, A.D. and Amorim, M.C.P. (2000) Spawning sounds of the male haddock, *Melanogrammus aeglefinus*. *Environmental Biology of Fishes* 59, 29–41.

Hemmingsen, W. and MacKenzie, K. (2001) The parasite fauna of the Atlantic cod, *Gadus morhua* L. *Advances in Marine Biology* 40, 2–60.

Hemre, G.-I., Taranger, G.L. and Hansen, T. (2002) Gonadal development influences nutrient utilisation in cod (*Gadus morhua*). *Aquaculture* 214, 201–209.

Hemre, G.-I., Karlsen, Ø., Mangor-Jensen, A. and Rosenlund, G. (2003) Digestibility of dry matter, protein, starch and lipid by cod, *Gadus morhua*: comparison of sampling methods. *Aquaculture* 225, 225–232.

Holdway, D.A. and Beamish, F.W.H. (1984) Specific growth rate and proximate body composition of Atlantic cod (*Gadus morhua* L.). *Journal of Experimental Marine Biology and Ecology* 81, 147–170.

Howell, B.R. (1984) The intensive rearing of juvenile cod, *Gadus morhua* L. In: Dahl, E., Danielssen, D.S., Moksness, E. and Solemdal, P. (eds) *The Propagation of Cod* Gadus morhua *L*. Flødevigen Rapportserie 1, Norway, pp. 657–675.

ICES (2005) Spawning and life history information for North Atlantic cod stocks. *ICES Cooperative Research Report 274*. ICES, Copenhagen, Denmark.

Imsland, A.K., Foss, A., Koedijk, R., Folkvord, A., Stefansson, S.O. and Jonassen, T.M. (2006) Short- and long-term differences in growth, feed conversion efficiency and deformities in juvenile Atlantic cod (*Gadus morhua*) startfed on rotifers or zooplancton. *Aquaculture Research* 37, 1015–1027.

Jobling, M. (1988) A review of the physiological and nutritional energetics of cod, *Gadus morhua* L., with particular reference to growth under farmed conditions. *Aquaculture* 70, 1–19.

Jobling, M. (2004) On-growing to market size. In: Moksness, E., Kjørsvik, E. and Olsen, Y. (eds) *Culture of Cold-water Marine Fish*. Blackwell Publishing, Oxford, UK, pp. 363–432.

Karlsen, Ø., Hemre, G.-I., Tveit, K. and Rosenlund, G. (2006a) Effect of varying levels of macronutrients and continuous light on growth, energy deposits and maturation in farmed Atlantic cod (*Gadus morhua* L.). *Aquaculture* 255, 242–254.

Karlsen, Ø., Norberg, B., Kjesbu, O.S. and Taranger, G.L. (2006b) Effects of photoperiod and exercise on growth, liver size, and age at puberty in farmed Atlantic cod (*Gadus morhua* L.). *ICES Journal of Marine Science* 63, 355–364.

Kjesbu, O.S. and Norberg, B. (2005) Stamfisk og gyting (Brood stock and spawning). In: Otterå, H., Taranger, G.L. and Borthen, J. (eds) *Oppdrett av torsk (Farming of Cod)*. Norsk Fiskeoppdrett AS, Bergen, Norway, pp. 39–47.

Kjesbu, O.S., Witthames, P.R., Solemdal, P. and Greer Walker, M. (1998) Temporal variations in the fecundity of Arcto-Norwegian cod (*Gadus morhua*) in response to natural changes in food and temperature. *Journal of Sea Research* 40, 303–321.

Knútsson, B. (1997) Þorskeldi á Íslandi: samanburður á arðsemi í strandeldi, kvíaeldi og fjarðaeldi (Farming of cod in Iceland: feasibility study of farming cod in a land-based facility, sea pens and fjordranching). MSc thesis, The University of Iceland, Reykjavík, 93 pp.

Kristiansen, T.S. (1999) Enhancement studies of coastal cod (*Gadus morhua* L.) in Nord-Trøndelag, Norway. In: Howell, B.R., Moksness, E. and Svåsand, T. (eds) *Stock Enhancement and Sea Ranching*. Fishing News Books, Blackwell Science, Oxford, UK, pp. 277–292.

Kristmundsson, Á., Eydal, M., Bambir, S.H. and Helgason, S. (2005a) *Loma* sp. infections of farmed cod, *Gadus morhua*, in Iceland (abstract). Cod farming in Nordic countries, a conference in Reykjavík, 6–8 September 2005, book of abstracts (http://www.sjavarutvegur.is/pdf/Ritverk2005/report1.pdf. Accessed 14 July 2009).

Kristmundsson, Á., Helgason, S., Bambir, S.H. and Eydal, M. (2005b) Natural outbreaks of *Aeromonas salmonicida* ssp. *achromogenes* among farmed Atlantic cod, *Gadus morhua*, in Iceland (abstact). Cod farming in Nordic countries, a conference in Reykjavík, 6–8 September 2005, book of abstracts (http://www.sjavarutvegur.is/pdf/Ritverk2005/report1.pdf. Accessed 14 July 2009).

Kurlansky, M. (1998) *Cod: A Biography of the Fish that Changed the World*. Penguin, Jonathan Cape, London, 304 pp.

Kvenseth, P.G. and Øiestad, V. (1984) Large-scale rearing of cod fry on the natural food production in an enclosed pond. In: Dahl, E., Danielssen, D.S., Moksness, E. and Solemdal, P. (eds) *The Propagation of Cod* Gadus morhua *L*. Flødevigen Rapportserie 1, Norway, pp. 645–655.

Lambert, Y. and Dutil, J.-D. (1997) Can simple condition indices be used to monitor and quantify seasonal changes in the energy reserves of Atlantic cod (*Gadus morhua*)? *Canadian Journal of Fisheries and Aquatic Sciences* 54 (Suppl. 1), 104–112.

Lambert, Y. and Dutil, J.-D. (2001) Food intake and growth of adult Atlantic cod (*Gadus morhua* L.) reared under different conditions of stocking density, feeding frequency and size-grading. *Aquaculture* 192, 233–247.

Lambert, Y., Dutil, J.-D. and Munro, J. (1994) Effects of intermediate and low salinity conditions on growth rate and food conversion of Atlantic cod (*Gadus morhua*). *Canadian Journal of Fisheries and Aquatic Sciences* 51, 1569–1576.

Le François, N.R., Savoie, A. and Siikavuopio, S. (2008) Gill metabolic and osmoregulatory responses of Atlantic cod juvenile (*Gadus morhua*) to chronic ammonia exposure. *Journal of Applied Ichthyology* 24, 112–115.

Leim, A.H. and Scott, W.B. (1966) *Fishes of the Atlantic Coast of Canada*. Fisheries Research Board of Canada, Ottawa, Bulletin 155.

Lie, Ø., Lied, E. and Lambertsen, G. (1988) Feed optimization in Atlantic cod (*Gadus morhua*): fat versus protein content in the feed. *Aquaculture* 69, 333–341.

Litvak, M.K. (1998) The development of haddock culture in Atlantic Canada. *Bulletin of the Aquaculture Association of Canada*, 98-1, 30–33.

Magnadóttir, B., Lange, S., Steinarsson, A. and Gudmundsdóttir, S. (2004) The ontogenic development of innate immune parameters of cod (*Gadus morhua* L.). *Comparative Biochemistry and Physiology Part B* 139, 217–224.

Martell, D.J., Kieffer, J.D. and Trippel, E.A. (2005) Effects of temperature during early life history on embryonic and larval development and growth in haddock. *Journal of Fish Biology* 66, 1558–1575.

Martin-Robichaud, D.J. and Berlinsky, D.L. (2004) The effects of photothermal manipulation on reproductive development in female haddock *Melanogrammus aeglefinus* L. *Aquaculture Research* 35, 465–472.

Martin-Robichaud, D.J. and Rommens, M. (2001) Assesment of sex and evaluation of ovarian maturation of fish using unltrasonography. *Aquaculture Research* 32, 113–120.

Meeren, T. van der and Ivannikov, V. (2001) Yngelproduksjon av gadoider: utvikling av intensiv oppdrettsmetode for torsk og hyse (Juvenile production of Gadoids: development of a farming method for cod and haddock). *Fisken og havet* 2-2001, 1–25.

Moe, H., Gaarder, R., Sunde, L.M., Borthen, J. and Olafsen, K. (2005) Rømmingssikker not for torsk (Sea cage which cod cannot escape from). Sintef Fiskeri og Havbruk Rapport SFH A054041, Trondheim, Norway, 59 pp.

Morais, S., Bell, J.G., Robertson, D.A., Roy, W.J. and Morris, P.C. (2001) Protein/lipid ratios in extruded diets for Atlantic cod (*Gadus morhua* L.): effects on growth, feed utilisation, muscle composition and liver histology. *Aquaculture* 203, 101–119.

Mørkøre, T. (2005) Produktkvalitet (Product quality). In: Otterå, H., Taranger, G.L. and Borthen, J. (eds) *Oppdrett av torsk (Farming of Cod)*. Norsk Fiskeoppdrett AS, Bergen, Norway, pp. 153–161.

Mørkøre, T. (2006) Relevance of dietary oil source for contraction and quality of pre-rigor filleted Atlantic cod, *Gadus morhua*. *Aquaculture* 251, 56–65.

Mutelet, A. (2003) Effet de la température sur la vitellogenèse du lieu jaune (*Pollachius pollachius*). MSc thesis, The University of Rennes I, France.

Myers, R.A., Hutchings, J.A. and Barrowman, N.J. (1996) Hypotheses for the decline of cod in the North Atlantic. *Marine Ecology Progress Series* 138, 293–308.

Norberg, B., Brown, C.L., Halldorsson, O., Stensland, K. and Björnsson, B.Th. (2004) Photoperiod regulates the timing of sexual maturation, spawning, sex steroid and thyroid hormone profiles in the Atlantic cod (*Gadus morhua*). *Aquaculture* 229, 451–467.

Ofimer (2006) Ventes sous criées: bilan 2005/2004 (http://www.ofimer.fr/99_up99load/2_actudoc/965d1_01.pdf, accessed 9 October 2006).

Olsen, A.B., Mikalsen, J., Rode, M., Alfjorden, A., Hoel, E., Straum-Lie, K., Haldorsen, R. and Colquhoun, D.J. (2006) A novel systemic granulomatous inflammatory disease in farmed Atlantic cod, *Gadus morhua* L., associated with a bacterium belonging to the genus *Francisella*. *Journal of Fish Diseases* 29, 307–311.

Omnes, M.H., Severe, A., Barone, H., Suquet, M., Buchet, V., Le Roux, A.,Gaignon, J.L., Fostier, A. and Fauvel, C. (2002). Growth and reproductive performances of juveniles and sexually mature pollack *Pollachius pollachius* in different conditions. In: Vandercoilden, L., Aspelagh, L. and Joncheere, H. (eds) *Seafarming Today and Tomorrow*. European Aquaculture Society, Special Publication No 32, pp. 394–395.

Otterå, H. and Folkvord, A. (1993) Allometric growth in juvenile cod (*Gadus morhua*) and possible effects on cannibalism. *Journal of Fish Biology* 43, 643–645.

Otterå, H., Kristiansen, T.S., Svåsand, T., Nordeide, J.T., Nævdal, G., Borge, A. and Pedersen, J.P. (1999) Enhancement studies of Atlantic cod (*Gadus morhua* L.) in an exposed coastal area in Western Norway. In: Howell, B.R., Moksness, E. and Svåsand, T. (eds) *Stock Enhancement and Sea Ranching*. Fishing News Books, Blackwell Science, Oxford, UK, pp. 257–276.

Otterlei, E., Nyhammer, G., Folkvord, A. and Stefansson, S.O. (1999) Temperature- and size-dependent growth of larval and early juvenile Atlantic cod (*Gadus morhua* L.): a comparative study of Norwegian coastal cod and northeast Arctic cod. *Canadian Journal of Fisheries and Aquatic Sciences* 56, 2099–2111.

Pálsson, Ó.K. (1994) A review of the trophic interactions of cod stocks in the North Atlantic. *International Council for the Exploration of the Sea Marine Science Symposia* 198, 553–575.

Pálsson, Ó.K. (2003) A length-based analysis of haddock discards in Icelandic fisheries. *Fisheries Research* 59, 437–446.

Pavlov, D., Kjørsvik, E., Refsti, T. and Anderson, Ø. (2004) Brood stock and egg production. In: Moksness, E., Kjørsvik, E. and Olsen, Y. (eds) *Culture of Cold-water Marine Fish*. Blackwell Publishing, Oxford, UK, pp. 129–203.

Peck, M.A., Buckley, L.J., O'Bryan, L.M., Davies, E.J. and Lapolla, A.E. (2004) Efficacy of egg surface disinfectants in captive spawning Atlantic cod *Gadus morhua* L. and haddock *Melanogrammus aeglefinus* L. *Aquaculture Research* 35, 992–996.

Person-Le Ruyet, J., Buchet, V., Vincent, B., Le Delliou, H. and Quéméner, L. (2006) Effects of temperature on the growth of pollack (*Pollachius pollachius*) juveniles. *Aquaculture* 251, 340–345.

Planque, B. and Frédou, T. (1999) Temperature and the recruitment of Atlantic cod (*Gadus morhua*). *Canadian Journal of Fisheries and Aquatic Sciences* 56, 2069–2077.

Potts, G.W. (1986) Predator–prey interactions between pollack and sand-eels. *Progress in Underwater Science* 11, 69–78.

Purchase, C.F. and Brown, J.A. (2001) Stock-specific changes in growth rates, food conversion efficiencies and energy allocation in response to temperature change in juvenile Atlantic cod. *Journal of Fish Biology* 58, 36–52.

Puvanendran, V. and Brown, J.A. (2002) Foraging, growth and survival of Atlantic cod larvae reared in different light intensities and photoperiods. *Aquaculture* 214, 131–151.

Quéméner, L., Suquet, M., Mero, D. and Gaignon, J.L. (2002) Selection method of new candidates for finfish aquaculture: the case of the French Atlantic, the Channel and the North Sea coasts. *Aquatic Living Resources* 15, 293–302.

Quéro, J.C. and Vayne, J.J. (1997) *Les poissons de mer des pêches françaises*. Delachaux and Niestlé, Lausanne-Paris.

Refstie, S., Førde-Skjærvik, O., Rosenlund, G. and Rørvik, K.-A. (2006) Feed intake, growth, and utilisation of macronutrients and amino acids by 1- and 2-year-old Atlantic cod (*Gadus morhua*) fed standard or bioprocessed soybean meal. *Aquaculture* 255, 279–291.

Rideout, R.M., Trippel, E.A. and Litvak, M.K. (2004a) Relationship between sperm density, spermatocrit, sperm motility and spawning date in wild and cultured haddock. *Journal of Fish Biology* 65, 319–332.

Rideout, R.M., Trippel, E.A. and Litvak, M.K. (2004b) Predicting haddock embryo viability based on early cleavage patterns. *Aquaculture* 230, 215–228.

Rideout, R.M., Trippel, E.A. and Litvak, M.K. (2004c) The development of haddock and Atlantic cod sperm cryopreservation techniques and the effect of sperm age on cryopreservation success. *Journal of Fish Biology* 65, 299–311.

Rosenlund, G. and Skretting, M. (2006) Worldwide status and perspective on gadoid culture. *International Council for the Exploration of the Sea Journal of Marine Science* 63, 194–197.

Rosenlund, G., Karlsen, Ø., Tveit, K., Mangor-Jensen, A. and Hemre, G.-I. (2004) Effect of feed composition and feeding frequency on growth, feed utilization and nutrient retention in juvenile Atlantic cod, *Gadus morhua* L. *Aquaculture Nutrition* 10, 371–378.

Russell, F.R.S. (1976) *The Eggs and Planktonic Stages of British Marine Fishes.* Academic Press, London.

Schmidt-Nielsen, K. (1979) *Animal Physiology: Adaptation and Environment.* Cambridge University Press, Cambridge, UK, 559 pp.

Scott, W.B. and Scott, M.G. (1988) Atlantic fishes of Canada. *Canadian Bulletin of Fisheries and Aquatic Sciences* No 219, University of Toronto, Toronto, Canada, 731 pp.

Shelbourn, J.E., Clarke, W.C., McBride, J.R., Fagerlund, U.H.M. and Donaldson, E.M. (1992) The use of 17 α-methyltestosterone and 3,5,3′-triiodo-L-thyronine for sterilizing and accelerating the growth of zero-age coho salmon smolts (*Oncorhynchus kisutch*). *Aquaculture* 103, 85–99.

Sin, F.Y.T. (1997) Transgenic fish. *Reviews in Fish Biology and Fisheries* 7, 417–441.

Skjæraasen, J.E., Salvanes, A.G.V., Karlsen, Ø., Dahle, R., Nilsen, T. and Norberg, B. (2004) The effect of photoperiod on sexual maturation, appetite and growth in wild Atlantic cod (*Gadus morhua*). *Fish Physiology and Biochemistry* 30, 163–174.

Sobel (1996) *Melanogrammus aeglefinus. 2006 IUCN Red List of Threatened Species* (accessed 11 May 2006). Listed as Vulnerable (VU A1d+2d v2.3).

Solberg, C., Willumsen, L., Amble, S., Johanessen, T. and Sveier, H. (2006) The effects of feeding frequencies on seasonal changes in growth rate and chemical composition of farmed cod (*Gadus morhua*). *Aquaculture Nutrition* 12, 157–163.

Solemdal, P., Dahl, E., Danielsen, D.S. and Moksness, E. (1984) The cod hatchery in Flødevigen – background and realities. In: Dahl, E., Danielssen, D.S., Moksness, E. and Solemdal, P. (eds) *The Propagation of Cod* Gadus morhua *L.* Flødevigen Rapportserie 1, Norway, pp. 17–45.

Statistical Series (2007) Report on Fisheries 2007:3 (14 May), Hagstofa Íslands, Reykjavík, Iceland, 28 pp.

Steinarsson, A. (2004) Framleiðsla þorskseiða (Hatchery production of cod). In: Björnsson, B. and Gunnarsson, V.I. (eds) *Þorskeldi á Íslandi (Cod Farming in Iceland).* Marine Research Institute, Report 111, Hafrannsóknastofnunin, Reykjavík, Iceland, pp. 41–86.

Steinarsson, A. and Björnsson, B. (1999) The effects of size and temperature on growth and mortality of cod larvae. *Journal of Fish Biology* 55 (Supplement A), 100–109.

Stoss, J., Hamre, K. and Otterå, H. (2004) Weaning and nursery. In: Moksness, E., Kjørsvik, E. and Olsen, Y. (eds) *Culture of Cold-water Marine Fish.* Blackwell Publishing, Oxford, UK, pp. 337–362.

Støttrup, J. (2002) Torskeopdræt-forskningsresultater og kundskab om torskeopdræt (Cod farming-results from research and knowledge about cod farming). Danmarks Fiskeriundersøgelser, Copenhagen, Denmark, 76 pp.

Suquet, M., Petton, B., Normant, Y., Dosdat, A. and Gaignon, J.L. (1996) First rearing attempts of pollack, *Pollachius pollachius. Aquatic Living Resources* 9, 103–106.

Suquet, M., Parfouru, D., Paquotte, P., Girard, S. and Gaignon, J.L. (2000) Method for selection of new marine fish species: the case of pollack (*Pollachius pollachius*). *Cahiers Options Méditerranéennes* 47, 127–134.

Suquet, M., Normant, Y., Gaignon, J.L., Quéméner, L. and Fauvel, C. (2005) Effect of water temperature on individual reproductive activity of pollack (*Pollachius pollachius*). *Aquaculture* 243, 113–120.

Svåsand, T., Jørstad, K.E., Otterå, H. and Kjesbu, O.S. (1996) Differences in growth performance between Arcto-Norwegian and Norwegian coastal cod reared under identical conditions. *Journal of Fish Biology* 49, 108–119.

Svåsand, T., Otterå, H.M. and Taranger, G.L. (2004) The status and perspectives for the species. In: Moksness, E., Kjørsvik, E. and Olsen, Y. (eds) *Culture of Cold-water Marine Fish*. Blackwell Publishing, Oxford, UK, pp. 433–474.

Taranger, G.L., Aardal, L., Hansen, T. and Kjesbu, O.S. (2006) Continuous light delays sexual maturation and increases growth of Atlantic cod (*Gadus morhua* L.) in sea cages. *International Council for the Exploration of the Sea Journal of Marine Science* 63, 365–375.

Tiwary, B.K., Kirubagaran, R. and Ray, A.K. (2004) The biology of triploid fish. *Reviews in Fish Biology and Fisheries* 14, 391–402.

Toppe, J., Aksnes, A., Hope, B. and Albrektsen, S. (2006) Inclusion of fish bone and crab by-products in diets for Atlantic cod, *Gadus morhua*. *Aquaculture* 253, 636–645.

Trippel, E.A. (2003) Paired mating as an alternative broodstock management strategy for haddock (*Melanogrammus aeglefinus*). In: Aiken, D.E. (ed.) *Early Rearing of Haddock: state of the art*. Aquaculture Association of Canada Special Publication 7, Saint Andrews, New Brunswick, Canada, pp. 23–34.

Trippel, E.A. and Neil, S.R.E. (2003) Effects of photoperiod and light intensity on growth and activity of juvenile haddock (*Melanogrammus aeglefinus*). *Aquaculture* 217, 633–645.

Trippel, E.A. and Neil, S.R.E. (2004) Maternal and seasonal differences in egg sizes and spawning activity of northwest Atlantic haddock (*Melanogrammus aeglefinus*) in relation to body size and condition. *Canadian Journal of Fisheries and Aquatic Sciences* 61, 2097–2010.

Trippel, E.A., Doherty, C.M., Wade, J. and Harmon, P.R. (1998) Controlled breeding technology for Haddock (*Melanogrammus aeglefinus*) in mated pairs. *Bulletin of the Aquaculture Association of Canada*, 98-3, 30–35.

Trippel, E.A., Castell, J.D., Neil, S.R.E. and Blair, T.J. (2000) Assessment of egg quality of haddock (*Melanogrammus aeglefinus*) in paired matings. In: Norberg, B., Kjesbu, O.S., Taranger, G.L., Andersson, E. and Stefansson, S.O. (eds) *Reproductive Physiology of Fish*. Proceedings of the 6th International Symposium on the Reproductive Physiology of Fishes, 4–9 July 1999, Bergen, Norway. Institute of Marine Research and University of Bergen, Bergen, Norway, pp. 405–407.

Trippel, E.A., Rideout, R.M., O'Reilly, O.T., Herbinger, C.M., Neil, S.R.E. and Hamilton, L. (2009) Communal spawning leads to high potential for inbreeding in gadoid aquaculture. *Aquaculture* 296, 27–35.

Tully, O. and Ceidigh, P.O. (1989) The ichthyoneuston of Galway Bay (west of Ireland): II. Food of post-larval and juvenile neustonic and pseudoneustonic fish. *Marine Ecology: Progress Series* 51, 301–310.

Tveite, S. (1971) Fluctuations in year-class strength of cod and pollack in southeastern Norwegian coastal waters during 1920–1969. *Fiskeridirektorates Skrifter, Serie Havundersøkelser* 16, 65–76.

Van Eeckhaute, L. and Brodziak, S. (2006) Assessment of Eastern Georges Bank Haddock for 2006. Transboundary Resource Assessment Committee Reference Document – 2006/06; 747.

Waiwood, K.G. (1994) Haddock. *Bulletin of the Aquaculture Association of Canada* 94-1, 16–21.

Watson, J.T., Sales, S., Cumming, G., Fitzsimmons, S.D., Walden, J., Arthur, G., Saravanan, S. and McEvoy, L.A. (2006) Comparison of performance of two size groups of farmed cod (*Gadus morhua* L.) juveniles following transfer to sea cages. *International Council for the Exploration of the Sea Journal of Marine Science* 63, 340–345.

14 The Snooks (Family: Centropomidae)

Chris Carter,[1] Brett Glencross,[2] Robin S. Katersky[1] and Michel Bermudes[2]

[1]*University of Tasmania, Launceston, Tasmania, Australia;* [2]*Department of Fisheries, Research Division, Government of Western Australia, North Beach, Australia*

14.1 General Introduction

The family Centropomidae (Snooks) (Fig. 14.1) has consisted of two genera, *Centropomus* and *Lates*, both of which contain species with aquaculture potential. Recently, a taxonomic reclassification has occurred for the genus *Lates*, moving it from the family Centropomidae to Latidae (Otero, 2004). The new family Latidae has three genera, including all *Lates* sp. as well as *Psammoperca waigiensis* (Waigieu sea perch, which is a commercially important fisheries species) and *Eolates* (fossil genus) (Otero, 2004). *Centropomus* species have a wide distribution around the Americas and *Lates* species around the Indo-west Pacific and African regions. *Centropomus* contains several species, including *C. undecimalis* (common snook) and the smaller *C. parallelus* (fat snook), which have been considered for aquaculture (Alvarez-Lajonchere *et al.*, 2002; Temple *et al.*, 2004). There are perhaps eight *Lates* species (Otero, 2004), of which *L. calcarifer* (barramundi/Asian seabass/giant sea perch) and several species of 'Nile perch' are the most well known with respect to aquaculture and fisheries (Greenwood, 1976; Harrison, 1991). Centropomidae and Latidae are characteristically tropical inshore euryhaline fishes.

Fat snook have been farmed extensively in grow-out ponds for over 300 years in north-eastern Brazil and attempts are being made to develop intensive aquaculture to take advantage of established markets that could be further developed (Temple *et al.*, 2004). However, the commercial viability of a more intensive form of aquaculture for both common and fat snooks will require considerable research to overcome the high mortality currently experienced during larval rearing (Alvarez-Lajonchere *et al.*, 2002). Barramundi has been farmed on small family-run farms as a relatively low-value species in Asia, but it is only since the 1980s that interest in developing larger-scale aquaculture has occurred in Thailand, Taiwan, Malaysia, Indonesia and Australia. There is now considerable interest in using indoor recirculation technology, particularly

 Finfish Aquaculture Diversification (eds N.R. Le François, M. Jobling, C. Carter and P.U. Blier)

Fig. 14.1. Centropomidae.

in Europe and the USA, and the species is on the verge of a dramatic increase in global aquaculture production.

Positive and negative attributes of barramundi as a species for aquaculture (Rimmer, 2003; Katersky and Carter, 2005) include:

Positive:

- It is a robust species throughout the production cycle and able to tolerate and perform well over a wide range of environmental conditions including temperature, salinity and water quality.
- There is a reliable supply of juveniles from well-developed techniques throughout Asia and Australia.
- Nutrition is well understood, weaning onto formulated feeds is straightforward, formulated feeds are readily available, well utilized and growth rates are high.
- Marketability is high due to the wide range of harvest sizes, high flesh quality, an established or increasing position as a high-quality restaurant fish and suitability for value-adding and market diversification.

Negative:

- Highly aggressive fish: it is cannibalistic in the hatchery and also develops high size variability, so needs frequent grading in some culture systems.

14.2 Barramundi, *Lates calcarifer*

The natural range encompasses the northern Indian and tropical western Pacific Oceans and it is found from Iran in the west to Australia in the east (Fig. 14.2). There has been limited exchange of individuals between river systems so that genetically distinct strains are recognized (Rimmer and Russell, 1998). It has now been widely exported outside of its natural range and barramundi can be found in closed aquaculture systems around the world in countries including

Fig. 14.2. World distribution of *Lates calcarifer*.

the USA (Massachusetts), England (New Forest), Netherlands and Israel (Rimmer, 2003; Segev *et al.*, 2003; Anon, 2006).

It is a euryhaline fish that inhabits fresh water, brackish water and marine coastal areas at different stages in its life cycle. There is a strong association between the life cycle of barramundi and the monsoonal seasons throughout the Indo-Pacific region (Barlow, 1981). Being a protandrous hermaphrodite, individuals first mature into males after 2–4 years in fresh water and then into females after 3–7 years in coastal waters. Fish less than 80 cm are generally males, while fish over 100 cm are usually females; most fish are female after 5 years (Tucker *et al.*, 2002). It is a catadromous species and reproduction and early life occurs in seawater. Spawning occurs on the incoming tide in the evening and for several days following the new and full moon. Barramundi are very fecund and females produce hundreds of thousands of eggs per kg on each spawning, large females of over 120 cm can produce over 40 million eggs. Females return to marine coastal seas and are joined by the males that migrated from rivers that year to spawn. The eggs and hatching larvae are pushed inshore with the high tide into flooded coastal feeding areas; they migrate upriver as the wet season ends and the wetlands dry. It is a highly aggressive and opportunistic carnivore that, depending on size, will take a wide range of aquatic, avian and terrestrial prey, including other barramundi.

This robust species can be cultured using intensive and extensive methods, indoors or outdoors, and can be grown over the full range of salinity (from fresh water to seawater), including inland brackish bore waters (Tucker *et al.*, 2002; Rimmer, 2003). Australia provides an excellent example of how different systems have been used to suit a range of geographic, local environmental, economic and market drivers. Open freshwater ponds, small cages held in freshwater ponds, raceways, large sea cages or indoor recirculation systems are used across the country from temperate southern Australia to the tropical north of the country. Smaller farms using freshwater ponds or recirculation systems tend to supply local markets with plate-size fish. Larger farms using sea cages grow fish to 2–3 kg for fillets or as whole 'banquet' fish.

14.2.1 Farming of barramundi

Barramundi aquaculture provides an interesting case of how a highly adaptable species can be grown successfully under many conditions. Although aquaculture production remains relatively small, there is great potential for increased global production. In 1984, production was 1726 t, by 1995 it reached 20,000 t, mainly from Taiwan (54%), Thailand (21%), Malaysia (12%) and Indonesia (10%), and stabilized for several years (Rimmer and Russell, 1998; Rimmer, 2003). Recently, production has increased and was 29,899 t in 2004. Thailand was the first commercial producer of barramundi with 5 t in 1963; their industry has now grown to over 14,000 t and in 2004 accounted for nearly half (48%) of global barramundi production, whereas Taiwan's production has decreased to 16%. Australia's production increased from 258 t and 1.3% in 1995 to 1567 t in 2004 and contributed 5% of the global barramundi production (FAO, 2006).

14.2.2 Broodstock management and husbandry

From the early practices of strip spawning running-ripe wild broodstock, the management of barramundi broodstock has evolved to improve supply reliability, egg quality and seed performance by reducing stress and incorporating genetic selection and health management programmes. Mature adults can be sourced from the wild or from cultured stocks, with wild broodstock requiring a period of at least 6 months to adapt to captivity before spawning is attempted (Kungvankij, 1987). Care must be taken during capture, transportation and on arrival to minimize stress and handling injuries. Sedatives can be used to minimize transportation and handling stress, and fish with injuries have wounds swabbed with antiseptic on arrival to minimize the risk of infection (Schipp, 1991, 2006). Barramundi can be spawned in tanks or strip spawned and this can be achieved with or without hormonal manipulation. Ambient conditions are very important for maturation and spawning; spawning will occur regularly under summer photoperiod and temperatures.

Wild and cultured broodstock can be held in cages (10–100 m^2, 2 m deep) and concrete or fibreglass tanks (20–100 m^3, 2 m deep) in either fresh or saltwater and stocked at a density of one fish per m^3 (Kungvankij, 1987; Garrett and O'Brien, 1994; Tucker *et al.*, 2002). Conditions have been manipulated to obtain continuous year-round spawning; monthly gonad maturation and spawning were achieved for 15 successive months by maintaining constant photoperiod (13 h light), temperature (28–29°C) and salinity (30–36‰) (Garrett and O'Brien, 1994). In South-east Asia (e.g. Thailand), breeding in the wild extends from April to August (Kungvankij, 1987). In Malaysia and Thailand, spawning is obtained year round at 27–34°C (Tucker *et al.*, 2002). Fish held in fresh water must be transferred to seawater (28–32 ppt) prior to breeding to enable final gonad maturation to take place; this simulates the migration from growing to spawning grounds (Kungvankij, 1987). Barramundi need to be cannulated regularly for sexing (sex-reversal can occur rapidly and deplete the number of males) and to assess gonadal development throughout the conditioning phase (see details of the ovarian biopsy technique in Garcia, 1989a). Males usually release milt on handling when close to spawning. Broodstock are fed once a day with fresh baitfish, mullet and squid at about 1–2% of biomass and overfeeding should be avoided as it is reported to reduce spawning success (Ruangpanit, 1987; Schipp, 2006). Feed can be supplemented with a vitamin mix to prevent deficiencies (Rimmer, 2006).

Spawners should be at least 3 years old and 4–5 kg in body weight. The sex ratio in spawning tanks (7–200 m^3) is 1:1 to 1:3 female to male (Kungvankij, 1987; Schipp, 2006). Two methods are used to induce spawning in barramundi. Hormone treatment is preferred in Australia and is used in Asia, where the alternative option of environmental manipulation is also common (Rimmer, 2006). Environmental manipulations aim to simulate the tidal regime of lower estuary spawning grounds (Kungvankij, 1987). Feeding is stopped 1 week prior to the new or full moon and fish conditioned in seawater will display prespawning behaviour 3 days prior to spawning (male and female swim together near the surface). At the start of the new or full moon, the water level is dropped in the middle of the day to increase temperature to 30–32°C. This is followed by rapid addition of fresh seawater to the tank, which simulates the rising tide and reduces the temperature to 27–28°C. If the fish do not spawn spontaneously on the night following manipulation, tank water drops can be repeated for 3 days to achieve spawning. Three to five spawns can be obtained on consecutive days by repeating the above steps and the same spawners can be induced during the full and new moon for 5–6 months (Kungvankij, 1987).

A range of hormones (human chorionic gonadotropin, carp pituitary, barramundi pituitary and gonadotropin-releasing hormone analogue) has been used to induce spawning in seabass (Tucker *et al.*, 2002). The luteinizing hormone-releasing hormone analogue (LHRHa) administered in a slow-release cholesterol pellet is now the preferred option to minimize handling stress while maximizing egg production and quality. Females are ready to be implanted when oocytes reach 400–450 μm. LHRHa at a dose rate of 19–75 μg/kg will induce spawning in 30–38 h (Garcia, 1989b; Garrett and O'Brien, 1994). Spawning can occur within 8–10 h of injection when oocytes are more advanced (500–550 μm) (Garrett and O'Connell, 1991). Induced fish spawn repeatedly, usually at dusk, at 24 h intervals over a 3–5 day period. If required, males can be induced with 25 μg LHRHa/kg (Schipp, 2006). Hormonal induction of spawning can be carried out effectively at any time during the lunar cycle (Garcia, 1992). Females induced to spawn produce 250,000 eggs/kg body weight (Tucker *et al.*, 2002). Fertilization is external and fertilized eggs float to the surface. Barramundi eggs are 0.74–0.80 mm in diameter and are collected in a 300 μm mesh net placed either inside (e.g. airlift collectors) or outside (over flow collectors) the spawning tank. For fish spawned in cages, cages are lined with a fine mesh ('hapa') (Rimmer, 2006).

Fertilized eggs (0.74–0.80 mm in diameter) generally hatch within 1 day (12–17 h) at temperatures of 27–30°C (Rimmer, 2006). At hatching, larvae are on average 1.5 mm in total length and have a yolk sac and oil globule, which are fully depleted after about 6 days at optimum temperatures of 28–30°C. Typically, larvae first-feed after 2–3 days (2.6 mm) and take around 25 days to metamorphose into juveniles (17 mm). There are many larval rearing approaches ranging from intensive to extensive. Intensive procedures are carried out in tanks using a rotifer and brine shrimp sequence, rotifers from day 2 to day 8–15 and brine shrimp from day 8–10. Stocking densities of 10–40 fish/l are used in growing the larvae to about day 21 (10 mm). Circular tanks of up to 10 m^3 and with a conical base are preferred for good water circulation and drainage (Rimmer, 2003; Kolkovski *et al.*, 2004). Barramundi are visual feeders and dark tank walls are used to provide sufficient contrast to highlight their prey. Less intensive procedures may also include a green water phase, for example *Nannochloropsis oculata* or *Tetraselmis* spp. can be added to the culture tanks to improve water quality and perhaps enhance larval nutrition.

Extensive procedures are carried out in earthen ponds relying on natural plankton blooms initiated by fertilizing the marine or brackish pond water. Although pond size may vary between 0.05 and 1.0 ha, they should be relatively shallow and are more effective when aeration is used. Stocking densities range between 400,000 and 900,000 fish/ha and 2-day-old larvae are stocked. Feeding follows a sequence including rotifers and copepod nauplii, copepodites, adult copepods and ending with benthic preys. The juveniles are harvested when they reach over 25 mm in length around 21 days. With an average survival of 40%, pond culture provides a relatively inexpensive method of large-scale production that is estimated to be 40–64% of the cost of intensive culture (Rimmer, 2003).

The nursery phase starts when the fish are 10–25 mm in length and, again, a variety of systems are used. Transfer to fresh water is possible at 10 mm and nursery operations, for example in smaller Australian farms, use small cages in freshwater ponds. In Asia, nursery cages or ponds are used and may be coastal, brackish or fresh water. Barramundi are weaned on to compound feeds at this stage; specialist weaning diets are recommended over crumbles and small sizes of grower diets (Rimmer, 2003). In Asia, during the nursery phase, minced trash fish may be used in decreasing amounts over a 21-day weaning period. After 30–45 days, fish reach 5–10 cm and are ready for grow-out.

Weaning barramundi from live feeds to commercial diets occurs around the time of metamorphosis. Co-feeding brine shrimp with a commercial diet should not start more than 3 days before stomach differentiation (18 days posthatch) and continue past metamorphosis (Curnow *et al.*, 2006a). This has been shown to improve growth by 25–30% and reduce the mortality rate during weaning (Curnow *et al.*, 2006a). Recently, it has been recommended that weaning protocols for barramundi should be specific to the commercial diets used to avoid a long weaning period and the increased occurrence of cannibalism (Curnow *et al.*, 2006b). Cannibalism is particularly important in larval and nursery rearing; it is a major cause of mortality, which is minimized through regular size grading.

14.2.3 On-growing to market size

Several production technologies are used throughout the Asia-Pacific region for the culture of barramundi. In marine and estuarine waters, cage culture is the most common production method. Cages range from small wooden cages (2 m × 2 m × 2 m) from which nets are suspended, used in some South-east Asian nations, to large modular steel systems (16 m × 16 m × 8 m) using heavy gauge galvanized steel nets, which were used in northern Australia. Use of Polar Cirkel™ cage systems, as in Atlantic salmon farming, has also gained some acceptance. In fresh water, ponds, cages (Fig. 14.3) and cages in ponds have all been used to varying degrees. Pond construction for barramundi culture varies between farms, with no specific design constraints being consistent, other than a minimum depth of 2 m and use of some aeration technology (e.g. paddle wheels) being commonplace. The majority of production throughout the Asia-Pacific region is in estuarine waters, though a significant number of farms also exist in freshwater lakes and pond systems.

A range of feed types is used in on-growing and nutritional requirements are relatively well understood (Boonyaratpalin and Williams, 2002; Glencross, 2006). In many South-east Asian nations, fish are still fed on a trash-fish diet. In Australia, barramundi are fed exclusively on extruded pellets. Formulated pellet feed specifications vary among feed suppliers. Generally, small pellets for small fish have low to moderate energy density (15–17 kJ/g) and are high in protein

Fig. 14.3. Freshwater cage farm on an inland lake (B. Glencross).

(50–55%); larger pellets, for fish larger than 200 g, are lower in protein (~ 45%) and higher in energy density (17–20 kJ/g). Pellet feeds are manufactured using principally fishmeals and fish oils, but a significant proportion of the diets now include other key ingredients such as wheat, lupin kernel meals, meat meals and poultry meals (Boonyaratpalin and Williams, 2002; Glencross, 2006).

Feed management for barramundi is typically undertaken on a manual basis, though some larger production systems are using automated feeding technology (Fig. 14.4). Fish are generally fed to apparent satiety as determined using primarily visual cues or by means of underwater sensors in automated systems. Feeding of small fish usually occurs several times a day, with larger fish needing to be fed only once daily (Williams and Barlow, 1999). Feed management guidelines based on empirical feed intake and growth data have been published. Others have been developed based on bioenergetic energy demand models and are being used by some production companies (Glencross *et al.*, 2007). However, the majority of producers in the Asia-Pacific region still feed to satiety manually using visual cues.

Harvesting practices vary depending on specific operators but typically involve nets being seined through either the pond or cage (Fig. 14.5) to corral the fish, which are then brailed out into an ice slurry. The fish are killed by immersion in this ice slurry, which decreases their metabolic rate rapidly until death is achieved. The fish are kept in the slurry until processing and/or packaging. The presence of geosmin and 2-methylisoborneol (MIB) taint has been noted from some Australian freshwater production systems, which gives a muddy-flavour taste to the flesh (Glencross *et al.*, 2007). Systems to purge the MIB taint from the fish are relatively simple and primarily require the fish to be

Fig. 14.4. Feeding using blower on a sea cage (B. Glencross).

Fig. 14.5. Harvesting from a sea cage (B. Glencross).

kept in untainted water for a period of 24–72 h depending on the extent of the initial taint problem (Howgate, 2004; Glencross *et al.*, 2007).

Routine health management of barramundi broodstock held in seawater is required, particularly during cooler periods, to treat and prevent diseases and parasites such as 'white spot' (*Cryptocaryon irritans*) and monogenean flatworms. Wild broodstock must be quarantined on arrival and screened by polymerase chain reaction (PCR) for the vertically transmitted viral nervous necrosis (VNN) from the causative agent *Nodavirus* sp. (Azad *et al.*, 2006). VNN can cause significant mortality during larval rearing and the nursery phase, particularly in intensive systems and in larvae reared under suboptimal conditions (Azad *et al.*, 2005). Several diseases are known to afflict barramundi during on-growing: *Streptoccocus iniae*, *Cytophaga johnsonae*, columnaris disease, bacterial haemorrhagic septicaemia, *Vibrio* sp., lymphocysitis, ciliated protozoans (*C. irritans*, *Chilodonella* sp., *Ichthyopthirius multifiliis*, *Trichodina* and *Amyloodinium*), as well as fungi (*Saprolegnia*), myxosporidians, flatworms and parasitic crustaceans with bacterial diseases being the most common cause of mortality (Owens, 2003; Rimmer, 2003). *S. iniae* is a bacterial pathogen linked to freshwater runoff and is becoming one of the main diseases in estuarine sea-cage culture (Owens, 2003). Infection generally occurs after a handling or temperature stress. Recently, the first commercial vaccine (Norvax-SI) produced by Intervet for barramundi is now being used in Indonesia against *S. iniae* (Delamare-Deboutteville *et al.*, 2006). Columnaris disease, which presents as fin and tail rot, is seen primarily on fingerlings that have undergone a rapid temperature change or handling stress (Anderson and Thomas, 1995). Improvements in farm practices and decreasing the stocking densities can prevent a reoccurrence. Bacterial haemorrhagic septicaemia occurs in freshwater-reared fish and has similar clinical signs to *Vibrio* sp. (in seawater), causing bacterial infections in the internal organs and skin ulcers (Anderson and Thomas, 1995). Lymphocystis is an iridovirus which results in cauliflower-like growths on the fish. Mortality is low and with stress reduction, fish usually recover. The greater impact is in the loss of market value if fish are presenting clinical signs at the time of harvest. Parasitic sea lice (*Argulus* sp.) have been reported in barramundi farming systems, although rare, mild infestations have been known to result in fish losses. White spot (*I. multifiliis*) is treated by lowering the salinity to below 10‰ for at least 3 days (Schipp and Pearce, 1991). Shorter freshwater bathing (i.e. 2 h) is also effective in the treatment of flatworms. Disease management in Australia usually comprises strict quarantine and disease testing prior to translocation of any animals to limit disease spread; antibiotics may be prescribed if further treatment is required.

14.2.4 Commercialization

There are three market sizes of barramundi produced in the Indo-Pacific region. In South-east Asia, small plate-size (~ 400 g) and banquet-size (~ 1 kg) fish predominate production, but considerable size variation can exist within each product class. In Australia, production standards are generally more tightly

Fig. 14.6. Retail display of plate-size barramundi on ice (C. Carter).

controlled, with a small plate-size fish having to satisfy a weight grade of 400–500 g (Fig. 14.6). The greatest number of fish in Australia is produced for this sector, but a large volume of fish is produced for the fillet market. For the fillet market, it is more common to target production of a 3 kg or larger fish. For a 'banquet' fish, a target weight of 1 kg is usual, but there is more leniency for several hundred grams above and below this weight as the fish are usually destined for the live market and are sold on location by their specific weight. To produce a plate-size fish typically takes 6–9 months, a banquet fish takes 8–12 months to grow and a 3 kg fillet-size fish takes 18–24 months. All production time periods are heavily dependent on environmental and management influences such as water temperature, feed type, feeding management and fish health. Around the world, barramundi have been grown in several European countries, including England and the Netherlands, as well as the USA. A premium market based around plate-size fish is being developed.

14.2.5 Future perspectives

As with most marine species, there is considerable scope and need for research to upscale commercial hatchery production and includes mapping the barramundi genome and developing breeding programmes; broodstock management

(Frost *et al.*, 2006; Wang *et al.*, 2006; Zhu *et al.*, 2006), particularly for small farms; reducing live feeds and rapid weaning on to pelleted feeds (Curnow *et al.*, 2006b); and generally increasing the supply of fry. The current rapid increase in the global demand for fry requires technology transfer from experimental or small-scale operations to develop large-scale commercial practices. Hatchery production is currently limiting farm production and impacting on uniform market supply and penetration, and clearly important if market demand is to continue. New ventures in Europe and North America are based on recirculation technology and there will be continuing requirements to develop the technology as production efficiencies are sought. Understanding the physiology of barramundi in relation to nutrition and environmental factors (e.g. Glencross and Felsing, 2006; Katersky and Carter, 2007a,b), principally temperature and water quality, will allow greater precision in feed formulation and feed management. The use of lights and manipulation of photoperiod may offer some potential for improved growth. Emphasis is also needed on harvesting methods and research needed in relation to product quality, investigation of rested harvest for example. Fish health issues will require attention, including immune function and vaccine development (Delamare-Deboutteville *et al.*, 2006). Variety in how barramundi are farmed and local requirements require continued research, technology development and commercialization of the research.

References

Alvarez-Lajonchere, L., Cerqueira, V.R., Silva, I.D., Araujo, J. and dos Reis, M. (2002) Mass production of juveniles of the fat snook *Centropomus parallelus* in Brazil. *Journal of the World Aquaculture Society* 33, 506–516.

Anderson, I.G. and Thomas, A. (1995) Recent advances in the bacterial disease of barramundi. *Proceedings Australian Barramundi Farming Workshop*. 15 and 16 June 1995, Walkamin, Queensland, Australia.

Anon. (2006) New Forest Barramundi. Cited from Sunday Telegraph (21/05/2006). *Austasia Aquaculture* August/September, 56.

Azad, I.S., Shekhar, M.S., Thirunavukkarasu, A.R., Poornima, M., Kailasam, M., Rajan, J.J.S., Ali, S.A., Abraham, M. and Ravichandran, P. (2005) Nodavirus infection causes mortalities in hatchery produced larvae of *Lates calcarifer*: first report from India. *Diseases of Aquatic Organisms* 63, 113–118.

Azad, I.S., Jithendrana, K.P., Shekhara, M.S., Thirunavukkarasua, A.R. and Penab, L.D. de la (2006) Immunolocalisation of nervous necrosis virus indicates vertical transmission in hatchery produced Asian sea bass (*Lates calcarifer* Bloch) – a case study. *Aquaculture* 255, 39–47.

Barlow, C.G. (1981) Breeding and larval rearing of *Lates calcarifer* (Bloch) (Pisces: *Centropomidae*) in Thailand. New South Wales State Fisheries Report. New South Wales Department of Fisheries, Sydney, Australia.

Boonyaratpalin, M. and Williams, K. (2002) Asian sea bass, *Lates calcarifer*. In: Webster, C.D. and Lim. C. (eds) *Nutrient Requirements and Feeding of Finfish for Aquaculture*. CAB International, Wallingford, UK, pp. 40–50.

Curnow, J., King, J., Bosmans, J. and Kolkovski, S. (2006a) The effect of reduced *Artemia* and rotifer use facilitated by a new microdiet in the rearing of barramundi *Lates calcarifer* (BLOCH) larvae. *Aquaculture* 257, 204–213.

Curnow, J., King, J., Partridge, G. and Kolkovski, S. (2006b) Effects of two commercial microdiets on growth and survival of barramundi (*Lates calcarifer* Bloch) larvae within various early weaning protocols. *Aquaculture Nutrition* 12, 247–255.

Delamare-Deboutteville, J., Wood, D. and Barnes, A.C. (2006) Response and function of cutaneous mucosal and serum antibodies in barramundi (*Lates calcarifer*) acclimated in seawater and freshwater. *Fish and Shellfish Immunology* 21, 92–101.

FAO Fishery Information, Data and Statistics Unit (FIDI) (2006) Fishery Statistical Collections. *FIGISDataCollection*.FAO–Rome(http://www.fao.org/figis/servlet/static?dom=collection&xml=global-aquaculture-production.xml, accessed 6 October 2006).

Frost, L.A., Evans, B.S. and Jerry, D.R. (2006) Loss of genetic diversity due to hatchery culture practices in barramundi (*Lates calcarifer*). *Aquaculture* 261, 1056–1064.

Garcia, L.M.B. (1989a) Development of an ovarian biopsy technique in the sea bass, *Lates calcarifer* (Bloch). *Aquaculture* 77, 97–102.

Garcia, L.M.B. (1989b) Dose-dependent spawning response of mature female sea bass, *Lates calcarifer* (Bloch), to pelleted luteinizing hormone-releasing hormone analogue (LHRHa). *Aquaculture* 77, 85–96.

Garcia, L.M.B. (1992) Lunar synchronization of spawning in sea bass, *Lates calcarifer* (Bloch): effect of luteinizing hormone-releasing hormone analogue (LHRHa) treatment. *Journal of Fish Biology* 40, 359–370.

Garrett, R.N. and O'Brien, J.J. (1994) All-year-round spawning of hatchery barramundi in Australia. *Austasia Aquaculture* 8, 40–42.

Garrett, R.N. and O'Connell, M.R.J. (1991) Induced breeding in barramundi. *Austasia Aquaculture* 5(8), 10–12.

Glencross, B.D. (2006) Nutritional management of barramundi, *Lates calcarifer* – a review. *Aquaculture Nutrition* 12, 291–309.

Glencross, B. and Felsing, M. (2006) Influence of fish size and water temperature on the metabolic demand for oxygen by barramundi, *Lates calcarifer* (Bloch), in freshwater. *Aquaculture Research* 37, 1055–1062.

Glencross, B.D., Percival, S., Jones, J.B. and Hughes, J. (2007) Sustainable development of barramundi cage aquaculture at Lake Argyle. Final Report to The Fisheries Research and Development Corporation, Canberra.

Greenwood, P.H. (1976) A review of the family *Centropomidae* (Pisces, Perciformes). *Bulletin of the British Museum (Natural History), Zoology* 29, 1–81.

Harrison, K. (1991) The taxonomy of east-African nile perch, *Lates* spp (Perciforms, *Centropromidae*). *Journal of Fish Biology* 38, 175–186.

Howgate, P. (2004) Tainting of farmed fish by geosmin and 2-methyl-iso-bomeol: a review of sensory aspects and of uptake/depuration. *Aquaculture* 234(1–4), 155–181.

Katersky, R.S. and Carter, C.G. (2005) Growth efficiency of juvenile barramundi, *Lates calcarifer*, at high temperatures. *Aquaculture* 250, 775–780.

Katersky, R.S. and Carter, C.G. (2007a) High growth efficiency occurs over a wide temperature range for juvenile barramundi *Lates calcarifer* fed a balanced diet. *Aquaculture* 272, 444–450.

Katersky, R.S. and Carter, C.G. (2007b) A preliminary study on growth and protein synthesis of juvenile barramundi, *Lates calcarifer* at different temperatures. *Aquaculture* 267, 157–164.

Kolkovski, S., Curnow, J. and King, J. (2004) Intensive rearing system for fish larvae research. I. Marine fish larval rearing system. *Aquaculture Engineering* 31, 295–308.

Kungvankij, P. (1987) Induction of spawning of sea bass (*Lates calcarifer*) by hormone injection and environmental manipulation. In: Copland, J.W. and Grey, D.L. (eds) *Management of Wild and Cultured Sea Bass/Barramundi (Lates Calcarifer)*. ACIAR Proceedings No 20. Australian Centre for International Agricultural Research, Canberra, pp. 120–122.

Otero, O. (2004) Anatomy, systematics and phylogeny of both recent and fossil latid fishes (Teleostei, Perciformes, Latidae). *Zoological Journal of the Linnean Society* 141, 81–133.

Owens, L. (2003) Barramundi. In: Lucas, J.S. and Southgate, P.C. (eds) *Aquaculture: Farming Aquatic Animals and Plants*. Blackwell Publishing, Oxford, UK, pp. 364–381.

Rimmer, M.A. (2003) Barramundi. In: Lucas, J.S. and Southgate, P.C. (eds) *Aquaculture: Farming Aquatic Animals and Plants*. Blackwell Publishing, Oxford, UK, pp. 364–381.

Rimmer, M.A. (2006) Cultured aquatic species information programme – *Lates calcarifer*. Cultured Aquatic Species Fact Sheets, FAO, Rome (http://www.fao.org/figis/servlet/static?dom=culturespecies&xmml=Lates_calcarifer.xml, accessed 1 September 2006).

Rimmer, M.A. and Russell, D.J. (1998) Aspects of the biology and culture of *Lates calcarifer*. In: De Silva, S.S. (ed) *Tropical Mariculture*. Academic Press, San Diego, California, pp. 449–476.

Ruangpanit, N. (1987) Developing hatchery techniques for sea bass (*Lates calcarifer*): a review. In: Copland, J.W. and Grey, D.L. (eds) *Management of Wild and Cultured Sea Bass/Barramundi (Lates Calcarifer)*. Australian Centre for International Agricultural Research Proceedings No 20, Canberra, pp. 132–135.

Schipp, G. (1991) Care of newly captured barramundi broodstock. Department of Primary Industry and Fisheries, Northern Territory, Australia. Fishnote No 8.

Schipp, G. (2006) Barramundi farming in the Northern Territory. Darwin Aquaculture Centre, Department of Primary Industry, Fisheries and Mines (DPIFM), Northern Territory, Australia.

Schipp, G. and Pearce, M. (1991) Diseases of barramundi broodstock. Department of Primary Industry and Fisheries, Northern Territory, Australia. Fishnote No 9.

Segev, R., Shoval, S., Ron, B., Peduel, A., Barki, A., Harpaz, S. and Karplus, I. (2003) Introduction of the barramundi (*Lates calcarifer*) to aquaculture farms in the Negev and Arava regions in Israel. *Israeli Journal of Aquaculture/Bamidgeh* 55, 228–229.

Temple, S., Cerqueira, V.R. and Brown, J.A. (2004) The effects of lowering prey density on the growth, survival and foraging behaviour of larval fat snook (*Centropomus parallelus* poey 1860). *Aquaculture* 233, 205–217.

Tucker, J.W. Jr, Russell, D.J. and Rimmer, M.A. (2002) Barramundi culture: a success story for aquaculture in Asia and Australia. *World Aquaculture* 33(4), 67–72.

Wang, C.M., Lo, L.C., Zhu, Z.Y. and Yue, G.H. (2006) A genome scan for quantitative trait loci affecting growth-related traits in an F1 family of Asian seabass (*Lates calcarifer*). *BMC Genomics* 7, Article No. 274.

Williams, K.C. and Barlow, C.G. (1999) Dietary requirement and optimal feeding practices for barramundi (*Lates calcarifer*). Project 92/63, Final Report to Fisheries R&D Corporation, Canberra.

Zhu, Z.Y., Lin, G., Lo, L.C., Xu, Y.X., Feng, F., Chou, R. and Yue, G.H. (2006) Genetic analyses of Asian seabass stocks using novel polymorphic microsatellites. *Aquaculture* 256, 167–173.

15 The Temperate Basses (Family: Moronidae)

Malcolm Jobling,[1] Stefano Peruzzi[1] and Curry Woods[2]

[1]*University of Tromsø, Tromsø, Norway;* [2]*Department of Animal and Avian Sciences, Animal Science Center, University of Maryland, College Park, USA*

15.1 General Introduction

The six species of temperate basses are representatives of a small family (Moronidae) (Fig. 15.1) of perciform fish (Nelson, 1994). Four of the species, representatives of the genus *Morone*, occur in North America. Two of the four North American species occur in brackish water and marine coastal areas and the other two are confined to fresh water. The natural distribution of the temperate basses within the genus *Morone* is along the eastern seaboard of the North American continent. Their natural geographic range encompasses the Atlantic Ocean and the Gulf of Mexico and the watercourses that drain into them. The original range of the striped bass, *M. saxatilis*, was from the St Lawrence River in Canada to northern Florida and along the Gulf coast from western Florida to Louisiana (Fig. 15.2). The St Lawrence River stock is thought to be extinct, but there is evidence of striped bass spawning grounds in the Miramichi River and Shubenacadie River systems in Canada. The Apalachicola–Chattahoochee–Flint river drainage system supports populations in the Gulf of Mexico. Striped bass have been introduced to the Pacific Coast and the fish is now widespread from Washington State to Baja, Mexico. Striped bass and/or its hybrids can be found in all of the 48 states of the contiguous US. In addition, striped bass, white bass, *M. chrysops*, and their hybrids have been introduced to other countries, including Mexico, China, Taiwan, Russia, Israel, France, Germany, Italy and Portugal.

The two European species, genus *Dicentrarchus*, occur in the coastal waters of the eastern Atlantic, the Mediterranean and Black Seas and they may also enter fresh water (Pickett and Pawson, 1994). The European seabass, *D. labrax*, has a more extensive distribution in the coastal waters off Europe and North Africa (Fig. 15.3) than does its congener, the spotted seabass, *D. punctatus*. Nevertheless, the distribution of the European seabass is relatively restricted. It is, for example, seldom found in great numbers north of

 Finfish Aquaculture Diversification (eds N.R. Le François, M. Jobling, C. Carter and P.U. Blier)

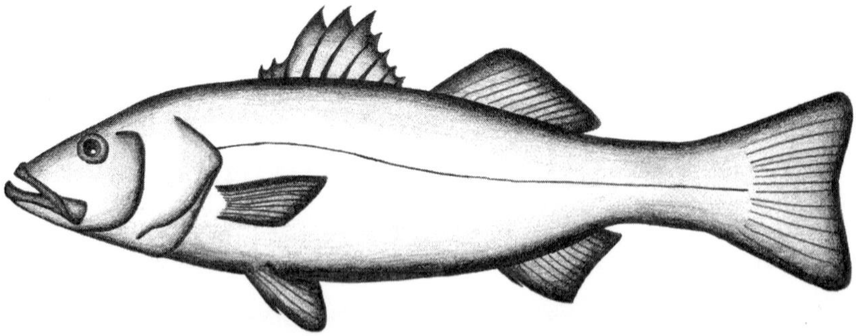

Fig. 15.1. Moronidae.

the UK. The European seabass is prized by sport fishermen and is also exploited in small-scale commercial fisheries throughout its range. It is highly regarded as a food fish, fetches quite high prices on international markets and appears on the menus of restaurants throughout Europe. The European seabass is currently being cultivated in several countries, particularly in the Mediterranean region.

There are several parallels that can be drawn between the European seabass and its North American relatives, particularly the striped bass. Both are typically inshore species that enter estuaries and are capable of surviving in brackish water. They are also both predators that feed mainly on small pelagic fish, and both are renowned as sporting fish. They are both regarded as good eating, there are limited commercial fisheries for both species and some quantities of both are cultivated for the table.

15.2 European Seabass, *Dicentrarchus labrax*

The European seabass is neither abundant nor particularly widely distributed (Fig. 15.3) and although the species forms the basis for commercial fisheries, the catches are neither constant nor large. The fish are taken by gill netting, in seines and trawls and on long-lines. Market requirements are usually for fresh, whole fish often destined for the restaurant trade. This places daily landings and rapid marketing of wild-caught fish at a premium and precludes at-sea processing and preservation. It also encourages the use of gears that lead to minimum damage to the fish (Pickett and Pawson, 1994). A combination of an established market, high prices and a demand that often exceeded supply promoted the development of the commercial culture of seabass during the latter years of the 20th century. Today, economically viable culture is well established and commercial producers can guarantee regular supplies of high-quality fish at a relatively stable price. Prices are usually lower during the summer and autumn as new generations of fish achieve market size and they remain low until farmers have cleared their unsold stocks. Stable supplies and prices cannot be rivalled by the commercial fishery, which depends on a seasonal and somewhat

Fig. 15.2. Distribution of wild populations of *Morone saxatilis*.

Fig. 15.3. Natural distribution of *Dicentrarchus labrax*.

unpredictable supply. There is, however, a market differentiation between wild-caught and farm-produced seabass, with prices paid for wild fish being much higher than those paid for farmed seabass. The biology and commercial exploitation of the European seabass have been described in detail by Pickett and Pawson (1994) and descriptions of the methods used for commercial culture of the species are also available (Nash and Novotny, 1995; Moretti *et al.*, 1999; Stickney, 2000; Theodorou, 2002).

15.2.1 Seabass biology

The European seabass frequents inshore coastal waters, occurs in estuaries and brackish-water lagoons and sometimes ventures upstream into fresh water. Although they can survive and grow in low-saline water, the fish are not able to reproduce in such environments. Spawning takes place in estuaries and coastal areas where the salinity approaches that of full-strength seawater. In addition to being euryhaline, the seabass is also eurythermal. The seabass can survive at temperatures from 2°C to 32°C and tolerates temperatures within the range 5–28°C very well, but spawning and egg development usually occur at 9–15°C. Spawning takes place during the winter, between late November and March in the Mediterranean basin. The start of the spawning season is delayed in the northern parts of the geographic range and spawning may occur until June along some parts of the Atlantic coastline of Europe.

During the spawning season, a female may produce 250,000–500,000 eggs/kg body weight. The small (1.1–1.4 mm diameter) eggs contain oil droplets, are neutrally buoyant and remain suspended in the water column; that is, seabass eggs are pelagic and form part of the plankton. The rate of development of the eggs depends on temperature; at 9°C, the eggs hatch after about 9 days and the time to hatch decreases to about 4 days at 15°C. The newly hatched seabass are 4–4.5 mm in length and they rely on the yolk sac for nutrition for the first few days. Mouth opening and first-feeding occur at about 14 days posthatch at 9°C and this is reduced to 5 days at 15°C. At the same time, the gas bladder becomes visible as a reflective ovoid bubble, so the young fish are capable of maintaining buoyancy despite having metabolized the low-density oil droplets that were present in the yolk sac.

The youngest juveniles are found close to shore from a few weeks after hatching. After about 2–3 months, they migrate to warm estuaries and relatively shallow inshore waters, such as bays, backwaters and harbours. These are the nursery areas for the juvenile seabass; the young seabass congregate, form shoals and feed mostly on small crustaceans. As the seabass increase in size, the diet is expanded to include small fish and larger crustaceans. The fish move further offshore as they grow and offshore migration is most pronounced during the winter months. The seabass will be sexually mature by the time they reach an age of 4–5 years; males may mature at 2–3 years of age, at a length of 25–30 cm, whereas females usually mature later (3–4 years of age) and at larger size (32–40 cm in length) (Pickett and Pawson, 1994).

15.2.2 Farming of seabass

Cultivation of the European seabass using extensive methods has been carried out for many years, mostly in Italy, France and Spain. Traditional cultivation methods are based on the rearing of wild hatchlings and juveniles that enter tidal lagoons and inlets, coastal ponds and marsh pools on tidal flows. The entrances to the lagoons, ponds and pools are first open during the spring and early summer to allow entry of the young fish. The entrances are then closed off with screens or barriers to prevent the escape of the fish. The fish are reliant on natural prey organisms for food and the rate of growth depends on food availability and temperature. In some cases, lagoons and coastal ponds are actively stocked with wild-caught juveniles. Canals and ditches may be dug to improve control over water exchange with the open sea and there may also be fertilization of the lagoons and ponds to increase biological productivity. Fish production in these modified systems is somewhat higher than in the traditional extensive systems. The seabass are harvested after 2–3 years, with 3 years being most usual in extensive systems. By this time, the fish may be 400–500 g in weight. Harvesting is usually carried out by seine netting, often following partial draining of the lagoon, pond or pool to reduce the water depth.

Interest in more intensive forms of rearing developed during the 1970s. Success with spawning seabass in captivity was achieved during the late 1960s. By the late 1970s, techniques for the mass production of juveniles were sufficiently well developed to provide the numbers required for commercial production. In addition, high market prices for seabass encouraged investment in farming. This resulted in seabass becoming the first non-salmonid species to be adopted for commercial marine culture in Europe. The seabass is currently of considerable importance in Mediterranean aquaculture; producing countries include Greece, Turkey, Italy, France, Spain, Croatia and Egypt. Italy is the main importer of farmed seabass. Farmed seabass are usually marketed as portion fish (250–300 and up to 450 g). The time required to achieve market size is 20–24 months in the cooler regions of the western Mediterranean, but is reduced to 14–15 months in warmer areas, such as along the African coastline. As production of farmed seabass increased, the market gradually became saturated, prices declined and profit margins were reduced. As such, the seabass farming industry can be considered to be relatively mature and to be meeting current market demand; lower prices may give some increase in demand in traditional markets and open new ones, but any major expansion in production would require a greater focus on market development, coupled with product diversification.

Intensive cultivation, involving the on-growing of hatchery-raised juveniles in enclosures, sea cages, tanks or raceways, is currently the commonest way for producing farmed seabass. Some companies specialize in hatchery production of juveniles for sale to on-growers, whereas others restrict their operations to the on-growing of juvenile fish to market size. Examples of complete integration are also common, where a single company owns both hatchery and on-growing units. The description of the methods used in the intensive culture of seabass presented here is based on accounts given in Nash and Novotny (1995),

Moretti *et al.* (1999), Stickney (2000) and Theodorou (2002), supplemented with information from other cited sources and from unpublished and restricted-access reports and experience documents.

15.2.3 Broodstock management and hatchery operations

Hatchery broodstock will be made up of mature fish of different ages. The European seabass is dioecious and gonochoristic; that is, the sexes are separate and there is no sex-reversal following sexual differentiation, but intersex individuals, i.e. with intratesticular oocytes, are sometimes observed. Male fish reach sexual maturity earlier than the females. This means that male broodstock will usually be younger (age 2–4 years) than the females (5–8 years). In wild European seabass, gametogenesis is initiated during the early autumn under conditions of reducing day-length and falling water temperatures, and the spawning season is of limited duration (late November–March). Consequently, reliance on natural cycles of reproductive development and spawning within a hatchery would result in the facility lying fallow for much of the year. In some cases, this is avoided by adopting multi-species production, in which different species with partially overlapping spawning seasons are used, e.g. seabass and gilthead seabream, *Sparus aurata*. In single-species seabass hatcheries, the spawning season can be extended by using multipopulation broodstocks, i.e. broodfish from Mediterranean and Atlantic populations. It is also usual that some form of photothermal manipulation is employed to extend the length of, or phase-shift, the spawning season. This enables production of 'out-of-season' eggs and ensures that juveniles can be made available to on-growers almost year round.

Reproductive development and spawning occur readily at temperatures of 12–15°C, but gametogenesis is inhibited at temperatures above 21–22°C and spawning may not occur when the water temperature is above 18°C or below 9°C. Control of the water temperature in the broodstock holding tanks is required to promote correct ovarian and testicular development and to ensure the production of gametes of good quality. Manipulation of the photoperiod is required if the aim is to extend the length of the spawning season or to induce a phase-shift in the onset of gametogenesis and influence the time of spawning. Several types of photothermal manipulation can be applied to promote out-of-season reproductive development and spawning:

- Fish are held under conditions of compressed cycles of photoperiod and temperature. This method is effective but can result in the production of gametes of reduced quality if not applied correctly.
- Fish are held under constant day-length conditions and are induced to commence reproductive development by brief exposure to periods of 'long' and then 'short' days.
- Groups of fish are held on 12-month photothermal cycles, but the groups are phase-shifted in relation to each other. For example, if there are four groups of broodstock and three of these are phase-shifted by 3, 6 and 9 months with respect to the natural annual cycle, the hatchery will have groups of fish that are ready to spawn at every season of the year.

It is relatively easy to determine the reproductive condition of broodstock males because they become running ripe, i.e. readily release milt when the belly is gently palpated, when ready to spawn. In females, however, maturation stage has to be ascertained by examination of oocytes removed from the ovary by means of a catheter; this is because post-vitellogenic females do not usually ovulate spontaneously and become running ripe. Females whose ovaries contain oocytes with a diameter greater than 650 μm are in post-vitellogenic condition and can be selected for spawning purposes. Final oocyte maturation, ovulation and spawning are induced by hormonal treatment of the selected females; the hormonal treatment triggers the final stages of egg maturation and ovulation and spawning then follow in sequence. The triggering of final maturation is usually induced by injecting the females with analogues of gonadotropin-releasing hormone (GnRHa). Injections are either administered as two low doses (5 and then 10 μg GnRHa/kg body weight) given a few hours (4–6 h) apart, or as a larger dose (10–15 μg GnRHa/kg body weight) given as a single intramuscular injection. Following injection, the females are transferred to spawning tanks, where they are stocked along with males at a sex ratio of 2 males:1 female; spawning will usually occur 54–72 h after transfer of the fish to the spawning tanks.

Some hours after hormonal treatment and transfer of the chosen broodfish to spawning tanks, the females release ripe eggs that are fertilized naturally by sperm from the males; this natural fertilization makes gamete stripping and artificial fertilization, as carried out with some other farmed species, salmonids for example, unnecessary. The small, pelagic eggs are then collected for incubation; eggs are usually collected as they are flushed from the spawning tanks in the water outflow. The egg collectors consist of fine-meshed net cones or mesh-screened boxes.

Egg incubation and larval rearing techniques vary somewhat from hatchery to hatchery, according to local experience, but the rearing techniques are based on principles determined by the biological characteristics of the species. One problem faced during incubation is that of keeping the eggs in suspension. This can be achieved by several means. For example, eggs may be incubated in a floating, cylindrical, nylon mesh net incubation basket with a conical base. Egg incubation baskets vary in size, e.g. with capacities 50–500 l, and egg densities will usually be over 5000 eggs/l (and up to about 12,000 eggs/l) during incubation. The eggs are held in suspension by arranging an upward flow of air or water from the base of the cone. The incubation baskets are often placed in tanks that are later used for larval rearing. The salinity of the water also plays a role in keeping the eggs suspended in the water column. The eggs are neutrally buoyant in seawater (above 33‰), so egg incubation is carried out in seawater rather than in water of reduced salinity.

As an alternative to being placed in incubation baskets, fertilized eggs can be stocked directly into cylindroconical tanks (500–1500 l) or, less commonly, in larger flat-bottomed tanks (5000–10,000 l) that are also used for larval rearing. In these cases, egg densities are lower than when eggs are held in specially designed incubators. Within the tanks, the eggs are kept in suspension either by the flow dynamics of the water in the tank and/or using an aeration system. Cylindroconical tanks are generally favoured because they are easy to use and maintain. They are easy to clean and wastes can be removed via a tap in the base of the conical bottom of the tank. The dynamics of water flow within the tank can be controlled by adjusting the supply of water and/or air entering at the base of the cone and water quality is maintained due to the continuous flow-through of water that leaves the tank via a mesh screen filter.

The eggs hatch after 5–9 days, depending on temperature; an incubation temperature of 13–17°C is usual. If floating incubation baskets have been used, rather than having larval rearing tanks doubling as incubators, the hatchlings are transferred gently to the larval rearing tanks. These tanks have a continuous flow of seawater to ensure that water quality is maintained at a level that promotes high survival of the hatchlings. Important developments in larval culture took place during the late 1980s and 1990s; these included elucidation of environmental requirements, the investigation of effects of tank colour and design and the introduction of technical innovations, such as surface skimmers, that resulted in increased survival (Chatain, 1997).

The temperature in the larval rearing tanks is initially the same as under egg incubation, but is increased gradually over time to about 20°C at 15–20 days posthatch and then to 23–25°C towards the end of the larval rearing stage. Larval rearing is generally completed at about 45 days posthatch. Light intensity is kept low at first (below 100 lux; egg incubation and early rearing until the time of start-feeding may be carried out in complete darkness), but is increased to about 500 lux at first-feeding (on day 10). Light intensity is increased further to 1200–1500 lux from around day 15 to day 20 and is then held at this level until the end of the larval rearing phase. Likewise, photoperiod is increased as time progresses; often from 12L:12D at 10 days posthatch to 16L:8D or 24L:0D by day 20 after hatching. Adjustments in salinity are also often undertaken, with salinity being reduced to about 25‰ over a 3-day period around the time of start-feeding.

Initially, the hatchlings are nourished by the nutrients present in the yolk sac, but as the yolk becomes exhausted, the fish are provided with exogenous feed, most often live feed organisms. Food is usually introduced at about day 10 after hatching. When methods of larval rearing of seabass were being developed, rotifers, *Brachionus plicatilis*, were commonly used for initial feeding, although other live prey organisms of suitable size were also used. An alternative method of start-feeding involved the direct introduction of nauplii of the brine shrimp, *Artemia salina*, and this is now common practice. At the same time, the larval rearing tanks may be supplied with algal cultures (e.g. *Nannochloropsis* sp.), giving the green-water method of larval rearing. In addition, attempts are continuing to develop inert microfeeds that are suitable as start-feeds. Some producers are using microfeeds directly, thereby circumventing the need for the production of live prey organisms. Several descriptions of the methods used for cultivation of live prey organisms for fish hatchlings and juveniles are available (e.g. Nash and Novotny, 1995; Moretti *et al.*, 1999; Stickney, 2000; Stottrup and McEvoy, 2002; Olsen, 2004) and there are a number of sources that provide information about the development of microfeeds for the culture of fish and crustacean larvae (e.g. Stickney, 2000; Cahu and Zambonino Infante, 2001; Langdon, 2003). At present, start-feeding of seabass is often based on live prey with inert, dry feeds of 80–200 µm being introduced from around day 20 onwards. Feed particle sizes are increased to 160–300 µm at about day 30; this represents the start of the weaning process, although weaning is not complete until much later (around day 60).

Seabass reared by these methods increase in size from 4–4.5 mm at hatch to 15–18 mm at 40–45 days. By this time, the fish will weigh 40–50 mg. Between day 40 and 45 posthatch, the fish are transferred to larger tanks for the completion of weaning and the continuation of growth to the juvenile stage. Following transfer to the larger tanks, the transition from live prey to inert, dry feed is undertaken gradually. The provision of live food is phased out gradually over a period of about 10 days and the supply of dry, weaning feed is increased progressively. On completion of weaning, the juveniles are held in the rearing units until they achieve the size at which they can be transferred to on-growing systems. The size at which the juveniles are transferred to the on-growing units depends on the type of system used and on the details of the production cycle employed by individual on-growing farms (Fig. 15.4).

Despite the maturity of the seabass aquaculture industry, most farmers rely on semi-domesticated broodstock for egg production and apply very simple breeding protocols to obtain cumulative genetic gains over generations. As such, the application of directed selective breeding programmes is still in its

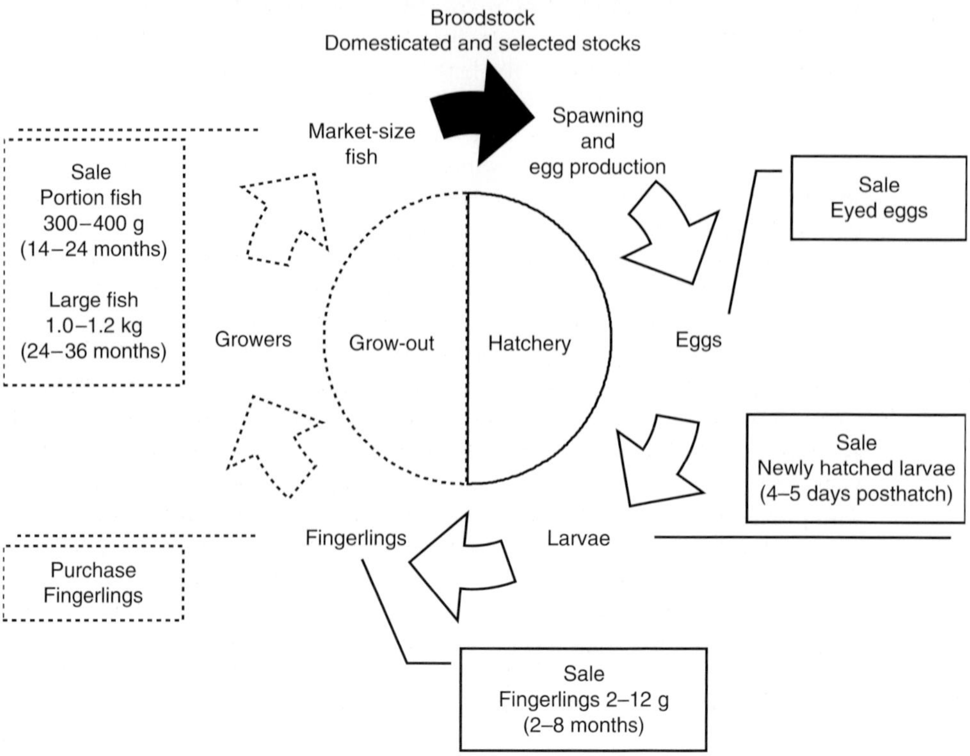

Fig. 15.4. Production cycle for farming European seabass, *Dicentrarchus labrax*.

infancy. Nevertheless, the study of the hereditability of major economic production traits, such as growth, disease resistance, flesh and carcass quality, has indicated considerable potential for genetic improvement through selective breeding. For example, the potential genetic gain for body weight may be of the order of 25–30% per generation and gains in growth at 6 months may reach almost 50% when comparisons are made between selected and wild or domesticated stocks of seabass (Chatain, personal communication).

Attempts to improve the performance of cultured species by hybridization, or cross-breeding between strains, with the aim of inducing heterosis for commercially valuable traits, have met varying success. Recently, hybridization trials have been carried out between the two European bass species, *D. labrax* and *D. punctatus*, and triploidization techniques have been applied to induce reproductive sterility. Culture performance of the triploid hybrid offspring has been investigated and results indicate that the triploid hybrid seabass have some advantages over the parental stocks. The technology used to produce these triploid hybrid seabass is currently under patent evaluation (French – FR2879401 – and European – WO2006067356).

15.2.4 On-growing to market size

On-growing of European seabass is carried out mostly in sea cages, although land-based on-growing units are not uncommon. The land-based systems may use either tanks or raceways as the rearing units and may be single-pass (flow-through), partial reuse or recirculation systems. Stocking densities are highest in land-based units (up to 40–50 kg/m^3) and may reach a maximum of about 20 kg/m^3 in cages. When on-growing is carried out onshore, the water must be pumped to the tanks or raceways. There is usually some form of pre- and post-treatment to ensure good quality inflowing water and to prevent excessive waste materials reaching the environment in the effluent. The water source may be brackish water pumped from a nearby lagoon, bore-water from an underground source, industrial cooling-water or water taken directly from the sea.

Juveniles for on-growing are supplied at a minimum size of 1.5–2.5 g, but most producers prefer to use juveniles of a larger size (*c.*5 g) for stocking into their on-growing units. The juveniles offered for sale are usually certified to be parasite- and pathogen-free and will also have been vaccinated against vibriosis (serotypes 1 and 2). In some instances, a sample of fish will be X-ray photographed to demonstrate that the fish do not have internal deformities. The fish usually reach a size of about 300 g in 16–18 months and 400–450 g in 18–24 months, but some fish may be retained for longer until they reach a body size of 1 kg or more. The latter are usually destined for the restaurant trade. Feeding may be carried out by hand, by computer-controlled automatic feeders that dispense feed as 2–3 meals/day, or on-demand feeding systems can be used. Self-feeders have been used to good effect in a number of land-based on-growing units and interactive feeding systems may be used at cage sites. Feeds are usually extruded dry pellets containing 43–50% protein and 12–25% fat (Kaushik, 2002).

The fish are held without feeding for a short period before harvesting; feed may be withheld for 1–12 days, depending on water temperature. For harvesting, the fish are first confined to a restricted area of the rearing unit and are then removed either using a dip-net or by pumping. The fish are transferred to plastic or stainless steel tubs and are usually killed by asphyxiation in chilled water, prepared by mixing seawater with ice slurry. Once killed, the fish are usually packed on ice, transported rapidly to market and are sold fresh within a couple of days (maximum 4–5 days). A small amount of farmed seabass is sold frozen and some is further processed and individually packaged for sale in specialist markets. Larger bass are supplied to hotels, restaurants and seafood specialty outlets.

One important problem encountered in the intensive commercial production of European seabass is the occurrence of high percentages of male fish (typically about 75% of the population). High proportions of the fish develop as males when hatchlings and early juveniles are reared at 19–22°C, rather than the lower temperatures (*c.*12–15°C) encountered in the wild. The development of male-biased populations is undesirable because the males grow slower than the females and males also mature earlier than the females. In culture, females will normally mature at 3 years of age, whereas most males mature at an age of 2 years, but some males mature precociously during their first year of life. This means that a substantial proportion of the male fish may mature before reaching harvest size, typically achieved after 15–24 months in culture, depending on rearing method and geographic area. Given this problem, the production of monosex female populations of seabass would be desirable, to take advantage of the sexual dimorphism in growth rates and avoid early sexual maturation. Although sex manipulation of seabass has been achieved by exposing the fish to exogenous sex steroid hormones, efforts to produce female monosex seabass populations using crosses involving sex-reversed individuals have not met with success. Some of the difficulties arise because the mechanisms responsible for sex determination and differentiation in this species are still not known with certainty (Piferrer *et al.*, 2005).

Although the seabass is sturdy and robust, it may be subject to parasite infestation and diseases during the production cycle (Rigos and Troisi, 2005). Woo *et al.* (2002) provide an overview of the diseases and parasites encountered in cage-cultured fish and Rigos and Troisi (2005) provide information on the use of antibacterial agents in the Mediterranean region. Under some circumstances, farm populations may be decimated by a disease outbreak, leading to serious losses of production. Outbreaks of disease are most likely to occur when rearing conditions are suboptimal, so a poor culture environment is often a major contributory factor to such outbreaks. Farmed seabass are subject to a number of viral and bacterial diseases (Table 15.1) (Rigos and Troisi, 2005) and may also be attacked by a relatively wide range of protistan parasites. The protistans are generally not species-specific, so may have several fish species as hosts. Some of the metazoan parasites display narrow host specificity, and have a relatively benign relationship with their seabass host, but other metazoan parasites found on seabass may use several fish species as hosts and cause more serious damage when they infest the farmed fish.

Table 15.1. Examples of diseases and parasites of farmed European seabass, *Dicentrarchus labrax*.

	Disease symptoms	Treatment
Viral diseases:		
Viral encaphalitis (Nodavirus)	Nervous disorder; whirling movement and loss of balance	No effective therapy
Bacterial diseases:		
Vibriosis (*Vibrio* spp.)	Haemorrhages, dark skin and skin ulcers and lesions	Vaccination and/or treatment with antibiotics (in medicated feed)
Pasteurellosis (*Photobacterium damselae*)	Acute septicaemia; enlarged spleen and gill necrosis	Antibiotic treatment in medicated feed and/or vaccination
Mycobacteriosis (*Mycobacterium marinum*)	Superficial ulcers; enlarged kidney and spleen with whitish nodules	Antibiotic treatment may be attempted, but may be of limited effect
Myxobacteriosis (*Flexibacter maritimus*)	Skin ulcers and necrotic lesions, fin rot and gill necrosis	Antibiotic treatment
Protistan parasites:		
Amyloodiniosis (*Amyloodinium ocellatus*)	'Dusty' appearance to skin	Copper sulphate bath treatment Placing fish in fresh water for a few minutes dislodges the parasites
Cryptocaryonosis (*Cryptocaryon irritans*)	Small white 'blisters' on the skin; increased mucus production	
Myxosporean infections (*Sphaerospora* spp.)	Infestations of the organs of the visceral cavity (intestine, gall bladder, swimbladder, etc.)	
Metazoan parasites:		
'Sea lice' (copepods and isopods)	Sites of infestation include the gills, mouth and surface of the skin	
Diplectanid monogeneans	Gill necrosis	

15.2.5 Concluding comments

During the latter years of the 20th century, the farming of European seabass became a success story in Mediterranean aquaculture, with production increasing to around 50,000t over a 15-year period. When farmed seabass first appeared on the market, prices were high and there were good profit margins for producers. As production increased, the relatively small regional market became saturated, there was a fall in price and the market also began to differentiate between farmed and wild fish, with prices for wild fish being much higher. Penetration into new markets was slow to develop, but sales are increasing in northern Europe. In addition, the development of novel products for sale in Mediterranean countries was hampered by consumer conservatism; the main market is for fresh, portion-size, whole fish and demand for processed or value-added products such as fillets, cutlets and pre-packaged products with a longer shelf life has been low. As such, growth of the European seabass industry had slowed by the turn of the century and the consolidation representative of a maturing industry came more to the fore.

15.3 Striped Bass, *Morone saxatilis*, and Striped Bass Hybrids

The culture of striped bass was started in the late 19th century to enhance coastal stocks for commercial and recreational fisheries. Stock enhancement was later expanded to include inland reservoirs (Stickney, 1996). Hormonal induction of spawning was achieved in 1962 and this enabled increased production of fish for stocking reservoirs (Stevens, 1966). By the end of the 20th century, private aquaculture firms were producing approximately 200 million juveniles each year, with the majority being sold in the USA. Crossing of striped bass and white bass was undertaken in 1965 (Logan, 1968) and this led to examination of the feasibility of hybrid striped bass aquaculture (Woods *et al*., 1981).

Most early aquaculture ventures of striped bass failed within 5 years (Woods, 2005), but commercial viability was achieved following a significant shift in production from striped bass to hybrids. The hybrid cross between female white bass and male striped bass, called sunshine bass, dominates production today. Production has grown steadily over the past two decades and currently exceeds 10,000t. The majority of hybrid bass producers in the USA raise their fish in freshwater ponds, but some use recirculation systems and others raise their fish in cages held in brackish water. Most producers of table fish purchase juveniles for on-growing from hatchery producers. At 18–24 months of age, marketable hybrid striped bass are harvested by seine and immediately packed whole on ice for shipment to market. Little additional processing is usually carried out until the fish reach the consumer.

A synopsis of striped bass biological data has been compiled by Setzler *et al*. (1980) and the culture of striped bass and its hybrids has been described in detail by Harrell (1997).

15.3.1 Striped bass biology

The striped bass and its hybrid crosses with the white bass have a silvery colour, a white abdomen and a black to olive-grey back (Kohler, 2000), and all have lateral, horizontal stripes on the body. The hybrids of white bass and striped bass are more tolerant of extremes in temperature and dissolved oxygen than either of the parental species, but both the striped bass and its hybrids are eurythermic and euryhaline.

Striped bass and white bass are considered to be dioecious, group-synchronous spawners, although hermaphroditism may occur in striped bass. They are iteroparous and spawn annually for several years once mature (Sullivan *et al.*, 1997). Males mature earlier than females. In the wild, male striped bass usually mature by age 2, while females require 4 or more years. White bass males mature by age 1 and females by 3–4 years of age (Stickney, 2000). Both species spawn during the spring under conditions of increasing day-length and water temperature. The time of peak spawning varies with latitude, and water temperatures most suitable for egg and larval development are 16–17°C (Fay *et al.*, 1983) and 14–18°C (Ruelle, 1971) for striped bass and white bass, respectively.

15.3.2 Farming striped bass and its hybrids

The culture of striped bass and its hybrids involves four distinct phases: the hatchery phase, rearing to a 25–75 mm juvenile, rearing juveniles for a full year where they may attain lengths of up to 250 mm, and on-growing of juveniles to market size (Brewer and Rees, 1990). Culture requirements for hybrid bass are similar to those of striped bass. The primary difference between striped bass and sunshine bass is that the hybrid is smaller at hatch and has a smaller mouth gape. Thus, the hybrids require special consideration with regard to first-feeding.

15.3.3 Broodstock management and hatchery operations

Striped bass and white bass broodstock will usually comprise mature fish of varying ages and domesticated striped bass males will usually be used for the *in vitro* fertilization of white bass eggs to produce sunshine bass. Currently, there are few efforts to domesticate white bass and broodstock females either are acquired ripe or are brought into the hatchery some weeks or months prior to spawning. Some form of photothermal manipulation is often employed to extend the length of the spawning season. This requires systems with environmental controls, but enables production of juveniles for on-growing at various latitudes and with different growing seasons. Nevertheless, juveniles are not currently available to on-growing producers on a year-round basis.

Wild, acclimated or domesticated broodfish are usually induced to spawn and protocols to spawn both striped bass and white bass have been described (Harrell *et al.*, 1990; Kohler *et al.*, 1994). The gametes are stripped from anaesthetized (75 mg/l MS-222, buffered with sodium bicarbonate) adults and fertilization is accomplished *in vitro*. It is relatively easy to determine the reproductive condition of striped bass males because spermiating males readily release milt when the belly is palpated gently. Striped bass spermatozoa can now be cryopreserved effectively (He and Woods, 2004), making it possible for hatcheries to bank spermatozoa for later use. Post-vitellogenic female striped bass and white bass do not usually ovulate spontaneously, so the stage of oocyte maturation must be ascertained by examination of oocytes removed from a catheterized lobe of the ovaries. Females whose ovaries contain oocytes with a diameter greater than 650 μm are in a post-vitellogenic condition and can be used for spawning. Final oocyte maturation, ovulation and spawning are induced by hormonal treatment of the selected females; the hormonal treatment triggers the final stages of egg maturation and ovulation and spawning then follow in sequence. The triggering of final maturation is usually induced by injecting or implanting the females with human chorionic gonadotropin (HCG) or mammalian analogues of gonadotropin-releasing hormone, D-Ala^{6}-Pro^{9}-Net-LHRH (mGnRHa). mGnRHa may be administered as two, intramuscular injections of relatively low dosages (5 and then 10 μg mGnRHa/kg body weight) given a few hours (4–6 h) apart, as a larger single dose (10–15 μg mGnRHa/kg body weight) or in time-released implants containing 20 μg mGnRHa/kg body weight (Woods and Sullivan, 1993; Mylonas *et al.*, 1998). If the broodfish are injected with HCG, the females should be administered intramuscular injections of 275–330 IU/kg body weight (Stickney, 2000). Following injection or implantation, females are transferred to spawning tanks, where they are held to check for ovulation and/or spawning. HCG-injected bass should be monitored every 2 h, starting 16 h post-injection, and ovulation usually occurs 24–36 hours post-injection. If striped bass females are administered mGnRHa, they may be placed in circular tanks with males and allowed to tank-spawn. Two to three males per female is best and spawning will usually occur 54–72 h after transfer of the female to the spawning tank. Once the individual spawns (volitional or manually strip-spawned) and fertilization complete, the broodfish are generally removed from the spawning tank and returned to a larger maintenance tank or pond.

Broodfish feeds have been formulated to have a digestible energy of 13,398 J/g and balanced amino acid compositions: arginine 1.5, histidine 0.7, isoleucine 1.1, leucine 2.0, lysine 2.3, methionine + cystine (total sulphur amino acids, TSAA) 1.0, phenylalanine 1.0, threonine 1.1, tryptophan 0.3 and valine 1.3 (Small *et al.*, 2000).

There are significant gaps in knowledge about population genetics, the molecular basis of hybrid vigour (Garber and Sullivan, 2006) and additive genetic variation within the genus *Morone* (Harrell, 1997). As such, initial attempts at domestication and selective breeding were based on trial and error. The most promising avenue for stock improvement appears to be recurrent selection based on progeny testing. This approach uses DNA markers, initially to track pedigrees and control the rate of inbreeding and later to construct a linkage map for identification and exploitation of qualitative trait loci. Garber and Sullivan (2006) summarize selective breeding techniques, including those based on molecular markers.

The management of striped bass broodstock uses molecular genetic markers (Rexroad *et al.*, 2006). Detailed genetic analyses of striped bass and white bass have been initiated using repeat-enriched striped bass DNA libraries (Couch *et al.*, 2006). Microsatellite markers have been used to develop a cost-effective DNA pooled genotyping technique for use with the large number of loci and large numbers of individuals (Skalski *et al.*, 2006). In combination, these techniques should facilitate further development of linkage maps for use in selective breeding.

Information about environmental requirements of striped bass and its hybrids is available (Tomasso, 1997). Striped bass usually spawn in fresh water; water of low salinity (2–5‰) is beneficial to the survival of eggs and larvae and eggs hatch best when salinity is below 10‰. Optimum spawning temperature is 16.7–19.4°C, but spawning may occur at temperatures as low as 12.8°C. The optimal temperature for egg incubation is approximately 18°C and eggs hatch after 48 h at that temperature (Bayless, 1972). Striped bass eggs and larvae may survive temperatures ranging from 12.8 to 23.9°C, but mortalities may occur at the extremes and there may be total mortality of larvae at temperatures below 10°C and above 26°C. Dissolved oxygen is another important factor in striped bass hatcheries, with survival of eggs decreasing when dissolved oxygen falls below 5 mg/l. Larval striped bass should be cultured in water with a pH of 7.0–8.5.

The larvae depend on nutrients deposited in the yolk sac for the first 5–6 days and exogenous feeding typically begins on day 7 posthatch. The fish are normally stocked into fertilized ponds with plankton blooms when they are ready to feed, with most culturists using ponds for production of juveniles (Harrell, 1997). At 2–10 days after hatch, the fish are stocked into fertilized ponds at 250,000–500,000/ha and the succession of phytoplankton and zooplankton in the ponds is monitored. In fresh water, cladocerans and cyclopoid copepods are important prey for fish that are less than 30 mm in total length (Woods *et al.*, 1985). In brackish water, the small fish usually consume calanoid and harpacticoid copepods (Harrell and Bukowski, 1990).

For larval striped bass (Fig. 15.5) and their hybrids held in intensive aquaculture systems, successful production requires the use of live food during the first 2 weeks of exogenous feeding. Brine shrimp, *Artemia salina*, are offered

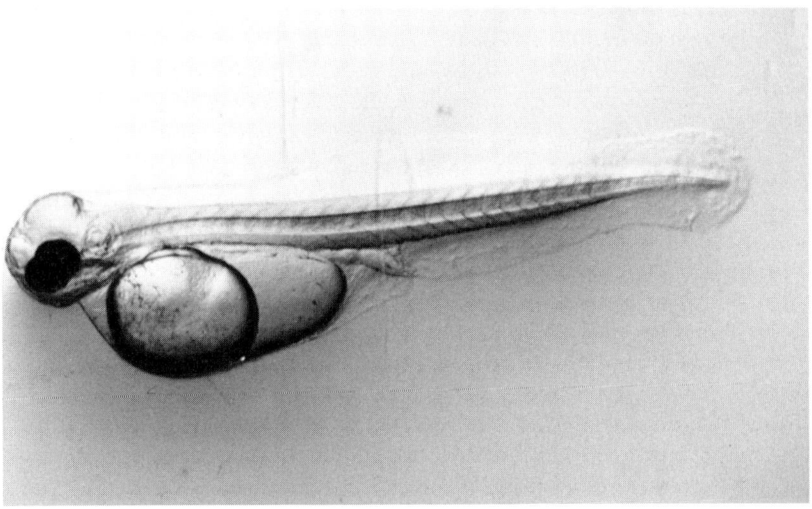

Fig. 15.5. One-day-old larval striped bass.

to striped bass. Sunshine bass, with smaller mouths than striped bass, are offered small zooplankton, such as rotifers, for several days to 1 week and are then offered brine shrimp. Due to the fact that both rotifers and brine shrimp are deficient in some essential fatty acids that striped bass and its hybrids are not able to synthesize (Harrell and Woods, 1995), it is routine to enrich these prey prior to feeding them to the fish.

In intensive systems, it is desirable to wean larvae from live prey to dry feeds as early as possible. Weaning can begin as early as 2 weeks posthatch, but dry feed should not be fed exclusively until 16 days posthatch, when the fish have their full complement of digestive enzymes and a developed stomach. Dry feeds fed to the small fish normally contain 45–55% protein. Some information is available about the nutritional requirements of fish that are in the larval and juvenile stages (Gatlin, 1997).

The juvenile rearing phase begins when the hatchery phase ends, that is, on completion of larval metamorphosis. Requirements for juvenile rearing have been collated and described (Harrell, 1997). If the fish are to be grown in ponds, an understanding of the dynamics of the nutrient–phytoplankton–zooplankton interactions is important, whereas if the fish are held in tanks, attention must be given to the feeds offered and the impact the feeds may have on water quality. The small juveniles are typically raised in 0.25–1.0 ha ponds designed specifically for rapid dewatering. After 30–60 days of rearing, the zooplankton in the ponds will not usually provide adequate food and the fish may become cannibalistic, so the fish should be provided with supplemental feeds at this time. They should then be harvested for grading and restocking for continued growth. Fine-mesh seines are used to move the fish toward a basin or harvest kettle, where fresh water is added to keep the oxygen level high and the water cool.

The second phase of juvenile rearing extends until the fish are 1 year old. The juveniles are stocked at 37,500–50,000 fish/ha (Carlberg *et al.*, 2000). In temperate pond culture, growth will be reduced during winter. To augment growth during the first year, culturists typically feed extruded feeds, with protein levels of 40% or higher, twice daily. This gives increased pond yields, at food conversion ratios of approximately 2:1. Survival during this stage of rearing should be greater than 85%. Juveniles may be stocked when they are 1–3 g and a bar grader can be used to remove potential cannibals. There are publications that describe culture techniques for rearing of striped bass and their hybrids semi-intensively in ponds (Smith *et al.*, 1990; Kelly and Kohler, 1996) and intensively in tanks and raceways (Nicholson *et al.*, 1990).

15.3.4 On-growing to market size

Grow-out occurs in the second growing season, during which the striped bass and its hybrids usually reach marketable size. At the start of grow-out, the fish weigh 90–225 g. Fish are harvested before the end of their second year in intensive tank systems and almost always before the beginning of the third growing season in pond systems. Production is conducted at high densities, with stocking rates being 7500–10,000 fish/ha. Production ponds are larger than those used for juvenile production; they range between 0.5 and 4 ha, but most are 2–2.5 ha (Harrell, 1997).

Striped bass and their hybrids often grow rapidly at temperatures close to 30°C, but best food conversion occurs at temperatures closer to 20°C (Tomasso, 1997). Temperatures between 18 and 32°C may be suitable for rearing juvenile striped bass, but growth is slowed at lower temperatures. Adult striped bass prefer cooler water temperatures (20–23°C) than the juveniles and often become stressed and go off feed when temperature exceeds 25°C. Oxygen concentrations below 4.5–4.0 mg/l result in reductions in food consumption and growth and 5 mg/l should be considered the minimum permissible level for intensive culture of striped bass. Juveniles and adults can survive pH values as low as 6.0, whereas the upper pH limit for striped bass is approximately 10.

Approximately half of cultured hybrid bass are produced in tanks, where it takes 8–13 months to reach market size. The other half is produced in ponds, where it takes 18–24 months to produce a marketable size fish. Striped bass and their hybrids are most often sold at weights ranging from 0.5 to 2 kg. Most of the fish are sold whole, but about 5% are processed into fillets.

During harvesting, the fish are often seined from the production ponds or tanks and are transferred directly to a live-hauling car using a hydraulically operated harvesting basket. These fish, to be sold live, must be handled carefully as any stress or physical damage occurring during harvesting can have significant negative impacts on product quality. Transporting fish on ice is the predominant alternative to live-hauling. Freshly caught fish that are iced immediately and held in ice can maintain a high quality for 8–9 days and be edible for up to 2 weeks (Rawles *et al.*, 1997) (Fig. 15.6).

Fig. 15.6. Fresh market-sized striped bass on ice.

Infectious diseases of the genus *Morone* are described by Plumb (1997). Striped bass and their hybrids respond negatively to improper handling, shipping and the chronic and acute stress associated with poor water quality or other environmental anomalies. These stressors often lead to secondary infections. In view of the fact that there is a lack of approved therapeutic drugs for striped bass and hybrid bass, it is imperative that culturists employ good management practices to maintain the health of their fish. Infectious agents that cause morbidity include viruses, bacteria, fungi, protozoa and metazoan parasites. Important viral diseases of striped bass include lymphocystis, infectious pancreatic necrosis and the striped bass aquareovirus. The most severe bacterial diseases are caused by *Flavobacterium columnare*, *Aeromonas hydrophila*, *Vibrio anguillarum*, *Edwardsiella tarda*, *Photobacterium damsela* subspecies *pisicida*, *Enterococcus* spp., *Streptococcus* spp. and *Mycobacterium marinum*. Secondary fungal infections of importance include *Saprolegnia parasitica*. The most severe parasitic diseases include the protozoans, *Ichthyophthirius multifiliis* and *Amyloodinium ocellatum*. In some cases where estuarine surface waters are used to culture striped bass, attacks by leeches can be a problem and these attacks may be lethal (Woods *et al*., 1990). Larval trematodes (white and yellow grubs), nematodes such as *Philometra* spp. and *Goezia* spp., cestodes and parasitic crustaceans such as *Lernaea* spp. and *Ergasilis* spp. are not uncommon metazoan parasites of striped bass and its hybrids. If left untreated, these parasites can give rise to serious morbidity and/or mortality within the population (Plumb, 1997).

15.3.5 Concluding comments

Striped bass and hybrid bass aquaculture has potential to give significant profit and return on investment. However, due to uncertainties about production costs and market prices, there are concerns within the investment community and this has limited involvement with this species (Lipton and Gempeshaw, 1997).

One of the problems faced by the industry throughout its existence, and repeatedly cited as a cause for failure of commercial ventures, is the lack of domesticated broodstock for genetic improvement (Woods *et al*., 1992; Carlberg *et al*., 2000). During the mid-to-late 1980s, a collapse in the Atlantic coast striped bass fishery led to reductions and bans on fishing, making it difficult to acquire broodstock for aquaculture production. The industry is still dependent on wild, gravid broodfish captured from public waters. Development of better control methods for spawning of broodfish, coupled with better methods for assessment of gamete quality could improve juvenile production efficiency. Industry expansion is also constrained due to the fact that striped bass and hybrid bass can still only be produced on a commercial scale during a few months of each year.

Nutritional studies have been carried out to improve gamete quality and it is also possible to control reproduction in captive striped bass broodstocks (Woods and Sullivan, 1993; Sullivan *et al*., 1997; Mylonas *et al*., 1998).

Progress has also been made in demonstrating exploitable genetic differences among geographically separated populations of striped bass (Leclerc *et al.*, 1996; Wirgin *et al.*, 1997; Roy *et al.*, 2000). The past decade has seen some improvement in production efficiency and there has been an initiation of efforts directed towards domestication (Woods *et al.*, 1999) and genetic improvement of stocks (Rexroad *et al.*, 2006)

Expansion of the industry is, however, constrained by relatively high production costs. The high production costs are related to a number of factors: the requirement for expensive, high-protein feeds, high feed conversion ratios and the susceptibility of the fish to environmental stress and subsequent disease. The challenge is to lower the costs of production significantly, which would allow increased production volume and expansion into new markets. In order to achieve improvements in production efficiency, domesticated broodstock must first be developed following evaluation of strains for important characteristics such as growth rate, disease resistance and tolerance to the environmental conditions found in aquaculture (Carlberg *et al.*, 2000). Improvements in domesticated stocks would then require the initiation of selective breeding programmes that targeted commercially important characters.

References

Bayless, J.D. (1972) *Artificial Propagation and Hybridization of Striped Bass,* Morone saxatilis *(Walbaum)*. South Carolina Wildlife and Marine Resources Department, Columbia, South Carolina.

Brewer, D.L. and Rees, R.A. (1990) Pond culture of phase I striped bass fingerlings. In: Harrell, R.M., Kerby, J.H. and Minton, R.V. (eds) *Culture and Propagation of Striped Bass and Its Hybrids*. Striped Bass Committee, Southern Division, American Fisheries Society, Bethesda, Maryland, pp. 99–120.

Cahu, C. and Zambonino Infante, J. (2001) Substitution of live food by formulated diets in marine fish larvae. *Aquaculture* 200, 161–180.

Carlberg, J.M., Van Olst, J.C. and Massingill, M.J. (2000) Hybrid striped bass: an important fish in US aquaculture. *Aquaculture Magazine* 26, 26–38.

Chatain, B. (1997) Development and achievements of marine fish-rearing technology in France over the last 15 years. *Hydrobiologia* 358, 7–11.

Couch, C.R., Garber, A.F., Rexroad, C.E. III, Abrams, J.M., Stannard, J.A., Westerman, M.E. and Sullivan, C.V. (2006) Isolation and characterization of 149 novel microsatellite DNA markers for striped bass, *Morone saxatilis*, and cross-species amplification in white bass, *Morone chrysops*, and their hybrid. *Molecular Ecology Notes* 6, 667–669.

Fay, C.W., Neve, R.J. and Pardue, G.B. (1983) *Species Profiles: Life Histories of Coastal Fishes and Invertebrates (Mid-Atlantic). Striped Bass*. FWS/OBS-82/11, TR EL-82/11. US Fish and Wildlife Service, Division of Biological Services and Army Corps of Engineers, Washington, D.C.

Garber, A.F. and Sullivan, C.V. (2006) Selective breeding for the hybrid striped bass (*Morone chrysops*, Rafinesque × *M. saxatilis*, Walbaum) industry: status and perspectives. *Aquaculture Research* 37, 319–338.

Gatlin, D.M. III (1997) Nutrition and feeding of striped bass and hybrid striped bass. In: Harrell, R.M. (ed.) *Striped Bass and Other Morone Culture*. Elsevier Science, Amsterdam, pp. 235–252.

Harrell, R.M. (1997) *Striped Bass and Other Morone Culture*. Elsevier Science, Amsterdam.

Harrell, R.M. and Bukowski, J. (1990) The culture, zooplankton dynamics and predator–prey interactions of Chesapeake Bay striped bass, *Morone saxatilis* (Walbaum), in estuarine ponds. *Aquaculture and Fisheries Management* 21, 195–212.

Harrell, R.M. and Woods, L.C. III (1995) Comparative fatty acid composition of eggs from domesticated and wild striped bass (*Morone saxatilis*). *Aquaculture* 133, 225–233.

Harrell, R.M., Kerby, J.H. and Minton, R.V. (1990) Striped bass and hybrid striped bass culture: the next twenty-five years. In: Harrell, R.M., Kerby, J.H. and Minton, R.V. (eds) *Culture and Propagation of Striped Bass and Its Hybrids*. Striped Bass Committee, Southern Division, American Fisheries Society, Bethesda, Maryland, pp. 253–261.

He, S. and Woods, L.C. III (2004) Changes in motility, ultrastructure and fertilization capacity of striped bass *Morone saxatilis* spermatozoa following cryopreservation. *Aquaculture* 236, 667–686.

Kaushik, S.J. (2002) European sea bass, *Dicentrarchus labrax*. In: Webster, C.D. and Lim, C.E. (eds) *Nutrient Requirements and Feeding of Finfish for Aquaculture*. CAB International, Wallingford, UK, pp. 28–39.

Kelly, A.M. and Kohler, C.C. (1996) Sunshine bass performance in ponds, cages, and indoor tanks. *The Progressive Fish-Culturist* 58, 55–58.

Kohler, C.C. (2000) Striped bass and hybrid striped bass culture. In: Stickney, R.R. (ed.) *Encyclopedia of Aquaculture*. Wiley-Interscience, New York, pp. 898–907.

Kohler, C.C., Sheehan, R.J., Habicht, C., Madison, J.A. and Kayes, T.B. (1994) Habituation to captivity and controlled spawning of white bass. *Transactions of the American Fisheries Society* 123, 964–974.

Langdon, C. (2003) Microparticle types for delivering nutrients to marine fish larvae. *Aquaculture* 227, 259–275.

Leclerc, G., Diaz, M. and Ely, B. (1996) Use of PCR-RFLP assays to detect genetic variation at single-copy nuclear loci in striped bass (*Morone saxatilis*). *Molecular Marine Biology and Biotechnology* 5, 138–144.

Lipton, D.W. and Gempeshaw, C.M. II (1997) Economics and marketing. In: Harrell, R.M. (ed.) *Striped Bass and Other Morone Culture*. Elsevier Science, Amsterdam, pp. 315–328.

Logan, H.J. (1968) Comparison of growth and survival rates of striped bass and striped bass white bass hybrids under controlled environments. *Proceedings of the Southeastern Association of Game and Fish Commissioners* 21, 260–263.

Moretti, A., Fernandez-Criado, M.P., Cittolin, G. and Guidastri, R. (1999) *Manual on Hatchery Production of Seabass and Gilthead Seabream*. FAO, Rome.

Mylonas, C.C., Woods, L.C. III, Thomas, P. and Zohar, Y. (1998) Endocrine profiles of female striped bass (*Morone saxatilis*) in captivity, during post-vitellogenesis and induction of final oocyte maturation via controlled-release GnRHa-delivery systems. *General and Comparative Endocrinology* 110, 276–289.

Nash, C.E. and Novotny, A.J. (1995) *Production of Aquatic Animals: Fishes*. Elsevier, Amsterdam.

Nelson, J.S. (1994) *Fishes of the World*. John Wiley and Sons, New York.

Nicholson, L.C., Woods, L.C. III and Woiwode, J.G. (1990) Intensive culture techniques for the striped bass and its hybrids. In: Harrell, R.M., Kerby, J.H. and Minton, R.V. (eds) *Culture and Propagation of Striped Bass and Its Hybrids*. Striped Bass Committee, Southern Division, American Fisheries Society, Bethesda, Maryland, pp. 141–158.

Olsen, Y. (2004) Live food technology of cold-water marine fish larvae. In: Moksness, E., Kjørsvik, E. and Olsen, Y. (eds) *Culture of Cold-water Marine Fish*. Blackwell Science, Oxford, UK, pp. 73–128.

Pickett, G.D. and Pawson, M.G. (1994) *Sea Bass: Biology, Exploitation and Conservation*. Chapman and Hall, London.

Piferrer, F., Blázquez, M., Navarro, L. and González, A. (2005) Genetic, endocrine, and environmental components of sex determination and differentiation in the European sea bass (*Dicentrarchus labrax* L.). *General and Comparative Endocrinology* 142, 102–110.

Plumb, J.A. (1997) Infectious diseases of striped bass. In: Harrell, R.M. (ed.) *Striped Bass and Other Morone Culture*. Elsevier Science, Amsterdam, pp. 271–314.

Rawles, D.D., Fernandes, C.F., Flick, G.J. and Ammerman, G.R. (1997) Processing and food safety. In: Harrell, R.M. (ed.) *Striped Bass and Other Morone Culture*. Elsevier Science, Amsterdam, pp. 329–365.

Rexroad, C., Vallejo, R., Coulibaly, I., Couch, C., Garber, A., Westerman, M. and Sullivan, C. (2006) Identification and characterization of microsatellites for striped bass from repeat-enriched libraries. *Conservation Genetics* 7(6), 971–982.

Rigos, G. and Troisi, G.M. (2005) Antibacterial agents in Mediterranean finfish farming: a synopsis of drug pharmacokinetics in important euryhaline fish species and possible environmental implications. *Reviews in Fish Biology and Fisheries* 15, 53–73.

Roy, N.K., Maceda, L. and Wirgin, I. (2000) Isolation of microsatellites in striped bass *Morone saxatilis* (Teleostei) and their preliminary use in population identification. *Molecular Ecology* 9, 827–829.

Ruelle, R. (1971) Factors influencing growth of white bass in Lewis and Clark Lake. *Transactions of the American Fisheries Society Special Publications* 8, 411–423.

Setzler, E.M., Boynton, W.R., Wood, K.V., Zion, H.H., Lubbers, L., Mountford, N.K., Frere, P., Tucker, L. and Mihursky, J.A. (1980) *Synopsis of Biological Data On Striped Bass,* Morone saxatilis *(Walbaum)*. National Marine Fisheries Service, Washington, D.C.

Skalski, G.T., Couch, C.R., Garber, A.F., Weir, B.S. and Sullivan, C.V. (2006) Evaluation of DNA pooling for the estimation of microsatellite allele frequencies: a case study using striped bass (*Morone saxatilis*). *Genetics* 173, 863–875.

Small, B.C., Soares, J.H. Jr and Woods, L.C. III (2000) Optimization of feed formulation for mature female striped bass. *North American Journal of Aquaculture* 62, 290–293.

Smith, T.I.J., Jenkins, W.E. and Minton, R.V. (1990) Production of advanced fingerling and sub-adult striped bass and striped bass hybrids in earthen ponds. In: Harrell, R.M., Kerby, J.H. and Minton, R.V. (eds) *Culture and Propagation of Striped Bass and Its Hybrids*. Striped Bass Committee, Southern Division, American Fisheries Society, Bethesda, Maryland, pp. 121–139.

Stevens, R.E. (1966) Hormone-induced spawning of striped bass for reservoir stocking. *The Progressive Fish-Culturist* 28, 19–28.

Stickney, R.R. (1996) *Aquaculture in the United States – A Historical Survey*. John Wiley and Sons, New York.

Stickney, R.R. (2000) *Encyclopedia of Aquaculture*. Wiley-Interscience, New York.

Stottrup, J. and McEvoy, L. (2002) *Live Feeds in Marine Aquaculture*. Blackwell Science, Oxford, UK.

Sullivan, C.V., Berlinsky, D.L. and Hodson, R.G. (1997) Reproduction. In: Harrell, R.M. (ed.) *Striped Bass and Other Morone Culture*. Elsevier Science, Amsterdam, pp. 11–73.

Theodorou, J.A. (2002) Current and future technological trends of European seabass–seabream culture. *Reviews in Fisheries Science* 10, 529–543.

Tomasso, J.R. (1997) Environmental requirements and non-infectious diseases. In: Harrell, R.M. (ed.) *Striped Bass and Other Morone Culture*. Elsevier Science, Amsterdam, pp. 253–270.

Wirgin, I., Maceda, L., Stabile, J. and Mesing, C. (1997) An evaluation of introgression of Atlantic coast striped bass mitochondrial DNA in a Gulf of Mexico population using formalin-preserved museum collections. *Molecular Ecology Notes* 6, 907–916.

Woo, P.T.K., Bruno, D.W. and Lim, L.H.S. (eds) (2002) *Diseases and Disorders of Finfish in Cage Culture*. CAB International, Wallingford, UK.

Woods, L.C. III (2005) Striped bass and hybrid striped bass culture. In: Kelly, A.M. and Silverstein, J. (eds) *Aquaculture in the 21st Century*. American Fisheries Society, Bethesda, Maryland, pp. 339–372.

Woods, L.C. III and Sullivan, C.V. (1993) Reproduction of striped bass (*Morone saxatilis*) brood stock: monitoring maturation and hormonal induction of spawning. *Aquaculture and Fisheries Management* 24, 213–224.

Woods, L.C. III, Kerby, J.H., Huish, M.T., Rickards, W.L. and Gafford, G.M. (1981) Circular tank for intensive culture of hybrid striped bass. *The Progressive Fish-Culturist* 43, 199–200.

Woods, L.C. III, Lockwood, J.C., Kerby, J.H. and Huish, M.T. (1985) Feeding ecology of hybrid striped bass (*Morone saxatilis* × *M. chrysops*) in culture ponds. *Journal of the World Mariculture Society* 16, 71–81.

Woods, L.C., McCarthy, M.A., Kraeuter, J.N. and Sager, D.R. (1990) Infestation of striped bass, *Morone saxatilis*, by the leech *Myzobdella lugubris*. In: Perkins, F.O. and Cheng, T.C. (eds) *Pathology in Marine Science*. Academic Press, San Diego, California, pp. 277–282.

Woods, L.C. III, Bennett, R.O. and Sullivan, C.V. (1992) Reproduction of a domestic striped bass brood stock. *The Progressive Fish-Culturist* 54, 184–188.

Woods, L.C. III, Hallerman, E.M., Douglass, L. and Harrell, R.M. (1999) Variation in growth rate within and among stocks and families of striped bass, *Morone saxatilis. North American Journal of Aquaculture* 61, 8–12.

16 Seabreams and Porgies (Family: Sparidae)

Malcolm Jobling and Stefano Peruzzi

University of Tromsø, Tromsø, Norway

16.1 General Introduction

The seabreams and porgies (family Sparidae, Fig. 16.1) are primarily marine and brackish water fishes (a few species occur in fresh water) found around the world in tropical, warm temperate and temperate seas. In general, the seabreams and porgies occur in coastal waters, where they form groups or swim in loose shoals. They are most abundant in the tropical Atlantic, Indian and western Pacific Oceans, but some species are found in the eastern Pacific. This family of perciform fishes comprises 29 genera (e.g. *Archosargus*, *Boops*, *Chrysophrys*, *Dentex*, *Diplodus*, *Lagodon*, *Lithognathus*, *Pagellus*, *Pagrus*, *Rhabdosargus*, *Sarpa* and *Sparus*) with over 100 species in total (Nelson, 1994). Most species reach moderate size, but some can grow to a maximum of 40kg or more. They are important food and game fish throughout their distributional range and several species are cultivated for human consumption.

Cultivation of seabreams and porgies is carried out mostly in South-east Asia and in the Mediterranean region and commercial production is concentrated on two species (Ikenoue and Kafuku, 1992; Nash and Novotny, 1995; Stickney, 2000; Rigos and Troisi, 2005). The two major species are the red seabream, *P. major*, which is the main culture species in South-east Asia, particularly Japan, and the gilthead seabream, *S. aurata*, which dominates seabream production in the Mediterranean. The gilthead seabream is usually marketed as a portion fish, weighing 300–350g. This size is reached after 16–20 months in the cooler parts of the Mediterranean, but production may be reduced to about a year in the warmer waters that occur along the North African coastline.

There is also cultivation of several other species. Aquaculture-related activities have been, or are being, carried out on over 15 species of seabreams and porgies (http://www.fao.org/figis). These activities encompass commercial production of low-to-moderate volumes of fish for sale, the testing of pilot-scale

Fig. 16.1. Sparidae.

commercial systems and R & D and feasibility studies. For example, several species within the genus *Acanthopagrus*, including the black porgy (*A. schlegeli*), yellowfin seabream (*A. latus*) and sobaity (*A. cuvieri*), have attracted attention in Japan, Taiwan and Kuwait. In addition, the goldlined seabream, *R. sarba*, and crimson seabream, *Evynnis japonica*, have aroused interest in some Asian countries, such as Korea, Taiwan and Hong Kong. In the countries of the Mediterranean region, the culture potential of several *Diplodus* species has been examined, as have the common seabream, *P. pagrus*, and the common dentex, *D. dentex*.

16.2 Gilthead Seabream, *Sparus aurata*

In common with most other seabreams and porgies, the gilthead seabream occurs in warmer waters. The gilthead seabream is found in the Mediterranean and Black Seas and in the eastern Atlantic Ocean from southern UK and the Bay of Biscay in the north to the coasts of Senegal and Ghana in the south (Fig. 16.2). The species is encountered only rarely in north European waters. The gilthead seabream is fished commercially and it appears regularly in the fish markets of Mediterranean countries; the fish may be sold fresh or frozen. Within the Mediterranean, the gilthead seabream is just one representative of the over 25 species of the Sparidae that occur there. Most species occur near the coast and any deep-bodied silvery fish seen there is likely to be some kind of seabream. Several descriptions of the biological characteristics and life cycle of the gilthead seabream, and information about cultivation techniques, are available (e.g. Nash and Novotny, 1995; Moretti *et al.*, 1999; Stickney, 2000; Theodorou, 2002).

The gilthead seabream is a shallow-water fish that does not usually venture into water deeper than about 30 m, although adults can be found at 50 m

Fig. 16.2. Natural distribution of *Sparus aurata*.

and sometimes even deeper. The fish tolerates, and may even prefer, brackish water, so it is commonly found in coastal lagoons and estuarine areas. The fish usually occur in small groups over rocky, mud or sand bottoms and they may congregate in large numbers in brackish water during the spring and summer. Gilthead seabream feed mostly on crustaceans and molluscs that occur in or on the bottom sediments. The fish may bury their heads partially into the substrate to remove their prey, which they then crush using their strong molariform teeth.

Although the gilthead seabream occurs in shallow coastal, estuarine and brackish lagoon waters during most of the year, it moves into deeper water during the late autumn. It spends the winter in deeper water. The fish is not very tolerant of low temperatures and there may be considerable die-offs during cold winters. Spawning occurs during the winter. The spawning season commences in late November and continues until early April, but most spawning occurs when day-lengths are at their shortest. The adults return to shallower water after spawning and are joined later by the young-of-the-year juveniles.

The gilthead seabream is a protandrous hermaphrodite, meaning that the fish undergoes a sex change from male to female during the course of its life. The fish first matures as a male and, at the end of a spawning season, may then start to change sex and develop ovaries. Sexual maturity (as males) usually occurs when the fish are 2–3 years of age and 20–30 cm in length. The sex change to female occurs one or more years later, when the fish are 35–40 cm in length. The sex change occurs gradually over the spring and summer and, during early autumn, the fish may complete development into a functional female. Alternatively, the ovarian tissue may degenerate and be reabsorbed, such that the fish continues as a male. The proportion of males that complete sex-reversal is dependent on the sex ratio in the population, a deficit of females leading to an increase in the numbers of males that undergo the sex change. The presence of large numbers of small, young fish, i.e. potential males, increases the proportion of older males that change sex to become females. On the other hand, the presence of a large number of older females inhibits sex-reversal of the younger fish and they continue to be males. If they fail to change sex at the end of one spawning season, the males do not lose their ability for sex-reversal and they may change sex in a subsequent year if placed in an environment that is conducive to the change.

Prior to spawning, the fish undergo changes in body colour and behaviour and there is active courtship behaviour. The females spawn their eggs in batches. Cohorts of eggs undergo final maturation and ovulation in sequence and are spawned on a daily basis. A female can produce 20,000–80,000 eggs/day for a period of up to 4 months, so females can lay 1–3 million eggs during a spawning season. The small, transparent eggs (0.95–1 mm diameter) contain a single large oil droplet, are fertilized externally and are pelagic. Hatching occurs after 65–70 h at 16–18°C, but egg development is inhibited at both low (below 12°C) and high (above 28°C) temperatures. The newly hatched fish are about 2.7 mm in length and they reach 3.5–4 mm by the time the yolk sac has been absorbed at 4–5 days posthatch. At this stage, the eyes are pigmented, the mouth is developed and the young fish commence exogenous feeding on small prey organisms. Inflation of the gas bladder occurs when the fish are 4–5mm in length. Shortly afterwards, there is clearer differentiation of the digestive system, median fin development commences once the fish reach a length of 14–18mm and the other definitive characters are developed once the juveniles are about 30mm in length. The time required for completion of early development and the timing of metamorphosis is very much dependent on temperature and food availability. The transition to the juvenile phase should, however, be complete within 80–90 days posthatch.

Following metamorphosis, the small juvenile fish recruit to shallow inshore areas, where they feed and grow throughout the spring and summer months. The juvenile fish move offshore into deeper water with the approach of winter. The fish then return to shallower water the following spring and repeat the seasonal cycles of onshore–offshore migration until they mature as 2–3-year-old fish. The fish continue these seasonal onshore–offshore migrations as mature adults.

16.2.1 Farming of seabream

Traditionally, gilthead seabream were cultivated extensively in coastal lagoons and saltwater ponds. Limited production volumes were accompanied by high market prices for the farmed fish. Research into intensive culture of the species was given higher priority during the 1980s and culminated in the development of methods for large-scale hatchery production of juveniles. This gave commercial farms access to large numbers of juvenile fish that could be stocked into lagoons, raised in sea cages or held in onshore tanks and raceways for on-growing. Intensive cultivation of gilthead seabream spread rapidly throughout many Mediterranean countries, e.g. Italy, Greece, Spain, France, Turkey and Croatia, and also to countries bordering the Red Sea and Persian Gulf. Cultivation in sea cages became the most popular method for the on-growing of juveniles (stocked at 5–10g) to market size (300–400g) because of the simplicity of the method and the high economic margins it gave. The rapid increase in the expansion of cage farming of gilthead seabream, with the resulting increase in production volume, resulted eventually in a fall in the market price of the fish. Thus, there was a rapid transition from an industry of low volumes and high economic returns to one of high volumes and low margins. As such, the traditional markets became saturated, there seemed to be limited prospects for expansion and production volume reached a plateau around the turn of the century.

The rapid expansion of the farming of gilthead seabream was not accompanied by an equivalent development of diagnostic and veterinary support services and disease outbreaks and parasite infestations often caused major production losses for farmers. Health management remains a matter of major concern for producers, as farmed stock may be subjected to attack by a range of viruses, bacteria, protistan and metazoan parasites (Table 16.1) (Rigos and Troisi, 2005). A general overview of the characteristics, diagnosis and treatment of diseases and parasites encountered in cage-farming of fish is provided by Woo *et al*. (2002).

Table 16.1. Examples of major diseases and parasites that may attack gilthead seabream, *Sparus aurata,* in culture.

	Disease symptoms	Treatment
Viral diseases:		
Lymphocystis (iridovirus)	Tumour-like growths on the body surface	No effective therapy; infected fish may recover within a few weeks
Bacterial diseases:		
Vibriosis (*Vibrio* sp.)	Haemorrhages, dark skin and skin lesions	Vaccination and/or treatment with antibiotics (in medicated feed)
Pasteurellosis (*Photobacterium damselae*)	Acute septicaemia; enlarged spleen and gill necrosis	Antibiotic treatment in medicated feed and/or vaccination
Protistan parasites:		
Amyloodiniosis (*Amyloodinium ocellatus*)	'Dusty' or 'velvet' appearance to gills and skin	Copper sulphate bath treatment
Enteritis (*Myxidium leei*)	Distended abdomen; fluid-filled gut	No effective treatment
Monogenean parasites:		
Endoparasitic 'gut' worms	Gut haemorrhage and inflammation	
Diplectanid monogeneans	Gill necrosis	

Although extensive and semi-intensive culture of gilthead seabream is still practised, production volumes are low compared with those produced in intensive culture. Extensive culture initially was based on the migration of wild juveniles into coastal ponds and lagoons, but this has now generally been replaced by the stocking of these waterbodies with wild-caught or hatchery-reared juveniles (2–3 g) in spring (April–May). The fish feed on the natural prey organisms that occur within the ponds and lagoons and, under these conditions, may reach market size (*c*.350 g) in about 20 months. Production within ponds and lagoons may be intensified by fertilization or supplemental feeding to increase the availability of food. The time between stocking and harvest can be reduced by stocking the ponds and lagoons with larger juveniles held in the hatchery for some months prior to stocking. A variant of the more intensive form of culture involves the enclosure of limited areas of lagoons using large net barriers, the stocking of the enclosures with hatchery-reared juveniles and the provision of supplemental feed in addition to the prey organisms that occur naturally within the enclosure.

Intensive on-growing methods are reliant on a reliable supply of hatchery-raised juveniles that can be used for stocking the rearing units, irrespective of whether they are enclosures, cages placed in the sea or onshore tanks and raceways (Fig. 16.3). In addition, producers that farm using extensive methods are increasingly reliant on supplies of hatchery-raised fish for stocking. The account of intensive cultivation of gilthead seabream given here is based largely

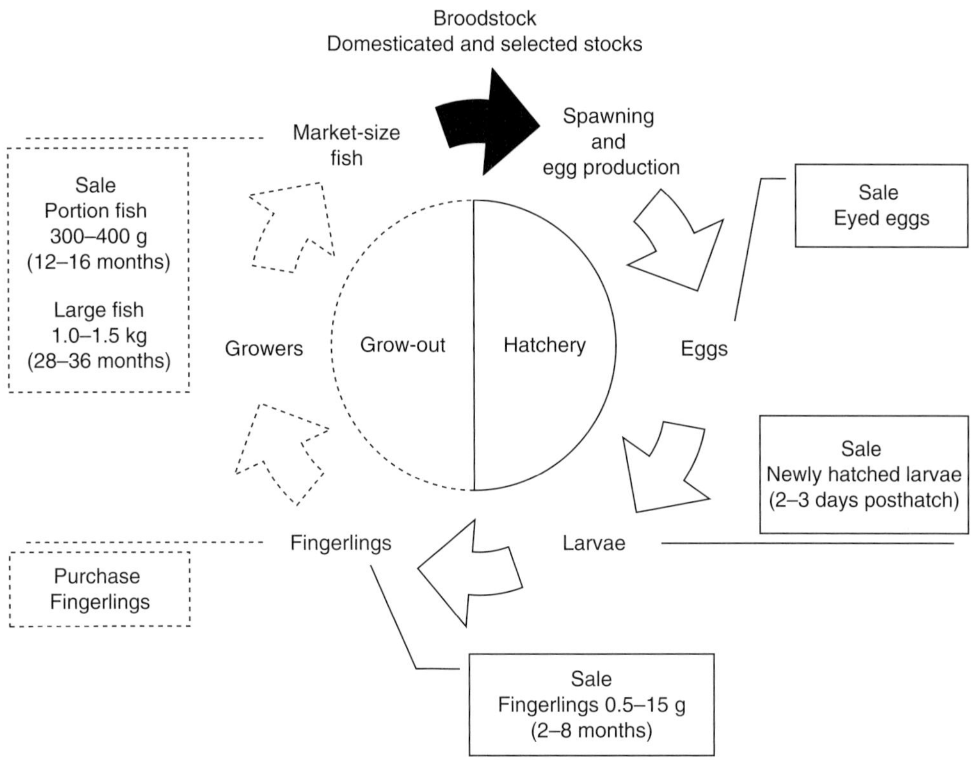

Fig. 16.3. Production cycle of gilthead seabream, *Sparus aurata*.

on information in Nash and Novotny (1995), Moretti *et al.* (1999), Stickney (2000) and Theodorou (2002), combined with supplementary material taken from other cited sources, unpublished reports and restricted-access experience documents.

16.2.2 Broodstock management and hatchery operations

The key points in the hatchery production of juveniles for stocking into on-growing units are the holding and management of the mature broodstock fish, the control of reproduction and spawning, effective incubation of eggs and control over the feeding and early growth of the fish from the time of hatching until they have undergone metamorphosis and have become juveniles (Fig. 16.3).

Hatchery broodstock will be made up of fish of different ages and, since the gilthead seabream is protandrous, the mature male broodstock fish will be younger (age 2–3 years) than the females (4–6 years old). In the wild, the spawning season extends over no more than a few months but, in intensive culture, it is usual for the broodstock to be subjected to environmental manipulation, i.e. photothermal treatments, to extend the length of the spawning season or modify the time at which the fish come into spawning condition. Photothermal manipulation requires the holding of broodstock in tanks equipped with water-cooling and heating systems and where there is also the possibility to regulate lighting conditions. Reproductive development occurs readily at temperatures within the range 14–18°C, but gametogenesis is inhibited at temperatures above 24°C and spawning may not occur when the water temperature is above 20°C or below 13°C. This means that temperature control is required to ensure that broodstock produce gametes of good quality. Further, photoperiod manipulation is required if the aim is to extend the length of the spawning season or to induce a phase-shift in the timing of spawning.

Males that are ready to spawn are easy to identify because they release milt when their belly is palpated gently. In females, however, maturation stage is usually ascertained by examination of oocytes removed from the ovary by means of a catheter. This examination is carried out because post-vitellogenic females may not complete final maturation and ovulate spontaneously. Females whose ovaries contain oocytes with a diameter greater than 500 μm are in post-vitellogenic condition and can be used as spawners. The selected females may then be transferred to spawning tanks along with running-ripe males. Groups comprising two or three females and similar numbers of males will usually give successful spawning and a high rate of egg fertilization; when spawning populations are larger, best results are obtained when the sex ratio is adjusted to 1 male:2 females. The fish may spawn spontaneously when held together in small mixed-sex groups, so hormonal induction of final maturation, ovulation and spawning is not always required. If, however, it is important that all the female broodstock spawn, and/or the aim is to ensure synchrony in spawning, then the females should be given hormonal treatments to trigger final maturation and induce spawning. Spawning may be induced using either intraperitoneal slow-release implants of gonadotropin-releasing hormone analogue (GnRHa) or by giving two injections, 4–6 h apart, of a low dose (1 μg/kg body weight) of GnRHa.

Gilthead seabream eggs float in seawater and can be harvested from the spawning tanks using automatic egg collectors. These may be either overflow collectors placed outside the tank or airlift collectors that float within the tank. The former consist of fine-screened containers that receive a gentle flow of surface water, containing eggs, from the tanks. The main tank outlet, which is situated below the water surface, is screened to prevent loss of eggs via this route. The overflow egg collectors have sufficient water depth to allow the eggs to remain floating. Water circulation within the collectors is maintained by a combination of the flow of water that passes through them and gentle aeration provided by air-diffusers. The airlift egg collector is placed in the spawning tank itself; it is a screened floating box fitted with small airlifts that transfer the surface water, containing eggs, from the tank to the egg collector. The eggs are retained within the collector by the screens when the water passes through the mesh and returns to the tank.

Once they have been removed from the collectors, the fertilized eggs are usually given a surface disinfection with iodophor solution prior to being transferred to incubators. The disinfection reduces the chance of pathogen transmission to the eggs and larvae. The incubators are usually cylindroconical tanks of 100–250 l capacity. An aeration system placed close to the bottom of the tank ensures the water circulation is adequate to keep the eggs in suspension. A second aeration system placed close to the surface outlet screen prevents the eggs from impinging on the outlet screen and clogging it. There is a gentle inflow of water that ensures that water quality is maintained, aids in prevention of water stratification and assists in keeping the eggs afloat. Eggs are usually stocked at 6000–15,000 eggs/l and they will hatch after about 65–70 h at 17–18°C. Eggs can also be incubated directly in the tanks that are to be used for larval rearing, but then the stocking density will be lower (1000–10,000 eggs/l).

The hatchlings are 2.5–3 mm in length and after about 4 days, they are ready to start feeding. The earliest stages of rearing are usually carried out in round 6–10 m^3 tanks provided with either seawater or water of slightly lower salinity (25‰). Temperature will be maintained initially at about 18°C. Initial stocking density may be 150–200 fish/l. The photoperiod will usually be set at 16L:8D and light intensity at the water surface will be within the range 1000–3000 lux. As time progresses, water temperature is raised gradually to about 24°C, a 20L:4D or continuous light regime is introduced and light intensity is reduced to 500–1000 lux during the latter stages of larval rearing. Initially, there is little or no renewal of water and circulation is maintained by aeration from the tank bottom. Later, an inflow of water is introduced to ensure that the water in the rearing tanks is of sufficiently high quality to promote survival of a large proportion of the fish.

Protocols for the production of larval seabream are generally based on live feed organisms and the first food offered to the hatchlings will usually be rotifers, *Brachionus plicatilis*; rotifers are chosen due to a combination of being of suitable size and because of the relative ease with which they can be cultured in large numbers. After 10–11 days, the rotifers will be supplemented with, and eventually replaced by, brine shrimp, *Artemia salina*, nauplii and metanauplii. Neither rotifers nor brine shrimp are ideal foods for the young fish and both are enriched routinely with essential fatty acids and vitamins to improve their nutritional value. In addition, microalgae (e.g. *Nannochloropsis* spp., *Chlorella* spp., *Isochrysis galbana* and *Dunaliella tertiolecta*) may be used in rotifer production. Algae may also be added directly to the larval rearing tanks. This is known as the green-water production method. Information about the cultivation of live prey organisms for fish hatchlings and juveniles is available from several sources (e.g. Ikenoue and Kafuku, 1992; Nash and Novotny, 1995; Moretti *et al.*, 1999; Stickney, 2000; Stottrup and McEvoy, 2002; Olsen, 2004). Towards the end of the larval rearing period, inert feed is introduced; starting at particle sizes of 80–200 μm and then increasing to 150–350 μm as the fish progress through metamorphosis. Although larval rearing protocols are based on live prey organisms, continued efforts are being made to develop formulated microfeeds that can replace live feeds at progressively earlier stages of larval rearing (Stickney, 2000; Langdon, 2003). Metamorphosis occurs at around 32–35 days and the fish are held in the larval rearing tanks for an additional 10 days or so before being transferred to nursery tanks for weaning on to dry feeds.

The nursery tanks are larger (10–25 m^3) than the larval rearing tanks and the fish are held in these tanks from about day 45 posthatch until they are about 2–5 g and are ready for transfer to on-growing units (Fig. 16.3). Towards the end of the nursery phase, the stocking density in the tanks may be 20 kg/m^3 or so. During nursery rearing, the temperature is usually 18–22°C and the salinity of the water (20–25‰) is below that of seawater. Live feed may be supplied for the first few days after transfer of the fish to the nursery tanks, but live feed is soon replaced by dry feed as weaning progresses. Dry, weaning feeds are initially particulates of 150–350 µm in size and particle size is increased gradually as the fish grow. The feed is usually distributed using automatic feeders programmed to deliver feed at regular intervals; regular feed delivery is required to reduce the risk of high levels of cannibalism. In addition, size grading will be required if a marked size heterogeneity develops within the tanks.

Buyers of fingerlings (2–15 g fish) will be provided with documentation relating to source and quality. The documentation will also provide information about whether the fish have been given any special treatments during the nursery phase. Special treatments may include the provision of feeds with additional vitamins, probiotics or immunostimulants and vaccination against certain pathogens.

Selective breeding of gilthead seabream is being carried out in the main producing countries. There appears to be significant additive genetic variation for traits of economic importance such as body weight, external colour and spinal deformities and there may be up to 23% growth advantage for first-generation selected fish in comparison with controls (Thorland *et al.*, 2006). Nevertheless, despite the awareness of the benefits of genetics as a tool for enhancing productivity and competitiveness, the seabream aquaculture industry is still at a very early stage of developing domesticated and selectively bred stocks. As is the case for other cultured fish species, key decisions for the instigation of selective breeding programmes relate to the uncertainties about the potential long-term returns on investment in a labile market, even though cost–benefit projections may appear attractive.

In the farming of some types of fish, hybrids have been shown to display growth, quality or survival superiority compared to the parental species, or hybrids express particular combinations of traits advantageous for production purposes. Sterile diploid and triploid hybrids have been developed between several species of seabreams and porgies, including the gilthead seabream, *S. aurata*, the red seabream, *P. major*, the red porgy, *P. pagrus*, and the common dentex, *D. dentex* (Gorshkov *et al.*, 2002, and references therein). Most of the reciprocal crossings failed to show superior performance with few production benefits and hybridization has encountered limited commercial success.

16.2.3 On-growing to market size

On-growing of gilthead seabream may be carried out in land-based tanks or raceways, or in sea cages; on-growing in sea cages is the most common method in the Mediterranean region. Floating cages are used in sheltered sites, whereas semi-submersible cages are the rearing unit of choice at very exposed sites. Tanks and raceways vary considerable in volume (200–3000 m^3), depending on the demands of production at particular farms and on the type of rearing strategy employed.

On-growing in sea cages is simpler and, generally, more economical than on-growing in land-based tank and raceway systems. In cage rearing, there is, for example, no need to pump water and there are no costs for aeration and for the treatment of incoming water. It is, however, not possible to control temperature and other water quality parameters in cages. Stocking densities are usually lower in cages (10–15 kg/m^3) than in tanks and raceways (15–50 kg/m^3). This means that the productivity of cages is generally lower than in tanks and raceways and the production cycle is usually longer. Juveniles stocked into cages at 10 g will usually need about 1 year to reach the harvest size of 350–400 g, whereas juveniles stocked at 5 g will grow to 350–400 g in about 16 months. By comparison, under favourable temperature conditions (18–26°C), seabream stocked into tanks and raceways as 5 g juveniles might reach market size in less than 1 year. Although most gilthead seabream are harvested, size graded and marketed as portion-size fish, some are cultured to sizes of up to 1.5 kg. The smaller fish are distributed through shops and supermarkets, whereas the larger fish have greater appeal for caterers and restaurants. Raising the fish to larger size is also carried out to increase the range of products made available to the market, as the larger fish can be processed to be sold as fillets (Fig. 16.3).

During on-growing, the fish will be provided with dry pellet feeds, given either by hand or by centrally-controlled automatic feeders that dispense feed at intervals. Feeding activity decreases during the winter months and may cease if temperature falls below 10°C. Reduced feeding during winter increases the time required to complete the production cycle. Low winter temperatures may also lead to the appearance of a pathological condition termed winter syndrome or winter disease (Tort *et al.*, 1998, 2004). The low temperatures seem to induce immunological suppression and a range of physiological disturbances. Affected fish are characterized by abnormal swimming behaviour, abdominal distension and the formation of lesions in several internal organs. There may be considerable die-off of affected fish if they are handled or subjected to other forms of stress.

Work on the nutritional requirements of the species is ongoing and feed formulations are amended as new information becomes available. At present, most of the commercial feeds used in the farming of seabream are relatively high-energy extruded pellets, with 45–50% protein, around 20% fat and 20% carbohydrate. The major limitation on including fat at higher concentrations is the effect on the body composition of the fish; excessive amounts of fat in the fish reduce their appeal to consumers and result in a lower market price (Koven, 2002). Attempts are also being made to develop special winter feeds to reduce the incidence and alleviate the symptoms of winter syndrome (Tort *et al.*, 2004).

Feed is withdrawn some time before the fish are harvested, the length of time depending on temperature; one day without feeding may be sufficient for the fish to evacuate their gut at 25°C, but at lower temperatures, the fish may be held without feed for 48–72h, or perhaps longer. For harvesting, the fish are first confined to a restricted area of the rearing unit and are then removed by dip-netting or by pumping for slaughter. The fish may be killed by putting them in tubs containing iced water saturated with carbon dioxide; acute exposure of the fish to low temperature results in death due to a thermal shock. Alternative methods are under investigation in an attempt to reduce the stressful effects of harvesting and slaughter on the fish. Following harvesting and slaughter, the fish are processed as rapidly as possible and packed on ice for transport to fish markets and other sales outlets.

16.2.4 Concluding comments

The great expansion of farming of gilthead seabream in the Mediterranean region took place during the latter part of the 20th century and, by the turn of the century, the industry had transformed from one of low volumes and high profit margins to one of high production volumes and minimal profits. In the period 1999–2004, annual production volumes were of the order 70,000–95,000t (http://www.fao.org/figis). The industry is in transition; developing towards maturity, but still being an industry that is in need of more efficient production systems and more sophisticated methods for developing and marketing its products. Further, there is also a need for diagnostic and veterinary services and for the development of effective disease treatments (Rigos and Troisi, 2005). Most of the traditional markets in Mediterranean countries are saturated; for expansion to occur, new markets must be sought and developed and the potential for product diversification explored more systematically.

References

Gorshkov, S., Gorshkova, G., Hadani, A., Gordin, H. and Knibb, W. (2002) Chromosome set manipulations and hybridization experiments in gilthead seabream (*Sparus aurata*). II. Assessment of diploid and triploid hybrids between gilthead seabream and red seabream (*Pagrus major*). *Journal of Applied Ichthyology* 18, 106–112.

Koven, W. (2002) Gilt-head sea bream. In: Webster, C.D. and Lim, C.E. (eds) *Nutrient Requirements and Feeding of Finfish for Aquaculture*. CAB International, Wallingford, UK, pp. 64–78.

Ikenoue, H. and Kafuku, T. (1992) *Modern Methods of Aquaculture in Japan*. Elsevier, Amsterdam.

Langdon, C. (2003) Microparticle types for delivering nutrients to marine fish larvae. *Aquaculture* 227, 259–275.

Moretti, A., Fernandez-Criado, M.P., Cittolin, G. and Guidastri, R. (1999) *Manual on Hatchery Production of Seabass and Gilthead Seabream*. FAO, Rome.

Nash, C.E. and Novotny, A.J. (1995) *Production of Aquatic Animals: Fishes*. Elsevier, Amsterdam.

Nelson, J.S. (1994) *Fishes of the World*. John Wiley and Sons, New York.

Olsen, Y. (2004) Live food technology of cold-water marine fish larvae. In: Moksness, E., Kjørsvik, E. and Olsen, Y. (eds) *Culture of Cold-water Marine Fish*. Blackwell Science, Oxford, UK, pp. 73–128.

Rigos, G. and Troisi, G.M. (2005) Antibacterial agents in Mediterranean finfish farming: a synopsis of drug pharmacokinetics in important fish species and possible environmental implications. *Reviews of Fish Biology and Fisheries* 15, 53–73.

Stickney, R.R. (2000) *Encyclopedia of Aquaculture*. Wiley-Interscience, New York.

Stottrup, J. and McEvoy, L. (2002) *Live Feeds in Marine Aquaculture*. Blackwell Science, Oxford, UK.

Theodorou, J.A. (2002) Current and future technological trends of European seabass-seabream culture. *Reviews in Fisheries Science* 10, 529–543.

Thorland, I., Papaioannou, N., Kottaras, L., Refstie, T., Papasolomontos, S. and Rye, M. (2006) Family based selection for production traits in gilthead sea bream (*Sparus aurata*) and European sea bass (*Dicentrarchus labrax*) in Greece. Book of Abstracts. IAGA IX, International Symposium on Genetics in Aquaculture, Montpellier, France 26–30 June 2006, pp. 104.

Tort, L., Padrós, F., Rotllant, J. and Crespo, S. (1998) Winter syndrome in the gilthead sea bream *Sparus aurata*. Immunological and histopathological features. *Fish and Shellfish Immunology* 8, 37–47.

Tort, L., Rotllant, J., Liarte, C., Acerete, L., Hernández, A., Ceulemans, S., Coutteau, P. and Padros, F. (2004) Effects of temperature decrease on feeding rates, immune indicators and histopathological changes of gilthead sea bream *Sparus aurata* fed with an experimental diet. *Aquaculture* 229, 55–65.

Woo, P.T.K., Bruno, D.W. and Lim, L.H.S. (2002) *Diseases and Disorders of Finfish in Cage Culture*. CAB International, Wallingford, UK.

17 Tilapia (Family: Cichlidae)

Kevin Fitzsimmons[1] and Wade O. Watanabe[2]

[1]*Environmental Research Lab, University of Arizona, Tucson, USA;*
[2]*Center of Marine Science, University of North Carolina at Wilmington, Wilmington, USA*

17.1 General Introduction

The Cichlidae (Fig. 17.1) are a family of between 1400 and 2000 species in 220 genera that are native to Africa and South and Central America. Cichlids are perciform fishes and are important as food fish, ornamentals and as the primary lacustrine fishes in most African lakes. Cichlids have been heavily studied by evolutionary biologists examining the species clusters that have developed in the Rift Lakes of East Africa. Although cichlids are found most commonly in lakes and ponds, they are also seen in rivers and even coastal estuaries (Fig. 17.2). For the most part, they are omnivores; however, many species have evolved specialized feeding habits as part of the speciation seen in the Rift Lake species clusters.

Cichlids represent one of the most popular groups of aquarium fishes including the freshwater angelfishes, Oscars, discus and many of the Rift Lake species. The tilapias, primarily the genus *Oreochromis*, are the second most important group of farmed fishes. In 2006, almost 2,400,000 Mt were produced from farms in over 100 countries (Table 17.1). It is estimated that an additional 600,000 Mt were harvested from the wild (Fig. 17.3).

Cichlids, and especially tilapia, are used frequently in integrated aquaculture–agriculture. Multiple use of water for fish and plant crops, with nutrient recycling, increases farming efficiency, which must be achieved if we are to feed increasing populations with limited resources. The oldest and most diversified integrated systems come from China, the Indian subcontinent and South-east Asia. It is probably not a coincidence that this very efficient form of agriculture is found in the most heavily populated regions of the planet. It is generally accepted that the healthiest human diets are those high in complex carbohydrates, vegetables and fish. This corresponds to the rice, vegetable and fish diets common throughout Asia. Likewise, these farms are highly efficient at producing large amounts of food with minimal inputs of non-renewable

Fig. 17.1. Cichlidae.

resources. This healthy diet and effective use of natural resources contributed to the development of some of the world's oldest civilizations, supports much of the world's current population and indicates a direction for reforming non-sustainable agricultural practices.

17.2 Tilapia, *Oreochromis* sp.

Tilapia, several closely related cichlid species, are the world's second most important farmed food fish after the carps. People in Africa have caught these fish in the wild for millennia. Paintings from Egypt depict tilapia being held in pens for the Pharaohs. Tilapia is a common name that is now applied to several genera and species of fish that were formerly classified in the genus *Tilapia*, within the family Cichlidae. In the reclassification scheme developed by Trewavas (1983), the several hundred species of *Tilapia* were split into three genera, *Oreochromis*, *Sarotherodon* and some remained as *Tilapia*. The *Oreochromis* are maternal mouthbrooders, the *Sarotherodon* are paternal mouthbrooders and the *Tilapia* are substrate spawners. The species that are most commonly reared in aquaculture are in the genus *Oreochromis*. These include the Nile tilapia, *O. niloticus*, the Mozambique tilapia, *O. mossambicus*, the blue tilapia, *O. aureus*, and *O. urolepis hornorum*, sometimes called the Wami River tilapia. These species will all hybridize readily in captivity. There are now many strains of the parent species, along with many hybrid strains, available to growers.

Domestication of the tilapias began in the 1950s and 1960s with development in several countries. Tilapia have been important to aquaculture because of the ease with which they can be bred in captivity and the wide variety of water conditions in which the fish will grow. Various strains can be grown in water varying in salinity from fresh water to full strength seawater (35‰). They will grow in water ranging from acidic (pH of 5) to alkaline (pH of 9). Tilapia can survive low dissolved oxygen (< 2 mg/l) and high ammonia levels

Fig. 17.2. World distribution of *Oreochromis* sp.

Table 17.1. Global list of significant tilapia producers.

Producing country	Yearly production (Mt in 2006)
China	1,110,000
Egypt	250,000
Indonesia	200,000
Philippines	185,000
Thailand	130,000
Mexico	100,000
Brazil	100,000
Taiwan	72,581
Colombia	38,656
Ecuador	36,000
Honduras	30,000
Vietnam	25,000
Cuba	20,000
Costa Rica	20,000
Malaysia	15,000
USA	9,000
Others	40,000
GLOBAL TOTAL	2,381,237

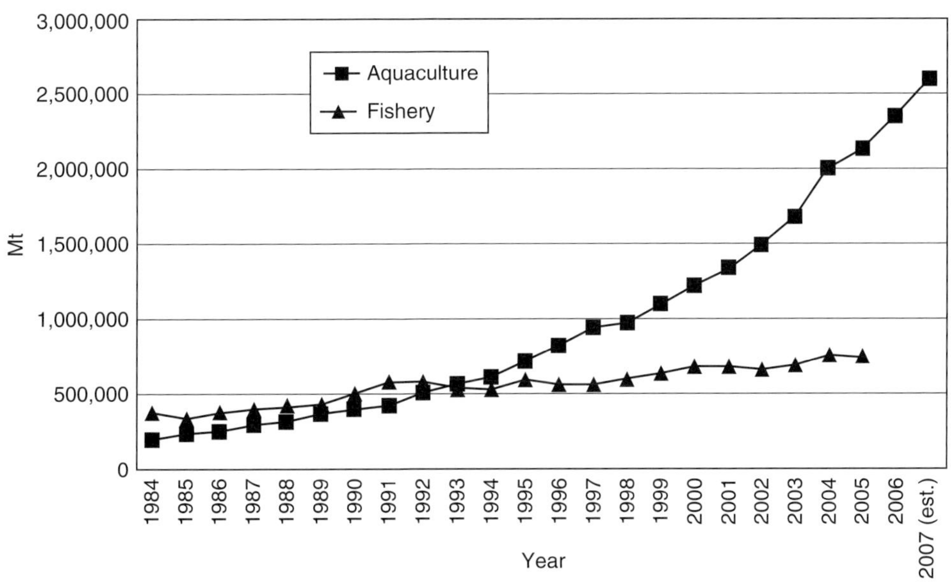

Fig. 17.3. Global tilapia supply.

(50 mg/l) for longer periods than most other fish. Consequently, they can be grown in densities greater than virtually any other kind of fish. These characteristics make them ideal for aquaculture.

Another characteristic that facilitates selective breeding and domestication is their reproductive behaviour. The tilapias used in aquaculture are maternal mouthbrooders. A female lays her eggs in a simple nest prepared by the male, the male fertilizes the eggs, then the female picks the eggs up and incubates them in her mouth. Even after eggs hatch, fry will remain in the mother's mouth. Once the fry are free-swimming, they will return to her mouth for protection. Females can produce several hundred to several thousand young per spawn. The high level of parental care allows breeders to raise thousands of young quickly for directed selection or for stocking into production units. Another advantage is that the adults become sexually mature in less than 6 months, when they are still a fraction of their potential size. This is an additional advantage for selective breeding, allowing several generations to be produced in the time it takes other fish to reach maturity. The drawback to this high potential for reproduction is that tilapia introduced to new (exotic) locations can spread quickly and impact native fish populations. Likewise in production ponds without predators, tilapia can overpopulate, producing large numbers of small, stunted fish. This can present a serious problem for aquaculturists who are attempting to rear a large size fish for market. Several methods are used to avoid overpopulation and stunting and are described below.

Eggs of tilapia are relatively large and fry are hardy and omnivorous. Fry feed readily on a variety of foods including periphyton and phytoplankton, zooplankton and powdered feed. This allows the culturist to manipulate spawning further by removing the young from the female and raising them independent of the mother. Removal of fry will encourage the female to begin eating again, she eats little while brooding, and be ready to spawn in a few weeks. Sex of fry can be manipulated in several ways. Undifferentiated sexual organs of juvenile tilapia can be induced to produce phenotypic all-male or all-female populations. Males grow more rapidly and crops of primarily males will avoid problems associated with unwanted spawning. There are several methods and reasons for this 'sex-reversal', covered in detail below.

Another reason that tilapia are prized as aquaculture species is because they are herbivorous or omnivorous, depending on the species. In nature, tilapia graze on algae, higher plants, detrital matter and/or small invertebrates. This makes it easy to grow the fish in ponds with minimal inputs of feed or fertilizer in extensive aquaculture. If semi-intensive systems are used to generate greater production from a facility, fertilizers can be used to produce algae and zooplankton. In intensive production, feeds containing primarily plant proteins can be fed. Consuming herbivorous fish is a more economical and ecologically efficient transfer of energy and protein to human consumers than using carnivorous fish that require fish or other animal proteins in their diets. This has made tilapia one of the preferred fishes for environmental groups opposed to the use of fishmeal in aquaculture.

17.2.1 Broodstock management and hatchery operations

From a practical point, there are three major issues regarding tilapia reproduction. First, control of reproduction for purposes of domestication and other directed breeding. Second, elimination of excessive breeding to grow a controlled number of fish in a rearing unit. Third, all or predominantly male fish are preferred in production systems.

Typically, fish breeders isolate young male and female broodstock. When the fish have reached the desired size, they can be stocked into a spawning unit. Males will exhibit species-specific spawning coloration. Due to the large number of hybridization events in the past, these colour patterns will be intermediate in hybridized strains. The optimal temperature for breeding is between 20 and 35°C, depending on species. Breeding will occur at a temperature higher than 20°C and spawning will occur at a temperature higher than 22°C.

Typically, a male will excavate a nest from the bottom sediment. In many culture settings, however, breeders are stocked into tanks in which there is a solid surface, where the male will scrape away the periphyton, leaving a cleared area delineating the nest. A female lays her eggs in the nest, the male fertilizes the eggs and then the female picks her eggs up and incubates them in her mouth.

Many culturists will collect fertilized embryos from the female at this point. If a female can be netted with minimal disturbance, she will continue to hold the eggs and a wash bottle can be used to rinse eggs from the buchal cavity. Eggs collected in this manner are typically hatched in a MacDonald jar, or various round-bottomed containers with an overflow spout. Eggs are ovoid or pear-shape, they measure between 2.8 and 4.3 mm and are negatively buoyant. The optimal temperature range for egg incubation is 25–32°C, as for newly hatched larve. Hatched fry swim up and out and are accumulated in a screened tray. One of the concerns encountered when incubating eggs is the prevalence of *Saprolegnia* fungal infections. Fungus tends to grow first on unfertilized or dead eggs, spreading to adjacent viable embryos. The most effective treatment is to remove infected eggs carefully by hand on a daily basis. Keeping eggs in motion is also effective, as the fungus cannot spread easily to eggs in motion. There are several chemicals that are known to be effective for controlling fungus, including potassium permanganate, formalin and salt. Regulations vary on how these compounds are used. After the embryos hatch, fry can survive 8–10 days on the yolk sac before they begin feeding on phytoplankton, periphyton or powdered feeds. Survival rate from hatching is usually between 80 and 90%.

Many small-scale culturists prefer to allow the female to continue to incubate the eggs. Even after eggs hatch and absorb the yolk sac, free-swimming fry will still return to the mother's mouth for protection. Tilapias of the genera *Tilapia* and *Sarotherodon* are more likely to be nest builders and care for eggs laid in the nest. However, they also invest in the care of the young by producing large eggs and then protecting the embryos and fry in the nest. Females produce several hundred to several thousand young per spawn. The high level of parental care allows breeders to raise thousands of young quickly for directed selection or for stocking into production units.

Precocious juveniles can become sexually mature in less than 6 months, when they may weigh less than 50 g. This can be an additional advantage for selective breeding, allowing many generations to be produced in the time it takes other fish to reach maturity. However, most of the sophisticated large-scale breeding programmes developing stocks for international-scale breeding programmes use fully grown broodstocks to determine desirable growth characteristics through the production cycle. The drawback to precocious sexual development and a high potential for reproduction is that tilapia released in exotic locations can spread quickly and impact native fish populations. Likewise in ponds with no predators, tilapia can overpopulate and end up with large numbers of stunted, unmarketable fish.

Effective hatcheries are a prerequisite for establishment of an aquaculture industry. Hatchery success or failure is subject to the technique and skill of the breeder, the quality of the fish, the food and the water available. Simple pond spawning is sufficient in many locations. A shallow pond is built and the males and females are stocked and allowed to spawn at random. Typically, three or four females will be stocked for each male. Seine nets are used on a regular basis to capture the young fish, which are then moved to another pond, nursery tank or hapa net (fine meshed nets that are suspended from a frame in a body of water). Young fish, usually less than 1 g and only a few days old, will be collected and placed in the hapa for initial feeding and protection from predators. A variation used in large hatcheries and for selective breeding programmes is to place the spawners in tanks or hapas suspended in ponds. After the fish spawn, the eggs or fry are collected directly from the mouth of the female, or the young can be collected as soon as they are swimming away freely from the mother. Eggs and fry removed from a brooding parent can be reared in artificial systems described above.

Development of genetically male tilapia, sex-reversed fry and sophisticated selective breeding programmes using dozens of family lines have contributed to a rapid improvement in the growth rates and harvest size of farmed tilapia. Domestication of tilapia is still in the earliest stages and already we are witnessing tremendous improvements, with more to follow in the near future.

One technique developed to generate all-male populations is the use of hybrids. Certain hybrid crosses, *O. aureus* × *O. mossambicus* and *O. aureus* × *O. urolepis hornorum*, result in a skewed sex ratio favouring males. The technique used most commonly to produce all-male populations is to sex-reverse fry. Newly hatched fry have undifferentiated gonads. By including a hormone in the feed, or by immersion in a solution containing a hormone, fry can be induced to develop morphologically as male or female. The normal technique is to feed methyltestosterone to fry for 28 days, which is sufficient to induce most, if not all, to develop testes. The small number that do not reverse will develop as females or hermaphrodites.

Another approach involves genetically male tilapia (Mair *et al.*, 1997). In this process, selected fry are fed a feminizing hormone, oestrogen, yielding an all-female population. Genetically male but phenotypic female fish are then bred to normal males, yielding a brood of fry with a normal distribution of 1/4 XX, 1/2 XY and 1/4 YY progeny. The YY males can then be found by progeny testing and, once identified as such, can be sold to other hatcheries and bred to selected females. The YY males bred with normal XX females should yield 100% XY or normal males. The primary benefit of this technique is that the production fish are all male and have never been treated with any hormone. Although the fish have been trademarked as Genetically Male Tilapia by

FishGen of the UK, they are not genetically engineered. Transgenic tilapia have been produced in the UK and Cuba (Martinez *et al.*, 1999), but have not been released outside the lab.

17.2.2 On-growing to market size

Tilapia are reared in a greater variety of production systems and environments than any other aquaculture product (Watanabe *et al.*, 2002). The optimal temperature for on-growing is between 25 and 30°C, depending on species. Using extensive aquaculture methods, the fish can be grown in small ponds or lakes with no additional inputs, depending solely on the primary productivity of the system. Yield per hectare may be small, but so is investment. More intensive systems use increasingly greater inputs. The acadja systems, developed in western Africa, incorporate stakes or poles driven into the bottom mud of ponds. The stakes provide substrate for the attachment of algae and bacteria, which tilapia will graze, increasing productivity without fertilizing.

Fertilizer is an additional input that increases fish yield greatly. Input of organic or inorganic fertilizers increases production of algae and then invertebrates and bacteria that graze on or decompose algae, respectively, with tilapia grazing on algae, invertebrates and bacteria. A good fertilization programme can increase the tilapia yield of a pond from several hundred kg/ha/year to several thousand kg/ha/year. Egna and Boyd (1997) provide a thorough review of pond dynamics and aquaculture.

Cage culture is a still more intensive method of rearing tilapia. Harvest densities can reach 169 kg/m^3 (Carro-Anzalotta and McGinty, 1986). Cages can be constructed out of simple bamboo poles and nets or sophisticated materials including steel, plastic and composites. By increasing the density of fish and keeping them concentrated, the farmer has better control over feeding, can reduce unwanted reproduction and can simplify harvest. Cages are especially useful for producers who use public or communal waters including village ponds, lakes, bays or irrigation systems. In many countries, cage culture operations provide jobs for thousands as members of cooperatives or company employees, with fish produced for domestic consumption and international trade.

Intensive flow-through pond or raceway farms have been built in many countries. Ponds of 1 ha or less, raceways or tanks are built with complete exchanges of water measured in hours. Supplemental aeration may be provided with paddle wheels or air injection. These farms are especially attractive in areas that can recover aquaculture effluent water for field crop irrigation. Wastes from the fish provide nutrients for field and tree crops and the producer does not need to worry about polluting the environment.

Other forms of integrated tilapia culture have been developed in Asia based on traditional carp culture. In these systems, agricultural wastes from a farm are used to fertilize ponds. Afterwards, pond water and nutrient-rich sediments are used in vegetable gardens. In rice-growing areas from South-east Asia across the Indian subcontinent to Egypt, tilapia are often grown in rice paddies. The fish help to control insects, aquatic weeds and algae that compete with the rice for fertilizers. Tilapia can then be harvested with the rice, yielding an additional edible crop from the same field. The channels used to drain water from the field even provide a convenient area in which to capture fish.

Integration of tilapia culture with hydroponic vegetable production is a high technology version of an integrated system. Hydroponic beds can be used to filter wastes from the tilapia and water can be returned to the fish or discharged (Rakocy and Hargreaves, 1993). As plants thrive on the nitrogenous wastes of the fish and the roots of the plants support bacteria that will filter the water further, this ecological system has become popular both as an efficient food production system and as a teaching tool in many schools. In Europe and North America, greenhouses that support year-round fish and plant production can be used to supply live tilapia, fresh herbs and vegetables.

Markets for live tilapia have led to the development of the most intensive of all aquaculture systems. Highly engineered systems recirculate all the water using a variety of physical and biological filtration systems to maintain water quality and retain heat. Fish are reared in concrete, fibreglass or plastic lined tanks. These systems represent the most intensive systems, requiring large investments of capital, technology and rearing skill. The cost of production is high, but market prices for live fish justify the investment (Watanabe *et al.*, 2002).

Much of the world's current population and economic growth is concentrated in arid and semi-arid zones. In those regions, multiple use of water for aquaculture and plant crop production is desirable for environmental, economic and even socio-political reasons. The seemingly incongruous concept of growing fish in a desert is directly related to the value of the limited water resource and the need to make the most efficient use of that resource. Tilapia are especially well suited to being grown in irrigation production systems (Mires *et al.*, 1990; Redding and Midlen, 1990; Fernando and Halwart, 2000) or in drainage waters collected after irrigation (Siddiqui *et al.*, 1991).

Lieberman and Shilo (1989) reported that 60% of the fish cultured in Israel were grown in irrigation reservoirs, with the water being used during the summer to irrigate field crops including citrus, grains and vegetables. Experimental production of tilapia in cages in irrigation canals demonstrated that yields of 40 kg/m^3 could be achieved with estimates of 100 Mt/year from a 1 ha field of cages placed in flowing canal water (Ishak, 1982, 1986). Shelton (Survey of aquaculture resources/activities in the Delta Region of Egypt, 1989) reported that an intensive tilapia culture facility was being constructed in Egypt to examine the use of aquaculture effluents for irrigation of citrus and bananas. Al-Jaloud *et al.* (1993) used tilapia farm effluent to grow wheat in Saudi Arabia. A 50% reduction in chemical nitrogen application could be achieved by irrigating with effluent containing 40 mg N/l.

Aquaculture effluents can be significant pollutants in the environment if they are not managed correctly (Boyd and Tucker, 1995; Goldburg and Triplett, 1997). In September 2002, the US Environmental Protection Agency published draft guidelines to regulate discharges from aquaculture facilities. In many cases, appropriate land application is the preferred method for disposal of effluents. Usually, these aquaculture effluents are considered to be a form of nitrogen fertilizer and are regulated under best management practices

(Fitzsimmons, 1992). Land application not only reuses water and nutrients, it also reduces apprehension over the introduction of diseases, pests and exotics to the natural environment. Land application is typically limited only by the capacity of the soil to absorb water or by the amount of nitrogen delivered per surface area.

Water used in intensive recirculating tilapia culture systems has been reported to have total ammonia levels of up to 19.2 mg/l, nitrate levels of up to 181 mg/l (Lightner *et al.*, 1988), phosphates up to 53 mg/l and potassium up to 150 mg/l (Nair *et al.*, 1985). Olsen *et al.* (1993) reported that water from an intensively stocked pond contributed 6.8 kg total nitrogen/ha when used to irrigate cotton for one season, representing 15% of the needed nitrogen. Nutrients from tilapia culture effluents are readily available to plants (Roy *et al.*, 1990) and the organic matter is especially useful for improving soils.

Researchers and farmers have tested several methods of incorporating tilapia culture into existing irrigation systems (Redding and Midlen, 1990; Budhabhatti and Maughan, 1993). Milstein *et al.* (1989) presented information on the limnology of reservoirs used for fish cage culture and crop irrigation. Their research demonstrated the feasibility of cage culture of tilapia and carp in irrigation reservoirs and documented the fish contributions of nitrates to irrigation water and reduction of pond water pH. Lieberman and Shilo (1989) recommended methods of improving water quality in reservoirs used for fish production. Their recommendations include siting of reservoirs to take advantage of prevailing winds, mechanical mixing and siting irrigation intakes to remove worst-quality water for fish culture but best-quality for irrigation and fertilization of field crops.

Several farmers in the western USA use geothermal waters for aquaculture and then field crop irrigation. This procedure provides the added benefit of maintaining preferred water temperatures for tilapia year round and providing warmth to plants to protect them from frosts and/or to encourage early germination and growth. These systems use modified irrigation ditches as raceways for intensive production of tilapia and catfish. A problem that may arise with geothermal waters is the salinization of soils from compounds in the water. Tilapia production in irrigation systems is likely to be the system of choice for integrated aquaculture–agriculture in arid regions. It will allow irrigated agriculture not only to survive but also to thrive using a sustainable methodology that will increase productivity, increase profitability and increase the amount of food available to a needy world population. Tilapia can be cultured either in dedicated facilities or directly in irrigation structures.

Tilapia production is scattered widely across the USA and many different production systems are used. These range from simple pond systems and extensive culture techniques to closed recirculating systems with controlled environments and the most intensive aquaculture techniques yet developed. Some of the highest fish densities of any aquaculture system are found on tilapia farms using liquid oxygen, microscreens, fluidized bed biofilters and UV sterilizers. Production densities over 100 kg/m^3 water are not uncommon.

Commercial tilapia producers in the USA primarily rear fish for the live markets. The demand for live tilapia is relatively stable across the year and producers need to use production systems that allow for frequent harvests. There are several broad groupings of recirculating systems for tilapia production (Malone and De Los Reyes, 1997; Watanabe *et al.*, 2002).

Many of the world's major tilapia producing countries are also major shrimp producers. When many of the shrimp farms were hit with disease problems, they looked to tilapia culture as an alternative crop. Suresh and Kwei-Lin (1992) and Watanabe *et al.* (1997a,b, 2002, 2006) report on several strains and production systems that can be used to rear tilapia in saline waters. Several techniques of polyculture of tilapia and shrimp have been tested. These include simultaneous polycultures, with tilapia and shrimp being grown in the same pond at the same time, and sequential polycultures, with tilapia and shrimp being grown in the same pond at different times or in the same water at different times. There seem to be several ecological and economic reasons why polyculture of tilapia and shrimp has been successful. The most important is the beneficial impact tilapia seem to have on microbial communities. The presence of yellow fluorescent vibrio bacteria has been noted as an indicator of poor water quality for shrimp farming and attendant disease problems. Rearing tilapia in a shrimp pond, or in water before it enters the shrimp pond, appears to decrease the proportion and/or number of yellow vibrio and to increase the proportion and/or number of green fluorescent vibrio which do not appear to be pathogenic. Tilapia culture also has been reported to condition water so that populations of green algae and diatoms that are most beneficial for shrimp predominate (Cruz *et al.*, 2008).

Tilapia raised in marine waters seem to be especially susceptible to parasites and bacterial infection. A lack of natural defences to marine and brackish water pathogens and parasites, along with the physiological stress of culture in salinities beyond what the fish normally encounters, can result in high levels of mortality if fish are not treated. *Amyloodinium ocellatum*, a ciliate, *Neobenedenia melleni*, a monogenean, *Caligus* spp., a copepod, and *Vibrio*, a bacterial infection, represent the most severe disease problems in saltwater culture. Careful filtration, or use of water from seawater wells, is the best way to avoid these parasites. In the case of infected fish, the most effective treatment appears to be freshwater baths. Some farms use a freshwater dip or bath as a preventative measure, others will wait for a problem to appear and then treat the fish with a freshwater dip.

17.2.3 Nutrition

In nature, members of the genus *Oreochromis* are all omnivores, feeding on algae, aquatic plants, small invertebrates, detrital material and associated bacterial films. Individual species may have preferences between these materials and are more or less efficient grazing on these foods, depending on morphological and physiological differences and life stages. Each of the commonly cultured species is somewhat opportunistic and will utilize any and all of these feeds when they are available. This provides an advantage to farmers because the fish can be reared in extensive situations that depend on the natural productivity of a waterbody or in intensive systems where fish will still take advantage of any bacterial or algae growth in the culture unit. This characteristic helps to lower feed costs compared to virtually any other fish.

In extensive aquaculture, tilapia act as primary consumers and detritivores, able to grow by eating algae and detrital matter (Diana *et al.*, 1991). Juveniles of all the *Oreochromis* species are efficient filter feeders of phytoplankton. Fry use gill rakers for filtering phytoplankton. In some species, filter feeding continues to be important, but all the tilapias become more omnivorous as they grow. Tilapia have small teeth in their jaws used for scraping algae and bacterial films. The fish have small stomachs and long convoluted intestines that allow for assimilation of vegetative materials. Juvenile tilapia are also capable of feeding on a higher trophic level, consuming small invertebrates and even small tilapia fry. In hatchery settings, it is not uncommon to observe 2–6 cm juveniles eating young fry. There is a large body of research that examines the growth of tilapia in pond systems that are fertilized with organic (Green *et al.*, 1990; Knud-Hansen *et al.*, 1991, 1993; Green, 1992) and inorganic fertilizers (Green *et al.*, 1989).

In intensive systems, tilapia have the advantage that they can be fed a prepared feed that includes a high percentage of plant proteins. Carnivorous fish require fishmeal or other animal proteins in their diets, which in general are more expensive than plant proteins. This characteristic is also important to those who are concerned about the use of fishmeal in aquaculture. Higher growth rates can be achieved by increasing the protein content of diets by including fish and animal proteins. However, there are many nutritional studies that substitute plant proteins supplemented with specific amino acids to reach the production levels that can be achieved with fishmeal.

Complete diets are used in systems that cannot provide dependable nutrition from natural productivity. This would include intensive recirculating systems, cages placed in water with low productivity and even heavily stocked ponds that do not provide enough nutrition for all the fish in the system. Supplemental diets will provide only portions of the nutritional demands of the fish, with the assumption that they will get most of the nutrients from the growing system. Supplemental diets are usually much less expensive than complete diets and usually high in carbohydrates. Some simple supplemental diets serve a dual purpose of fertilizing the pond as well as increasing productivity. Considerable research has been conducted on complete diets and on fertilization programmes for natural and man-made waterbodies. Development of supplemental diets directed specifically to provide limiting nutrients is a growing area of research.

Tilapia exhibit their best growth rates when they are fed a balanced diet that provides a proper mix of protein, carbohydrates, lipids, vitamins, mineral and fibre. El-Sayed and Teshima (1991) and Lim and Webster (2006) provide excellent reviews that examine details of tilapia nutrition. The nutritional requirements are slightly different for each species and, more importantly, vary with life stage. Fry and fingerling fish require a diet higher in protein, lipids, vitamins and minerals and lower in carbohydrates, as they are developing muscle, internal organs and bone with rapid growth. Subadult fish need more calories from fat and carbohydrates for basal metabolism and a smaller percentage of protein for growth. Of course, the absolute amount the fish is eating will still be increasing as the fish is much larger.

Tilapia are not naturally high in omega 3 fatty acids. However, the fatty acid profile in the edible portions of the fish can be determined by adjusting dietary fatty acid levels and by altering environmental conditions (Fitzsimmons *et al.*, 1997; Huang *et al.*, 1998; Chou *et al.*, 2001). Consumers are increasingly aware of the beneficial aspects of omega 3 fatty acids in their diet and products with higher omega 3 content may bring a higher price due to their higher nutritional value (Karapanagiotidis, 2002).

Vitamins and minerals are critical to proper nutrition in tilapia and considerable research has been conducted to determine these requirements (Roem *et al.*, 1990; El-Sayed and Teshima, 1991; Watanabe *et al.*, 1997a,b). Commercial premixes are available which allow feed makers to purchase a whole group of micronutrients rather than attempting to determine how much is available from the productivity of the system and the other ingredients

A tremendous amount of research has been conducted to examine the use of various agricultural by-products for tilapia diets (El-Sayed, 1999). Because of the ability of tilapia to digest a wide variety of vegetative products, plant proteins can be substituted for expensive fishmeal or other animal-based proteins (Viola *et al.*, 1988). Plants such as duckweed (Fasakin *et al.*, 1999), Azolla (Naegel, 1997), algae (Nwachukwu, 1997), soybean (Shiau *et al.*, 1990) and coffee pulp (Ulloa Rojas and Weerd, 1997; Ulloa Rojas, 2002) have been tested.

This is meant to be only a short introduction to tilapia nutrition. Specific nutritional needs vary by species, age of fish, production system and salinity. A wealth of information is available and feed manufacturers have developed considerable expertise. Tilapia nutrition is critical to further increases in efficiency and profitability for both the small producer growing for personal consumption and the large producer involved in international trade.

17.2.4 Impacts of introductions and farming

Tilapia are considered exotic species in many of the countries into which they have been introduced (over 100, on all continents except Antarctica) and are subject to legal restrictions (Pullin *et al.*, 1997). Most of the early introductions were for insect control or aquatic weed control (McIntosh *et al.*, 2003). The agencies responsible for these introductions dispersed the fish widely. In most countries, introductions of tilapia for farming purposes came later, or else fish already established in local water bodies were domesticated. Many populations of tilapia are now so well established they are a permanent part of the fish community. However, there are some steps that aquaculture operations can take to mitigate any additional harm. The eventual goal should be to develop fully domesticated strains of tilapia that will have little chance of surviving outside a culture setting, in much the same manner as most domestic farm animals. The industry is well on its way with tilapia. Red strains of fish are an important step. Red tilapia are found in domesticated populations only and they have very little chance of surviving in the wild. Predation is high from birds, fish and humans because they are so visible in the water. Strains that have been bred to have very large fillets and a more rounded body form are also unlikely to survive outside a farm. Finally, all-male populations, developed from hybrids, sex-reversal or genetically male parentage, are less likely to be able to establish a breeding population off-farm. All of these techniques should be considered as contributing to the reduction of the ability of tilapia to impact native communities.

Tilapia's spread to countries around the world has been accompanied by environmental externalities; negative impacts on the ecosystem outside the farm. Environmental impacts of tilapia can be grouped loosely into two major categories. First is the impact of feral populations of tilapia on native fishes. Introductions of tilapia around the world frequently occur in concert with severe human impacts on local aquatic systems. These impacts often help the tilapia to thrive at the expense of native species. Tilapia may compete for resources with native fish, or they may just thrive in the altered conditions.

The second environmental issue is the nutrient enrichment of local waters from intensive farming of tilapia. Intensively fed fish generate faecal waste and leave uneaten food. Nitrogen and phosphorus dissolved in the effluent and the biological oxygen demand of decaying organic matter can impact the receiving water. Wastewater from processing plants can also impact receiving waters if they are not treated sufficiently. Conventional water treatment plants, constructed wetlands and irrigation of crop plants are suggested as methods of reducing negative impacts from eutrophication caused by fish farm effluents.

Conservation of genetic variability of wild and domesticated tilapia stocks is also of importance. There are several instances where this is important. First, tilapia stocks have been moved repeatedly and allowed to interbreed with extant populations. In some cases, this has led to a decrease in genetic diversity and 'contamination' of endemic populations. The loss of this diversity becomes important because whole genomes may be lost. Some of this genetic variability may be important as a genetic reservoir of material that may be useful for future conservation or breeding efforts.

Several of the most common strains of tilapia came from very small founder stocks. These fish have a high degree of introgression and may be subject to genetic bottlenecks. Development of domesticated stocks that have been selected for certain culture conditions from an adequate breeding population would be the best way to avoid this problem. The Genetically Improved Farmed Tilapia (GIFT) programme is probably the best example of such a genetic improvement effort (Ponzoni *et al.*, 2005; Ng and Hanim, 2007). Maintaining diversity in feral and captive stocks is important both for the fish in their native environment and for those in captivity.

There are several environmental impacts associated with the discharge of wastewaters from tilapia farms. The most obvious is eutrophication. To reduce this impact, farmers use nutrient-dense diets. These are diets designed to minimize wastes by delivering a feed that provides the exact ratio of nutrients for growth. Maximizing the food conversion ratio is closely related. Use of effluents for irrigation and fertilization of field crops is an alternative action. Saltwater culture of tilapia is growing quickly and methods of treating saline effluent are also available. Effluents can be used to irrigate and fertilize halophytic plants used for commercial purposes (Brown *et al.*, 1999) or they can be used for seaweed culture, especially *Gracilaria* (Nelson *et al.*, 2001).

17.2.5 Diseases

Tilapia are hardy fishes but are still susceptible to opportunistic pathogens. When subjected to environmental or physiological stress, their defence mechanisms are less able to withstand infections. Plumb (1997) and Shoemaker *et al.* (2006a,b) provide thorough reviews of tilapia diseases.

17.2.5.1 Viral

Viral infections of tilapia (Chen *et al.*, 1985; Avtalion and Shlapobersky, 1994) have been isolated instances and have not had an impact on commercial tilapia production.

17.2.5.2 Bacterial

Streptococcus iniea has been the primary disease problem for commercial tilapia farming. It has been reported from recirculating culture systems and very intensive cage or flow-through systems (Shoemaker and Klesius, 1997; McNulty *et al.*, 2003). Poor water quality is almost always a contributing factor. Symptoms include haemorrhages at the base of fins, exophthalmia and darkening of skin, lack of appetite and erratic swimming behaviour. Mortality rates can be very high in affected populations, although the fish will respond to antibiotics. Several vaccines have been developed, with encouraging results (Klesius *et al.*, 2006). Haemorrhagic septicaemia is a condition caused by one or more bacteria, including *Aeromonas hydrophila*, *Edwardsiella tarda*, *Pseudomonas fluorescens* and *Vibrio* spp. Recent findings (Cruz *et al.*, 2008) suggest that Gram-positive bacteria, or their metabolites, may reduce populations of *Vibrio in vitro* and in production systems.

17.2.5.3 Mycotic

In coldwater conditions, tilapia are susceptible to *Saprolegnia* spp. fungal infections. Tilapia species have varying levels of cold tolerance, but in all cases, as temperatures drop, the fish will be more likely to be stricken. Fungal infection from cold water is a chronic condition seen in fish held just above the lethal limit. The fungus appears first on body areas that have minimal production of body slime or where the body slime may have been rubbed off from handling.

Dorsal surfaces of the head and along the caudal peduncle are the most frequent sites, with white cottony colonies (mycelium) being visible. Salt and potassium permanganate treatments will slow the fungal growth, but the primary cure is elevated water temperatures. *Saprolegnia* also attacks eggs in artificial incubation systems. Typically, dead eggs will become infected and then the fungus will spread to live eggs. Maintaining warm temperatures, keeping eggs in motion and prompt removal of all dead and infected eggs are the best control measures. In extreme cases, iodine or potassium permanganate may be used to remove fungus.

17.2.5.4 Parasites

Tilapia are susceptible to most of the common warmwater protozoan and crustacean parasites. A common sign that parasites are affecting the fish is scratching or flashing.

17.2.5.5 Flagellates

Ichthyobodo necator is found primarily on smaller fish that have been stressed in the hatchery or nursery system. *Piscinoodinium pillulare* is another flagellate that is found most frequently on the gills. A 25 ppm formalin bath is the most frequent method used to control flagellates.

17.2.5.6 Ciliates

Ich or White Spot, caused by *Ichthyophthirius multifiliis*, can be identified by the large white cysts that form on the skin of infected fish. Without treatment, lesions will form eventually and fish will die. *Chilodonella* and *Tricodina* are additional ciliates that attack stressed fish, especially in water with large amounts of organic matter. Both ciliates are most common in the gills and skin of the fish. Improved water quality and formalin baths are most effective for control.

17.2.5.7 Saltwater parasites

The rapid expansion of tilapia culture into brackish and marine waters has been accompanied by infections from marine parasites. *A. ocellatum*, a ciliate, *N. melleni*, a monogenean, and *Caligus* spp., a copepod, represent the most severe disease problems in saltwater culture. In each case, the most effective treatment appears to be freshwater baths. Some farms use a freshwater dip or bath as a preventative measure, others will wait for a problem to appear and then treat the fish with a freshwater dip.

17.2.6 Production and markets for Tilapia

During the 1990s, tilapia products became an important commodity in the international seafood trade. Tilapia farming began with fish being introduced around the world by development agencies to feed the rural poor and has now developed into a highly domesticated livestock product with sales exceeding two billion dollars a year. The description of tilapia as the aquatic chicken becomes more appropriate every day. Tilapia, as in chicken farming, can be successful on any scale, from subsistence farmers with a few essentially feral fish in a pond, to multinational corporations rearing highly domesticated fish with vertically integrated farms, processing plants and markets in many countries. Tilapia have been domesticated faster and to a greater extent than any other group of fish. The hatchery technology is relatively simple and because of the ease with which tilapias can be hybridized, and the several species that readily hybridize, they have a large genetic base. Tilapia probably surpassed salmonids in production volume in 2004 and may eventually equal the carps.

Millions of small farmers in over 100 countries supplement their diet with occasional meals of farmed tilapia. Many more utilize tilapia for small enterprises including family-operated restaurants. The overall economic value is difficult to assess, but some studies are available in specific regions (Neira and Engle, 2001; Engle, 2006). Commercial size is 200–250g for developing country domestic markets and 750–1200g for processing and international trade. Those sizes can be reached in 4 and 12 months, respectively. Then, flesh yield is between 33–43%.

In the mid-1980s, the only tilapia product found in international trade was whole frozen tilapia grown and exported from Taiwan. There have been tremendous increases in the number of producing countries who are exporting tilapia and in the quantity and quality of the processed fish. In the 1990s, high-quality, low-cost frozen fillets from Indonesia, Taiwan and Jamaica and fresh fillets from Costa Rica, Jamaica and Colombia opened a floodgate of demand. This was followed by equal increases in production from China, Thailand, Vietnam, Brazil, Ecuador and Honduras as major producers.

World production of farmed tilapia was approximately 2,381,237Mt in 2006. China is the world's major producer and consumer of tilapia. The Chinese mainland's production in 2006 was 1,110,000Mt and Taiwan produced another 72,581Mt (Fig. 17.4). Other Asian countries produced an additional 555,000Mt. The USA is the world's major importer of tilapia. The imports for 2006 were 158,254Mt with an import value of US$482,742,515. The products were divided between frozen whole fish, frozen fillets and fresh fillets. These products represent a live weight of 359,295Mt. Linked with 2006, domestic production of 9000Mt sets the 2006 US consumption of live weight fish at 369,295Mt, or 814 million pounds. Considering that US consumers prefer filleted products, the actual consumption in 2005 (the last year with a complete record) was 1lb per capita (Table 17.2). This made tilapia the fifth most popular seafood in the USA. Consumption in 2006 and 2007 increased considerably, further raising per capita consumption. Although China is the largest consumer of tilapia, the per capita consumption is still less that the USA. The EU is increasing its consumption of tilapia rapidly, importing frozen products from Asia and fresh fillets from Africa, the Caribbean and Latin America. Tilapia have already become one of the most important farm-raised fishes and is increasingly taking its place as a major item in the international seafood trade.

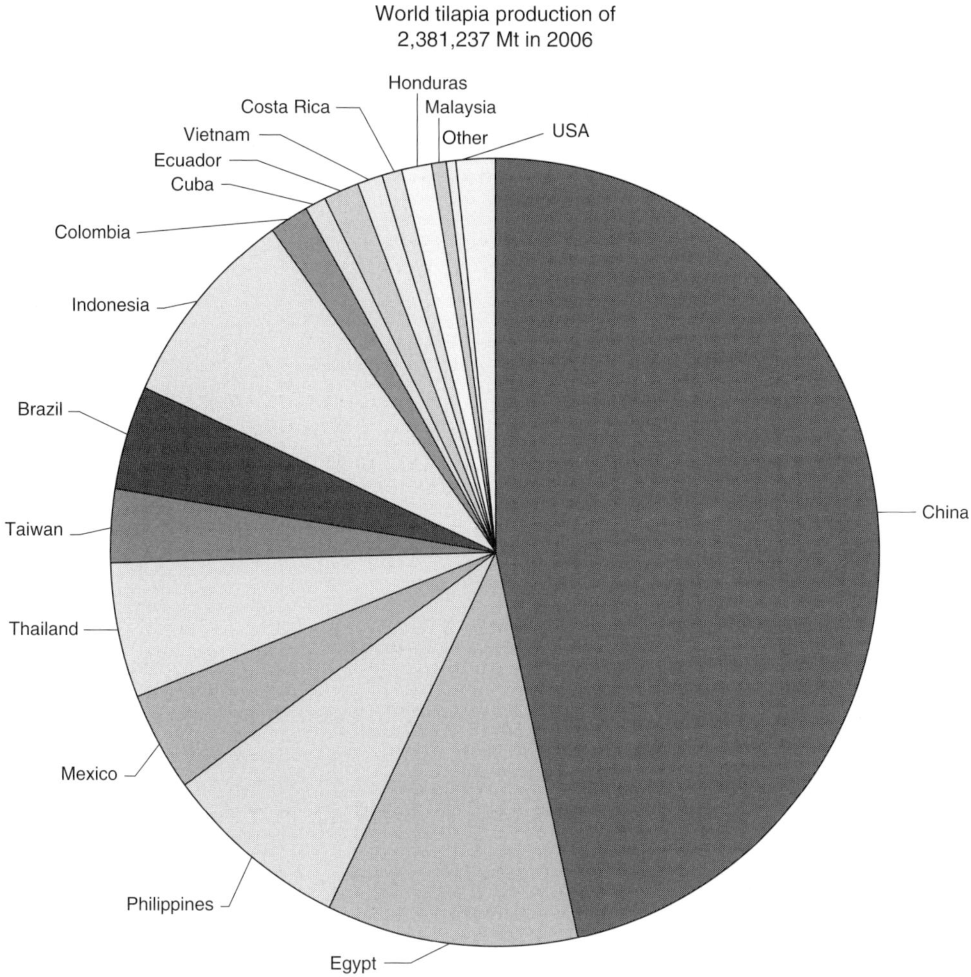

Fig. 17.4. Distribution of tilapia aquaculture.

The tilapia industry is currently in the middle stages of the market developments previously seen in the salmon and shrimp industries. Commodity prices that were dependent on wild catches and seasonal availability have been overtaken by the year-round availability and quality of farm-raised product. Rapid expansion of demand for and production of tilapia products has had the effect of maintaining prices to the grower and consumer in most nations. However, the costs of most inputs (feed, energy, labour) for farmers continue to rise. Rapid improvements in technology, feeds, genetics and experience levels of farmers have improved productivity enough to remain profitable. Continued improvement in productivity will be required for the industry to continue its remarkable growth and consistency of prices.

Table 17.2. Growth of tilapia in the US market (per capita consumption/lbs) (statistics from National Fisheries Institute).

2000		2001		2002		2003		2004		2005		2006	
Tuna	3.5	Shrimp	3.4	Shrimp	3.7	Shrimp	4.0	Shrimp	4.2	Shrimp	4.1	Shrimp	4.4
Shrimp	3.2	Tuna	2.9	Tuna	3.1	Tuna	3.4	Tuna	3.4	Tuna	3.1	Tuna	2.9
Pollock	1.6	Salmon	2.0	Salmon	2.0	Salmon	2.2	Salmon	2.2	Salmon	2.4	Salmon	2.0
Salmon	1.5	Pollock	1.2	Pollock	1.1	Pollock	1.7	Pollock	1.7	Pollock	1.5	Pollock	1.6
Catfish	1.1	Catfish	1.1	Catfish	1.1	Catfish	1.1	Catfish	1.1	Catfish	1.0	Tilapia	1.0
Cod	0.8	Cod	0.6	Cod	0.7	Cod	0.6	Tilapia	0.7	Tilapia	0.8	Catfish	0.97
Clams	0.5	Clams	0.5	Crabs	0.6	Crabs	0.6	Cod	0.6	Crabs	0.6	Crabs	0.66
Crabs	0.4	Crabs	0.4	Clams	0.5	Tilapia	0.5	Crabs	0.6	Cod	0.6	Cod	0.51
Flatfish	0.4	Flatfish	0.4	Tilapia	0.4	Clams	0.5	Clams	0.5	Clams	0.4	Clams	0.44
Scallops	0.3	Tilapia	0.4	Flatfish	0.3	Scallops	0.3	Scallops	0.3	Scallops	0.3	Scallops	0.31
Tilapia	0.3												

References

Al-Jaloud, A., Hussain, G., Alsadon, A., Siddiqui, A.Q. and Al-Najada, A. (1993) Use of aquaculture effluent as a supplemental source of nitrogen fertilizer to wheat crop. *Arid Soil Research and Rehabilitation* 7, 233–241.

Avtalion, R.R. and Shlapobersky, M. (1994) A whirling disease of tilapia larvae. *Israeli Journal of Aquaculture* 46, 102–104.

Boyd, C.E. and Tucker, C.S. (1995) Sustainability of channel catfish farming. *World Aquaculture* 26(3), 45–53.

Brown, J., Glenn, E., Fitzsimmons, K. and Smith, S. (1999) Halophytes for the treatment of saline aquaculture effluent. *Aquaculture* 175, 255–268.

Budhabhatti, J. and Maughan, O.E. (1993) Production of channel catfish, *Ictalurus punctatuts*, in flowing agricultural waters in Arizona. *Journal of World Aquaculture Society* 24(4), 493–498.

Carro-Anzalotta, A.E. and McGinty, A.S. (1986) Effects of stocking density on growth of *Tilapia nilotica* cultured in cages in ponds. *Journal of World Aquaculture Society* 17, 52–57.

Chen, S.N., Kou, G.H., Hedrick, R.P. and Fryer, J.L. (1985) The occurrence of viral infections of fish in Taiwan. In: Ellis, A.E. (ed.) *Fish and Shellfish Pathology*. Academic Press, London, pp. 313–319.

Chou, B., Shiau, S. and Hung, S. (2001) Effect of cod liver oil on growth and fatty acids of juvenile hybrid tilapia. *North American Journal of Aquaculture* 63, 277–284.

Cruz, P., Andalecio, M., Bolivar, R. and Fitzsimmons, K. (2008) Tilapia-shrimp polyculture in Negros Island, Philippines: a review. *Journal of the World Aquaculture Society* 39(6), 713–725.

Diana, J.S., Lin, C.K. and Schneeberger, P.J. (1991) Relationship among nutrient inputs, water nutrient concentrations, primary production, and yield of *Oreochromis niloticus* in ponds. *Aquaculture* 92, 323–341.

Egna, H.S. and Boyd, C.E. (1997) *Dynamics of Pond Aquaculture*. CRC Press, Boca Raton, Florida.

El Sayed, A.F.M. (1999) Alternative dietary protein sources for farmed tilapia, *Oreochromis* spp. *Aquaculture* 179, 149–168.

El Sayed, A.F.M. (2001) Tilapia culture in Arabia. *Global Aquaculture Advocate* 4(6), 46.

El Sayed, A.F.M. and Teshima, S.I. (1991) Tilapia nutrition in aquaculture. *Reviews in Aquatic Sciences* 5(3–4), 247–265.

Engle, C. (2006) Marketing and economics. In: Lim, C. and Webster, C. (eds) *Tilapia: Biology, Culture, and Nutrition*. Haworth Press, Binghamton, New York, pp. 619–644.

Fasakin, E.A., Balogun, A.M. and Fasuru, B.E. (1999) Use of duckweed, *Spirodela polyrhiza* L. as a protein feedstuff in practical diets for tilapia, *Oreochromis niloticus* L. *Aquaculture Research* 30, 313–318.

Fernando, C.H. and Halwart, M. (2000) Possibilities for the integration of fish farming into irrigation systems. *Fisheries Management and Ecology* 7, 45–54.

Fitzsimmons, K. (1992) Extending the value of aquaculture effluents through sustainable agriculture practices. In: *National Livestock, Poultry, and Aquaculture Waste Management*. Publication 03-92, American Society of Agricultural Engineers, St Joseph, Michigan, pp. 344–346.

Fitzsimmons, K., Dickenson, G., Brand, C. and Davis, J. (1997) Effect of reducing dietary lipid levels on growth of hybrid tilapia in an intensive recirculating system. *Progressive Fish Culturist* 59, 290–296.

Goldburg, R. and Triplett, T. (1997) *Murky Waters: Environmental Effects of Aquaculture in the United States*. Environmental Defense Fund, New York.

Green, B. (1992) Substitution of organic manure for pelleted feed in tilapia production. *Aquaculture* 101, 213–222.

Green, B.W., Phelps, R.P. and Alvarenga, H.R. (1989) The effect of manures and chemical fertilizers on the production of *Oreochromis niloticus* in earthen ponds. *Aquaculture* 76, 37–42.

Green, B.W., Teichert-Coddington, D.R. and Phelps, R.P. (1990) Response of tilapia production and economics to varying rates of organic fertilization in two tropical American countries. *Aquaculture* 90, 279–290.

Huang, C., Huang, M. and Hou, P. (1998) Effect of dietary lipids on fatty acid composition and lipid peroxidation in sarcoplasmic reticulum of hybrid tilapia, *Oreochromis niloticus* × *O. aureus*. *Comparative Biochemistry and Physiology* 120B, 331–336.

Ishak, M.M. (1982) *Aquaculture Egypt – Development of Fish Farming in Egypt*. International Development Research Centre Rep. 1 (phase 1). Central Lab for Aquaculture Research, Department of Agriculture, Abbassa, Sharkia, Egypt.

Ishak, M.M. (1986) *Development of Fish Farming in Egypt (Cage and Pen Culture)*. Institute of Oceanography and Fisheries and the International Development Research Centre Rep. 4 (phase 2). Central Lab for Aquaculture Research, Department of Agriculture, Abbassa, Sharkia, Egypt.

Karapanagiotidis, I. (2002) Studies to optimize the polyunsaturated fatty acid composition of tilapias for human consumption in South-east Asia. *Stirling University Aquaculture News* 28, 6–7.

Klesius, P., Evans, J., Shoemaker, C. and Pasnik, K. (2006) Vaccines to prevent *Streptococcus iniae* and *S. agalactiae* disease in tilapia, *Oreochromis niloticus*. In: Contreras-Sanchez, W. and Fitzsimmons, K. (eds) *Tilapia, Sustainable Aquaculture for the New Millenium*. Proceedings of the Seventh International Symposium on Tilapia in Aquaculture. World Aquaculture Society, Baton Rouge, Louisiana.

Knud-Hansen, C., Batterson, T., McNabb, C., Harahat, I., Sumanatadinata, K. and Eidman, M. (1991) Nitrogen input, primary productivity and fish yield in fertilized ponds in Indonesia. *Aquaculture* 94, 49–63.

Knud-Hansen, C., Batterson, T. and McNabb, C. (1993) The role of chicken manure in the production of Nile tilapia, *Oreochromis niloticus* (L). *Aquaculture and Fisheries Management* 24, 483–493.

Lieberman, O.S. and Shilo, M. (1989) Water quality improvement of deep, dual-purpose reservoirs. *Israeli Journal of Aquaculture* 41(1), 34–37.

Lightner, D., Redman, R., Mohney, L., Dickenson, G. and Fitzsimmons, K. (1988) Major diseases encountered in controlled environment culture of Tilapias in fresh- and brackish water over a three-year period in Arizona. In: Pullin, R.S.V., Bhukaṣwan, T., Tonguthai, K. and Maclean, J.L. (eds) *The Second International Symposium on Tilapia in Aquaculture*. ICLARM, Manila, pp. 111–116.

Lim, C. and Webster, C. (2006) *Tilapia: Biology, Culture, and Nutrition*. Haworth Press. Binghamton, New York.

McIntosh, D., King, C. and Fitzsimmons, K. (2003) Tilapia for biological control of Giant Salvinia. *Journal of Aquatic Plant Management* 41, 28–31.

McNulty, S.T., Klesius, P.H., Shoemaker, C.A. and Evans, J.J. (2003) Hematological changes in Nile Tilapia (*Oreochromis niloticus*) infected with *Streptococcus iniae* by nare inoculation. *Journal of the World Aquaculture Society* 34(3), 418–422.

Mair, G.C., Capili, J.B., Skibinski, D.O.F., Beardmore, J.A., Pascaul, L.P., Abucay, J.S., Danting, J.C.S. and Reyes, R.A. (1997) Sex ratios and growth performance of crossbred genetically male tilapia using YY males in *Oreochromis niloticus* L. In: Fitzsimmons, K. (ed.) *Proceedings of the Fourth International Symposium on Tilapia in Aquaculture*. Northeast Regional Agricultural Engineering Service, Ithaca, New York, pp. 262–271.

Malone, R.F. and De Los Reyes, A.A. (1997) Categories of recirculating systems. In: Timmons, M. and Losordo, T. (eds) *Advances in Aquacultural Engineering* III. Publication No. NRAES 105. Northeast Regional Agricultural Engineering Service, Ithaca, New York, pp. 197–208.

Martinez, R., Arenal, A., Estrada, M.P., Herrera, F., Huerta, V., Vazquez, J., Sanchez, T. and Fuente, J.N.A. de la (1999) Mendelian transmission, transgene dosage and growth phenotype in trangenic tilapia (*Oreochromis hornorum*) showing ectopic expression of homologous growth hormone. *Aquaculture* 173, 271–283.

Milstein, A., Feldlite, M., Mozes, N. and Avnimelech, Y. (1989) Limnology of reservoirs used for fish farming and crop irrigation, with integrated free and cage fish culture. *Israeli Journal of Aquaculture* 41(1), 12–22.

Mires, D., Amit, Y., Avnimelech, Y., Diab, S. and Cochaba, M. (1990) Water quality in a recycled intensive fish culture system under field conditions. *Israeli Journal of Aquaculture* 42(4), 110–121.

Naegel, L.C. (1997) Azolla meal as supplemental feed ingredients for tilapias. In:. Fitzsimmons, K. (ed.) *Proceedings of the Fourth International Symposium on Tilapia in Aquaculture*. Northeast Regional Agricultural Engineering Service, Ithaca, New York, pp. 20–30.

Nair, A., Rakocy, J.E. and Hargreaves, J.A. (1985) Water quality characteristics of a closed recirculating system for tilapia culture and tomato hydroponics. In: *Proceedings of the 2nd International Conference on Warm Water Aquaculture*. Brigham Young University, Hawaii, pp. 223–254.

Neira, I. and Engle, C. (2001) Markets for tilapia (*Oreochromis* spp.) in Nicaragua: a descriptive analysis of restaurants, supermarkets, and stands in open markets. Sixth Central American Symposium on Aquaculture at Tegucigalpa, Honduras, 22–24 August 2001. Pond Dynamics/ Aquaculture Collaborative Research Support Program, Tegucigalpa, Honduras.

Nelson, S.G., Glenn, E.P., Conn, J., Moore, D., Walsh, T. and Akutagawa, M. (2001) Cultivation of *Gracilaria parvispora* (Rhodophyta) in shrimp-farm effluent ditches and floating cages in Hawaii: a two-phase polyculture system. *Aquaculture* 193, 239–248.

Ng, W.K. and Hanim, R. (2007) Performance of genetically improved Nile tilapia compared with red tilapia fed diets containing two protein levels. *Aquaculture Research* 38, 965–972.

Nwachukwu, V.N. (1997) Tilapia nutrition through substrate enhancement in ponds: a cheap sustainable, and environmentally friendly feeding method. In: Fitzsimmons, K. (ed.) *Proceedings of the Fourth International Symposium on Tilapia in Aquaculture*. Northeast Regional Agricultural Engineering Service, Ithaca, New York, pp. 5–57.

Olsen, M.W., Fitzsimmons, K.M. and Moore, D.W. (1993) Surface irrigation of cotton using aquaculture effluent. In: Wang, J.K. (ed.) *Techniques for Modern Aquaculture*. ASAE Publication, St Joseph, Michigan, 02-93, pp.159–165.

Plumb, J.A. (1997) Infectious diseases of tilapia. In: Costa-Pierce, B.A. and Rakocy, J.E. (eds) *Tilapia Aquaculture in the Americas, Volume 1*. World Aquaculture Society, Baton Rouge, Louisiana, pp. 212–228.

Ponzoni, R., Hamzah, A., Tan, S. and Kamaruzzaman, N. (2005) Genetic parameters and response to selection for live weight in the GIFT strain of Nile tilapia (*Oreochromis niloticus*). *Aquaculture* 247, 203–210.

Pullin, R.S., Palmares, M.L., Casal, C.V., Dey, M.M. and Pauly, D. (1997) Environmental impacts of tilapias. In: Fitzsimmons, K. (ed.) *Proceedings of the Fourth International Symposium on Tilapia in Aquaculture*. Northeast Regional Agricultural Engineering Service, Ithaca, New York, pp. 554–572.

Rakocy, J.E. and Hargreaves, J.A. (1993) Integration of vegetable hydroponics with fish culture: a review. In: Wang, J.K. (ed.) *Techniques for Modern Aquaculture*. ASAE Publication, St Joseph, Michigan, 02-93, pp. 112–136.

Redding, T.A. and Midlen, A.B. (1990) *Fish Production in Irrigation Canals. A Review*. FAO Fisheries Technical Paper No. 317, Rome.

Roem, A.J., Stickney, R.R. and Kohler, C.C. (1990) Vitamin requirements of blue tilapias in a recirculating water system. *Progressive Fish Culturist* 52, 15–18.

Roy, B., Das, D.N. and Mukhopadhyay, P.K. (1990) Rice–fish–vegetable integrated farming: towards a sustainable ecosystem. *Naga, The ICLARM Quaterly* 13(4), 17–18.

Shiau, S.Y., Lan, S.F., Yu, S.L., Lin, A.L. and Kwok, C.C. (1990) Defatted and fullfat soybean meal as partial replacement for fishmeal in tilapia (*Oreochromis niloticus* × *O. aureus*) diets at low protein level. *Aquaculture* 86, 401–407.

Shoemaker, C.A. and Klesius, P.H. (1997) Streptococcal disease problems and control – a review. In: Fitzsimmons, K. (ed.) *Tilapia Aquaculture*. Northeast Regional Agricultural Engineering Service, Ithaca, New York, pp. 671–682.

Shoemaker, C.A., Xu, D.H., Evans, J.J. and Klesius, P.H. (2006a) Parasites and diseases. In: Lim, C. and Webster, C. (eds) *Tilapia: Biology, Culture, and Nutrition*. Haworth Press, Binghamton, New York, pp. 561–582.

Shoemaker, C.A., Lim, C., Yildirim-Aksoy, M., Welker, T.L., Phillip, H., Klesius, P.H. and Evans, J.J. (2006b) Growth response and acquired resistance of Nile Tilapia *Oreochromis niloticus* following infection or vaccination with *Streptococcus iniae*. In: Contreras-Sanchez, W. and Fitzsimmons, K. (eds) *Tilapia, Sustainable Aquaculture from the New Millennium – Proceedings of the Seventh International Symposium on Tilapia in Aquaculture*. American Tilapia Association, Aquaculture CRSP, Veracruz, Mexico, 389 pp.

Siddiqui, A.Q., Howlander, M.S. and Adam, A.A. (1991) Management strategies for intensive culture of Nile Tilapia (*Oreochromis niloticus* L.) in tanks using drainage water in Al-Hassa region of Saudi Arabia. *Arab Gulf Journal of Scientific Research* 9(2), 149–163.

Suresh, A.V. and Kwei-Lin, C. (1992) Tilapia culture in saline waters: a review. *Aquaculture* 106, 201–226.

Trewavas, E. (1983) *Tilapiine fishes of the genera* Sarotherodon, Oreochromis *and* Danakilia. British Museum of Natural History, London.

Ulloa Rojas, J.B. (2002) Use of coffee pulp as feed ingredient for tilapia culture. PhD dissertation, Wageningen University, The Netherlands.

Ulloa Rojas, J.B. and Weerd, H. (1997) The growth and feed utilization of *Oreochromis aureus* fingerlings fed diets with various coffee pulp levels. In: Fitzsimmons, K. (ed.) *Proceedings of the Fourth International Symposium on Tilapia in Aquaculture*. Northeast Regional Agricultural Engineering Service, Ithaca, New York, pp. 40–49.

Viola, S., Arieli, Y. and Zohar, G. (1988) Animal protein-free feeds for hybrid tilapia (*Oreochromis niloticus* × *O. aureus*) in intensive culture. *Aquaculture* 75, 115–125.

Watanabe, T., Kiron V. and Satoh, S. (1997a) Trace mineral in fish nutrition. *Aquaculture* 151, 185–207.

Watanabe, W.O., Wicklund, R.I., Olla, B.L. and Head, W.D. (1997b) Saltwater culture of the Florida red and other saline tolerant tilapias: a review. In: Costa-Pierce, B.A. and Rakocy, J.E. (eds) Tilapia *Aquaculture in the Americas, Volume 1*. World Aquaculture Society, Baton Rouge, Louisiana, pp. 54–141.

Watanabe, W.O., Losordo, T.M., Fitzsimmons, K. and Hanley, F. (2002) Tilapia production systems in the Americas: technological advances, trends and challenges. *Reviews in Fisheries Science* 10, 465–498.

Watanabe, W.O., Fitzsimmons, K. and Yang, Y. (2006) Farming of tilapia in saline waters. In: Webster, C. and Lim, C. (eds) *Tilapia: Biology, Culture, and Nutrition*. Haworth Press, Binghamton, NY, pp. 347–447.

18 Drum-fish or Croakers (Family: Sciaenidae)

JUAN P. LAZO,[1] JOAN G. HOLT,[2] CHRISTIAN FAUVEL,[3] MARC SUQUET[4] AND LOIC QUÉMÉNER[5]

[1]*Department of Aquaculture, Center for Scientific Research and Higher Education of Ensenada, Ensenada, México;* [2]*Marine Science Institute, The University of Texas at Austin, Port Aransas, USA;* [3]*Ifremer, Station expérimentale d'Aquaculture, Palavas les flots, France;* [4]*Ifremer, Laboratoire LALR, Station Expérimentale de Palavas, Palavas, France;* [5]*Ifremer, Centre de Brest, Plouzané, France*

18.1 General Introduction

The family Sciaenidae (Fig. 18.1) includes about 70 genera and 270 species (Nelson, 1994) that occur in temperate and tropical regions of the world. Sciaenids are well represented in the Indo-Pacific (approximately 65 species; Leis and Trnski, 1989), the Caribbean (17 genera; Randall, 1983) and in the temperate waters of the Atlantic and Pacific Oceans. Two additional species are found in the lakes of the Amazon flood plain. Thirty-four species occur in North America, including one freshwater species, and five are species associated with Caribbean coral reefs. The common name given to sciaenids, drums or croakers, is derived from the sounds they emit while vibrating muscles adjacent to their well-developed air bladders that act as resonating chambers. Sciaenids are carnivorous fishes usually found in shallow coastal and estuarine waters and many species are of considerable commercial value. Spawning takes place in or near estuaries, reefs, or in shelf waters, usually in the evening (Holt *et al.*, 1985). The larvae are transported to shallow coastal areas or to estuaries, where they settle into nursery habitats. Juveniles generally exploit different habitats within the estuaries until they reach sexual maturity (at 1–6 years of age).

Habitat degradation and overharvesting have led to a decrease in the population size of sciaenids in many areas. For example, the totoaba (*Totoaba macdonaldi*) is a threatened species endemic to the Gulf of California, Mexico. Ongoing conservation studies include attempts to spawn and rear totoaba for possible restocking (True *et al.*, 1997). Several other sciaenids are cultured for replenishment of natural populations that are under pressure, including red drum (*Sciaenops ocellatus*) (McEachron *et al.*, 1995; Serafy *et al.*, 1999), spotted seatrout (*Cynoscion nebulosus*) (Robert Vega, Texas Parks and Wildlife, personal

 Finfish Aquaculture Diversification (eds N.R. Le François, M. Jobling, C. Carter and P.U. Blier)

Fig. 18.1. Sciaenidae.

communication) and white seabass (*Atractoscion nobilis*) (Drawbridge, 2002). Other sciaenids are cultivated for food production and have good potential for farming, including red drum, white seabass, meagre (*Argyrosomus regius*) and yellow croaker (*Pseudosciaena crocea*) (WanShu and Yong, 2003), black drum (*Pogonias cromis*) (Henderson-Arzapalo *et al.*, 1994), Japanese croaker (*Nibea japonica*) (Furuya, 1995; Liao *et al.*, 1995) and the orange mouth corvina (*C. xanthulus*) (Van Olst and Carlberg, 1987). Global interest in culturing sciaenids developed after red drum were spawned successfully and reared under controlled conditions during the 1970s (Arnold *et al.*, 1979). Following years of research, it is clear that many species of sciaenids are well suited to captivity and amenable to cultivation, which has provided the interest and incentive to apply similar culture and rearing techniques to other species in the family.

18.2 Red Drum, *Sciaenops ocellatus*

The distribution of red drum ranges from Cape Cod in the north-western Atlantic Ocean to Tuxpan, Mexico in the Gulf of Mexico (Simmons and Breuer, 1962) (Fig. 18.2). Red drum or redfish, as this species is also commonly known, usually inhabit coastal and estuarine waters. They have a characteristic red-orange colour that can vary from grey to red-bronze and one or more black spots near the base of the caudal fin (Fig. 18.3). Red drum are eurythermal and euryhaline, carnivorous and of considerable commercial value. Early juveniles feed primarily on bottom-dwelling invertebrates and later stages feed on fish, shrimp and crabs. Juveniles are found in bays and estuaries until sexually mature, except in their more northerly range, where they move offshore or south in the winter to warmer waters. Red drums are not tolerant of cold temperatures and death results below 10°C.

Sexually mature adults of 3–5 years of age (4–5 kg) emigrate to offshore waters and spawning takes place in shallow coastal waters during the evening from August to November (Holt *et al.*, 1985; Murphy and Taylor, 1990). The maximum age documented is 56 years (1250 mm fork length, FL) for males and 52 years (1346 mm FL) for females (Ross *et al.*, 1995).

Fig. 18.2. World distribution of *Sciaenops ocellatus*.

Fig. 18.3. Adult red drum.

18.2.1 Farming of red drum

Aquaculture of red drum began in the mid-1970s, when adult fish were induced to spawn in the laboratory using photoperiod and temperature manipulations and larval rearing techniques were developed (Arnold *et al.*, 1979; Holt *et al.*, 1981b). Interest in culturing this species arose due to severe reduction in population size along the Texas coast as a result of environmental changes and heavy harvesting by the gill net fishery (Caldwell and Carr, 2000). The spawning and rearing technologies were upscaled rapidly for mass production by the Texas Parks and Wildlife Hatchery and used to produce juveniles for stocking into local estuaries. A ban on commercial exploitation of red drum was implemented concurrently (Matlock, 1984), increasing the interest in the potential of aquaculture practices as a source of food production.

Adult wild-caught red drum adapt well to captivity and are easily conditioned to spawn in the laboratory without the use of hormones by simulating temperature and photoperiod regimes found under natural conditions (Arnold *et al.*, 1979). Larvae and juveniles feed readily and grow quickly, reaching one pound (453g) in about 9 months under favourable conditions. Infestations of the parasite *Amyloodinium ocellatus* are frequently encountered but can be treated easily with copper. Red drum are produced commercially in several states along the Gulf of Mexico and south-eastern Atlantic in the USA, as well as in Taiwan, China, Mexico and several other countries in Latin America (Lutz, 1999). In the USA, red drum have been cultured in ponds, indoor raceways, geothermal regions, power plants, brackish aquifers and net pens in coastal waters of the Gulf of Mexico.

18.2.2 Broodstock management and hatchery operations

Temperature and photoperiod are the most important environmental factors controlling gonadal spawning and recrudescence in red drum and can be manipulated to induce spawning. Condensing the natural annual cycle of temperature and photoperiod into a 180-day period induces spawning under the natural spawning conditions of 12–13h light and 24–28°C temperatures. In addition, prolonged spawning (up to 22 months) can be accomplished by maintaining red drum under constant spawning conditions (Arnold, 1988). The timing of egg production in the laboratory can be controlled by inducing spawning events of short duration, reducing the temperature for a couple of weeks and later increasing the temperature once again (Thomas *et al.*, 1995). As a result of these successful manipulations, it is feasible to have eggs out of season year round from one or two tanks of broodstock. Precocial spawning of captive fish is also possible. Second generation red drum have been induced to spawn at 19 months of age (2.9kg) when reared under constant warm temperatures (26°C) and a regime of 12h light:12h dark (Arnold, 1991). Thus, the process of egg production is induced and maintained easily in captivity, contributing to the popularity of this species in aquaculture.

Broodstock are maintained in circular fibreglass tanks that vary in size from 10 to 17 m^3 and have a minimum height of 1.5m. Generally 4–6 fish of 1:1 sex ratio are maintained in a single tank and fed frozen penaeid shrimp and fish at 2% body weight/day. They will also accept commercial pelleted food readily (19mm floating pellets are used at the University of Texas Marine Science Institute, Port Aransas, Texas). Egg production is 1–1.5 × 10^5/kg female per spawn; they are multiple spawners with an estimated seasonal (4 month) production of 15 ×10^6 eggs (Arnold, 1988).

Spawning in red drum occurs in the evening and eggs are collected overnight in 500 micron nitex bags or special collecting baskets in the overflow from the spawning tank (Holt *et al.*, 1990). Red drum eggs (*c.*1.0mm diameter) are positively buoyant and spherical, with a single oil globule. They can be counted volumetrically averaging 1000 eggs/ml. Egg development time is a function of temperature, but most eggs hatch within 24h (2.2mm) and exogenous feeding generally initiates within 48–72h (Holt *et al.*, 1981b). Larvae (Fig. 18.4) are grown initially on rotifers for the first 10 days and *Artemia* thereafter to metamorphosis, but they can be weaned to commercial larval diets at 7–10 days posthatch precluding the use of *Artemia* (Holt, 1993). Complete elimination of live prey in red drum larvae reared in the lab has been successful and suggests that replacement diets for live prey is feasible (Lazo *et al.*, 2000, 2002). Live prey is enriched with microalgae or commercial products with highly unsaturated fatty acids (HUFA) to fulfil the HUFA requirement (Brinkmeyer and Holt, 1998). Optimum temperature and salinity for growth of larvae is 27°C and 25–30‰ (Holt *et al.*, 1981a).

Larvae are reared in high density, recirculated-water tanks ranging in volume from a few hundred to thousands of litres. Generally, circular fibreglass tanks with dark coloured walls and bottom are used. Initial densities range from 10 to 30 larvae/l and survival through the larval stage can be > 50%. Oxygen

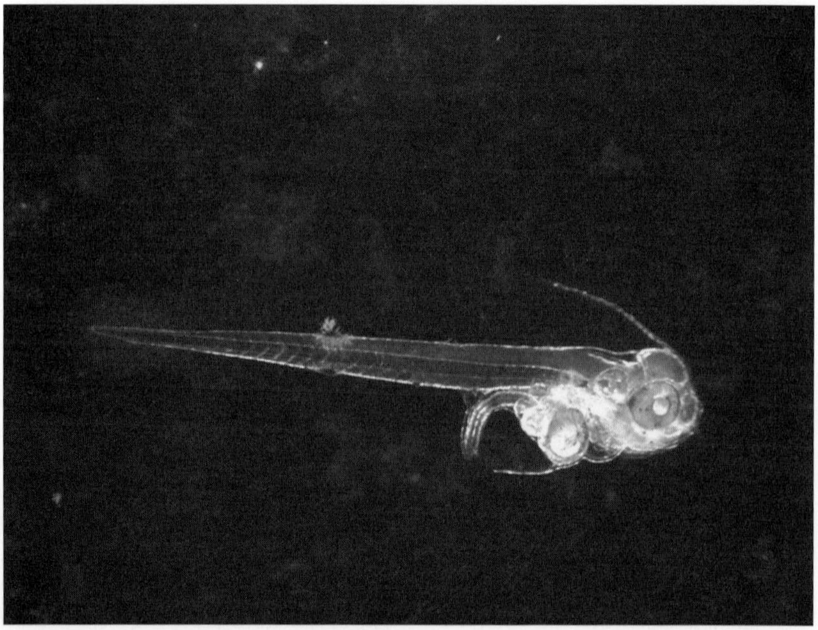

Fig. 18.4. One-day-old red drum larvae.

requirements are not unusual and maximum ammonia (1 ppm) and nitrite (10 ppm) levels (Holt and Arnold, 1983) are controlled easily by biofiltration. Growth in the warm temperatures in which they are spawned is fast, with fish reaching the juvenile stage (25 mm) in less than a month.

Red drum larvae for stock enhancement are generally reared in ponds with natural blooms of plankton induced by adding fertilizer (Colura *et al.*, 1990). Methods for pond culture (discussed in Holt, 2005) are well developed and production estimates for the Texas Parks and Wildlife stocking programme are 20–30 million fingerlings/year.

18.2.3 On-growing to market size

Commercial grow-out of red drum is typically performed in earthen ponds, raceways, cages and net pens (Matlock, 1990; Miller, 1995). In the USA, red drum is commonly cultured in earthen ponds and sometimes in intensive raceway culture systems using either flow-through or recirculating water systems (Gatlin, 2000). Harvest size is usually 1–2 kg in North America and yearly production levels are in the range of 900–1000 Mt. In countries such as China, cage culture in embayments is the preferred method, with production levels reaching 3000 Mt (Holt, 2005). Typical rearing densities range from 0.5 to 2.2 kg/m^2 (for 1–1.5 kg fish) (Hopkins, 1990; Davis, 1991; Thaker *et al.*, 1991).

It takes 12–18 months to reach commercial size, depending on rearing temperature, which seems to be the most critical environmental parameter for adequate growth. Optimal rearing temperatures are in the range of 24–30°C, with 28°C providing the highest growth rate without significantly affecting survival (Neill, 1987). One of the main problems encountered during pond culture of red drum is their low tolerance to cold temperatures, for mortality occurs below 10°C (Craig *et al.*, 1995). Low temperature exposure particularly has been an issue in outdoor pond culture operations in the southern USA during winter. Several overwintering techniques have been evaluated to overcome this problem, including dietary manipulations and the use of thermal refuges, although they have met with limited economic success (Gatlin, 2000).

Although red drum can tolerate oxygen levels below 2.5 mg/l for a limited time, the recommended levels are near or above 5 mg/l. It is a good practice to maintain unionized ammonia levels in the water below 0.3 mg/l for early juveniles and below 1.0 mg/l during the grow-out period (Holt and Arnold, 1983; Holt, 2005). Red drum can be raised in fresh water if hardness is at least 100 mg/l (Neill, 1987). As fish grow, there is an increase in the tolerance to low salinities and the grow-out phase can be performed in waters with salinities as low as 4 ppt.

Red drum juveniles accept dry diets readily. The type of diets most commonly used are high energy extruded floating diets or slow sinking pellets, with a nutrient content of 35–45% protein, 7–11% lipid and 40 kJ/g diet total energy. Typical feeding rates vary from 2 to 7% of body weight, depending on the size of the fish (Gatlin, 2002). Feed particle size ranges from 0.25 mm for a 0.5 g fish to > 9 mm for a > 500 g fish (Gatlin, 2002). During the grow-out period, susceptibility to disease is low. The most commonly encountered problem during culture of larvae and early juveniles is infection with the dinoflagellate parasite of the genus *Amyloodinium*. This parasite is usually treated with fresh water baths or chelated copper compounds (Johnson, 1988).

On reaching commercial size, harvesting is commonly performed by using a seine in earthen ponds or fish pumps in raceway systems. Farm-raised red drum are sold to wholesale seafood markets and brokers as whole fish on ice. Larger fish can be processed and sold as fillets. Red drum fillets are white, flaky, mild and lacking in a strong fish taste, which is appropriate for the US market. Fillet characteristics are as follows: 77.5% water, 19.2% protein and 1.25% lipid (Miller, 1995). Yield is 34% in the case of skin-on fillets and 28% for skinless fillets. Price per kg of whole fish depends on presentation, but gutted whole fish can range from US$4.32 to US$7.41/kg. Skinless fillet can reach prices around US$10.00/kg fillet (wholesale). Red drum can also be commercialized as live fish for sport fishing and for seeding of golf course lakes. In Texas, there is an ongoing production of juveniles for stock enhancement programmes.

The nutritional requirements of red drum have been reviewed recently by Gatlin (2002). Protein requirements have been estimated to be around 35–45%, depending on size, culture system, protein quality and energy level in the diets, among others factors (Daniels and Robinson, 1986; Serrano *et al.*, 1992). A recent study concluded that red drum reared in seawater required at least 44% crude protein for maximum growth and feed efficiency (Thoman *et al.*, 1999). In addition, feedstuffs of plant origin may provide an adequate source of protein for red drum. For example, soybean meal can provide up to 90% of the protein in the diet without affecting performance, as long as 10% of the protein is derived from fishmeal (McGoogan and Gatlin, 1997).

If lipids are provided as marine fish oils, including them at levels of 7–11% of the diet result in adequate growth and survival if protein levels are around 40% (Gatlin, 2002). Red drum seem to have a limited capacity to chain-elongate and desaturate short chain fatty acids to produce polyunsaturated fatty acids (PUFAs) and therefore linolenic acid is usually included in the feed to meet n-3 PUFA requirements (Lochmann and Gatlin, 1993). Juvenile red drum require 10% of the total lipids as n-3 HUFA, mainly docosahexaeonic acid (DHA) and eicosapentaenoic acid (EPA) in a ratio of 2:1 (Lochmann and Gatlin, 1993; Brinkmeyer and Holt, 1998).

As in most marine fish, red drum use dietary protein and lipids preferentially as energy instead of carbohydrates. Although carbohydrate requirements have not been established, red drum can be fed moderate levels of carbohydrates if they are provided in soluble form in the diet (Daniels and Robinson, 1986; Williams and Robinson, 1988). Diets with 35–45 kJ/g diet of total dietary energy (estimated digestible energy 15 kJ/g diet) provide maximum growth and survival (Serrano *et al.*, 1992). If excess energy is available in the diet, red drum will tend to deposit lipid in the visceral cavity (intraperitoneal fat) and muscle, which can lead to poor dress-out percentages, reduce shelf life and is undesirable for the 'health conscious' consumer. Alternative lipid sources such as medium chain triglycerides have been evaluated to reduce this problem, with limited success (Davis *et al.*, 1999).

Red drum diets do not usually contain more than 7% fibre (Gatlin, 2000). Red drum requires 15 vitamins for adequate growth and survival. Although vitamin- and mineral-specific requirements have not been fully established at this time, warmwater fish formulations are typically used successfully (NRC, 1993).

In summary, practical diet formulations for red drum typically contain 40% crude protein from fish and soybean meal, 7% lipids mainly from fish oil, 25% carbohydrate from wheat, corn or rice, no more than 7% total fibre and the recommended warmwater fish vitamin and mineral premixes.

Red drum is a disease-resistant species that is very suitable for aquaculture practices. Although susceptibility to disease is low, red drum are infected occasionally by some protozoan parasites, bacteria and viruses. Under culture situations, the most common disease-causing organisms are the dinoflagellate, *A. ocellatus*, and bacteria such as *Vibrio* (Gatlin, 2000). *Amyloodinium* is usually only a problem during the larval and early juvenile stages, when density of organisms in culture systems is high. Infection by *Amyloodinium* affects the gills, causing fusion of the lamellae, which reduces oxygen uptake and leads to mortality due to suffocation. The most common treatment involves the use of copper compounds such as copper sulphate or a freshwater bath, if possible. Free copper concentration in the water should be kept around 0.3 ppm for several hours to kill the parasite effectively (Johnson, 1988, 1990; Hawke, 1991).

18.2.4 Conclusion

The red drum is a hardy, fast-growing, disease-resistant species that can tolerate a wide range of salinities and relatively high temperatures. Due to a ban on commercial fishing in the USA, an increase in support for aquaculture was implemented that led to the development of sound ecological and economical culture techniques to overcome many of the problems associated with marine fish culture. Current annual production of red drum worldwide is estimated to be around 45,000 Mt, with China producing over 94% of total production (FAO, 2007). Other countries besides the USA which produce red drum are Israel, Martinique and Mexico. The total value of this production is around US$55 million. Today, diets developed specifically for red drum grow-out have eliminated much of the waste load in recirculating systems. Nevertheless, problems are still associated with overwintering conditions, where high mortalities occur due to this species' low tolerance to cold temperatures. Offshore cage culture or land-based recirculating systems may resolve the overwintering issues. Further research is needed in the areas of broodstock nutrition, egg and larval quality, weaning diets, feed efficiencies and disease prevention and control.

18.3 Meagre, *Argyrosomus regius*

The genus *Argyrosomus* is represented presently by eight species. The vernacular names for *A. regius* are meagre, drum, jew, jewfish, croaker, salmon-basse, sea-sheep (English), corvina (Spanish and Portuguese), ombrevis (Dutch), maigre (French), adlerfisch (German), bocca d'oro (Italian) and sariagiz, which means yellow mouth (Turkish).

The fisheries of common meagre, *A. regius*, are restricted to the East Atlantic and Mediterranean areas. Total catches of around 10,000 t/year are limited and present high yearly variations with high dispersion of fish size. Such characteristics dictate large price fluctuations, which indicate consumer interest in meagre. Moreover, the species exhibits the major features of a good candidate for aquaculture development. These are flesh quality and flavour, fast growth at mild temperatures and market size obtained in a reasonable period. Meagre in the Mediterranean has developed over recent years, supported by a limited number of hatcheries located in France.

The meagre (Fig. 18.5) has an elongated, laterally depressed body. The fish can weigh up to 103 kg (Maigret and Ly, 1986) and reach 230 cm in length (Quéro and Vayne, 1987). Externally, the species is characterized by large silver scale flanks becoming purple bronze near the dorsal fins. The lateral line is perfectly visible and extends conspicuously to the caudal fin. The second dorsal fin is longer than the first (Quéméner, 2002).

Fig. 18.5. Meagre.

In the Atlantic, the species is seen from Iceland to Senegal (Quéro and Vayne, 1987) but the usual distribution limit is the Bay of Biscay (France). The meagre is also widely distributed in the Mediterranean and Black Seas and may reveal a Lessepsian species[1] since it is also found in the Red Sea and occasionally in the Indian Ocean. (Quéro, 1989) (Fig. 18.6).

A. regius is a benthopelagic amphibiotic fish living in coastal waters in a depth range from 15 to 80 m, and sometimes to a depth of 200–300 m (Quéro and Vayne, 1997; Riede, 2004). It also occurs in surface waters chasing cephalopods or shoals of mugilids and clupeids. The species shows reproduction migration and congregates inshore and in shallow waters of the ocean shelf during spring. Adults and juveniles move along the shore or offshore–onshore to find their thermic preferenda (Chao and Trewavas, 1990). The temperature which determines migration to higher locations should be close to 13–14°C, while the optimum for growth is estimated to be 17–21°C and the maximum 23°C (Tixerant, 1974).

Although widely spread, meagre is seen to migrate and spawn into only three different areas. In the Atlantic Ocean, spawning aggregations have been observed in the bay of Lévrier in Mauritania (Tixerant, 1974) and the Gironde estuary in France (Quéro and Vayne, 1993). In the Mediterranean, reproduction occurs in the Nile Delta (Arraes, 1994). The environmental and climatic conditions of these areas are different and the determining features for meagre reproduction vary in relation to these different areas. These are temperatures from 17 to 23°C, salinity from 10 to 37‰ and a varying photoperiod due to latitude differences and climatic season. This suggests segregation of discrete populations (Quéméner, 2002). The different strains may show variations in otolith shape (Tixerant, 1974).

The age of maturity is assumed to be 4–5 years in the wild, with males around 70 cm long and females 85 cm long. Tixerant (1974) reported a relative fecundity of 800,000 eggs (990 μm diameter) for a 1.2 m-long female when the water temperature was between 17 and 23°C.

[1] Lessepsian migration is the ongoing migration of marine species across the Suez Canal, usually from the Red Sea to the Mediterranean Sea, more rarely in the opposite direction. Named after Ferdinand de Lesseps, the engineer in charge of the construction. *Source*: http://en.wikipedia.org/wiki/Lessepsian_migration.

Fig. 18.6. World distribution of *Argyrosomus regius*.

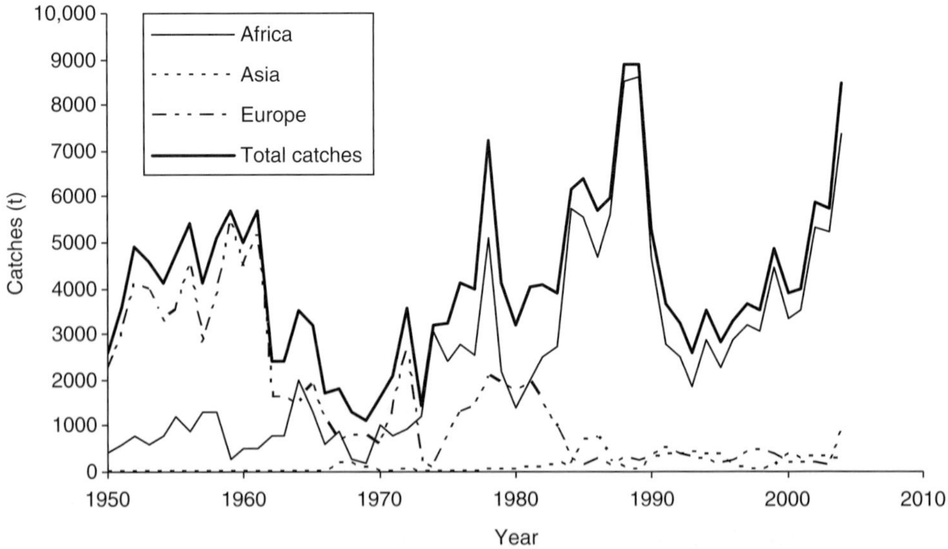

Fig. 18.7. Variations of landings of meagre from 1950 to the present. *Source*: FAO/FIGIS, 2006.

The wild meagre growth curve was established using the von Bertalanffy model by the following equation: TL (total length) = $210(1\text{-}e^{-0.089(t\text{-}to)})$ based on catches from Mauritania. This model provides interesting growth evaluation for aquaculture, with fish weighing around 2 kg after 3 years using the relation weight = $0.00826 \times TL^{3.059}$ (Tixerant, 1974). According to this author, meagre less than 30 cm chase mainly mysidacae, shrimps and small demersal fishes. Then the feeding regime is comprised of mainly cephalopods and pelagic fishes.

The total annual catches of *A. regius* have increased up to 10,000 t from the 1950s to the present, although it has remained a poor contribution of the total world or even European finfish capture. Figure 18.7 shows that the increase of capture is characterized by high annual variations of landings, so that annual recruitment may be very dependent on climatic conditions. North-east Atlantic fisheries were the major provider of meagre until the early 1960s, but they collapsed and are now compensated by a rapid increase in fisheries yield from Central East Atlantic (Morocco and Mauritania). According to Quéméner (2002), the Mediterranean and North-east Atlantic yield only 10% of total meagre fisheries landings each.

18.3.1 Farming of meagre

The production of meagre by aquaculture is recent and increasing rapidly up to about 1000 t/year in restricted areas of southern France, Corsica and Italy (Fig. 18.8). However, information is scarce about meagre biology in aquaculture conditions due to the small size of the meagre industry. One French farm (Les Poissons du Soleil Inc) with hatchery, nursery and on-growing has succeeded in

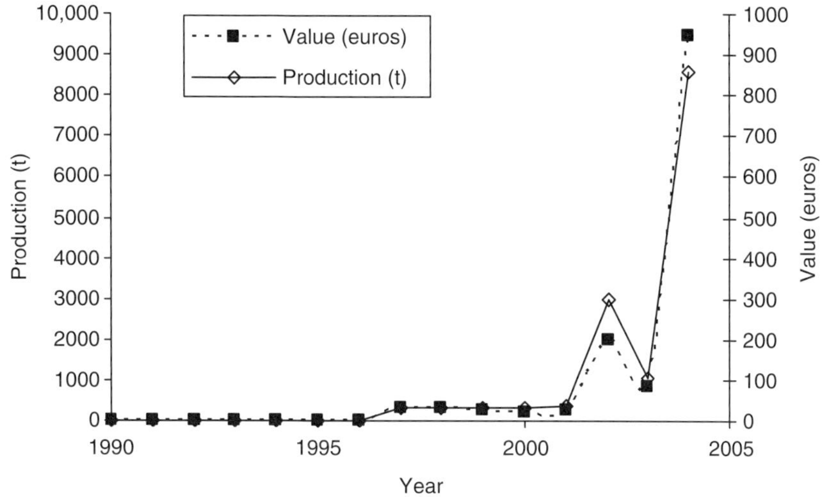

Fig. 18.8. Meagre aquaculture production surge from 1990 to the present (FAO/Figis, 2006).

the artificial reproduction and rearing of meagre. This operation is actually producing over 3 million fry/year in recirculated systems that also supply French and Italian on-growing farms (Poli *et al.*, 2003) and one Spanish offshore cage farm operation (San Pedro del Pinatar, south-east Spain). The Spanish facility is composed of 18 offshore cages, 16 m in diameter and 15 m net depth (approximately 3000 m^3/cage), with a total production of 400 t/year of gilthead (*Sparus aurata*: see Chapter 16) and meager (Aguado-Giménez *et al.*, 2007). Under normal rearing densities, the fish can achieve 1.2–1.5 kg after 2 years in the cages.

18.3.2 Broodstock management and hatchery operations

Wild-caught meagre adapt quite well to captivity, so it is possible to set up broodstocks with high genetic variability. However, it is difficult to evaluate precisely the age of puberty of the meagre. The set-up of broodstock from progeny born in captivity shows that the first reproduction occurs in 5-year-old females, as is the case in wild meagre. Under adequate combinations of external factors, the meagre reproductive season can be shifted to allow a longer and better scheduling of fry production. During late spring, ovulation, egg laying and fertilization occur spontaneously in tanks and the embryos can be collected quickly by a net trap situated at the surface outflow. The application of hormonal GnRHa treatment during the reproductive season is used successfully to trigger the process to ensure accurate scheduling. Since the treated females are generally allowed to spawn in a communal broodstock tank, there is no information about individual fecundity and ovulation rhythm, but 'tank spawning' occurs within 3 days after heterologous GnRHa stimulation and can last for several days without any means of knowing which fish produce eggs. At present, only two hatcheries located in France maintain broodstock of about 20 females whose global fecundity covers the current industry supply needs of fry.

The males produce sperm for at least 3 months covering the female spawning time. Sperm is collected easily by a gentle pressure of the male abdomen. Mean milt concentration is about 30 10^9 spz/ml, with no significant variations along the season. Sperm can be stored for several hours over ice without any loss of activity potential. The initial motility triggered by seawater concerns 80–100% of the gametes and the progressive movement lasts 1.5–3 min (Fauvel, unpublished).

The following data are drawn from Tixerant (1974), since current features of cultured meagre larvae are not available. Meagre egg diameter is around 900–1000 μm. The eggs float and usually show a unique oil droplet of 250 μm diameter. Embryonic development is temperature dependent and leads to hatching at around 750 degree days. The survival of starved larvae does not exceed 96 h at 25–30°C. Current knowledge indicates there is no major difficulty in larval rearing of meagre. The size of hatching larvae and their growth by yolk resorption implies the use of rotifers as a first feed and the rapidity of growth makes it necessary to provide *Artemia* in large quantities to prevent cannibalism, which is a characteristic of the species. The technology for meagre larval rearing and weaning has improved steadily and it is presently compatible with economic requirements and is profitable in current market conditions. Useful information may be drawn from rearing trials published by Battaglene and Talbot (1994) on a very proximate species, the mulloway, *A. hololepidotus*.

18.3.3 On-growing to market size

The meagre is adapted to the classic rearing conditions used for seabass and seabream in shallow waters (cages) as well as onshore (tanks). In the northern Mediterranean, meagre can reach 1.2 kg in less than 24 months in semi-intensive cage culture (final charge: 50 kg/m^3). The species is eurythermic and euryhaline, with significant growth at 17°C, but shows high growth rate during summer when water temperatures are around 21°C. As was the case for rearing facilities, the meagre is fed successfully using the same pellets as for other Mediterranean marine species, with food conversion between 1.5 and 2, depending on rearing conditions.

According to Bykov (2000), meagre flesh is well liked by consumers. Its characteristics are comparable to those of other marine fish of the same area. (Table 18.1). The low rate of lipids and their balance in muscle gives the species

Table 18.1. Composition of meagre flesh (Bykov, 2000).

	%
Water	75
Proteins	19.5
Lipids	2.3
Ashes	1.4

a high dietetic value and a positive image for the consumers (Table 18.1). Moreover, the fillet yield when processing is comparable to other fish, like seabass for example (Table 18.2) (Bykov, 2000; Poli *et al.*, 2001a). The recent work from Poli *et al.* (2001b and 2003) provides interesting information on the species' long shelf life and the morphological, marketable and nutritional traits of farmed meagre.

Available market information on meagre is poor. The species is known and appreciated in limited areas of South France and seems better known in Italy (Montfort, 2006). The market price of wild-caught meagre in France (€2–9/kg) is highly dependent on fish landing (Fig. 18.9), but it is much lower than wild seabass prices (€7–17/kg), for example. But meagre and seabass produced by aquaculture currently show similar prices around €7/kg, probably due to the homogeneity of the product and the regularity of market supply.

Table 18.2. Proportions of different body parts of meagre of 1.8–4.1 kg from North Africa fisheries and comparisons with seabass (*Dicentrarchus labrax*) and Atlantic salmon (*Salmo salar*).

	Meagre (%)	Salmon (%)	Seabass (%)
Head	18–24	18–24	19–30
Body (without fins)	61–64	75	55–59
Skin	1.2–2	5	–
Viscera	9–17	10–17	9.3–23
Scales	0.5–3	–	–
Fillet yield	46.5	52.5	37

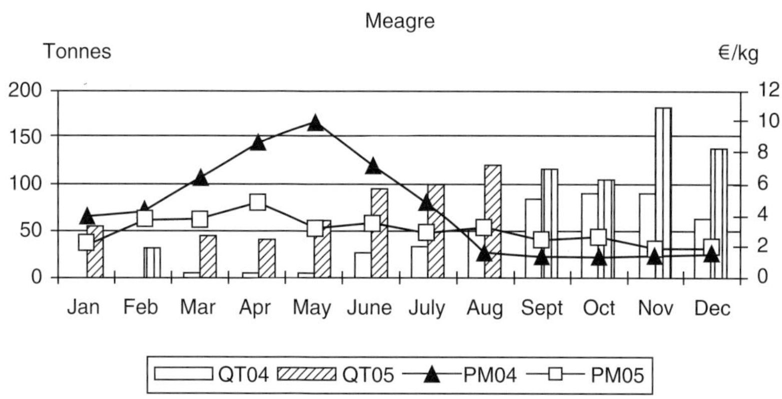

Fig. 18.9. An example of variations of price in relation to availability of meagre. The figure compares monthly average prices (PMn) and landed quantities (QTn) between 2004 and 2005. Poor captures during the first month of 2004 resulted in a high market price of meagre, while an increase of landing depreciates the market (drawn from OFIMER, 2006).

However, the increase of global supply through the combination of meagre aquaculture and wild harvested meagre may influence the market price positively by increasing the number of people who taste the fish, thereby influencing its market acceptability and demand.

18.3.4 Concluding comments

The adoption of general marine fish rearing technology has been sufficient to reach the present status of meagre aquaculture. The limitation of its development is related to the competition with well-farmed seabass in a cultured fish market. However, the meagre can grow faster and reach a larger size, so it may occupy a different market segment. Such considerations should direct research into economics and marketing to provide an identity for farmed meagre and forecast its market evolution to increasing production and gaining more consumers. If this is resolved, meagre can meet the problems of supply regularity in high-quality fry production and the associated production costs. Such bottlenecks would be solved by a better control of reproduction and satisfaction of nutritional needs, both in terms of quality and quantity. The study of environmental determinants of reproductive activity and larva behaviour would sustain the progress of production successfully. The rapid growth in current rearing conditions under a classic marine fish commercial diet does not justify a large research effort on the growth phase. Finally, since availability of wild-caught broodstock is restricted, it may be profitable to study the potential somatic and germinal effects of consanguinity on progenies.

18.3.5 Conclusion

Meagre aquaculture is quite recent and has not reached its maturity. Rearing techniques and production are sufficient to provide fry for current needs. Better knowledge of potential market share may be a prerequisite for further developmental effort. As meagre does not show peculiarities in classic rearing conditions, an increase of production may raise domestication problems, the resolution of which can be drawn from other aquaculture research.

References

Aguado-Giménez, F., Marin, A., Montoya, S., Marin-Guirao, L., Piedecausa, A. and Garcia-Garcia, B. (2007) Comparison between some procedures for monitoring offshore cage culture in western Mediterranean Sea: sampling methods and impact indicators in soft substrata. *Aquaculture* 271, 357–370.

Arnold, C.R. (1988) Controlled year-round spawning of red drum *Sciaenops ocellatus* in captivity. In: Arnold, C.R., Holt, G.J. and Thomas, P. (eds) *Red Drum Aquaculture: Proceedings of a Symposium on the Culture of Red Drum and Other Warm Water Fishes. Contributions in Marine Science. Supplement to Volume 30*. University of Texas at Austin, Texas, USA, pp. 65–70.

Arnold, C.R. (1991) Precocious spawning of red drum. *Progressive Fish Culturist* 53, 50–51.

Arnold, C.R., Bailey, W.H., Williams, T.D., Johnson, A. and Lasswell, J.L. (1979) Laboratory spawning and larval rearing of red drum and southern flounder. *Proceedings of the 1977 Annual Conference Southeastern Association of Fish and Wildlife Agencies* 31, 437–440.

Arraes, R. (1994) *Analyse des captures de maigre Argyrosomus regius (Poissons, Perciformes, Sciaenidae) dans le golfe de Gascogne de 1987 à 1993*. Mémoire de DEA (MSc thesis), Université Pierre et Marie Curie, Paris VI, 50 p.

Battaglene, S.C. and Talbot, R.B. (1994) Hormone induction and larval rearing of Mulloway, *Argyrosomus hololepidotus* (Pisces: Sciaenidae). *Aquaculture* 126, 73–81.

Brinkmeyer, R. and Holt, G.J. (1998) Highly unsaturated fatty acids in diets of red drum (*Sciaenops ocellatus*) larvae. *Aquaculture* 161, 253–268.

Bykov, V.P. (2000) *Chemical Composition and Processing Properties of Marine and Ocean Fishes: Handbook*. M. VNIRO, Moscow.

Caldwell, S. and Carr, F. (2000) *Changes of Tides*. Coastal Conservation Association, Houston, Texas.

Chao, L.N. and Trewavas, E. (1990) *Sciaenidae*. In: Quéro, J.C., Hureau, J.C., Karrer, C., Post, A. and Saldanha, L. (eds) *Check-list of the Fishes of the Eastern Tropical Atlantic* (CLOFETA), Volume 2. JNICT, Lisbon, SEI, Paris and UNESCO, Paris, pp. 813–826.

Colura, R.L., Bumguardner, B.W., Henderson-Arzapalo, A. and Gray, J.D. (1990) Culture of red drum fingerlings. Report No 22, Management Data Series. Texas Parks and Wildlife Department, Coastal Fisheries Branch, Austin, Texas.

Craig, S.R., Neill, W.H. and Gatlin, D.M. III (1995) Effects of dietary lipid and environmental salinity on growth, body composition and cold tolerance of juvenile red drum (*Sciaenops ocellatus*). *Fish Physiology and Biochemistry* 14, 49–61.

Daniels, W.H. and Robinson, E.H. (1986) Protein and energy requirements of juvenile red drum (*Scieanops ocellatus*). *Aquaculture* 53, 243–252.

Davis, D.A., Lazo, J.P. and Arnold, C.R. (1999) Response of juvenile red drum (*Sciaenops ocellatus*) to practical diets with medium chain triglycerides. *Fish Physiology and Biochemistry* 21, 235–248.

Davis, J.T. (1991) *Red Drum: Production of Food Fish*. Publication No 322. Southern Regional Aquaculture Center, Stoneville, Mississippi.

Drawbridge, M.A. (2002) The role of aquaculture in the restoration of coastal fisheries. In: Costa-Pierce, B. (ed.) *Ecological Aquaculture, The Evolution of the Blue Revolution*. Blackwell Science, Osney Mead, Oxford, UK, pp. 314–336.

FAO (2007) Red drum (*Sciaenops ocellatus*): cultured aquatic species, fact sheets (http://www.fao.org/fishery/culturedspecies/Sciaenops_ocellatus/en, accessed 11 May 2007).

Furuya, K. (1995) Current condition and subject of marine culture in Japan. In: Maim, K.L. and Rosenfeld, C. (eds) *Culture of High-value Marine Fishes in Asia and the United States*. Oceanic Institute, Honolulu, Hawaii, pp. 219–230.

Gatlin, D.M. III (2000) Red drum culture. In: Stickney, R.R. (ed.) The *Encyclopedia of Aquaculture*. John Wiley and Sons, New York, pp.736–743.

Gatlin, D.M. III (2002) Red drum, *Sciaenops ocellatus*. In: Webster, C.D. and Lim, C.E. (eds) *Nutrient Requirements and Feeding of Finfish for Aquaculture*. CAB International, New York, pp. 147–158.

Hawke, J.P. (1991) An overview of the diseases of cultured hybrid striped bass and red drum. Louisiana Aquaculture Conference 1991, Louisiana Cooperative Extension Service and Louisiana Agricultural Experiment Station, LSU Agricultural Center, Baton Rouge, Louisiana, pp. 52–54.

Hernderson-Arzapalo, A., Colura, R. and Maciorowski, A.F. (1994) A comparasion of black drum, red drum and their hybrids in saltwater pond culture. *Journal of the World Aquaculture Society* 25, 289–296.

Holt, G.J. (1993) Feeding larval red drum on microparticulate diets in a closed recirculating water system. *Journal of the World Aquaculture Society* 24, 225–230.

Holt, G.J. (2005) Red drum aquaculture. In: Kelly, M. and Silverstein, J. (eds) *Aquaculture in the 21st Century. American Fisheries Society Symposium* 46, 457–463.

Holt, G.J. and Arnold, C.A. (1983) Effects of ammonia and nitrite on growth and survival of red drum eggs and larvae. *Transactions of the American Fisheries Society* 112, 314–318.

Holt, G.J., Holt, S.A. and Arnold C.R. (1985) Diel periodicity of spawning in sciaenids. *Marine Ecology Progress Series* 27, 1–7.

Holt, G.J., Arnold, C.A. and Riley, C.M. (1990) Intensive culture of larval and post-larval red drum. In: Chamberlain, G.W., Midget, R.J. and Haby, M.G. (eds) *Red Drum Aquaculture*. Texas A&M Sea Grant College Program No TAMU-SG-90-603. College Station, Texas, pp. 53–56.

Holt, J., Godbout, R. and Arnold, C.R. (1981a) Effects of temperature and salinity on egg hatching and larval survival of red drum, *Sciaenops ocellata. Fisheries Bulletin* 79, 569–573.

Holt, J., Johnson, A.G., Arnold, C.R., Fable, W.A. Jr and Williams, T.D. (1981b) Description of eggs and larvae of laboratory reared red drum, *Sciaenops ocellata. Copeia* 4, 751–756.

Hopkins, J.S. (1990) Semi-intensive grow-out in saltwater. In: Chamberlain, G.W., Miget, R.J. and Haby, M.G. (eds) *Red Drum Aquaculture*. Texas A&M Sea Grant College Program, College Station, Texas, pp. 170–177.

Johnson, S.K. (1988) Control of diseases in red drum. In: Arnold, C.R., Holt, G.J. and Thomas, P. (eds) *Red Drum Aquaculture: Proceedings of a Symposium on the Culture of Red Drum and Other Warm Water Fishes. Contributions in Marine Science. Supplement to Volume 30*. University of Texas at Austin, Texas, USA, pp. 113–130.

Johnson, S.K. (1990) Recognition and control of diseases common to grow-out aquaculture of red drum. In: Chamberlain, G.W., Midget, R.J. and Haby, M.G. (eds) *Red Drum Aquaculture*. Texas A&M Sea Grant College Program Technical Report TAMU-SG-90-603. College Station, Texas, pp. 113–130.

Lazo, J.P., Dinis, M.T., Holt, G.J., Faulk, C. and Arnold, C.R. (2000) Co-feeding microparticulate diets with algae: toward eliminating the need of zooplankton at first feeding in larval red drum (*Sciaenops ocellatus*). *Aquaculture* 188, 339–351.

Lazo, J.P., Holt, G.J. and Arnold, C.R. (2002) Towards the development of suitable microdiets for substitution of live prey in the rearing of red drum (*Sciaenops ocellatus*) larvae: Applications of studies on digestive physiology. *Fisheries Science* 68 (Supplement 1), 888–891.

Leis, J.M. and Trnski, T. (1989) *The Larvae of Indo-Pacific Shorefishes*. New South Wales University Press, Sydney, Australia.

Liao, C., Su, M.S. and Chang, S.L. (1995) A review of the nursery and grow-out techniques of high value marine fishes in Tawain. In: Maim, K.L. and Rosenfeld, C. (eds) *Culture of High-value Marine Fishes in Asia and the United States*. Oceanic Institute, Honolulu, Hawaii, pp. 121–137.

Lochmann, R.T. and Gatlin, D.M. III (1993) Evaluation of different types and levels of tryglicerides, singly and in combination with different levels of n-3 highly unsaturated fatty acids ethyl esters in diets of juvenile red drum (*Sciaenops ocellatus*). *Aquaculture* 114, 113–130.

Lutz, G. (1999) Red drum: a re-emerging aquaculture species. *Aquaculture Magazine* 25, 1–3.

McEachron, L.W., McCarty, C.E. and Vega, R.R. (1995) Beneficial uses of marine fish hatcheries: Enhancement of red drum in Texas coastal waters. *American Fisheries Society Symposium* 15, 161–166.

McGoogan, B.B. and Gatlin, D.M. III (1997) Effects of replacing fish meal with soybean meal in diets for red drum *Sciaenops ocellatus. Journal of the World Aquaculture Society* 28, 374–385.

Maigret, J. and Ly, B. (1986) *Les poissons de mer de Mauritanie*. Sciences Naturelles, Compiègne, France.

Matlock, G.C. (1984) A basis for the development of a management plan for red drum in Texas. Doctoral dissertation, Texas A&M University, College Station, Texas.

Matlock, G.C. (1990) Preliminary results of red drum stocking in Texas. In: Sparks, A. (ed.) *Marine Farming and Enhancement: Proceedings of the 15th US–Japan Meeting on Aquaculture*, 22–23 October 1986, Kyoto, Japan. Marine Farming and Enhancement, US Department of Commerce, pp. 11–15.

Miller, W.W. (1995) *Red drum. Aquaculture Curriculum Guide*. National Council for Agricultural Education, USDA, Iowa.

Monfort M.C. (2006) FAO markets and marketing of aquaculture, finfish in Europe, focus on the Mediterranean Basin. For FAO Fisheries Division, April 2006, *www.marketing-seafood.com*.

Murphy, M.D. and Taylor, R.G. (1990) Reproduction, growth, and mortality of red drum *Sciaenops ocellatus* in Florida waters. *Fishery Bulletin* US 88, 531–542.

Neill, W.H. (l987) Environmental requirements of red drum. In: Chamberlain, G.W., Miget, R.J. and Haby, M.G. (eds) *Manual on Red Drum Aquaculture*. Texas Agricultural Extension Service and Sea Grant College Program, College Station, Texas, pp. IVl–IV8.

Nelson, J.S. (1994) *Fishes of the World*, 3rd edn. John Wiley and Sons, New York.

NRC (National Research Council) (1993) *Nutrient Requirements of Fish*. National Academy Press, Washington, DC.

Poli, B.M., Parisi, G., Mecatti, M., Lupi, P., Iurzan, F., Zampacavallo, G. and Gilmozzi, M. (2001a) The meagre (*Argyrosomus regius*), a new species for Mediterranean aquaculture. 1. Morphological, merchantable and nutritional traits in a commercial wide size-range. *European Aquaculture Society Special Publication* 29, 209–210.

Poli, B.M., Parisi, G., Mecatti, M., Lupi, P., Iurzan, F., Zampacavallo, G. and Gilmozzi, M. (2001b) The meagre (*Argyrosomus regius*), a new species for Mediterranean aquaculture. 2. Freshness involution and flesh dietetic traits in large commercial-size fish. *European Aquaculture Society Special Publication* 29, 211–212.

Poli, B.M., Parisi, G., Zampacavallo, G., Iurzan, F., Mecatti, M., Lupi, P. and Bonelli, A. (2003) Preliminary results on quality changes in reared meagre (*Argyrosomus regius*): body and fillet traits and freshness changes in refrigerated commercial-size fish. *Aquaculture International* 11, 301–311.

Quéméner, L. (2002) *Le maigre commun (*Argyrosomus regius*). Biologie, pêche, marché et potentiel aquacole*. Editions Ifremer, Plouzané, France.

Quéro, J.C. (1989) Le maigre, *Argyrosomus regius* (Asso, 1801) (Poissons, Sciaenidae) en Méditerranée occidentale. *Bulletin de la Société Zoologique de France* 114(4), 81–89.

Quéro, J.-C. and Vayne, J.-J., (1987) Le maigre, *Argyrosomus regius* (Asso 1801) (Pisces, Perciformes, Sciaenidae) du Golfe de Gascogne et des eaux plus septentrionales. *Revue des Travaux des Pêches Maritimes* 49(1–2), 35–66.

Quéro, J.C. and Vayne, J.J. (1993) Nouvel indice sur les pérégrinations du maigre. *Annales de la Société des Sciences Naturelles de la Charente-Maritime*, 8(2), 127–128.

Quéro, J.C. and Vayne, J.J. (1997) *Les poissons de pêche des mers francaises*. Delachaux et Niestlé SA, Lausanne-Paris, France.

Randall, J.E. (1983) *Caribbean Reef Fishes, Revised Edition*. T.F.H. Publications, Neptune City, New Jersey.

Riede, K. (2004) Global register of migratory species – from global to regional scales. *Final Report of the R&D-Projekt 808 05 081*. Federal Agency for Nature Conservation, Bonn, Germany.

Ross, J.L., Stevens, T.M. and Vaughan, D.S. (1995) Age, growth, mortality, and reproductive biology of red drums in North Carolina waters. *TAFS* 124, 37–54.

Serafy, J.E., Ault, J.S., Capo, T.R. and Schultz, D.R. (1999) Red drum, *Sciaenops ocellatus* L., stock enhancement in Biscayne Bay, FL, USA: assessment of releasing unmarked early juveniles. *Aquaculture Research* 30(10), 737–750.

Serrano, J.A., Nematipour, G.R. and Gatlin, D.M. III (1992) Dietary protein requirements of the red drum (*Sciaenops ocellatus*) and relative use of dietary carhohydrates and lipid. *Aquaculture* 101, 283–291.

Simmons, E.G. and Breuer, J.P. (1962) A study of redfish, *Scianops ocelleta* (Linnaeus), and black drum, *Pegonias cromis* (Linnaeus). *Publications of the Institute of Marine Science University of Texas* 8, 184–211.

Thacker, S.G., Griffin, W., Rahman, A. and Lambregts, J. (1991) Modeling the financial risk of intensive redfish *Sciaenops ocellatus* aquaculture in Texas. *Journal of the World Aquaculture Society* 22, 42–43.

Thoman, E., Davis, A. and Arnold, C.R. (1999) Evaluation of growout diets with varying protein and energy levels for red drum (*Sciaenops occelatus*). *Aquaculture* 176, 343–353.

Thomas, P., Arnold, C.R. and Holt, G.J. (1995) Red drum and other sciaenids. In: Bromage, N.R. and Roberts, R.J. (eds) *Broodstock Management and Egg and Larval Quality*. Blackwell Scientific Publishers, Oxford, UK, pp. 118–137.

Tixerant, G. (1974) Contribution à l'étude de la biologie du maigre ou courbine *Argyrosomus regius* (Asso) = *Sciaena aquila* (Lacép.) sur la cote mauritanienne. Thèse d'université d'Aix-Marseille, France.

True C.D., Loera, A.S. and Castro, N.C. (1997) Acquisition of broodstock of *Totoaba macdonaldi*: field handling, decompression, and prophylaxis of an endangered species. *Progressive Fish-Culturist* 59, 246–248.

Van Olst, J.C. and Carlberg, J.M. (1987) The Feasibility of Corvina Aquaculture for Population Enhancement and Food Production. Phase I report. NSF/IST-87069, National Technical Information Service, San Diego, USA.

WanShu, H. and Yong, Z.Q. (2003) Review of captive bred species and fry production of marine fish in China. *Aquaculture* 227, 305–318.

Williams, C.D. and Robinson, E.H. (1988) Response of red drum to various dietary levels of menhaden oil. *Aquaculture* 70, 107–120.

19 The Wolffishes (Family: Anarhichadidae)

Nathalie R. Le François,[1,2] Helge Tveiten,[3] Laura C. Halfyard[4] and Atle Foss[5]

[1]*Biodôme de Montréal, Montréal, Québec, Canada and* [2]*Université du Québec à Rimouski, Rimouski, Québec, Canada;* [3]*Nofima Marine, Tromsø, Norway;* [4]*Marine Institute, Memorial University of Newfoundland, St John's, Newfoundland, Canada;* [5]*Akvaplan-Niva A/S, Bergen, Norway*

19.1 General Introduction

The Anarhichadidae (Order: Perciformes Suborder: Zoarcoidei) (Anderson, 1994) (Fig. 19.1) is a small family of primarily demersal marine fish species found in the North Atlantic and the North Pacific Oceans and inhabiting shallow to deep cold waters. This family includes two Pacific species (wolf-eel, *Anarrhichthys ocellatus* Ayres, and Bering wolffish, *Anarhichas orientalis* Pallas) and three Atlantic species (spotted wolffish, *A. minor* Olafsen; common or Atlantic wolffish, *A. lupus* L.; and northern wolffish, *A. denticulatus* Krøyer). Figures 19.2 and 19.3 show the world distribution of Atlantic and spotted wolffish, respectively. In the northern Atlantic, the distribution of the three species overlaps, but each species displays distinct depth preferences. The most coastal of the species is the Atlantic wolffish (1–500 m, found mostly from 18–110 m), followed by the spotted wolffish (25–590 m but most common from 100–400 m) and, in the deepest depth range, the Northern wolffish (60–900 m). Wolffishes have compressed and moderately elongated bodies (torpedo-like) reaching 120–145 cm, with a blunt head and heavy jaws with a dentition adapted to their distinctive feeding niche, i.e. bottom-living crustaceans, echinoderms and invertebrates (Albikovskaya, 1982). The swimming activity level of this family of fish can be qualified as sluggish. Wolffish display very limited migrations that are associated mostly with spawning events. Various aspects of the biology of the Atlantic wolffish have been investigated in the north-west Atlantic regions (Albikovskaya, 1982; Keats *et al.*, 1985; Templeman, 1984, 1986), off Iceland (Jonsson, 1982; Gunnarsson *et al.*, 2006) and in the north-east Atlantic Ocean and the Barents Sea (Barsukov, 1959; Falk-Petersen and Hansen, 1991).

Some species are of commercial importance in the fisheries, especially those from the Atlantic Ocean and the Barents Sea (*A. lupus* and *A minor*).

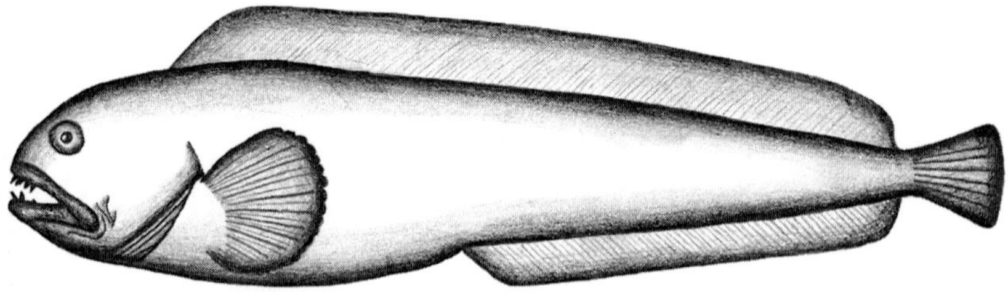

Fig. 19.1. Anarhichadidae.

The Northern wolffish (*A. denticulatus*), contrary to the Atlantic and spotted wolffish (*A. minor*), is not regarded as a commercial species (Simpson and Kulka, 2002). The absence of large concentrations of wolffishes qualifies their fishery in Canadian waters as accidental or as by-catch. Much of the present global market supply of wolffish is still dependent on wild stocks. In Canada, the Species At Risk Act (SARA) presently prohibits wild harvest of the northern and spotted wolffish, placing them on the 'threatened' conservation status due to a 93% decrease in stocks since the early 1980s; the common wolffish is on the 'special concern' list (DFO, 2006, www.sararegistry.gc.ca). In the USA the Atlantic wolffish stocks are considered depleted and experiencing low to moderate levels of threats. However, these threats are not sufficient to put them at risk of extension now or in the foreseeable future. The species is however retained on the Species of Concern List Management Plan (K. Damon-Randall, NOAA, USA, personal communication). Wolffishes belong to a group of fish that are cryptic, that do not congregate throughout most of their lives and are difficult to observe in their natural habitat.

Wolffishes have attributes and biological characteristics that make them highly suitable for intensive farming (Brown *et al.*, 1995; Le François *et al.*, 2002; Foss *et al.*, 2004). Pavlov and Moksness (1994) compared the culture of wolffish to commercial salmon farming, with similar husbandry and systems technology transfer and accordingly they share many desirable aquaculture traits. Research into wolffish aquaculture was first initiated by Norway and Russia (Pavlov and Novikov, 1986; Moksness *et al.*, 1989) and focused initially on the Atlantic wolffish. Studies have since shown the superior growth of the spotted wolffish, *A. minor* (Moksness, 1994), and actual commercialization strategies are now based primarily on this species. A recent study (Imsland *et al.*, 2008) and reports from Barsukov (1959) suggest that wolffish species hybridize in nature. Characterization of the reciprocal hybrids of *A. lupus* and *A. minor* as aquaculture alternatives is currently being assessed (Gaudreau *et al.*, 2009). This chapter will discuss primarily the spotted wolffish (Fig. 19.4) since the commercial potential for this species (growth and behaviour) outweighs that of the common wolffish and other species in this family significantly. However, relevant research on the common wolffish will be provided.

Fig. 19.2. World distribution of *Anarhichas lupus*.

Fig. 19.3. World distribution of *Anarhichas minor*.

Fig. 19.4. Spotted wolffish (photograph by Marc Lajoie, MAPAQ).

19.2 Spotted Wolffish, *Anarhichas minor*, and Atlantic Wolffish, *A. lupus*

The Atlantic wolffish is believed to be adapted to a wider range of environmental conditions. Ovarian growth takes place over several months starting in May–June and spawning occurs in September–October (Keats *et al.*, 1985; Tveiten and Johnsen, 2001). As the spawning season approaches, the fish appear to migrate to cold water (Jonsson, 1982; Pavlov and Novikov, 1993) and egg survival is reduced if oocyte maturation and ovulation occur at high temperatures (Tveiten *et al.*, 2001). During summer and autumn, wolffish form pairs (Keats *et al.*, 1985; Pavlov and Novikov, 1986) and the fish spawn in nesting holes at depths from 15 to 300 m (Pavlov and Novikov, 1993). Fertilization is considered to be internal and probably occurs in the few hours following copulation. Internal fertilization in fishes is associated generally with reduced milt and egg production, unusual sperm motility characteristics, a long ovarian maturation period, a large egg size, a long incubation period, paternal care and precocious ontogeny at hatching. Large fertilized eggs (5–6 mm) are released in ovarian fluid (Johannessen *et al.*, 1993; Pavlov, 1994a). The female wraps herself around the eggs, which forms a spheroid egg mass. Later, the male protects the egg mass throughout incubation, which may last for 9–10 months (Keats *et al.*, 1985; Pavlov and Novikov, 1993). In captivity, males do not usually display normal spawning behaviour and the females release unfertilized eggs. Methods for artificial fertilization of wolffish eggs have been developed (Pavlov, 1994a,b). Comparatively, not much is known of the spawning events of the spotted wolffish in the wild due to the difficulties involved in their observation.

Both species of wolffish are not sufficiently abundant on the Canadian coast to be of economic importance. No directed fisheries exist and catches are usually taken as a by-catch from offshore trawlers or long-line fisheries. Commercial fishery of mostly *A. lupus* takes place in the Barents Sea and the banks off the north of Norway, and the Icelandic, the Russian and the Faroe Island coasts.

19.2.1 Farming of spotted wolffish

Norwegian wolffish culture has led global production with broodstock development; a pilot commercial farm was in operation in Tomma (Nordland) until 2007. It was established in 1999 and produced 100–150t in 2005. The company had its own broodstock, juvenile production, on-growing facilities, where shallow raceways were used, and on-site processing facilities. The species is also evaluated at the pilot-scale level in Iceland and Canada. Both countries have established wild-caught broodstock and have produced juveniles for on-growing initiatives.

19.2.2 Broodstock management and husbandry

Wolffish males have small testes and a low gonadosomatic index (0.10–0.30), produce a small volume of ejaculate (range 0.2–10.6ml; mean 1.9ml) (Moksness and Pavlov, 1996) and have low sperm concentrations (range 5–1198 × 10^6 spz/ml) (Johannesen *et al.*, 1993; Pavlov, 1994a; Tveiten and Johnsen, 1999). Unlike the sperm of most teleosts, that of both common (Moksness and Pavlov, 1996) and spotted (Kime and Tveiten, 2002) wolffish is motile on stripping and can retain activity for at least 2 days. The sperm loses motility when exposed to seawater.

In Atlantic wolffish, maximum sperm concentration is registered during peak ovulation (Moksness and Pavlov, 1996) and it appears that sperm concentration does not decrease if stripping is carried out at 12–14 day intervals. In spotted wolffish, milt volume, sperm density and motility is decreased significantly if stripping is carried out at 3–4 day intervals. In captive male wolffish, the number of sperm per ejaculate is, however, often insufficient to fertilize all the eggs from one female. This indicates that under culture condition, where appropriate environmental and social cues may be lacking, sperm production may be impaired. Today, low sperm/milt production during the breeding period is one constraint to the development of wolffish cultivation. To circumvent this problem, gonadotropin-releasing hormone analogue (GnRHa) treatment, which is shown to be effective in increasing sperm/milt production in several other fish species, has been used. GnRHa treatment did not result in any significant changes in sperm volume, activity or density. However, steroid treatment (11-KT and 17,20β-P) increased milt volume 2–3 times compared to control fish without affecting sperm density. Thus, sex steroid treatment may be a method to circumvent, at least in part, the problem of low milt and sperm production.

Recent studies identified a commercial semen preservation media (Cryo-Fish©, IMV Technologie, France) as promising (Le François *et al.*, 2008: *A. lupus*; Gunnarsson *et al.*, 2009: *A. minor*). Both studies recognize the need for further experiments in order to define optimum conditions for cryopreservation. The natural presence of antifreeze proteins in the semen of both species (*A. lupus* > *A. minor*) was confirmed possibly by conferring some degree of freezing protection in both types of gametes (eggs and sperm cells) (Le François *et al.*, 2008).

Maturing wolffish females are easily recognized by their large rounded belly. A couple of days before egg release, there is an additional increase in abdominal size, probably related to the increased production of ovarian fluid. Ovulation is indicated by the opening of the genital pore. Unfertilized eggs are stripped when the opening is 6–8 mm. Stripping prior to or after this stage may result in eggs of low developmental capacity or the release of unfertilized eggs to the water, respectively. Eggs are fertilized by undiluted sperm from at least 2–3 males. To ensure high fertilization, eggs and milt are incubated for 2–3 h, with gentle mixing every half hour. Tveiten and Johnsen (1999) found that the fertilization rate was generally between 80 and 90% when sperm–egg ratios were in excess of $1.5{:}10^4$. In Atlantic wolffish, the fertilization rate is also influenced by the contact time between gametes and Pavlov (1994a,b) reported that more that 2 h is required for high fertilization. If the sperm–egg ratio is low, contact time may be increased up to 7 h. In spotted wolffish, using egg and sperm of apparently good quality, 100% fertilization may, however, be achieved within 1 h. A subjective assessment of wolffish milt quality (i.e. sperm motility) prior to fertilization is carried out easily using a binocular microscope to avoid/discard poor quality milt.

After transfer to seawater, eggs are incubated in upwelling incubators at temperatures of 6–8°C until hatch, which occurs after 900–950 degree days (D°). Eggs incubated at 6°C display higher survival during initial feeding and subsequent growth (Hansen and Falk-Petersen, 2001a). Mortality during incubation is generally highest during the first couple of weeks and low or insignificant after the early eye stage (*c.*240 D°) in good egg batches (Falk-Petersen *et al.*, 1999; Hansen and Falk-Petersen, 2001a). Due to the long incubation period, eggs should be disinfected once per month with glutaraldehyde (150 ppm for 5 min) for the first two-thirds of the incubation period (Hansen and Falk-Petersen, 2001b). Less toxic disinfection agents that would meet health agency regulations, such as Citrox© (bioflavonoids with natural fruit acids) or Perosan© (peracetic acid and hydrogen peroxide), should be developed (I.-B. Falk-Petersen, personal communication).

Responses to temperature changes during reproductive development differ between species, but temperature tolerance seems to be lowest in species inhabiting environments of low and stable temperature, such as wolffishes. In Atlantic wolffish, exposure to elevated temperature (> 8°C) during vitellogenesis resulted in shifts in the timing of ovulation (4–5 weeks delay), a significantly lower egg survival (Tveiten and Johnsen, 2001) and a lower survival and growth rate of the newly hatched larvae (Lamarre *et al.*, 2004). The largest eggs were produced by the fish kept at 8°C (Tveiten and Johnsen, 1999). Effects on egg survival were more pronounced when fish were exposed to different temperatures during final oocyte maturation and ovulation than during ovarian growth (Tveiten *et al.*, 2001). A temperature within the range 4–12°C did not have any clear effects on male reproductive performance (Tveiten and Johnsen, 1999; Tveiten *et al.*, 2001).

It would be advantageous to be able to manipulate maturation and spawning time to produce a year-round supply of eggs and fry and also to delay maturation until after harvest size, avoiding reduced flesh quality. In Atlantic wolffish kept at two different light–dark regimes (18L:6D and 6L:18D) for a year, milt volume and sperm concentration were not affected significantly (Moksness and Pavlov, 1996). Preliminary experiments showed a significant influence on the timing of ovulation. In spotted wolffish, both sexes respond to photoperiod manipulation. When compared to a control broodstock (natural photoperiod), spawning after 18 months in the manipulated group resulted in a lower percentage of fertilized eggs, a lower relative fecundity and smaller eggs, while no difference in egg survival was detected.

For wolffish, no detailed study regarding the effect of broodstock fish nutrition on egg quality has been carried out. However, there appears to be a potential to improve egg quality by manipulating the composition of the broodstock fish diet, since Tveiten *et al.* (2004) found that embryonic survival was correlated to fatty acid composition of the unfertilized egg. Savoie *et al.* (2008a) strongly suggest that poor digestive enzyme activity (trypsin deficiency) is not the main agent responsible for poor survival, but rather the failure to initiate feeding. Feeding stimulants such as protein hydrolysates (PH) and egg quality issues (incubation protocols, broodstock nutrition) should be targeted for further investigations aimed at improving survival (see Savoie *et al.*, 2006, 2008b: protein hydrolysate inclusion).

In contrast to most marine fish species of interest in aquaculture, newly hatched wolffishes are quite large (20–25 mm) (Fig. 19.5), do not go through any metamorphosis or yolk sac resorption phase and accept dry feeds readily without the need for live feed (rotifers or *Artemia*). The general practice for intensive commercial production for the spotted wolffish is to provide only a

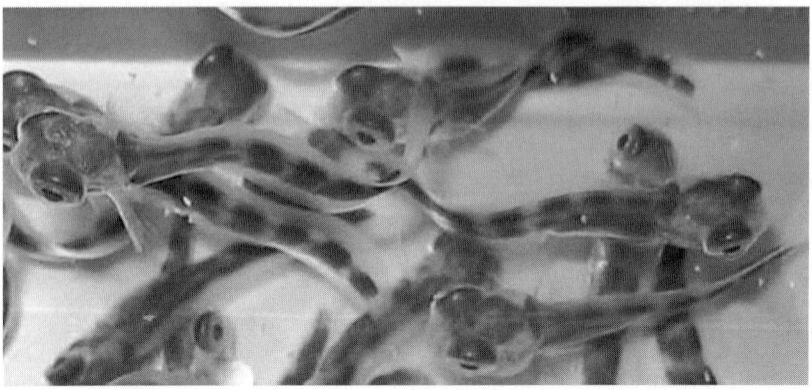

Fig. 19.5. Spotted wolffish juveniles shortly after hatching (photograph by A. Savoie, UQAR).

dry diet for first feeding and to avoid *Artemia* production. Adapted feeds are available but have at this stage a limited availability outside Norway. Marine fish feeds developed for cod or halibut may be used, as similar nutrient requirements for wolffish have been observed (Strand *et al.*, 1995; Halfyard *et al.*, 2000). First-feeding and on-growing rearing generally are conducted in low-level raceways of increasing volume as growth occurs, such as described in Strand *et al.* (1995). Key parameters are low depth, low current velocity and the use of a floating feed. A recent study by Savoie *et al.* (2006, 2008b) incorporating 20% PH improved the survival of newly hatched spotted wolffish significantly. The feed that contained 87% of the peptides below 1900 Da (molecular weight) yielded the best survival rates, 82.7 ± 5.79% compared to 67.3 ± 3.5% for the control without PH.

Falk-Petersen *et al.* (1999) studied newly hatched spotted wolffish larvae reared at 4, 6 and 8°C and growth was found to increase with increasing temperature during the first 48 days after hatching. Hansen and Falk-Petersen (2002) reared newly hatched larvae at ambient temperature (2.9–4.5°C) and at 6, 8, 10 and 12°C. During the first 63 days, growth rates increased with increasing temperature (SGRs from 1.8 %/day at ambient temperature to 3.1 and 4.7%/day at 6 and 12°C, respectively). In a second experiment, fish were reared at 8, 10, 12, 14 and 16°C for 30 days to detect any potential detrimental effects of temperature and all groups were then reared at a constant temperature of 8°C for an additional 33 days. Overall, when considering both growth and survival, the optimal temperature during the weaning phase was calculated to be 10.3°C (Hansen and Falk-Petersen, 2002). Savoie *et al.* (2006) obtained similar SGRs in weight per day: 4.01 and 4.88% at 8 and 12°C, respectively. McCarthy *et al.* (1998) and Lamarre *et al.* (2007), working with the newly hatched Atlantic wolffish, reported growth rates of 5.5% and 5.4% (0–30 and 0–50 days posthatch, respectively) at 14 and 9–10°C, respectively. Recent studies by Lamarre *et al.* (2009) and Lamarre *et al.* (in press), indicate that the spotted wolffish growth potential (i.e. maximum growth rate at a given fish mass) is equivalent to many other cold-water species of aquaculture interest. Given that aquaculture protocols and requirement of that species in terms of nutrition and environmental needs are still to be defined in details, this species can be considered as an attractive alternative species under cold environments.

Very few conclusive studies on light and photoperiod requirement have been conducted on different wolffish species and their life stage (Foss *et al.*, 2004). Currently, low light intensities are generally used for on-growing of juveniles and broodstock, whereas long day-lengths (LD18:6) are used during husbandry.

19.2.3 On-growing to market size

Due to the benthic behaviour of this fish and its social tolerance to high stocking densities, raceway or shallow tank systems similar to those used for flatfish species have been proposed. However, problems with water quality maintenance at higher densities need to be addressed. This section will discuss other options being investigated for wolffish culture. One hyperintensive concept consists of a shallow raceway system where shallow raceways are stacked in racks. The water is reused between levels and this system effectively would reduce the overall logistic needs with respect to buildings and water supply systems. Also, through its compactness and extended automation, it could simplify the operation in the production process (Øiestad, 1999). Another option might be a flow-through system, which may be hyperoxygenated prior to the single passage through the fish tanks. Such systems may also be used in conjunction with recirculation aquaculture system (RAS) technology to intensify the production system further and maximize the use of water (see Langan and Couturier, Chapter 24, this volume). Recently, Imsland *et al.* (2007) compared growth performances between shallow raceways and conventional circular tanks (without shelves). Overall growth rates and food conversion efficiency were 14% and 17% higher, respectively, in shallow raceways.

The possibility of rearing wolffish in sea cages was investigated recently by Mortensen *et al.* (2007). The systems tested were flat-bottom net cages with shelves, modified from similar cage systems commonly used for sea-based culture of Atlantic halibut. Results from the studies indicated that juveniles could be transferred to the net cages at a size of approximately 20 g and, from that size until the fish reached 300 and 450 g, fish grew equally well compared to fish held in conventional land-based systems.

Moksness (1994) compared growth of wild-caught common and spotted wolffish fry. The spotted wolffish obtained weights more than four times higher than that of the common wolffish (1.58 kg and 0.37 kg, respectively) 2 years after hatching. A similar difference between the two species was demonstrated by Monsaas *et al.* (unpublished data, referenced in Foss *et al.*, 2004), where growth between the species was compared in juveniles reared at three different temperatures around the respective species expected $T_{opt}G$ (6, 8 and 10°C for spotted wolffish and 8, 10 and 12°C for common wolffish). After approximately 6 months, spotted wolffish (all groups) had obtained weights 4–6 times higher than that of the common wolffish. Recent investigations reveal sex differences in the growth rate of both *A. lupus* and *A. minor* (Dupont-Cyr *et al.*, submitted).

Generally, fish show a temperature optima for growth ($T_{opt}G$) and survival which may change with age and size. Imsland *et al.* (2006) reared spotted wolffish for approximately 7 months at 4, 6, 8 and 12°C. They calculated a $T_{opt}G$ of juvenile spotted wolffish in the size range 135–380 g, dropping from 7.9°C for 130–135 g to 6.6°C for 260–380 g juveniles.

Tolerance to low salinities was investigated by Foss *et al.* (2001) and Le François *et al.* (2004b). These studies revealed that both species were strong osmoregulators, which makes possible their cultivation at reduced salinities with expected benefits in productivity. Identification of the optimal rearing densities at a given size are still needed, but work from Jonassen (2002) and current studies conducted by Tremblay-Bourgeois *et al.* (in press) suggest that optimal rearing densities within the size range of 50–160 g lie between 30 and 40 kg/m^2, which approximates to densities higher than 200 kg/m^3. Foss *et al.* (2004)

suggested that stocking densities of 200–300 kg/m^3 will provide acceptable growth rates in both juvenile (70–200 g) and near market-sized fish (1.5–3 kg). Evaluation of density induced stress response of spotted wolffish revealed that the species is very tolerant to crowding (Tremblay-Bourgeois *et al.*, in press).

Espelid (2002) reviewed the disease risks for commercial wolffish farming and found them to be very low. There have been recorded incidences of the following diseases in wolffish: *Vibrio anguillarum*, atypical furunculosis, *Aeromonas salmonicida*, *Trichodina*, *Costia* and *Pleistophora*. Treatments have included: *Trichodina* (formalin bath); *Vibrio* sp., *Flexibacter* sp. (tetracycline and Chloramine-T).

A Canadian market study, mirroring on a previous European study (Richardsen and Johansen, 2002) was conducted in major Canadian cities and showed high consumer and chef ratings of the product (Laflamme *et al.*, 2005). Canadian chefs preferred smaller market size 2–2.5 kg instead of 5 kg fish reported in the European study. Overall, both the Norwegian and Québec market studies had positive impressions of wolffish. However, the product is not well known and requires the initiation of a marketing strategy aimed at long-term brand building. Respondents appreciated the freshness and long shelf life of the product. Other commercial avenues include the potential for fish leather (see Ingram and Dixon, 1994) and residual by-products (blood, skin, viscera, mucus) (Le François *et al.*, 2004a; Desjardins *et al.*, 2006, 2007; Desrosiers *et al.*, 2008) (see Blier and Le François, Chapter 25, this volume). Both market studies (European and North American) and Johnson and Halfyard (2002) identified the need to settle on a marketable name that would avoid confusion in the consumer market.

19.2.4 Concluding comments

During the past decade, spotted wolffish has emerged as a promising candidate for coldwater aquaculture and is now looked upon with growing interest in Norway, Iceland and Canada. Through committed research, a complete production cycle has been developed for the species in less than 10 years. Despite the reported robustness and rapid growth characteristics of wolffishes, one remaining hindrance to increased juvenile production is the high variability in egg and larval quality. Proper conditioning of broodstock (nutrition, environmental temperature, reduced stress levels), the development of domestic broodstock, genetic selection and breeding programmes should allow the production of better gamete and larval quality at hatching. Research groups generally agree that the major phases have been researched but, as in all emerging species, there is a general need for optimization of the production cycle, i.e. mainly light intensity, photoperiod, density and broodstock nutrition.

References

Albikosvkaya, L.K. (1982) Distribution and abundance of Atlantic wolffish, spotted wolffish and Northern wolffish in the Newfoundland area. *NAFO Scientific Council Studies* 3, 29–32.

Anderson, M.E. (1994) Systematics and osteology of the Zoarcidae (Teleostei: Perciformes). *Ichthyological Bulletin of the J.L.B. Smith Institute of Ichtyology* 60, 1–120.

Barsukov, V.V. (1959) The wolffish (*Anarhichadidae*). Fauna SSSR. Moscow N.S. 73. (Translation by Smithsonian Institute, 1972). NTIS, Springfield, Virginia.

Brown, J.A., Helm, H. and Muir, J. (1995) New candidate species for aquaculture. In: Boghen, A.G. (ed.) *Coldwater Aquaculture in Atlantic Canada*. Tribune Press, Sackville, Nebraska, pp. 341–362.

Desjardins, M., Le François, N.R., Fletcher, G.L. and Blier, P.U. (2006) Seasonal modulation of plasma antifreeze protein levels in Atlantic (*Anarhichas lupus*) and spotted wolffish (*A. minor*). *Journal of Experimental Marine Biology and Ecology* 335, 142–150.

Desjardins, M., Le François, N.R., Fletcher, G.L. and Blier, P.U. (2007) High antifreeze protein levels in Atlantic wolffish (*Anarhichas lupus*) make them ideal candidates for potentially ice-laden waters. *Aquaculture* 272, 667–674.

Desrosiers, V., Blier, P.U. and Le François, N.R. (2008) Trypsin-like enzyme from Atlantic wolfish (*Anarhichas lupus*) viscera: purification and characterization. *Journal of Aquatic Food Product Technology* 17(1), 11–26.

Desrosiers, V., Le François, N.R., Tveiten, H. and Blier, P.U. (2008) Ontogenesis of catabolic and energy metabolism capacities during the embryonic development of spotted wolffish (*Anarhichas minor*). *Comparative Physiology and Biochemistry Part B* 150(2), 200–206.

DFO (2006) www.sararegistry.ca (accessed 3 May 2006)

Dupont-Cyr, B.-A., Le François, N.R., Roy, R.L., Vandenberg, G.W. and Tveiten, H. (submitted) Sex difference in the growth rate of two species of wolffishes (*Anarhichas minor and A. lupus*). *Aquaculture.*

Espelid, S. (2002) Susceptibility of spotted wolffish to infectious diseases and use of immunoprophylaxis. *Bulletin of the Aquaculture Association of Canada* 102(2), 37–38.

Falk-Petersen, I.-B. and Hansen, T.K. (1991) *Reproductive Biology of Wolffish* Anarhichas lupus *from North Norwegian Waters*. International Council for the Exploration of the Sea, CM 1991/G.14.

Falk-Petersen, I.B. and Hansen, T.K. (2003) Early ontogeny of the spotted wolffish (*Anarhichas minor* Olafsen). *Aquaculture Research* 34, 1059–1067.

Falk-Petersen, I.B., Hansen, T.K., Fieler, R. and Sunde, L.M. (1999) Cultivation of the spotted wolffish *Anarhichas minor* (Olafsen) – a new candidate for cold-water fish farming. *Aquaculture Research* 30, 711–718.

Foss, A., Evensen, T.H., Imsland, A.K. and Oiestad, V. (2001) Effects of reduced salinities on growth, food conversion efficiency and osmoregulatory status in the spotted wolffish. *Journal of Fish Biology* 59, 416–426.

Foss, A., Imsland, A.K., Falk-Petersen, I.B. and Øiestad, V. (2004) A review of the culture potential of spotted wolffish *Anarhichas minor* Olafsen. *Reviews in Fish Biology and Fisheries* 14, 277–294.

Gaudreau, C., Le François, N.R., Ditlecadet, D., Tveiten, H. and Blier, P.U. (2009) Characterization of the early stages of the wolffish hybrid *Anarhichas minor* × A. lupus: conservation and aquaculture applications. *Aquatic Living Resources* 22, 371–377.

Gunnarsson, A., Hjörleifsson, E., Thórarinsson, K. and Marteinsdóttir, G. (2006) Growth, maturity and fecundity of wolffish *Anarhichas lupus* in Icelandic waters. *Journal of Fish Biology* 68, 1158–1176.

Gunnarsson, S., Sigurdsson, S., Thorarensen and Imsland, A.K. (2009) Cryopreservation of sperm from spotted wolffish. *Aquaculture International* 17, 385–389.

Halfyard, L.C., Parrish, C.C. and Jauncey, K. (2000) Nutritional requirements of Atlantic Wolffish *Anarhichas* lupus L. and *A. minor* Olafsen: Two New Coldwater Aquaculture Species. In: Shahidi, F. (ed.) Seafood in Health and Nutrition – Transformation in Fisheries and

Aquaculture: Global Perspectives. ScienceTech Publishing Company, St John's, Canada, chapter 35, pp. 449–476.

Hansen, T.K. and Falk-Petersen, I.B. (2001a) The influence of rearing temperature on early development and growth of spotted wolffish *Anarhichas minor* (Olafsen). *Aquaculture Research* 32, 369–378.

Hansen, T.K. and Falk-Petersen, I.B. (2001b) Effects of egg disinfection and incubation temperature on early life stages of spotted wolffish. *Aquaculture International* 9, 333–344.

Hansen, T.K. and Falk-Petersen, I.B. (2002) Growth and survival of first-feeding spotted wolffish (*Anarhichas minor* Olafsen) at various temperature regimes. *Aquaculture Research* 33, 1119–1127.

Imsland, A.K., Foss, A., Sparboe, L.O. and Sigurdsson, S. (2006) The effect of temperature and fish size on growth and feed efficiency ratio of juvenile spotted wolffish *Anarhichas minor. Journal of Fish Biology* 68, 1107–1122.

Imsland, A.K., Foss, A., Gunnarsson, S., Sparboe, L.O., Oiestad, V. and Sigurdsson, S. (2007) Culture of juvenile spotted wolffish, *Anarhichas minor* in shallow raceways compared to conventional raceway tanks. *Journal of World Aquaculture Society* 38(1), 154–160.

Imsland, A.K., Stensland, K., Johansen, T., Le François, N.R., Lamarre, S.G., Naevdal, G. and Foss, A. (2008) Genetic population structure of the spotted wolffish *Anarhichas minor* in the North Atlantic. *Open Marine Biology Journal* 2, 7–12.

Ingram, P and Dixon, G. (1994) Fish skin leather: an innovative product. *Journal of the Society of Leather technologies and Chemists* 79, 103–106.

Johannessen, T., Gjøsæther, J. and Moksness, E. (1993) Reproduction, spawning behaviour and captive breeding of the common wolffish *Anarhichas lupus* L. *Aquaculture* 115, 41–51.

Johnson, H. and Halfyard, L.C. (2002) Marketing issues and opportunities for farmed wolffish in the United States and Europe. *Bulletin of the Aquaculture Association of Canada* 102(2), 11–12.

Jonassen, T.M. (2002) Effects of photoperiod, stocking density and diet on growth in young spotted wolffish (*Anarhichas minor* olafsen). *Aquaculture International* 10, 411–420.

Jonsson, G. (1982) Contribution to the biology of catfish (*Anarhichas lupus*) at Iceland. *Rit Fiskideildar* 6, 2–26.

Keats, D.W., South, G.R. and Steele, D.H. (1985) Reproduction and egg guarding by Atlantic wolffish (*Anarhichas lupus*: Anarhichidae) and ocean pout (*Macrozoarces americanus*: Zoarcidae) in Newfoundland waters. *Canadian Journal of Zoology* 63, 2565–2568.

Kime, D.E. and Tveiten, H. (2002) Unusual motility characteristics of sperm of the spotted wolffish. *Journal of Fish Biology* 61, 1549–1559.

Laflamme, J., Michaud, J.-C., Lévesque, M. and Le François, N.R. (2005) Potentiel commercial et technico-financier de l'élevage du loup de mer au Québec. Final Report (http://www.sodim.org/pdf/AutresEspeces/710,15_Potentiel_comm_loup_tach.pdf, accessed 7 December 2006).

Lamarre, S., Le François, N.R., Falk-Petersen, I.-B. and Blier, P.U. (2004) Can digestive and metabolic enzyme activity levels predict growth rate and survival of newly hatched Atlantic wolffish? *Aquaculture Research* 35, 608–613.

Lamarre, S.G., Le François, N.R., Falk-Petersen, I.-B. and Blier, P.U. (2007) The digestive and metabolic enzyme activity profiles of a non-metamorphic marine fish species: effect of feed type and feeding level. *Canadian Journal of Fisheries and Aquatic Sciences* 64, 849–856.

Lamarre, S.G., Le François, N.R., Dreidzic, W. and Blier, P.U. (2009a) Protein synthesis is lowered while 20s proteasome activity is maintained following acclimation to low temperature in juvenile spotted wolffish (*Anarhichas minor,* Olafsen). *Journal of Experimental Biology* 212, 1294–1301.

Lamarre, S.G. Blier, P.U., W. Dreidzic and Le François, N.R. (in press). White muscle 20s proteasome activity is negatively correlated to growth rate at low temperature in the spotted wolffish (*Anarhichas minor;* Olafsen). *Journal of Fish Biology.*

Le François, N.R., Lemieux, H. and Blier, P.U. (2002) Biological and technical evaluation of the potential of marine and anadromous fish species for cold-water mariculture. *Aquaculture Research* 33, 95–108.

Le François, N.R., Desjardins, M. and Blier, P.U. (2004a) Enhancement of profitability perspectives of an aquaculture production by the extraction of high commercial value biomolecules: evaluation of the potential of wolffish cultivation. In: Shahidi, F. and Simpson, B.K. (eds) *Seafood Quality and Safety: Advances in the New Millennium*. Science Tech Publishing, St John's, Newfoundland, Canada, pp. 61–69.

Le François, N.R., Lamarre, S.G. and Blier, P.U. (2004b) Tolerance growth and haloplasticity of the Atlantic wolffish (*Anarhichas lupus*) to low and intermediate salinities. *Aquaculture* 236, 659–675.

Le François, N.R., Lamarre, S.G., Tveiten, H., Blier, P.U. and Bailey, J. (2008) Sperm cryoconservation in *Anarhichas* sp., endangered cold-water aquaculture species with internal fertilization. *Aquaculture International* 16(3), 273–279.

McCarthy, I., Moksness, E. and Pavlov, D.A. (1998) The effects of temperature on growth rate and growth efficiency of juvenile common wolffish. *Aquaculture International* 6, 207–218.

Moksness, E. (1994) Growth rates of the common wolffish, *Anarhichas lupus* L., and spotted wolffish, *A. minor* Olafsen, in captivity. *Aquaculture and Fisheries Management* 25, 363–371.

Moksness, E. and Pavlov, D.A. (1996) Management by life cycle of wolffish, *Anarhichas lupus* L., a new species for cold-water aquaculture; a technical paper. *Aquaculture Research* 27, 865–883.

Moksness, E., Gjosaeter, J., Reinert, A. and Fjallstein, I.S. (1989) Start-feeding and on-growing of wolffish (*Anarhichas lupus*) in the laboratory. *Aquaculture* 77, 221–228.

Mortensen, A., Toften, H. and Aas, K. (2007) Cage culture of spotted wolffish (*Anarhichas minor* Olafsen). *Aquaculture* 264, 475–478.

Øiestad, V. (1999) Shallow raceways as a compact resource-maximising farming procedure for marine fish species. *Aquaculture Research* 30, 1–10.

Pavlov, D.A. (1994a) Fertilization in the wolffish, *Anarhichas lupus*: external or internal? *Journal of Ichthyology* 34, 140–151.

Pavlov, D.A. (1994b) Maturation and artificial fertilization of eggs of captive common wolffish, *Anarhichas lupus* L. from the White Sea. *Aquaculture and Fisheries Management* 25, 891–902.

Pavlov, D.A. and Moksness, E. (1994) Production and quality of eggs obtained from wolffish (*Anarhichas lupus* L.) reared in captivity. *Aquaculture* 122, 295–312.

Pavlov, D.A. and Novikov, G.G. (1986) On the development of biotechnology for rearing of the White Sea wolffish, *Anarhichas lupus marisalbi*. I. Experience on obtaining mature sex products, incubation of eggs and rearing of the young fish. *Journal of Ichthyology* 26, 95–106.

Pavlov, D.A. and Novikov, G.G. (1993) Life history and peculiarities of common wolffish (*Anarhichas lupus*) in the White Sea. *ICES Journal of Marine Science* 50, 271–277.

Richardsen, R. and Johansen, J.A. (2002) Cultured wolffish (*Anarhichas minor*) – a potential species for exclusive restaurants? *Bulletin of the Aquaculture Association of Canada* 102(2), 13–16.

Savoie, A., Le François, N.R., Cahu, C., Blier, P.U. and Andreassen, P.U. (2006) Do protein hydrolysates improve survival and growth of newly-hatched spotted wolffish (*Anarhichas minor*), a non-metamorphic aquaculture fish species. *Aquaculture* 261, 782–788.

Savoie, A., Le François, N.R., Cahu, C. and Blier, P.U. (2008a) Metabolic and digestive enzyme activity profiles of newly hatched spotted wolffish (*Anarhichas minor* Olafsen): effect of temperature. *Aquaculture Research* 39, 382–389.

Savoie, A., Le François, N.R., Lamarre, S.G. and Cahu, C. (2008b) Dietary protein hydrolysates and trypsin inhibitor effects on the digestive capacities and performance of newly-hatched wolffish. Aquaculture America 2008, 9–12 February, Lake Buena Vista, Florida.

Simpson, M.R. and Kulka, D.W. (2002) États des stocks de trois espèces de loup de mer (*Anarhichas lupus*, *A. minor* et *A. denticulatus*) dans les eaux terreneuviennes. Ministère des Pêches et des Océans, Document de recherche 2002/078 (in French).

Strand, H.K., Hansen, T.K., Pedersen, A., Falk-Petersen, I.B. and Øiestad, V. (1995) First feeding of common wolffish on formulated dry feed diets in a low water-level raceway system. *Aquaculture International* 3, 1–10.

Templeman, W. (1984) Migration of wolffishes, *Anarhichas* sp. from tagging in the Newfoundland area. *Journal of Northwest Atlantic Fishery Science* 5, 93–97.

Templeman, W. (1986) Some biological aspects of Atlantic wolffish (*Anarhichas lupus*). *Northwest Atlantic Fishery Science* 7, 57–65.

Tremblay-Bourgeois, S., Le François, N.R., Roy, R.L., Benfey, T.J. and Imsland, A.K. (in press) Effect of rearing density on the growth and welfare indices of juvenile spotted wolffish, *Anarhichas minor* (Olafsen). *Aquaculture Research*.

Tveiten, H. and Johnsen, H.K. (1999) Temperature experienced during vitellogenesis influences ovarian maturation and the timing of final maturation in common wolffish (*Anarhichas lupus* L.). *Journal of Fish Biology* 55, 809–819.

Tveiten, H. and Johnsen, H.K. (2001) Thermal influences on temporal changes in plasma testosterone and oestradiol-17β concentrations during gonadal recrudescence in female common wolffish. *Journal of Fish Biology* 59, 171–178.

Tveiten, H., Solevåg, S.E. and Johnsen, H.K. (2001) Holding temperature during the spawning season influences final maturation and egg quality in common wolffish (*Anarhichas lupus* L.). *Journal of Fish Biology* 58, 374–385.

Tveiten, H., Jobling, M. and Andreassen, I. (2004) Influence of egg lipids and fatty acids on egg viability, and their utilization during embryonic development of spotted wolf-fish, *Anarhichas minor* Olafsen. *Aquaculture Research* 35, 152–161.

20 The Tunas (Family: Scombridae)

Chris Carter,[1] Barbara Nowak[1] and Steven Clarke[2]

[1]*University of Tasmania, Launceston, Tasmania, Australia;* [2]*South Australian Research and Development Institute (SARDI), Henley Beach, Australia*

20.1 General Introduction

There are 15 genera of scombrid tuna (Fig. 20.1) and about 50 species of these epipelagic marine fishes. Perhaps the most remarkable of which are the large *Thunnus* sp., which have an impressive array of adaptations for their pelagic lifestyle, involving large migrations between cold and tropical waters and deep dives (Collette *et al*., 2001). There are eight reported *Thunnus* sp., including the giant Atlantic (ABT, *T. thynnus*), Pacific (PBT, *T. orientalis*) and southern (SBT, *T. maccoyii*) bluefin tuna. Although recent molecular evidence questions whether the Atlantic and Pacific bluefin tuna are separate species (Chow *et al*., 2006), in this review, partly due to their separation in aquaculture, they will be treated individually. These three species will be the focus of this chapter and general aspects of their physiology are outlined in this section; several comprehensive collections and reviews on key aspects of tuna biology provide further details (Block and Stevens, 2001; Fromentin and Powers, 2005). Bigeye (*T. obesus*) and yellowfin (*T. albacares*) tuna are also being considered for aquaculture, mainly in Central and South America (Smart and Sylvia, 2006), South-east Asia and Oceania, but will not be discussed further.

The dorsal surface is blackish-blue, with silvery lower sides and ventral surface; they are superbly adapted for efficient sustained forward high-speed swimming in a straight line (Fig. 20.2). Body shape is thickened towards the anterior and a narrow caudal peduncle leads to a classic-shaped lunate caudal fin. Other fins are short, robust and retract into grooves on a smooth, strong elastic skin. Cone-shaped locomotory muscle blocks act against a complex system of tendons, septa, bones and skin to transfer considerable locomotory force to the caudal fin (Westneat and Wainwright, 2001). Tuna are obligate ram ventilators and suffocate rapidly if prevented from swimming (Brill and Bushnell, 2001). Minimum cruising speeds for *Thunnus* sp.

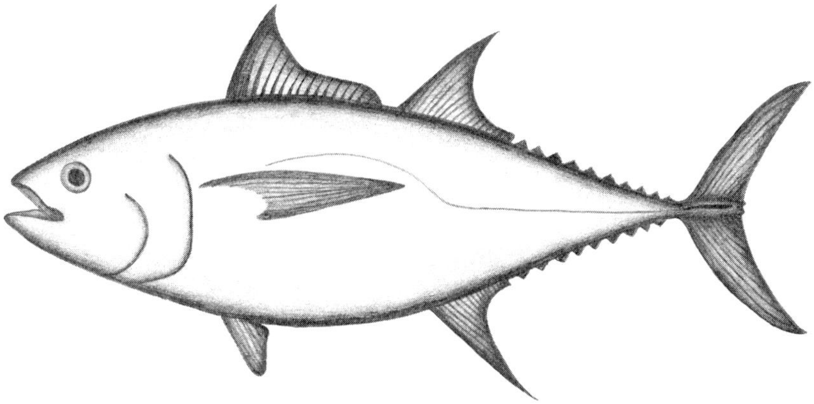

Fig. 20.1. Scombridae.

Fig. 20.2. Southern bluefin tuna swimming in a sea cage (C.G. Carter).

are about 1 body length (BL)/s (≡5.4 km/h for 150 cm fish), although lower for very large individuals (0.3 BL/s, ≡3.2 km/h for 300 cm fish) (Altringham and Shadwick, 2001).

Tuna are opportunist feeders and take a wide range of prey, mainly from schooling fish, cephalopod and crustacean species. For much of their life tuna swim together in schools; they may first locate prey via a chemical trail and they have large nasal rosettes that are very sensitive to low concentrations of small molecules such as amino acids. Once in range, they locate prey visually, feeding is frenetic and the water boils in a 'feeding frenzy'. Prey is usually swallowed whole. Tuna compete with other members of the school for food; individuals need to feed rapidly and take as much prey as possible in a limited time, so the stomach wall is elastic and the stomach has a large capacity (van Barneveld *et al.*, 1997).

20.2 Farming of Tuna, *Thunnus* sp.

Aquaculture industries have developed around the on-growing of wild-caught juvenile tuna, principally Atlantic bluefin tuna (ABT) in the Mediterranean, Pacific bluefin tuna (PBT) in Japan and Mexico and Southern bluefin tuna (SBT) in Australia (Carey *et al*., 1984; Kaji *et al*., 1996; Carter *et al*., 1998; Sawada *et al*., 2005). Sea cage-based operations generally have been successful at increasing the value of stock rapidly through relatively fast weight gain (Table 20.1), increasing the per kg value through improved condition and controlling the time of harvest in relation to demand, as well as managing foreign currency exchange risk. The majority of the bluefin tuna are marketed in Japan for sashimi and sushi, where they command a high price. Grading

Table 20.1. Growth of bluefin species held in cages on farms or experimentally under semi-commercial conditions.

Weight/age	Water temperature (°C)	Feed	Final weight (kg)	Growth (SGR %/day)	Final K	Reference
ABF						
0–1 y		W	4	2.3		Fromentin and Powers, 2005
1–10 y		W	150	0.1		
10–20 y		W	300	0.03		
32 kg		F	63	0.29		Aguado-Gimenez *et al*., 2006
219 kg			255	0.06		
1–2 y 6.4 kg	Increased 12 to 25 (mean 18.3)	F	28.3	0.29	2.33	Ticina *et al*., 2007
NBF						
Larvae 0–30 dph	23–26	L		7.68		Sawada *et al*., 2005
0.3 kg 100 dph		F	0.9	2.19	1.41	
SBT						
20–26 kg	Decreased	F	33.7	0.21	2.38	Carter *et al*., 1998
1–3 y	17 to 12	S	25.2	0.04	2.08	
		S	20.8	0.01	1.85	
27 kg 1–4 y	Decreased 20 to 16	S	33.8	0.28	2.13	Glencross *et al*., 2002b
30 kg		F		81 g/day		Glencross *et al*., 2002a
		S		81 g/day		
		E		21 g/day		

Note: K, condition factor = 100%*(weight [g]/length [cm]^3); y, years old; dph, days posthatch. Feed: E, formulated diet in form of extruded pellet; F, whole or part fish; L, larval live feeds; S, formulated diet in form other than extruded; W, wild prey. SGR, specific growth rate.

determines price and the grading of an individual fish relies on external characteristics and visual inspection of the 'tail-cut' and thickness of the belly flaps (Starling and Diver, 2005). In 2004, annual aquaculture production reported by FAO for these three species was 6958, 4030 and 517 t for ABT, SBT and PBT, respectively, with the majority coming from Spain (92% ABT) and Australia (100% SBT) (FAO, 2006). Sawada *et al.* (2005) suggested reasons that motivated research into closure of the life cycle were the variable quality of wild-caught juveniles and, as with tuna on-growing from the wild, further reductions to the size of fishery quota. Breeding from hatchery-raised first generation offspring has been achieved recently with PBT (Sawada *et al.*, 2005; Sakamoto *et al.*, 2006). Propagation programmes are now at different stages of development for SBT and ABT (Elizur *et al.*, 2006; Garcia, 2006; Mylonas *et al.*, 2006).

Factors that will determine whether tuna are suitable for aquaculture based around a closed life cycle are more complex than those which have made sea cage on-growing of juveniles profitable. Broodstock management and larval rearing on a commercial basis is a huge challenge for ABT and SBT industries. Attributes that make tuna suitable for on-growing include:

- Schooling at high densities in surface waters at predictable locations facilitates capture.
- Wild-caught tuna are robust fish that cope well with capture and towing from the ocean to inshore farms, adapt well to the confines of sea cages, are active feeders and grow quickly.
- There is an established high-value market for quality fresh and frozen tuna in Japan, the market accepts farmed tuna readily and there are increasing markets outside of Japan (China, Europe and the USA).

20.3 Atlantic Bluefin Tuna, *Thunnus thynnus*

Atlantic bluefin tuna (ABT) inhabits the temperate waters of the entire North Atlantic and adjacent seas so that its range extends from the equator to north of Norway and from the Gulf of Mexico to the Black Sea, via the Mediterranean (Collette *et al.*, 2001) (Fig. 20.3).

There is a long history of tuna fishing in the Mediterranean and evidence that it was happening 9000 years ago. Traditional fish traps are found in many countries including Spain, Morocco and Tunisia (Farwell, 2001; Fromentin and Powers, 2005). However, it was in Canada in the 1970s that early attempts at on-growing ABT were made. Large adults (350 kg plus) were held after natural spawning in 50 m × 15 m deep pounds constructed of heavy netting and fed baitfish (mackerel and herring) to increase their weight and condition (Carey *et al.*, 1984). Currently, on-growing is located in the Mediterranean, it started in the mid-1990s and the majority of ABT are produced in Spain, Croatia, Malta, Italy, Tunisia and Libya (FAO, 2006). Many farms on-grow 2–4-year-old wild-caught fish over one season in sea cages (Farwell, 2001; Ticina *et al.*, 2004) and the similarity to SBT farming is to be expected due to close links

Fig. 20.3. World distribution of *Thunnus thynnus*.

between the two industries. There has also been some long-term holding and on-growing of smaller fish for as long as 3 years (Ticina *et al*., 2004). The very large size of some farmed ABT, which can reach 500 kg plus (Aguado-Gimenez and Garcia-Garcia, 2005a,b), is one significant difference between SBT and poses further technical challenges. Significant research has been undertaken on ABT propagation over the past 3 years, with captive fish having been successfully hormone induced to spawn and fertilized eggs collected and grown through to 5-day-old larvae before being discarded.

Wild ABT can attain a very large size (300 cm, 900 kg). Being endothermic, ABT maintain body temperatures in waters ranging between 3 and 30°C, they dive to 1000 m and migrate large distances to distinct spawning grounds located in the warm waters of the Mediterranean and Gulf of Mexico (Fromentin and Powers, 2005). A capture-based fattening industry has developed for ABT relying on the capture of migrating wild fish. Maturity is reached in smaller fish in the east (110–120 cm, 30–35 kg) than in the west (200 cm, 150 kg). Spawning may occur only every second or third year and reproductive physiology and behaviour are similar to the other large tuna: asynchronous oocyte development and multiple batch spawning every 1–2 days. Several spawning grounds have been identified, required water temperatures are above 24°C and this places spatial and temporal limits on the location of spawning grounds. The development of captive broodstock management and spawning methods is essential for the support of a true and sustainable aquaculture industry for this species. Research is progressing; for example, Corriero *et al*. (2007) studied the effects of treatment with GnRHa on the reproductive maturation of ABT.

Eggs are about 1 mm diameter and they hatch to 3–4 mm yolk-sac larvae within 2 days. They are opportunistic feeders throughout the life cycle: larvae feed on zooplankton, mainly copepods, juveniles on crustaceans, fish and cephalopods (and may take octopus, salps, crabs and sponges) and adults mainly on fish (herring, anchovy, sardine and mackerel). Growth rates in the wild and on farms appear comparable to other tuna (Table 20.1). Baitfish are the main feed used in the industry and information on its utilization is appearing in the literature (Ticina *et al*., 2004; Aguado-Gimenez *et al*., 2006). As has been suggested previously, research into farmed tuna is difficult and often requires the development of special techniques and approaches to nutrition (Carter *et al*., 1998, 1999). Aguado-Gimenez *et al*. (2006) presented an analysis of nutrient budgets for growing medium- (32 kg) and large- (219 kg) sized ABT: they showed that medium ABT with a daily feed intake of 4.3% body weight (BW) grew at 0.29%/day specific growth rate (SGR), a food conversion ratio (FCR) of 15.4 and nitrogen and phosphorous retention efficiencies of 6.5 and 10.5%, respectively. Large fish with a daily feed intake of 1.5% BW grew at 0.06%/day SGR, FCR of 24.9 and nitrogen and phosphorous retention efficiencies of 4 and 8%, respectively. Nevertheless, over about 8 months, the medium and large tuna grew to nearly 196% and 116% their original size, respectively.

High mortalities of captive ABT reared in sea cages in the Mediterranean (Adriatic) Sea have been reported on a number of occasions; some were most likely due to pasteurellosis (Mladineo *et al*., 2006). Pasteurellosis has a seasonal occurrence, peaking usually in June and deteriorating slowly within 3–5 months (Mladineo *et al.*, 2006). It was suggested that other risk factors, that is, poor diet (including long storage), sudden rise of water temperature, lack of current and forest fires, contributed to the outbreak (Mladineo *et al.*, 2006). A range of parasites can be present in tuna, including nematodes, helminths, copepods and microsporidians (Mladineo, 2006; Nowak *et al.*, 2006); however, none has been considered a significant risk to the fish.

20.4 Southern Bluefin Tuna, *Thunnus maccoyii*

The southern bluefin tuna (SBT) is thought to have one population with a single spawning ground and an immense circumpolar range in temperate waters of the southern hemisphere (Sund *et al.*, 1981; Grewe *et al.*, 1997) (Fig. 20.4). This includes the coasts of Australia and New Zealand, the South Atlantic and Indian Oceans (Gunn and Block, 2001; Safina, 2001). The 'Oka' ground, to the north-west of Australia and south of Java, is the only known spawning ground (Sund *et al.*, 1981; Safina, 2001). After spawning in these tropical Indian Ocean waters, the fertilized eggs hatch and pelagic larvae drift southwards. One- to two-year-old juveniles inhabit waters off western Australia, older juveniles migrate southwards into the Great Australian Bight and by 6 years of age, the majority of surviving fish have moved into deeper waters and can range as far south as 30°S in winter and 40°S in warmer months (Jenkins *et al.*, 1991). Juveniles 'spike' dive to depths greater than 400 m in the winter, but usually inhabit the upper 100 m (Gunn and Block, 2001).

Sexual maturity is reached after between 8 and 14 years. Mature SBT are abundant in the 'Oka' spawning ground during summer and there may be two peaks in activity, September–October and February–March (Young, 2001). Larger individuals spawn near to the surface, where water temperatures are around 30°C. There is no sexual dimorphism and it is very difficult to sex adult tuna using external characteristics (Schaefer, 2001). Females are multiple spawners, having asynchronous ovaries (that contain oocytes in various stages of development). SBT grow to a maximum length of 200 cm and a weight of 200 kg and can live for 20–40 years (Kailola *et al.*, 1993). They are opportunistic carnivores and school to feed at the surface on schooling or aggregated prey such as sardines, crustaceans and squid (Young *et al.*, 1997).

A large decline in catches and the reduction in quotas, set by CCSBT (Commission for the Conservation of Southern Bluefin Tuna) in the 1980s, stimulated the successful industry based at Port Lincoln, South Australia (Clarke *et al.*, 1997). This is the only location that SBT are farmed and current practices are the result of extensive experience. The first aquaculture production recorded by FAO was 335 t in 1992 and stabilized at around 4000 t from 2001 to 2004 (FAO, 2006). The industry is based around on-growing wild-caught fish and there is little published information available concerning SBT broodstock or approaches to larval rearing. This situation will change rapidly with current commercially driven propagation research and development (Elizur *et al.*, 2006).

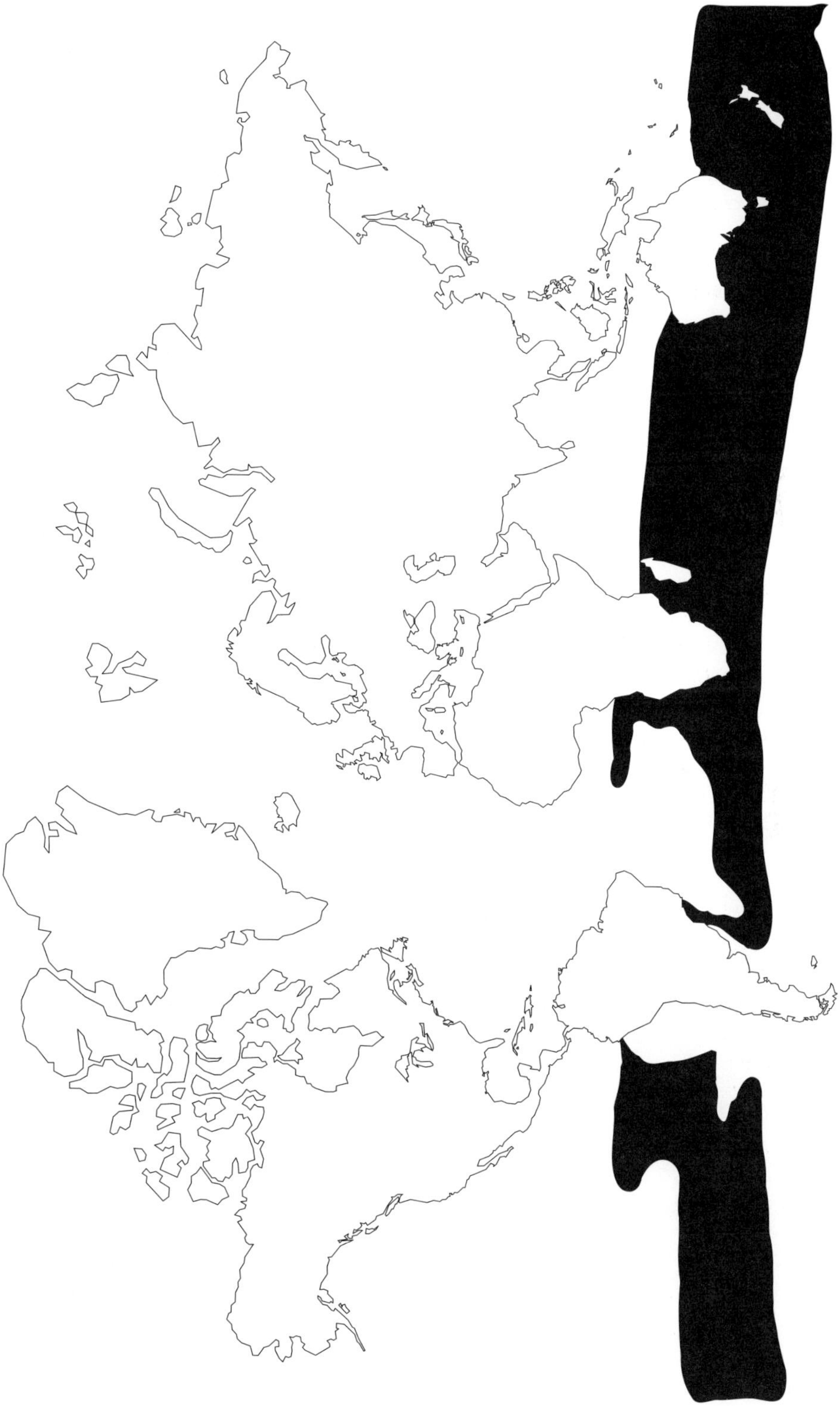

Fig. 20.4. World distribution of *Thunnus maccoyii*.

Several publications include details of SBT farming methods (Clarke *et al.*, 1997; Carter *et al.*, 1998; Farwell, 2001; Glencross *et al.*, 2002a). Juvenile SBT schools are caught in the Great Australian Bight by purse-seine in December–March and towed back in specially designed sea cages. They are transferred to moored cages for 4–8 months on-growing over the summer and early autumn. SBT are usually 10–20 kg at catch and approximately double their weight on the farm. Floating ring cages with typical diameters of 40 m and low stocking densities of around 2 kg/m^3 are used. Sardine species have provided the majority of commercial feed and are very well accepted by SBT; they elicit a 'feeding frenzy' and a high level of feeding activity that results in high feed intake and promotes substantial growth. In some situations, vitamin premix is used to coat the sardines immediately prior to feeding (Fig. 20.5). There is considerable variation in availability of baitfish, many different sources and species are used, although in recent years the Australian sardine, *Sardinops neopilchardus*, from local sources has dominated. The nutrient composition of baitfish can vary substantially, depending on species, season and storage methods. This has resulted in an innovative approach in deciding on the best ratio between the available baitfish for feeding to the SBT: *Formu-bait* is a formulation package that, based on the desired nutrient intake parameters, calculates the ratios of different baitfish that should be fed (van Barneveld and Ellis, unpublished data).

SBT are harvested with the use of 'crowd' nets to contain the desired number of fish within a smaller and shallower area and divers pull SBT individually to the harvest boat (Fig. 20.6), where they are killed quickly by removal of the brain core and a wire is inserted down the spinal cord to destroy the central nervous system. The fish are bled through cuts into the lateral line that cut the

Fig. 20.5. Sardines coated with white vitamin premix.

Fig. 20.6. Divers handling southern bluefin during harvesting.

Fig. 20.7. Processing southern bluefin tuna prior to shipping.

lateral artery. Fish to be taken ashore for processing are first placed in a salt ice bath on board the vessel and subsequently gilled and gutted, cleaned and either packed fresh, chilled for transport by road to a suitable airport and then air-freighted, or 'snap frozen' to −80°C for storage and subsequent transport by sea to Japan (Fig. 20.7). Alternatively, some SBT are harvested directly on to 'mother ships' adjacent to the farms and processed and deep frozen on board.

Farming increases the value of the fishery catch due to the increased fish weight and condition and the ability to optimize marketing.

In 1998, and after several years of development (Table 20.1), compound feeds based on simple processing technology were shown to produce comparable growth and FCR to baitfish (Glencross *et al.*, 2002a). At this time, growth rates (SGR) of around 0.2–0.4%/day and FCR of 4–7 were typical for 30kg fish and analysis suggested SBT should be fed at approximately 10g protein/day/kg biomass and 300kJ/day/kg biomass (Glencross *et al.*, 2002a). Feed intake is governed by water temperature and nutritional status: daily feed intake of 3.6% BW was recorded at water temperatures of around 20°C and decreased to minimal levels at below 15°C (Glencross *et al.*, 2002b). Compound feeds were an important advance (Fig. 20.8) and were followed by commercial moist feeds, made using cold-extrusion technology; these were not 'shelf stable' and had to be transported under refrigeration. Future advances will be made on production of more stable compound extruded feeds. Technology is also required to reduce the reliance on hand feeding (Fig. 20.9). Industry has yet to adopt compound feeds due to their relatively high cost and it should be noted that they provide an important alternative nutritional strategy if baitfish become unavailable. Compound feeds are an integral component of planned propagation strategies, for on-growing with hatchery-raised SBT.

Infections by the ciliate, *Uronema nigricans*, were linked in the past to SBT mortalities (Munday *et al.*, 1997, 2003); however, improvements in husbandry and improved location of sites seemed to have almost eliminated this problem (Nowak, 2004; Nowak *et al.*, 2006). Another parasite currently linked to potential tuna health problems is *Caligus chiastos* (Haywood *et al.*, 2008).

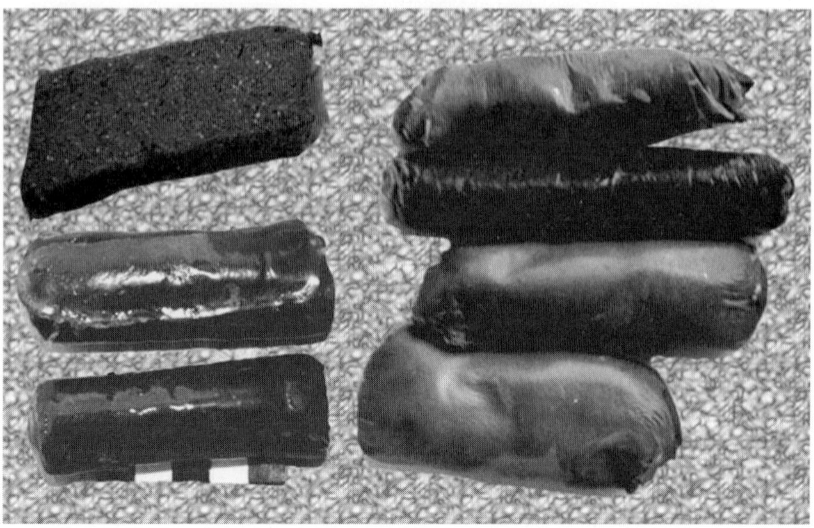

Fig. 20.8. Selection of non-extruded feeds (scale bar = 5cm) used in the early development of feeds for southern bluefin tuna (C. Carter).

Fig. 20.9. Hand feeding pelleted feed to southern bluefin tuna.

20.5 Pacific Bluefin Tuna, *Thunnus orientalis*

The Pacific bluefin tuna (PBT) inhabit the waters of the north Pacific from Japan, where it spawns, to California, although it has sometimes been caught south of the equator (Ward *et al.*, 1995) (Fig. 20.10). Key aspects of its natural history and biology are included in reviews on tuna (Block and Stevens, 2001).

On-growing of wild-caught PBT is practised in Japan (the islands of Shikoku and Kyushu) and Mexico (Baja, California) (Farwell, 2001) and aquaculture, based on a closed life cycle, is closest to being commercialized with this species of tuna due to recent successes of the propagation programme at Kinki University (Sawada *et al.*, 2005; Sakamoto *et al.*, 2006). The PBT programme at Kinki University has an associated business and 1000 fish have been sold between 2004 and 2006 (Sawada *et al.*, 2006). Mexico has eight farms, which produced 3800 t in the 2004/05 season, and this amount is predicted to increase with the increased stocking of wild fish (Smart and Sylvia, 2006).

In Japan, 1-year-old fish are caught and on-grown in sea cages over 2 years to reach about 40 kg from 200–300 g initial weight. In Mexico, the approach is similar to that used for SBT in Australia: older larger PBT are on-grown over 3–9 months from an initial weight of 15–50 kg (Farwell 2001; Smart and Sylvia, 2006). Initial weight is influenced by season and location; fish can be caught from July to November, larger fish are caught in the north and smaller in the south. Stocking densities are typically low (< 5 kg/m^3) and locally caught baitfish are fed in both regions.

The life cycle of PBT was closed in Japan following successful larval production from broodstock that were themselves grown in culture from larvae (Sawada *et al.*, 2005; Sakamoto *et al.*, 2006). Considerable development will

Fig. 20.10. World distribution of *Thunnus orientalis*.

be required, both because of the inherent difficulties of rearing marine fish larvae and juveniles and also due to the complexity of upscaling facilities and methods from experimental to commercial. There may be key differences in nutritional and health status between wild-caught and domestic broodstock that could result in poor larval quality and a low success rate in the production of viable juveniles from the second generation of cultured larvae (Sawada *et al.*, 2005). However, given current levels of knowledge on rearing marine fishes, the rapid increase in data on tuna, as well as the resources being allocated, it is likely that technical issues will be overcome. Reproductive biology (Mourente *et al.*, 2002; Chen *et al.*, 2006; Sawada *et al.* 2007) and cryopreservation procedures are being studied (Gwo *et al.*, 2005).

Typical larval rearing would occur at around 25°C, > 33‰ to full strength salinity, pH 8 and under a 12:12 photoperiod and result in a high hatch rate (> 95%) (Takashi *et al.*, 2006). The mouth is open by 2 days posthatch (dph) and rotifers are fed from 3 dph; swimming ability develops at about 10 dph. An overlapping sequence of rotifers, *Artemia* and fish larvae are fed to 30 dph, when the PBT reach 40 mm total length (Sawada *et al.*, 2005). Mortality in the first year can reach 99% and three factors have been identified (Miyashita *et al.*, 2000; Takashi *et al.*, 2006): initial mortality due to poor feeding, adhesion to the water surface and contact with the tank floor occurs up to day 10 posthatch; cannibalism from post-flexion and through the juvenile stages; and high-speed collisions with the sides of tanks and sea cages. As in other marine fish hatcheries, betanodavirus infections can be a problem. Juvenile production can fail as a result of viral nervous necrosis caused by betanodoviruses (Sugaya *et al.*, 2005). A genetic polymorphism study of betanodaviruses in cultured and wild fish suggested that this was a vertically transmitted disease (from broodstock); however, wild fish present around hatchery could become a risk factor in the future. Close proximity of tuna cages to red seabream and yellowtail aquaculture has been suggested to be a risk factor (Munday *et al.*, 2003). Infection with the scuticociliate, *U. nigricans*, can cause mortalities in hatcheries (Munday *et al.*, 2003).

Recent progress in propagation and husbandry, coupled with economic analyses, suggest tuna aquaculture has considerable potential for growth, particularly to replace the declining domestic catch (Hidaka, 2006). The Japanese market for tuna is approximately 0.5 Mt and is considerably greater than the current aquaculture supply of less than 30,000 Mt in 2003 (Hidaka, 2006). Research challenges include the development of compound feeds that are economic, have limited environmental impacts and produce tuna that meet consumer demands; mass production of viable eggs and larvae; selective breeding; effective slaughter methods; and the development of appropriate holding facilities for different stages of the life cycle.

References

Aguado-Gimenez, F. and Garcia-Garcia, B. (2005a) Changes in some morphometric relationships in Atlantic bluefin tuna (*Thunnus thynnus thynnus* Linnaeus, 1758) as a result of fattening process. *Aquaculture* 249, 303–309.

Aguado-Gimenez, F. and Garcia-Garcia, B. (2005b) Growth, food intake and feed conversion rates in captive Atlantic bluefin tuna (*Thunnus thynnus* Linnaeus, 1758) under fattening conditions. *Aquaculture Research* 36, 610–614.

Aguado-Gimenez, F., Garcia-Garcia, B., Hernandez-Lorente, M.D. and Cerezo-Valverde, J. (2006) Gross metabolic waste output estimates using a nutritional approach in Atlantic bluefin tuna (*Thunnus thynnus*) under intensive fattening conditions in western Mediterranean Sea. *Aquaculture Research* 37, 1254–1258.

Altringham, J.D. and Shadwick, R.E. (2001) Swimming and muscle function. In: Block, B.A. and Stevens, E.D. (eds) *Tuna: Physiology, Ecology and Evolution*. Academic Press, San Diego, California, pp. 314–344.

Block, B.A. and Stevens, E.D. (eds) (2001) *Tuna: Physiology, Ecology and Evolution*. Fish Physiology Series Volume 19. Academic Press, San Diego, California, 368 pp.

Brill, R.W. and Bushnell, P.G. (2001) The cardiovascular system of tunas. In: Block, B.A. and Stevens, E.D. (eds) *Tuna: Physiology, Ecology and Evolution*. Academic Press, San Diego, California, pp. 79–120.

Carey, F.G., Kanwisher, J.W. and Stevens, E.D. (1984) Bluefin tuna warm their viscera during digestion. *Journal of Experimental Biology* 109, 1–20.

Carter, C.G., Seeto, C.G., Smart, A., Clarke, S. and van Barneveld, R.J. (1998) Correlates of growth in farmed juvenile southern bluefin tuna *Thunnus maccoyii* (Castelnau). *Aquaculture* 161, 107–119.

Carter, C.G., Bransden, M.P., van Barneveld, R.J. and Clarke, S.M. (1999) Alternative methods for nutrition research on the southern bluefin tuna, *Thunnus maccoyii*: *in vitro* digestibility. *Aquaculture* 179, 57–70.

Chen, K.-S., Crone, P. and Hsu, C.-C. (2006) Reproductive biology of female Pacific bluefin tuna *Thunnus orientalis* from south-western North Pacific Ocean. *Fisheries Science* 72, 985–994.

Chow, S., Nakagawa, T., Suzuki, N., Takeyama, H. and Matsunaga, T. (2006) Phylogenetic relationships among *Thunnus* species inferred from rDNA ITS1 sequence. *Journal of Fish Biology* 68, 24–35.

Clarke, S., Smart, A., van Barneveld, R. and Carter, C. (1997) The development and optimisation of manufactured feeds for farmed southern bluefin tuna. *Austasia Aquaculture* 11, 59–62.

Collette, B.B., Reeb, C. and Block, B.A. (2001) Systematics of the tunas and mackerels (Scombridae). In: Block, B.A. and Stevens, E.D. (eds) *Tuna: Physiology, Ecology and Evolution*. Academic Press, San Diego, California, pp. 1–33.

Corriero, A., Medina, A., Mylonas, C.C., Abascal, F.J., Deflorio, M., Aragon, L., Bridges, C.R., Santamaria, N., Heinisch, G., Vassallo-Agius, R., Belmonte, A., Fauvel, C., Garcia, A., Gordin, H. and De Metrio, G. (2007) Histological study on the effects of treatment with gonadotropin-releasing hormone agonist (GnRHa) on the reproductive maturation of captive-reared Atlantic bluefin tuna (*Thunnus thynnus* L.). *Aquaculture* 272, 675–686.

Elizur, A., Stehr, H., Stehr, M., Deichmann, M. and Stone, S. (2006) Clean seas tuna – a vision for southern bluefin tuna. In: *Australasian Aquaculture 2006, Innovation in Aquaculture*, 27–30 August 2006, Adelaide. Conference CD.

FAO (2006) Fishery Statistical Collections c2002. In: *FIGIS Data Collection*. FAO, Rome (http://www.fao.org./figis/servlet/static?dom=collectionandxml=global-aquaculture-production.xml, accessed 6 October 2006).

Farwell, C.J. (2001) Tunas in captivity. In: Block, B.A. and Stevens, E.D. (eds) *Tuna: Physiology, Ecology and Evolution*. Academic Press, San Diego, California, pp. 391–412.

Fromentin, J.M. and Powers, J.E. (2005) Atlantic bluefin tuna: population dynamics, ecology, fisheries and management. *Fish and Fisheries* 6, 281–306.

Garcia, A. (2006) General overview of the EU bluefin tuna research problems and issues. In: Sakamoto, W., Miyashita, S. and Nakagawa, Y. (eds) *Ecology and Aquaculture of Bluefin Tuna*. Kinki University Press, Fisheries Laboratory, Kinki University, Wakayama, Japan, pp. 68–71.

Glencross, B., Carter, C., Gunn, J., van Barneveld, R., Rough, K. and Clarke, S. (2002a) Southern bluefin tuna, *Thunnus maccoyii*. In: Webster, C.D. and Lim, C. (eds) *Nutrient Requirements and Feeding of Finfish for Aquaculture*. CAB International, Wallingford, UK, pp. 159–171.

Glencross, B.D., Clarke, S.M., Buchanan, J.G., Carter, C.G. and van Barneveld, R.J. (2002b) Temporal growth patterns of farmed juvenile southern bluefin tuna, *Thunnus maccoyii* (Castelnau) fed moist pellets. *Journal of the World Aquaculture Society* 33, 138–145.

Grewe, P.M., Elliott, N.G., Innes, B.H. and Ward, R.D. (1997) Genetic population structure of southern bluefin tuna (*Thunnus maccoyii*). *Marine Biology* 127, 555–561.

Gunn, J. and Block, B.A. (2001) Advances in acoustic, archival, and satellite tagging of tunas. In: Block, B.A. and Stevens, E.D. (eds) *Tuna: Physiology, Ecology and Evolution*. Academic Press, San Diego, California, pp. 167–224.

Gwo, H.-H., Weng, T.-S., Fang, L-S. and Lee, Y.-H. (2005) Development of cryopreservation procedures for semen of Pacific bluefin tuna *Thunnus orientalis*. *Aquaculture* 249, 205–211.

Hayward, C.H., Aiken, A. and Nowak, B.F. (2008) An epizootic of *Caligus chiastos* on farmed southern bluefin tuna (*Thunnus maccoyii*) off South Australia. *Diseases of Aquatic Organisms* 79, 57–63.

Hidaka, T. (2006) Comparative analysis of blue-fin aquaculture: comparison of products and value chain. In: Sakamoto, W., Miyashita, S. and Nakagawa, Y. (eds) *Ecology and Aquaculture of Bluefin Tuna*. Kinki University Press, Fisheries Laboratory, Kinki University, Wakayama, Japan, pp. 61–67.

Jenkins, G.P., Young, W.J. and Davis, T.L.O. (1991) Density dependence of larval growth of marine fish, the southern bluefin tuna *Thunnus maccoyii*. *Canadian Journal of Fisheries and Aquatic Sciences* 48, 1358–1363.

Kailola, P.J., Williams, M.J., Stewart, P.C., Reichelt, R.E., McNee, A. and Grieve, C. (1993) *Australian Fisheries Resources*. Bureau of Resource Sciences, DPIE and the Fisheries Research and Development Corporation, Canberra.

Kaji, T., Tanaka, M., Takahashi, Y., Oka, M. and Ishibashi, N. (1996) Preliminary observations on development of Pacific bluefin tuna *Thunnus thynnus* (Scombridae) larvae reared in the laboratory, with special reference to the digestive system. *Marine and Freshwater Research* 47, 261–269.

Miyashita, S., Sawada, Y., Hattori, N., Nakatsukasa, H., Okada, T., Murata, O. and Kumai, H. (2000) Mortality of northern bluefin tuna *Thunnus thynnus* due to trauma caused by collision during growout culture. *Journal of the World Aquaculture Society* 31, 632–639.

Mladineo, I. (2006) Microsporidia sp. in Atlantic bluefin tuna (*Thunnus thynnus*). *Bulletin of European Association of Fish Pathologists* 26, 153–156.

Mladineo, I., Miletic, I. and Bocina, I. (2006) *Photobacterium damselae* subsp. *piscicida* outbreak in cage-reared Atlantic bluefin tuna *Thunnus thynnus*. *Journal of Aquatic Animal Health* 18, 51–54.

Mourente, G., Megina, C. and Diaz-Salvago, E. (2002) Lipids in female northern bluefin tuna (*Thunnus thynnus thynnus* L.) during sexual maturation. *Fish Physiology and Biochemistry* 24, 351–363.

Munday, B.L., O'Donoghue, P.J., Watts, M., Rough, K. and Hawkseford, T. (1997) Fatal encephalitis due to the scutociliate *Uronema nigricans* in sea caged, southern bluefin tuna *Thunnus maccoyii*. *Diseases of Aquatic Organisms* 30, 17–25.

Munday, B., Sawada, Y., Cribb, T. and Hayward, C. (2003) Diseases of tunas, *Thunnus* spp. *Journal of Fish Diseases* 26, 187–206.

Mylonas, C.C., Bridges, C., Belmonte, A., Gordin, H., Garcia, A., Fauvel, C., Rosenfeld, H., Medina, A., Corriero, A., Demetrio, G., Vassallo-Agius, R. and Zohar, Y. (2006) Induction of ovulation and production of viable larvae of bluefin tuna *Thunnus thynnus* using GnRHa implants. In: *Australasian Aquaculture 2006, Innovation in Aquaculture*, 27–30 August 2006, Adelaide, Australia. Conference CD.

Nowak, B.F. (2004) Assessment of health risks to southern bluefin tuna under current culture conditions. *Bulletin of European Association of Fish Pathologists* 24, 45–51.

Nowak, B., Mladineo, I., Aiken, H., Bott, N. and Hayward, C. (2006) Results of health surveys of two species of farmed tuna: southern bluefin tuna (*Thunnus maccoyii*) in Australia and northern bluefin tuna (*Thunnus thynnus*) in the Mediterranean. *Bulletin of European Association of Fish Pathologists* 26, 38–42.

Safina, C. (2001) Tuna conservation. In: Block, B.A. and Stevens, E.D. (eds) *Tuna: Physiology, Ecology and Evolution*. Academic Press, San Diego, California, pp. 413–459.

Sakamoto, W., Miyashita, S. and Nakagawa, Y. (2006) *Ecology and Aquaculture of Bluefin Tuna: Proceedings of the Joint International Symposium of Bluefin Tuna, 2006*. Kinki University Press, Wakayama, Japan, 140 pp.

Sawada, Y., Okada, T., Miyashita, S., Murata, O. and Kumai, H. (2005) Completion of the Pacific bluefin tuna *Thunnus orientalis* (Temminck et Schlegel) life cycle. *Aquaculture Research* 36, 413–421.

Sawada, Y., Kurata, M., Okada, T., Miyashita, S., Murata, O. and Kumai, H. (2006) Juvenile production technology of the Pacific bluefin tuna in Japan. In: *Australasian Aquaculture 2006, Innovation in Aquaculture*, 27–30 August 2006, Adelaide, Australia. Conference CD.

Sawada, Y., Seoka, M., Kato, K., Tamura, T., Nakatani, M., Hayashi, S., Okada, T., Tose, K., Miyashita, S., Murata, O. and Kumai, H. (2007) Testes maturation of reared Pacific bluefin tuna *Thunnus orientalis* at two-plus years old. *Fisheries Science* 73, 1070–1077.

Schaefer, K.M. (2001) Reproductive biology of tunas. In: Block, B.A. and Stevens, E.D. (eds) *Tuna: Physiology, Ecology and Evolution*. Academic Press, San Diego, California, pp. 225–270.

Smart, A. and Sylvia, P.C. (2006) Current status of bluefin tuna farming in North America. In: *Australasian Aquaculture 2006, Innovation in Aquaculture*, 27–30 August 2006, Adelaide, Australia. Conference CD.

Starling, E. and Diver, G. (2005) *The Australian Tuna Handling Manual: A Practical Guide for Industry*. Seafood Services, Ascot, Australia, 80 pp.

Sugaya, T., Mori, K., Nishioka, T., Masuma, S., Oka, M., Okinaka, Y. and Nakai, T. (2005) Genetic polymorphisms of betanodaviruses detected in diseased juveniles of bluefin tuna *Thunnus orientalis*. EAFP 12th International Conference, 11–16 September, Diseases of Fish and Shellfish Abstracts, Copenhagen, pp. 129.

Sund, P.N., Blackburn, M. and Williams, F. (1981) Tunas and their environment in the Pacific ocean: a review. *Oceanography and Marine Biology Annual Review* 19, 443–512.

Takashi, T., Kohno, H., Sakamoto, W., Miyashita, S., Murata, O. and Sawada, Y. (2006) Diel and ontogenetic body density change in Pacific bluefin tuna, *Thunnus orientalis* (Temminck and Schlegal), larvae. *Aquaculture Research* 37, 1172–1179.

Ticina, V., Grubisic, L. and Katavic, I. (2004) Sampling and tagging of live bluefin tuna in growth-out floating cages. *Aquaculture Research* 35, 307–310.

Ticina, V., Katavic, I. and Grubisic, L. (2007) Growth indices of small northern tuna (*Thunnus thynnus*, L.) in growth-out cages. *Aquaculture* 269, 538–543.

van Barneveld, R.J., Smart, A.R., Clarke, S.M., Carter, C.G., Davis, B.J., Tivey, D.R. and Brooker, J.D. (1997) Nutritional management of sea-caged southern bluefin tuna (*Thunnus maccoyii*). In: Corbett, J.L., Choct, M., Nolan, J.V. and Rowe, J.B. (eds) *Recent Advances in Animal Nutrition in Australia 1997*. University of New England, Armidale, Australia, pp. 88–97.

Ward, R.D., Elliott, N.G. and Grewe, P.M. (1995) Allozyme and mitochondrial DNA separation of Pacific northern bluefin tuna, *Thunnus thynnus orientalis* (Temminck and Schlegel), from southern bluefin tuna, *Thunnus maccoyii* (Castelnau). *Marine and Freshwater Research* 46, 921–930.

Westneat, M.W. and Wainwright, S.A. (2001) Mechanical design for swimming: muscle, tendon, and bone. In: Block, B.A. and Stevens, E.D. (eds) *Tuna: Physiology, Ecology and Evolution*. Academic Press, San Diego, California, pp. 272–311.

Young, J.W., Lamb, T.D., Le, D., Bradford, R.W. and Whitelaw, A.W. (1997) Feeding ecology and interannual variations in diet of southern bluefin tuna, *Thunnus maccoyii*, in relation to coastal and oceanic waters off the eastern Tasmania, Australia. *Environmental Biology of Fishes* 50, 275–291.

Young, P.C. (2001) *Southern Bluefin Tuna: Its Propagation in Australia*. Fisheries Research and Development Corporation, Deakin, Australia.

21 The Flatfishes (Order: Pleuronectiformes)

ALBERT K. IMSLAND

Akvaplan-Niva A/S, Iceland and Department of Biology, University of Bergen, Bergen, Norway

21.1 General Introduction

Flatfishes are laterally compressed benthic fishes of the order Pleuronectiformes, including flounders, soles and tonguefishes. Recent estimates of flatfish species diversity include that of Nelson (1994) with approximately 570 species in 123 genera and about 11 families. Hensley (1997) estimated flatfish diversity at 570–620 species, whereas Munroe (2005) recognized 716 species of flatfish (669 named and 47 recognized but not described), placed into 123 genera. Of marine eutoleostean fishes, flatfishes rank as the third most diverse Order both in number of species and genera; only the Perciformes and Scorpeeniformes have greater diversity (Munroe, 2005). Flatfishes are widely distributed and are found circumglobally in cold, temperate and tropical seas in depths from the intertidal zone to the continental slope. Many important food fish are in this order, including the flounders, soles, turbot, plaice and halibut.

The most important species for commercial fisheries, and for aquaculture, are representatives of the families Scophthalmidae (including turbot, megrim and brill), Pleuronectidae (including Atlantic and Pacific halibut, dab, lemon sole, plaice, Greenland halibut and winter flounder), Solidae (including common sole and Senegal sole), Rhombosoleidae (including greenback flounder) and Paralichthyidae (including summer flounder and Japanese flounder). The fisheries for flatfish are located primarily in the northern hemisphere in both the Atlantic and Pacific Oceans, where representatives from the Pleuronectidae, Scophthalmidae, Soleidae and Paralichthyidae dominate the catch. South temperate regions also support commercial fisheries in other groups, e.g. the Rhombosoleidae (Munroe, 2005). Aquaculture production of flatfishes has been increasing steadily during the past 30 years and had reached 60,000 t by the turn of the century (FAO, 2007). The largest aquaculture production comes from the Japanese flounder and other flounder species in Japan and Korea (mostly for sea ranching), turbot, Atlantic halibut and, recently, sole. There is farming interest for local species in many places, including north-east USA (summer flounder), Australia (greenback flounder) and Canada (winter flounder).

Several of the flatfish species have attributes and biological characteristics that make them highly suitable for intensive farming, although the number of positive attributes varies from species to species. The characteristics possessed by flatfishes that are deemed desirable in a cultured species can be summarized as follows:

- Large, established markets. World catches of flatfishes have declined from 1.5 to 1.0 Mt/year in the period from 1980 to 2004. Unmet market demand can only be fulfilled with an increase in culture.
- High market prices. Globally farmed flatfishes are commanding high market prices (€8–15/kg).
- The timing of spawning can be controlled using thermal and photoperiod manipulations to give 'out-of-season' eggs and enable a continuous cycle of production, for example, in turbot and, recently, Atlantic halibut.
- Several species adapt well to farm conditions and can be fed dry, pellet feeds throughout the grow-out phase. The fish tolerate a moderate degree of crowding and handling and are moderately resistant to disease, even though several disease problems have been recognized (Mulcahy, 2002).
- Some species grow quite quickly to a relatively large body size, for example turbot reaches market size (1–2 kg) in 14–24 months.
- Large body size is an advantage when it comes to processing the harvested fish; it is easier to process large fish than small fish and the range of products that can be produced from large fish is also greater. The meat yield of flatfish is high, with the fillet representing 50–60% of the body mass in turbot and Atlantic halibut.
- The meat quality of flatfish is good; it has characteristics that are generally appealing to a broad range of potential consumers. Flatfishes are well known as food fish and they have good acceptability in the market.

Flatfishes are among the most popular and most valuable fishes used for human consumption, displaying a wide variety of developmental patterns and a diversity of life histories. As a group, they display a nearly global occurrence in marine habitats (Munroe, 2005). The four species discussed in this chapter are, however, representative of flatfishes from the North Atlantic Ocean, ranging from the Artic-boreal (Atlantic halibut, winter flounder) to subtropical (soles) areas. Flatfishes occupy diverse bathymetric environments from shallow marine and freshwater areas to deep-water habitats. All flatfishes begin life as pelagic, bilaterally symmetrical fishes. During larval development, flatfishes undergo an ontogenetic metamorphosis where one eye migrates from one side of the head to the other (Munroe, 2005). Depending on the species, either the right (e.g. Atlantic halibut, winter flounder, soles) or left (e.g. turbot, Japanese flounder) eye migrates. As a group, flatfishes are the only vertebrates to deviate so radically from a bilaterally symmetrical body plan.

Many flatfish species have been, and still are considered to be, promising candidates for intensive farming, but only two species, the Japanese flounder and turbot, have been fully commercialized (Howell and Yamashita, 2005). Other species of flatfishes are on the verge of commercialization, including Atlantic halibut and soles, whereas others, for example, winter flounder, are still only on the test trial stage. In Europe, the relative uniformity of finfish aquaculture limits its potential for expansion and makes the industry very vulnerable to market fluctuations. Successful diversification requires the introduction of new species whose growth can be optimized while costs are controlled (cost effectively). Many members of the order Pleuronectiformes have, due to high market prices and declining and/or variable catches, gained considerable interest as candidates for marine aquaculture, e.g. soles (Houghton *et al.*, 1985; Imsland *et al.*, 2004; Fig. 21.1), Japanese flounder, *Paralichtys olivaceus* (Howell and Yamashita, 2005), summer flounder, *P. dentatus* (Nardi, 1996), Atlantic halibut (Imsland and Jonassen, 2002; Fig. 21.2) and turbot (Lavens and Remmerswaal, 1994; Imsland and Jonassen, 2002; Fig. 21.3). Elsewhere in the world, many of the species from these two families (i.e. Scophthalmidae and Pleuronectidae) are highly priced and seen as aquaculture candidates, including the winter flounder *Pseudopleuronectes americanus* (Fig. 21.4).

Fig. 21.1. Common sole, *Solea solea* (photo by Albert Imsland).

Fig. 21.2. Atlantic halibut, *Hipppoglossus hippoglossus* (photo by Karin Pittman, University of Bergen, Norway).

Fig. 21.3. Turbot, *Scophthalmus maximus* (photo by Lars Olav Sparboe, Akvaplan-niva).

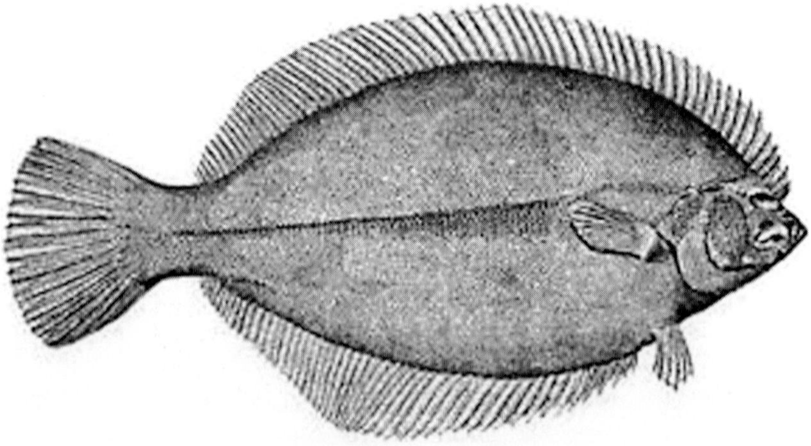

Fig. 21.4. Winter flounder, *Pseudopleuronectes americanus.*

21.2 Turbot, *Scophthalmus maximus*

Turbot is a marine demersal carnivorous flatfish of the Scophthalmidae family (Fig. 21.5). It is relatively abundant in Europe, from Iceland (66°N) and western Norway in the north to Morocco (30°N) in the south (Fig. 21.6). It is also

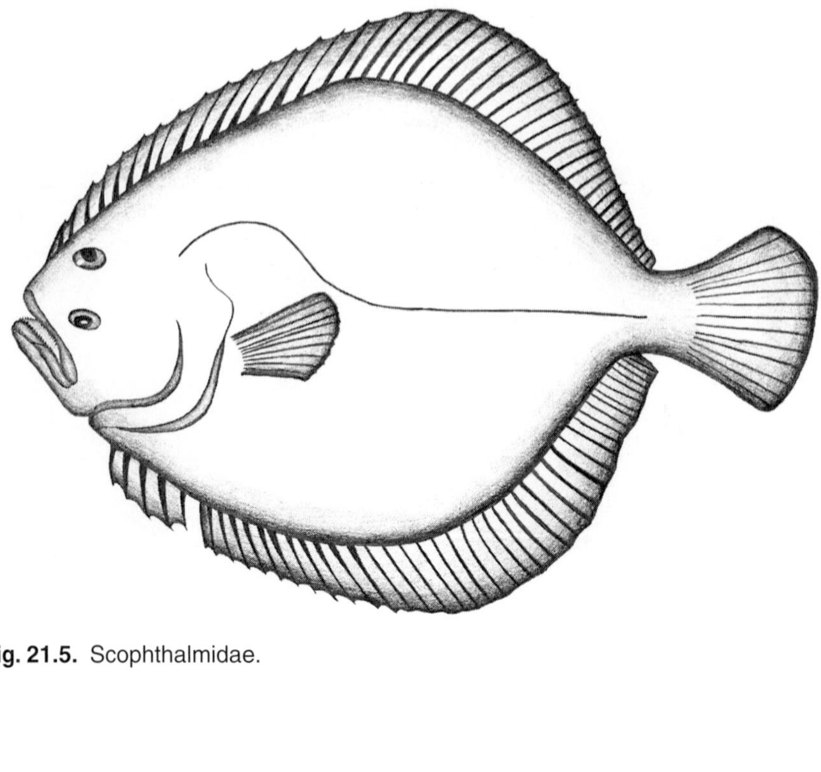

Fig. 21.5. Scophthalmidae.

Fig. 21.6. World distribution of *Scophthalmus maximus.*

abundant in the Mediterranean Sea as far as Turkey. In the Black Sea, a closely related species (*Scophthalmus maximus maeoticus*) can be found. Turbot is caught on sand, gravel or mixed bottoms at a depth of 20–70 m. The total annual catches vary, but have remained less than 10,000 Mt for the past 5 years (Fig. 21.7). The main fishery nations are the UK, the Netherlands, France, Spain, Belgium and Denmark. Turkey catches about 300 t annually of Black Sea turbot. Turbot fishing is generally intensified in the period from March to June, when the sea temperature increases, and approximately 65% of turbot is caught in this period.

The studies of Aneer and Westin (1990), Déniel (1990), Iglesias and Rodríguez-Ojea (1994) and Bergstad and Folkvord (1997) indicate that turbot do not undergo long migrations but are a stationary species. Different spatial distribution between juveniles and adults is seen, as only large fish migrate to colder areas (Aneer and Westin, 1990; Iglesias and Rodríguez-Ojea, 1994), which may be partly explained by decreasing temperature sensitivity and a downshift in temperature optimum with size (Imsland *et al.*, 1996, 2001a, 2006a), whereas this might also be a strategy to reduce predation risk by hiding from enemies. Lack of long migration, together with the fact that this species is found in different environments (among others, different salinities), makes it reasonable to believe that turbot in European waters belong to more than one population. Imsland *et al.* (2003a) suggested that turbot in north European waters consisted of more than one population. Recent studies using polymorphic DNA microsatellites have also indicated genetic differentiation between turbot populations (M.Ö. Stefánsson, Marine Research Institute, Reykjavík, Iceland, personal communication). DNA microsatellite studies have also indicated genetic differentiation between wild and farmed turbot populations (Coughlan

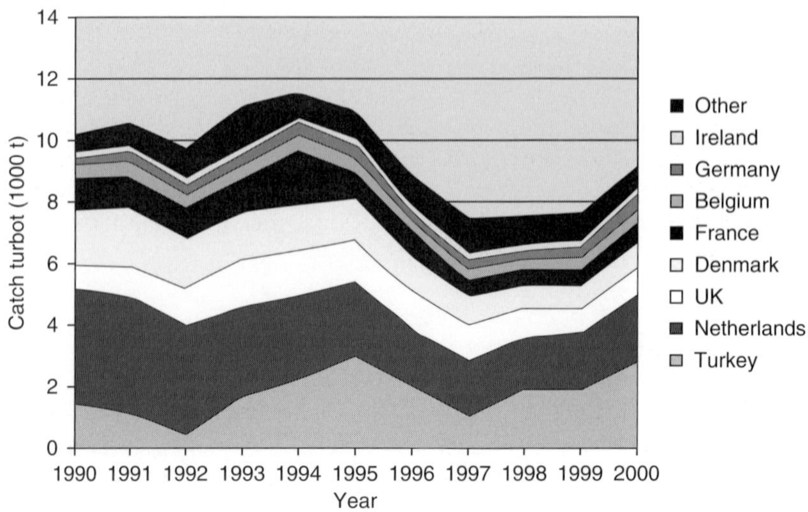

Fig. 21.7. Development in annual catches of turbot during 1990–2000 (*Source*: Norwegian Seafood Export Council (NSEC)). Turbot caught by Turkey is predominantly the subspecies *Scophthalmus maximus maeoticus* (Black Sea turbot).

et al., 1998) and it was presumed that this was caused by genetic drift in the hatcheries and that this drift could cause a considerable loss of genetic heterogeneity over a period of only a few generations.

21.2.1 Farming of turbot

Trials on turbot rearing have been going on for a century. As early as 1892, Holt (1892) reported successful artificial fertilization of turbot eggs from the North Sea. Dannevig (1895) also reported successful hatching of turbot larvae at Dunbar marine hatchery in Scotland. Malard (1899) and Anthony (1910) succeeded in getting naturally fertilized embryos from captive spawners kept in large tanks. In his paper, Anthony (1910) described the methods used by him and his co-workers in 1907 to produce turbot fry. He used a 50 l barrel with a continuous aeration system and a temperature of 15–20°C to hatch the eggs he sampled from naturally spawning turbots and fed the larvae with natural plankton. Anthony (*op. cit.*) then asked 'What is left to be done in the culture of the turbot?' In spite of his optimism and promising results, the fact remains that no further research on turbot rearing was conducted or published until the 1970s.

Following the initial commercialization of turbot farming in the UK and France during the 1980s (Jones *et al.*, 1981), the emerging industry became centred on northern Spain, owing to favourable water temperatures for on-growing. Problems of oversupply were encountered in the early 1990s, associated with a proliferation of grant-assisted Spanish farms (Shields, 2001). The industry has subsequently consolidated and output has risen gradually to approximately 6000 t in Europe in 2005. Accurate numbers for production in other parts of the world are difficult to obtain, but are estimated around 3000 t in 2005 (400 t in Chile, 2500 t in China). The farmed product has gained increasingly in standing in the discerning Spanish and French markets, due in part to feed improvements and to the greater availability of larger-sized fish. Consolidation of production has taken place both in the on-growing and hatchery sectors. The Norwegian-owned company, Stolt Sea Farm, and its Spanish subsidiary, Prodemar, dominate the on-growing sector. Juvenile supply is dominated by the company, France Turbot, which produced approximately 5 million intensively reared turbot in 2001.

21.2.2 Broodstock management and hatchery operations

Turbot spawn naturally during the summer months and its natural fecundity is very high. Females produce more than 1 million eggs/kg in each spawning season. At turbot hatcheries, the spawning season can be manipulated using photoperiod and light to produce an almost year-round supply of eggs. Turbot eggs are about 1 mm in diameter. The period of incubation before hatch is dependent on temperature, with eggs taking approximately 6 days at 14°C to hatch. The survival rate of turbot from fertilization to hatching varies considerably and averages at about 20% at hatcheries. Commercial size is around 1.5–2 kg. Productivity of turbot culture depends mostly on fry quality, rearing temperatures and on the control of the main pathologies affecting captive individuals (vibriosis and furunculosis). Environmental factors favouring optimal growth in turbot are now reasonably well known (see review by Imsland and Jonassen, 2002). The production of turbot is today carried out almost entirely in land-based facilities. In Spain and France, ambient seawater is used but in other countries, turbot is often reared in recirculation systems.

Maturation of the broodfish is determined mainly by photoperiod and occurs between June and August. The optimal temperature for spawning is between 13 and 14°C and it is possible to achieve year-round spawning by manipulation of the photoperiod. This is used by the major juvenile producers (France Turbot, Stolt Sea Farm) in order to produce high-quality juveniles throughout the year. Data indicate that female turbot do not necessarily spawn every season (Stoss and Røer, 1993; Leclercq, 1994; Imsland *et al.*, 2003b). Leclercq (1994) noted that the number of spawning females in a given group is around 70% but, for young stocks, it drops to 50%. Further, it was noted that within a group, the number of females that spawned might vary from one season to another without apparent reason.

Broodstocks are based on both wild and farmed individuals. Breeders are maintained in concrete or cement squared tanks (density: 3–6 kg/m^3) and fed on moist pellets. Turbot do not spawn spontaneously in captivity, thus gametes must be hand-stripped. As in the wild, females exhibit significantly higher growth rates than males and reach sexual maturity earlier (Imsland *et al.*, 1997a). Males produce poor sperm in terms of both quality and quantity compared to other marine teleosts, but females can produce 5–10 million eggs. Spawning can be obtained all year round, modifying rearing temperatures and day–night rhythms (Stoss and Røer, 1993; Leclercq, 1994; Imsland *et al.*, 2003b). Hormone treatment can also be used to manage advanced spawning in broodstock and to obtain egg production all year round (Mugnier *et al.*, 2000).

Mature fish are stripped and the eggs hatched in small silos (30 l). At 16°C, the eggs will hatch in 4–5 days. Methodology for turbot juvenile production is similar to that used for many other marine fish species. Fry production is mainly intensive, although good results have been obtained using extensive production methods in Norway and Denmark (Støttrup *et al.*, 1998). The duration of incubation for the 1–1.2 mm diameter eggs ranges from approximately 2450 degree hours at 10°C to 1800 degree hours at 20°C, equivalent to 5–6 days at 14°C (Iglesias *et al.*, 1991). Leclercq (1994) stated a preferred stocking density of 3000 eggs/l and an incubation temperature of 13°C. Newly hatched larvae range in length from 2.7 to 3.1 mm (Person-Le Ruyet, 1989) and rely on endogenous reserves for the following 2–3 days.

Larval culture may be semi-intensive or intensive. In semi-intensive systems, larvae are cultured at low density (2–5 larvae/l) in a large volume (50 m^3), while in intensive culture, larval density is higher (15–20/l) and tank volume is 20–30 m^3. In both systems, the rearing temperature is 18–20°C. Person-Le Ruyet (1989) and Shields (2001) described a hatchery rearing protocol that formed the basis of the intensive turbot production technique. Contemporary descriptions of the technique are similar to the published report of Minkoff and Broadhurst (1994), with a somewhat shortened duration of the live feeding stage, more stable and higher rates of survival and better quality of the resultant fry.

Newly hatched turbot larvae are stocked at densities of *c*.30–40/l into continuously illuminated rearing tanks, several cubic metres in volume. During yolk resorption, the water temperature is elevated from the initial egg incubation level, in preparation for start-feeding. Person-Le Ruyet (1989) reported a preferred rearing temperature of 18–19°C, while Riaza and Hall (1993) used a broader temperature range (16–21°C) in a semi-intensive rearing system. Lighting levels in the range 1000–4000 lux have been reported by Minkoff and Broadhurst (1994) and Planas (1994), although optimal settings will depend on system configuration, in particular related to the quantity and type of microalgae present. The newly hatched larvae feed from their vitelline reserves; mouth opening occurs on day 3. Larvae are first fed on rotifers, *Brachionus plicatilis*, and later given *Artemia* nauplii and metanauplii. They are weaned on to dry diets at the end of the first month posthatching. Phytoplankton is sometimes added to the culture medium. Various commercial feeds are used at the weaning stage. A schematic overview of the production cycle of turbot is shown in Fig. 21.8.

21.2.3 On-growing to market size

Juvenile turbot are nursed in square or circular tanks (10–30 m^3) with open-circuit pumped seawater. Aeration systems are usually used to maintain the water at oxygen saturation. Juveniles are fed with dry pelleted feed, introduced manually or automatically. The weight range varies between 5 and 10 g and 80 and 100 g during the prefattening period (duration 4–6 months). Turbot are reared either in onshore tanks (the most common technique for this species) or flat-bottomed cages. Commercial onshore tanks are square or circular cement tanks (25–100 m^3), with open-circuit pumped seawater. Aeration or oxygenation systems are normally used to maintain the water at oxygen saturation. Feeding consists of extruded pellets, introduced manually or automatically.

Production of juvenile turbot has increased fast in the last few years and there are plans for further expansion. Up to 2000, the main part of the production was for the Spanish market, with a relatively stable demand. Since 2000, there has been a surge in demand for juvenile turbot (mainly due to production in China). As a result of intensive studies in turbot fry rearing, the production of turbot juveniles has increased rapidly throughout Europe during the past two decades. The first metamorphosed juveniles were obtained in the early 1970s and in 2002, approximately 8 million turbot juveniles were produced. Over 90% of these were produced by intensive hatcheries.

In juvenile turbot, the growth rate is influenced significantly by temperature, following a pattern typical of most fish species (cf. Imsland and Jonassen,

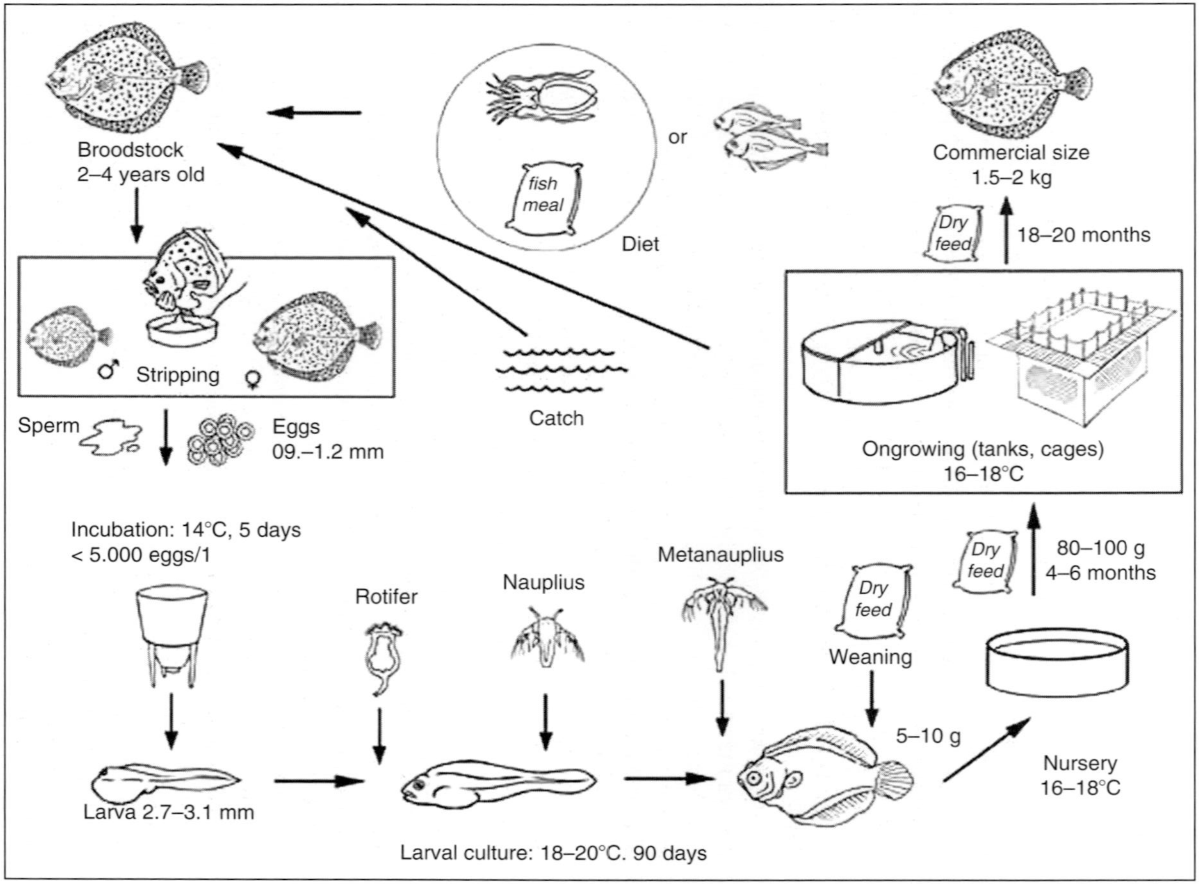

Fig. 21.8. Production cycle of turbot, illustrated schematically (www.fao.org/figis).

2002). Fish typically show a rapid increase in relative growth rate as the temperature rises, passing through a peak at optimum temperature ($T_{opt}G$) and falling rapidly at temperatures beyond $T_{opt}G$ (cf. Imsland *et al.*, 1996, 2000, 2006b). A common finding in studies examining the relationship of temperature and size on growth is that $T_{opt}G$ shifts to lower temperatures as fish increase in size. The findings of different temperature optima for different size classes, together with the downward trend of the $T_{opt}G$ with size, can be summarized in what we call the 'stepwise temperature hypothesis'. Instead of using constant rearing temperatures, one uses specific 'temperature steps' where the fish are reared at optimum temperatures defined for each size class, that is, the temperature should be lowered following changes in fish size, mimicking a mechanism suggested for wild turbot (Aneer and Westin, 1990; Iglesias and Rodríguez-Ojea, 1994). The $T_{opt}G$ for juvenile turbot is highly size dependent and drops rapidly in the first 6–8 months of the juvenile period. For juveniles < 50 g, $T_{opt}G$ is reported to be from 20 to 22°C (Imsland *et al.*, 2000, 2001a); for 50–100 g juveniles, $T_{opt}G$ is reported to be around 19°C (Burel *et al.*, 1996; Imsland *et al.*, 1996); and for juveniles > 100 g, $T_{opt}G$ is reported to be 16–17°C (Imsland *et al.*, 2006a).

Applying optimal rearing temperatures, a 2 kg fish can be produced in 18–22 months. However, the majority of aquaculture production of turbot is now in land-based flow-through systems in Spain, which are based on the ambient temperature cycle at the Spanish coast. During the second summer (the normal production cycle is based on hatching in June/July and dry feed weaned juveniles in August/September), ambient temperatures become too high for optimal growth. The fish loose appetite and are usually slaughtered during this period. The majority of the produced fish is therefore from 1–2 kg. Best prices are, however, achieved for fish > 2 kg. An obvious strategy for expanding turbot production would be to aim at producing fish in the 2–4 kg size range. This requires environmental control but, by applying state-of-the-art land-based technology, it should be obtainable. Commercial turbot feeds are available, with a current (2006) cost of €900/t. Typical food conversion efficiency (FCE) found in experimental trials is 0.8–1.2 (Imsland *et al.*, 2001a,b).

In turbot culture, photoperiod manipulation is used to produce eggs and sperm on a year-round basis (Fores *et al.*, 1990; Stoss and Røer, 1993). Studies on the effects of abrupt change in photoperiod on maturation and growth in turbot are, however, scarce. Imsland *et al.* (2003b) exposed female turbot (6 years old) to continuous light from the spring equinox (previously reared on simulated natural photoperiod). They reported that exposure to continuous light reduced the proportion of maturing females, spawning was delayed for 4 weeks, individual spawning frequency was lower, egg production was reduced by 90% and growth was significantly higher (Fig. 21.9). Delayed maturity in turbot exposed to continuous light during the juvenile stage has been documented for first-time spawners (age 2 years, size range 1500–2000 g; Imsland *et al.*, 1997b). However, it is clear that the timing of exposure to continuous light is critical for subsequent maturation for turbot, as in other flatfish species (Norberg *et al.*, 2001).

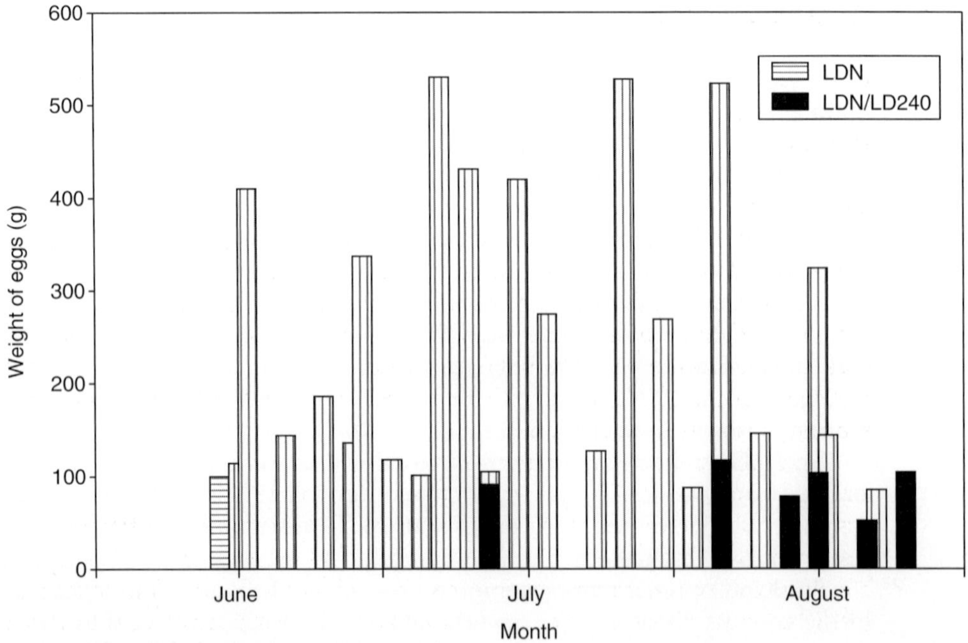

Fig. 21.9. Individual spawning frequency and timespan for mature females in both experimental groups. Each triangle (LDN group) and square (LND/LD24:0 group) shows time of successful stripping of eggs from the fish.

Studies have revealed that growth of turbot can be improved when reared in water iso-osmotic to blood (Gaumet *et al.*, 1995; Imsland *et al.*, 2001a, 2002) and that salinity and temperature will have an interactive effect on growth and feed conversion efficiency. Imsland *et al.* (2001a) found the optimal temperature – salinity combination for growth of turbot < 150 g to be 21.8 ± 0.9°C and 18.5 ± 0.8‰, and the optimal temperature – salinity combination for FCE was found to be 18.3 ± 0.6°C and 19.0 ± 1.0‰.

Farmed turbot may be subject to attack by a variety of disease organisms, but turbot does not appear to be more susceptible to disease outbreaks than other commonly cultured marine species. Bacterial diseases encountered in turbot farming include furunculosis (with *Aeromonas salmonicida* as the causative agent) and streptococcosis (with *Streptococcus parauberis* as the causative agent). The fish may be subject to attack by *Vibrio anguillarum*, the agent causing vibriosis, but vaccines that provide protection against this pathogen are available. Farmed turbot may also be attacked by several skin and gill micro- and macroparasites. Herpesvirus (*Herpesvirus scophthalmi*) is reported in turbot in the UK, Denmark and Norway (Hellberg *et al.*, 2002). Turbot is also susceptible to infections by an aquatic birnavirus, infectious pancreatic necrosis virus (IPNV), and several outbreaks of disease have been reported (Mortensen *et al.*, 1990; Vazquez Branas *et al.*, 1994). Aquatic birnaviruses are widespread in the marine environment (Bergh *et al.*, 2001) and other, less virulent, birnavirus strains have been reported for turbot (Novoa *et al.*, 1993).

21.2.4 Commercialization

Currently, fisheries supply approximately 55% of the market for turbot in Europe. The annual catch fluctuates as a result of differences in year class strength. It is difficult to predict how turbot landings will develop in the long term. Considering the discussion on the sustainability of fisheries and increasing pressure to close parts of the North Sea from fisheries, it is obvious that landings are more likely to decrease than increase. The gap between total landings and market demand can only be filled by aquaculture.

It should be possible to expand the current market (approximately 15,000 t) for turbot. Compared to other aquaculture species, such as salmon and seabass, the increase in production volume has been slow. This is due partly to the fact that almost all production is land-based, compared to sea pen culture in seabass, seabream and salmon.

The production of turbot is today carried out almost entirely in land-based facilities. In Spain and France, ambient seawater is used, but in other countries, turbot is often reared in recirculation systems. The advantage of recirculation systems is the ability to control the water temperature so that optimal growth conditions for different size classes can be achieved. Farmed turbot can be marketed from about 0.7 kg (1 year) to 3 kg (> 2 years) or more, with larger fish commanding higher prices. Prices have dropped from €14/kg in 1991 to €8/kg in 2001, but have remained remarkably stable over the past 10 years. For the larger fish, the price is around €11–12/kg or more. The market is in expansion as some farmers have started up production in the Far East (China, South Korea) and South America (Chile), and there are plans for expanding production in Europe (including most major European farmers). It is, however, impossible to foresee price development but, based on trends in other farmed species, the price may drop somewhat before reaching a stable point. Based on information from Norway and Spain, the current production cost is approximately €5–6/kg. Similar numbers have been estimated for production in Iceland (Oddgeirsson *et al.*, 2000). With the current market price of €8–14/kg, there is at the present time a positive profit margin of €3–6/kg for the production of turbot, depending on size.

21.2.5 Future perspectives

High prices and promising market forecasts are positive factors for the culture potential of turbot. There are also other factors that must be considered when appraising the culture potential of turbot, including enough space for building land-based culture facilities, good quality seawater and the possibility of lower food costs (using domestic feed sources and higher FCE due to the control of culture temperature). Production of turbot in northern Europe requires heated recirculation facilities as the optimum rearing conditions with respect to growth rates and feed conversion ratios are approximately 14–22°C. By this method, it would be possible to tailor production with respect to size classes. With the introduction of specially designed optimum recirculation systems, the cost of production would be lowered considerably and could be similar, or even lower, than the cost of producing in on-growing facilities in southern Europe (see also Oddgeirsson *et al.*, 2000). Overall, turbot remains one of the most promising aquaculture candidates in land-based culture in both south and north Europe. Rapid expansion is also taking place in other parts of the world (Asia).

21.3 Atlantic Halibut, *Hippoglossus hippoglossus*

Atlantic halibut is part of the Pleuronectidae family (Fig. 21.10) and is distributed in parts of the Arctic Ocean, in the northern part of the Atlantic Ocean, occasionally as far south as the Bay of Biscay and New York on the eastern and western side of the Atlantic, respectively (Haug, 1990) (Fig. 21.11). Immature and mature halibut occupy different habitats as coastal areas of 30–60 m depth serve as nursery areas, whereas the mature fish leave these nursery areas and migrate, often to very distant areas (Godø and Haug, 1988). The halibut congregate for spawning in winter on well-defined deepwater spawning grounds. Spawning occurs over a soft clay or mud bottom in 300–700 m depth. This deep incubation environment is dark, cool (5–7°C) and relatively sterile. The larvae hatch after 30–40 days and metamorphose from a two-sided 'fish shape' to a one-sided horizontal 'flatfish shape' after another 90 days. This rather complex deepwater natural life cycle has to be mimicked in culture and, until recently, the main obstacle for further progression of halibut culture has been the stable production of good quality larvae and juveniles. Large improvements in recent years indicate that these problems are close to being solved and on-growing of halibut may expand rapidly in the next 5–10 years.

Total annual catches of Atlantic halibut from the northern Atlantic have declined and are now close to 3500 Mt (Fig. 21.12). In some areas, extensive fishery has led to severe depletion of stocks. Today the main fishery nations are Canada, Norway, Iceland and the Faroe Islands. Total landings of halibut in

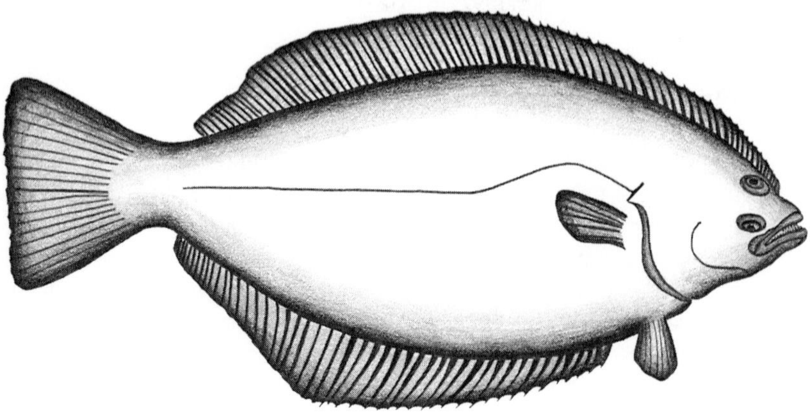

Fig. 21.10. Pleuronectidae.

Fig. 21.11. World distribution of *Hippoglossus hippoglossus*.

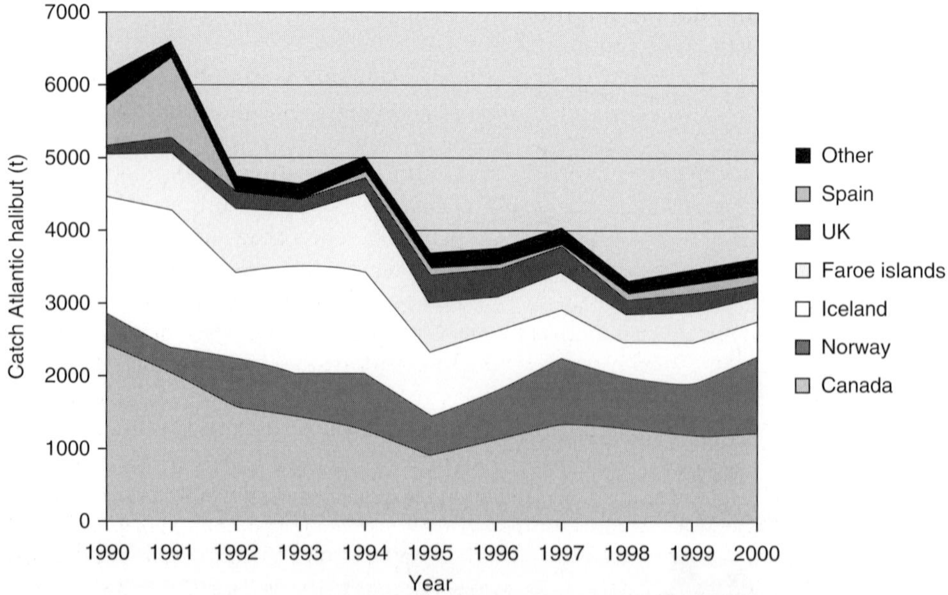

Fig. 21.12. Landings of Atlantic halibut during the period 1990–200 (*Source*: NSEC, ICES.).

Iceland have declined from 4000 Mt in 1965 and 2500 Mt in 1992 to 650 Mt in 2001. Due to the depletion of breeding and fishery stock and dramatically reduced catch per unit effort during the last decade, the Icelandic Marine Research Institute suggested a total fishery ban in Icelandic waters in 2002–2003.

Fevolden and Haug (1988) and Foss *et al.* (1998) found that Atlantic halibut from the Faroes–Iceland–Greenland region were similar in terms of allele frequencies, whereas halibut from northern Norway deviated significantly from those from Faroes–Iceland–Greenland (Foss *et al.*, 1998). However, it is still unclear whether Atlantic halibut from coastal areas off east Canada are separate from halibut in Icelandic and Norwegian waters. Tagging experiments with Atlantic halibut have revealed annual migrations between feeding and spawning grounds (Devold, 1938; Godø and Haug, 1988; Stobo *et al.*, 1988). This suggests that adult Atlantic halibut return to the same spawning grounds for several successive years so that different spawning groups may exist in the north Atlantic.

Pacific and Atlantic halibut are considered close substitutes in catch statistics, although two separate species. The total global wild catch was > 40,000 t in 1998, of which Pacific halibut made up more than 90%. Of the total catches, the North American market consumed about 75% and the EU markets 25%. The largest geographic markets in Europe are the UK, Sweden and Norway. In these markets, fresh Atlantic halibut is a high-priced product for which restaurants and the catering market constitute the prime customer segments. For the European market, Atlantic halibut is sold mainly as fresh fish and frozen products.

21.3.1 Farming of Atlantic halibut

The first known experiments on the artificial fertilization of halibut eggs were carried out by Rollefsen (1934) at Trondheim's Biological Station, Norway. Eggs were obtained from captive fish living in the station aquarium and Rollefsen was able to incubate eggs through the embryonic period (Haug, 1990). Norwegian hatching and rearing experiments with Atlantic halibut recommenced in 1974. Until 1982, all eggs used in experiments were obtained from females captured with gill nets on the spawning grounds (Haug, 1990). Since then, there has been considerable focus on Atlantic halibut as a new promising marine fish species for aquaculture in Norway. The development of a complete production line, which gave 70,000 metamorphosed juveniles in 1990, was considered as the first breakthrough for the industry. However, several problems with the upscaling of production are still giving a small and unpredictable supply of juveniles, which has been a limiting factor for the development and optimization of the on-growing phase. The first commercially farmed halibut were harvested in 1993 (11 t) and in 1999, production reached approximately 310 t. Estimated production in 2006 is approximately 1 m juveniles and 2000 t of harvested fish. Suitable rearing systems and knowledge about the environmental and genetic factors affecting growth are essential for an optimal halibut production and the further development of the industry.

Farmed halibut can be marketed from about 3–4 kg (currently 4 years or more), with larger fish commanding higher prices. Prices dropped from NOK80/kg in 1997 to NOK63/kg in 1999, but seem to have stabilized at around NOK70–75/kg since 1999. For the larger fish, the price is around NOK90/kg or higher. The market is in expansion as Norwegian halibut farmers have started to upscale their production. The concurrent volume and price increase may indicate that the total supply of Atlantic halibut does not satisfy market demand and that the market should be able to absorb a high volume of farmed halibut.

21.3.2 Broodstock management and hatchery operations

Juvenile halibut may achieve very high growth rates at temperatures around 14°C. Optimal temperatures for growth in halibut are reduced as the fish grow larger. The on-rearing phase can be divided into three temperature-dependent phases: 1–100 g (11–14°C), 100–500 g (9–12°C) and > 500 g (7–11°C). Growth can be enhanced with the use of extended photoperiods and this may also be used to lower the frequency of precocious males. Halibut can be grown at high stocking densities (up to 100 kg/m^2), but the recommended stocking density is between 25–75 kg/m^2. Although farming of Atlantic halibut has proven to be commercially viable in some countries, there are several problems that may complicate successful upscaling of the production. One major problem that hinders the establishment of halibut farming in some regions relates to the requirement of the species for seawater of appropriate temperature. High water temperatures may increase the risk of disease outbreaks in juvenile fish

(Imsland *et al.*, 2002) in broodstock or in fish that are close to market size (Bleie, 2003). During the egg incubation period, the temperature should be between 5 and 7°C. This means that the successful holding of broodstock and hatchery operations are only possible in locations where suitable seawater temperatures are available on an all-year basis.

Broodstocks of Atlantic halibut normally are maintained in cylindrical tanks (diameter 5–15 m, 1 m depth, light protected, 34‰ salinity and temperature around 8°C). Increase in temperature during the spring and transitions in water temperature in general affect spawning rhythm and egg quality in a negative way. The broodfish are normally fed fresh fish at least three times each week. Atlantic halibut is a batch spawner and each female releases several batches of eggs (4–16 per season, 3–4-day intervals). Appropriate photoperiod and temperature manipulations may be used to enable continuous egg production, and this method is implemented throughout the industry (Mangor-Jensen *et al.*, 1998; Olsen *et al.*, 1999). Halibut broodstock are held mostly in indoor tanks of varying dimensions, provided with artificial lighting. Photomanipulation of spawning can be achieved by using 12-month out-of-phase photoperiods. Spawning performance varies according to water temperature, with improved fecundity and egg viability among stocks receiving water at a temperature of 6°C during the spawning season (Brown *et al.*, 1995).

Natural spawning of halibut in large holding tanks has been used for the commercial production of eggs in Norway in some instances (Mangor-Jensen *et al.*, 1998). Successful natural egg production was reported by Holmefjord and Lein (1990) for stocks held at Sunndalsøra, western Norway. More than 30 l of good quality eggs were collected from fish in 1 m deep × 10 m diameter tanks during the 1989 spawning season. Although this method of egg production has limitations because of the large tank facilities needed and the production of two-parent groups, it may regain favour in the future as a result of the lower input of labour involved and the higher amount of good quality eggs than achieved by stripping.

Atlantic halibut has a multiple spawning strategy with several successive egg batches during a single season or up to 40% of body mass. Fry production is mainly intensive, although good results have been obtained in Norway using extensive production methods. The greatest advantage of the intensive method is the possibility of achieving a year-round production. Due to this fact, and technical improvements during the larval and weaning stages, over 90% of all cultured halibut fry are produced by the intensive method. Eggs are obtained by stripping; at least 90% of the stripped eggs are normally fertilized and 75–80% of the fertilized eggs hatch.

Halibut eggs are transparent and naturally buoyant at a salinity of 32–34 ‰. Eggs are normally reared in upwelling incubators, with plankton net to prevent them from being drained out at the overflow. The optimum water temperature limits are relatively narrow at 5–7°C. The eggs hatch approximately 82 day degrees after fertilization when incubation temperatures are kept within these limits (Mangor-Jensen *et al.*, 1998; Bergh *et al.*, 2001). The eggs hatched in large silos (250 l or larger). At 6°C, the eggs will hatch after ≈ 14 days. After hatching, the larvae are very small (6–7 mm) and poorly developed. During the next 35–45 days (depending on temperature), the larvae obtain their nutrition from the yolk sac and are very vulnerable to any disturbance in the water column. At the end of this stage, the larvae are moved to start-feeding tanks. Methodology for halibut juvenile production is similar to that used for many other marine fish species. Larvae are first fed *Artemia* nauplii and later receive *Artemia* metanauplii. They are weaned on to dry diets at approximately 90 days posthatch. It is essential that fry is given feed of a quality that mimics the composition of the food they would eat in the wild. In particular, it is important to have a high ratio of the correct fatty acids, i.e. high DHA:EPA ratio, achieved by using specially developed emulsified oils. This will improve dramatically the frequency of correct pigmentation and metamorphosis.

The duration of the yolk sac stage of halibut is approximately 35–45 days at 6–8°C (280 day degrees). The fragile yolk-sac larvae are very sensitive to both physical and microbial conditions and must be handled with great care. The production of larvae for first-feeding is undertaken in large (normally 5–10 m^3) silo-shaped, flow-through systems in complete darkness (Bolla and Holmefjord, 1988; Mangor-Jensen *et al.*, 1998). The stocking density is 1–20 larvae/l and survival ranges between 50–70% of the initial stocked population in extensive realistic experimental trials, which have also shown fairly good reproducibility for replicate treatments (10% failures). The results of production trials in commercial operations are still variable. High frequency of jaw deformation has been, and is still, a major problem in the use of flow-through silos. The production costs for this stage are relatively low (US$30/1000 larvae; Olsen *et al.*, 1999) if survival remains high. The process can easily be partly automated.

The fragility of the prolonged halibut yolk-sac larval stage has been obvious since the initial rearing trials in the mid-1980s (Opstad and Raae, 1986; Mangor-Jensen *et al.*, 1998). Halibut yolk-sac larvae have been shown to be sensitive to a range of environmental parameters including temperature, salinity, light, water flow and mechanical stress. Therefore, this developmental phase has represented a major bottleneck in the establishment of a reliable halibut production method and yolk-sac larvae have been the subject of much research (see review by Mangor-Jensen *et al.*, 1998), including morphological and histological studies of organogenesis, behavioural studies and studies of microbial factors.

A breakthrough for a reliable year-round production of halibut fry was reached in 1996 in Iceland in the company, Fiskey. The key was the improved nutritional content (enriching) of *Artemia* nauplii by using special emulsified oils developed from marine resources, with marine oil as one of the main components in the enrichment formula. The development of this enrichment mix is an important reason for Fiskey's strong position during the past decade, as this company has dominated the production of halibut fry.

21.3.3 On-growing to market size

Recent development shows that production of halibut juveniles will take place in intensive land-based systems at high densities. An effective exploitation of such systems calls for detailed knowledge on the impact of key rearing factors on fundamental production characteristics such as growth performance, FCE and fish welfare. Earlier studies (Imsland and Jonassen, 2002) have indicated clearly that temperature and photoperiod are the most important environmental factors suitable for promoting accelerated growth in juvenile Atlantic halibut, while natural catabolites from fish metabolism, especially ammonia, will probably be the most important limiting factors for growth of young fish reared at high densities (Fivelstad *et al.*, 1991).

Estimated temperature optimum for maximum growth ($T_{opt}G$) of juvenile halibut fed in excess suggest the $T_{opt}G$ for juvenile halibut in the size range 5–100g is between 12 and 15°C (Hallaråker *et al.*, 1995; Björnsson and Tryggvadóttir, 1996; Jonassen *et al.*, 1999) (Fig. 21.13). In nature, juvenile halibut settle in nursery areas along the coast at 20–60m, where the water temperature varies from 7–8°C in July to 2.5–3.5°C in March and May, and they stay there for 3–4 years (Haug, 1990). The very few observations on the growth of wild juvenile halibut indicate that a size of 44–70mm can be reached 5–7 months after

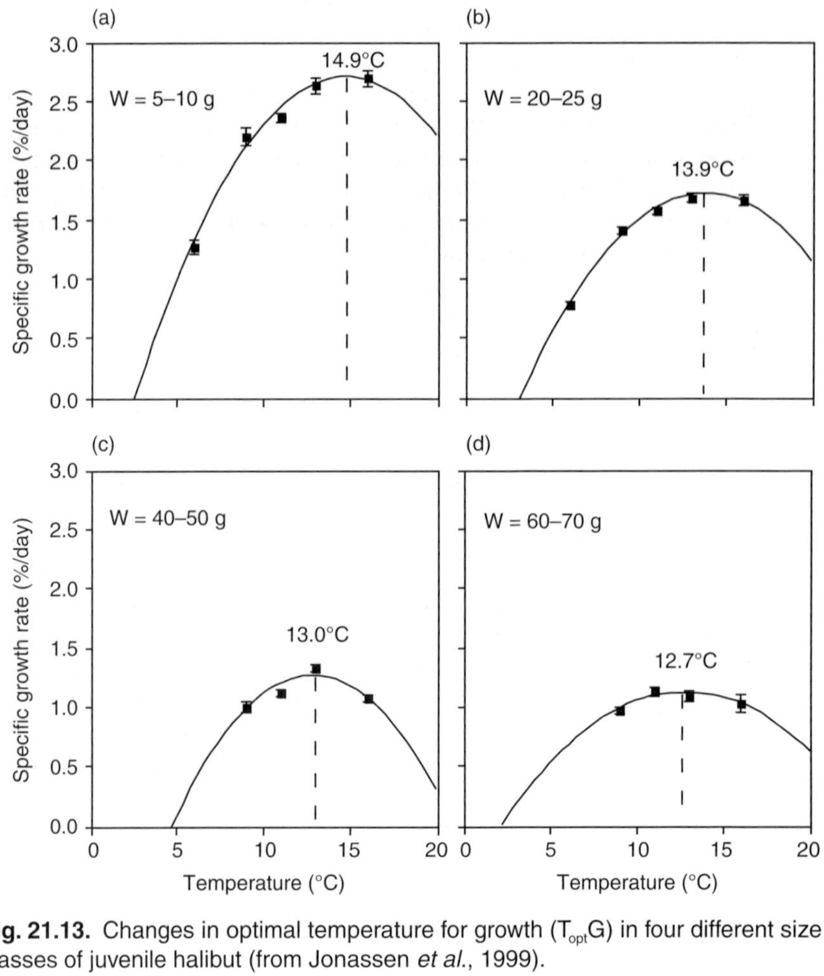

Fig. 21.13. Changes in optimal temperature for growth ($T_{opt}G$) in four different size classes of juvenile halibut (from Jonassen *et al.*, 1999).

hatching and, after 2 years, they can reach *c*.25 cm (Haug, 1990). Compared with the high optimum temperature for growth and good overall growth rates in halibut (Hallaråker *et al.*, 1995; Björnsson and Tryggvadóttir, 1996; Jonassen *et al.*, 1999), this demonstrates that wild juvenile halibut never experience natural temperature conditions where the intrinsic growth potential can be exploited.

The optimal temperature for FCE (T_{opt}FCE) is generally lower than T_{opt}G and this is more pronounced in juveniles (5–500 g) than in larger halibut (Björnsson and Tryggvadóttir, 1996). Studies on turbot, Atlantic halibut and Atlantic cod indicate, however, that the 'temperature-step' rearing method is far superior to using constant temperatures, as demonstrated by Imsland *et al.* (1996, 2006b). The possible benefit of 'temperature steps' has been verified experimentally in one case with Atlantic halibut. Aune *et al.* (1997) reared Atlantic halibut at two constant temperatures (constant 11°C and 14°C), and fish were transferred from either 11°C to 14°C (F11:14) or from 14°C to 11°C (F14:11). The authors demonstrated an 18% higher weight gain in only 3 months in the downward temperature step group.

Today, no systematic studies on the effects of abrupt change in photoperiod on maturation and growth in both sexes of Atlantic halibut have been made. However, there are indications that exposure to continuous light during the juvenile stage might have great impact on subsequent growth and maturity. Imsland and Jonassen (2003, 2005) reared halibut at four different photoperiod regimes in an experiment lasting 2 years and found that juveniles subjected to continuous light exhibited faster growth than those experiencing a natural photoperiod or a constant short day. Differences in maturity proportions between the different photoperiod regimes were also noted. This indicates a large phenotypic plasticity for the onset of maturation in halibut and that environmental manipulations during the juvenile stage can affect the onset of maturation. Maturation seems to be linked to differences in growth in month 19–25 post-hatch or approximately 1 year prior to first maturation in males. Based on these findings, there is ongoing study where manipulation of rearing conditions in this period is changed in order to enhance growth and reduce maturation (Imsland *et al.*, unpublished data). Current data indicate that applying continuous light from May to November during the second and third year may result in 20% growth improvement compared to control (Imsland *et al.*, unpublished data).

For halibut, no systematic studies on the effect of low and intermediate salinities on growth physiology in juveniles and adult exist. Unpublished data from a land-based farm in Iceland indicate that large production advantages can be gained by rearing juvenile halibut at low salinities. There are also indications that culture of halibut larvae may be improved by rearing the larvae at 15–20‰ (Opstad and Rust, 2004). For turbot (Imsland *et al.*, 2001a, 2002), rearing juveniles at intermediate salinities (15‰) improved growth and feed conversion efficiency by 10–25% compared to full salinity. In a recent study (Imsland *et al.*, unpublished data), results of similar magnitude were found for juvenile Atlantic halibut (20–100 g).

As with other domesticated species, disease problems have been experienced in halibut culture (Bergh *et al.*, 2001). In economic terms, the most important losses have been suffered at the larval and juvenile stages. The most important infections are caused by nodaviruses, causative agents of viral encephalopathy and retinopathy (VER), which are the major reason why Norway's production of halibut fry has been level since 1995 (Grotmol *et al.*, 1997). An aquatic birnavirus, infectious pancreatic necrosis virus, is also an important agent of mortality (Biering *et al.*, 1994; Biering and Bergh, 1996). *V. anguillarum*, *Flexibacter ovolyticus* and atypical *A. salmonicida* are the major bacterial pathogens (Hansen *et al.*, 1992; Bergh *et al.*, 2001). The protozoan parasites recorded include *Ichthyobodo* sp., the microsporidium *Enterocytozoon* sp. and *Trichodina hippoglossi* and the metazoan pathogens include myxozoans, helminths, *Entobdella hippoglossi*, *Lepeophtheirus hippoglossi* and other parasitic copepods (Nilsen *et al.*, 1995; Bergh *et al.*, 2001). Experimental vaccines have been tested against *V. anguillarum* and atypical *A. salmonicida*, with good results (Ingilæ *et al.*, 2000). A recombinant vaccine against nodaviruses is under development. A few trials have been carried out on non-specific immunostimulants, but no such treatment is currently available. A number of efficacy and pharmacokinetic trials with various antibacterial agents have also been published (see review in Bergh *et al.*, 2001).

21.3.4 Commercialization

The production of halibut has, until now, been almost entirely in land-based facilities. It is foreseen that the juvenile and part of the on-growing phase will remain land-based but, in Norway, it is anticipated that the majority of the on-growing production will take place in sea pens. The halibut market is now in rapid expansion, whereas prices have remained stable during the last few years. It is, however, impossible to foresee price development, but based on trends in other farmed species, the price may drop somewhat before reaching a stable point. Based on experience from Norway, the current production cost is approximately €4–7/kg. With current prices around €8–10/kg, this gives a positive profit margin for halibut production of €3–5/kg, depending on size. Production costs may even be lower for full-scale production in sea pens, thereby increasing profitability.

21.3.5 Future perspectives

High prices and promising market forecasts are positive factors for the culture potential of halibut. There are also other factors that must be considered when appraising the culture potential of Atlantic halibut, including enough space for building land-based culture facilities, good quality seawater and the possibility of lower food costs (using domestic feed sources and higher FCE due to the control of culture temperature). The demand for recirculation systems and the need for technology development at all stages of the culture cycle are but two of the most obvious negative factors. It is important to optimize rearing during the on-growing phase by the use of recirculation systems, as the optimal temperature for growth and FCE varies from 15°C for early juveniles to 7–9°C for > 1 kg fish.

21.4 Winter Flounder, *Pseudopleuronectes americanus*

The winter flounder or blackback, *P. americanus*, is part of the Pleuronectidae family (Fig. 21.10) and is distributed in the north-west Atlantic from Labrador to Georgia. Abundance is highest from the Gulf of St Lawrence to Chesapeake Bay (Bigelow and Shroeder, 1953) (Fig. 21.14). Winter flounder may attain sizes up to 64 cm total length. Winter flounder are one of the most stationary of fishes, displaying a very limited seasonal migration. Fish stay over winter in inshore areas. During summer, the larger fish move offshore to deeper waters. Although a given population usually remains fairly stationary (Saucerman and Deegan, 1991), there is evidence of wide-scale movement of some individuals, perhaps in search of food or to avoid high temperatures (Goldberg *et al.*, 2002). Genetics indicate separate groups of winter flounder north of Cape Cod, east and south of Cape Cod and on Georges Bank (Crivello *et al.*, 2004). At least groups are recognized for assessment purposes: Gulf of Maine, southern New England – Middle Atlantic and Georges Bank.

Winter flounder are eurythermal, euryhaline and extremely hardy; they possess antifreeze proteins that allow them to withstand temperatures below −1.0°C (Fletcher, 1977; Litvak, 1999). Researchers working on antifreeze proteins have often used winter flounder as a model species. Winter flounder genes have also been used to create a transgenic line of cold-tolerant Atlantic salmon (Fletcher *et al.*, 1988). This hardiness and resistance to cold temperatures will allow for expansion of the number of sites suitable for culture along the Canadian coast, making winter flounder an attractive species for Atlantic Canada. It also adapts well to a cage environment and fish grown in captivity are much heavier than wild fish of similar lengths (Litvak, 1994). According to Lee and Litvak (1996), wild young-of-the-year and laboratory-reared juveniles can be weaned easily on to commercial diets. Nevertheless, mass production of fry is not yet possible and remains a main constraint in commercial production.

Winter flounder has also gained some interest in restocking programmes in the Atlantic coast of Canada and USA. Fairchild *et al.* (2005) reported evaluation of possible release sites of winter flounder in the Great Bay Estuary, New Hampshire, USA.

21.4.1 Farming of winter flounder

There is little information specific to the development of winter flounder for aquaculture (Litvak, 1999). Development of new species for aquaculture can be a challenge. In most of these species, the major bottleneck has been survival to the juvenile stage. Until now, protocols for mass rearing of winter flounder have not been developed and early attempts were unsuccessful (Sawyer and Hoornbeek, 1980; Litvak, 1999).

Fig. 21.14. World distribution of *Pseudopleuronectes americanus*.

21.4.2 Broodstock management and husbandry

Both male and female winter flounder normally reach sexual maturity at 3 years of age. The fecundity (number of eggs produced each year) increases with body size, with smaller females producing about 500,000 and larger females around 1,500,00 eggs/year. In New England, reproduction occurs in estuaries from January to May, with peak activity during February and March when the water temperatures are the coldest of the year, ranging from 0 to 4°C. Evidence suggests that specific individuals return for many years to the same site to spawn (Crivello *et al.*, 2004). When kept in captivity, stress-related physiological disturbances and survival of wild-caught broodstock are improved when the fish is reared in iso-osmotic water (15‰) (Plante *et al.*, 2002).

Unlike the floating eggs of all other local flatfish, eggs of the winter flounder clump together in masses on the bottom. Eggs, usually laid on clean sand, hatch 15–18 days after being released. Winter flounder has high fecundity but spawn small eggs (Ben Khemis *et al.*, 2000). These small eggs (700–800 μm diameter) produce poorly developed larvae (Laroche, 1981) with limited energy reserves. The appropriateness of food supply from the onset of feeding is therefore the key to successful juvenile production. Development and time of hatching can be manipulated by temperature regime (Buckley, 1982). Eggs take approximately 7–14 days to hatch when incubated at 12 to 8°C. Winter flounder hatch as pelagic larvae with bilateral symmetry and have a total length of approximately 2.4–3.2 mm. Metamorphosis occurs after 360–400 degree days, when larvae reach about 7.8 mm in size (Ben Khemis *et al.*, 2003). Larvae exhaust their yolk supply in 4, 6, 9 and 13 days after hatching when reared at 10, 7, 5 and 2°C, respectively (Buckley, 1982; Litvak, 1999).

Litvak (1999) described the larval rearing system currently in use for winter flounder. Larvae are generally stocked at 20/l (based on the cod model from Norway). Upwelling cylindroconical tanks were used to rear the larvae (Litvak, 1999). Larvae were reared in the system at 14–16°C and they tended to reach metamorphosis at approximately 26–33 days posthatch (Lee and Litvak, 1996). Photoperiod treatment gives positive results, as larvae reared under a continuous light regime experienced close to a fivefold increase in survival (48%) over the natural photoperiod (10%) and also grew significantly faster (Litvak, 1999). It took larvae grown in continuous light 4 fewer days to reach metamorphosis than larvae grown under ambient photoperiod (> 10% faster). Although photoperiod was found to be very important, light intensity (light treatments of 5, 110 and 200 lux) did not affect growth or survival in winter flounder (Litvak, 1999).

Litvak (1999) fed the larvae in green water (with *Isochrysis galbana*) and larvae were fed enriched rotifers (5/ml twice daily) and *Artemia* (1–2/ml twice daily). Winter flounder larvae are usually fed rotifers (Mercier *et al.*, 2004) and algae during the first few weeks after feeding has been initiated. The timing of the switch to *Artemia* is monitored to match larval gape size. Usually, *Artemia* is added at 18–21 days posthatch (14–16°C). Larvae and recently metamorphosed fish (6.5–7.5 mm standard length) are removed from larval rearing tanks between 26 and 33 days (depending on temperature). The cylindroconical tank allows for easy removal of juveniles. A plexi-glass box, fitted with a water-tight lid and a screen filter overflow valve, is attached directly to the bottom of the cylindroconical tank. Metamorphosed fish are flushed gently through the bottom of the tank and captured in the plexi-glass box. The box is then placed into the juvenile tank and fish are released.

21.4.3 On-growing to market size

Weaning protocols for laboratory-reared winter flounder have been developed (Litvak, 1999; Ben Khemis *et al.*, 2003), but there is still improvement of this weaning protocol. Hebb *et al.* (1997, 2003) show that winter flounder require a high percentage of protein in their diet. Their results suggest that winter flounder do not appear to have the capability to spare protein. However, preliminary results suggest that winter flounder are able to utilize a significant proportion of plant protein in their diet, which will dramatically reduce feed costs.

Wild winter flounder do not grow particularly fast; they require approximately 3.5 years to reach 1 kg. However, larvae reared in the laboratory do grow quickly and growth rates of hatchery-reared juveniles appear to be almost an order of magnitude higher than those of their wild cousins (Litvak, 1999). Growth rates under culture conditions suggest that a dramatic decrease in time to market size, compared to its wild counterpart, is possible. The development of protocols for growing winter flounder in a culture situation has reached the stage that it can and has been transferred to industry. Currently, there is enough known about winter flounder that we are ready for a pilot-stage attempt for hatchery, on-growing and grow-out production (Litvak, 1999). However, optimization of key culture aspects for winter flounder in culture is still unsolved, including: stocking density, temperature, salinity and photoperiod regimes for the on-growing stage.

Relatively little is known about the disease susceptibility of winter flounder as it is only recently being looked on with interest for aquaculture. At present, no vaccine programme has been initiated and most disease outbreak reports are limited to wild fish. Of bacterial infections, winter flounder may be subject to attack by *V. anguillarum*, causing vibriosis (Levin *et al.*, 1972) and fin rot disease (Murchelano, 1975). Winter flounder is susceptible to infections by a nodavirus strain causing viral encephalopathy and retinopathy (VER) (Olivier, 2002). Farmed winter flounder may also be attacked by several micro- and macroparasites, including *Tricodina murmanica* and *Gyrodactylus pleuronecti* (Barker *et al.*, 2002).

21.4.4 Commercialization

Winter flounder is a hardy fish, possessing antifreeze proteins that allow it to withstand temperatures below −1.0°C. Their resistance to low temperature provides an opportunity for aquaculture development in some of the colder waters along the Atlantic Canadian coast previously deemed unsuitable for fish

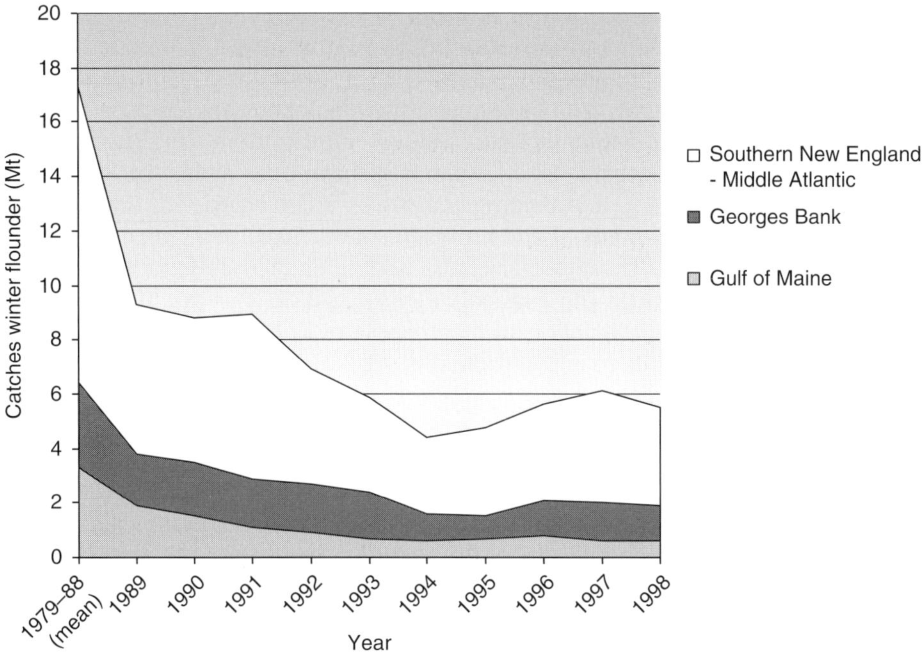

Fig. 21.15. Commercial landings of winter flounder during the period 1988–1998 (Nitschke *et al.*, 2000).

culture. The production of winter flounder is, however, still in its infantry, but it is most likely that possible future culture will almost entirely be in land-based facilities. As in other flatfish species, it is foreseen that the juvenile and part of the on-growing phase will remain land-based, but in Canada it is anticipated that part of the on-growing production could take place in sea pens (Litvak, 1999). There has been a decline in commercial availability of winter flounder for over 20 years. Landings were at an all time low in the 1990s (Litvak, 1999) and total winter flounder landings in 1998 was 5500 t (Nitschke *et al.*, 2000), among the lowest on record (Fig. 21.15). Winter flounder is sold fresh dressed or filleted. It has the thickest fillet of all the small flounders found north of Cape Cod (Litvak, 1999). Flounder market value depends on availability, size, location of catch and name given by broker but, given the decline in the commercial catch, there is a market possibility for this species.

21.4.5 Future perspectives

At present, it is not known what will be the production cost of winter flounder in culture. If we assume comparable production costs as for other cultured flatfishes, e.g. turbot and halibut (US$5–8/kg), it is questionable if winter flounder culture can be developed into a viable industry. The current price

(September 2006) for winter flounder is between US$10 and US$12/kg (fresh fillet), which is perhaps too low to support possible production costs. However, if it is possible to rear the fish in scale production in sea pens, production costs may even be lowered, thereby increasing the possibility for a net profitability of winter flounder culture. Furthermore, this species is currently part of the coastal fisheries (landings have been stable for 3 successive years (DFO, 2004)), it is commercialized as a flatfish species among others (no product differentiation) and it presents a relatively small maximal size and a low flesh yield. These factors have been identified as impediments to its aquaculture relevance (Le François *et al.*, 2002), whereas Motnikar *et al.* (2006) present an evaluation of the economics of land-based winter flounder cultivation that leads to a non-profitable verdict.

21.5 Sole, *Solea solea* and *Solea senegalensis*

Sole farming involves two species: *Solea solea* (L., 1758) and *S. senegalensis* (Kaup, 1858), which are part of the *Soleidae* family (Fig. 21.16). The world distribution of *S. solea* is shown in Fig. 21.17. These species are closely related and very similar (Dinis *et al.*, 1999) and only specialists are able to distinguish them. According to current scientific opinion, the species are so similar that experimental results yielded for one species can be applied directly to the other species. There are, however, great differences in market values at different European markets. From an economical point of view, the suitability for culture of one of the two species therefore depends on the part of Europe where the market is located. In general, *S. solea* is highly appreciated and priced accordingly in Western and Northern Europe. The market value of *S. senegalensis* is much lower in this part of Europe. In Southern Europe, the opposite is the case. Although pilot production of juveniles is conducted in Spain and Portugal, there is currently very limited production of market-size fish.

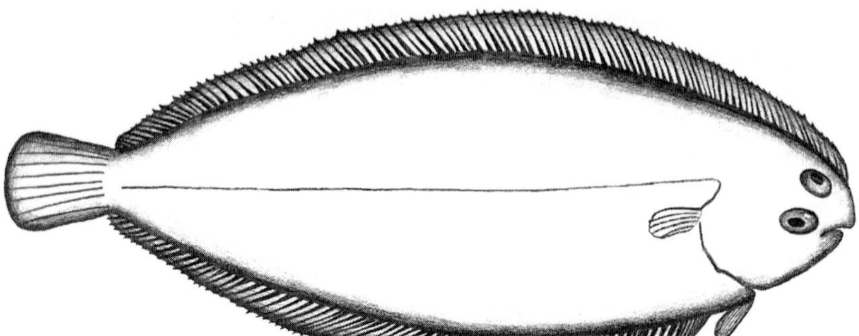

Fig. 21.16. Soleidae.

Fig. 21.17. World distribution of *Solea solea*.

In order to provide an independent insight into the systematics of the *Solea* genus, Tinti and Piccinetti (2000) studied the molecular genetics of Atlanto-Mediterranean *Solea* species using sequence analysis of two mtDNA genes (Fig. 21.18). Phylogenetic relationships were assessed applying different methods of analysis at the generic level of differentiation. Samples from seven taxa of *Solea* were taken in the Adriatic Sea, Gulf of Cádiz and Gulf of Taranto. Within the genus *Solea*, four sister lineages have evolved that actually correspond to the species *S. solea (vulgaris)*, *S. senegalensis*, *S. kleini* and *S. lascaris*. This pattern of taxonomy agrees fully with that proposed by Ben-Tuvia (1990) in a study investigating the morphological features of these species. Both Tinti and

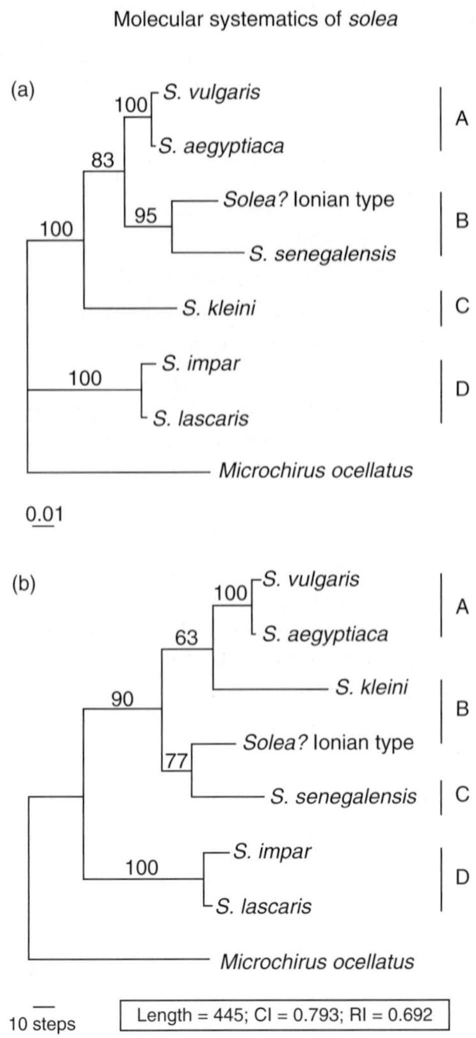

Fig. 21.18. Maximum-likelihood (a) and maximum-parsimony (b) phylogenetic tree for *Solea* taxa (from Tinti and Piccinetti, 2000).

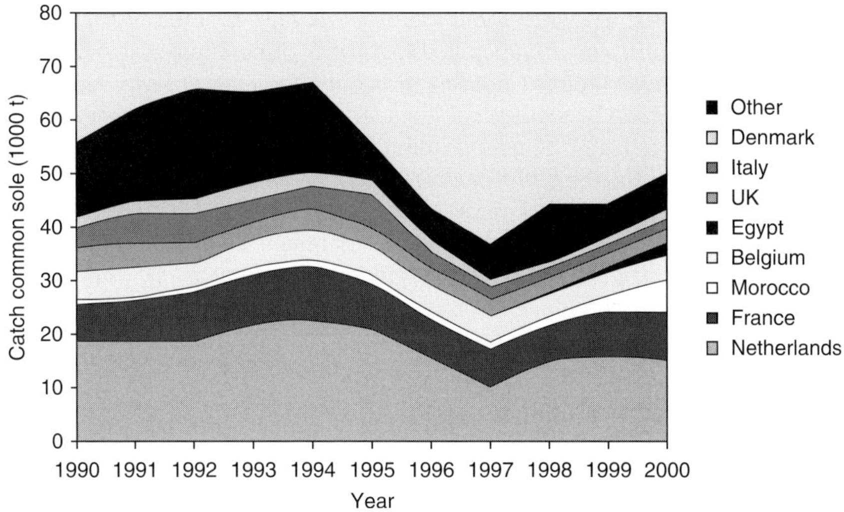

Fig. 21.19. Annual catch of sole during the period 1990–2000, sorted by nations (*Source*: NSEC).

Piccinetti's (2000) and Ben-Tuvia's (1990) studies support the phylogentic relatedness of *S. solea* and *S. senegalensis*, being found as closest sister lineages in most reconstructions. At the same time, both types of approach (i.e. molecular genetics and morphological traits) give diagnostic differences that lead consistently to their taxonomic separation at the specific rank.

In Europe, scientific and technical interests have focused on high-value native species whose biological cycle can be reproduced using currently available breeding techniques. From this point of view, both sole species appear as credible candidates for marine culture. The current market price in the EU is around €8.5–16.5/kg, depending on size and season, for *S. solea* and around €8–14/kg for *S. senegalensis*. Currently, consumer demand for both sole species is provided by the fisheries. In 1998, the total landing of sole in Europe was 45.586 t (Anon., 1999) (Fig. 21.19). Annual catches fluctuate as a result of differences in year class strength and it is hard to predict how sole landings will develop in the long term. However, considering the discussion on the sustainability of the fisheries and increasing pressure to close down fishing in parts of the North Sea (producing approximately 50% of the total landing for *S. solea*), it is obvious that landings are more likely to decrease than increase. The little information that can be found regarding wild catches of *S. senegalensis* indicate that catches are declining for this species (Anon., 1999, 2004a,b).

21.5.1 Farming of sole

Thirty years ago, sole was already considered one of the most interesting and promising species for marine fish farming in Europe (Howell, 1997). However, the species never did become a commercial success, as technological and disease problems hampered the development of commercial sole culture. The main problem was the occurrence of a disease called black patch necrosis (BPN) (Bernadet *et al.*, 1990). It is now known that BPN is contributed to by poor nutrition in the natural diets fed to the sole (Baynes and Howell, 1993). Another restraint of sole farming in Western and Northern Europe used to be water temperature control. Even the more northern of the two species (*S. solea*) requires relatively warm water (~ 20°C) for optimal growth in the juvenile and on-growing stage. As a result, suitable sites (e.g. near power plants) were rare. Today, recirculation technology is fully established and freely available. This means that optimal growth conditions for sole can be realized year round, even in temperate areas. This, together with recent advances in feed technology surrounding weaning and on-growing, have fuelled a renewed interest in sole as an aquaculture species (Howell, 1997; Dinis *et al.*, 1999). A considerable number of scientific reports have been published over the past three decades, and several laboratory- and pilot-scale experiments have been conducted in recent years.

21.5.2 Broodstock management and hatchery operations (*S. solea*)

To summarize, the environmental conditions used for broodfish in published studies are: sex ratio of 0.5–3 males to each female, moderate densities (0.6–3.0 kg/m^3; Devauchelle *et al.*, 1987), large tanks (> 10 m^3), with light intensity of 20–1500 lux, temperature should be kept between 8 and 12°C and allowed to fluctuate with ambient fluctuations. Broodfish are fed *ad libitum* with fresh mollusc and polychaetes at an average rate of about 10% body weight per week (Baynes *et al.*, 1993). The fish are allowed to spawn naturally and fertilized eggs are collected in the water column. Temperatures are allowed to follow the annual cyclic regime where the difference between maximum and minimum temperature should be around or less than 12°C. First signs of spawning are correlated to raising the temperature in spring and this temperature rise may be an important environmental determinant for spawning. The use of photoperiod for manipulation of spawning has not been studied systematically for sole. Fecundity varies between 10 and 140 eggs/kg and a fertilization rate between 20 and 80%. No indication of correlation between fertilization rates and sex ratios was found (Baynes *et al.*, 1993).

Devauchelle *et al.* (1987) reported that the best incubation performances (defined as max % hatching and min % of deformed newly hatched larvae) for *S. solea* eggs was 13–15°C, which was slightly higher than the optimal temperature for spawning found by the same authors (8–12°C). Optimal salinity range for eggs and embryos is reported to be from 20 to 35‰ (Fonds, 1979; Devauchelle *et al.*, 1987). The size of the eggs ranges from 1.0 to 1.6 mm and has been reported to decrease during the spawning season (Baynes *et al.*, 1993; Houghton *et al.*, 1985). Fonds (1979) incubated sole eggs at five different temperatures between 10 and 22°C and five different salinities between 20 and 50‰. High survival and normal development until hatching was observed at temperatures from 10 to 16°C and salinities from 20 to 40‰. At 22°C, no viable larvae hatched and at 19°C, many embryos were abnormal. Survival was near 100% at 10°C. Incubation time (i.e. from fertilization to start of feeding) was highly dependent on the incubation temperatures and was 27.5, 19.7 and 15 days at 10, 13 and 16°C, respectively. The optimal temperature for successful development of the eggs was lower than the optimal temperature for growth of the larvae, as maximum growth rates were found at 19 and 22°C.

Howell (1997) reported that rearing the larvae through metamorphosis presented few problems, with survival rates being consistently in excess of 70% in small-scale laboratory systems. The larvae can be reared on a diet of freshly hatched *Artemia* nauplii without prior enrichment with algae or proprietary 'booster' diets. The larvae have also been reared on a diet of rotifers, offered either as the exclusive food source (Howell, 1973) or in combination with *Artemia* nauplii (Fuchs, 1982), but survival was not enhanced by the availability of the smaller food organism in either study. According to Howell (1997), the relative ease with which larvae can be reared may, in part, reflect the consistent quality of the fertilized eggs produced by captive stocks. Because these stocks are the product of natural spawning, they are not subject to varying quality caused by overripening, as is often evident in fish where the gametes are stripped manually. Dietary requirement for (*n*-3) HUFA in *S. solea* is found to be less stringent than in many other marine species (Howell and Tzoumas, 1991), so that enhancing the lipid content of *Artemia* is not a prerequisite for high larval survival as long as *Artemia* strains rich in eicosepentaenoic acid, 20:5 (*n*-3), are used. Quantitative requirements for specific nutrients, particularly lipids, during the larval stages remain unknown.

Weaning for *S. solea* larvae has been started at different days posthatch, from 10 dph (Gatesoupe and Luquet, 1982) to 25–40 dph (Bromley, 1977). Studies indicate that a mixture of inert and live food may increase the weaning attractants in the dry feed (Cadena-Roa *et al.*, 1982; Métailler *et al.*, 1983). New studies (Howell, 1997; Day *et al.*, 1999) have demonstrated that young sole can be weaned on to commercially prepared formulated feeds with high survival and growth approaching those attainable on live foods.

21.5.3 Broodstock management and hatchery operations (*S. senegalensis*)

The reproduction of *S. senegalensis* captivity has been the subject of research since the early 1980s (Rodriguez, 1984; Dinis, 1986, 1992; Bedoui, 1995). Bedoui (1995) reared *S. senegalensis* at an experimental scale. The rearing was conducted from broodstock to juveniles of 2 months. The broodstock was collected from the wild and acclimatized in a raceway with a sand bottom. Natural spawning was obtained at 18°C during a 3-month period (April–June). The pelagic egg size ranged from 0.99 to 1.02 mm. The incubation was performed in stagnant water and lasted 42 h at 19°C. Dinis and Reis (1995, cited in Dinis *et al.*, 1999) described the broodstock management and larval rearing of *S. senegalensis* obtained in a pilot research project in Portugal since 1994. After 7 months in captivity, a wild broodstock spawned naturally and the spawning season lasted from March until June, at temperatures ranging from 16.5 ± 0.5°C to 22 ± 1.0°C and salinities from 30 to 35‰. Batches of eggs with 100% fertilization presented viability ranging from 90 to 100%. Newly hatched larvae had a total length of 2.6 ± 0.1 mm and were reared in 200 l fibreglass cylindroconical tanks.

Incubation of *S. senegalensis* eggs has been performed using temperatures of 19°C (Cañavate and Fernandez-Díaz, 1999) and 18–21°C (Dinis *et al.*, 1999). Various types of experimental units have been used for incubation: cylindroconical 300 l tanks with gentle aeration and continuous upwelling at 0.5 l/min (Cañavate and Fernandez-Díaz, 1999), 150 l fibreglass cylindroconical tanks (Dinis, 1992) and 500 l fibreglass cylindroconical tanks (Vázquez *et al.*, 1994).

Dinis *et al.* (1999) summarized culture experience with *S. senegalensis* in Portugal and Spain in the period 1993–1997. Natural spawning of broodstock in captivity was accomplished and was the only way viable eggs were obtained. The broodstock feed regime was based on squid, *Loligo vulgaris*, and was supplemented with polychaetes, *Hediste diversicolor*, during final maturation. Temperature played a very important role in the onset and duration of the spawning period, with egg emission stopping below 16°C. Observed duration of the spawning period ranged from 4 to 6 months. The total weight of the eggs collected daily during the spawning season ranged from 0 to 180 g for a broodstock of 15 fish. Egg fertilization rates varied between 20 and 100% and the percentage of viable eggs (percentage of fertilized eggs hatching) was 72.1 ± 26.5%. Variations in egg size between batches were detected, with egg size tending to decline during the spawning season. A similar trend is found for *S. solea* (Baynes *et al.*, 1993).

In conclusion, the available literature indicates clearly that the broodstock management and natural spawning of *S. senegalensis* in captivity can be achieved successfully (Dinis *et al.*, 1999) and that spawning occurs normally in broodstock fed on squid supplemented with polychaetes. Stocking density in the maturation tanks should be 1–1.5 kg/m^2 and temperature should be kept above 16°C, as emission stops below that temperature. Recent data (Dinis *et al.*, 2003) indicate that broodstock density might be up to 5 kg/m^2 and that the annual temperature cycle triggers sole maturation. In nature, the onset of spawning is related to the rise in temperature during spring (March–June). Portuguese studies indicate that *S. senegalensis* can be found in coastal and estuary areas where the temperature rises from approximately 16 to 25°C during this period of the year (Cabral and Costa, 1999; Dinis *et al.*, 1999; Cabral, 2000). Salinity should be kept constant around 33–35‰ and the fish are reared under simulated natural photoperiod (LDN). In other cultured flatfish species, a change in the photoperiod is the key environmental signal used to manipulate and control maturation (Imsland *et al.*, 1997b, 2003b; Norberg *et al.*, 2001) but at the present time, there are no published experimental studies to verify or contradict this for either *S. senegalensis* or *S. solea*.

21.5.4 On-growing of *S. solea* to market size

Temperature is a rate-controlling factor for all chemical processes in poikilotherms. However, comprehensive growth data for sole are lacking and there are very few studies that have tried to quantify the effects of temperature (Irvin, 1973; Fonds, 1976). Irvin (1973) monitored the growth rate of hatchery-reared juvenile *S. solea* of an initial mean total length of about 5 cm at five temperatures ranging from 11 to 27°C for 12 weeks. The fish were fed *ad libitum* on an oligochaete worm. The fish showed an approximately linear increase in growth from 9 to 23°C and a drop in growth after that (Fig. 21.20). Fonds (1976) worked with wild-caught *S. solea* of a larger initial size (12–13 cm) and followed their growth for over a year at temperatures ranging from 10 to 25°C. The fish were fed daily with fresh chopped mussel, *Mytilus edulis*, or live lugworm, *Arenicola marina*. In his study, he found that the fish grew slower, as they were larger than the fish in Irvin's (1973) study, but both experiments showed little increase in growth rates above 20°C and indicated that the optimum temperature for growth (i.e. $T_{opt}G$) was between 20–25°C. Howell (1997) extrapolated the data from these two studies and found that fish of

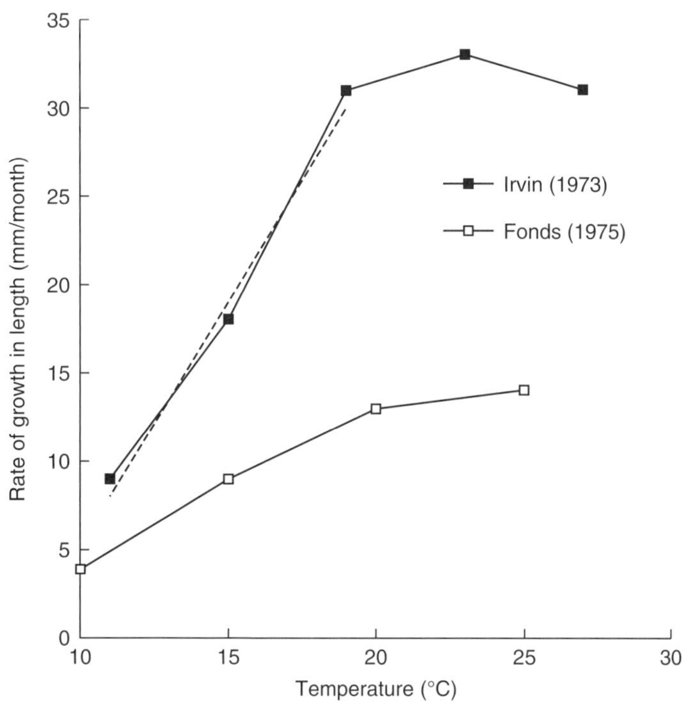

Fig. 21.20. The relationship between rate of increase in length and temperatures for juvenile *S. solea* calculated from Irvin (1973) and Fonds (1976). For Irvin's data the regression of growth rate on temperatures from 11 to 19°C is shown (dotted line) (from Howell, 1997).

about 5 cm might reach minimum market size of 24 cm (125 g) at temperatures close to optimum in less than 300 days. Day *et al*. (1997) reared weaned juveniles for an 18-month period at an average temperature of 16.5°C, achieving a final average weight of 133 g ± 40 SD (217.5 mm ± 19.1 SD). The study shows that intensive on-growing of sole may be performed in sand-free tanks, as a near market size of 22 cm may be obtained in 18 months with no mortalities. The growth rates reported in this study are similar to those reported by Danielssen and Gulbrandsen (1989), who reared sole on chopped blue mussel for the first 150 days, followed by pellets containing blue mussel and squid.

Early work on the culture of sole, *S. solea*, revealed that the species appeared to be extremely vulnerable to disease (see discussion in Baynes and Howell, 1993). The most common and devastating of these is black patch necrosis (BPN), a condition first described by McVicar and White (1979) and later confirmed as being caused by the bacterium, *F. maritimus* (Bernadet *et al*., 1990). BPN was reported to be highly infectious, but was found to be both prevented and controlled by providing a sand substrate in the rearing tanks (McVicar and White, 1979). Although it has been widely held that a sand substrate is essential for successful culture of juvenile sole, later research has shown that this is not necessarily so and that sole may not be as vulnerable to disease as these early trials suggested. Baudin-Laurencin (1986) investigated the susceptibility of *S. solea* to vibriosis caused by a specific *V. anguillarum* strain, *V. anguillarum* 408, and found that juvenile sole seemed fairly resistant to *V. anguillarum* 408, at least if the level of virulent bacteria was not very high. The first occurrence of viral nervous necrosis (VNN) in *S. solea* was documented recently (Starkey *et al*., 2001). This disease is characterized by the development of a vacuolating encephalopathy and retinopathy associated with arrays of virus-like particles in infected neurons. It has been associated with high mortalities in farmed Atlantic halibut in Norway (Grotmol *et al*., 1997) and has the potential to cause severe economic loss in the farming of marine species (Starkey *et al*., 2001).

21.5.5 On-growing of *S. senegalensis* to market size

Dinis *et al*. (1999) investigated the growth of juvenile *S. senegalensis* during an experimental on-growing trial. Two types of earthen ponds were stocked with unweaned juveniles. Annual temperature in the ponds fluctuates between 15°C (January) and 24°C (July). One pond (1000 m^2) was equipped with nets at the water entrance to exclude predators and competitors. No food was supplied and the fish were fed on naturally occurring prey only. A total of 2000 fish were stocked in this pond in late July. One year later, the pond was harvested and 20% of stocked fish were captured. After 1 year, the fish had obtained an average total length of 16.6 ± 2.1 cm and weighed 40.3 ± 2.5 g. Fish were also stocked in a pond with *Sparus aurata*, at a density of two sole juveniles/m^2. In this pond, the fish were fed with pellets. After 1 year, the fish had obtained a total length of 35.3 ± 1.8 cm and weighed 456.1 ± 3.6 g, with a survival of 8%. Studies of benthic fauna in commercial seabream ponds, carried out by Pousão-Ferreira *et al.* (1995), reported a high occurrence of polychaetes, which may explain the good growth obtained in the former study.

The effect of different feeding frequencies and night versus day feeding has been studied in juvenile (1530 g) *S. senegalensis* (Engrola *et al.*, 2002). Higher feeding frequencies seem to reduce growth heterogeneity, but have little effect on mean growth rates. The light regime does not seem to have a major effect on feed intake or mean growth, but in general tend to show a slight improvement under illuminated conditions. Juvenile *S. senegalensis* (10–25 g) seem to be little affected by stocking densities up to 4.5 kg/m^2 (Engrola *et al.*, 2002). Still, the smaller fish in a batch seem to grow better if graded and reared at lower densities. *S. senegalensis* has an apparent good adaptation capacity to different diet compositions and feeding regimes. However, suboptimal feed formulations or inadequate feeding practices may impose a toll in terms of growth heterogeneity, nitrogen waste output or flesh quality. Further studies are required regarding optimal rearing technology and conditions, feeding behaviour and nutritional requirements.

In general, less is known about diseases in *S. senegalensis* than in *S. solea*. In Spain, there have been outbreaks of a viral disease with high mortality affecting *S. senegalensis*. This disease is characterized by dark coloration, hyperactivity, erratic swimming and abnormal behaviour. Rodriquez *et al.* (1997) isolated and characterized a birnavirus, named solevirus, from the skin and internal organs of moribund and dead soles. A serological comparison of solevirus with reference strains of IPNV (infectious pancreatic necrosis virus) show that the virus is clearly related to the Sp serotype, which is the most common serotype in Spain (Perez-Prieto *et al.*, 2001). In February 2001, an outbreak associated with moderate mortalities (20%) in populations of *S. senegalensis* cultured in a farm in south Spain was observed (Zorrrilla *et al.*, 2003). Bacteria isolated from the outbreak were identified as *V. harveyi* and *V. parahaemolyticus*. For both species, vaccination with sublethal doses of extracellular products reduced mortality by 32–37% (*V. parahaemolyticus*) and 76–83% (*V. harveyi*) compared to unvaccinated fish. Zorrilla *et al.* (1999) reported the first description of pasteurellosis affecting *S. senegalensis* cultured in the south-west of Spain. Microbiological analysis of these fish revealed the presence of a bacterial colony in all organs examined, which was biochemically and serologically characterized as *Photobacterium damsela* ssp. *piscicida*. This bacterium is the causative agent of pasteurellosis, a disease which provokes massive mortalities in cultures of several marine fish species.

21.5.6 Commercialization

Research on sole culture has increased over the past decade and has extended our current knowledge of the species and promoted renewed interest in the aquaculture of both species. As sole, yet again, is looked on as a promising aquaculture candidate, it is important to study the species in its natural environment, as optimal conditions in culture will try to mimic the optimal preferences of the species in nature. It seems clear that there are several biotic and abiotic mechanisms that act as structural forces in these species (see review by Imsland *et al*., 2004) and recent studies support the phylogenetic relatedness of *S. solea* and *S. senegalensis*, being found as closest sister lineages in most reconstructions. At present, sole is caught by commercial fisheries in many countries in Europe, but landings show strong fluctuations between years and seasons (Anon., 2004a,b). Moreover, there is a considerable pressure on exploitation due to overfishing, which makes a continuous supply of sole in the future uncertain, based on fishery. This makes sole an interesting species for aquaculture. Production of sole usually requires heated recirculation facilities as the optimum rearing conditions with respect to growth rates and feed conversion ratios are approximately 20–25°C. Studies on sole have indicated clearly that, in contrast to other cultured marine fishes, fingerling production is not the bottleneck in proceeding towards commercial culture. This is mastered at the laboratory- and pilot-scale, but experience with upscaling is still lacking.

21.5.7 Future perspectives

In sole culture, there are still major problems with feeding and on-growing systems, which are due to the specific feeding behaviour of this species. Sole require on-growing facilities that achieve a compromise between self-cleaning capacities, feed residence time and feed distribution. There is currently no commercial feed available specifically for sole that suits both the needs of the species and the farming technique. Furthermore, specialized sole on-growing systems are currently non-existent. High prices and promising market forecasts are positive factors for the culture potential of sole. Demand for recirculation systems, slow growth, import of all culture material (broodfish, eggs and juveniles) and the need for technology development at all stages of the culture cycle are the most obvious negative factors.

References

Aneer, G. and Westin, L. (1990) Migration of turbot (*Psetta maxima* L.) in the northern Baltic proper. *Fisheries Research* 9, 307–315.

Anon. (1999) *Fishery Statistics Capture Production Yearbook*. FAO, Rome, Italy.

Anon. (2004a) *Fishery Statistics Capture Production Yearbook*. FAO, Rome, Italy.

Anon. (2004b) *ICES Catch Database*, International Council for the Exploration of the Sea, Denmark, 2004 (http://www.fishbase.org/report/ICES/ICESSearchMenu.cfm, accessed 6 February 2008).

Anthony, T. (1910) The cultivation of the turbot. *Proceedings of the Fourth International Fisheries Congress, Washington, 1908*. US Bureau of Fisheries Bulletin, Washington, DC, pp. 859–870.

Aune, A., Imsland, A.K. and Pittman, K. (1997) Growth of juvenile halibut (*Hippoglossus hippoglossus* L.) under a constant and switched temperature regime. *Aquaculture Research* 28, 931–939.

Barker, D.E., Cone, D.K. and Burt, M.D.B. (2002) *Trichodina murmanica* (Ciliophora) and *Gyrodactylus pleuronecti* (Monogenea) parasitizing hatchery-reared winter flounder,

Pseudopleuronectes americanus (Walbaum): effects on host growth and assessment of parasite interaction. *Journal of Fish Diseases* 25, 81–89.

Baudin-Laurencin, F. (1986) Sensitivity of the sole (*Solea solea*) to vibriosis. In: Vivares, C.P., Bonami, J.R. and Jaspers, E. (eds) *Pathology in Marine Aquaculture*. European Aquaculture Society, Special Publication No. 9, Bredene, Belgium, pp. 345–350.

Baynes, S.M. and Howell, B.R. (1993) Observations on the growth, survival and disease resistance of juvenile common sole, *Solea solea* (L.), fed *Mytilus edulis* L. *Aquaculture and Fisheries Mangagement* 24, 95–100.

Baynes, S.M., Howell, B.R. and Beard, T.W. (1993) A review of egg production by captive sole, *Solea solea* (L.). *Aquaculture and Fisheries Mangagement* 24, 171–180.

Bedoui, R. (1995) Rearing of *Solea senegalensis* (Kaup, 1958) in Tunisia. *Cahiers Options Méditerranéenes* 16, 31–39.

Ben Khemis, I., de la Noue, J. and Audet, C. (2000) Feeding larvae of winter flounder *Pseudopleuronectes americanus* (Walbaum) with live prey or microencapsulated diet: linear growth and protein, RNA and DNA content. *Aquaculture Research* 31, 377–386.

Ben Khemis, I., Audet, C., Fournier, R. and de la Noue, J. (2003) Early weaning of winter flounder (*Pseudopleuronectes americanus* Walbaum) larvae on a commercial microencapsulated diet. *Aquaculture Research* 34, 445–452.

Ben-Tuvia, A. (1990) A taxonomic reappraisal of the Atlanto-Mediterranean soles *Solea solea*, *S. senegalensis* and *S. lascaris*. *Journal of Fish Biology* 36, 947–960.

Bergh, Ø., Nilsen, F. and Samuelsen, O.B. (2001) Diseases, prophylaxis and treatment of the Atlantic halibut *Hippoglossus hippoglossus*: a review. *Diseases of Aquatic Organisms* 48, 57–74.

Bergstad, O.A. and Folkvord, A. (1997) Dispersal of tagged juvenile turbot *Scophthalmus maximus* on the Norwegian Skagerrak coast. *Fisheries Research* 29, 211–215.

Bernadet, J.F., Campbell, A.C. and Buswell, J.A. (1990) *Flexibacter maritimus* is the agent of 'black patch necrosis' in Dover sole in Scotland. *Diseases of Aquatic Orgamisms* 8, 233–237.

Biering, E. and Bergh, Ø. (1996) Experimental infection of Atlantic halibut, *Hippoglossus hippoglossus* L., yolk-sac larvae with infectious pancreatic necrosis virus: detection of virus by immunohistochemistry and in situ hybridization. *Journal of Fish Diseases* 19, 405–413.

Biering, E., Nilsen, F., Rødseth, O.M. and Glette, J. (1994) Susceptibility of Atlantic halibut *Hippoglossus hippoglossus* to infectious pancreatic necrosis virus. *Diseases of Aquatic Organsims* 20, 183–190.

Bigelow, H.B. and Shroeder, W.C. (1953) Fishes of the Gulf of Maine. *Fisheries Bulletin of the Fish and Wildlife Service* 53, Woods Hole Oceanographic Institution, Woods Hole, Massachusetts.

Björnsson, B. and Tryggvadóttir, S.V. (1996) Effect of size on optimal temperature for growth and growth efficiency of immature Atlantic halibut (*Hippoglossus hippoglossus* L.) *Aquaculture* 142, 33–42.

Bleie, H. (2003) Health situation in Norwegian cultured stocks. Annual Report of the Norwegian Marine Institute, pp. 57–60 (in Norwegian) (http://www.imr.no/__data/page/3887/2.9_Marine_arter__Helsesituasjonen_potensielle_problemer_og_forebyggende_tiltak.pdf, accessed 6 February 2008).

Bolla, S. and Holmefjord, I. (1988) Effect of temperature and light on development of Atlantic halibut larvae. *Aquaculture* 74, 355–358.

Bromley, P.J. (1977) Methods of weaning juvenile hatchery reared sole (*Solea solea* (L.)) from live food to prepared diets. *Aquaculture* 12, 337–347.

Brown, N.P., Bromage, N.R. and Shields, R.J. (1995) The effect of spawning temperature on egg viability in the Atlantic halibut (*Hippoglossus hippoglossus*). In: Goetz, F.W. and Thomas, P. (eds) *Proceedings of the Fifth International Symposium on the Reproductive Physiology of Fish*. University of Texas, Austin, Texas, pp. 181.

Buckley, L.J. (1982) Effects of temperature on growth and biochemical composition of larval winter flounder *Pseudopleuronectes americanus*. *Marine Ecology Progress Series* 8, 181–186.

Burel, C., Person-Le Ruyet, J., Gaumet, F., Le Roux, A., Sévère, A. and Boeuf, F. (1996) Effects of temperature on growth and metabolism in juvenile turbot. *Journal of Fish Biology* 49, 678–692.

Cabral, H.N. (2000) Distribution and abundance patterns of flatfishes in the Sado Estuary, Portugal. *Estuaries* 23, 351–358.

Cabral, H. and Costa, M.J. (1999) Differential use of nursery areas within the Tagus Estuary by sympatric soles *Solea solea* and *Solea senegalensis*. *Environmental Biology of Fishes* 56, 389–397.

Cadena-Roa, M., Huelvan, C., Le Borgne, Y. and Metailler, R. (1982) Use of rehydratable extruded pellets and attractive substances for the weaning of sole (*Solea vulgaris*). *Journal of World Mariculture Society* 13, 246–253.

Cañavate, J.P. and Fernández-Díaz, C. (1999) Influence of co-feeding larvae with live and inert diets on weaning the sole *Solea senegalensis* onto commercial dry feeds. *Aquaculture* 174, 255–263.

Coughlan, J.P., Imsland, A.K., Galvin, P.T., FitzGerald, R.D., Nævdal, G. and Cross, T.F. (1998) Microsatellite DNA variation in wild populations and farmed strains of turbot *Scophthalmus maximus* from Ireland and Norway. *Journal of Fish Biology* 52, 916–922.

Crivello, J.F., Danilla, D., Lorda, E., Keser, M. and Roseman, E.F. (2004) The genetic stock structure of larval and juvenile winter flounder larvae in Connecticut waters of eastern Long Island Sound and estimations of larval entrainment. *Journal of Fish Biology* 65, 62–76.

Danielssen, D.S. and Gulbrandsen, K.E. (1989) Growth rate of turbot (*Scophthalamus maximus* L.) and sole (*Solea solea* L.) based on dry pellets. In: De Pauw, N., Jaspers, E., Ackefors, H. and Wilkins, N. (eds) *Aquaculture: A Biotechnology in Progress*, Volume 1. European Aquaculture Society, Bredene, Belgium, pp 631–636.

Dannevig, H.C. (1895) On the hatching operation at Dunbar Marine Hatchery. *Reports of the Fisheries Board of Scotland* 13, 123–132.

Day, O.J., Howell, B.R. and Jones, D.A. (1997) The effect of dietary hydrolysed fish protein concentrate on the survival and growth of juvenile Dover sole, *Solea solea* (L.), during and after weaning. *Aquaculture Research* 28, 911–921.

Day, O.J., Howell, B.R., Aksnes, A. and Nygård, E. (1999) Recent advances in the weaning of sole, *Solea solea* (L.). *Abstracts of Contributions Presented at the International Conference Aquaculture Europe 1999*. European Aquaculture Society, Oostede, Belgium, Special Publication No. 27, 40–41.

Déniel, C. (1990) Comparative study of growth of flatfishes on the west coast of Brittany. *Journal of Fish Biology* 37, 149–166.

Devauchelle, N., Alexandre, J.C., Le-Corre, N. and Letty, Y. (1987) Spawning of sole (*Solea solea*) in captivity. *Aquaculture* 66, 125–147.

Devold, F. (1938) The North Atlantic halibut and net fishing. *Fiskeridirektoratets Skrifter Serie Havundersøkelser* 5, 5–47.

DFO (2004) Winter flounder in the southern Gulf of St Lawrence (Div. 4T). DFO Science Stock Status Report 2004/005.

Dinis, M.T. (1986) Quatre *soleidae* de l'estuaire du Tage. Reproduction et croissance. Essai d'élevage de *Solea senegalensis* Kaup. Thèse de Doctorat – Sciences Naturelles, Université de Bretagne Occidentale, Brest, Bretagne, France.

Dinis, M.T. (1992) Aspects of the potential of *Solea senegalensis* Kaup for aquaculture: larval rearing and weaning to an artificial diet. *Aquaculture and Fisheries Management* 23, 515–520.

Dinis, M.T. and Reis, J. (1995) Culture of *Solea* spp. In: *Marine Aquaculture Finfish Species Diversification*. CIHEAM-IAMZ, Zaragoza, Spain, pp. 9–19.

Dinis, M.T., Ribeiro, L., Soares, F. and Sarasquete, C. (1999) A review on the cultivation potential of *Solea senegalensis* in Spain and in Portugal. *Aquaculture* 176, 27–38.

Dinis, M.T., Soares, F., Ribeiro, L., Engrola, S., Cação, P., Aragão, C., Pousão-Ferreira, P. and Conceição, L.E.C. (2003) Broodstock management and larval rearing of Senegalese sole (*Solea senegalensis*). *Book of Abstracts of the World Aquaculture 03*, 19–23 May 2003, Salvador, Brazil. World Aquaculture Society, Bahia Convention Center, Salvador, Brazil, 164 pp.

Engrola, S., Conceição, L.E.C., Pereira, R. and Dinis, M.T. (2002) Growth and diet utilisation of sole, *Solea senegalensis* reared under different feeding regimes. *Book of Abstracts of the 10th International Symposium on Nutrition and Feeding in Fish*, 2–7 June 2002, Rhodes, Greece. Nutrition and Production of Fish and Shellfish Committee of the International Union of Nutritional Sciences (INUS), Rhodes, Greece, 330 pp.

Fairchild, E.A., Fleck, J. and Howell, H.W. (2005) Determining an optimal release site for juvenile winter flounder *Pseudopleuronectes americanus* (Walbaum) in the Great Bay Estuary, NH, USA. *Aquaculture Research* 36, 1374–1383.

FAO (2007) Yearbook of fishery statistics summary tables (ftp://ftp.fao.org/fi/stat/summary/default.htm, accessed 4 June 2008).

Fevolden, S.E. and Haug, T. (1988) Genetic population structure of Atlantic halibut, *Hippoglossus hippoglossus. Canadian Journal of Fisheries and Aquatic Sciences* 45, 2–7.

Fivelstad, S., Berghem, A. and Tyvold, T. (1991) Studies of limiting factors governing the waterflow requirement for atlantic salmon (*Salmo salar* L.) in landbased seawater systems. *Aquacultural Engineering* 10(4), 237–249.

Fletcher, G.L. (1977) Circannual cycles of blood plasma freezing point and Na+ and Cl– concentrations in Newfoundland winter flounder (*Pseudopleuronectes americanus*), correlation with water temperature and photoperiod. *Canadian Journal of Zoology* 55, 789–795.

Fletcher, G.L., Shears, M.A., King, M.J., Davies, P.L. and Hew, C.L. (1988) Evidence for antifreeze protein gene transfer in Atlantic salmon (*Salmo salar*). *Canadian Journal of Fisheries and Aquatic Sciences* 45, 352–357.

Fonds, M. (1976) The influence of temperature and salinity on growth of young sole *Solea solea* L. *10th European Symposium on Marine Biology* 1, 109–125.

Fonds, M. (1979) Laboratory observations on the influence of temperature and salinity on development of the eggs and growth of the larvae of *Solea solea* (Pisces). *Marine Ecology Progress Series* 1, 91–99.

Fores, J., Iglesias, J., Olmedo, M., Sanchez, F.J. and Peleteiro, J.B. (1990) Induction of spawning in turbot (*Scophthalmus maximus* L.) by a sudden change in the photoperiod. *Aquaculture Engineering* 9, 357–366.

Foss, A., Imsland, A.K. and Nævdal, G. (1998) Population genetics studies of Atlantic halibut in the North Atlantic Ocean. *Journal of Fish Biology* 52, 901–905.

Fuchs, J. (1982) The production of juvenile sole (*Solea solea*) under intensive conditions. 1. The first month of rearing. *Aquaculture* 26, 321–337.

Gatesoupe, F.J. and Luquet, P. (1982) Weaning of the sole (*Solea solea*) before metamorphosis. *Aquaculture* 26, 359–368.

Gaumet, F., Boeuf, G., Severe, A., Le Roux, A. and Mayer-Gostan, N. (1995) Effects of salinity on the ionic balance and growth of juvenile turbot (*Scophthalmus maximus*). *Journal of Fish Biology* 47, 865–876.

Godø, O.R. and Haug, T. (1988) Tagging and recapture of Atlantic halibut (*Hippoglossus hippoglossus*) in Norwegian waters. *Journal du Conseil International pour l'Exploration de la Mer* 44, 169–179.

Goldberg, R., Phelan, B., Pereira, J., Hagan, S., Clark, P., Bejda, A., Calabrese, A., Studholme, A. and Able, K.W. (2002) Variability in habitat use by young-of-the-year winter flounder, *Pseudopleuronectes americanus*, in three northeastern US estuaries. *Estuaries* 25, 215–226.

Grotmol, S., Totland, G.K., Thorud, K. and Hjeltnes, B.K. (1997) Vacuolating encephalopathy and retinopathy associated with a nodavirus-like agent: a probable cause of mass mortality of cultured larval and juvenile Atlantic halibut *Hippoglossus hippoglossus. Diseases of Aquatic Organisms* 29, 85–97.

Hallaråker, H., Folkvord, A. and Stefansson, S.O. (1995) Growth of juvenile halibut (*Hippoglossus hippoglossus*) related to temperature, light period and feeding regime. *Netherlands Journal of Sea Research* 34, 139–147.

Hansen, G.H., Bergh, Ø., Michaelsen, J. and Knappskog, D. (1992) *Flexibacter ovolyticus* sp. nov., a pathogen of eggs and larvae of Atlantic halibut, *Hippoglossus hippoglossus* L. *International Journal of Systematic Bacteriology* 42, 451–458.

Haug, T. (1990) Biology of the Atlantic halibut, *Hippoglossus hippoglossus* (L., 1758). *Advances in Marine Biology* 26, 1–70.

Hebb, C.D., Castell, J.D., Anderson D.M. and Batt, J. (1997) Nutritional studies on growth and protein utilization during the juvenile stage of winter flounder (*Pleuronectes americanus*). *Bulletin of the Aquculture Association of Canada* 97, 45–47.

Hebb, C.D., Castell, J.D., Anderson, D.M. and Batt, J. (2003) Growth and feed conversion of juvenile winter flounder (*Pleuronectes americanus*) in relation to different protein-to-lipid levels in isocaloric diets. *Aquaculture* 221, 439–449.

Hellberg, H., Koppang, E.O., Torud, B. and Bjerkas, I. (2002) Subclinical herpesvirus infection in farmed turbot *Scophthalmus maximus. Diseases of Aquatic Organisms* 49, 27–31.

Hensley, D.A. (1997) An overview of the systematics and biogeography of the flatfishes. *Journal of Sea Research* 37, 187–194.

Holmefjord, I. and Lein, I. (1990) Natural spawning of Atlantic halibut (*Hippoglossus hippoglossus*) in captivity. *International Council for the Exploration of the Sea, Copenhagen, C.M.* F:74 (in mimeo).

Holt, E.W.L. (1892) *Rhombus maximus*, Linn. (the turbot). *Journal of the Marine Biology Association of the United Kingdom* 2, 399–404.

Houghton, R.G., Last, J.M. and Bromley, P.J. (1985) Fecundity and egg size of sole (*Solea solea* (L.)) spawning in captivity. *Journal du Conseil International pour l'Exploration de la Mer* 42, 162–165.

Howell, B.R. (1973) Marine fish culture in Britain VIII. A marine rotifer, *Brachionus plicatilis* Muller, and the larvae of the mussel, *Mytilus edulis* L., as foods for larval flatfish. *Journal du Conseil International pour l'Exploration de la Mer* 32, 385–394.

Howell, B.R. (1997) A re-appraisal of the potential of the sole, *Solea solea* (L.), for commercial cultivation. *Aquaculture* 155, 359–369.

Howell, B.R. and Tzoumas, T.S. (1991) The nutritional value of *Artemia* nauplii for larval sole, *Solea solea* (L.), with respect to their (n-3) HUFA content. In: Lavens, P., Sorgeloos, P., Jaspers, E. and Ollevier, F. (eds) *Larvi '91 – Fish and Crustacean Larviculture Symposium*, Gent, Belgium. European Aquaculture Society, Special Publication, No. 15, pp. 63–65.

Howell, B.R. and Yamashita, Y. (2005) Aquaculture and stock enhancement. In: Gibson, R.N. (ed.) *Flatfishes – Biology and Exploitation*. Blackwell Publishing, Oxford, UK, pp. 347–371.

Iglesias, J. and Rodríguez-Ojea, G. (1994) Fitness of hatchery-reared turbot, *Scophthalmus maximus* L., for survival in the sea: first year results on feeding, growth and distribution. *Aquaculture and Fisheries Management* 25, 179–188.

Iglesias, J., Rodriguez-Ojea, G. and Peleteiro, J.B. (1991) Effect of light and temperature on the development of turbot eggs (*Scophthalmus maximus* L.). *International Council for the Exploration of the Sea* CM 1991, F:51 (in mimeo).

Imsland, A.K. and Jonassen, T.M. (2002) Regulation of growth in turbot (*Scophthalmus maximus* Rafinesque) and Atlantic halibut (*Hippoglossus hippoglossus* L.): aspects of environment × genotype interactions. *Reviews in Fish Biology and Fisheries* 11, 71–90.

Imsland, A.K. and Jonassen, T.M. (2003) Effects of exposure to extended photoperiods on long-term growth and age at first maturity in turbot and Atlantic halibut. *Aquaculture International* 11, 463–475.

Imsland, A.K. and Jonassen, T.M. (2005) The relation between age at first maturity and growth of Atlantic halibut (*Hippoglossus hippoglossus* L.) reared under four different light regimes. *Aquaculture Research* 36, 1–7.

Imsland, A.K., Sunde, L.M., Folkvord, A. and Stefansson, S.O. (1996) The interaction between temperature and size on growth of juvenile turbot. *Journal of Fish Biology* 49, 926–940.

Imsland, A.K., Folkvord, A., Grung, G.L., Stefansson, S.O. and Taranger, G.L. (1997a) Sexual dimorphism in growth and maturation of turbot, *Scophthalmus maximus* (Rafinesque 1810). *Aquaculture Research* 28, 101–114.

Imsland, A.K., Folkvord, A., Jónsdóttir, Ó.D.B. and Stefansson, S.O. (1997b) Effects of exposure to extended photoperiods during the first winter on long-term growth and age at first maturity in turbot (*Scophthalmus maximus*). *Aquaculture* 159, 125–141.

Imsland, A.K., Foss, A., Stefansson, S.O. and Nævdal, G. (2000) Hemoglobin genotypes of turbot (*Scophthalmus maximus*): variation in growth and optimal temperature for growth. *Fish Physiology and Biochemistry* 23, 75–81.

Imsland, A.K., Foss, A., Gunnarsson, S., Berntssen, M., FitzGerald, R., Bonga, S.W., van Ham, E., Nævdal, G. and Stefansson, S.O. (2001a) The interaction of temperature and salinity on growth and food conversion in juvenile turbot (*Scophthalmus maximus*). *Aquaculture* 198, 353–367.

Imsland, A.K., Foss, A. and Stefansson, S.O. (2001b) Variation in food intake, food conversion efficiency and growth of juvenile turbot from different geographic strains. *Journal of Fish Biology* 59, 449–454.

Imsland, A.K., Foss, A., Bonga, S.W., van Ham, E. and Stefansson, S.O. (2002) Comparison of growth and RNA/DNA ratios in three populations of juvenile turbot reared at two salinities. *Journal of Fish Biology* 60, 288–300.

Imsland, A.K., Scanu, G. and Nævdal, G. (2003a) New variants of the hemoglobin of turbot: possible use in aquaculture and population genetics studies. *Sarsia* 88, 55–64.

Imsland, A.K., Dragsnes, M. and Stefansson, S.O. (2003b) Exposure to continuous light inhibits maturation in turbot (*Scophthalmus maximus*). *Aquaculture* 219, 911–919.

Imsland, A.K., Foss, A., Conseição, L.E.C., Dinis, M.T., Kamstra, A., Schram, E., Daanbar, D., Rema, P. and White, P. (2004) A review of the culture potential of *Solea solea* and *S. senegalensis*. *Reviews in Fish Biology and Fisheries* 13, 379–409.

Imsland, A.K., Wergeland, T., Jonassen, T. and Stefansson, S.O. (2006a) Does malpigmentation improve growth in juvenile turbot (*Scophthalmus maximus*) and halibut (*Hippoglossus hippoglossus*). *Aquaculture Research* 37, 306–312.

Imsland, A.K., Foss, A., Folkvord, A., Stefansson, S.O. and Jonassen, T.M. (2006b) The interrelation between temperature regimes and fish size in juvenile Atlantic cod (*Gadus morhua*): effects on growth and feed conversion efficiency. *Fish Physiology and Biochemistry* 31, 347–361.

Ingilæ, M., Arnesen, J.A., Lund, V. and Eggset, G. (2000) Vaccination of Atlantic halibut *Hippoglossus hippoglossus* L., and spotted wolffish *Anarhichas minor* L., against atypical *Aeromonas salmonicida*. *Aquaculture* 183, 31–44.

Irvin, D.N. (1973) The growth and survival of Dover sole, *Solea solea*, and some observations on the growth of the plaice *Pleuronectes platessa*, considered at various temperatures. PhD thesis, University of Liverpool, Liverpool, UK.

Jonassen, T.M., Imsland, A.K. and Stefansson, S.O. (1999) The interaction of temperature and size on growth of juvenile Atlantic halibut. *Journal of Fish Biology* 54, 556–572.

Jones, A., Brown, J.A.G., Douglas, M.T., Thompson, S.J. and Whitfield, R.J. (1981) Progress towards developing methods for the intensive farming of turbot (*Scophthalmus maximus* L.)

in cooling water from a nuclear power station. In: Tiews, K. (ed.) *Proceedings of World Symposium on Aquaculture in Heated Effluents and Recirculation Systems*. Technical Paper. FAO, Paris, pp. 481–496.

Laroche W.A. (1981) Development of larval smooth flounder, *Liopsetta putnami*, with a redescription of development of winter flounder, *Pseudopleuronectes americanus* (family Pleuronectidae). *Fisheries Bulletin* 78, 897–909.

Lavens, P. and Remmerswaal, R.A.M. (1994) *Turbot Culture: Problems and Prospects*. European Aquaculture Society, Oostende, Belgium.

Leclercq, D. (1994) Turbot broodstock management: a key point to mid-term progress of the turbot industry. In: Lavens, P. and Remmerswaal, R.A.M. (eds) *Turbot Culture: Problems and Prospects*. European Aquaculture Society, Oostende, Belgium, pp. 3–13.

Lee, G.W.Y. and Litvak, M.K. (1996) Weaning of metamorphosed winter flounder (*Pleuronectes americanus*) reared in the laboratory, comparison of two commercial artificial diets on growth, survival and conversion efficiency. *Aquaculture* 144, 251–263.

Le François, N.R., Lemieux, H. and Blier, P.U. (2002) Biological and technical evaluation of the potential of marine and anadromous fish species for cold-water mariculture. *Aquaculture Research* 33, 95–108.

Levin, M.A., Wolke, R.E. and Cabelli, V.J. (1972) *Vibrio anguillarum* as a cause of disease in winter flounder (*Pseudopleuronectes americanus*). *Canadian Journal of Microbiology* 18, 1585–1592.

Litvak, M.K. (1994) Winter flounder. *Bulletin of the Aquaculture Association of Canada* 94, 4–8.

Litvak, M.K. (1996) Development of winter flounder for aquaculture (*Pleuronectes americanus*), state of the art. *Bulletin of the Aquaculture Association of Canada* 96, 24–26.

Litvak, M.K. (1999) The development of winter flounder (*Pleuronectes americanus*) for aquaculture in Atlantic Canada: current status and future prospects. *Aquaculture* 176, 55–64.

McVicar, A.H. and White, P.G. (1979) Fin and skin necrosis of cultivated Dover sole, *Solea solea*. *Journal of Fish Diseases* 2, 557–562.

Malard, A.E. (1899) Sur le développement et la pisciculture du turbot. *Les Comptes rendus de l'Académie des sciences, Paris* 129, 181–183.

Mangor-Jensen, A., Harboe, T., Shields, R.J., Gara, B. and Naas, K.E. (1998) Atlantic halibut, *Hippoglossus hipppoglossus* L., larvae cultivation literature, including a bibliography. *Aquaculture Research* 29, 857–886.

Mercier, L., Audet, C., de la Noue, J., Parent, B., Parrish, C.C. and Ross, N.W. (2004) First feeding of winter flounder (*Pseudopleuronectes americanus*) larvae: use of *Brachionus plicatilis* acclimated at low temperature as live prey. *Aquaculture* 229, 361–376.

Métailler, R., Cadena-Roa, M. and Person-Le-Ruyet, J. (1983) Attractive chemical substances for the weaning of Dover sole (*Solea vulgaris*): qualitative and quantitative approach. *Journal of World Mariculture Society* 14, 679–684.

Minkoff, G. and Broadhurst, A.P. (1994) Intensive production of turbot, *Scophthalmus maximus*, fry. In: Lavens, P. and Remmerswaal, R.A.M. (eds) *Turbot Culture: Problems and Prospects*. European Aquaculture Society, Oostende, Belgium, pp.14–31.

Mortensen, S.H., Hjeltnes, B., Rødseth, O.M., Krogsrud, J. and Christie, K.E. (1990) Infectious pancreatic necrosis virus, serotype N1, isolated from Norwegian halibut (*Hippoglossus hippoglossus*), turbot (*Scopthalmus maximus*) and scallops (*Pecten maximus*). *Bulletin of the European Association of Fish Pathologists* 10, 42–43.

Motnikar, S., Légaré, O., Vaillancourt, R., Cyr, C. and Lauzier, L. (2006) Atelier sur l'élevage de poissons marins au Québec – 16th and 17th November 2004, Ministère de l'Agriculture, des Pêcheries et de l'Alimentation du Québec, 196 pp. Compte rendu No. 23.

Mugnier, C., Guennocc, M., Lebegue, E., Fostier, A. and Breton, B. (2000) Induction and synchronisation of spawning in cultivated turbot (*Scophthalmus maximus* L.) broodstock by implantation of a sustained-release GnRH-a pellet. *Aquaculture* 181, 241–255.

Mulcahy, M.F. (2002) Diseases of flatfish. *Bulletin of the European Association of Fish Pathologists* 22, 86–94.

Munroe, T.A. (2005) Systematic diversity of the Pleuronectiformes. In: Gibson, R.N. (ed.) *Flatfishes – Biology and Exploitation*. Blackwell Publishing, Oxford, UK, pp. 10–41.

Murchelano, R.A. (1975) The histopathology of fin rot disease in winter flounder from the New York Bight. *Journal of Wildlife Diseases* 11, 263–268.

Nardi, G. (1996) Culture of summer flounder, *Paralichthys dentatus* at Greatbay Aquafarms. *Journal of Shellfish Research* 15, 458.

Nelson, J.S. (1994) *Fishes of the World,* 3rd edition. John Wiley and Sons, New York.

Nilsen, F., Ness, A. and Nylund, A. (1995) Observations on an intranuclea microsporidian in lymphoblasts from farmed halibut larvae (*Hippoglossus hippoglossus* L.). *Journal of Eukaryotic Microbiology* 42, 131–135.

Nitschke, P., Brown, R. and Hendrickson, L. (2000) Status of the fisheries resources off northeastern United States – winter flounder (http://www.nefsc.noaa.gov/sos/spsyn/fldrs/winter, accessed 15 January 2008).

Norberg, B., Weltzien, F.A., Karlsen, Ø. and Holm, J.C. (2001) Effects of photoperiod on sexual maturation and somatic growth in male Atlantic halibut (*Hippoglossus hippoglossus* L.). *Comparative Biochemistry and Physiology* 129B, 357–365.

Novoa, B., Figueras, A., Puentes, C.F., Ledo, A. and Toranzo, A.E. (1993) Characterization of a birnavirus isolated from diseased turbot cultured in Spain. *Diseases of Aquatic Organisms* 15, 163–169.

Oddgeirsson, M., Steinarsson, A. and Björnsson, B. (2000) Mat á arðsemi sandhverfueldis á Íslandi (Evaluation of economic feasibility of turbot culture in Iceland). *Marine Research Institute Special Series* (in Icelandic). Marine Research Institute, Reykjavik.

Olivier, G. (2002) Disease interactions between wild and cultured fish – perspectives from the America Northeast (Atlantic Provinces). *Bulletin of the European Association of Fish Pathologists* 22, 103–109.

Olsen, Y., Evjemo, J.O. and Olsen, A. (1999) Status of the cultivation technology for production of Atlantic halibut (*Hippoglossus hippoglossus*) juveniles in Norway/Europe. *Aquaculture* 176, 3–13.

Opstad, I. and Raae, A.J. (1986) Physical stress on halibut larvae. *International Council of the Exploration of the Sea, Copenhagen,* C.M. F:19, (in mimeo).

Opstad, I. and Rust, M.B. (2004) Effect of salinity of acutely stressed halibut (*Hippoglossus hippoglossus* L.) larvae. *Aquaculture Research* 35, 1286–1291.

Perez-Prieto, S., Garcia-Rosado, E., Rodriguez, S., Castro, S. and Borrego, J.J. (2001) Antigenic properties and experimental transmission to several fish species of a marine birnavirus isolated from sole (*Solea senegalensis*). *Veterinary Microbiology* 82, 11–25.

Person-Le Ruyet, J. (1989) The hatchery rearing of turbot larvae (*Scopthalmus maximus,* L.). *Cuadernos da área de ciencias marinas* 3, 57–91.

Planas, M. (1994) R&D on production systems. In: Lavens, P. and Remmerswaal, R.A.M. (eds) *Turbot Culture: Problems and Prospects*. European Aquaculture Society, Oostede, Belgium, pp. 54–73.

Plante, S., Audet, C., Lambert, Y. and de la Noue, J. (2002) The effects of two rearing salinities on survival and stress of winter flounder broodstock. *Journal of Aquatic Animal Health* 14, 281–287.

Pousão-Ferreira, P., Machado, M. and Fonseca, L. (1995) Marine pond culture in southern Portugal: present status and future perspectives. *Cahiers Options Méditerranéenes* 16, 21–30.

Riaza, A. and Hall, J. (1993) Large-scale production of turbot. In: Reinertsen, H., Dahle, L.A., Jorgensen, L. and Tvinnereim, K. (eds) *Fish Farming Technology*. A.A. Balkema, Rotterdam, The Netherlands, pp. 147–148.

Rodriguez, R.B. (1984) Biologia e cultivo de *Solea senegalensis* Kaup en Golfo de Cadiz. PhD Thesis, Universidad de Sevilla, Spain.

Rodriguez, S., Vilas, M.P., Gutierrez, M.C., Perez-Prieto, S.I., Sarasquete, M.C. and Rodriguez, R.B. (1997) Isolation and preliminary characterization of a birnavirus from the sole *Solea senegalensis* in southwest Spain. *Journal of Aquatic Animal Health* 9, 295–300.

Rollefsen, G. (1934) The eggs and larvae of the halibut (*Hippoglossus vulgaris*). *Det Kongelige Norske Vitdenskabers Selskab Forhandlinger* 7, 20–23.

Saucerman, S.E. and Deegan, L.A. (1991) Lateral and cross-channel movement of young-of-the-year winter flounder (*Pseudopleuronectes-americanus*) in Waquoit Bay, Massachusetts. *Estuaries* 14, 440–446.

Sawyer, P.J. and Hoornbeek, F.W. (1980) Pilot attempt to grow winter flounders commercially. New Hampshire Fish and Game Department, Completion Report for Project 3-275-R for the Period 15 February 1977 – 30 June 1980.

Shields, R.J. (2001) Larviculture of marine finfish in Europe. *Aquaculture* 200, 55–88.

Starkey, W.G., Ireland, J.H., Muir, K.F., Jenkins, M.E., Roy, W.J., Richards, R.H. and Ferguson, H.W. (2001) Nodavirus infection in Atlantic cod and Dover sole in the UK. *Veterinary Record* 146, 179–181.

Stobo, W.T., Neilson, J.D. and Simpson, P.G. (1988) Movements of Atlantic halibut (*Hippoglossus hippoglossus*) in the Canadian North Atlantic. *Canadian Journal of Fisheries and Aquatic Sciences* 45, 485–491.

Stoss, J.H. and Røer, J.E. (1993) Controlled reproduction in turbot (*Scopthalmus maximus*). In: Reinertsen, H., Dahle, L.A., Jørgensen L. and Tvinnereim, K. (eds) *Fish Farming Technology*. A.A. Balkema, Rotterdam, The Netherlands, pp. 29–34.

Støttrup, J.G., Shields, R.J., Gillespie, M., Gara, M.B., Sargent, J.R., Bell, J.G., Henderson, R.J., Tocher, D.R., Sutherland, R., Naess, T., Mangor Jensen, A., Naas, K., van der Meeren, T., Harboe, T., Sanchez, F.J., Sorgeloos, P., Dhert, P. and Fitzgerald, R. (1998) The production and use of copepods in larval rearing of halibut, turbot and cod. *Bulletin of the Aquaculture Association of Canada* 98(4), 41–45.

Tinti, F. and Piccinetti, C. (2000) Molecular systematics of the Atlanto-Mediterranean *Solea* species. *Journal of Fish Biology* 56, 604–614.

Vázquez Branas, M., Morales, J.C. and Estepa, A. (1994) A sandwich ELISA to detect VHSV and IPNV in turbot. *Aquaculture International* 2, 206–217.

Zorrilla, I., Balebona, M.C., Moriñigo, M.A., Sarasquete, C. and Borrego, J.J. (1999) Isolation and characterization of the causative agent of pasteurellosis, *Photobacterium damsela* ssp. *piscicida*, from sole, *Solea senegalensis* (Kaup). *Journal of Fish Diseases* 22, 167–172.

Zorrilla, I., Arijo, S., Chabrillon, M., Diaz, P., Martinez-Manzanares, E., Balebona, M.C. and Moriñigo, M.A. (2003) *Vibrio* species isolated from diseased farmed sole, *Solea senegalensis* (Kaup), and evaluation of the potential virulence role of their extracellular products. *Journal of Fish Diseases* 26, 103–108.

III Market and Economic Analysis

22 Marketing New Species

James A. Young

Department of Marketing, University of Stirling, Scotland, UK

22.1 Introduction

Marketing has a pivotal role to play in the development and potential success of new farmed fish species. This importance is likely to grow as the number of new species, and the many product derivatives therefrom, expands and competes within an increasing range of food product alternatives in international retail and foodservice sectors. Aquaculture accounts for almost 50% of global fish consumption (FAO, 2007) and, combined with ongoing pressure on wild-captured supplies, it is widely accepted that still greater reliance will be placed on existing and new farmed species to satisfy increased future demand. Markets for fish are likely to be boosted through a combination of population growth, greater demand from increasingly wealthy sections of emergent societies and more favourable changes towards fish from existing consumers for reasons such as health, convenience and other perceived benefits. When these trends are coupled with the fact that fish is currently the world's most valuable internationally traded food (Anderson, 2003; Lem, 2007), it seems axiomatic that marketing will assume a more prominent position.

However, the historical use made of marketing within aquaculture has a somewhat varied track record, which may be explained for a number of reasons. Conflicts of interests and divergent interpretations of the relative importance of technical, production-oriented factors and market-oriented demands have been common. A widespread pattern has been for new species to be considered only when the technical capacity to overcome production challenges has been achieved. The focus of technical attention has, of course, often been narrowed by price signals from the market to those species commanding high unit values and experiencing consistent shortfalls in supply. Prevailing prices have tended to dominate new product development (NPD) decisions over the signals from elsewhere in the market and more proactive marketing strategies. In addition, individual company characteristics and decision-making

(eds N.R. Le François, M. Jobling, C. Carter and P.U. Blier)

processes have led to different interpretations on the timing and use made of marketing in developing and launching new species.

The traditionally narrow perspective of most aquaculture firms is unfortunate because, arguably, it has resulted in a tardy realization of the potential benefits of new species. When compared to other major protein sources such as poultry, red meat, etc., fish has a significant comparative advantage in its ability to serve as a product in itself, as whole fish, or to provide the raw material for transformation into numerous product variants. These may present as fresh, chilled, frozen, canned, smoked and other forms and provide numerous further variations via fillets, steaks and cutlets, along with many other attributes and accompaniments (Muir and Young, 1999), or even high-value biomolecules originating from fish by-products (see Chapter 25, this volume). This potential ubiquity and diversity of individual new species thus suggests that when considering 'new species', implicit consideration should also be given to the various additional products that each might support. Such a collective approach of integrating new species and derivative new products is adopted in this chapter.

As a consequence of the historical tendency to focus primarily on the whole fish, many farmed species have experienced rapid growth resultant from a variety of technical, species-led innovations embracing gains from the hatchery, through husbandry and feeding regimes and improved cost structures in production. But in many such instances, insufficient attention to the vagaries of the market has contributed to cyclical movements of prices, profits and thus industry structure, its spatial distribution and performance. In many cases, fish farms have created a situation wherein they are forced to sell what they have produced without having first considered whether this is what consumers actually want to buy. The instability that results is clearly neither conducive to the long-run viability of aquaculture firms nor desirable for such a key component of future world food supplies. Marketing cannot provide any guarantee of success and should not be seen alone as some panacea. It can nevertheless assist in reducing the risks associated with new species and can contribute positively to the ongoing viable operation of the firms concerned along the entire value chain, from the points of deciding to produce to those of post-consumption. Importantly, marketing can make a significant contribution to the overall benefits to those communities that engage with aquaculture.

Within the scope of the book, this chapter is concerned primarily with exploration of some of the key ways in which marketing can contribute to the process of new species development in aquaculture. Given the potential disciplinary diversity of the readership, and the very ubiquity of the term 'marketing' (which commonly leads to confusion with selling, advertising and other misnomer activities), this is approached by first clarifying what is meant here by marketing and the critical characteristics of the aquatic foods environment in which marketing is undertaken. A marketing perspective on new species, or product, development is then introduced that illustrates its holistic function, which could hardly be assumed from a narrower science or technology background. The subsequent discussion and conclusions suggest some of the ways in which greater integration of marketing into the process of developing new farmed species, and related products, might enhance the chances of success and enable more effective responses to emergent challenges.

22.2 What is Marketing?

The basis of the marketing concept is to reconcile the needs and wants of customers. To do this, marketers embark on a four-stage process, which begins with gaining an understanding of what consumers want. This task is achieved using a variety of marketing research tools which typically gather secondary (published) data and primary data that are specific to the investigation in hand. As we shall see, this stage underpins decisions about what new products (including species) an organization might consider developing. Once an understanding of what consumers want, who may form part of discrete market segments or be spread among wider boundaries, marketers engage with other professions to create appropriate products which may incorporate related services. In aquatic marketing, the term 'product' may be determined by many factors, such as the species, farmed or captured from wild fisheries; how it is presented, such as whole fish, fillets, skin-on or off, boneless, flesh colour (feed influences), texture, seasonal availability and freshness, etc. In addition, other attributes such as quality marks, information pertaining to its provenance, certification of its compliance with different criteria set by various certifiers, organic status (see Chapter 26, this volume) and other factors may be considered integral parts of the offering to the market. The marketer's aim is to ensure that the overall bundle of attributes, including price, will be perceived by consumers to have 'value' and that the product offered will meet and, wherever possible, exceed the expectations of customers – while at the same time generating a profit.

The creation of products, part of NPD, is typically time- and resource-consuming, uncertain in outcome and involves a number of different disciplinary inputs. Many products fail because of the time lag and unfolding events between understanding trends in the market and getting a product to the launch phase of NPD. Problems may stem from the introduction of alternative competing products in the interim time period, price competition and the possibility that consumers' preferences change over time. New species commonly present particular problems because of the time to trial and generate solutions, none of which is helped by the innate constraints of the biological product cycle. Unlike many other products, new parts or solutions can seldom be generated and tried instantly. And even once an interim solution is developed, the end product can be marketed only after the growth cycle of the fish has been completed. The combination of these risks and potential rewards has led to the adoption of a complex screening process at this stage of creation, which is explored in more detail later. Assuming, for the interim, that the desired values can be created, some efforts will then be required on their communication.

Communication is necessary once values have been created to ensure that individuals (who may be individual consumers or organizations) are aware of and know about the product. Communication of values may be done in a number of different ways using the tools of the promotion mix: advertising, publicity, personal selling, sales promotion, direct sales, e-marketing and other information and communications technology (ICT)-based emergent techniques. Each has their own advantages and disadvantages in different settings and more general discussions are to be found in a number of non-fish specific sources (Shimp, 1997; Fill, 2006). Historically, advertising and other high-cost promotion tools have been used to a comparatively limited extent with new species, and more generally when fish is compared to other foods (Tveteras *et al.*, 2006). While some generic promotions have been undertaken for established farmed species such as salmon (Kinnucan and Myrland, 2000) and others (Scholderer and Grunnert, 2001), by definition the limitations of any new species market combined with the commonly small scale of firms involved tend to constrain the viability of any high-profile promotions.

It should be noted, however, that the emergence of greater access to ICT-based techniques has created a number of potential opportunities for fish farming. The geographical separation of the points of fish production and fish consumption, as evident from the pattern of international trade, has benefited from the ability to communicate with a much wider, yet more targeted, audience on an effectively instantaneous and much more cost-effective basis. Given the rapid endemic perishability of many fish products and their potentially short shelf life, the ability to convey fast and accurate signals to the market is important. As we shall see, however, there can also be downsides to the greater transparency of news on market and product developments.

The final stage is to ensure that producers and all other elements of the value chain deliver to the consumer. This does not mean just ensuring the physical presence of the product but, more importantly, delivering consumers' expectations of the values which have been understood, created and communicated. This must be done consistently so that consumers will develop trust in the claims that are made and so be encouraged, post-consumption, to respond positively to future communications. One recent and novel illustration of this was the launch of the world's first organic cod in 2006 from a fish farm in Shetland, located at the confluence of the North Sea and the Atlantic Ocean. The launch made much of its provenance from pristine seas, being ethically farmed and with emphasis on its sustainability at a time when supplies of North Sea cod from capture fisheries were receiving a lot of media attention about overexploitation. The organic farmed status was emphasized and the adoption of a brand name, 'No Catch' (see http://www.nocatch.co.uk/) did set out its clear position in a number of supermarket chains in the UK and elsewhere in Europe. This case is of particular interest as an example of a firm which deviated from the traditional pattern of new species launch. It is neatly illustrative of the need to ensure consistency of the product message along the chain from production to consumption.

Following the above four-part sequence: understand, create, communicate and deliver, the end objective is to enable the mutual satisfaction of producers and consumers, along with all other intermediaries in the chain. In its simplest form, marketing can be seen as part of the exchange process, as shown in

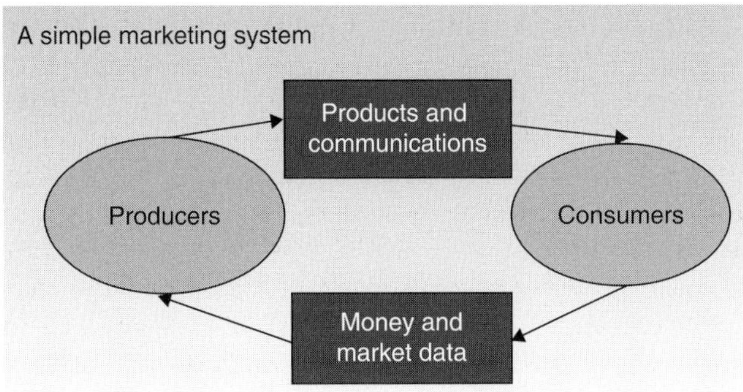

Fig. 22.1. A simple marketing system and the exchange process.

Fig. 22.1, wherein producers and customers will generate mutual gain from the products and prices paid through a series of transactions. These processes of exchange will manage competitively the supply chain in which producers operate, while collecting data to understand the market better and communicate about their products and services. This simple marketing system drives all the other processes of exchange which take place in markets. From a marketing perspective, the aim is to create value in the fish products offered by proactively understanding customers' expectations and making such products available in the market at a price which is competitive with other substitute products.

The creation and delivery of desired values is the central core of the marketing approach. Hence, a distinction may be made between an organization that is market-led compared with one that is a product-led, only selling what it makes. As noted earlier, this latter tendency has been common in aquaculture organizations as they have strived first to solve technical problems, often with neither proactive nor simultaneous concern for the market they are hoping to supply. The simple marketing system lies at the core of the exchange process undertaken in a wide variety of different marketing organizations, each of which has its particular influences within its own business or marketing environment. It may be argued that the fish marketing environment, including both cultured and captured products, poses a number of demanding challenges to its constituents, which are made all the more difficult when new products are added to the scene. An overview of this background environment is presented in the following section so that the more specific challenges of launching new species can be better appreciated.

22.3 The Fish Marketing Environment

In the fish marketing environment, shown in Fig. 22.2, a number of additional forces operate and combine to produce a complex and challenging set of tasks for the marketer to solve. Critical factors include, but are not restricted to, the fast-changing, dynamic nature of the fish business environment, coupled with

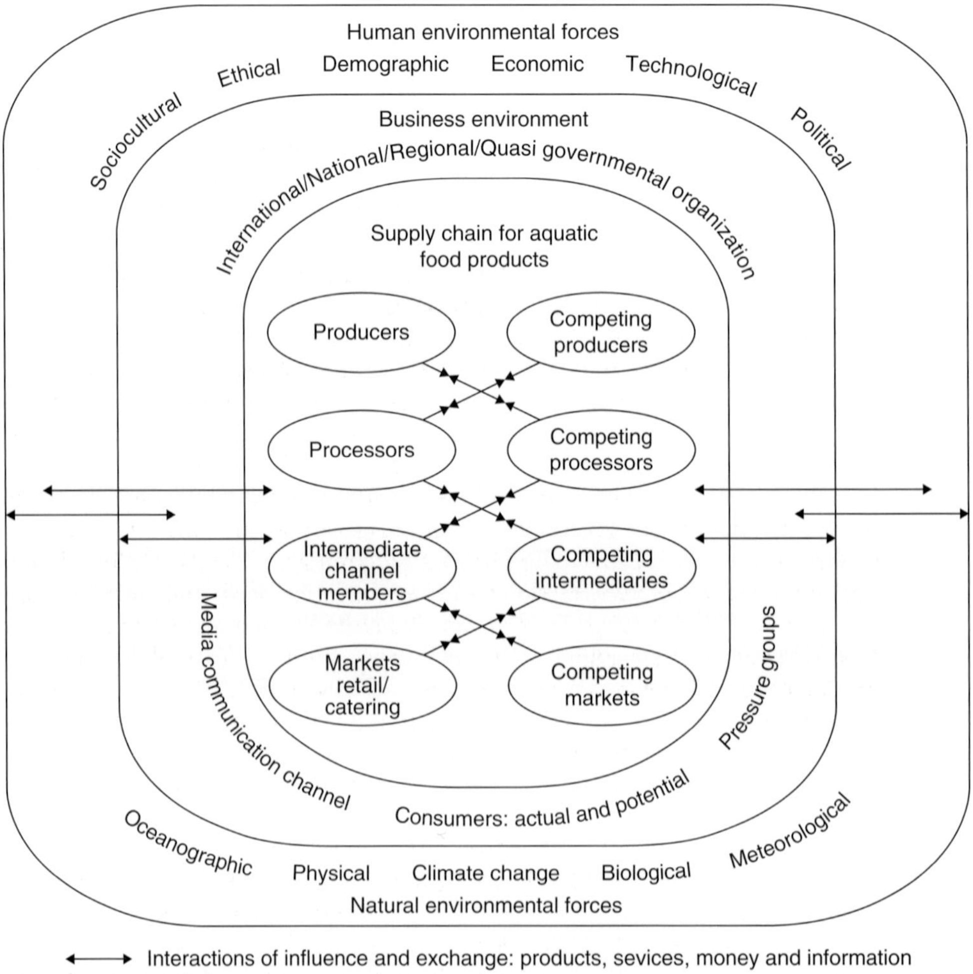

Fig. 22.2. The fish marketing environment. *Source*: Young, 1987.

a highly perishable product with uncertain supply availability and attendant price fluctuations. Farmed products may compete with wild-capture fisheries supplies and the many uncertainties therein. Although there may be greater scope for planning and control over farmed product supplies, individual firms commonly have only limited power in the market and their performance will thus tend to be determined largely by aggregate supplies. Farmed products can also be susceptible to diseases, especially with new species where treatments have yet to be developed; they may also be subject to the vagaries of the natural environment, for example, escapees through storm damage or mortality from algal blooms or predation (Cherry and Zimmerman, 2007; Intrafish, 2007). Consequently, despite detailed production planning, the availability of raw material and its price, on which downstream channel members and investments depend, can alter quite markedly from one year to another, and often within shorter periods with little or no warning. Such uncertainty clearly has a potentially significant impact on profitability, industry stability and propensity to invest in new species.

At the macro scale, other natural and human environmental factors are shown to impact on the inner business environment and core supply chain. Space constraints preclude discussion of each of the factors identified in Fig. 22.2, but it is suggested that readers consider the potential implications and interdependence of these phenomena in the context of their own and other fish marketing environments. While some change drivers such as climate change may pervade much wider platforms for contemporary debate, others such as the emergence of ethical considerations in how fish are farmed and the related workforces treated may be currently more localized, yet worthy of future note. Critically, the impacts of these forces tend to be volatile, unpredictable and present both challenges and opportunities. One illustration of the turbulence of the fish farming sector is to be found in the EU seabass and seabream industry where, following rapid and classic production-led expansion through the 1990s, events culminated in a price crisis in the early years of the 21st century (report produced for the European Commission DG XIV, 2004). In an attempt to assist the ailing sector, the EU adopted a policy of encouraging species diversification through financial assistance to those investing in new species. Whether the best time to invest in new species is at a time of downturn in profitability is debatable and questions may be raised about alternative mechanisms to encourage aquaculture innovation at this supranational level.

The scope for intervention and supporting measures also extends to the postharvest infrastructure of the sector, where various schemes encouraging processing plants and other channel structures have been implemented with varying degrees of success. Fundamentally, the processes of exchange and transfer between different parties along the supply chain are potentially problematical because of the rapid rate at which fish can spoil and the often constrained period available for decision making. The need for specialist infrastructure, such as chilled and cold chains to ensure fish preservation at controlled temperatures and appropriate handling throughout, present logistical constraints and incur costs. Because of the small volumes typically associated with new species, attempts are often made to share logistics with existing established channels. Clearly, this may present conflicts of interests and, in some locations, presents a significant barrier to entry into a new sector. Apart from these and other supply-side issues, market demand exhibits its own vagaries.

Individual consumers in different markets across the globe are shown from consumption data (FAO, 2007) to have their own fish product preferences. These too are constantly subject to a changing composition of competing alternative fish and non-fish foods. Within different population groups, determined by age, sociocultural influences, stage in the life cycle, willingness and ability to purchase, changing personal tastes and many other determinants, consumers will exhibit varied preferences for fish at different points in time. Gaining some working understanding of these potential target markets and some evaluation of their prospective viability is difficult, even where firms have access to good quality market information. The labour skills, financial and time resources to invest in such market analysis can place considerable demands on individual firms, whatever their size. This task is all the more difficult in the case of new products, or new markets for existing products, where, by definition, much less information is likely to be available. Where new farmed species are being considered, individuals commonly face still greater knowledge gaps.

Many firms involved in new species tend to be small innovative organizations, sometimes just individuals, intent on pioneering a new sector but with limited resources to allocate to all the demands of their enterprise. While the case for understanding the market in advance may be accepted in theory, in practice many entrepreneurs become enmeshed in the immediate demands of current difficulties and are less able to maintain a clear vision of the longer term perspective. This focus on intervening bottlenecks at the expense of market analysis and planning has been noted in emerging industries elsewhere (Porter, 1980). Given the biological growth cycle which must be gone through before a product can be placed on the market, this tendency is understandable. Even once some market presence is achieved, the revenue stream is likely to be small in relation to the costs which have been incurred and poses the risk of having insufficient cash flow to maintain intended targets. Such was allegedly the case with the aforementioned 'No Catch' organic cod farm, which was placed in administration at the time of writing when it exceeded the limits of its investors (Cherry, 2008).

Apart from the day-to-day, and sometimes novel, husbandry, demands of new species firms typically encounter a number of barriers in gaining and making use of market information. By definition, there is often little precedent practice available to learn from and actions undertaken with other species may not be applicable with the new one in question. In a study of farming blue mussels in Norway, Otteson and Gronhaug (2004) noted the difficulties this could bring, along with the potential danger of information tending to be skewed to portray a more favourable situation, which might then be used to attract additional resources (see also Chapter 23, this volume). Issues were also identified about the scope for cooperation among innovating firms. While at one level there may be a willingness and enthusiasm to share experience, such as that evident currently among cod farmers (Borch, 2008), there is an inevitable point at which co-pioneers become competitors. On a pragmatic level, accurate determination of the onset of this changed status may also be critical in deciding how one responds to information made available.

Having established a brief overview of the business environment in which fish marketing decisions are made, the next section examines some of the more specific challenges of launching new species and how this process might be approached; always assuming it is not abandoned beforehand.

22.4 Marketing Decisions on New Species

Observation and analysis of the food marketing environment helps the fish marketer to understand and so manage factors both internal and external to the organization. The range and importance of influential factors is often difficult to determine, not least because of the rapid pace of change and the scope for impacts to cross over from other food and non-food sectors. For example, over the past few decades, numerous markets have been disrupted with food scares, causing consumers to respond differently to market information and changing their food purchase patterns. To remain successful, organizations need to maintain an ongoing evaluation of their marketing variables in relation to the internal strengths and weaknesses of their corporate resources and the external opportunities and threats identified in the market. Ultimately, the aim is to identify where and in what way the firm can utilize its comparative advantage to maximum benefit by reconciling their immediate business demands and longer term planning needs.

Short- and long-term opportunities and threats, coupled with an evaluation of the enterprise's strengths and weaknesses, should be undertaken on an ongoing basis against competitors and market trends. This SWOT (strength, weakness, opportunity and threat) analysis is often the starting point for identifying new strategic directions and the catalyst for bringing new species and products into the business or targeting new markets. For example, the rapid expansion of pangasius throughout many European markets resulted from the Vietnamese exporters responding to the threat to their sector from the imposition of increased tariffs in their, then, main market – the USA (Young, 2006).

Many other models are available to help determine strategy and marketing planning which are not specific to fish, but may of course provide lessons for consideration (see Kotler, 1997; McDonald, 2002). Typically, these invoke analysis of current and potential new products against performance in current and new markets. It should be appreciated that what constitutes a new product, or species, in one market might not be new in another; and that there may be debate about when a product is new or simply an extension of an existing offering. Thus, the relative competitive position of the firm, in terms of products and markets, can result from many different perspectives and will almost certainly change over time. Critically, these analytical tools will seldom provide some prescription as to what must be done, but rather will identify options. Different interpretations may be made of the product/market environments and these provide still further scope in how the firm chooses to control its marketing variables in targeted market segments. Organizations thus need to be alert to their weaknesses and strengths, current and emergent, so that ongoing planning can be made for change. Indeed, in a marketing environment which is constantly changing, and at a faster pace, this demands continuous review of the product range.

Despite the logic of the need for change, there are many reasons why organizations may resist modifications to their range of products. Familiarity and established expertise with the associated value chain compared to the risks and uncertainty of new product areas and their technologies explain some of the reluctance to alter offerings to the market (see Chapter 23, this volume). Similarly, knowledge of markets and long-standing relationships with customers may discourage interest in developing new channels. Any form of NPD involves creativity and an element of risk. Generally, the more challenging the project and the less the experience in launching in a new product sector, as might be the case with a new species, the higher will be the risk of failure. None the less, market demands do change, as well as what consumers value; so, too, must the products that deliver those values. In addition to this underlying driver to review the current product range constantly, the ability to satisfy new and emergent desires through the characteristics of new species might also present a logical case for product range revision. For example, the health attributes attached to omega-3 fatty acids have encouraged consumption of certain oily fish products, including salmonids, and might influence both the product range and communication of these advantages accordingly.

22.5 Planning for New Species and Products

Given the risks and potential benefits that are attached to developing new species and their related products, firms ideally might aim to achieve a balanced portfolio of products, old and new. This process can be explained by considering the notion of the product life cycle (PLC), shown below in Fig. 22.3, which provides some indication of the pattern of sales and profits that might be expected over time.

While the PLC is not intended to be used as a means of forecasting product success or otherwise, it does provide a reflection of the general stages that products might be expected to pass through from their initial conception as an idea, R & D, market launch and subsequent sales on the market through phases of growth, maturity and subsequent decline and possible death. At each of these stages, a range of marketing actions would be taken, in line with competing products and their own situation in the market. Each phase of the PLC will thus tend to result in varying levels of sales and profitability. Consequently, firms generally try to manage their product range to achieve some spread over the PLC rather than have all products at the same stage. In particular, the proactive firm would aim to secure an income stream from the existing profitable products sufficient to finance R & D for the new. Newborn firms though, which clearly tend to focus on new products, will not enjoy this luxury, which places them in a more vulnerable position during the early years; a status confirmed by the high number of failures in the first 5 years (Craig and Hart, 1992).

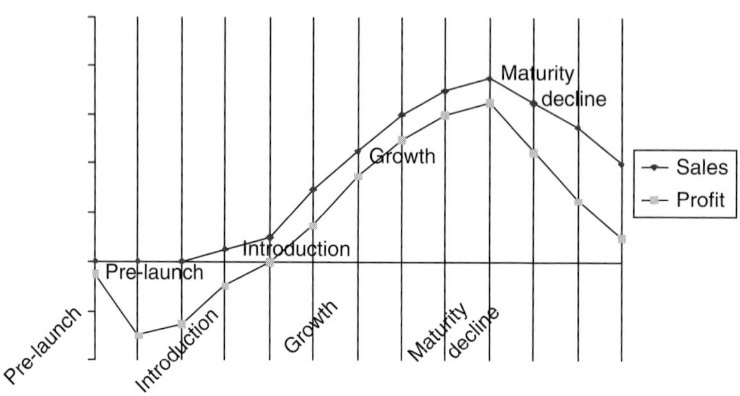

Fig. 22.3. The product life cycle (PLC) of new species.

In some cases, products may move from launch, through a rapid growth phase to maturity within a short time period, whereas other species may take a much longer time to achieve market penetration goals. Some launches may fail almost immediately, possibly because of a new entrant to the market. In other instances, firms may attempt to reposition the product on the PLC; for example, repackaging to increase appeal and effect a shift from decline to growth phases. Differing perceptions of the species and the product derivatives in different markets illustrate the inherent uncertainty that marketers of new species need to contend with. For example, while farmed tilapia has long since attained a significant level of acceptance in the North American markets, its adoption in Europe has been at a much slower pace (Muir and Young, 2002). Even where attributes surrounding the new species are entirely consistent with emergent market trends, many other factors can intervene and disrupt predicted adoption.

Generating a stream of new products is critically dependent on having a systematic process in place which responds to, *inter alia*, identified changes in the market, technical innovations and emergent opportunities perceived. Much has been written on the ways in which new products can be identified (Fuller, 1994; Hart, 1996), but one common variant is to scrutinize prospective ideas according to a sequential screening process, such as shown below in Fig. 22.4.

The essence of the above approach is to identify only those new product ideas which hold greatest scope for commercial success. From the initial range of ideas, possibly generated through different routes such as market gap analysis, observing competitors, food and aquaculture firms elsewhere, in-house think-tanks, brainstorming and other techniques, those considered to have least chance of success are discarded or possibly stored for a later review. Elimination continues via a systematic series of assessments, which span a range of disciplinary perspectives so that the full context of the product can be gauged. At each stage, the commitment to the product will increase and at some stages,

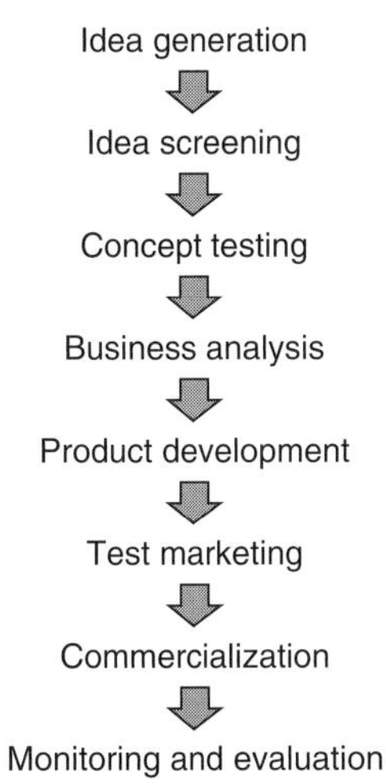

Fig. 22.4. The new product development (NPD) screening process.

such as product development, can invoke an exponential rise in costs (Baker and Hart, 1999). While this screening process is liable to consume a lot of resources, it should result in greater efficiency by deleting product ideas which are more likely to fail before they are commercialized. Some sharing of costs, and potential benefits, may be enabled through joint cooperation with other channel members, typically upstream or downstream, such as feed companies, processors or retailers. These arrangements can improve access to specialist expertise and resources whose costs could not be justified as a stand-alone function. For example, the UK retailer Marks & Spencer linked with Scottish Sea Farms to launch a discrete 'Lochmuir' salmon brand based on its innovative feed and husbandry regime (Cherry, 2006).

The planning of new products may be undertaken in a variety of ways and, unsurprisingly, there is no single standard technique which can be hailed as a certain route to success. Some firms may opt not to innovate but instead imitate or follow the lead of others. Being an innovator ahead of competitors demands the ability to secure commercial confidentiality and restrict market intelligence about planned activities; these may be undermined by stages such as test marketing which, by definition, will reveal forthcoming product launches. Despite the costs and risks of launching new products, some firms use NPD as a means of reinforcing their status as an innovative firm, leading new developments in the market. For example, in the highly competitive and mature market for salmon, the Loch Duart brand of Scottish salmon has established a unique market position based on the quality and husbandry of the fish (Loch Duart, 2008). Such a position can help gain a competitive advantage in newly established segments which may be more financially lucrative, especially during initial periods with limited competition. However, as noted earlier, markets do change and the length of the various stages of the PLC, and the extent of profitability therein, cannot be predicted.

A further challenge to aquaculture value chains concerns the impact of additional species launched on to the market. Hitherto, a succession of species has been introduced in different markets around the world and it is possible to trace their generally cyclical performance for reasons earlier discussed. Typically, the launch of each has been accompanied by an initial period of high prices, made possible because of the comparative novelty and limited supply of the new product. However, as firms approach the fifth decade of farmed fish markets, the number of species available inevitably has increased, and at a faster pace. As noted before, when the number of different product variants that may be supported by each species is considered, the real increase in choice available to consumers is even more substantial. In such circumstances, doubt may be cast about the continuation of the earlier existent 'honeymoon' period of relatively high profitability (Muir and Young, 1999). So far, the price positions of some new entrants have shown a capacity to withstand high price tags, notably the launches of farmed cod and halibut; however, in other instances, the increased international trade in products has encouraged market penetration using quite different tactics of highly competitive prices on entry, such as pangasius and some tilapia. These different approaches to the market themselves suggest the emergence of a more complex picture.

The complexity of the market for new species is also likely to be increased through the imposition and adoption of various certification schemes on the existing range of established products and their subsequent application to new product launches. Space precludes discussion of the multitude of market segments in which products increasingly are differentiated according to various criteria such as quality, provenance, sustainability, animal welfare, ethics, organic status and otherwise. In some cases, quality marks like Label Rouge for salmon and seabass have become long-established points of differentiation (Mariojouls and Roheim Wessels, 2002) but, with the increased proliferation of further criteria, especially over the past 3 years, greater doubt must exist as to the ability and inclination of fish consumers to absorb these issues accurately into their (frequent) food choice decisions. It thus remains to be seen whether these criteria will provide a meaningful basis for more specific product selection and encourage trial of new products, or simply become part of the already complex myriad of messages with which producers attempt to communicate with their audiences.

22.6 Conclusion

This chapter has emphasized the role of marketing in the development of new species, but it should be clear that this contribution could be effective only if undertaken in conjunction with the various other disciplinary inputs to NPD. Organizations that incorporate the marketing function in isolation and without due attention to other contributory components of the product are arguably equally at risk of failing to deliver what the market wants, as are those who remain in the mould of product orientation. This holistic perspective is likely to become all the more important as new farmed species stretch the boundaries of traditional acceptance. Already, new species, and their related products, in many markets have been introduced from outside conventional supply chains to enable delivery of additional attributes. Some species formerly sourced in tropical environments only are also being produced and consumed in the main temperate market areas with attached green credentials, which some may debate. Other new entrants openly challenge, for some times at least, those available from conventional capture fisheries. The evolving market for farmed species is consequentially more complex.

Clearly, the incorporation of new species in the market creates many additional marketing opportunities for fish products to compete in the wider market for foods. But, as has been noted, the greater and more subtle forms of differentiation of the fish products launched has led to the emergence of a very diverse product range within which it is almost certainly becoming more difficult for consumers to assimilate and respond rationally to the subtle variants on offer. The resultant confusion inevitably makes it difficult for further new products launched to claim credibility and capture new market share, which in turn is likely to depress the profitability and propensity to innovate further still. Yet, despite the risks attached to NPD, ignoring it is guaranteed to result in failure later, if not sooner.

References

Anderson, J.L. (2003) *The International Seafood Trade*. CRC Press LLC and Woodhead Publishing Ltd, Cambridge, UK, 222 pp.

Baker, M. and Hart, S. (1999) *Product Strategy and Management*. Prentice Hall, Upper Saddle River, New Jersey.

Borch, O.J. (2008) Future strategic challenges within cod farming. Paper presented to *Go for Cod Conference*, 13–15 February, Tromso, Norway.

Cherry, D. (2006) Grown to order. *Intrafish*, 7 September 2006 (http://www.intrafish.no/global/news/article115244.ece, accessed 3 May 2008).

Cherry, D. (2008) No catch execs: bank sunk us. *Intrafish* 19 March 2008 (http://www.intrafish.no/global/news/article167117.ece, accessed 5 May 2008).

Cherry, D. and Zimmerman, R. (2007) Escapes cost Norway's fish farmers millions in 2006. *Intrafish* 19 April 2007 (http://www.intrafish.no/global/news/article132590.ece, accessed 30 April 2008).

Craig, A. and Hart, S. (1992) Where to now in new product development research. *European Journal of Marketing* 26, 1–46.

European Commission DG XIV (2004) Study of the market for aquaculture-produced seabass and seabream species. Report. Stirling Aquaculture, University of Stirling (http://ec.europa.eu/fisheries/publications/studies/aquaculture_market_230404.pdf, accessed 30 April 2008).

FAO (2007) *The State of the World Fisheries and Aquaculture 2006*. Food and Agriculture Organization of the United Nations, Rome.

Fill, C. (2006) *Marketing Communications: Engagments, Strategies and Practice*, 4th edn. Prentice Hall, Upper Saddle River, New Jersey.

Fuller, G.W. (1994) *New Food Product Development*. CRC Press, Boca Raton, Florida.

Hart, S. (1996) *New Product Development: A Reader*. Dryden Press, Fort Worth, Texas.

Intrafish (2007) Northern Ireland's only salmon farm wiped out by jellyfish (http://www.intrafish.no/global/news/article148079.ece, accessed 22 November 2007).

Kinnucan, H. and Myrland, O. (2000) What are the results of the Pan-European salmon marketing campaigns? Presented to *Seafood Opportunities in the New Millenium Aquavision Conference*, May 2000, Stavanger, Norway.

Kotler, P. (1997) *Marketing Management: Analysis, Planning, Implementation, and Control*, 9th edn. Prentice Hall, Upper Saddle River, New Jersey.

Lem, A. (2007) Seafood production and international trade: global trends. Paper presented to *Marketing of Seafood Products: Trends and Challenges Workshop*, 12–16 November 2007, Zaragoza, Spain.

Loch Duart (http://www.lochduart.com/tastethedifference.htm, accessed 3 May 2008).

McDonald, M. (2002) *Marketing Plans: How to Prepare Them, How to Use Them*, 5th edn. Butterworth Heinemann, Oxford, UK.

Mariojouls, C. and Roheim Wessels, C. (2002) Certification and quality signals in the aquaculture sector in France. *Marine Resource Economics* 17(2), 153–162.

Muir, J.F. and Young, J.A. (1999) Strategic issues in new species development for aquaculture. In: *Elsevier Biofutures Series. XXXIII International Symposium on New Species for Mediterranean Aquaculture*. Elsevier/Biofutur, Cambridge, UK, pp 85–96.

Muir, J.F. and Young, J.A (2002) Tilapia: both fish and fowl? *Journal of Marine Resource Economics* 17(2), 163–173.

Otteson, G.G. and Gronhaug, K. (2004) Perceived opportunities and pursued strategies in an emerging industry: the case of Norwegian blue mussel farming. *Aquaculture Economics and Management* 8(1/2), 19–39.

Porter, M.E. (1980) *Competitive Strategy*. The Free Press, New York.

Scholderer, J. and Grunnert, K.G. (2001) Does generic advertising work? A systematic evaluation of the Danish campaign for fresh fish. *Aquaculture Economics and Management* 5, 253–271.

Shimp, T.A. (1997) *Advertising, Promotion, and Supplemental Aspects of Integrated Marketing Communications*, 4th edn. The Dryden Press Fort Worth, Texas.

Tveteras, R., Anderson, J., Asche, F., Kinnucan, H., Myrland, O. and Young, J.A. (2006) Marketing: whose job? Facing the challenges from the global food market. Presentation to *Seafood Opportunities in the New Millenium Aquavision Conference*, Stavanger, Norway.

Young, J.A. (1987) Marketing in a dynamic environment: an overview of the UK fish processing industry. *Food Marketing* 3(1), 144–161.

Young, J.A. (2006) Technical: Positioning pangasius on the global market – a case of diversification through adversity? Paper presented to *Remaining Competitive in a Global Economy Session WAS Aquaculture America*, 13–16 February 2006, Las Vegas, Nevada.

23 Diversification Pays: Economic Perspectives on Investment in Diversified Aquaculture

James R. Wilson[1] and Bruno Archer[2]

[1]Université du Québec à Rimouski, Gestion des Ressources Maritimes, Rimouski, Québec, Canada; [2]Centre Aquacole de la Côte-Nord (CACN), North-Shore Aquaculture Center (NSAC), Sept-Îles (Québec), Canada

23.1 Introduction

This chapter lays out some ideas that might be useful in considering the question of investment in diversified aquaculture development. The first question, and arguably the most important one, is what we mean by *diversification*. This book is aimed mainly at the question of species diversification. However, economists understand the word in the context of other important related notions that cannot be ignored. These are the notions of *global economic efficiency*, *opportunity cost*, *resource scarcity* and *risk*. Therefore, to an economist, diversification is a strategy that a public or private manager may use, by taking several different (and occasionally conflicting) 'positions' with respect to a market or markets in order to maximize expected gains *and* to minimize exposure to risk. In the private trading of common stock and other assets, diversified portfolio management has developed into both a science as well as an art. In the case of public investments, different strategies of public investment are often the objects of intense public debate, because the positions public managers take with regard to various investments can have a direct impact on the welfare of a population. Global economic efficiency requires that we address questions about what is being given up and what is being obtained through the attribution of scarce resources to different projects.

In applying these notions to aquaculture, we may well ask ourselves if private investors in the Province of Québec (Canada) have taken positions in the aquaculture industry as part of their own investment strategies. If we observe that they have not, then it is entirely reasonable to ask why not. It is not very sound economic reasoning to say that since private investors are not investing 'enough' in diversified aquaculture, the public sector should step in. This is because, as we explain later, the public sector may be in no better a position to make these investments than those specialists who make a living by investing. We may also ask ourselves if more public investment in aquaculture is justified, given the other public investment 'positions' to consider in the economy. And if so, then what are the economic arguments for such an investment? Still further, assuming that public investments in aquaculture development are called for, are these public investments diversified enough across those sectors that touch aquaculture, such as factor markets, logistics and R & D? In other words, does the 'mix' of investment in the technical knowledge of production, marketing, trade and logistics contribute to the well-being of the citizenry? Finally, assuming that public investments in aquaculture are properly diversified in this way, might public managers consider the possibility of diversifying their development efforts by exploring the feasibility of producing different species, or different product forms of species already produced? Some of these questions would apply equally to a private investor as well.

The point of this discussion is that there are many different levels of diversification and many different 'positions' a private or public manager might take with respect to the sector in order to be economically efficient. Therefore, any discussion on diversified investment must, by the very nature of the term, consider a broad range of alternative investments; not just alternative positions with respect to choice of species, but alternative positions with respect to other possible investments in the economy, alternative investments in different aspects of the production problem or alternative investments in public works supporting that production, like education and roads. Diversified investment issues in aquaculture are best discussed within the backdrop of public and private portfolio management. These ideas from the economics perspective allow us to place the issue of investment in diversified aquaculture in proper perspective.

It is also important to be clear at the outset about some other terms we use in this discussion. The term *investment* means the conscious decision to place scarce resources in one endeavour or another. Investment can mean everything from buying a piece of an existing company, to building a company oneself, to preserving natural or man-made infrastructure. Disinvestment is when you get out of an endeavour. It is usually associated with taking a 'cash position' or investing in something else. Killing a tank of fish at a certain time and selling them means that you are disinvesting in that tank of live fish and taking (for the time being) a cash position. These decisions are (and most economists would strongly argue *should*) be guided by the investor's perceptions of *opportunity cost*. The perception of opportunity cost is the result of having to choose between investments having different expected pay-offs. All things equal, the choice of the most attractive option is compared to the next best option. The difference in the two is the net gain (or loss) of the impending decision. If the differences are small or non-existent, the investor will 'stay put' at the status quo. Almost all rational investment decisions boil down to these basic ideas.

Next, we should say something about the role of uncertainty and risk in investment decisions. Uncertainty can be divided into two types: *weak uncertainty* and *strong uncertainty*. When modern economists (see, for example, Dequech, 2006) speak of weak uncertainty, they mean that there is a probability distribution attached to an event. If the investor is capable of perceiving these probabilities and forming preferences, then this implies that *risk* is quantifiable as the welfare position of the investor associated with the expected value of a certain negative (or positive) outcome. Risk is therefore a combined concept of weak uncertainty and the preferences and aversions of an investor for different states of nature.

For example, a salmon aquaculture firm in eastern Canada may have historical information on the probability that 100% of a stock of Atlantic salmon, *Salmo salar*, might escape from a marine cage. Let us say that this probability has been historically calculated at 1%. If the consequences of such an escape means the loss of US$1,000,000 of product, then the expected value of the loss (US$10,000) is a measure of exposure to the risk of a cage failure and the escape of the stock. Note that these numbers are purely hypothetical and serve only as an example to illustrate the idea of weak uncertainty and the risk arising from this uncertainty.

To deal with that risk, a manager who is *risk averse* may pay up to US $10,000 in insurance premiums to be indemnified against that event. The amount he or she is actually willing to pay as an insurance premium is related to personal *risk aversion*, but is also a measure of the money (decrease in economic welfare) that he or she is willing to give up to be indemnified *with certainty* against the risk of loss. Therefore, when a person says, 'I run the risk of going broke raising marine fish', the words 'going broke' is that event (often quantifiable) multiplied by the probability associated with profits falling at or below the 'broke' level. These insights form the basis of a vast literature in investment and finance. Buying insurance is but one way of dealing with uncertainty and risk.

We turn our discussion now to *strong uncertainty*. The economist, Frank Knight, in the seminal book 'Risk Uncertainty and Profit' (1921),[1] introduced this idea because many investments and decisions were made *without* any knowledge of the statistical distribution of an outcome. There is therefore no information on mean or variance that can be used to calculate expectations. The occasional marine toxic algae bloom occurring in the eastern part of Canada can be considered a good example of a strongly uncertain event that can affect blue mussel production, *Mytilus edulis*.[2] When it occurs, producers can lose their entire stock or, at the very least, are prevented from harvesting and selling their product until the bloom passes and the toxins are flushed out of the animals. This poses a problem none the less, because producers often cannot wait to sell animals at a later date, because the animals may become too large for the market. To take another example, some species of *Ascidiacea* are also increasingly present in the eastern part of Canada and are now reaching waters surrounding the province of Québec. These organisms have a high growth and propagation rate and can hurt producers by competing for space and filter-feeding activities on the aquaculture structures for blue mussel production. The mussel producer in this case confronts strong uncertainty, because he or she cannot predict or evaluate the probability of the occurrence, although

[1] It is noteworthy that this book is still part of the required reading list for advanced studies in finance and economics in North America, despite the publication date.

[2] Another one is marine ducks that can attack and destroy a crop of mussels.

there is some evidence (Locke *et al.*, 2007) suggesting that the event is becoming more frequent through time.

There is often strong uncertainty concerning the competitive environment in which some aquaculture producers work. Non-regulated competitive environments and liberalized international trade may lead to marketing problems at a local level, especially when high volumes from several sources suddenly hit markets. This is why agricultural economists are often interested in estimating farm gate and consumer demand correctly, as well as the *price elasticity of demand* for different agricultural products. This tells us something about the degree of risk associated with producing in such markets. Since the quantity demanded of most products is related negatively to price, gluts in the supply of the fresh market for blue mussels, for example, will lead to declines in price, and subsequently to declining profits for all mussel producers. Without quantitative knowledge of farm gate and consumer demand, price effects are more uncertain.

Despite these problems with strong uncertainty, there are those investors who thrive on using their entrepreneurial 'sense of smell' to make decisions, even in the face of strong uncertainty. These people might be called *entrepreneurs* or *venture capitalists*. Most investors operate partly with weak uncertainty problems, but strong uncertainty plays a central role in many investment strategies.

It could be argued that much of the motivation for technological change towards more intensive use of real capital and man-made components, rather than technologies that are intensive in 'natural technologies' and 'natural capital', is to avoid exposure to strong uncertainty arising from the use of natural processes. Much of this uncertainty may also arise from the relatively high costs of information acquisition and control when using extensive technologies. The less control one has over an aquaculture production technology, the greater the likelihood that dealing with strong uncertainty plays a predominant role in decision making. Therefore, from an investment standpoint, the choice between intensive and extensive technologies may well have to do with the costs of managing uncertainty.

Another important part of this discussion has to do with the place of private entrepreneurs and venture capitalists in the development of a diversified aquaculture sector and the proper place of public investments in the sector. While no one disputes the importance of public investment in new technologies and methods of food production such as aquaculture, there is the question of how to structure this investment in order to maximize the economic benefits to the taxpayer, who ultimately finances such investment. These discussions usually also involve education and extension, technological diffusion and research in R & D.

23.2 Economic Reasons for Diversification

The first reason for public or private types of investment in diversified aquaculture might be for the purposes of managing risk and uncertainty.

Paul Samuelson provides a general proof that 'diversification pays' (Samuelson, 1967). However, close attention must be paid to the assumptions.

Samuelson was speaking of a private investor facing many different investment possibilities, each having uncertain pay-offs, but which are nevertheless measurable because of historical data on the statistical distribution of the pay-off. Assuming knowledge about the degree of risk averseness of the investor, it can be shown that the private investor is better off by investing in a portfolio of activities aimed at striking a balance between minimizing exposure to variance and maximizing the expected value of the pay-off. Thus, if an investor had prior knowledge of the means and variances of pay-offs of different types of culture of different species, the problem of how much to invest in the different activities would be no different than forming a diversified portfolio of investments. However, and importantly, the Samuelson proof unambiguously shows that the investor *would never* decide to invest in only one species (for example), unless the probability distributions *and* the pay-offs between the species were exactly identical; a near impossibility.

In the case of strong uncertainty, the recommendations of Samuelson still apply; a diversified portfolio, even based on subjective information on the moments of an undefined distribution, is better than putting all eggs in one basket.

However, this general result does have to be qualified. One important qualification is that there cannot be different costs of *learning* when making a diversified investment. Lockwood (1998) has made the point that in the case of aquaculture in the 1980s, large conglomerates 'diversified' into aquaculture without sufficient knowledge of the management and organizational aspects of that sector. This resulted in the failure of many of these branches. If diversification implies learning new management techniques, then the costs of diversification may outweigh the benefits.

The case of the public investor is more complex and bears some explanation. Arrow and Lind (1970) have argued that the implications of uncertainty for public investment remain controversial. Despite the date of this paper, this controversy continues to this day and both sides of the debate are reflected in current arguments for and against public investment in many sectors, including aquaculture. As we have said earlier, most individuals are averse to risk and will therefore not value assets with uncertain pay-offs at their expected (or statistical) values. This is well-known behaviour for individual investors. However, at issue is whether it is appropriate to treat public investments differently, because the perception of risk by a government, or its actual exposure to risk, may be different from that of the private sector (Arrow, 1971, p. 239). There are arguments that might favour the opinion that the role of risk in a public investment decision should be less, because they are exposed to less of it (Samuelson and Vickrey, 1964). However, the dominant opinion of economists who have studied the economics of risk bearing (Arrow, 1971) is that risk should be discounted in public investments in the same way as in private investments. Extending this observation to the case at hand, it is invalid to think that the public sector is somehow in a better position to bear risk of investments in R & D in aquaculture, or that the scale size of government reduces its exposure to risk or uncertainty. As we argue below, although the public sector may have certain scale advantages for the treatment of aquaculture investment, superior position with regard to risk bearing is not one of them.

On the other hand, there are excellent economic arguments for public involvement in investment in diversification based on notions of scale size, epidemiology and the divergence between private perceptions of opportunity and profit versus social exposure to risk of disease and food security. Weitzman (2000) argues that individual producers are strongly motivated to use standard species and techniques for food production leading to widespread monoculture over large areas, creating what are called *network externalities*. In many cases, this diffusion has been aided by universities and governments through extension activities, leading to a negative relationship between the levels of agricultural subsidies and plant biodiversity (Di Falco and Perrings, 2005). Weitzman argues that as a technology diffuses through a geographic area, each producer contributes to what he calls an increasing 'disease externality risk', which exposes the sector to possible collapse. This is because when a disproportionate part of the energy flow-through in an ecosystem is occupied by one species, the pathogen and parasite biomass specific to that species, which is determined endogenously, grows as well, providing for more opportunities for mutation, adaptation and successful attack on the host population. In this regard, monoculture aquaculture development could be considered as an accelerator for the spread of local and invasive species. Sea lice propagation helped along by densely developed salmonid farms is a good example of this mechanism.

Thus, there is a role for public investment in preserving or encouraging what Weitzman calls 'ecological entropy', or biodiversity. These investments might be viewed as insurance premiums against widespread collapse of one monoculture species, which could have severe negative impacts on the welfare of humans if the sector is large. Public investments of this type could be considered effective hedges against the strong uncertainty associated with these events. Therefore, in this particular case, public investment in diversification pays and Weitzman's work draws a clear line between the issue of investing in diversity and the motivation of the individual taxpayer to finance investment in such diversity. This link between the taxpayer and the risky event to be avoided is an important condition in order to establish the efficiency of a public investment in sectors like aquaculture.

Another economic argument for diversified aquaculture has to do with the efficient use of space and ecosystems. Different forms of polyculture involving aquaculture have been practised for millennia. There is a revival of interest in what has come to be known as multi-trophic aquaculture. Examples would be the culture of various seaweeds along with fish and shellfish species. These practices have been surveyed by Chopin *et al*. (2001). Although this literature has sought to attract attention to the sustainability features of such culture, efficient use of nutrients available in the water through developing multiple outputs by the same firm may also be economically efficient. Diversification aimed at creating these types of efficiencies leads naturally to multiple input/multiple output conceptions of production. These types of operations are frequently encountered in agricultural production and there is no reason to believe that the evolution of aquaculture will be any different.

The economics of diversification and efficiencies thereof extend to the output side as well. Increased interest in biomolecules has led to markets for nutritional supplements for human and animal consumption and recovery of biomaterials for medicinal and pharmacological markets. All of these cases are examples of diversification as well.

23.3 The Political Economics of Aquaculture Diversification

Economists study the behaviour of humans, especially as they distribute scarce resources. This extends also to the behaviour of groups of humans, both as private actors but also as members of the public sector. In this regard, what economists have to say about the political economics of public investment is pertinent to our discussions on diversified aquaculture.

There are three considerations under the rubric of 'political economics' that might be mentioned. These are: (i) the implications of public investment in diversified aquaculture from the standpoint of the efficiency of investment and the social returns on capital; (ii) the interactions between private and public investors; and (iii) the scale and dynamic implications of public investment in the development of factor markets, notably in human capital, land and biotic resources.

We take the last case first. Choices of species to cultivate often imply the organization of research and development agendas, which will affect input markets. Investment in human capital through education and training remains one of the most productive areas of public investment. For example, Mingat and Tan (1996) estimate that from 1960 to 1995, the annual social rate of return on tertiary education for OECD countries is more than 10%. However, R & D programmes also fix human capital on career pathways. These developed human resources can exert influence on collective decision making and may provide impetus to certain parts of sector development, even though other elements of development may not be in place to take full advantage of the expertise created. These human resources may become public employees and may stay in the public sector for decades. The exertion of influence to fund research agendas having uncertain outcomes publicly may lead to a perpetuation of public funding in certain sectors that may be unjustified from a strict efficiency standpoint and may result eventually in public investments yielding low social returns on investment.

Considerable space in the economics literature is devoted to the relative efficiency of public investment in infrastructure and R & D and the impact this investment has on the behaviour of private investors. As we have argued above, the rationality of public investment in large-scale infrastructure or potentially large-scale ideas or technologies is hard to deny. However, as we look at different elements of a public investment portfolio (roads, telecommunications, police, education, health care, investments in agriculture and in aquaculture), there are differing social returns to the tax dollar invested. As we mention above in the introduction, the opportunity costs in terms of foregone returns in other sectors is important in any analysis. This is why economists have spent a considerable amount of intellectual effort in understanding the many ways in which public investment can go wrong through mismanagement and perverse economic incentives.

Shapiro's argument (1973) in this regard is pertinent. He argued that public managers tend to select investment projects that benefit pressure groups of constituents rather than the general citizenry. These pressure groups then help the growth of the agency. Such growth benefits public managers, either in the form of direct monetary compensation or in other forms. Agency size increases through investments in projects with low or negative rates of return. This is possible because transaction costs preclude a successful opposition by disaffected taxpayers, who are large in number and who do not always have free or easy access to public information. In this way, Shapiro argues, inefficient public investments are a rule rather than an exception. It is not difficult to see how such an outcome would work against more diversified investments.

There may be other less serious problems with the contractual relationships implied by public investment in R & D. Universities often take the lead in extension activities. However, the contract structure of university workers, especially those involved in more fundamental research, may not be well defined when it comes to applied research and extension activities. In these cases, researchers may end up working in areas that, although important, have little to do with the needs of the industry. Yet, industry leaders have no easy way to exert influence on the research agendas of public employees. Part of this problem is rectified through the creation of extension appointments at universities, where extension and outreach to the industry is an element of performance for the faculty member.

In addition, part of this literature discusses the possibility of 'crowding out' or eviction of private capital by public spending. The argument goes that it is possible for public management to create circumstances where the 'smart money' is crowded out and invested elsewhere. While these ideas are not conclusive or universally accepted, it is interesting to note that, at least in the Province of Québec, in some cases the public sector content of the investment initiative in aquaculture can appear large, compared to that of the private sector. It is difficult, however, to prove causality in these cases.

There are a number of explanations for these results. First, the aquaculture sector is characterized as risky (Lockwood, 1998) and this perception of risk has been reinforced by a number of well-publicized failures, especially among conglomerates who have attempted to diversify themselves *too broadly* into aquaculture, without having the organizational or technical skills to actually manage an aquaculture firm. Lockwood (1998) as well as Jarvinen (2000) point out that aquaculture is constrained by the biology of the animals being grown and that the relatively high rates of return expected by traditional venture capitalists (50% or more a year) can far exceed the rates of return that can be anticipated in aquaculture. Since competition for venture capital is fierce in other sectors, it is not surprising that the aquaculture sector suffers a relative lack of funding. In addition, Jarvinen (2000) points out that emerging marine aquaculture faces special problems arising from the nature of marine property rights, finite period leasing, technological change and uncertainty and catastrophic risk. All of these problems may act to limit the market for venture capital for this sector to private sources and 'angels'; private, non-institutional venture capitalists whose opportunity cost for capital is relatively modest and who can therefore afford to be more patient.

There is one other implication of public investment in diversified aquaculture that should be mentioned. This has to do with the possible effects of public investment in diversified aquaculture on the distribution and use of productive factors. The important factors would seem to be human capital, access to land/sea areas, energy and feed. Decisions to invest in a species undoubtedly fix human capital in certain vocations, which, because of path dependencies, cannot be changed easily. Since know-how resides in the human, and is usually transmitted through a combination of the written and oral word as well as 'learning by doing', there needs to be a critical mass of resident know-how for the aquaculture sector to develop. The transformation of 'old technologies' and human capital that used to be associated with collapsed fisheries may be important, but education and re-training, especially in a diversified economy, is not a trivial issue.

Second, the slow development of sea tenure (or the lack of it) in some countries where resources are available is arguably the most important reason for slow development of the sector. However, rapid development of sea tenure can lead to unresolved external effects between tract-holders, especially in monoculture situations. As an example, the dense development of salmonid farms in eastern and western Canada may result in unresolved externalities related to the use of coastal resources and the area related to fish production in competition with other uses of these same natural goods.

Increased public investment in aquaculture could be expected to lead to increased demands for energy, and this demand would be met largely through the price system, albeit with rising costs related to relative scarcity of energy. Feed is another issue. An important part of protein sources for feed still comes from wild-caught fish. However, there is a growing discussion among economists, biologists and policy makers over the impact that aquaculture development has on the input markets for fishmeal and, as a consequence, the growing pressures on open access fishing leading to overexploitation of fish resources as feed and as broodstock. Hannesson (2003) has argued that open access fishing of wild stocks worldwide eventually will limit both aquaculture production and wild fish production as well. Willman (2005), on the other hand, contends that there appears to be enough evidence of substitution effects with other protein sources, but that overfishing of forage fish is still the result of inefficient management of marine fisheries. This is an ongoing debate among marine resource economists.

23.4 A Case Study: Application to the Province of Québec

In the forgoing discussion, we enumerated several reasons why a diversified approach to aquaculture development, at either the private or the public level, might be justified. Further, we argued that there was certainly a place for private investors and possibly a place for public investors, in the right context. Then we provided some cautions regarding the political economics of public investment. We turn our attention now to some observations on Québec.

The province of Québec could be characterized as a slow starter in the marine aquaculture sector. The sector so far has developed the production of blue mussels and scallops. The freshwater sector is based mainly on salmonid production, which supplies the stocking activities of artificial ponds and watercourses in managed wildlife areas.

The development of marine salmonid culture for the table market faces important constraints, which are thought to affect their competitive position compared to others using more conventional sea pen technologies in more temperate climates. In Québec, salmonid production is therefore a land-based activity using internal holding tanks or external closed ponds. Because of both climatic and governmental restrictions, marine fish production will likely use a land-based approach coupled with biofiltration technologies. These technologies imply a higher cost of operation.

Environmental issues (in particular access to water resources and waste management) are an important topic in the freshwater aquaculture sector of this province. These are not yet issues in the marine sector, because no commercial marine fish production is yet present in Québec, in part because of governmental restrictions.

Much of this slow start-up therefore has to do with climate and lack of venture capital. Add to this a justified caution on the part of the Québec and Canadian governments about the sustainability issues surrounding some marine aquaculture development. Another reason may have to do with federal–provincial relationships, as well as the extensive administrative requirements associated with aquaculture start-up. All of these things might explain the seeming reticence of private investors to enter this sector.

In this section, we will use the Québec experience in aquaculture to raise a few issues related to investment decisions in diversified aquaculture. These issues may be related. First, if indeed the markets for private venture capital in Canada are relatively efficient, why do we NOT see more private venture capital in the aquaculture industry of this province? Second, if diversified investments are such a good thing, as we have argued in this paper, why do we NOT see a more diversified strategy on the part of public managers in Québec? We take up these questions in the next two sections.

23.4.1 Private venture capital markets and diversified aquaculture

To put this discussion into perspective, let us step out of aquaculture and consider private investment in mining industries. When mining companies in Canada obtain exploration rights, the costs of doing so are relatively small in comparison to the expected pay-offs. Further, the property rights aspects of the contract are relatively clear. Still further, the administrative procedures for obtaining access to Crown lands for mineral exploration are relatively clear. Yet, in the aquaculture sector, the competitive environment for species such as blue mussels may make it such that the expected pay-off is not assured *compared to* the costs of becoming established. For other species, such as wolffish, which seem to have promise, the likely technologies are more intensive, but there is still an important amount of uncertainty around being able to control different aspects of production, which are still largely exogenous and dependent on public managers, such as feed and energy costs and genetic sourcing from wild populations for the development of strains. In both cases, the costs of contracting and development may be larger than the expected pay-off to the Canadian investor, especially when compared to alternative investments in other economic sectors outside aquaculture. This is partly due to the important amount of uncertainty associated with controlling different aspects of production.

Whether or not private investors invest in different sectors may be related to a number of issues. Obviously, the key variable is the opportunity cost of the capital invested: why invest in a venture with a 15% net return per year when you can invest in one with a considerably higher net return per year, say, 25%? The fluidity of venture capital markets and the availability of other investment options may be one reason why aquaculture investment in Québec seems comparatively unattractive to investors. One way in which governments attract private investment to sectors like aquaculture has been to provide access to public goods at well below the scarcity value of the public good, essentially giving the investor a production subsidy. The efficiency, equity and sustainability implications of such policy are obviously debatable, but if public managers in resource-rich countries like Canada wish to develop a sector rapidly, they might be tempted to do this.

Not all venture capitalists are alike and some may have specialized knowledge of a sector, which allows them to control uncertainties better than other investors might. This brings us to another important consideration, which is whether investors from other countries might be willing to invest in aquaculture in Québec, even before Canadian investors are willing or ready and even though the administrative costs seem comparatively high at home. Further, is foreign direct investment in aquaculture in Québec a good thing or a bad thing?

In countries that have well-functioning tax laws and which charge for the use of their public goods at or near their scarcity value, encouraging foreign direct investment in aquaculture may be the surest way to encourage both diversity and development. In this circumstance, the standard gains from trade arguments apply. The only question, which is not a trivial one, is whether the governmental agencies involved in the attribution of permits to use submerged lands and other public goods will make the effort to form efficient contracts and transfer public goods in a way which would be advantageous to the Canadian taxpayer. As the economic history in forestry, fisheries and even aquaculture in Canada has shown, raising such concerns is neither trivial nor alarmist.

23.4.2 Public investment strategies

Our arguments up to now have been that regardless of whether we encourage private or public investment (or both), we would do well collectively to have a more diversified investment portfolio. Private investors usually do this as a matter of principle and good sense. Yet, we do not tend to see public managers engaged in this type of behaviour, at least in the aquaculture sector in Québec. We have already provided some reasons why public managers might be reluctant to diversify. Here are some other reasons taken from the Québec experience. Public mangers in Québec have been mainly involved in supporting the development of the culture of invertebrate filter feeders (mainly blue mussels and scallops in aquaculture or open-water culture), possibly because the extensive nature of the technologies were thought to be less costly and easier to learn. Public funding has therefore been concentrated and very focused over a long period of time. Although much was learned about alternative technologies, the returns on investment have been slow in coming.

Our central thesis (diversification pays) is in opposition to these practices of public agencies. These agencies, at least in Québec, target specific projects, possibly because it is felt that an extensive type of production, even assuming strong uncertainty, is a better bet than investing in technologies that are more intensive. Yet, after many years of public investment, this strategy may not have borne fruit. During this time of focused investment, when 'all the eggs' were put in one basket, other promising aquaculture sectors may have been ignored. The presence of strong uncertainty suggests that aquaculture diversification is preferable. Even if the presence of strong uncertainties suggests that aquaculture diversification is preferable, the complex problem of defining development objectives could be perceived as a rational justification for a more focused investment policy. Yet, in the case of Québec, this diversification has been slow to occur.

We have discussed briefly why and how focused investment policy might have taken place, both generally and in Québec. It is, in part, a rational response to the complex problem of defining development objectives. A declining scallop industry in the Isles de la Madeleine, along with a build-up in biological expertise, led to projects aimed at trying to manage scallop enhancement and harvests cooperatively. However, although there were tremendous gains in biological and technical knowledge, other knowledge related to property rights and contracting for public goods use, marketing, distribution and trade issues probably did not keep pace with this knowledge development.

In the case of Québec's blue mussel industry, the relatively late arrival of Québec, coupled with the competitive environment in mussel aquaculture in eastern Canada (aided possibly by lower management costs due to somewhat easier provincial–federal relations elsewhere), caused managers in Québec to focus attention on this sector, even though the competitive and physical/climatic aspects of the sector made these investments in Québec difficult from the start and clearly prevented them from looking for diversification opportunities.

Again, the existence of fixed human capital in biology may have played a role in perpetuating public investment in the sector. Production subsidies also tended to fix entrepreneurial capacity in mussel production, justifying further public investments. However, there are indications that development in aquaculture production is more rapid than the development in other important parts of the production chain, such as marketing and distribution. Indeed, a productive sector was created, but it could be argued that the sector was largely an artifact of public investment and somewhat isolated from other important elements necessary for development, logistics and marketing being but two examples.

Even though we are able to explain why these things might have happened, and some of the reasons are certainly understandable, 'focused public investment' is still not necessarily a good strategy for using scarce public funds in an industrial sector such as marine aquaculture. In this sector as others, exogenous events worldwide will impose important pressures on the home sector to perform at world standards, which can expose new and marginal firms to high risks of failure. 'Focused private investment' under these circumstances would be ill advised as well; most investors allow themselves only a few speculative investments (usually at bargain prices) out of a large portfolio of more solid investments.

'Path-dependent' public investment may contribute to reducing the diversification process and, although the competitiveness of potential new firms that receive attention may well be enhanced, examples from Québec suggest that the opposite effect might occur as well. Further, focused public investment may discourage entrepreneurs who do have good ideas but who do not 'fit the vision' of the public managers.

Focused public investment can also have the effect of *adversely selecting* private partners that participate in the development of the sector. The various subsidies implied by some public investment programmes could be expected to attract investors whose objectives may not be the generation of their own profit but the capture of publicly generated rents and other benefits. Such behaviour has nothing to do with the generation of new wealth. In practical terms, such investors may have less skill related to the perception and management of risk, compared to their colleagues elsewhere. Some may also come to the sector without the entrepreneurial capacity normally needed in competitive firms elsewhere. It is sometimes argued that injections of public funds to producers are necessary, so that they may learn how to become competitive. This type of 'infant industry argument'[3] can lead to a sector composed of firms that can survive only through the constant injection of public funds. The protection of unfit managers may have negative impacts on the development of the entire sector.

The two examples we cite here, the sea scallop and the blue mussel industry during the past 20 years, are two possible examples of how focused top-down public investment strategies can slow diversification and transfer of technologies that are advantageous in the marine coldwater aquaculture sector in North America. However, it is not the choice of species that is at issue. Rather, it is the way in which investments at the public level are carried out, the impact this has on the type of entrepreneur who enters the sector, the evolution of sea tenure and property rights and the costs of contracting for these

[3] The 'infant industry' argument in international trade theory says that focused public investment in one sector might result in the generation of a comparative advantage in that sector over time. Proponents of this argument cite Québec cases like Bombardier (commercial aviation and train technology) and Hydro-Québec (hydroelectricity), although the economic history of both sectors shows that they are special cases of natural comparative advantage, very large positive network externalities and imperfect competition. Most historical evidence for other sectors suggests that a sector so subsidized can become dependent on public funding and will underperform with respect to the world market for the products or services it produces (Krugman and Obstfeld, 2006).

rights, the focused nature of investment in the face of risk and uncertainty and the level of specialization and fixity in human capital, along with the impacts this specialization and fixity has on the public investment process. Moving to another species and doing the same type of focused investment in the same way could well yield the same results.

23.5 Concluding Remarks

This chapter has discussed some of the economic considerations associated with investment in diversified aquaculture. There are a number of reasons why public investment in diversification may be attractive. A more diversified aquaculture would lead to a greater variety of products available to consumers, possibly reducing pressure on wild stocks of species having similar qualities. Diversification research may lead to the more efficient use of productive factors, either by raising complementary species together or by specifically exploring production processes having multiple products from the same species. Minimizing exposure to risk for both the private investor and the public manager is probably the most important reason to study diversified strategies. There may be other benefits related to scale and network economies as well. Public investments of sufficient size in aquaculture may lead to a strengthening of other infrastructure necessary to develop an industry, with processing, marketing and distribution being possible examples.

However, there are other issues to consider as well. One other important issue is the sustainability of a proposed technology. There is a growing problem (and literature) on world overfishing, especially the overfishing of forage fish used for feeding livestock, fishing down the food chain and biodiversity loss in ecosystems resulting from non-selective fishing techniques (Pauly and Palomares, 2005). One possible difficulty with the development of cultivation techniques for carnivorous marine fish species may be the link with open access capture fisheries. The effects of open access fishing on the price of fish are well known and, if substitutes are not found, this could have negative impacts on the cost structure of firms growing carnivorous fish. Energy costs may well be an issue in the years to come as well. Food and energy costs can represent more than 60% of the production costs of marine carnivorous fish culture. It is worthwhile noting that the species developed and raised for food by humans through the ages have been mainly vegetarian and often grazers. With that in mind, diversification in aquaculture may well mean the development of species that are less costly energetically and less dependent on animal proteins coming from declining forage fish stocks.[4]

Biomass recovery and recycling of biomass in meal recipes may take place on a grand scale, thus reducing the dependence of aquaculture operations on wild-caught forage fish. Improvements in the quality of feed consequently may

[4] The growth in demand for these inputs is related to the rapid growth of the aquaculture sector, as well as the growth in other agricultural activities such as beef, pork and chicken production through the world.

improve conversion ratios, further reducing pressure on wild stocks. However, to the extent that aquaculture diversification leads to increased pressures on wild stocks fished in open access conditions, both as a form of food for farmed fish and as a source of genitors and genetic diversity for the aquaculture producer, such investments could result in non-sustainable aquaculture practices. This consideration may argue for more attention to be paid to investing in species whose natural food is farther down the food chain; an investment in plankton or plant-eating fish and of course filter-feeders, for example. Pauly's observations on capture fisheries may apply eventually to aquaculture as well.

There are also questions related to public finance and public economics. One has to do with the relative efficiency of public versus private investment. While there are excellent reasons for public investment for the provision of certain goods (roads, education, electric grids, biodiversity), these have to do mainly with the fact that private investors cannot or will not become involved, whereas the government can in certain cases. Here, we make an important distinction. Public investment in the provision of a good is justified when the nature of the good is such that the economic benefits are large but are dispersed so widely over the population and the networks of benefits are so varied and non-predictable and uncontrolled that private investors cannot capture the benefits of their investment and therefore realize a return. Therefore, because of these scale and network effects, the public sector is thought at times to be relatively more efficient at providing certain goods, called public goods.

Project promoters often invoke this argument to justify public investment in a number of sectors. However, there are cases where this argument is merely a subterfuge for an interest group to exert political or coercive power under conditions economists call 'information asymmetries', much as Shapiro (1973) described. This may occur when an interest group is able to obtain a disproportionate amount of funding through the legislature or the minister responsible for the portfolio without the public becoming aware.

We are not arguing that this has happened in Québec aquaculture development. However, the fact that the projects of interest groups often masquerade as public investments makes it necessary for economists who study sectors such as aquaculture to explain carefully why such investments are globally efficient for the economy. The 'public good' argument for public investment in a sector is less convincing the smaller the sector is and the more focused the interest group is. One way to start such an analysis is to understand better why private venture capitalists are less present in a sector like aquaculture (which seems to have promise) when other sectors, which are also risky, seem to have fewer problems attracting risk capital (pharmacology, mining, energy production, genetics, etc.). This may give important clues as to the changes that need to be made so that private *and* public investments are efficient.

Canada and Québec face many of the same constraints to aquaculture development that other latecomers have experienced, including some (like climate) that other contestants may not have. However, a fundamental constraint is the relatively slow development of a working sea tenure system, as well as the lack of cost-effective institutions and contracting that allow the transfer of property rights on public goods to entrepreneurs wishing to invest in aquaculture. This probably has an impact on the ability of entrepreneurs to access venture capital and to other markets for managing risk, such as crop insurance. As mentioned earlier, formation of human resources and education is necessary, but such formation may result not in diversification but in a tendency to focus on specific species. This tendency to focus on one or two species may be reinforced by a number of other considerations, such as not enough development capital to go around, climate and path dependence related to past strategic decisions.

Economies may need a strong economic base before embarking on marine aquaculture diversification projects, for the reasons we describe above. The marginal efficiency of public or private investment is related to gross output of the economy. As output grows in an economy and as money supply expands through monetary policy, investors (public and private) will search out ventures having increasingly uncertain or lower marginal expected pay-offs. It is no coincidence, therefore, that coastal regions that have experienced booms in their economy for one reason or the other are sometimes seen as leaders in diversified aquaculture. Nowhere is this more clearly the case than in Norway. Petroleum production in Norway arguably permitted the development of research and development in aquaculture. This tendency can also be seen in some states such as Alaska, where rehabilitation and enhancement of fisheries stocks (a venture possibly more risky even than aquaculture) was tied directly to the generation of oil revenues.

Researchers (seen, for example, in Sachs and Warner, 2001) have detected a negative relationship between natural resource endowments and the percentage rate of growth in economies, which they have called the 'resource curse'. This phenomenon might be explained in part by economies that 'soak up' the capital generated by large resource discoveries in investments that have progressively lower rates of return. The slow development of diversified aquaculture in places like Québec may be the result of a confluence of limited investment funds tied to the macroeconomics of the province, climate and a regulatory environment that is comparatively less favourable to aquaculture development.

Nevertheless, a diversified aquaculture sector can be an important motor for regional and national development and can have significant positive network impacts on the economy, where the benefits of public investments could well outweigh the costs.

In the cases we presented for Québec, these social returns on investment do not seem to be as convincing as we would have liked. Rather than repeating the errors of the past, a next step could be to have a fuller understanding of the barriers to more diversified investment. This will lead us to examine more carefully markets for risk management, alternative investment markets, market structure for fish feed, energy and other important factors of production including human capital and a re-examination of the informational, institutional and legal barriers to the evolution of sea tenure in the sector.

References

Arrow, K. (1971) *Essays in the Theory of Risk-bearing*. North-Holland Publishing Company, Amsterdam.

Arrow, K. and Lind, R.C. (1970) Uncertainty and the evaluation of public investment decisions. *American Economic Review* 60, 364–378.

Chopin, T., Buschman, A.H., Halling, C., Troell, M., Kautsky, N., Neori, A., Kraemer, G.P., Zertuche-Gonzalez, J.A., Yarish, C. and Neefus, C. (2001) Integrating seaweeds into marine aquaculture systems; a key toward sustainability. *Journal of Phycology* 37, 975–986.

Dequech, D. (2006) The new institutional economics and the theory of behaviour under uncertainty. *Journal of Economic Behavior and Organization* 59, 109–131.

Di Falco, S. and Perrings, C. (2005) Crop biodiversity, risk management and the implications of agricultural assistance. *Ecological Economics* 55, 459–466.

Hannesson, R. (2003) Aquaculture and fisheries. *Marine Policy* 27(2), 169–178.

Jarvinen, D. (2000) Federal and state support for aquaculture development in the United States. *Aquaculture Economics and Management* 4(3/4), 209–225.

Knight, F.H. (1921) *Risk, Uncertainty, and Profit*. Hart, Schaffner and Marx; Houghton Mifflin Company, Boston, Massachusetts (Internet version available through the Library of Economics and Liberty, http://www.econlib.org/library/Knight/knRUP.html, accessed June 2007).

Krugman, P. and Obstfeld, M. (2006) *International Economics: Theory and Policy*, 7th edn. Pearson Education Inc., Boston, Massachusetts.

Locke, A.J.M., Hanson, K., Ellis, M., Thompson, J. and Rochette, R. (2007) Invasion of the southern Gulf of St. Lawrence by the clubbed tunicate (*Styela clava* Herdman): potential mechanisms for invasions of Prince Edward Island estuaries. *Journal of Experimental Marine Biology and Ecology* 342, 69–77.

Lockwood, G. (1998) Capital markets and aquaculture failures. *Aquaculture Magazine* 24(6), 37–47.

Mingat, A. and Tan, J. (1996) *The Full Social Returns to Education: Estimates Based on Countries' Economic Growth Performance*. Human Capital Development Working Papers, World Bank, Washington, DC.

Pauly, D. and Palomares, M.-L. (2005) Fishing down marine food webs: it is far more pervasive than we thought. *Bulletin of Marine Science* 76(2), 197–211.

Sachs, J.D. and Warner, A.M. (2001) The curse of natural resources. *European Economic Review* 45, 827–838.

Samuelson, P.A. (1967) General proof that diversification pays. *The Journal of Financial and Quantitative Analysis* 2(1), 1–13.

Samuelson P.A. and Vickrey, W. (1964) Discussion. *American Economic Review (proc.)* 59,88–96.

Shapiro, D.L. (1973) Can public investment have a positive rate of return? *Journal of Political Economy* 81(2), 401–413.

Weitzman, M. (2000) Economic profitability versus ecological entropy. *The Quarterly Journal of Economics* February, 237–263.

Willman, R. (2005) Economic aspects in the relationship between capture fisheries. Paper presented at the Regional Workshop on Low Value and 'Trash Fish' in the Asia-Pacific Region, 7–9 June 2005, Hanoi, Vietnam.

IV Future Perspectives

24 Offshore and Recirculation Technologies

Richard Langan[1] and Michel Couturier[2]

[1]*Atlantic Marine Aquaculture Center, University of New Hampshire, Durham, New Hampshire, USA;* [2]*Chemical Engineering Department, University of New Brunswick, Fredericton, New Brunswick, Canada*

24.1 Introduction

Aquaculture systems are very diverse in their design and can be classified on a scale ranging from open to closed systems. Open system culture generally refers to fish farming in natural waterbodies and typically involves the confinement of fish in floating cages or net pens. This form of aquaculture relies on natural currents to replenish oxygen in the cages and to disperse fish wastes. In contrast, closed systems use tanks or raceways to confine the fish and rely on pumped water to bring oxygen to the fish and to carry fish wastes out of the rearing units. The outflowing water is then reconditioned and recirculated to the culture units. A partially closed system which recycles the majority of the water flowing through the rearing units is known as a recirculating system. Open systems built for offshore operation and recirculating aquaculture systems (RAS) suited for land-based aquaculture are discussed in this chapter.

24.2 Offshore Technologies

24.2.1 The context for offshore farming

Aquaculture has played an important role in filling the gap between seafood supply and demand. At its current rate of growth, aquaculture will not keep pace with demand due to political, environmental, economic and resource constraints on land-based and nearshore marine culture. The gap between supply and demand is projected to reach 40 Mt by 2030 (FAO, 2006). To close this gap, additional production strategies must be identified.

Farming in offshore marine waters has been identified as one potential option for increased production; however, wind and wave conditions in most of the world's oceans pose significant technical and operational challenges that will require a completely new engineering approach. Despite these challenges, there is sufficient rationale for pursuing the development of open ocean farming. Favourable features of open ocean waters, which include ample space, tremendous carry capacity, the potential to reduce some of the negative environmental impacts of coastal fish farming (Helsley and Kim, 2005; Ward *et al.*, 2006) and optimal environmental conditions for a wide variety of marine species (Ostrowski and Helsley, 2003; Ryan, 2004; Benetti *et al.*, 2006; Howell *et al.*, 2006), have encouraged many countries to engage in offshore development.

24.2.2 Current status of offshore finfish aquaculture

24.2.2.1 Defining offshore aquaculture

To discuss technology approaches to offshore aquaculture, it is important first to define 'offshore aquaculture'. For most, the term is generally accepted to mean farming in locations that are subjected to ocean waves. Clearly, a wide range of sea conditions fall under this broad definition and technology developed for some sites will not be appropriate for others. Ryan (2004) reported on a site classification system for marine waters developed in Norway that is based on significant wave height exposure (Table 24.1). While this classification method is instructive, the full range of conditions that occur at a particular site over several years must be known to design robust engineered systems.

Other environmental conditions also must be considered, including seabed characteristics and contour, current velocities, temperature profiles, dissolved oxygen and the occurrence of harmful algal blooms. Water depth is another consideration when locating offshore farms. While, in theory, it is possible to anchor these operations in very deep water, as is done with oil platforms, the cost and practicality of setting and inspecting anchors in deep waters is currently beyond the economic reach of the fish farming industry.

24.2.2.2 Engineering design and assessment

Initial attempts at offshore farming relied to a large extent on trial and error and, as a result, many failures occurred. Beginning in the early 1990s, several groups began to apply a more sophisticated engineering approach to cage (Lisac, 1996;

Table 24.1. Norwegian Aquaculture site classification (from Ryan, 2004).

Site class	Significant wave height (meters)	Degree of exposure
1	<.0.5	Small
2	0.5–1.0	Moderate
3	1.0–2.0	Medium
4	2.0–3.0	High
5	>3.0	Extreme

Loverich and Goudey, 1996) and mooring design (Fredricksson, *et al.*, 2004); assessment of the structural integrity of cage materials (DeCew *et al.*, 2005); and modelling the effects of hydrodynamic forcing on cages and netting (Lader and Fredheim, 2003; Swift *et al.*, 2006). An approach that includes numerical modelling, scale model testing and in-field measurement of line tensions and physical forcing on cage and mooring components has been shown to inform effectively the design, materials selection and integrity of offshore systems and to reduce the possibility of system failure (Fredricksson *et al.*, 2003).

24.2.2.3 Containment technologies

A wide array of containment technologies has been applied or proposed for use in exposed locations. These have been described in detail by Scott and Muir (2000) and more recently by Ryan (2004), though a number of new technologies have emerged in the concept or prototype phase since these documents were published. The above-mentioned authors parsed technologies into categories based on structural and operational properties, which can be divided into one of two main categories: (i) surface referenced or gravity cages that use steel, high density polyethylene (HDPE) or rubber collars to float the cage and net; and (ii) submersible cages. Further subdivision can be made based on how they are moored to the seafloor and whether they are of flexible or rigid design and construction. In addition, some amount of hybridization has blurred the distinction between the two categories and some older technologies and recent innovations defy categorization.

Given the wide range of sea conditions that fall under the definition of 'offshore aquaculture', no single cage technology can be considered ideal or even appropriate for use under all circumstances. Currently, the greatest production in exposed locations is achieved with gravity cages. Large rubber (e.g. Bridgestone, Dunlop) and HDPE (e.g. PolarCirkel, Fusion, Aqualine) collar cages are in use in high-energy sites in Ireland, Scotland, the Faroe Islands, Canada, the Mediterranean Sea, Australia and Mexico. The trend in recent years has been towards increasing larger diameter HDPE collar cages. The increased size results in greater flexibility in response to waves, as well as an enormous production volume that now approaches nearly 50,000 m^3.

There are a number of advantages to the use of HDPE and rubber collar gravity cages – including a relatively long history of use and operational familiarity at protected sites; their ability – in some circumstances – to use existing automated infrastructure such as air-piped centralized feeding systems; and their low cost relative to containment volume. However, there are also limitations to their use (Fredricksson *et al.*, 2007). These include structural failures, operational difficulties related to feeding, harvesting and fish monitoring in rough weather and increased maintenance to repair and replace system components due to excessive wear and tear. Surface conditions such as waves and high currents during storms can also compress cage volume and can have detrimental effects on the fish.

Cage manufacturers continue to make structural improvements and several companies are developing submersible versions of their cages. These adaptations are likely to result in more robust systems and the option to submerge during storms will expand the range of sites in which these systems can be operated. It is likely that in the near term, these technologies will continue to be used for offshore farming at suitable sites.

Variations on the gravity cage include the Farm Ocean, which uses a rigid steel umbrella-like frame for floatation. The SADCO Shelf uses a similar steel upper structure and can be operated in a submerged configuration. Due in part to high cost per volume, as well as some structural and operational issues, neither of these cages has achieved wide-scale adoption.

Fully submersible cages include the semi-rigid SeaStation from Ocean Spar and the relatively new, rigid construction AquaPod from Ocean Farm Technologies. Both are capable of operation in a submerged position at all times, which allows their use in very high-energy sites. Despite their demonstrated ability to withstand extreme sea conditions and provide a stable environment for fish, fully submersible cages have not achieved widespread use due primarily to their relatively small size and high cost. Farms that are currently using fully submersible systems tend to be small (2–8 cages) with relatively low production volumes of high-value niche species.

A number of new designs for submersible cages have emerged in recent years, such as the 40,000 m^3 Ocean Globe from Byks of Norway, but few of them have been built at full scale and virtually none have been tested in field situations. Innovation in submersible cage technology continues and several new designs are due for unveiling in the near future; however, reluctance by industry to embrace submerged technologies has hampered their development.

There are several cage technologies – some implemented and others in the design phase – that do not fall neatly into either the gravity cage or submersible categories. Ocean Spar developed a 20,000 m^3 anchor tension cage in the 1990s called the AquaSpar that used spar buoys rather than a floatation collar to provide buoyancy and create the cage volume. RefaMed Italy has developed a tension-leg mooring with a very small seafloor footprint to secure what is essentially an inverted gravity cage (Lisac, 1996) and a number of these cages are in use in the Mediterranean for bass and bream culture. The New Brunswick (Canada) company, Aquaculture Engineering Group (AEG), has developed a unique farming concept that integrates an array of modified gravity cages with an automated feeding system and current velocity deflector that is moored at a single point and designed for use in the high current open waters of the Bay of Fundy.

24.2.2.4 Operational systems

In addition to the challenge of developing sufficiently robust containment and mooring systems, offshore farming presents additional challenges for nearly all aspects of day-to-day farm operation. Methods and equipment developed for routine operations such as feeding, harvesting and monitoring at protected inshore sites cannot be transferred directly to the offshore environment. Development of alternative operational systems has not kept pace with cage development and farmers have struggled to integrate existing, as well as new and unproven, supporting technologies into offshore installations.

Of all offshore operations, feeding is probably the most important. Inshore approaches, which include dispensing feed by cannons from a service vessel or automated feeding with blowers mounted on centralized feed barges, are severely hampered by rough seas. An ideal feeding system for offshore aquaculture would be robust, remotely controlled, fully automated, able to accommodate the volume of food needed for a 2- to 3-week period and possess a hydraulic rather than pneumatic feed delivery system. It would also ideally be capable of wireless transmission of in-cage video, monitoring data or other information critical to farm operation. Though no system as described currently exists, some progress has been made. The Scottish company, Gael Force, has developed the Sea Cap, a concrete feed barge that has operated successfully in exposed locations for several years. The University of New Hampshire (UNH) has developed two small, remotely operated feeders that have been in use since 2001 (Rice *et al.*, 2003) and deployed a much larger centralized feeder in 2007 that has four separate feed silos and can dispense feed to four submerged cages (Turmelle *et al.*, 2006). Farm Ocean and SADCO also have integrated feeding systems into their cage designs.

Other routine offshore operations such as grading, harvesting, biofouling control and removal of mortalities are complicated by sea conditions and additional strategies must be developed to make these practices safer and more efficient.

24.2.2.5 Species cultivated in offshore cages

Fish cultivated in offshore cages include both temperate and warmwater species. In terms of volume, the leading species are Atlantic salmon, *Salmo salar*, which are produced in gravity cages in Ireland and Scotland, and seabass, *Dicentrarchus labrax*, and seabream, *Sparus auratus*, grown in gravity, tension leg and a few submersible cages throughout the Mediterranean (Ryan, 2004). Fattening of northern bluefin tuna, *Thunnus thynnus*, occurs in large gravity cages in exposed locations throughout the Mediterranean. Other tuna fattening operations that use large HDPE collar cages in exposed locations include southern bluefin tuna, *T. maccoyii*, in South Australia, Pacific bluefin, *T. thynnus orientalis*, and smaller quantities of yellowfin, *T. albacares*, and bigeye tuna, *T. obesus*, in Mexico. In the north-east USA, The University of New Hampshire operates an experimental offshore farm and has produced small quantities of Atlantic cod, *Gadus morhua*, haddock, *Melanogrammus aeglefinis*, and Atlantic halibut, *Hippoglossus hippoglossus*, in submerged SeaStation cages (Howell *et al.*, 2006).

In Hawaii, small volumes of Pacific threadfin, *Polydactylus sexfilis*, and amberjack, *Seriola rivoliana*, are being grown commercially in submerged SeaStation cages at offshore sites in Oahu and the Big Island. Another warmwater species of interest for offshore farming is cobia, *Rachycentron canadum*. A commercial offshore cobia farm off Culebra Island in Puerto Rico that uses submerged SeaStations and an AquaPod cage has been in operation since

2002. In South Korea, a commercial farm is producing parrotfish, *Oplegnathus faciatus*, and olive flounder, *Paralichythys olivaceus*, in submerged SeaStation cages off the south coast of Jeju Island.

A number of other species have been proposed for offshore culture. In the USA, California yellowtail, *Seriola dorsalis lalandi*, striped bass, *Morone saxatilis*, California halibut, *P. californicus*, and tuna, *Thunnus* sp., have been proposed as candidate species for offshore culture in southern California, while Florida pompano, *Trachinotus carolinus*, red drum, *Sciaenops ocellatus*, cobia and tunas are being considered for the Gulf of Mexico. It is likely that many more species will come into production as offshore farming technologies are further developed.

24.2.2.6 Environmental considerations

One of the anticipated benefits of offshore farming is the reduced impact to the seafloor and water column as a consequence of greater dispersion of organic and inorganic wastes. The limited amount of published data available indicates that this is indeed the case. Ryan (2004), Alston *et al.* (2005), Helsley and Kim (2005), Lee *et al.* (2006), Rapp (2006) and Ward *et al.* (2006) have reported that benthic and water column impacts are greatly reduced, if not more or less absent, at open ocean farm sites in Ireland, Hawaii, New Hampshire and Puerto Rico.

Data also indicate that offshore sites may offer other environmental benefits. Ryan (2004) reported lower incidence of sea lice at offshore salmon farms in Ireland and lower stress levels and better fish health have been attributed to the more stable salinity regimes, as well as the higher oxygen levels and reduced ammonium levels that result from the greater water exchange through offshore cages (Ryan, 2004; Benetti *et al.*, 2006; Howell *et al.*, 2006). Bricknell (2006) concludes that 'offshore aquaculture offers many opportunities to reduce disease interactions between wild and farmed fish'.

Fish escapement and the use of fishmeal and fish oil as feed ingredients have also been raised as potential environmental impacts of offshore farming; however, these issues nor their solutions are not specific to offshore sites and will not be addressed in this chapter.

24.2.3 Future prospects and challenges

Developments in offshore marine cage culture over the past two decades clearly indicate that offshore farming is feasible and can be conducted in an environmentally responsible manner. However, a number of technical and operational challenges must be addressed to achieve the high levels of production needed to fill the projected gap between seafood supply and demand. What has also become clear is that offshore aquaculture will be a technology-driven enterprise. Therefore, continued investment in R & D from public and private sectors will be needed to reach the level of efficiency required for economic viability. In particular, research should be focused on the development of highly mechanized and fully integrated offshore farming systems to achieve greater efficiency and ensure worker safety. Until 'turnkey' systems that are essentially autonomous are available and economic viability offshore farming can be demonstrated, expansion of this sector in the near future will be limited.

24.3 Recirculation Technologies

24.3.1 Recirculating aquaculture systems (RAS)

The three basic categories of land-based aquaculture systems are flow-through, reuse and recirculating systems. In flow-through systems, the culture water makes one pass through the system and is discharged (Fig. 24.1a). This is the simplest and cheapest fish culture system when sufficient quantities of good quality water are available and no discharge treatment is required. If sufficient water is not available, production levels can be maintained or increased by reusing the water in multiple rearing units (Fig. 24.1b). The water in serial-reuse systems is treated between culture units but is not used in the same unit twice. Treatment usually includes the removal of faecal matter, the addition of oxygen and the removal of carbon dioxide. The number of serial reuses is, however, limited by the accumulation of harmful ammonia. This barrier is removed in recirculating systems (Fig. 24.1c) by the addition of biofilters which oxidize ammonia to nitrate. The outflowing water in recirculating systems is reconditioned and recirculated to the culture units, except for a small portion which must be discharged with the waste solids to prevent nitrate and trace minerals from accumulating to harmful levels.

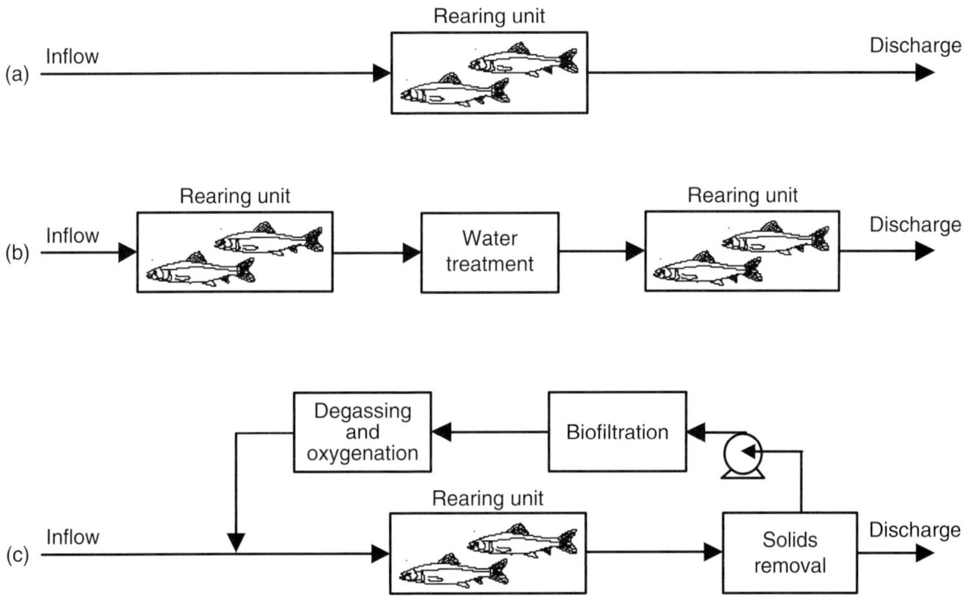

Fig. 24.1. Types of land-based aquaculture systems: (a) flow-through; (b) reuse; and (c) recirculating.

24.3.2 Potential of RAS for species diversification

Water recirculation is becoming a familiar strategy for reducing inflow and outflow volumes when water conservation is important or when environmental considerations are paramount. However, water recirculation alone does little to alleviate the environmental impact of aquaculture activities. Waste disposal remains an issue because recirculating systems generate as much waste (on a dry basis) per tonne of feed as other production systems. The main advantage of recirculating systems lies in their ability to produce lower volume and higher strength wastewater streams which can be treated at a lower cost with existing technologies.

By minimizing water consumption, recirculating systems not only benefit the environment but also open up exciting possibilities for aquaculture. It becomes feasible to use well water or municipal water rather than surface waters. These alternative water sources have smaller seasonal temperature variations and reduce the risk of disease entering the farm. It also becomes feasible to maintain the water at the optimum rearing temperature year round by heating or cooling the inflow water and housing the system in an insulated building. Heat retention is not the only benefit to be derived from keeping rearing units indoors; light manipulation regimes can be implemented, the potential negative visual effect of the equipment is limited and a protective barrier is inserted between the fish and the environment. By completely controlling the rearing environment in this manner, it is possible not only to mitigate the risks of outdoor aquaculture (disease transmission from wild fish or birds, fish escapes, temperature fluctuations, predation, air- and waterborne contaminants) but also to optimize fish growth on a year-round basis.

Since recirculating systems need only a fraction of the source water that flow-through systems require, they can be located where there is a limited water supply and in geographic locations with unfavourable climatic conditions. Consequently, recirculating systems can be built closer to the consumer and can provide a fresher product at a lower transportation cost. In principle, any species can be farmed anywhere in the world using a properly designed RAS. Furthermore, by employing a continuous production strategy involving year-round stocking and harvesting of fish, recirculating systems with temperature control can provide consistent volumes of quality products throughout the year. This allows recirculating systems to meet out-of-season market demands and to access niche markets that cannot be satisfied by outdoor systems with variable seasonal outputs.

As with all intensive aquaculture systems, the major operational expenses of recirculating systems are feed, labour and fingerling costs. But there are additional expenses that are specific to the tasks of recycling water and controlling the rearing environment. These include the capital and operating costs of the water reconditioning equipment, the cost of the protected space and the cost of the energy for recirculating and heating the water. Also, because recirculation systems are more complex than flow-through systems, their risk-related costs are perceived to be higher. The water reconditioning equipment is a life-support system and the risk of loosing production due to mechanical or electrical failure, poor water quality or operator error cannot be overlooked. On the other hand, water quality may be more consistent and reliable in a well-designed recirculating system than in some flow-through farms using surface waters. Furthermore, since optimum culture conditions can be maintained year round, fish grow faster and the productivity per unit volume of rearing space is higher than in outdoor farms. As a result, the same rearing volume can yield a higher annual production or, conversely, the same annual production can be achieved in a smaller rearing volume.

Other potential disadvantages of recirculating systems are related to product quality and disease eradication. Although the level of control provided by recirculation systems reduces significantly the risk of disease entering the farm, it does not eliminate the need for strict biosecurity measures. Once a disease is introduced, it can spread quickly through all components of the recycle system and may be difficult to eradicate. Disinfection of the system is especially difficult when surfaces are made of porous materials (e.g. concrete) or when components are not easily accessible to clean accumulated biosolids. Fish grown in recycle systems can also acquire a muddy taste, caused by the accumulation of geosmin in the flesh of the fish. The geosmin is produced by filamentous bacteria, *Actinomycetes*, which feed on dissolved organic solid and grow on the surfaces of the culture system. Off-flavour problems can be avoided by removing waste solids efficiently before they solubilize and by cleaning all surfaces frequently, including the inside walls of pipes. If these measures fail, the off-flavour can be removed by holding the fish in a flow-through system fed with clean water for several days before harvest.

24.3.3 Operation and design

The production capacity of RAS is limited by the level of feeding which can be sustained without degrading water quality to the point where fish become affected. Unless the consumed oxygen is replaced and the waste metabolites are removed, the quality of the water will degrade. Poor water quality is a significant stressor that makes fish more vulnerable to disease outbreaks. Each species has its own tolerance level; what is good quality for one species may be of marginal quality for another. The critical water parameters are temperature, suspended solids, pH, salinity, alkalinity and concentrations of dissolved oxygen, ammonia, nitrite and carbon dioxide. The water treatment equipment must be sized to maintain the water quality parameters within acceptable limits. The equipment typically includes: clarifiers and filters for removing solids, a biological filter for the nitrification of ammonia, an aeration column for carbon dioxide stripping, an oxygenation system and pumps for moving the water. The methodology for designing these units can be found in aquacultural engineering books (Timmons and Losordo, 1994; Lawson, 1995; Timmons and Ebeling, 2007).

The challenge in designing recirculating systems is to develop reliable systems while minimizing the cost of production. The cost of production is the total cost, inclusive of amortized capital costs, incurred in producing a unit weight of fish. The difference between the fish market price and the cost of production is a direct measure of profitability. To minimize the cost of production, it is usually necessary to maximize the percentage of water recycled, not only to reduce water use, but also to reduce the heat lost in the effluent and the volume of wastewater that must be processed. Energy costs can also be reduced by minimizing the energy for pumping water. Since the cost of pumping a given volume of water is directly proportional to the height to which the water is pumped, the design should minimize elevation differences and use efficient pumps. Substantial savings in both capital and labour costs can also be realized by shifting production into fewer but larger tanks (Timmons and Ebeling, 2007). The time to service a tank is nearly independent of diameter, whereas the tank's capital cost is roughly proportional to its volume raised to the power 2/3. This leads to economies of scale and tanks with diameters exceeding 10 m are becoming the norm rather than the exception. Another promising approach for reducing the cost of production is to simplify the design by using equipment that can accomplish more than one task, such as the multi-drain tank which acts as a fish rearing unit as well as a primary solids separator (Timmons *et al*., 1998) and the low-head moving bed biofilter, which provides nitrification as well as some aeration and carbon dioxide degassing (Couturier *et al*., 2006).

Large production levels are best achieved by replicating a successful recycle unit that has been optimized to take advantage of economies of scale. By spreading the production over several independent modules, this strategy minimizes the probability of a complete failure and allows for incremental expansions of the farm in step with the availability of funds or market demand.

24.3.4 Future prospects and challenges

As the aquaculture industry strives to improve control over the fish rearing environment and seeks to reduce its environmental impact, water recirculation is becoming the technology of choice for land-based aquaculture farms. Indoor fish production using water recirculation is progressively replacing traditional outdoor aquaculture methods because it has smaller water requirements, provides greater control over water temperature, allows the implementation of more stringent biosecurity measures, is less vulnerable to environmental disasters and is the most secure system to prevent fish escapement. Successful commercial RAS are currently being used to provide a secure environment for broodfish and to produce high-value species for niche markets. These include salmon smolts, live tilapia, ornamental fish, hybrid striped bass, sturgeon, eel, turbot and halibut. Further development is required to reduce the production cost of RAS so that they can also be used to produce lower-priced species for commodity markets. To achieve this goal, it will be necessary to increase the scale of the production systems, maximize stocking densities and use continuous production methods. The classical batch production approach, where all tanks are stocked with small fingerlings and harvested once the fish reach market size, greatly underutilizes the rearing capacity of the tanks for much of the grow-out period. The production capacity of the culture system can be nearly doubled with year-round fish stocking and harvesting (Watten, 1992). However, continuous production strategies require that the fish be graded frequently and that the water conditioning equipment be operated continuously at design capacity. This will necessitate greater automation of the equipment but has the prospect of improving feed conversion and making more efficient use of labour. Clearly, the development of recirculating aquaculture systems will need to continue with the objective of reducing their capital cost.

There is also the subsidiary aspect of fish welfare. It is economically desirable to operate RAS at stocking densities greater than 50 kg/m^3 in order to maximize production per unit volume. However, crowding of animals in intensive farming is viewed by some consumers as unethical, even if physiological indicators, such as growth and survival, suggest that the animals are fine. The aquaculture industry will thus have to reconcile economic realities and societal concerns in defining suitable stocking density standards.

References

Alston, D.E., Cabarcas, A., Capella, J., Benetti, D., Keene-Maltzof, S., Bonilla, J. and Cortes, R. (2005) Environmental and social impact of sustainable offshore cage culture production in Puerto Rican waters. Final Report for NOAA Grant No. NA16RG1611. National Oceanic and Atmospheric Administration, Silver Spring, Maryland.

Benetti, D., O'Hanlon, B., Brand, L., Orhun, R., Zink, I., Doulliet, P., Collins, J., Maxey, C., Danylchuk, A., Alston, D. and Cabarcas, A. (2006) Hatchery, on growing technology and environmental monitoring of open ocean aquaculture of Cobia (*Rachycentron canadum*) in the Caribbean. World Aquaculture Society, Abstract in the *Proceedings of Aquaculture 2006*, Florence Italy (http://www.was.org/meetings/AbstractData.asp?AbstractId=10424, accessed 7 November 2006).

Bricknell, I. (2006) Disease implications of offshore aquaculture. Breaking the disease cycle? In: *Offshore Mariculture 2006, Conference Programme*. 11–13 October 2006, Malta. Society for Underwater Technology, London.

Couturier, M., Ma, Z., Conroy, J., Zheng, Y. and Desbarats, A. (2006) A low-head biofilter with degassing and oxygenation capabilities. In: *Proceedings of the Sixth International Conference on Recirculating Aquaculture*. Virginia Tech, Roanoke, Virginia.

DeCew, J., Fredriksson, D.W., Bougrov, L., Swift, M.R., Eroshkin, O. and Celikkol, B. (2005) A case study of a modified gravity type cage and mooring system using numerical and physical models. *IEEE Journal of Oceanic Engineering* 30(1), 47–58.

FAO (2006) State of world aquaculture: 2006. FAO Fisheries Technical Paper No. 500. Food and Agriculture Organization of the United Nations, Rome.

Fredriksson, D.W., Swift, M.R., Tsukrov, I., Irish, J.D. and Celikkol, B. (2003) Fish cage and mooring system dynamics using physical and numerical models with field measurements. *Aquacultural Engineering* 27(2), 117–270.

Fredricksson, D., DeCew, W.J., Swift, M.R., Tsukrov, I., Chambers, M.D. and. Celikkol, B. (2004) The design and analysis of a four-cage grid mooring for open ocean aquaculture. *Aquacultural Engineering* 32, 77–94.

Fredricksson, D.W., DeCew, J.C., Tsukrov, I., Swift, M.R. and Irish, J.D. (2007) Development of large fish farm numerical modeling techniques with in-situ mooring tension comparisons. *Aquacultural Engineering* 36, 137–148.

Helsley, C.E. and Kim, J.K. (2005) Mixing downstream of a submerged fish cage: a numerical study. *IEEE Journal Of Oceanic Engineering* 30, 12–19.

Howell, W.H, Watson, W.H. and Chambers, M.D. (2006) Offshore production of cod, haddock and halibut. CINEMar/Open Ocean Aquaculture Annual Progress Report for the Period from 1/01/05 to 12/31/05. Final Report for NOAA Grant No. NA16RP1718 (http://ooa.unh.edu, accessed 31 January 2006).

Lader, P. and Fredheim, A. (2003) Modeling of net structures exposed to three-dimensional waves and current. In: Bridger, C.J. and Costa-Pierce, B.A. (eds) *Open Ocean Aquaculture: From Research to Commercial Reality*. World Aquaculture Society Publication, Baton Rouge, Louisiana, pp. 177–189.

Lawson, T.B. (1995) *Fundamentals of Aquacultural Engineering*. Chapman and Hall, New York.

Lee, H.W., Bailey-Brock, J.H. and McGurr, M.M. (2006) Temporal changes in the polychaete infaunal community surrounding a Hawaiian mariculture operation. *Marine Ecology Progress Series* 307, 175–185.

Lisac, D. (1996) Recent developments in open-sea cages: practical experience with the tension leg cage. In: Polk, M. (ed.) *Open Ocean Aquaculture: Proceedings of an International Conference*. 8–10 May 1996, Portland, Maine. New Hampshire/Maine Sea Grant College Program Report No. UNHMP-CP-SG-96-9, pp. 513–522.

Loverich, G.F. and Goudey, C. (1996) Design and operation of an offshore sea farming system. In: Polk, M. (ed.) *Open Ocean Aquaculture: Proceedings of an International Conference*. 8–10 May 1996, Portland, Maine. New Hampshire/Maine Sea Grant College Program Report No. UNHMP-CP-SG-96-9, pp. 495–512.

Ostrowski, A.C. and Helsley, C.E. (2003) The Hawaii offshore aquaculture research project: critical research and development issues for commercialization. In: Bridger, C.J. and Costa-Pierce, B.A. (eds) *Open Ocean Aquaculture: From Research to Commercial Reality*. The World Aquaculture Society, Baton Rouge, Louisiana, pp. 119–128.

Rapp, P. (2006) Measurement of the benthic loading and the benthic impact from an open-ocean fish farm in tropical waters. In: Bridger, C.J. and Costa-Pierce, B.A. (eds) *Open Ocean Aquaculture: From Research to Commercial Reality*. Preliminary Report for NOAA Grant No: NA040AR4170.The World Aquaculture Society, Baton Rouge, Louisiana, pp. 119–128.

Rice, G., Stommel, M. Chambers, M.D. and Eroshkin. O. (2003) The design, construction and testing of the University of New Hampshire feed buoy. In: Bridger, C.J. and Costa-Pierce, B.A. (eds) *Open Ocean Aquaculture: From Research to Commercial Reality*. The World Aquaculture Society, Baton Rouge, Louisiana, pp. 197–203.

Ryan, J. (2004) Farming the deep blue. Board Iascaigh Mhara and Marine Institute Technical Report, Dublin.

Scott, D.C.B. and Muir, J.F. (2000) Offshore cage systems – a practical overview. In: Muir, J. and Basurco, B. (eds) *Mediterranean Offshore Mariculture*. Serie B: Etudes et Recherches, No. 30 Options Mediterraneennes. CIHEAM (Centre International de Hautes Etudes Agronomiques Mediterraneennes), Zaragoza, Spain, pp. 79–89.

Swift, M.R., Fredriksson, D.W., Unrein, A., Fullerton, B., Patursson, O. and Baldwin, B. (2006) Drag force acting on biofouled net panels. *Aquacultural Engineering* 35(3), 292–299.

Timmons, M.B. and Ebeling, J.M. (2007) *Recirculating Aquaculture*. Cayuga Aqua Ventures, Ithaca, New York.

Timmons, M.B. and Losordo, T.M. (1994) *Aquaculture Water Reuse Systems: Engineering Design and Management*. Elsevier Science, Amsterdam.

Timmons, M.B., Summerfelt, S.T. and Vinci, B.J. (1998) Review of circular tank technology and management. *Aquacultural Engineering* 18, 51–69.

Turmelle, C., Swift, M., Celikkol, B., Chambers, M., DeCew, J., Fredriksson, D., Rice, G. and Swanson, K. (2006). Design of a 20-ton capacity finfish aquaculture feeding buoy. *Proceedings of the Oceans 2006*, MTS/IEEE Conference. Boston, Massachussetts.

Ward, L.G., Grizzle, R.E. and Irish, J.D. (2006) UNH OOA Environmental Monitoring Program, 2005. CINEMar/Open Ocean Aquaculture Annual Progress Report for the period from 1/01/05 to 12/31/05. Final Report for NOAA Grant No. NA16RP1718 (http://ooa.unh.edu, accessed 31 January 2006).

Watten, B.J. (1992) Modeling the effects of sequential rearing on the potential production of controlled environment fish-culture systems. *Aquacultural Engineering* 11, 33–46.

25 Valorization of Aquaculture By-products

Pierre U. Blier[1] and Nathalie R. Le François[1,2]

[1]Université du Québec à Rimouski, Rimouski, Québec, Canada; [2]Biodôme de Montréal, Montréal, Québec, Canada

25.1 Introduction

Within the agri-food sector, the aquaculture industry is generally considered as a high-risk venture for private investors and/or governmental agencies. The reasons behind this uncertainty are: (i) the significant investments required to set up a commercial aquaculture operation; (ii) the time needed to become profitable (i.e. return on investment); (iii) the risk associated with the rearing of live aquatic organisms, either by the direct impacts of environmental conditions and water quality or the reliability of the rearing systems and perhaps, to a higher degree (iv) the relative vulnerability of this type of activity to the economic conditions that dictate the achievement of profitability. Among them are production costs, dictated largely by feed and energy costs and the value of the output products on the traditional seafood market. As pointed out by Burbridge *et al.* (2001), a major problem faced by the aquaculture industry is the lack of a framework for its objective economic evaluation, resulting in a distorted and inconsistent analysis of the associated costs and benefits of expanding and diversifying.

In North America and Europe, most fish production is oriented toward carnivorous species in intensive rearing systems, which therefore implies high sensitivity to variation of feed price and quality and the energy costs of all aspects of operations. Given the consistent escalation of energy costs (a barrel of crude oil exceeded US$125 in 2008), it seems futile in the short term to expect any significant gain from energy savings, even if major improvements in operating costs were achieved.

Meanwhile, in the world market, the value of seafood production is still dictated by the high availability of low-value products originating from commercial fishery landings. For example, the price in US$ for cod, hake and pollack fillets did not change significantly from 1986 to 2006 (FAO, 2006), despite the collapse of the fisheries. At the same time, the increase of aquaculture production induced an unexpected depreciation of the value of some products. For example, the price of Norwegian fresh gutted salmon has decreased significantly in the past 20 years. To maintain profitability, the high-valued salmoniculture industry adopted a dual strategy. The first line of attack was aimed at significant improvement of fish growth performance through the optimization of nutrition and genetics. The second strategy was directed toward reducing the cost of rearing production technologies. Both efforts required substantial investments in research and development that, in the end, translated in significant enhancement of productivity (see Wilson and Archer, Chapter 23, this volume). In this chapter, we will suggest an alternative approach to increase the benefits from fish production by making use of otherwise discarded or undervalued resources from the fish farming industry: the by-products. This concept, also introduced briefly by Muir (2005), is considered in order to maximize profit to compensate for the narrow profit margins expected from traditional activities and subscribe to the concepts of eco-efficiency and sustainability of aquaculture operations. A series of activities intended to verify the application of this principle will be outlined briefly in this chapter.

The by-products from a standard fish farm can represent over 50% of the biomass production, depending on the species (Le François *et al.*, 2002). Not only are these by-products not recovered or transformed, but they usually represent a significant expenditure in the form of waste management. For example, the fillet yield of cod is around 42% of body weight, while this value reaches less than 35% for small flatfish species (Straus, 1991). More than half of what is produced is simply thrown away and lost to solid waste landfills. Arguably, this is an unsustainable practice and, in addition to ecological considerations, we suggest that this narrow and outdated perception of a unique output of production (transformed or untransformed flesh) is not profitable.

In the following chapter, we will consider the potential value of seafood by-products in general and put forward the importance of considering this added value at the beginning of the process of diversification when considering new species.

Marine biotechnology activities rely partly on fisheries for the supply of the biomass used for extraction processes. However, the seasonality and variation encountered in the quality and quantity of the biomass requires optimization of extraction processes. Thus, aquaculture can be considered as a potential source of biomass without the variability experienced when relying on fisheries as a main supplier of marine biomass. As an example, the by-products from two wolffish species of aquaculture interest will be presented with the aim of showing how increasing their utilization to produce value-added products will enable improvement of profitability and reduce waste. Screening for potential biomolecules will be outlined and will include antimicrobial polypeptides, antifreeze proteins and digestive enzymes. Processing methods to extract biomolecules with application in food, nutraceuticals and pharmaceuticals will be studied and optimized, characterization will be realized and the markets for these compounds reviewed. Finally, the bioeconomics of this opportunity for diversification within an aquaculture production cycle will be assessed. This approach offers a real opportunity since the biomolecule of interest that will be extracted originates from controlled conditions of production, providing the basis for optimized extraction processes and yields.

In support of these assumptions, we will evaluate the potential value of different extracts from fish farm by-products by first exploring the valorization activities of fishery by-products worldwide and then concentrating on a case study involving the coldwater-adapted wolffishes (*Anarhichas lupus* and *A. minor*) (see Chapter 19, this volume) to illustrate how these potential new products should be included in feasibility studies when considering new species for diversification.

25.2 Feed and Food Use of By-products

Fishery wastes have been used successfully in the nutritional and non-nutritional industry, giving a wide variety of products with a large spectrum/range of value categories. Of low value, fishmeals and fish oils contribute to the highest volume of production. These commodities offer the advantages of low-cost, simple production technologies and easy and direct access to markets with no need for complex marketing and positioning strategies. The other side of the coin is the low price and, consequently, the low benefit that can be expected. For example, fishmeal on the international market almost reached US$0.90/kg, while fish oil was detailed at US$0.75/kg in 2006 (Klinkhardt, 2006). The expected yield from raw material is approximately 22% for meal (FAO, 1986) and good extraction from fatty fish can generate 5% of oil (Chantachum *et al.*, 2000). From simple calculations, for 1 kg of flesh produced, one can expect approximately 1 kg of raw material, from which 50 g of oil and 220 g of meal could be extracted, representing on the international market a value of US$0.23 (according to the value of fishmeal and fish oil in 2006). If we consider the high level of investment required for the process and the cost of production, it is unlikely that this product would constitute the main strategy to add value to fish production. However, if there is a world shortage in long-chain omega-3 fatty acids, the value of fish oil for nutraceuticals from fishery or aquaculture may increase significantly. Production of fishmeal and fish oil appears to be economically viable only when the fish farm has close access to the processing plant or when small-scale equipment and technology are be available at an affordable price (Archer, 2001).

We suggest that a stronger economic strategy would be first to extract high-value biomolecules and then consider directing the other by-products through the appropriate low price–low technology channel. Despite the low profitability of such operations, these low-value products could at least partly reduce the cost and solve the environmental issues of waste management.

While fishmeal is still a cheap source of animal proteins (price per kg of protein), further processing of these proteins through enzymatic hydrolysis could be a valuable approach to obtain products commanding higher price. Fish protein hydrolysate (FPH) is actually used in the food industry, mainly for flavour and aroma properties (Kristinsson, 2007) and has also been proposed as a fish feed ingredient since larvae and juvenile fish usually possess low digestive capacity associated to a low level of proteolytic enzymes (Lemieux *et al.*, 2003; Lamarre *et al.*, 2004, 2007; Savoie *et al.*, 2008). In several species, inclusion of protein hydrolysates has been reported to have a positive effect on larval performance (Cahu and Zambonino Infante, 1995; Zambonino Infante *et al.*, 1997; Savoie *et al.*, 2006). Hydrolysis can be performed with proteolytic enzymes that are commercially available. Standard technologies from the food industry such as filtration and spray drying can then be adapted easily to dry the protein hydrolysate. The challenge in the production of FPH for nutritional use is to optimize the production conditions (enzyme sources and quantity, temperature, pH and duration) in order to minimize bitterness and ensure flavour development (Kristinsson, 2007). Fish proteins and fish protein hydrolysates also demonstrate good attributes for their use in human nutrition and as part of strategies against metabolic syndromes and type II diabetes (Lavigne *et al.*, 2001, Ouellet *et al.*, 2007). These authors observed increased insulin sensitivity following the inclusion of FPH in the diets of insulin-resistant rats and humans. These potential impacts on metabolic regulation associated with fish consumption could also be involved in the overall prevention of heart diseases (He *et al.*, 2004).

Other traditional products that can be extracted from aquaculture by-products are collagen and gelatin, as the demand for these products exploded due to consumer concern about the risks of bovine spongiform encephalopathy associated with collagen and gelatin of mammalian origin (Regenstein and Zhou, 2007). These molecules are used as functional ingredients in the food industry, but they can also be used by the pharmaceutical industry to manufacture edible capsules, tablets and pastilles. Products aimed at human health supplements usually command high prices, but necessitate the utmost standards of manufacturing, for which the financial resources of small companies are usually inadequate.

25.3 Specialized Products

25.3.1 Enzymes

Fish viscera contain extractable quantities and varieties of enzymes that could be of interest for different types of industrial processes. Because fish are ectotherms and have evolved in widely diverse aquatic environments (freshwater/seawater, cold/warm habitats), their enzymatic systems have evolved and adapted to a wide range of physico-chemical conditions which give them unique characteristics unparalleled by endothermic animal representatives or terrestrial plants. Fish enzymes are already used for fish processing applications in the food industry. Further examples and details of applications have been summarized by Shahidi and Janak Kamil (2001) (see Table 25.1).

Enzyme extractions rely on easily manageable technologies and depend on access to relatively sophisticated equipment: industrial centrifuges and different filtration and drying systems that can be purchased at standard food or dairy processing equipment retailers. The price that can be obtained for extracted enzymes will depend on purity and functional properties. Roughly, US$5–10/kg/dry weight (detailed price) can be realized for crude proteolytic enzymes if their activity range is within 400–1000 International Units/g. This could well represent, depending on the enzyme content of the available tissues and the

Table 25.1. Application of enzymes from fish.

Area of application	Examples
Selective tissue degradation	De-skinning of fish and aquatic invertebrates Purification of fish roe – 'caviar production' Removal of membrane from cod liver Removal of exoskeleton from shellfish Production of salted cod swim bladder
Fermentation and curing of fish	Production of fish sauce Production of fish silage Production of 'maatjes herring'
Production of hydrolysed products	Fish protein hydrolysate Flavour compounds
Extraction of pigments	Enzymatic recovery of pigments from shellfish waste
Coagulation of protein	Application of chymosin as a rennet substitute in cheese manufacturing Removal of oxidized flavour of milk
Waste management (viscosity reduction of stickwater)	Enzymatic treatment of stickwater
Other potential applications	Meat tenderizing Enzymatic extraction of fish oil from raw material Gene cloning technology Antibacterial enzymes Antioxidative enzymes Production of omega-3 fatty acid concentrates

Source: After Shahidi and Janak Kamil (2001).

extraction efficiency of the processes, a considerable financial gain from by-product processing activities. In a recent study, we estimated total proteolytic activities in herring and mackerel ranging from 10 to 550 International Units/g of viscera, depending on the species and the season.

25.3.2 Antifreeze proteins

In cold oceans, different fish species have developed strategies to survive at temperatures below their freezing point. Antifreeze proteins (AFP) have evolved in different taxonomic groups and five different classes of AFP have been discovered so far (Fletcher *et al.*, 2001). Commercial applications of AFP have been identified in the following areas: (i) protection of fish and plants against cold and freezing temperatures; (ii) cold protection of mammalian cells, tissues and organs; (iii) enhanced tumour cell destruction during cryosurgery; and (iv) longer shelf life for and better quality of frozen foods (Fletcher *et al.*, 1999). The major commercial use of AFP is to prevent ice crystal formation in frozen food and to extend its shelf life. While AFP content in fish plasma is relatively low, its high market price justifies its consideration for production that could improve significantly the profitability of rearing fish species that synthesize AFP naturally and, ultimately, at least for coldwater aquaculture, be included as a selection criteria when choosing new species for diversification both as a freeze-protection agent and for its high economical value.

25.3.3 Antimicrobial polypeptides

Antimicrobial polypeptides (AMP) display considerable potential as novel therapeutic agents or disinfectants for the treatment of systemic fungal, bacterial and viral infections, tumours, gastric and skin ulcers, oral cavity disease, ophthalmic disease, sterilization, production of sterile coatings used in cosmetics or laundry detergents, use as a food preservative, etc. (Tossi and Sandri, 2002). The potential for the application of AMP was recognized early due to the rapidity and broad spectrum of their activity. The marine environment is an abundant source of AMP but only very little of this rich resource has been studied. Most marine organisms rely heavily on antimicrobial components of their innate immune defences to fight pathogens (Patrzykat and Douglas, 2003). Several types of antimicrobial peptides have been isolated from fish (Douglas *et al.*, 2003). In teleosts, secretions from mucous gland cells protect the living epidermal surface of the skin (Shephard, 1994). This mucosal epithelial layer is the interface with an austere external environment and it provides a physical and biochemical barrier that is crucial as a first line of host defence (Ellis, 2001). The use of classical antimicrobial agents (chemical or synthesized products) in animal feed displayed persistence of those compounds in the derived products (poultry, beef, fish and pork), which caused increased toxicity, the development of allergies in humans and a growing resistance of the undesired microbes to their action. Public opinion is now more inclined to use antimicrobial agents of natural origin instead. Increased research efforts aimed at natural AMP sources are ongoing and a promising market is foreseen.

Antimicrobial polypeptides are found in various tissues and organs (skin, digestive tract, respiratory tract) of many species, such as *Pseudopleuronectes americanus* and *Squalus acanthias* (Moore *et al.*, 1993; Cole *et al.*, 1997). These natural peptides are frequently found in the mucus and exert a broad range of antimicrobial activity against pathogenous bacteria, parasites and encapsulated viruses (Cole *et al.*, 1997). The most known of these AMP are: cecropines, melittines, magainines and defensines (Hancock and Lehrer, 1998).

25.4 A Case Study: Wolffish, *Anarhichas lupus* and *A. minor*

For domestication and organoleptic attributes, wolffishes have been identified as one of the best species for diversification of marine aquaculture in eastern Canada (Le François *et al.*, 2002; Le François *et al.*, Chapter 19, this volume). Market surveys confirmed that the interest in this species was mainly from high-end restaurants (Laflamme *et al.*, 2005). Studies aimed at the characterization of biomolecules, or at the regulation of their production to evaluate their potential for diversification of output products from wolffish farming, were initiated (Le François *et al.*, 2004) (see Fig. 25.1). The first biomolecule that we investigated is AFP (type III) (Desjardins *et al.*, 2006, 2007). These studies showed that adult Atlantic wolffish could produce up to 15 mg AFP/ml plasma in winter. Knowing that the cycle of production is controlled partly by photoperiod and the highest concentrations are obtained at lower temperatures, we can consider the optimization of the environmental condition triggering and maintaining AFP production levels throughout the year. AFP can be sold on the market at close to US$5000/g (Dr Fletcher, A/F Protein Canada Inc, personal communication). Simple calculations and verification allowed us to estimate that a 5 kg Atlantic wolffish, which contained roughly 125 ml of extractable blood (25 ml/kg) would yield about 2/3 of plasma which, at an AFP concentration of 15 mg/ml, translated into a value of US$6 250/5 kg fish. These estimates do not, however, include extraction/purification and marketing costs.

In a second series of experiments, we investigated the extraction potential of trypsin, a digestive enzyme of known commercial value and of wide applications in the food industry (Desrosiers *et al.*, 2008). The major conclusion from this study was that trypsin derived from Atlantic wolffish displayed activity levels similar to bovine trypsin at high temperatures and shared the same wide range

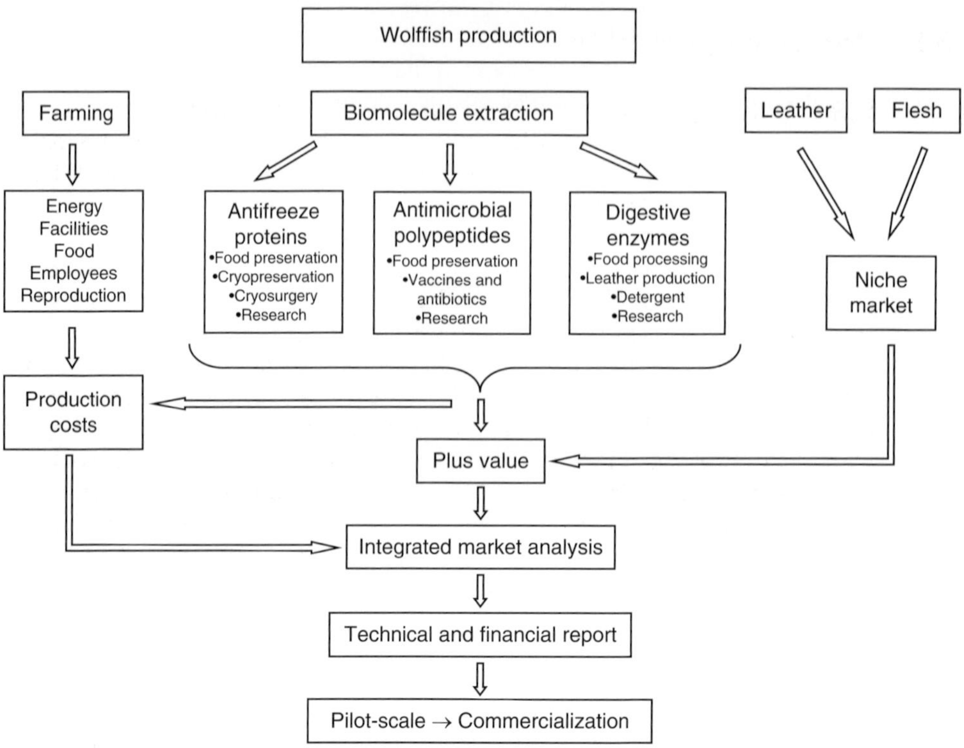

Fig. 25.1. Extraction potential of various biomolecules, case study of wolffish (adapted from Le François *et al.*, 2004).

of pH optima. The wolffish trypsin enzyme appears to be much more stable after heat treatment than bovine trypsin and other fish trypsin. It is also particularly stable in alkaline conditions and at high sodium concentrations. Furthermore, wolffish trypsin enzyme could be used where salt-tolerant enzymes are required in fermentation processes. Trypsin from wolffish could then be an appropriate substitute to bovine or most other mammal trypsin. This enzyme alone could hardly generate sufficient benefits to justify the required investments for enzyme purification, but it could be a valuable portion of a global process of different enzyme extraction and purification from farmed fish by-products.

In addition to biomolecule extraction, the combination of leather extraction and tanning activities in a commercial wolffish production is also worth considering (Le François *et al.*, 2004). For example, it was estimated that a raw spotted wolffish skin had a commercial value around CAN$5–10, depending on the size of the fish (L.O. Sparboe, personal communication). The attractiveness of fish leather worldwide is based essentially on the waterproof quality and durability of fish skins (Ingram and Dixon, 1994). The fish skin is used primarily for the production of high-quality fashion shoes, handbags and clothing. Grey *et al.* (2006) recently provided an analysis of the imports and exports of fish leather by the USA in relation to wild fish population exploitation, conservation and sustainability.

25.5 Concluding Remarks

By-products originating from fish farming can yield, once transformed, a wide spectrum of products with a wide range of commercial value and demanding varying levels of technological sophistication. Low value-added products likely necessitate lower investments but generally, to ensure high returns, must rely on a large volume of final products. High-value biomolecules, on the other hand, entail the processing of important quantities of raw material to yield a significant quantity of the targeted molecules. In both cases, the limiting steps of the exercise are close access to the know-how, the technologies and the markets. A clever strategy would be to take advantage of the presence of established food or biotechnology companies already having advanced expertise in extraction and purification techniques and having a good understanding of market expectations. In terms of economic development politics, it would be wise to synchronize efforts of aquaculture diversification (species and products) with investments already made in the development of an industry involved in the extraction of biomolecules from aquatic organisms. This could translate in localizing new farms closer to knowledge centres.

References

Archer, M. (2001) Fish Waste Production in the United Kingdom: The Quantities Produced and Opportunities for Better Utilisation. *Seafish Report No. SR537*. The Sea Fish Industry Authority, UK.

Burbridge, P., Hendrick, V., Roth, E. and Rosenthal, H. (2001) Social and economic policy issues relevant to marine aquaculture. *Journal of Applied Ichthyology* 17, 194–206.

Cahu, C.L. and Zambonino Infante, J.L. (1995) Maturation of the pancreatic and intestinal digestive function in sea bass (*Dicentrarchus labrax*) effect of weaning with different protein sources. *Fish Physiology and Biochemistry* 14, 431–437.

Chantachum, S., Benjaku, S. and Sriwirat, N. (2000) Separation and quality of fish oil from precooked and non-precooked tuna heads. *Foods Chemistry* 69, 289–294.

Cole, A.M., Weis, P. and Diamond, G. (1997) Isolation and characterization of pleurocidin, an antimicrobial peptide in the skin secretions of winter flounder. *Journal of Biological Chemistry* 272, 12008–12013.

Desjardins, M., Le François, N.R., Fletcher, G.L. and Blier, P.U. (2006) Seasonal modulation of plasma antifreeze protein levels in Atlantic (*Anarhichas lupus*) and spotted wolffish (*A. minor*). *Journal of Experimental Marine Biology and Ecology* 335, 142–150.

Desjardins, M., Le François, N.R., Fletcher, G.L. and Blier, P.U. (2007) High antifreeze protein levels in wolffish (*Anarhichas lupus*) make them an ideal candidate for culture in cold, potentially ice laden waters. *Aquaculture* 272, 667–674.

Desrosiers, V., Le François, N.R. and Blier, P.U. (2008) Trypsin-like enzyme from Atlantic wolffish (*Anarhichas lupus*) viscera: purification and characterization. *Journal of Aquatic Food Product Technology* 17, 11–26.

Douglas, S.E., Gallant, J.W., Liebscher, R.S., Dacanay, A. and Tsoi, S.C.M. (2003) Identification and expression analysis of hepcidin-like antimicrobial peptides in bony fish. *Developmental and Comparative Immunology* 27, 589–601.

Ellis, A.E. (2001) Innate host defense mechanisms of fish against viruses and bacteria. *Developmental and Comparative Immunology* 25, 827–839.

FAO (1986) *The production of fish meal and oil*. FAO Fisheries Technical Paper 142. FAO, Rome.

FAO (2006) *State of world aquaculture*. FAO technical paper 500. FAO, Rome.

Fletcher, G.L., Goddard, S.V. and Wu, Y. (1999) Antifreeze proteins and their genes: from basic research to business opportunity. *Chemtech* 29, 17–28.

Fletcher, G.L., Hew, C.L. and Davies, P.L. (2001) Antifreeze proteins of teleost fishes. *Annual Review of Physiology* 63, 359–390.

Grey, M., Blais, A.-M., Hunt, B. and Vincent, A.C.J. (2006) The USA's international trade in fish leather, from a conservation perspective. *Environmental Conservation* 33, 100–108.

Hancock, R.E.W. and Lehrer, R. (1998) Cationic peptides: a new source of antibiotics. *Trends in Biotechnologies* 16, 82–88.

He, K., Song, Y., Daviglus, M.L., Liu, K., Van Horn, L. and Dyer, A.R. (2004) Fish consumption and incidence of stroke: a meta-analysis of cohort studies. *Stroke* 35, 1538–1542.

Ingram, P. and Dixon, G. (1994) Fishskin leather: an innovative product. *Journal of the Society of Leather Technologies and Chemists* 79, 103–106.

Klinkhardt, M. (2006) Demand for fish meal and fish oil rising. *Eurofish Magazine* 6 (http://www.globefish.org/index.php?id=3531&easysitestatid=603325997, accessed 12–20 May 2008).

Kristinsson, H.G. (2007) Aquatic food protein hydrolysates. In: Shahidi, F. (ed.) *Maximising the Value of Marine By-products*. CRC Press, Boca Raton, Florida, pp. 229–248.

Laflamme, J., Michaud, J.C., Lévesque, M. and Le François, N.R. (2005) *Potentiel commercial et technico-financier de l'élevage du loup tacheté au Québec* (http://www.sodim.org/pdf/Potentiel_comm_loup_tach.pdf, accessed 1 July 2009).

Lamarre, S.G., Le François, N.R., Falk-Petersen, I. and Blier, P.U. (2004) Can digestive and metabolic enzyme activity levels predict growth rate and survival of newly hatched Atlantic wolffish (*Anarhichas lupus* Olafsen)? *Aquaculture Research* 35, 608–613.

Lamarre, S.G., Le François, N.R., Lemieux, H., Falk-Petersen, I.-B. and Blier, P.U. (2007) The digestive and metabolic enzyme activity profiles of a non-metamorphic marine fish species:

effects of feed type and feeding level. *Canadian Journal of Fisheries and Aquatic Sciences* 64, 849–856.

Lavigne, C., Tremblay, F., Asselin, G., Jacques, H.N.E. and Marette, A. (2001) Prevention of skeletal muscle insulin resistance by dietary cod protein in high fat-fed rats. *American Journal of Physiology – Endocrinology and Metabolism* 281(1 44-1), E62–E71.

Le François, N.R., Lemieux, H. and Blier, P.U. (2002) Biological and technical evaluation of the potential of marine and anadromous fish species for cold-water mariculture. *Aquaculture Research* 33, 95–108.

Le François, N.R., Desjardins, M. and Blier P.U. (2004) Enhancement of profitability perspectives of an aquaculture production by the extraction of high commercial value biomolecules: evaluation of the potential of wolffish cultivation. In: Shahidi, F. and Simpson, B.K. (eds) *Seafood Quality and Safety: Advances in the New Millennium*. Science Tech Publishing Company, St John's, Newfoundland, Canada, pp. 61–69.

Lemieux, H., Le François, N.R. and Blier, P.U. (2003) The early ontogeny of digestive and metabolic enzyme activities in two commercial strains of arctic charr (*Salvelinus alpinus* L.). *Journal of Experimental Zoology Part A: Comparative Experimental Biology* 299, 151–160.

Moore, K.S., Wehrli, S., Roder, H., Rogers, M., Forrest, J.N., McCrimmon, D. and Zasloff, M. (1993) *Squalamine: an aminosterol antibiotic from the shark*. Proceedings of the National Academy of Sciences of the United States of America 90, 1354–1358.

Muir, J. (2005) Managing to harvest? Perspectives on the potential of aquaculture. *Philosophical Transactions of the Royal Society* 360, 191–218.

Ouellet, V., Marois, J., Weisnagel, S.J. and Jacques, H. (2007) Dietary cod protein improves insulin sensitivity in insulin-resistant men and women: a randomized controlled trial. *Diabetes Care* 30, 2816–2821.

Patrzykat, A. and Douglas, S.E. (2003) Gone gene fishing: how to catch novel marine antimicrobials. *Trends in Biotechnology* 21, 362–369.

Regenstein, J.M. and Zhou, P. (2007) Collagen and gelatin from marine by-products. In: Shahidi, F. (ed.) *Maximising the Value of Marine By-products*. CRC Press, Boca Raton, Florida, pp. 279–303.

Savoie, A., Le François, N.R., Cahu, C., Blier, P.U. and Andreassen, I. (2006) Do protein hydrolysates improve survival and growth of newly-hatched spotted wolffish (*Anarhichas minor*), a non-metamorphic aquaculture fish species? *Aquaculture* 261, 782–788.

Savoie, A., Le François, N.R., Cahu, C. and Blier, P.U. (2008) Metabolic and digestive enzyme activity profiles of newly hatched spotted wolffish (*Anarhichas minor* Olafsen): effect of temperature. *Aquaculture Research* 39, 382–389.

Shahidi, F. and Janak Kamil, Y.V.A. (2001) Enzymes from fish and aquatic invertebrates and their application in the food industry. *Trends in Food Science and Technology* 12, 435–464.

Shephard, K.L. (1994) Functions for fish mucus. *Reviews in Fisheries Biology* 4, 401–429.

Straus, K. (1991) *The Seafood Handbook: Seafood Standards*. Seafood Business, Rockland, Maine.

Tossi, A., and Sandri, L. (2002) Molecular diversity in gene-encoded, cationic antimicrobial polypeptides. *Current Pharmaceutical Design* 9, 743–761.

Zambonino-Infante, J.L., Cahu, C. and Péres, A. (1997) Partial substitution of native protein by di- and tripeptides in diet improves sea bass (*Dicentrarchus labrax*) larvae development. *Journal of Nutrition* 127, 608–617.

26 Organic and Ecolabelling

James A. Young

Department of Marketing, University of Stirling, Scotland, UK

26.1 Introduction

The options of diversifying product offerings to the market through environmentally friendly (also referred to as ecolabelled or green) and organic labelling have become increasingly important and attractive to aquaculture producers over the relatively recent past. A number of factors can be cited to explain this movement, ranging from those driven by the traditional product focus of the sector through to consumer concerns within the wider markets for food, in some of which aquatic food products compete. The green market sectors have become increasingly dynamic and variable in their responses and uptake and it seems likely that as wider trends within the markets for fish continue, they will become more important. This is all the more so when one considers the status of aquaculture.

The comparatively recent emergence of aquaculture as a significant contributor to global fish and food markets has created a dynamic sector within which a number of changes have emerged and continue to evolve. It is now generally accepted that the contribution of aquaculture will increase (FAO, 2007) and questions arise about how this expanding volume of aquatic food products will appear on the market. In this chapter, the intention is to provide some insight into the emergent 'greening' of aquaculture marketing activities based on events primarily from the turn of the 21st century and the contemporary changes within the wider markets for foods. The chapter can only provide an overview of the organic and ecolabelling phenomena. The organic market is perhaps best considered as but one part of a wider continuum of environmentally friendly-driven markets ranging from conventional products through to evermore esoteric points of differentiation. Some would see organics as representing the purest form of environmentally friendly products; others would contend differently because adoption in some cases would result in reduced production, reduced food supply and thus possibly adverse repercussions in terms of nutrition, health, socio-economic impacts, etc. Given the overall constraints within this book, the focus is deliberately narrow, but pertinent to the underlying theme of species diversification.

26.2 Production Pressures on Diversification

Throughout its recent history of less than 40 years, aquaculture has tended to be production, rather than market, led. Advances in aquaculture science and technology have been the main determinants of species offered to the markets, often with scant and only reflexive regard for what the market might want. This production orientation has driven a series of profit cycles in different species whereby increased ability to produce has led to an expanded volume being placed on markets; but on markets without sufficient demand to absorb the increase in supply readily without corresponding price reductions. As prices have fallen in response to the increased quantities available, producers have tended to adopt a strategy of selling yet more in an effort to maintain total sales revenues and profits. Declining profits have resulted in periodic contractions in production capacity, whose more limited volume has, in turn, eventually met with higher market prices; thereby encouraging further expansion of production and repetition of the cycle.

While this train of events may appear foolhardy, possibly short-sighted, in retrospect, it is more understandable given the very limited market power of most individual producers and the contemporaneous actions of competing suppliers. All producers in the same geographical areas have tended to have near simultaneous access to similar advantages realized through technical progress and thus invariably have been at similar stages along the learning curve for the limited range of individual species launched. The cyclical erosion of profitability has tended to encourage greater focus on production costs and emphasis on the species with which they are most familiar. This sequence has long since been observed in agriculture and its manifestation as an aquatic crop cycle is entirely consistent with events witnessed in a number of aquaculture sectors, including salmon, seabass, seabream, shrimp and catfish. (Bjorndal, 1990; Anon., 2004; Globefish, 2007, 2008a).

Faced with a seemingly endless progression of peaks and troughs in market prices for individual species, many have considered diversification. While some producers have opted for strategies seeking to add value to the species produced, classically through routes in fish processing or incorporation of additional product attributes such as convenience, premium freshness, etc., more have tended to stick to their area of core competence: fish farming. Elsewhere, producers have diversified into new species, but the broad pattern identified earlier has tended to repeat, in part again due to contemporaneous access to newly available technology and similar interpretations of market signals. More significant perhaps is the likelihood for new species launched to enter into this aquatic crop cycle within shorter time periods in the future.

During the initial decades of aquaculture development, a comparatively small number of species has been brought to the market. While these have been selected primarily because of their amenity to scientific and technical capability, their ability to command a price level on the market thought sufficient to cover their R & D costs has also played some part in the decision. With only a limited number of species available to choose from, consumers are more liable to see merit in sampling new species and quite possibly being willing to pay higher price levels as they are launched on to the market. This 'honeymoon period' (Muir and Young, 1999) of higher prices and enhanced profitability is, however, liable to be increasingly short-lived as more species are launched. Each successive new entrant has to compete with a yet wider array of alternative aquatic products at price levels commonly depressed by their earlier cycle of rapid expansion. Given an increased choice, the propensity for prospective consumers to purchase at prices higher relative to alternative options is diminished.

Moreover, as the number of species available from aquaculture increases, it is less likely that each one will command a discrete position in the market. While it seems reasonable to expect that consumers will continue to harbour preferences for particular species, as reflected in the price differentials for wild captured supplies, so too is it likely that the tendency to the commoditization of fish, again evident in the wider market for fish, will embrace farmed supplies too. This increasing difficulty in identifying a unique species proposition to the market tends to focus emphasis on price and thus cost reduction to maintain profitability. Indeed, it may be conjectured that because market prices play some role in R & D decisions about which species to farm and launch, and that these values in turn reflect some engrained contexture of consumer preference for similar attribute combinations, there may be a greater inherent tendency for farmed products to be perceived as more similar, a more uniform commodity, than the natural greater diversity of product available from capture fisheries. If so, this accelerates the need for aquaculture producers to consider alternative routes to differentiation on the market other than simple species diversity.

26.3 The Role of Branding in Differentiation

In the market for fish, differentiation via branding is relatively uncommon. This reflects many of the basic characteristics of fish as a commodity wherein buyers often perceive there to be comparatively little difference between products. Historically, this perceived homogeneity has been promoted through the practice of fish typically being sold without packaging, labelling or other ready means of identifying its origins (Bjorndal *et al.*, 2001). Without the classic communication cues available through labelling and packaging at the point of sale, and supported by advertising and other messages in different media, individual fish products are effectively left to the abilities and whims of individual buyers to determine and decide on relative product attributes.

While buyers in some specialized markets are undoubtedly highly skilled and specific in their particular preferences between and within species, much consumer research suggests that many feel a lacking of knowledge and expertise in the fish purchase and post-purchase activities related to storage and preparation (Marshall, 1988; Brunso, 2003). To some extent, this uncertainty has accounted for the growth in fish being sold in more advanced forms of preparation such as fillets, steaks and other more convenient products. These products provide partial solutions to consumer concerns about the whole format of the natural raw material and reflect added values which have long since been built in to many other food products too, such as poultry. Given that all fish products compete with many other substitutes in the wider market for food, it seems quite logical that these would tend to follow patterns and trends evident elsewhere.

In the broader market for food, consumers have become increasingly concerned about the tendency for food to be intensively produced, with an increasing emphasis on low unit cost rather than other attributes. Undoubtedly, a significant factor in this process has been consumers' logical preference to pay as little as possible for products, coupled with the competitive drivers in the value chain that tend to put added pressure on the need to produce more at lower cost, as has also been noted in aquaculture. However, over the past 25 years, various food markets have been the subject of food scares whereby fundamental causes for concerns have been raised about the integrity and safety of the products on sale. Food scares have extended across a wide range of commodities and other ingredients ranging from salmonella in eggs and poultry, BSE (mad cow disease), dioxins in soft drinks and heavy metals and other contaminants in fish among others (Beardsworth, 1990; Harris and O'Shaughnessy, 1997).

These concerns have prompted many consumers to question the safety of the food being sold and their trust in the production process used to bring it to their table. Such fundamental concerns strike at the core of the relationship between food consumers and those supplying their required daily foods. Increased uncertainty and dissatisfaction with the supply chains that have evolved to provide intensively produced foods have led many to seek alternative sources which are considered to be more natural and safe products that can be traced and verified back to their origin. For many, the answer to these concerns was thought to lie in organic products and this gave rise to the emergence of a rapidly expanding market in the 1990s. Organics appeared to offer a discrete and identifiable point of demarcation in the market which would neatly delineate what consumers were looking for, albeit at some price. Buyers demonstrated their willingness to pay price premiums of 15–25% over conventional foods in the EU (Reithe and Tveterås, 1998), where the total market was estimated at US$4.5 billion in 1997 (Lohr, 1998). A similar, if somewhat smaller, trend was also evident on a wider scale, with global sales of organic food and drink reporting an increase of over 10% to US$23 billion by 2002 (http://www.organicconsumers.org/organic/070603_organic.cfm, Organic Consumers Association, 2003). Production of organic food was also encouraged by governmental incentives in some countries which, combined with organic price premiums, encouraged further supplies.

In the case of fish, there was arguably less background pressure to the organic route during this period of initial expansion. Historically, fish had escaped lightly from consumer concerns, although much more recently there has been a significant growth in the attention paid to the source of fish products and the impact of buying and consuming them. These trends have provoked a number of responses, which have included opportunities for producers to diversify through differentiation of their products. In aquaculture, some producers opted for the potentially problematic task of responding to market signals through a shift into organic production. Appreciation of the challenges that this route invoked requires consideration of both production and market-related phenomena. These are best approached from the production perspective, since organic products must meet certain stipulated criteria before they can be designated as such on the market.

26.4 Organic Differentiation

The foundation of organic production systems generally is rooted in ideological and non-economic rationality (Høgh-Jensen, 1998). At a simplistic level, the essence of organic has been noted as 'the cycling of nutrients and pesticide-free production with an agreed understanding between the producer and the consumer over how product is produced and what consumption will deliver' (Aarset and Young, 2004). However, on further inspection, the literature reveals greater ambiguity of understanding and interpretations of the organic concept (Guthman, 1998). This reflects, in part, the various interpretations of organic status promulgated by different certification organizations and the consequent variation in the standards manifest in the market (Aarset *et al*., 2004). Some indication of the scope for confusion is evident in the range of issues forwarded by the International Federation of Organic Agriculture Movements (IFOAM). This certification organization defines organic farming to: 'include(s) all agricultural systems that promote the environmentally, socially and economically sound production of food and fibres.... Organic agriculture dramatically reduces external inputs by refraining from the use of chemosynthetic fertilisers, pesticides and pharmaceuticals. Instead, it allows the powerful laws of nature to increase both agricultural yields and disease resistance.... Organic agriculture adheres to globally accepted principles, which are implemented within local social-economic, geoclimatical and cultural settings' (www. ifoam.org).

The status of food as 'organic' is determined by a number of different certification organizations, each with their own programmes (e.g. Debio, Norway; Soil Association, UK; Naturland, Germany; Krav, Sweden, and others). These apply criteria which producers must meet in order for their products to be marketed under their particular organic labels. Producers in turn pay a fee to the certification organizations, in part for the inspection process undertaken and maintained and by way of compensation for the benefits of increased price and market recognition and acceptance that should accrue. Certification bodies are not confined to the market of their headquarters; instead, they are free to operate internationally and thus in any market, a number of different organic labels may be present. Differences between certification organizations also exist in the detail of the criteria used to determine organic status.

The ambiguity of what constitutes an organic product is compounded when applied to aquatic food products. There is a fundamental divergence as to whether captured and/or cultured fish can be considered organic. In Europe, unlike North America, wild fish are generally not considered to be eligible for organic status since the inputs to their living environment cannot be controlled. For example, some might argue that wild fish may feed off overexploited stocks and thus cannot be considered to be sustainable; others might contend that the waters they swim in might be polluted, and again this would debar certification. Farming fish, however, affords the opportunity to exercise control over much of the husbandry process in terms of feeds, stocking density, treatment with pharmaceuticals, postharvest treatment, etc., and thus may be deemed eligible for organic status by some. However, the designation of farmed fish as organic is not uniform in its application to all species and some are thus permitted by some certification organizations, while being refuted by another. For example, the launch of 'organic' cod farmed in Shetland has been certified by IFOAM, whereas the main UK organic certification body, the Soil Association, has refused to incorporate cod, *inter alia*, because of the use of artificial light to manipulate the photoperiod (Olsen, 2007).

As might be appreciated, there is almost endless scope for debate over the interpretation of what is, and is not, permissible. A central problem undoubtedly lies in the relative recency of the 'emergence' of aquaculture. For some, fish are still perceived as wild species whose entrapment and enclosure within pens, ponds or cages represents a fundamental violation of the organic principal of natural behaviour. However, for the majority of consumers, the much earlier domestication of animals such as chickens and cattle provides no excuse for exclusion and they can be accepted readily as organic, subject to compliance with relevant criteria. But it is within these additional criteria, and especially when applied to fish, that there are many further contentious points of debate which invariably are resolved through the adoption of technical standards for implementation.

26.5 Market Impacts of Differentiation

However, when consumers are confronted with organic aquaculture products in the marketplace, research shows considerable confusion to prevail. This is perhaps not surprising given that fish, being the most widely traded commodity (Lem, 2007), often ends up in markets far removed from its source. In the case of organic products, the numerous certification bodies which may prevail in different markets can result in consumers coming into contact with a diverse range of labels, marks and other product communications. Consumers' understanding and interpretation of the organic message is mixed and undoubtedly is not helped by the varied specifications to be found between different certification organizations along the supply chain. Variables might include stocking density, feeds and their source, the use of chemicals and pharmaceuticals in the treatment of disease, environmental impact, slaughtering and welfare concerns about fish. In addition, other aspects such as food miles, the carbon footprint of production and distribution along the supply chain to the point of purchase and preparation and debates about 'greener' alternatives figure in the mindset of concerned consumers.

It is unclear as yet to what extent individual consumers pick up on the more subtle undertones behind the labels, whether they are inclined to research these differences and the importance they ascribe to any such variations. It might be expected that this would vary between and within groups over time and familiarity with the products concerned. In the longer term, as certification organizations converge on more uniform standards and product communications, and as consumers become increasingly familiar through greater exposure, it might be expected that some of this dissonance will subside. Until this is done, there is some risk that the perceived standing of organic products may be affected adversely by the different standards present. Variation in the required organic specification risks lowering the perceived standard of all products to the level of the lowest common denominator. Thus buyers, especially those who are new, will tend to form their opinions as to what constitutes organic according to the lowest standard encountered. This could mean that one product of relatively lower, but legitimate, standards could cannibalize sales of other organic products produced to higher specifications. Such variability could well promote ambiguity and heighten concerns about the authenticity of the generic organic product.

At the sectoral level, caution might also be raised about the potentially adverse impact of a differentiation strategy which seeks to promote one component of production as 'superior' to the rest. Inevitably, this promotes questions about the relative shortcomings of the other products. And given that non-organic products are likely to account for the greater proportion of output, any negative concerns raised may have a correspondingly large net negative impact on total revenue. In a global food market increasingly bombarded with adverse communications, such prospects pose a potentially significant threat. Organic producers too are liable to encounter challenges in communicating with the wider market, primarily because of the resources available to such smaller firms and the dispersed but small relative size of their target audience. In these circumstances, it would seem likely that producers will be all the more dependent on their products to deliver customer expectations and countermand any doubts they may encounter along the channel. Given these difficulties and the particular dilemmas that may face producers producing both organic and non-organic products, alternative recourse may be sought in focusing on the potentially less contentious environmentally friendly attributes of the product.

26.6 Ecolabelling

Rather like organic, 'ecolabelling' is not an exact and uniform product specification. Indeed, as noted earlier, it spans a continuum of product attributes which, in one way or another, are intended to communicate positive messages about the environmental/ecological impact of the product, its origins, route to market and consumption implications. Ecolabelled products are intended to communicate a mix of concerns for conservation, the environment, sustainable management, impacts on by-catch, use of critical resources, social impacts, contamination risks and food safety, as have already been noted by some authors (Wessells *et al.*, 1996; Wandel and Bugge, 1997; Wessells, 1998; Norberg and Myrland, 2003). In aquaculture, this again is a relatively recent feature and results from a combination of more specific adverse comments about farmed fish and wider spillover market trends in relation to other foods, including capture fisheries.

While there has been some evidence of concern about the environmental credentials of fish products with promotions such as tuna which avoid dolphin by-catches, the movement has gained much wider prominence over the relatively recent past. A prime factor in this trend of consumer concerns with the consequences of their fish purchases and consumption on the sustainability of fish stocks has been the formation of the Marine Stewardship Council (MSC). Established in 1997 by WWF and Unilever, the world's largest seafood buyer, it began promotion of an ecolabelling scheme for certification of wild-caught seafood products from sources deemed sustainable. Since gaining independence in 1999, the organization has become the most prominent of an expanding band of similar eco-focused organizations, with over 20 fisheries worldwide certified and a similar number in the process of certification, representing over 3 Mt of seafood annually and more than 300 seafood products (MSC, 2007). Further growth is certain given the lead of a number of retailers, especially in the UK and other parts of the EU, to place certified sustainability as a prerequisite for seafood procurement. This endorsement of an ecolabelling programme around the world will extend consumers' awareness and make them potentially more responsive to such generic product differentiation and other ecolabels established in the market.

With ecolabels becoming entrenched and mainstream in the market for fish, some farmed products have placed greater emphasis on their environmental credentials. Indeed, some products have long since established niche positions, often attached to wider quality marks, Scottish salmon's Label Rouge designation in the French market, for example (Young *et al.*, 2006). To date, the MSC has decided not to become involved in aquaculture (Holland, 2008) but, given trends in the contribution of farmed fish, the attached revenue streams and the similarities of much of the supply chain, this strategy could well change. Other organizations such as the Global Aquaculture Alliance (http://www.gaalliance.org/offi.html) have already become heavily involved in communicating a sustainable aquaculture dialogue and this is liable to become more common. This seems all the more probable through market interventions of retailers and other foodservice outlets demanding environmentally friendly attributes as part of their core offerings. Such movements, of course, raise wider issues about what constitutes an appropriate 'green' standard and the socio-economic impact of their adoption and imposition by non-governmental organizations.

The range of species being marketed under ecolabels is expanding and some are now being launched proactively with environmental attributes rather than being introduced retrospectively once products have reached maturity or some other post-launch phase to differentiate the product. Because of its status as a mature product and its long-term presence on the market, farmed salmon probably presents a good species case study. Having matured in the international food-industry complex, salmon has evolved from an industrial structure consisting of many small independent firms to one now dominated by a few transnational companies with production bases in a small number of countries. Around 90% of the international market for salmon is produced from only five countries (Globefish, 2008b). As noted, salmon farming has been the subject of an increasing amount of criticism, initially focused more on environmental degradation through pollution and visual amenity, impacts on wild salmon populations, animal welfare and the use of antibiotics. This has broadened to include the sustainability of a food production system dependent on the exploitation of other fish stocks for feed inputs and possible contaminants in the pelagic species used. Attempts to redress some of these issues, such as using substitute vegetable oils, have been criticized for unnatural diets, alleged poorer product quality and risks of mixing GMO products in the crops used. These concerns were highlighted in the 2004 *Science* case of alleged PCB and dioxin contamination of farmed salmon (Hites *et al.*, 2004), although the evidence of differentials between conventional and organic feeds remains debatable. Whatever the truth of such comments, their existence and circulation alone will tend to cast some further doubts, which may need more formal and accessible clarification. It is already evident that such issues are becoming increasingly important in the market, as illustrated by the decision of Marks & Spencer in the UK to launch an environmentally friendly brand, Loch Muir, based in part on the product being fed 'clean' feed (Cherry, 2007).

26.7 Conclusion

Differentiation of farmed fish along a continuum of ecolabelled products, including organic, has been shown to be increasingly prevalent. On the basis of trends in the relatively recent past, the foundations appear to have been laid for these attributes to remain a more permanent means of appeal to more discrete market segments. The evolution of delivery of these emergent consumer expectations has been somewhat chequered by the comparatively recent status of farmed fish. While this has led some buyers to reject the credibility of fish as organic, there appears to be much less contention about the more central notion of ecolabelled products. Since organic products may be regarded as a more refined and discrete component of the wider spectrum of ecolabelled products, wider acceptance of organic fish may well be anticipated in the future. This is also likely to be accelerated by the existing adoption of the organic concept by an increasing number and range of consumers.

During this interim period, it is also likely there will be ongoing dialogue about the agreed definition of organic fish. And with this debate comes the potentially adverse impact on the wider farmed fish market, as meanings are variously interpreted and customer expectations are only partially fulfilled at least some of the time. In the longer run, it is tempting to suppose that issues pertaining to uniformity, consistency and ambiguity will converge and greater agreement will prevail across the international market. While it is futile to speculate just how long this will take, when it does materialize, it is almost certain that yet further points of differentiation will have been introduced to the market.

References

Aarset, B. and Young, J.A. (2004) Aquatic foods: options for organic growth? *Aquatic Resources, Culture and Development* 1(1), 1–9.

Aarset, B., Beckmann, S., Bigne, J., Beveridge, M.C., Bjorndal, T., Bunting, M.J., McDonagh, P., Mariojouls, C., Muir, J.F., Prothero, A., Reisch, L.A., Smith, A., Tveteras, R. and Young, J.A. (2004) The European consumers' understanding and perceptions of the 'organic' food regime: the case of aquaculture. *British Food Journal* 106(2), 93–105.

Anon. (2004) Study of the market for aquaculture produced seabass and seabream species. Stirling Aquaculture, University of Stirling. Report produced for the European Commission DG XIV (see http://ec.europa.eu/fisheries/publications/studies/aquaculture_market_230404.pdf, accessed 29 April 2008).

Beardsworth, A.D. (1990) Trans-science and moral panics: understanding food scares. *British Food Journal* 92(5), 11–16.

Bjorndal, T. (1990) *The Economics of Salmon Aquaculture*. Blackwell Scientific Publications, Oxford, UK.

Bjorndal, T., Asche, F. and Young, J.A. (2001) Market interactions for aquaculture product. *Aquaculture Economics and Management* 5(5/6), 303–318.

Brunso, K. (2003) Consumer research on fish in Europe. In: Luten, J.B., Oehlenschlager, J. and Olafsdottir, G. (eds) *Quality of Fish from Catch to Consumer. Labelling, Monitoring and Traceability*. Wageningen Academic Publishers, Wageningen, Netherlands, pp. 335–344.

Cherry, D. (2007) M&S, Scottish Sea Farms strike deal for 'tailor-made' salmon. *Intrafish* 7 September (http://www.intrafish.no/global/news/article115193.ece, accessed 29 April 2008).

FAO. (2007) *The State of the World Fisheries and Aquaculture 2006*. Food and Agriculture Organisation of the United Nations, Rome.

Globefish (2007) Pangasius Market Report – November. Food and Agriculture Organisation of the United Nations, Rome (see http://www.globefish.org/index.php?id=4338, accessed 29 April 2008).

Globefish (2008a) Shrimp Monthly Market Report – March. Food and Agriculture Organisation of the United Nations, Rome (see http://www.globefish.org/index.php?id=4430, accessed 29 April 2008).

Globefish (2008b) Salmon Monthly Market Report – February. Food and Agriculture Organisation of the United Nations, Rome (see http://www.globefish.org/index.php?id=4420, accessed 29 April 2008).

Guthman, J. (1998) Regulating meaning, appropriating nature: the codification of California organic agriculture. *Antipode* 30(2), 135–154.

Harris, P. and O'Shaughnessy, N. (1997) BSE and marketing myopia communications: daisy and the death of the sacred cow. *Risk Decision and Policy* 2(1), 29–39.

Hites, R.A., Foran, J.A., Carpenter, D.O., Hamilton, M.C., Knuth, B.A. and Schwager, S.J. (2004) Global assessment of organic contaminants in farmed salmon. *Science* 303(5655), 226–229.

Høgh-Jensen, H. (1998) Systems theory as a scientific approach towards organic farming. *Biological Agriculture and Horticulture* 16(1), 37–52.
Holland, J. (2008) MSC eyes aquaculture. *Intrafish* 14 February (http://www.intrafish.no/global/news/article163169.ece, accessed 29 April 2008).
Lem, A. (2007) Seafood production and international trade: global trends. In: *Marketing of Seafood Products: Trends and Challenges*. Workshop. 12–16 November, Instituto Agronomico Mediterraneo de Zaragoza, Zaragoza, Spain.
Lohr, L. (1998) Implications of organic certification for market structure and trade. *American Journal of Agricultural Economics* 80(5), 1125–1129.
Marshall, D.W. (1988) Behavioural variables influencing the consumption of fish and fish products. In: Thompsen, D.M.H. (ed.) *Food Acceptability*. Elsevier, London, pp. 219–231.
MSC (2007) November (http://www.msc.org/html/content_458.htm, accessed 29 April 2008).
Muir, J.F. and Young, J.A. (1999) Strategic issues in new species development for aquaculture. In: Enne, G. and Greppi, G.F. (eds) *Elsevier Biofutures Series. XXXIII International Symposium on New Species for Mediterranean Aquaculture*. Elsevier, Amsterdam, pp. 85–96.
Norberg, H.M. and Myrland, O. (2003) The influence of collective trademarks in consumers' buying decision process. In: Luten, J.B., Oehlenschlager, J. and Olafsdottir, G. (eds) *Quality of Fish from Catch to Consumer. Labelling, Monitoring and Traceability*. Wageningen Academic Publishers, Wageningen, The Netherlands, pp. 413–428.
Olsen, K.E. (2007) Organic seafood the responsibility of authorities. *Intrafish* 20 February (http://www.intrafish.no/global/news/article128002.ece, accessed 20 February 2007).
Organic Consumers Association (2003) (see http://www.organicconsumers.org/organic/070603_organic.cfm, accessed 29 April 2008).
Reithe, S. and Tveterås, R. (1998) *Demand for Organic Salmon in the European Union*. Report No. 50. Stiftelsen for samfunns-og naeringslivforskning, Bergen, Norway.
Wandel, M. and Bugge, A. (1997) Environmental concern in consumer evaluation of food quality. *Food Quality and Preference* 8(1), 19–26.
Wessells, C. (1998) Assessment of the effectiveness of ecolabelling as a market-based approach to sustainable seafood production. In: *Proceedings of the 9th International Conference of the International Institute of Fisheries Economics and Trade*, 8–11 July 1998, Tromso, Norway. The International Institute for Fisheries Economics and Trade (IIFET), Corvallis, Oregon, pp. 317–321.
Wessells, C., Roheim, K.J. and Anderson, J.G. (1996) Seafood safety perceptions and their effects on consumption choices under varying information treatments. *Agricultural and Resource Economics Review* 25, 12–21.
Young, J.A., Sturrock, H.T. and Muir, J.F. (2006) Seafood certification and ecolabelling: a new wrapping on fisheries resource management? In: *Proceedings of the XIIIth Biennial IIFET Conference: Rebuilding Fisheries in an Uncertain Environment*, 11–14 July, Portsmouth, UK. The International Institute for Fisheries Economics and Trade (IIFET), Corvallis, Oregon.

27 The Future of Aquaculture: Insights from Economic Theory

GUNNAR KNAPP

University of Alaska Anchorage, Institute of Social and Economic Research, Anchorage, Alaska, USA

27.1 Introduction

What is the future of aquaculture? How much fish will be farmed? What prices will farmers receive? How does diversification in aquaculture occur? What effects do government policies, technological change and marketing have on the development of aquaculture?

All of these important questions about the future of aquaculture have to do with the economics of aquaculture: how economic factors interact with political and technological factors to affect the development of aquaculture. Economic theory provides a variety of useful insights into these questions. This chapter discusses how basic economic theory – particularly supply and demand models – may be applied to thinking about the future of aquaculture.

The models we discuss will be familiar to economists, although not necessarily their application to aquaculture. Our primary goal is to introduce these models to non-economists and to discuss some of the most useful insights they suggest about the future of aquaculture.

27.2 Supply and Demand Model

Supply and demand models are basic tools of economics taught in introductory economics courses. Supply and demand models use graphs of supply and demand functions to explain the quantity of a product sold in a market during a given period of time and what price it commands. The supply and demand functions are graphed with quantity on the horizontal axis and price on the vertical axis.[1]

[1] Our discussion here of supply and demand models, which can be deceptively complex, is far from comprehensive. A more detailed discussion of supply and demand models may be found in any introductory economics textbook (for example, Samuelson and Nordhaus, 1985; Gwartney and Stroup, 1990; McConnell and Brue, 2002).

(eds N.R. Le François, M. Jobling, C. Carter and P.U. Blier)

The supply function or 'supply curve' shows that volume sellers would be willing to supply at any given price. This curve is usually depicted as upward sloping, reflecting the fact that as price increases, sellers are willing to supply more.

Figure 27.1a shows a hypothetical supply curve for farmed fish of a given species in a given region during a given period of time. The sellers are fish farmers. For prices below price P^{min} – corresponding to the production cost for the lowest cost farmer – the quantity supplied would be zero. As the price increases, the quantity supplied increases as more farming operations at more remote sites would become profitable. If there is a limit to total production (for example, because the number of sites is limited or the government imposes a production quota), the supply curve will become vertical at the maximum production quantity Q^{max}, indicating that further price increases would not result in greater production.

Note that the shape of the supply curve for farmed fish depends on numerous factors such as the production technology, the costs of inputs such as feed and labour and how the industry is regulated. As these factors change, the shape of the supply curve changes. A shift outwards or downwards in the supply curve (indicating that the quantity supplied at any given price is higher) is referred to as an 'increase in supply'. A shift inwards or upwards in the supply curve (indicating that the quantity supplied at any given price is lower) is referred to as a 'decrease in supply'.

Note also that the shape of the supply curve depends on the period of time for which it represents quantity supplied. In a short period of time, if the price increases, farmers may not be able to increase production very much because of the time required to grow more juveniles and install new cages. Over a longer period of time, a similar price increase may result in a much larger increase in production. Thus, a 'short-run' aquaculture supply curve is likely to be steeper than a 'long-run' aquaculture supply curve.

The demand function or 'demand curve' shows that volume buyers would be willing to purchase at any given price. The demand curve is usually depicted as downward sloping, reflecting the fact that as price decreases, buyers demand more.

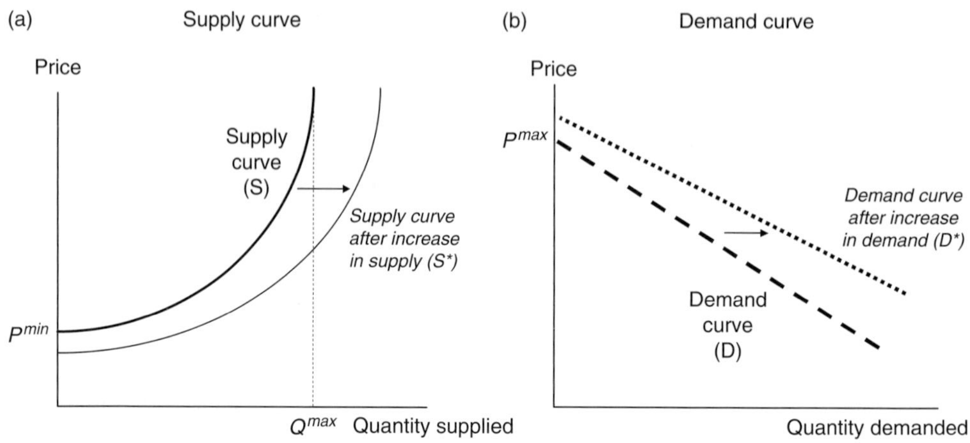

Fig. 27.1. Hypothetical supply and demand curves for farmed fish.

Figure 27.1b shows a hypothetical demand curve for farmed fish of a given species in a given region during a given period of time. The buyers might be (for example) retail stores and restaurants. For prices above P^{max} – the maximum price any buyer is willing to pay – the quantity demanded would be zero. This maximum price might be, for example, the price at which equivalent quality fish could be obtained from a different region, or the price at which consumers would choose to eat a different, lower cost species. As the price decreases, the quantity demanded increases.

Note that the shape of the demand curve for farmed fish depends on numerous factors such as consumer tastes, consumer income and (importantly) the price of the competing products available to buyers (such as other farmed or wild fish species and other proteins such as poultry or beef). As these factors change, the shape of the demand curve changes. A shift outwards or upwards in the demand curve (indicating that the quantity demanded at any given price is higher) is referred to as an 'increase in demand'. A shift inwards or downwards in the demand curve (indicating that the quantity demanded at any given price is lower) is referred to as a 'decrease in demand'.

Note also that the shape of the demand curve depends on the period of time for which it represents quantity demanded. In a short period of time, if the price decreases, demand may not change very much, because consumer tastes change only gradually. Over a longer period of time, a similar price decrease may lead to a much larger increase in quantity demanded. Thus, a 'short-run' aquaculture demand curve is likely to be steeper than a 'long-run' aquaculture demand curve.

As illustrated in Fig. 27.2, for any given supply and demand curves, there is an equilibrium price at which the quantity producers are willing to supply

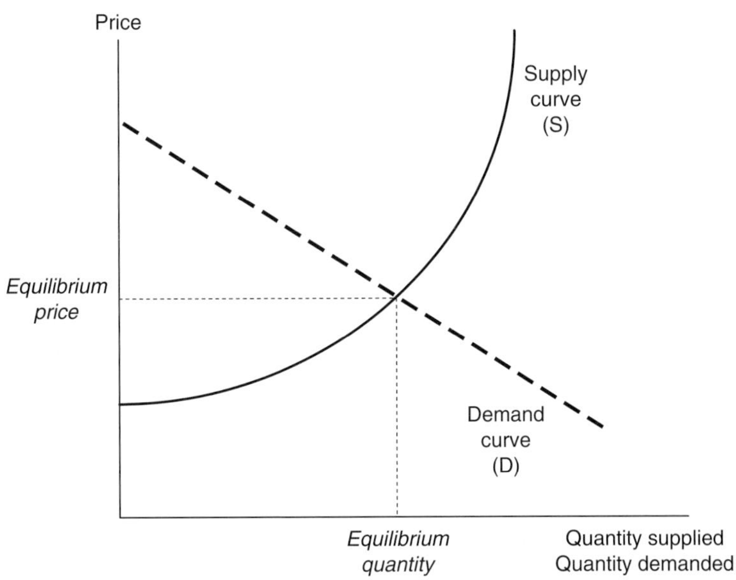

Fig. 27.2. Equilibrium price and quantity.

equals the quantity buyers demand. As prices rise (or fall), producers supply more (or less) and buyers demand less (or more) until the quantity supplied by producers equals the quantity demanded by buyers.

27.3 Applying Supply and Demand Modelling to Aquaculture

The utility of supply and demand modelling is in the insights it provides into the effects of economic factors which *shift* the supply or demand curves, and thus shift the equilibrium price and production. Figures 27.3a–d illustrate the effects of four kinds of shifts in the demand or supply curves on equilibrium aquaculture

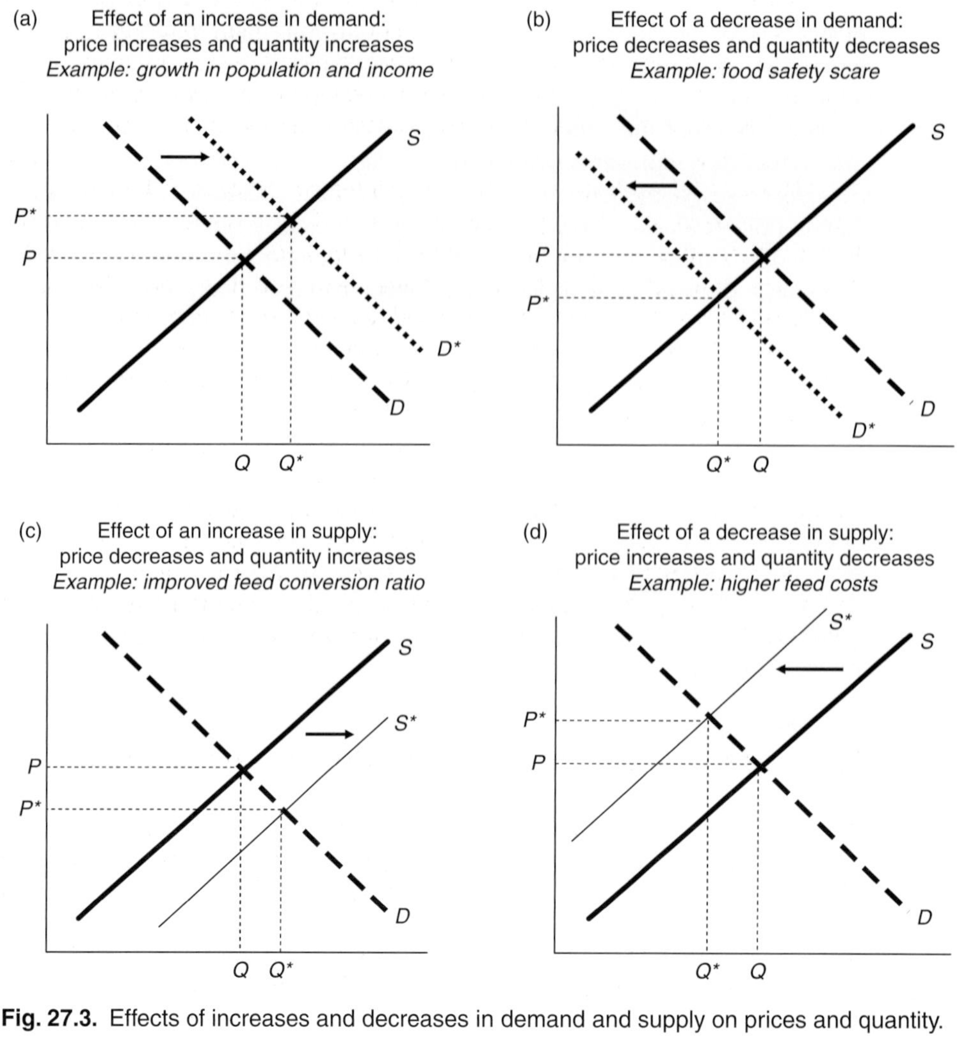

Fig. 27.3. Effects of increases and decreases in demand and supply on prices and quantity.

production quantity and price for a particular species in a particular geographic region. For simplicity, the supply and demand curves are depicted as lines; as discussed above, they are not necessarily linear and may become horizontal or vertical over parts of their range.

Figure 27.3a illustrates the effect of an outward or upward shift in the demand curve, or an *increase in demand*. Growth in population and income is an example of a factor which would cause an increase in demand for fish: at any given price, buyers would demand more fish. The effect of an increase in demand is to increase equilibrium price and quantity.

Figure 27.3b illustrates the effect of an inward or downward shift in the demand curve, or a *decrease in demand*. A food safety scare is an example of a factor which would cause a decrease in demand: at any given price, buyers would demand fewer fish. The effect of a decrease in demand is to decrease equilibrium price and quantity.

Figure 27.3c illustrates the effect of an outward or downward shift in the supply curve, or an *increase in supply*. An improvement in the feed conversion ratio is an example of a factor which would cause an increase in the supply of fish: at any given price, producers would be willing to supply more fish. The effect of an increase in supply is to decrease equilibrium price and increase equilibrium quantity.

Figure 27.3d illustrates the effect of an inward or upward shift in the supply curve, or a *decrease in supply*. An increase in feed costs is an example of a factor which would cause a decrease in the supply of fish: at any given price, producers would be willing to supply fewer fish. The effect of a decrease in supply is to increase equilibrium price and decrease equilibrium quantity.

Figures 27.3a–d are examples of how supply and demand analysis may be used to see how changes in a single factor affecting demand or supply might affect future aquaculture production quantity and price, *assuming nothing else changed*. In reality, many different factors affecting both demand and supply may change simultaneously. In using supply and demand analysis to explain past changes in quantity or price, or to project how quantity and price may change in the future, we need to consider the net shifts in the supply and demand curves resulting from multiple factors which may tend to shift the demand and supply curves in different ways. It is important to distinguish between the independent effects of any single factor and the net effects of multiple factors.

27.4 Modelling How Aquaculture May Change in the Future

Table 27.1 summarizes how a variety of factors may affect supply and demand for aquaculture products in the future. In general, a wide variety of factors are likely to increase demand, while relatively few are likely to reduce demand. Thus, it seems almost certain that demand for most aquaculture products will increase in the future: the demand curves will shift outwards.

In contrast, it is less certain how aquaculture supply may change in the future. As we discuss below, cost-reducing technological advances are likely to

Table 27.1. Potential effects of selected factors on aquaculture production and prices.

Selected factors which may shift supply or demand for farmed fish in the future	Effect on the demand or supply curves	Effects on equilibrium production and prices	
Growing world population Growing world income Marketing by aquaculture producers Reduced supply from wild fisheries Increased awareness of health of seafood Expanded distribution of seafood products	Increase in demand: demand curve shifts out	Higher production	Higher prices
Anti-aquaculture campaigns by aquaculture opponents	Decrease in demand: demand curve shifts in	Lower production	Lower prices
Changes in prices for other proteins	Uncertain effect on demand	Uncertain effect on production	Uncertain effect on prices
Cost-reducing technological advances in aquaculture	Increase in supply: supply curve shifts out	Higher production	Lower prices
Higher feed costs Higher energy costs Increased competition from other uses of farming sites	Decrease in supply: supply curve shifts in	Lower production	Higher prices
Changes in aquaculture regulations Changes in trade restrictions Changes in costs of capital Changes in costs of transportation	Uncertain effect on supply	Uncertain effect on production	Uncertain effect on prices

shift supply curves out. However, increases in costs of feed and energy, as well as increased competition from other uses of farming sites, will tend to shift supply curves in. The net effects on the future supply curves for aquaculture products – whether they are likely to shift outwards or inwards – are uncertain.

Figures 27.4a–d illustrate four potential net effects of future changes in aquaculture supply and demand. In each example, demand increases, as depicted by the shift outwards in the demand curve from D to D*. However, the net future change in equilibrium quantity and price depends on whether supply increases or decreases and how much supply changes relative to the increase in demand.

In Fig. 27.4a and b, supply increases, as depicted by the shift outwards in the supply curve from S to S*. If the increase in supply is large relative to the increase in demand (Fig. 27.4a), the price falls. If the increase in supply is small relative to the increase in demand (Fig. 27.4b), the price increases.

In Fig. 27.4c and d, supply decreases, as depicted by the shift inwards in the supply curve from S to S*. If the decrease in supply is small relative to the increase in demand (Fig. 27.4c), total production increases. If the decrease in supply is large relative to the increase in demand (Fig. 27.4d), total production decreases.

One of the most subtle and confusing aspects of supply and demand analysis is the difference between a change in the supply or demand curves and a change in equilibrium quantity. It is possible for equilibrium quantity to decrease, even if supply or demand increases (if the supply or demand curve shifts outwards).

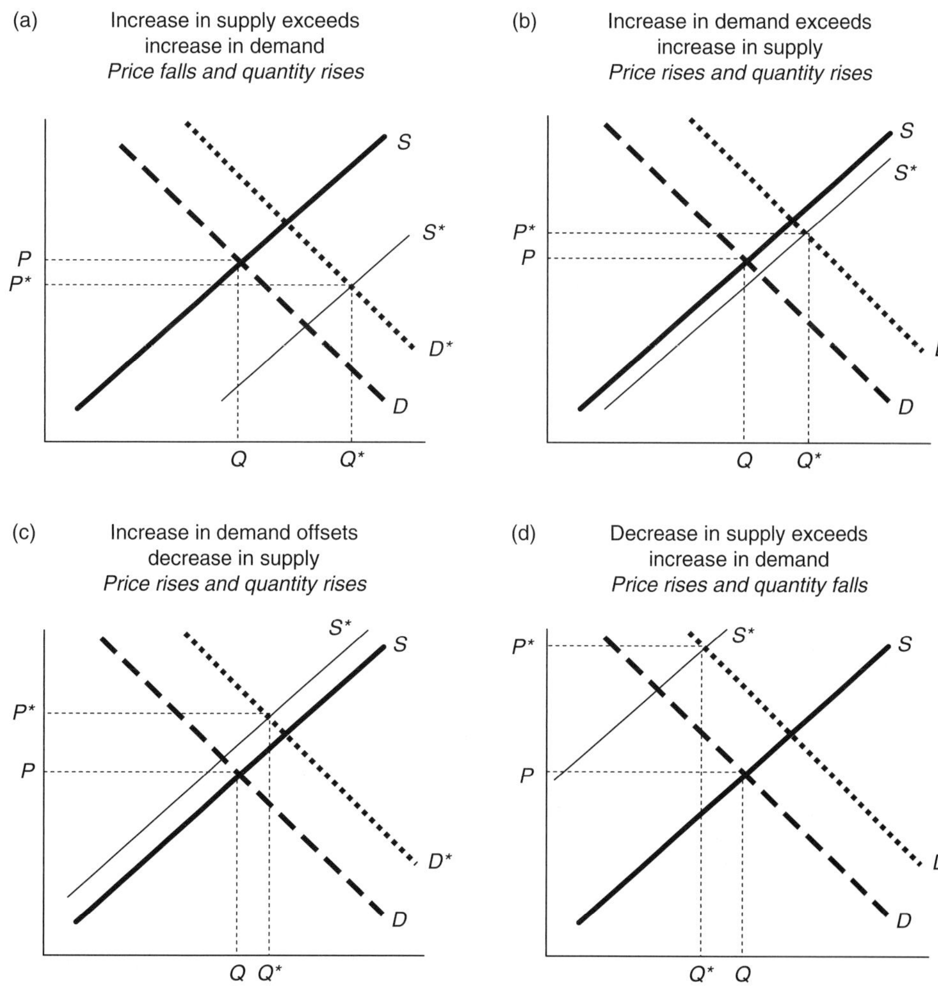

Fig. 27.4 Potential net effects of future changes in aquaculture supply and demand.

For example, in Fig. 27.4d, demand increases: at any given price, buyers are willing to purchase more. Nevertheless, the quantity actually purchased decreases – because suppliers are willing to supply less at any given price.

Even the simple supply and demand analysis illustrated in Fig. 27.4a–d can provide powerful and sometimes counter-intuitive insights. A key insight is that the fact that demand for aquaculture is likely to grow in the future does not necessarily mean that aquaculture production will increase. For example, if the costs of aquaculture production were to rise dramatically (for example, due to higher feed costs), this could drive prices up so much that consumers would consume less and production would fall.

More generally, future prices and production will be driven by changes in *both* demand and supply. We cannot explain past changes or predict future changes in prices and production by thinking only about demand or only about supply.

27.5 Technological Innovation in Aquaculture

Innovation – the development of new technology – is a fundamental driving factor in the growth and diversification of aquaculture. Aquaculture faces technological hurdles which must be overcome at each stage of development of a species from experimental production to large-scale commercial production (Table 27.2). Different technological hurdles become limiting factors to the expansion of aquaculture at different stages of development. A critical initial hurdle is the production of juveniles. Later hurdles for large-scale commercial production include mechanization to reduce labour costs and development of new product forms.

Note that technological hurdles are not limited to challenges encountered in farming fish but also include challenges encountered in transporting, processing and marketing fish. Put differently, to develop an aquaculture industry, it is not sufficient to be able to grow fish. It is also necessary to be able to transport

Table 27.2. Selected technological hurdles in the development of aquaculture for a species.

Stage of development	Hurdle
Experimental production	Juvenile production
Small-scale commercial production	Disease control
	Feed conversion efficiency
Large-scale commercial production	Live fish transportation
	Mechanization
Future opportunities and challenges	Alternative feeds
	Genetic improvement
	Offshore cage construction
	Closed containment systems

the fish to processing plants, process them into product forms that consumers want and transport the products to markets where consumers can buy them – at a cost less than what consumers are willing to pay.

We may model the effects of innovation as an increase in supply: an outward or downward shift in the supply curve. As was illustrated in Fig. 27.3c, the effect of innovation is to increase production and lower the price of a species.

Given the importance of innovation for aquaculture, it is useful to think about how and why innovation occurs. Innovation is driven by a combination of *basic research*, *applied research* and *experience*. Basic research in a wide range of fields such as genetics, immunology, nutrition, electronics and engineering expands the frontier of what is technically possible in aquaculture. Applied research addresses a particular challenge, such as how to grow juveniles, optimize feeding or design cages, utilizing knowledge gained from basic research. Experience in growing, transporting, processing and marketing fish provides opportunities to test and refine innovations.

In theory, both basic and applied research could be publicly or privately funded and conducted by public organizations (such as government laboratories, universities and demonstration farms) or private organizations (such as fish farming companies, feed producers and equipment manufacturers). In practice, the private sector is most likely to invest in applied research from which there is an earlier and greater expected rate of return on investment. How much the private sector invests in research depends on the expected timing, risk and rate of return from research investment – which are in turn affected by government aquaculture policy, economic conditions such as prices and costs and the scale and profitability of the industry.

We may draw several conclusions about the process of innovation. First, industry will invest in applied research to address technological hurdles in the order in which they become important. For example, industry will invest in research on juvenile production of a species before it invests in research on how to increase feed conversion efficiency. Thus, aquaculture innovation will tend to occur when it is needed, but not before it is needed.

Second, innovation will occur in response to changes in economic conditions such as prices and costs. If the price of a particular species rises, the rate of return from investment into research about that species is likely to increase, leading to more research investment and more innovation. If the cost of fish-based feeds increases, we may expect that research into alternative vegetable-based feeds will expand, because the expected cost savings from switching to alternative feeds will grow.

Third, the rate of innovation will increase, at least initially, as an industry grows in scale. The larger the scale of production, the greater the incentive to invest in research which could lower costs or increase prices. The more sites at which production is occurring and the more companies involved in production, the more experience gained annually and the more opportunity for experimenting with and testing innovations.

We may draw several policy conclusions from this brief discussion. First, public funding for basic research may have a greater impact on the rate of innovation in aquaculture than public funding for applied research. Basic research is a 'public good' which may be undersupplied by the private sector because private firms cannot capture as great a share of the benefits from advances in basic science. In contrast, private firms have an incentive to invest in applied research and are likely to have a better understanding than government of what kinds of applied research will provide the greatest return on investment.

Second, as discussed more below, the most effective way in which a government can contribute to innovation may be through efficient leasing and regulatory policies which increase the expected rate of return in aquaculture and the corresponding economic incentive for private industry to invest in research and production. Put differently, private industry will not invest in applied research on farming a species, or gain experience in farming a species, without a reasonable expectation of an opportunity to earn a profit from farming that species. Publicly funded research on aquaculture will not lead to aquaculture innovation and development unless public policy is favourable to aquaculture development.

Third, in regulating aquaculture, it is reasonable to expect that innovation can and will occur to address challenges and opportunities as they arise. Thus, the fact that a technology does not yet exist to address a regulatory goal does not necessarily mean that a particular type of farming should not be allowed. It may be more appropriate to make regulatory approval contingent on development of the needed technology.

The expectation of continuing technological innovation in aquaculture is not blind faith. It is supported by both theory and real-world experience in not only aquaculture but also numerous other industries – ranging from air travel to computing and telecommunications – in which almost unimaginable technological progress has occurred during the professional careers of readers of this chapter.

27.6 The Importance of Marketing for Aquaculture

An important insight provided by supply and demand analysis is the critical importance of marketing for the future development of aquaculture. Broadly defined, marketing includes any efforts by industry or government that have the effect of shifting the demand curve for aquaculture outwards, so that buyers demand more at any given price. Marketing may include, for example, efforts to inform consumers about the health benefits of eating fish; development of new product forms, selling aquaculture products in new geographic locations and types of stores, or simply newspaper or television advertisements for particular products (see Young, Chapter 26, this volume). By shifting the demand curve out, marketing has the effect of increasing both production quantity and price.

As illustrated in Fig. 27.5a and b, marketing is particularly important for those types of aquaculture for which cost-lowering technological innovation increases supply, as depicted by the shift outwards in the supply curve from S to S*. If there is only limited marketing (Fig. 27.5a), the price will fall, limiting the extent to which producers benefit from technological innovation and the extent to which they are able to expand production.

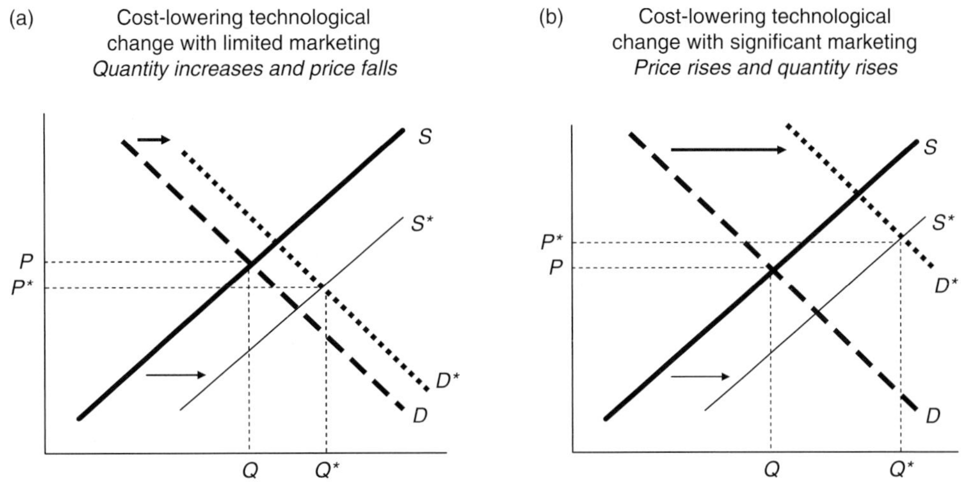

Fig. 27.5. The importance of marketing for aquaculture.

In contrast, if there is significant marketing (Fig. 27.5b), the price will rise. Producers will enjoy higher profits, benefiting from both lower costs and higher prices, and they will be able to expand production much more.

Put differently, future aquaculture production and profitability will be limited not only by what producers are able to produce but by what consumers are willing to buy and the prices they are willing to pay. How much farmers are able to expand demand will determine the extent to which technological improvements increase farmers' profits rather than simply driving down prices.

The importance of marketing is illustrated by the experience of the world salmon farming industry, which has seen a dramatic increase in production over the past two decades (Fig. 27.6). Part of this growth in salmon production is due to cost reductions, which are attributable both to economies of scale (cost savings from larger-scale production) as well as wide-ranging technological innovations such as genetic improvements, better feeding technology and reductions in losses from disease (Tveterås and Heshmati, 2002). Cost reductions have had the effect of shifting the farmed salmon supply curve outwards: for any given price, farmers have been able to supply ever-larger volumes.

However, the growth in world farmed salmon production would not have been possible without a corresponding dramatic growth in world demand, which allowed global consumption to more than double between 1996 and 2006 without a corresponding decline in prices. This growth in demand could not have occurred if salmon farmers had continued to sell only the same products in the same markets – which would have depressed prices below the costs of production. It was possible only because the salmon farming industry developed a wide variety of new products, such as convenient skinless boneless fillets, and introduced salmon to a wide variety of new markets, ranging from the American mid-west to Russia (Knapp *et al.*, 2007).

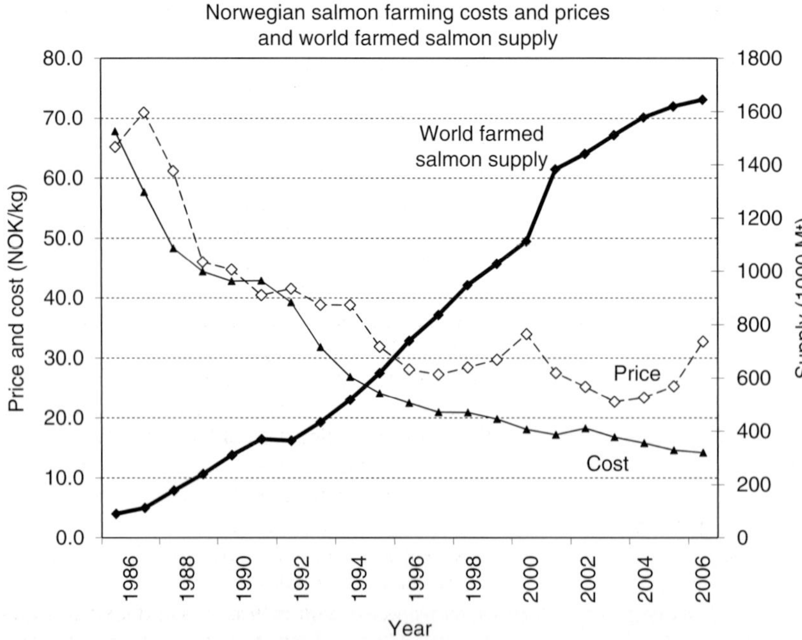

Fig. 27.6. Norwegian salmon farming costs and prices and world salmon supply. Sources: FAO (2008); Norwegian Directorate of Fisheries (2008); Statistics Norway (2007).

27.7 Effects of Government Policies on Aquaculture

We may next use supply and demand modelling to examine the effects of government policies on aquaculture. Clearly, government aquaculture policies are critically important for the development of the aquaculture industry. To see this, we need only contrast farmed salmon production in British Columbia with that in neighbouring south-east Alaska, two regions with very similar environmental conditions. In 2004, farmed salmon production in British Columbia was 56,300t (GSGislason and Associates, 2006), while farmed salmon production in Alaska was zero. The primary reason for this difference is that finfish farming is banned in Alaska (Alaska Statutes 16.40.210).

In the subsequent discussion, we use the term 'policy' to refer to all government policies which affect the aquaculture industry, directly or indirectly. We may group these broadly as leasing policies, regulatory policies and other policies (Table 27.3). The effects of government policies on the aquaculture industry may be modelled in terms of their effects on the aquaculture supply curve. Figure 27.7a depicts four potential supply curves corresponding to four different sets of policies. Curve S^{FEN} depicts a supply curve which results from government policies which are 'favourable', 'efficient' and 'revenue neutral'. We define these terms in the following discussion: basically they refer to policies which maximize aquaculture production without subsidizing it.

Table 27.3. Selected government policies affecting the aquaculture supply curve.

Category	Selected key issues
Leasing policies	Is there a process by which farmers may lease sites? How predictable is the process? How long does it take? How legally secure are sites? How flexible are permitted uses of sites? Can sites be transferred? What do sites cost?
Regulatory policies	What regulations does government impose on farmers? How costly are the regulations? What is the process for developing regulations? How stable and predictable are the regulations? What are the objectives of the regulations? How efficient are the regulations? Could the same objectives be achieved at lower cost?
Other policies	How is aquaculture taxed? What kinds of subsidies are available for the aquaculture industry? To what extent and in what ways does government support aquaculture research, education and marketing? What are trade policies towards farmed fish? What kinds of infrastructure (roads, ports, etc.) does government provide in areas with aquaculture potential?

Curve S^B depicts the supply curve that would result from a government ban on aquaculture, so that no production is possible at any price. Curve S^R depicts a supply curve that might result from restrictive policies that impose significant regulatory costs on farmers (increasing the cost of production for any given quantity compared with curve S^{FEN}) and that limits sites or stocking to a level at which the maximum possible production is quantity Q^R. Curve S^S depicts a supply curve that might result from policies that include government subsidies for aquaculture (reducing the cost of production for any given quantity compared with curve S^{FEN}). As the supply curve shifts out from S^B to S^S, the equilibrium production quantity rises and the equilibrium price falls.

It is useful to distinguish formally between three characteristics of government policy which affect the aquaculture supply curve: objectives, efficiency and subsidization. *Objectives* are what the government is trying to achieve through its aquaculture policies. There may be a trade-off between aquaculture development and other potential objectives of society. As a simple example, suppose fish farms reduce the beauty of the marine environment. We could characterize government policy objectives in terms of the relative government priority for aquaculture and scenery. As depicted in Fig. 27.7b, if aquaculture is a relatively high priority compared to scenery, the government may impose relatively few restrictions on aquaculture, resulting in the 'favourable' supply curve S^F. If aquaculture is a relatively low priority compared to scenery, government policies

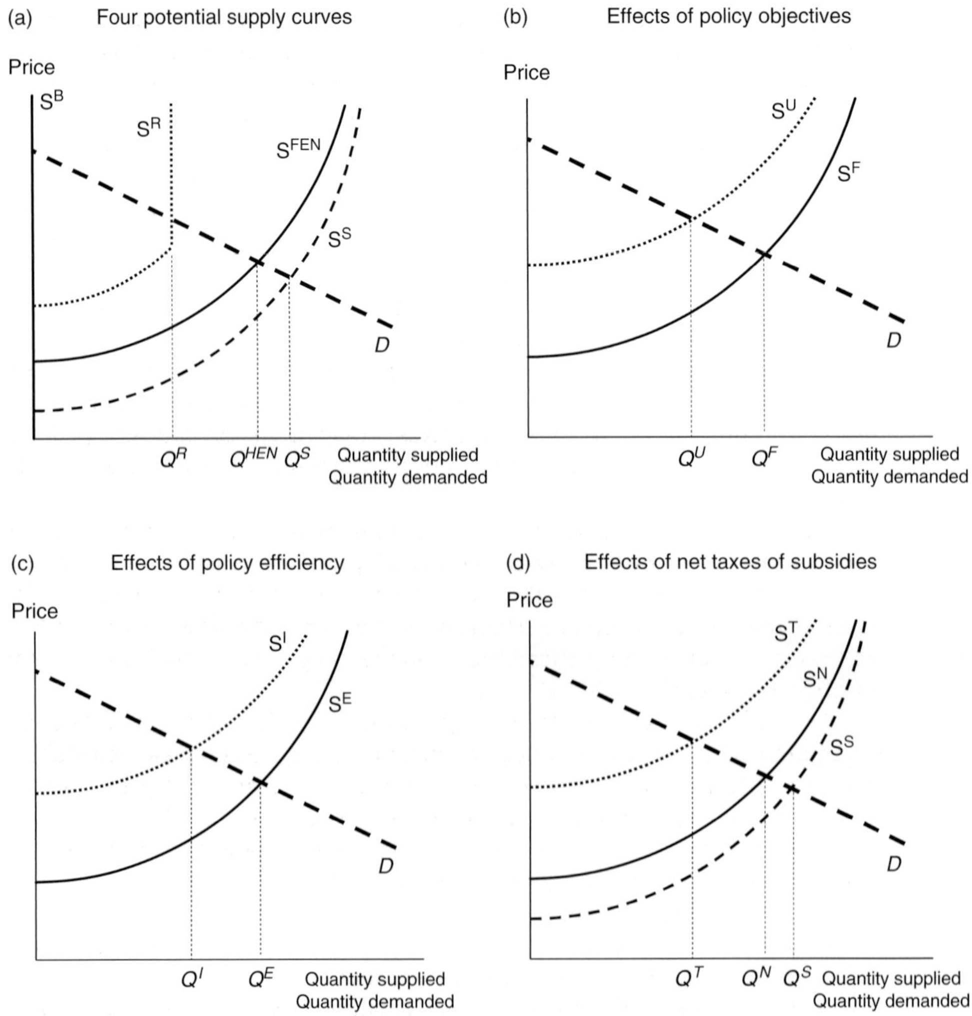

Fig. 27.7. Potential effects of government policies on the aquaculture supply curve.

may impose significant restrictions on aquaculture, resulting in the 'unfavourable' supply curve S^U.

Efficiency refers to the costs that government imposes on aquaculture in order to achieve a given set of policy objectives. For example, the policy objectives might include limiting escapes to a certain level. Requiring the use of a particular type of high-cost cage to achieve this objective would be a less efficient policy than allowing the use of a lower-cost cage which is equally good at preventing escapes. In Fig. 27.7c, supply curve S^E depicts the supply curve which would be most efficient or least expensive for achieving a given set of policy objectives, while supply curve S^I depicts an inefficient or higher-cost potential supply curve for achieving the same set of policy objectives.

Subsidization refers to the extent to which a government subsidizes or taxes aquaculture directly or indirectly, for a given set of policy objectives and a given level of policy efficiency. In Fig. 27.7d, supply curve S^T depicts a supply curve under which the government imposes net taxes on aquaculture, collecting more revenue from the industry than it spends in supporting the industry through services such as research, education, marketing, infrastructure or through direct subsidies. S^N depicts a 'revenue neutral' supply curve in which a government collects the same amount of revenue from aquaculture as it spends in supporting the industry. S^S depicts a supply curve under which a government provides net subsidies to aquaculture, collecting less revenue from aquaculture than it spends in supporting the industry.

All three characteristics of government policy – objectives, efficiency and subsidization – simultaneously determine how policy affects the supply curve. Put differently, how a government affects the aquaculture industry depends on the extent to which government objectives are to promote aquaculture compared with potentially conflicting objectives; how efficient government policy is at achieving its objectives; and the extent to which a government taxes or subsidizes aquaculture.

We may use our supply and demand framework to derive several insights about the implications of government policy for aquaculture development. First, as previously mentioned, government policy is critically important for aquaculture. In the extreme, a government can ban aquaculture entirely. Or, it can constrain aquaculture effectively to any level by limiting site leases and/or imposing regulatory costs and taxes.

Second, the potential effects of government policies on aquaculture are asymmetrical. A government has an unlimited ability to constrain aquaculture but only a limited ability to promote it. Even if a government places a high priority on aquaculture relative to potentially competing objectives and regulates aquaculture efficiently, at best it can only reduce regulatory costs to zero. A government can shift the supply curve out further by subsidizing aquaculture, but the costs of subsidization increase as the scale of production increases. Even the wealthiest countries have a limited ability to subsidize large-scale aquaculture (and it is questionable why they would wish to).

Third, the 'optimal' government aquaculture policy depends on the policy objectives. How a government should regulate aquaculture depends on the relative priority of aquaculture development compared with potentially conflicting objectives. Thus, Alaska's policy towards salmon aquaculture is not necessarily better or worse than British Columbia's. In theory, if the two regions have different policy objectives, each region's policy could be appropriate.

Fourth, regardless of a government's policy objectives, government aquaculture policies should be efficient. The government should consider carefully the costs to industry of aquaculture leasing, regulatory and other policies and whether it is possible to achieve the same objectives at lower cost to industry. An extensive body of economic literature suggests several basic economic principles for efficient aquaculture policy:

- *Policies should be clear and stable.* Regulatory uncertainty – the risk that planned farming investments will not be approved or that regulations may change and impose additional costs and/or delay – adds to the costs perceived by the aquaculture industry.

- *Policies should avoid unnecessary delay*. The longer the time from when an investment is made to when an economic return is realized, the lower the rate of return on the investment. To the extent possible, a government should respond rapidly to applications for leases and operating permits.
- *Site leases should be well defined and transferable*. Leases should be *well defined* so that farmers have a clear understanding of how and for what period of time they will be able to use a site. They should be *transferable* so that they will be operated by the most efficient farmers, who are able and willing to pay the most for the sites.
- *Policies should regulate outcomes rather than inputs*. If the goal of regulation is to achieve a certain outcome (such as maintaining water quality or limiting escapes), to the extent possible a government should allow industry to seek the most cost-effective way to achieve the outcome rather than mandating a particular way of achieving it.

Note that these principles do not require any less commitment to non-aquaculture policy objectives or any higher cost to government – but they can make a big difference to the costs incurred by fish farmers. Put differently, government indifference to regulatory efficiency has the potential to slow the development of aquaculture significantly.

27.8 Modelling Diversification of Aquaculture

Turning finally to the main theme of this book, how can we model diversification of aquaculture – the introduction of new species? If a species is not being produced, this suggests that the supply curve is above the demand curve for all quantities: no farmers are willing to produce any of the species for a price any consumers are willing to pay (Fig. 27.8a).

Production of the species may begin and expand for several different reasons. Technological change may lower the cost of production, shifting the supply curve down and out (Fig. 27.8b). Demand for the species may increase – perhaps because of an increase in the price of other, competing species (Fig. 27.8c).

As discussed above, for production of a species to grow significantly, both supply and demand need to increase (Fig. 27.8d). This will tend to occur once production of a species has become profitable, as both farmers and consumers gain experience with the species. Supply will tend to expand as farmers gain production experience and invest in applied research, leading to cost-lowering technological innovation. Demand will tend to expand as consumers become more familiar with the species and producers invest in marketing. Ultimately, the extent to which these processes continue will limit the extent to which production will grow.

Note that competition from other farmed species will be a major factor limiting diversification in aquaculture. The availability of other species limits the extent to which consumer demand for a new species can increase, particularly if other species are available at lower prices. Similarly, the option to produce other species limits the extent to which producers will wish to supply new

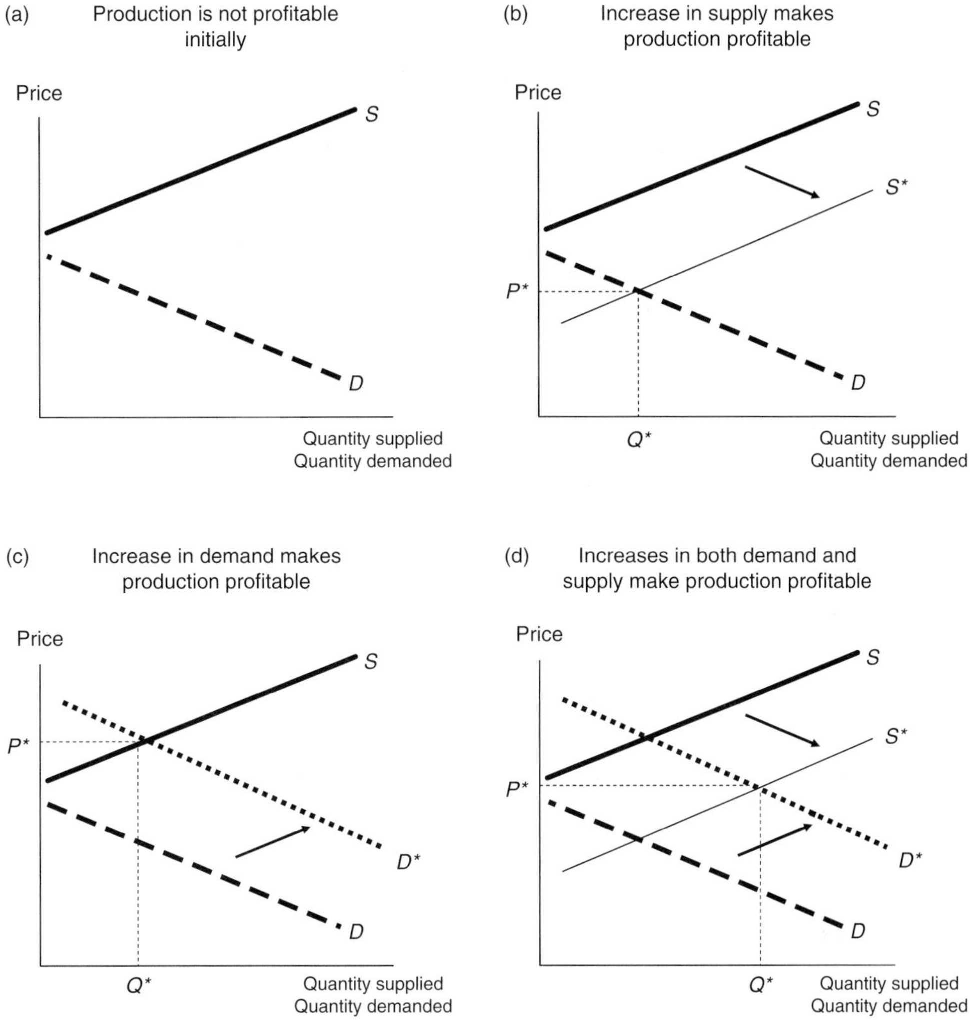

Fig. 27.8. Potential economic mechanisms for the introduction of new farmed species.

species, particularly if other species can be farmed more profitably. Ultimately, over time, the relative volumes of different species which are farmed, and the relative prices at which they are bought and sold, will tend towards an equilibrium at which neither producers nor consumers would benefit from a change in production. However, showing this requires more sophisticated economic models than the basic supply and demand models presented in this chapter.

Note that in the future, aquaculture will not necessarily become more diverse – as measured by the number of species being produced in significant volumes. In the USA, seafood consumption is becoming concentrated on fewer species: the top five species accounted for 75% of consumption in 2004, compared with 56% of consumption two decades earlier (Anderson, 2007).

By way of analogy, consider that of the dozens of animal species which have been domesticated by farmers, only a small number (particularly cows, pigs, sheep, chickens and turkeys) account for most of the world's meat production. Although many other species are farmed in relatively smaller volumes, it is not surprising that consumers have chosen to buy and farmers have chosen to produce a few species which can be raised most easily and cheaply. It is not unreasonable to think that the same ultimate trend may occur in aquaculture, with a few species being produced in very large volumes and other species being produced in much smaller volumes to supply the limited demands of niche markets (Anderson, 2006).

27.9 Conclusions

This chapter has illustrated how economic models may be applied to thinking about the future of aquaculture. Basic techniques of economic modelling, taught in introductory economics courses, provide a variety of useful insights into how and why aquaculture may change in the future. In conclusion, we summarize some of the most important of these insights.

Future changes in production and prices of farmed fish will be driven by changes in both demand and supply. Total world demand for farmed fish is likely to increase in the future, driven by growth in world population and income. Technological advances in aquaculture are likely to increase supply, while increases in the costs of aquaculture inputs such as feed may tend to reduce supply. The extent of future growth in total aquaculture production, and whether farmed fish prices tend to rise or fall, will be driven by the combined effects of all of these factors.

For particular species of fish in particular regions, demand and supply will also be affected by relative changes in demand and supply and how these affect prices for competing species and regions. Marketing will also be a critical factor determining the extent to which cost-reducing technological advances result in lower prices for producers or higher production and profits.

Government policies are critically important for the development of the aquaculture industry and will affect significantly where and what kind of aquaculture development occurs. The potential effects of government policies on aquaculture are asymmetrical: the government has an unlimited ability to constrain aquaculture, but only a limited ability to promote it. Of particular importance to aquaculture is the efficiency of government policies: the costs imposed on industry to achieve a given policy objective.

Innovation – the development of new technology – is a fundamental driving factor in the growth and diversification of aquaculture. Aquaculture innovation will tend to occur when it is needed, but not before it is needed, in response to economic incentives of prices and costs. A government can contribute to innovation through efficient leasing and regulatory policies which increase the expected rate of return in aquaculture and the corresponding economic incentive for private industry to invest in research and production.

Diversification of aquaculture – the introduction of new species – may be driven both by reductions in the costs of production and by expansion of demand. Ultimately, the extent to which these processes continue will limit the extent to which production of a new species will expand. Competition from other farmed species will be a major factor limiting diversification in aquaculture. In the future, aquaculture will not necessarily become more diverse – as measured by the number of species being produced in significant volumes. As has occurred in animal farming, a relatively small number of species – those which can be produced most easily and cost effectively – may be produced in very large volumes, with other species being produced in much smaller volumes.

References

Anderson, J.L. (2006) Personal communication.

Anderson, J.L. (2007) *Aquaculture and Fisheries: Complement or Competition*. Proceedings of Global Trade Conference on Aquaculture, 29–31 May, Qingdao, China. FAO Fisheries Division, Rome.

FAO (Food and Aquaculture Organization of the United Nations) (2008) FishStat Plus database (http://www.fao.org/fishery/statistics/programme/3,1,2/en, accessed 18 November 2008).

GSGislason & Associates Ltd (2006) *The Canadian Farmed Salmon Industry: Benchmark Analysis for the US Market*. Report prepared for Agriculture and Agri-Food Canada, Vancouver, British Columbia.

Gwartney, J.D. and Stroup, R.L. (1990) *Economics: Private and Public Choice*, 5th edn. Harcourt Brace Jovanovich, Orlando, Florida.

Knapp, G., Roheim, C. and Anderson, J. (2007) *The Great Salmon Run: Competition Between Wild and Farmed Salmon*. TRAFFIC North America. World Wildlife Fund, Washington D.C.

McConnell, C.R. and Brue, S.L. (2002) *Economics: Principles, Problems, and Policies*, 15th edn. McGraw-Hill Irwin, Boston, Massachusetts.

Norwegian Directorate of Fisheries (2008) Preliminary Results – 2006: Average Results for Norwegian Fish Farms (www.fiskeridir.no/fiskeridir/english/aquaculture).

Samuelson, P.A. and Nordhaus, W.D. (1985) *Economics*, 12th edn. McGraw-Hill, New York.

Statistics Norway (2007) *Fish Farming 2005* (http://www.ssb.no/nos_fiskeoppdrett, accessed 12 October 2007).

Tveterås, R. and Heshmati, A. (2002) Patterns of productivity growth in the Norwegian salmon farming industry. *International Review of Economics and Business* 49, 367–393.

Species factsheets: Appendix (1–24)

1. Technical sheet – Shortnose and Atlantic sturgeons (Acipenser brevirostrum and A. oxyrhynchus)

Text in *italic* refers only to *A. brevirostrum* and text in **bold** refers only to *A. oxyrhynchus*

GAMETES AND DEVELOPING EGGS	
Egg characteristics	Black to brown, *3.5 mm* and **2.5 mm** diameter and adhesive[1,2,3]
Milt and sperm characteristics	Sperms are comprised of an acrosome, head region, midpiece and single flagellum. The sperm body is *9.71* or **4.1** μm long and the length of the flagellum is 37 μm[4,5,6]
Egg yield (eggs/kg female)	*11,600*[1] eggs or **23,700**[2] eggs/kg fish
Incubation time	*120 degree days;* **73–105 degree days**[7]
Optimal T°C	18–20°C[7,8]
Fertilization success	Variable, usually high over 80%
Overall incubation success	70–100%
LARVAE–JUVENILE STAGE	
Larval size at hatching	*7–11 mm*[9] or **7 mm**[7]
Optimal T°C	17–23°C
Rearing units	Circular tanks
Time until first-feeding	*9–12 dph*[10]; **8–10 at 20°C**
Live feed needed	*Artemia*
Survival from hatching to weaning	Over 70%

continued

ON-GROWING STAGE	
Commercial size	Suggest > *10 kg* and **> 30** kg for caviar production, > 3 kg for meat
Years to reach commercial size	*5 (caviar) and 2 (meat)*, **less than 9 (caviar) and 2 (meat)**
Rearing units	Circular tanks and/or raceways
Rearing density	Up to 180 kg/m^3 (with oxygen injection) for shortnose
Susceptibility to disease	Low
Optimal T°C	20°C
Diet	Salmonid feeds[11]
REPRODUCTION	
Period	*Spring*, **summer**
Fertilization mode	External fertilization
Fertilization protocol	Hormones are used to stimulate ovulation, in pellet[12] or injected. Females usually produce eggs within 24 h. Eggs are expressed into a bowl. Milt is procured by gentle squeezing of the testes, diluted in freshwater and added to the eggs
Time to reach sexual maturity	*5 years*, **9 years**
Optimal T°C	*8°C–13°C*[3,9,10,13,14,15] **13.3–17.8°C**[2]
Control of reproduction	Hormones are used to stimulate ovulation[12]
COMMERCIALIZATION	
Product characteristics	Caviar *3.5 mm* or **2.5 mm**
Flesh and/or egg yield	70% (meat)
Prices	Other similar sturgeons: of up to US$150/ounce of caviar and US$4.80/kg (meat). Wild Atlantic sturgeon caviar from NB is sold at US$50/ounce
Other commercialization avenues	Flesh (males), caviar (females), smoked, sausages, mouse, roulade and jerky

RESEARCH AVENUES

- Development of diets formulated for sturgeon
- Development of techniques to obtain all-female populations or for early gender identification
- Stocking density and rearing temperature
- Development of a predictive relationship between plasma hormone production and egg developmental stage in order to know when to harvest the caviar[16,17,18]

[1] Dadswell, M.J. (1979) Biology and population characteristics of the shortnose sturgeon, Acipenser brevirostrum, LeSeur 1818 (Osteichthyes:Acipenseridae), in the Saint John river estuary, New brunswick, Canada. Canadian Journal of Zoology 57, 2186–2210.

[2] Scott, W.B. and Scott, M.G. (1988) Atlantic fishes of Canada. *Canadian Bulletin of Fisheries and Aquatic Sciences* 219, University of Toronto Press, Toronto, 731 pp.

[3] Kynard, B. (1997) Life history, latitudinal patterns, and status of the shortnose sturgeon, *Acipenser brevirostrum. Environmental Biology of Fishes* 48, 319–334.
[4] Dilauro, M.N., Kaboord, W.S. and Walsh, R.A. (1999) Sperm-cell ultrastructure of North American sturgeons. II. The shortnose sturgeon (*Acipenser brevirostrum*, LeSueur, 1818). *Canadian Journal of Zoology* 77, 321–330.
[5] Dilauro, M.N., Kaboord, W.S. and Walsh, R.A. (2000) Sperm-cell ultrastructure of North American sturgeons. II. The lake sturgeon (*Acipenser fulvescens* Rafinesque, 1817). *Canadian Journal of Zoology* 78, 438–447.
[6] Dilauro, M.N., Kaboor, W., Walsh, R.A., Krise, W.F. and Hendrix, M.A. (1998) Sperm-cell ultrastructure of North American sturgeons. I. The Atlantic sturgeon (*Acipenser oxyrhynchus*). *Canadian Journal of Zoology* 76, 1822–1836.
[7] Smith, T.I.J., Dingley, E.K. and Marchette, D.E. (1980) Induced spawning and culture of Atlantic sturgeon. *Progressive Fish Culturist* 42, 147–151.
[8] Litvak, Matthew, University of New-Brunswick, NB, Canada.
[9] Buckley, J. and Kynard, B. (1981) Spawning and rearing of shortnose sturgeon from the Connecticut River. *Progressive Fish Culturist* 43, 74–76.
[10] NMFS (1998) *Recovery Plan for the Shortnose Sturgeon (*Acipenser brevirostrum). Prepared by the Shortnose Sturgeon Recovery Team for the National Marine Fisheries Service, Silver Spring, Maryland, 104 pp.
[11] Hung, S.S.O., Storebakken, T., Cui, Y., Tain, L. and Einen, O. (1997) High-energy diets for white sturgeon, *Acipenser transmontanus* Richardson. *Aquaculture Nutrition* 3, 281–286.
[12] Flynn, S.R., Matsuoka, M., Reith, M., Martin-Robichaud, D.J. and Benfey, T.J. (2006) Gynogenesis and sex determination in shortnose sturgeon, *Acipenser brevirostrum* Lesuere. *Aquaculture* 253, 721–727.
[13] Buckley J. and Kynard, B. (1985) Yearly movements of shortnose sturgeon in the Connecticut River. *Transactions of the American Fisheries Society* 114, 813–820.
[14] Kieffer, M. and Kynard, B. (1993) Annual movements of shortnose and Atlantic sturgeon in the Merimack River, Massachussets. *Transactions of the American Fisheries Society* 122, 1088–1103.
[15] COSEWIC (2005) COSEWIC assessment and update status report on the shortnose sturgeon *Acipenser brevirostrum* in Canada. Committee on the Status of Endangered Wildlife in Canada. Ottawa (www.sararegistry.gc.ca/status/status_e.cfm, accessed March 2007).
[16] Cuisset, B., Pelissero, C., Le Menn, F., Nunez-Rodriguez, J. (1991) ELISA for Siberian sturgeon (*Acipenser baeri* Brandt) vitellogenin. In: Williot, P. (ed.) *Acipenser. Proceedings of the First International Sturgeon Symposium,* 3–6 October 1989, CEMAGREF, Bordeaux, France, pp. 107–111.
[17] Van Eenennaam, J.P. and Doroshov, S.I. (1998) Effects of age and body size on gonadal development of Atlantic sturgeon. *Journal of Fish Biology* 53, 624–637.
[18] Linares-Casenave, J., Kroll, K.J., Van Eenennaam, J.P. and Doroshov, S.I. (2003) Effect of ovarian stage on plasma vitellogenin and calcium in cultured white sturgeon. *Aquaculture* 221, 645–656.

2. Technical sheet – Wolffish (*Anarhichas minor and A. lupus*)

Text in *italic* and **bold** refers respectively to *A. minor* and *A. lupus* only

GAMETES AND DEVELOPING EGGS	
Egg characteristics	Diam. *5–6 mm*[1] or **3.7–6.5 mm**[2,3] demersal, sticky during the first 5–6 h of incubation[4]
Milt and sperm characteristics	Milt volume: ≈ *0.5–10 ml*, **1–4 ml**[5]; sperm concentrations: *5–1198 × 10[6] spz/ml* or **12–1198 × 10[8] spz/ml**[4]; *sperm motile at stripping, motile up to 48 h at low T°*; **motile many days, max. fertility within 10 h of stripping**[5]
Egg yield (eggs/kg female)	*1500–4000*[6,7,8], **1350–3430**[4]
Incubation time	800–1000 degree days[1]
Optimal T°C	≈ *6°C*[1,6,9], **5°C**[1]
Fertilization success	50–100%[1,6,7], **90–100%**[1,4,7], depends on egg and sperm quality
Overall incubation success	*0–80%*[1,6,10], **60–90%**[4]
LARVAE–JUVENILE STAGE	
Larval size at hatching	20–24 mm, 80–110 mg[9,11]
Optimal T°C	*10.3°C in the first 60 days*[9,12], **6.5°C (first 17 days), 11–14°C (next 4 months)**[4]
Rearing units	Shallow raceways[13] and combi-tank[14]
Time until first-feeding	0[1,2,4,15,16]
Live feed needed	These species readily feed on formulated feed at hatching[1,2,6,9,15]
Survival from hatching to weaning	*60–90%*[1,12], **60–100%**[17,18,19]
ON-GROWING STAGE	
Commercial size	2.5 kg in North America and 5 kg in Europe[20]
Years to reach commercial size	*3 years from hatching*[21], **0.37 kg in 2 years**[22]
Rearing units	Shallow raceways[13,23] sea cages[24,25]
Rearing density	*100% bottom-coverage, i.e. 25–40 kg/m² (50–200 g fish)*[9], *25–40 kg/m² (500 g fish)*[26]; *< 70 kg/m² (1.5–3 kg fish)*[9], **50 kg/m²**[27]
Susceptibility to disease	Very low[9,28]
Optimal T°C	*10°C (0–70 g)*[12]; *8°C (130 g) and 6.6°C (370 g)*[29]; *6°C (700 g)*[30]; *4°C (+4 kg)*[21], **9–11°C**[13]
Diet	Floating diet, 45–62% proteins[9], 15% fat[26], < 20% carbohydrates[31]

continued

REPRODUCTION	
Period	*Ovarian growth May–July, spawning commences during Oct–Dec*[7], **Oct–Feb**[4]
Fertilization mode	Internal fertilization[31,33,34]
Fertilization protocol	Egg (in ovarian fluid) and sperm is incubated for 2–3 h before transfer to seawater[4]
Time to reach sexual maturity	3–4 years under culture conditions (no differences between males and females)[1,4,34], 7–9 years in the wild: (females 1–2 years earlier than males)[35,36,37,38,39]
Optimal T°C	Ovarian growth: 6–8°C; spawning: 4°C[7,38]
Control of reproduction	Spawning can be shifted by photoperiod and temperature manipulation[7,14,38]
COMMERCIALIZATION	
Product characteristics	Rich and tasty fillet[1]; white flesh, good taste and texture, long shelf life and firm texture[40], no parasites or internal bones, 82–83% humidity, 1.3–1.5% fat, 14–15% proteins, 1% ash[41]
Flesh and/or egg yield	Flesh: *41–50%*, **38–45%**[20,22,41]
Price/kg of whole fish, gutted	Norway: US$6.40, Germany: US$7.40, France: US$7.40[40], US$12/kg[20], €8–10/kg[9]
Other commercialization avenues	Leather (www.sealeatherbuttons.com; www.upscaleleather.com; www.oceanleather.com; www.funkifish.com), biomolecules
RESEARCH AVENUES	
– Optimization of feed composition[1,42] – Quality of the sexual products (egg and sperm)[10] – Genetic characterization and selection of the broodstock[9] – Hybridization with *A. lupus*[8] – Scaling-up of egg and larval production[9] – Fish health[9] – Species-specific market research[9,42]	

[1] Falk-Petersen, I.B., Hansen, T.K., Fieler R. and Sunde, L.M. (1999) Cultivation of the spotted wolffish *Anarhichas minor* (Olafsen) – a new candidate for cold-water fish farming. *Aquaculture Research* 30(9), 711–718.

[2] Falk-Petersen, I.-B. and Hansen, T.K. (2001) Organ differentiation in newly hatched common wolffish. *Journal of Fish Biology* 59, 1465–1482.

[3] Halfyard, L.C., Parrish, C.C. and Jauncey, K. (2000) Nutritional requirements of Atlantic wolffish *Anarhichas lupus* L. and *A. minor* Olafsen: two new coldwater aquaculture species. In: Shahidi, F. (ed.) *Seafood in Health and Nutrition.* ScienceTech Publishing Company, St-John's, Canada, pp. 449–476.

[4] Moksness, E. and Pavlov, D.A. (1996) Management by life cycle of wolffish, *Anarhichas lupus* L., a new species for cold-water aquaculture: a technical paper. *Aquaculture Research* 27, 865–883.

[5] Pavlov, D.A., Knudsen, P., Emelyanova, N.G. and Moksness, E. (1997) Spermatozoon ultrastructure and sperm production in wolffish (*Anarhichas lupus*), a species with internal fertilization. *Aquatic Living Resources* 10(3), 187–194.
[6] Hansen, T.K. and Falk-Petersen, I.B. (2001) The influence of rearing temperature on early development and growth of spotted wolffish *Anarhichas minor* (Olafsen). *Aquaculture Research* 32(5), 369–378.
[7] H. Tveiten, Fiskeriforskning, Tromsø, Norway, 2007, personal communication.
[8] N.R. Le François, Université du Québec à Rimouski, Canada, 2007, personal communication.
[9] Foss, A., Imsland, A.K., Falk-Petersen, I.B. and Oiestad, V. (2004) A review of the culture potential of spotted wolffish *Anarhichas minor* Olafsen. *Reviews in Fish Biology and Fisheries* 14(2), 277–294.
[10] Tveiten, H., Jobling, M. and Andreassen, I. (2004) Influence of egg lipids and fatty acids on egg viability, and their utilization during embryonic development of spotted wolf-fish, *Anarhichas minor* Olafsen. *Aquaculture Research* 35(2), 152–161.
[11] McCarthy, I., Moksness, E. and Pavlov, D.A. (1998) The effects of temperature on growth rate and growth efficiency of juvenile common wolffish. *Aquaculture International* 6, 207–218.
[12] Hansen, T.K. and Falk-Petersen, I.B. (2002) Growth and survival of first-feeding spotted wolffish (*Anarhichas minor* Olafsen) at various temperature regimes. *Aquaculture Research* 33(14), 1119–1127.
[13] Oiestad, V. (1999) Shallow raceways as a compact, resource-maximizing farming procedure for marine fish species. *Aquaculture Research* 30(11–12), 831–840.
[14] I. Andreassen, Tomma Marinfisk A/S, Bodo, Norway, 2006, personal communication.
[15] Strand, H.K., Hansen, T.K., Pedersen, A., Falk-Petersen, I.-B. and Oiestad, V. (1995) First feeding of common wolffish on formulated dry feed in a low water-level raceway system. *Aquaculture International* 3, 1–10.
[16] Pavlov, D.A. (1995) Growth of juveniles of White Sea common wolffish, *Anarhichas lupus* L., in captivity. *Aquaculture Research* 26, 195–203.
[17] Ringo, E., Olsen, R.E. and Boee, B. (1987) Initial feeding of wolf fish (*Anarhichas lupus* L.) fry. *Aquaculture* 62, 33–43.
[18] Moksness, E., Gjoesaeter, J., Reinert, A. and Fjallstein, I.S. (1989) Start-feeding and on-growing of wolffish (*Anarhichas lupus*) in the laboratory. *Aquaculture* 77, 221–228.
[19] Wiseman, D.L. and Brown, J.A. (1996) Early growth and survival of larval striped wolffish (*Anarhichas lupus*): a behavioural approach. *Bulletin of the Aquaculture Association of Canada* 96, 12–13.
[20] Laflamme, J., Michaud, J.-C., Lévesque, M. and Le François, N.R. (2004) Potentiel commercial et technico-financier de l'élevage du loup de mer tacheté au Québec. Rapport présenté au MAPAQ, à la SODIM et à DÉC. GRM-UQAR-ADRA, 80 pp.
[21] Falk-Petersen, I.B. and Hansen, T.K. (2003) Early ontogeny of the spotted wolffish (*Anarhichas minor* Olafsen). *Aquaculture Research* 34(12), 1059–1067.
[22] Moksness, E. (1994) Growth rates of the common wolffish, *Anarhichas lupus* L., and spotted wolffish, *A. minor* Olafsen, in captivity. *Aquaculture and Fisheries Management* 25(4), 363–371.
[23] Imsland, A.K., Foss, A., Gunnarsson, S., Sparboe, L., Oiestad, V., Sigurdsson, S. (2007) Culture of juvenile spotted wolffish *Anarhichas minor* in shallow raceways compared to conventional tanks. *Journal of the World Aquaculture Society* 38(1), 154–160.
[24] Mortensen, A., Toften, H. and Aas, K. (2007) Cage culture of spotted wolffish (*Anarhichas minor* Olafsen). *Aquaculture* 264, 475–478.
[25] L.O. Sparboe, Akvaplan-Niva A/S, Tromsø, Norway, 2007, personal communication.
[26] Jonassen, T.M. (2002) Effects of photoperiod, stocking density and diet on growth in young spotted wolffish (*Anarhichas minor* Olafsen). *Aquaculture International* 10(5), 411–420.

[27] Fam, S.D. (1997) Food and feeding requirements of juvenile striped wolffish (*Anarhichas lupus*). MSc thesis, Aquaculture, Memorial University of Newfoundland, St-John's, NF, Canada.
[28] Espelid, S. (2002) Susceptibility of spotted wolffish to infectious diseases and use of immunoprophylaxis. *Bulletin of the Aquaculture Association of Canada* 102(2), 37–38.
[29] Imsland, A.K., Le Francois, N.R., Lamarre, S.G., Ditlecadet, D., Sigurdsson, S. and Foss, A. (2006) Myosin expression levels and enzyme activity in juvenile spotted wolffish (*Anarhichas minor)* muscle: a method for monitoring growth rates. *Canadian Journal of Fisheries and Aquatic Sciences* 63(9), 1959–1967.
[30] Lundamo, I. (1999) Vekst og overlevels hos flekksteinbit (*Anarhichas minor*). Effekter av tempreatur og fotoperiode. MSc thesis, The University of Tromso, Norway.
[31] Stefanussen, D., Lie, O., Moksness, E. and Ugland, K.I. (1993) Growth of juvenile common wolffish (*Anarhichas lupus*) fed practical fish feeds. *Aquaculture* 114, 103–111.
[32] Pavlov, D.A. and Moksness, E. (1996) Repeat sexual maturation of wolffish (*Anarhichas lupus* L.) broodstock. *Aquaculture* 139, 249–263.
[33] Pavlov, D.A. (1994) Fertilization in the wolffish, *Anarhichas lupus*: external or internal? *Journal of Ichthyology* 34, 140–151.
[34] Pavlov, D. and Moksness, E. (1994) Reproductive biology, early ontogeny, and effect of temperature on development in wolffish: comparison with salmon. *Aquaculture International* 2, 133–155.
[35] Barsukov, V.V. (1959) The wolffish (Anarhichadidae). Zool. Inst. Akad. Nauk SSSR, Fauna SSSR vol. 5 (5). In Russian. English translation by Indian National Scientific Documentation Centre, New Delhi, 1972.
[36] Templeman, W. (1986) Some biological aspects of Atlantic wolffish (*Anarhichas lupus*) in the Northwest Atlantic. *Journal of the Northwest Atlantic Fishery Science* 7, 57–65.
[37] Shevelev, M.S. (1988) Ontogenetic stages in spotted wolffish (*Anarhichas minor* Olafsen) from the Barents Sea. International Council for the Exploration of the Sea, C.M. G31.
[38] Tveiten, H. (2002) Management of wolffish broodstock. *Bulletin of the Aquaculture Association of Canada* 102(2), 25–28.
[39] Pavlov, D.A. and Novikov, G.G. (1993) Life-history and peculiarities of common wolffish (*Anarhichas lupus*) in the White Sea. *ICES Journal of Marine Science* 50(3), 271–277.
[40] Richardsen, R. and Johansen, J.A. (2002) Cultured wolffish (*Anarhichas minor*) – a potential species for exclusive restaurants? *Bulletin of the Aquaculture Association of Canada* 102(2), 13–16.
[41] Coulombe, N. and Côté, E. (2001) Évaluation des rendements et évaluation sensorielle du loup de l'Atlantique et tacheté. Ministère de l'Agriculture, des Pêcheries et de l'Alimentation, Direction Générale des Pêches et de l'Aquiculture Commerciales, Direction de l'Innovation et des Technologies. *Rapport d'Analyse* 2001/15.
[42] Le Francois, N.R., Lemieux, H. and Blier, P.U. (2002) Evaluation of the potential of marine and anadromous fish for cold-water mariculture in Québec. *Bulletin of the Aquaculture Association of Canada* 102(2), 9–10.

3. Technical sheet – Meagre (*Argyrosomus regius*)

GAMETES AND DEVELOPING EGGS	
Egg characteristics	Floating; diameter 990 μm; lipid droplet of 250 μm diameter[1]
Milt and sperm characteristics	Milt volume 0.1–1.5 ml per stripping independent on male size Unmotile at stripping, activated by seawater, onward motility up to 1.5 min concentration: 20 10^9 spz/ml[2]
Egg yield (eggs/kg female)	200,000 after hormonal supply[2]
Incubation time	24–48 h[1]
Optimal T°C	19–25°C[1]
Fertilization success	95%[2]
Overall incubation success	
LARVAE–JUVENILE STAGE	
Larval size at hatching	2.2–2.4 mm[1]
Optimal T°C	N/A
Rearing units	N/A
Time until first-feeding	Need for rotifers at 24 h posthatch[1]
Live feed needed	Rotifers and *Artemias*[1]
Survival from hatching to weaning	N/A
ON-GROWING STAGE	
Commercial size	0.8–1.2 kg and > 3 kg for fillets or slices[3]
Years to reach commercial size	2 years and 3 years from hatching
Rearing units	Cages or ponds[3]
Rearing density	1–7 kg/m^3 (3–20 g), then 50 individuals of 0.1–1 kg/m3[3]
Susceptibility to disease	Resistant to bacterial diseases[3], no problem of handling, rapid cicatrization[2]
Optimal T°C	17–21°C
Diet	45–48% protein, 20–24% lipid, extruded pellets[3]
REPRODUCTION	
Period	Natural: late spring on specific zone after migration. Duration: probably sequential spawner but no confirmed data[2]
Fertilization mode	External
Fertilization protocol	Spontaneous mating in tanks; possibility of artificial insemination[2]
Time to reach sexual maturity	Similar between wild and domestic: 4–5 years
Optimal T°C	17–22°C[3]
Control of reproduction	Possibility to shift reproductive season by temperature and photoperiod management. Use of GnRH analogue to programme sharp ovulation peak[2]

continued

COMMERCIALIZATION	
Product characteristics	Appreciated flesh, long shelf life, sold as whole fish (0.6–2 kg), fillet or slices (2–5 kg); can be smoked[3], 2.06–2.93% fat[4]
Flesh and/or egg yield	44–46.5%[4,5]
Price/kg of whole fish, gutted	Europe: US$8/kg for fish 0.6–1 kg; US$9–16/kg for fish > 2 kg[3]
Other commercialization avenues	

RESEARCH AVENUES
– Economic sciences and marketing studies to forecast the market in case of production increase – Improvement of all rearing phases and particularly gametogenesis and genital activity control in order to secure hatchery supply if upscaling – Genetic characterization of stocks in order to prepare possible development of selection programmes

[1] Tixerant, G. (1974) Contribution à l'étude de la biologie du maigre ou courbine *Argyrosomus regius* (Asso) = *Sciaena aquila* (Lacép.) sur la cote mauritanienne. MSc thesis, Université d'Aix-Marseille, France.
[2] Battaglene, S.C. and Talbot, R.B. (1994) Hormone induction and larval rearing of Mulloway, *Argyrosomus hololepidotus* (Pisces: Sciaenidae). *Aquaculture* 126, 73–81.
[3] V. Usache, Palavas, France, 2001, personal communication.
[4] FAO (2006) *Argyrosomus regius* (http://www.fao.org/fi/website/, accessed 1 March 2007).
[5] Poli, B.M., Parisi, G., Zampacavallo, G., Iurzan, F., Mecatti, M., Lupi, P. and Bonelli, A. (2003) Preliminary results on quality and quality changes in reared meagre (*Argyrosomus regius*): body and fillet traits and freshness changes in refrigerated commercial-size fish. *Aquaculture International* 11(3), 301–311.

4. Technical sheet – Milkfish (*Chanos chanos*)

GAMETES AND DEVELOPING EGGS	
Egg characteristics	Diameter 1.10–1.25 mm with positive buoyancy at salinity greater than 30‰[1]
Milt and sperm characteristics	N/A
Egg yield (eggs/kg female)	Multiple spawning; total fecundity from less than a million to six million[2]
Incubation time	At 28°C and 35‰, hatching time is 28 h
Optimal T°C	23–30°C[3]
Fertilization success	Natural spawning and only fertilized eggs were collected
Overall incubation success	Greater than 50%, usually 80% or better[1]
LARVAE–JUVENILE STAGE	
Larval size at hatching	3.7 mm[3]
Optimal T°C	25–30°C
Rearing units	Outdoor commercial hatchery used pond size from 200 m² to 2000 m²[2,4]
Time until first-feeding	2nd day after hatching[5,6]
Live feed needed	Rotifers and *Artemias*
Survival from hatching to weaning	30% or higher[6–7]
ON-GROWING STAGE	
Commercial size	For baitfish market: 160–220 g, for food market: 300–600 g
Years to reach commercial size	7 months from hatching to 600 g[1]
Rearing units	Shallow-water system, deepwater system or sea-cage system
Rearing density	Pending on system from less than 1/m² to 2/m² or 30–60/m³ in cage[1,11]
Susceptibility to diseases	Very low
Optimal T°C	25–30°C[1]
Diet	Benthic algae or/and floating formulated feed
REPRODUCTION	
Period	May to August in Taiwan; year-round in the Philippines and Indonesia[1,8]
Fertilization mode	Natural fertilization
Fertilization protocol	Fertilizer eggs were transferred to hatcheries for incubation at embryonic stage
Time to reach sexual maturity	Fish at the size of 3 kg or more, male of 5-y-o and female of 6-y-o had better chance of mature in captivity[1]
Optimal T°C	27°C and higher[1]
Control of reproduction	Maturation process can be accelerated by hormonal control. Photoperiod and temperature were not used to adjust spawning season[9,10]

continued

COMMERCIALIZATION	
Product characteristics	Fish were sold as fresh or frozen form. Value-added products were processed to overcome the disadvantage as bony fish
Flesh and/or egg yield	Flesh: 65%
Price/kg of whole fish, gutted	About US$$2/kg for whole fish
Other commercialization avenues	N/A
RESEARCH AVENUES	
– Expansion of market – Stability of market price – Lower production cost – Transfer of milkfish culture technology to food shortage country	

[1] Lee, C.-S. (1995) Aquaculture of milkfish (*Chanos chanos).* TML Aquaculture Series No. 1. Tungkang Marine Laboratory, Taiwan Fisheries Research Institute, Taiwan, 141 pp.
[2] Bagarinao, T.U. (1991) Biology of milkfish *Chanos chanos* (Forsskal). Aquaculture Department, South-east Asian Fisheries Development Center, Philippines, 94 pp.
[3] Walsh, W.A., Swanson, C. and Lee. C.-S. (1991) Effects of development, temperature, and salinity on metabolism in eggs and yolk-sac larvae of milkfish, *Chanos chanos* (Forsskal*). Journal of Fish Biology* 39, 115–125.
[4] Chang, S.L., Su, M.S. and Liao, I.C. (1993) Milkfish fry production in Taiwan. In: Lee, C.-S., Su, M.S. and Liao, I.C. (eds) *Finfish Hatchery in Asia: Proceedings of Finfish Hatchery in Asia '91*, Tungkang Marine Laboratory, Tungkang, Taiwan, pp. 157–171.
[5] Liao, I.C. and Chen, T.I. (1979) Report on the induced maturation and ovulation of milkfish (*Chanos chanos*) reared in tanks. *Proceedings of the World Mariculture Society* 10, 317–331.
[6] Eda, H., Murashige, R., Eastham, B., Wallace, L., Bass, P., Tamaru, C.S. and Lee, C.S. (1990) Survival and growth of milkfish (*Chanos chanos*) in the hatchery. I. Feeding. *Aquaculture* 89, 233–244.
[7] Lee, C.-S., Leung, P.-S. and Su, M.-S. (1997) Bioeconomic evaluation of different fry production systems for milkfish (*Chanos chanos*). *Aquaculture* 155, 367–376.
[8] Liao, I.C. and Chen, T.I. (1984) Gonadal development and induced breeding of captive milkfish in Taiwan. In: Juario, J.V., Ferrari, R.P. and Benitez, L.V. (eds) *Advances in Milkfish Biology and Culture.* Island Publishing House Inc., Manila, Philippines, pp. 41–52.
[9] Lee, C.-S., Tamaru, C.S., Banno, J.E. and Kelley, C.D. (1986) Influence of chronic administration of LHRH- analogue and/or 17α-methyltestosterone on maturation in milkfish, *Chanos chanos. Aquaculture* 59, 147–159.
[10] Lee, C.-S., Tamaru, C.S. and Weber, G.M. (1987) Studies on the maturation and spawning of milkfish, *Chanos chanos*, in a photoperiod-controlled room. *Journal of the World Aquaculture Society* 18(4), 253–259.
[11] Marte, C.L., Cruz, P. and Flores, E.E.C. (2000) Recent developments in freshwater and marine cage aquaculture in the Philippines. In: Liao I.C. and Lin, C.K. (eds) *Cage Aquaculture in Asia.* Asian Fisheries Society, Manila, Philippines and World Aquaculture Society-South-east Asian Chapter, Bangkok, pp. 83–96.

5. Technical sheet – European whitefish (*Coregonus lavaretus*)

GAMETES AND DEVELOPING EGGS	
Egg characteristics	Diameter 2–3.5 mm[1,2]
Milt and sperm characteristics	N/A
Egg yield (eggs/kg female)	Gonadosomatic index 15–24[2,3]
Incubation time	256–313 degree days from 3 to 7°C[4,5]
Optimal T°C	3–7°C[5]
Fertilization success	N/A
Overall incubation success	Variable, commercial producers work with estimates 30–80%[6]
LARVAE–JUVENILE STAGE	
Larval size at hatching	12–14 mm[1,2,7]
Optimal T°C	16–20°C[8,9]
Rearing units	Intensive; plastic tanks. Extensive; natural food ponds and to some extent in illuminated net cages[2,9,10,11]
Time until first-feeding	3–5 days after hatching (dry diets)[12]
Live feed needed	None
Survival from hatching to weaning	Variable, commercial producers work with estimates of 60–80%[6]
ON-GROWING STAGE	
Commercial size	0.8–1.0 kg in Scandinavia[13]
Years to reach commercial size	2–3 years from hatching in Scandinavian temperature conditions[2]
Rearing units	Earth ponds, plastic tanks and sea cages (brackish water)[2]
Rearing density	Commercial producers work with estimates of 20–30 kg/m³[3]
Susceptibility to disease	Quite low[2,14,15]
Optimal T°C	17–19°C[2]
Diet	Dry sinking diet, 34–40% protein[16,17], in practical diets fat level is around 25%

continued

REPRODUCTION	
Period	Ovarian growth is initiated during August–September. Spawning commences during November–December. Timing is influenced by prevailing temperature[2,3]
Fertilization mode	External fertilization[2]
Fertilization protocol	Eggs (in ovarian fluid) and sperm are mixed and a small amount of fresh water is then added. After some time, the eggs are washed in fresh water to remove excess milt and transferred to incubation units[2]
Time to reach sexual maturity	3–4 years in culture conditions[2]
Optimal T°C	N/A
Control of reproduction	N/A
COMMERCIALIZATION	
Product characteristics	White flesh with mild taste. Fat: muscle 8.0–9.0% and carcass 10.5–12.7%[18,19]
Flesh and/or egg yield	60%[18]
Price/kg of whole fish, gutted	Farmed fish; US$6.70–8.00/kg (in Finland)[13]
Other commercialization avenues	Eggs
RESEARCH AVENUES	
– Selective breeding programme – Alternative dietary raw materials – Improving the market quality of production	

[1] Rojas-Beltran, R., Champigneulle, A., Gillet, D. and Le Rouille, N. (1992) Influence of egg source, initial characteristics, and rearing conditions on the growth, survival, and development of *Coregonus lavaretus* L. larvae initially fed a dry diet. *Polish Archives of Hydrobiology* 39, 683–691.

[2] Koskela, J., Määttä, V., Vielma, J., Rahkonen, R., Forsman, L., Setälä, J. and Honkanen, A. (2002*) Siian kasvatus ruokakalaksi (Table fish farming of whitefish).* Finnish Game and Fisheries Research Institute, F.G. Lönnberg, Helsinki (in Finnish).

[3] Rösch, R. (2000) Gonadosomatic index (GSI) of female whitefish (*Coregonus lavaretus*) in Lake Constance. *Limnologica* 30(2), 193–196.

[4] Braum, E. (1964) Experimentelle untersuchungen zur ersten nahrungsaufnahme und biologie an jungfischen von Blaufelchen (*Coregonus wartmanni* Bloch), Weissfelchen (*Coregonus fera* Jurine) und Hechten (*Esox lucius* L.) *Archiv fur Hydrobiologie* 28, 183–244.

[5] Eckmann, R. (1987) A comparative study on the temperature dependence of embryogenesis in three coregonids (*Coregonus* spp.) from Lake Constance. *Schweizerische Zeitschrift fuer Hydrologie* 49(3), 353–362.

[6] Mr Lankinen, Savon Taimen, Rautalampi commune, Finland, 2007, personal communication.

[7] Kolman, K. and Luczynski, M. (1986) Survival and growth rates of whitefish (*Coregonus lavaretus* L.) larvae after experimentally delayed hatching. *Aquaculture* 51, 207–210.

[8] Koskela, J. and Eskelinen, U. (1992) Growth of larval European whitefish (*Coregonus lavaretus*) at different temperatures. *Polish Archives of Hydrobiology* 39, 677–682.

[9] Rösch, R. (1995) Rearing of coregonid (*Coregonus* sp.) larvae in tanks: a review. *Advances in Limnology* 46, 293–300.

[10] Mamcarz, A. (1995) Rearing of coregonid (*Coregonus* sp.) larvae in illuminated cages: a review. *Advances in Limnology* 46, 287–292.

[11] Anon. (2006) Vesiviljely 2005. (Aquaculture 2005) Riista- ja kalatalouden tutkimuslaitos. SVT Maa-, metsä- ja kalatalous, Helsinki, 16 pp. (in Finnish, Swedish and English).

[12] Rösch, R. (1992) Food intake and growth of larvae of *Coregonus lavaretus*: effect of diet and light. *Polish Archives of Hydrobiology* 39, 671–676.

[13] Kankainen, M., Setälä, J. and Kause, A. (2007) Kasvatetun siian tuoteominaisuuksien taloudellisen arvon laskenta (The economical value of different product traits in farmed whitefish). *Kalaraportteja* 414 (in Finnish).

[14] Lönnström, L.-G., Rahkonen, R., Lunden, T., Pasternack, M., Koskela, J. and Gröndahl, A. (2001) Protection, immune response and side-effects in European whitefish (*Coregonus lavaretus* L.) vaccinated against vibriosis and furunculosis. *Aquaculture* 200, 271–284.

[15] Rahkonen, R., Vennerström, P., Rintamäki-Kinnunen, P. and Kannel, R. (2000) *Terve kala. Tautien ennaltaehkäisy, tunnistus ja hoito. (Healthy fish, Prevention, identification and treatment of disease).* Finnish Game and Fisheries Research Institute, Nykypaino, Helsinki, 140 pp. (in Finnish).

[16] Ruohonen, K., Koskela, J., Vielma, J. and Kettunen, J. (2003) Optimal diet composition for European whitefish (*Coregonus lavaretus*): analysis of growth and nutrient utilisation in mixture model trials. *Aquaculture* 225, 27–39.

[17] Dabrowski, K., Czesny, S. and Matusiewicz, M. (2002) Coregonids. In: Webster, C.D. and Lim, C.E. (eds) *Nutrient Requirements and Feeding of Finfish for Aquaculture.* CAB International, Wallingford, UK, pp. 230–244.

[18] Koskela, J., Rahkonen, R., Forsman, L., Norrdahl, O. and Lönnström, L.G. (2001) Siika ruokakalaviljelyssä – kahden siikakannan ja kantaristeymän vertailu. Abstract: Whitefish in aquaculture – comparison of two stocks and their hybrids. Finnish Game and Fisheries Research Institute, *Kalantutkimuksia – Fiskundersökningar* 171, 24 pp.

[19] Koskela, J., Jobling, M. and Savolainen, R. (1998) Influence of dietary fat level on feed intake, growth and fat deposition in the whitefish *Coregonus lavaretus. Aquaculture International* 6, 95–102.

6. Technical sheet – European seabass (*Dicentrarchus labrax*)

GAMETES AND DEVELOPING EGGS	
Egg characteristics	Diameter 1.1–1.2 mm, pelagic. Spherical and transparent, they present generally one oil droplet of 400–440 μm diameter. Buoyant at salinity > 37‰
Milt and sperm characteristics	Up to 60 × 10^9 spz/ml decreasing significantly at the end of the reproductive season. Motility at activation ≈ 100%. Duration of sperm motility ≈ 40 s[1,2,3]
Egg yield (eggs/kg female)	≈ 200,000–300,000[4]
Incubation time	Hatching starts approximately 72 h after fertilization at 13–14°C
Optimal T°C	13–14°C
Fertilization success	90–95%
Overall incubation success	75–85%
LARVAE–JUVENILE STAGE	
Larval size at hatching	4–4.5 mm
Optimal T°C	16–22°C
Rearing units	Round fibreglass tanks of 6–10 m^3 in a semi-closed or flow-through water system[4,5]
Time until first-feeding	Commonly 3–5 days posthatching. May be increased to 10–12 by keeping the larvae in complete darkness and at temperature of 16°C[4,5]
Live feed needed	Rotifers and *Artemias*, partial replacement of live preys with artificial microparticulate diets under test
Survival from hatching to weaning	40 ± 10% (end of weaning)
ON-GROWING STAGE	
Commercial size	Commonly 450 g. Other sizes: 200–300 g (small) 600–800 g (large) depending on origin and time of the year
Years to reach commercial size	2–3 years
Rearing units	Sea cages, raceways or ponds
Rearing density	20–30 kg/m^3 (cages) to 40–50 kg/m^3 (raceways)
Susceptibility to diseases	Virus (nodavirus), bacteria (*Photobacterium damselae* and *Tenacibaculum maritimum*, formerly *Flexibacter maritimus*)[4,6]
Optimal T°C	22–25°C
Diet	Pressed or extruded: protein (47–52%), fat (11–18%)

continued

REPRODUCTION	
Period	January–February (Mediterranean Sea), April–May (North-east Atlantic Ocean), May–June (North Atlantic Ocean)
Fertilization mode	External fertilization
Fertilization protocol	Optimal sperm/egg ratio: 70,000 spz/egg
Time to reach sexual maturity	Males reach maturity at 1–2 years (50–300 g), females at 3 years (> 700 g)
Optimal T°C	Gametogenesis: with decreasing temperatures from 17°C to 13°C; spawning: 13°C (Mediterranean Sea), 10–11°C (Atlantic Ocean)
Control of reproduction	Spawning period can be shifted by photoperiod and temperature manipulation
COMMERCIALIZATION	
Product characteristics	White, firm and tasty flesh. Good shelf life, 19.8% proteins, total lipids 10.6%, 1.2% ash, 68.3% moisture[7]
Flesh and/or egg yield	Fillet yield: 35–40%[8]
Price/kg of whole fish, gutted	Commonly traded as whole fresh fish. Spain, Greece, Italy and France: €6–8/kg with occasional market lows resulting from increased supplies
Other commercialization avenues	Development of an EU organic label still pending. Organic certification issued by 'Agriculture Biologique' (AB) currently recognized by the EU
RESEARCH AVENUES	
– Fish nutrition and immunology – Fish welfare and product quality – Selective breeding, sterilization and hybridization (intra-specific)	

[1] Fauvel, C., Savoye, O., Dreanno, C., Cosson, J. and Suquet M. (1999) Characteristics of sperm of captive seabass in relation to its fertilization potential. Journal of Fish Biology 54(2), 356–369.

[2] Zohar, Y., Billard, R. and Weil, C. (1984) Reproduction of the sea bream and the sea bass: sexual cycle and induction of spawning. In: Barnabé, G. and Billard R. (eds) *L'aquaculture du bar et des sparidés*. INRA, Paris, pp. 3–24.

[3] Sorbera, L.A., Mylonas, C.C., Zanuy, S., Carillo, M. and Zohar, Y. (1996) Sustained administration of GnRHa increases milt volume without altering sperm counts in the sea bass. *Journal of Experimental Zoology* 276, 361–368.

[4] Moretti, A., Pedini Fernandez-Criado, M., Cittolin, G. and Guidastri, R. (1999) *Manual on Hatchery Production of Seabass and Gilthead Seabream*, Vol. 1. Food and Agriculture Organization of the United Nations, Rome.

[5] Moretti, A., Pedini Fernandez-Criado, M., and Vetillart, R. (2005) *Manual on Hatchery Production of Seabass and Gilthead Seabream*, Vol. 2. Food and Agriculture Organization of the United Nations, Rome.

[6] Breuil, G., Bonami, J.R., Pepin, J.F. and Pichot, Y. (1991) Viral infection (picorna-like virus) associated with mass mortalities in hatchery-reared sea bass (*Dicentrarchus labrax*) larvae and juveniles. *Aquaculture* 97, 109–116.
[7] Orban, E., Di Lena, G., Nevigato, T., Casini, I., Santaroni, G., Marzetti, A. and Caproni, R. (2002) Quality characteristics of sea bass intensively reared and from lagoon as affected by growth conditions and the aquatic environment. *Journal of Food Science* 67(2), 542–546.
[8] Peruzzi, S., Chatain, B., Saillant, E., Haffray, P., Menu, B. and Falguiere, J.-C. (2004) Production of meiotic gynogenetic and triploid sea bass, *Dicentrarchus labrax* L.: 1. Performances, maturation and carcass quality. *Aquaculture* 230, 41–64.

7. Technical sheet – Atlantic cod (*Gadus morhua*)

GAMETES AND DEVELOPING EGGS	
Egg characteristics	Pelagic eggs, diameter 1.2–1.6 mm[1,2]
Milt and sperm characteristics	Milt 12–17% of whole weight.[4] Sperm concentration: 7–20 × 10^9 spz/ml.[3] Sperm motility: mean speed 150 μm/s.[32] Successful cryopreservation.[4] Commonly, 1–5 ml of sperm used to fertilize 100 ml of ova[5]
Egg yield (eggs/kg female)	300,000–900,000[6]
Incubation time	80–100 degree days at 7–8°C[7,8]
Optimal T°C	7–8°C[8]
Fertilization success	20–50% when stripped, 90–100% when eggs collected from the tank[5]
Overall incubation success	10–30% when stripped[5]
LARVAE–JUVENILE STAGE	
Larval size at hatching	4–5 mm, 0.04–0.07 mg dry wt[9,10,11] (dry weight ≈ 10% of wet weight)
Optimal T°C	Increasing from 9.7 to13.4°C with body size from 0.07 to 0.25 mg dry weight[12]
Rearing units	Dark, circular tanks, flat bottom, slow-rotating bottom scraper, adjustable light level, subsurface feeders (intensive production)[8]
Time until first-feeding	3–5 days[10]
Live feed needed	Enriched rotifers and *Artemias*[13,14]
Survival from hatching to weaning	10–20% (in intensive production)[8]
ON-GROWING STAGE	
Commercial size	3–5 kg[15]
Years to reach commercial size	2–3 years from hatching[15,16]
Rearing units	Sea-cages[17]
Rearing density	Optimal 20–25 kg/m^3 in sea cages,[18] 40–50 kg/m^3 in land-based tanks[19]
Susceptibility to disease	Susceptible to atypical furunculosis and vibriosis under stressful conditions[20]
Optimal T°C	Optimal T° decreases with weight (w) of fish ($T_{opt.G} = 15.57 - 0.8426 \ln(w)$)[16]
Diet	Slow sinking feed, 50–60% proteins, 10–20% fat[21]

continued

REPRODUCTION	
Period	In nature, ovarian growth: October–November, spawning: March–April[2]
Fertilization mode	External ferilization[22]
Fertilization protocol	Sperm diluted with seawater (1:100) and mixed with ripe eggs for 10 min before transfer to seawater[5,8]
Time to reach sexual maturity	Normally 2 years in captivity[17]
Optimal T°C	Ovarian growth: 6–10°C; spawning: 7–8°C[8]
Control of reproduction	Spawning time can be shifted by photoperiod manipulation[23]
COMMERCIALIZATION	
Product characteristics	White, lean fillet; 79–88% water, 1% fat, 10–19% protein[24]
Flesh and/or egg yield	40–50% of whole weight (50–60% of gutted weight)[25]
Price/kg of whole fish, gutted	US$2.5–3.0/kg whole fish (US$3–4/kg of whole gutted fish)[15,26,27]**
Other commercialization avenues	Gonads, liver, head, etc., amount to 10% of total revenue of fish[28]

RESEARCH AVENUES

- Selective breeding[29]
- Intensive larval production[8,30]
- Growth rate (temperature and weight,[16] water quality and density,[19] photoperiod[31])
- Feed composition[32]
- Fish health, vaccines[20]

[1] Kjesbu, O.S. and Norberg, B. (2005) Stamfisk og gyting (Brood stock and spawning). In: Otterå, H., Taranger, G.L. and Borthen, J. (eds) *Oppdrett av torsk (Farming of Cod).* Norsk Fiskeoppdrett AS, Bergen, pp. 39–47.

[2] Brander, K. (ed.) (2005) Spawning and life history information for North Atlantic cod stocks. ICES Cooperative Research Report 274.

[3] Rakitin, A., Ferguson, M.M. and Trippel, E.A. (1999) Spermatocrit and spermatozoa density in Atlantic cod (*Gadus morhua*): correlation and variation during the spawning season. *Aquaculture* 170, 349–358.

[4] Rideout, R.M., Trippel, E.A. and Litvak, M.K. (2004) The development of haddock and Atlantic cod sperm cryopreservation techniques and the effect of sperm age on cryopreservation success. *Journal of Fish Biology* 65, 299–311.

[5] A. Steinarsson, Marine Research Institute, Mariculture Laboratory, Grindavík, Iceland, 2007, personal communication.

[6] Kjesbu, O.S., Solemdal, P., Bratland, P. and Fonn, M. (1996) Variation in annual egg production in individual captive Atlandic cod (*Gadus morhua*). *Canadian Journal of Fisheries and Aquatic Sciences* 53, 610–620.

[7] Galloway, T.F., Kjørsvik, E. and Kryvi, H. (1998) Effect of temperature on viability and axial muscle development in embryos and yolk sac larvae of the Northeast Arctic cod (*Gadus morhua*). *Marine Biology* 132, 559–567.

[8] Steinarsson, A. (2004) Framleiðsla þorskseiða (Hatchery production of cod). In: Björnsson, B. and Gunnarsson, V.I. (eds) *Þorskeldi á Íslandi (Cod Farming in Iceland)*. Marine Research Institute, Report 111, Reykyavik, Iceland, pp. 41–86.
[9] Marteinsdottir, G. and Steinarsson, A. (1998) Maternal influence on size and viability of Icelandic cod *Gadus morhua* eggs and larvae. *Journal of Fish Biology* 52, 1241–1258.
[10] Kjørsvik, E., Pittman, K. and Pavlov, D. (2004) From fertilisation to the end of metamorphosis-functional development. In: Moksness, E., Kjørsvik, E. and Olsen, Y. (eds) *Culture of Cold-water Marine Fish*. Blackwell Publishing, Oxford, UK, pp. 204–278.
[11] Meeren, T. van der, Pedersen, J.P. and Kolbeinshavn, A.-G. (2005) Yngelproduksjon-larvefase (Juvenile production-larval phase). In: Otterå, H., Taranger, G.L. and Borthen, J. (eds) *Oppdrett av torsk (Farming of Cod)*. Norsk Fiskeoppdrett AS, Bergen, pp. 85–111.
[12] Steinarsson, A. and Björnsson, B. (1999) The effects of size and temperature on growth and mortality of cod larvae. *Journal of Fish Biology* 55 (Supplement A), 100–109.
[13] Olsen, Y. (2004) Live food technology of cold-water marine fish larvae. In: Moksness, E., Kjørsvik, E. and Olsen, Y. (eds) *Culture of Cold-water Marine Fish*. Blackwell Publishing, Oxford, UK, pp. 73–128.
[14] Reitan, K.I. (2005) Produksjon av levende fôr (Production of live feed). For og ernæring (Feed and nutrition). In: Otterå, H., Taranger, G.L. and Borthen, J. (eds) *Oppdrett av torsk (Farming of Cod)*. Norsk Fiskeoppdrett AS, Bergen, pp. 121–142.
[15] Engelsen, R., Asche, F., Skjennum, F. and Adoff, G. (2004) New species in aquaculture: some basic economic aspects. In: Moksness, E., Kjørsvik, E.and Olsen,Y. (eds) *Culture of Cold-water Marine Fish*. Blackwell Publishing, Oxford, UK, pp. 487–515.
[16] Björnsson, B., Steinarsson, A. and Árnason, T. (2007) Growth model for Atlantic cod (*Gadus morhua*): effects of temperature and body weight. *Aquaculture* 271, 216–226.
[17] Svåsand, T., Otterå, H.M. and Taranger, G.L. (2004) The status and perspectives for the species. In: Moksness, E., Kjørsvik, E. and Olsen, Y. (eds) *Culture of Cold-water Marine Fish*. Blackwell Publishing, Oxford, UK, pp. 433–474.
[18] Gunnarsson, V.I., Björnsson, B., Kjartansson, D., Grétarsson, E.H., Stefánsson, G.W., Karlsson, H., Pétursson, H., Pálsson, J.Ö., Elíasson, K., Gudmundsson, R., Ingvason, Ó.M., Sigurdsson, S., Halldórsson, S., Haraldsson, S. and Ólafsson, Th. (2006) Þorskeldiskvóti: yfirlit yfir föngun og áframeldi þorsks á árinu 2004 (Quota for on-growing of cod in Iceland: a report on the capture and on-growing of cod in the year 2004). *Marine Research Institute Report* 124, Reykyavik, Iceland, 76 pp.
[19] Björnsson, B. and Ólafsdóttir, S.R. (2006) Effects of water quality and stocking density on growth performance of juvenile cod (*Gadus morhua* L.). *ICES Journal of Marine Science* 63, 326–334.
[20] Bricknell, I.R., Bron, J.E. and Bowden, T.J. (2006) Diseases of gadoid fish in cultivation: a review. *ICES Journal of Marine Science* 63, 253–266.
[21] Rosenlund, G., Karlsen, Ø., Tveit, K., Mangor-Jensen, A. and Hemre, G.-I. (2004) Effect of feed composition and feeding frequency on growth, feed utilization and nutrient retention in juvenile Atlantic cod, *Gadus morhua* L. *Aquaculture Nutrition* 10, 371–378.
[22] Pavlov, D., Kjørsvik, E., Refsti, T. and Anderson, Ø. (2004) Brood stock and egg production. In: Moksness, E., Kjørsvik, E. and Olsen, Y. (eds) *Culture of Cold-water Marine Fish*. Blackwell Publishing, Oxford, UK, pp. 129–203.
[23] Norberg, B., Brown, C.L., Halldorsson, O., Stensland, K. and Björnsson, B.Th. (2004) Photoperiod regulates the timing of sexual maturation, spawning, sex steroid and thyroid hormone profiles in the Atlantic cod (*Gadus morhua*). *Aquaculture* 229, 451–467.
[24] Holdway, D.A. and Beamish, F.W.H. (1984) Specific growth rate and proximate body composition of Atlantic cod (*Gadus morhua* L.). *Journal of Experimental Marine Biology and Ecology* 81, 147–170.

[25] Gunnarsson, V.I. and Jóakimsson, K.G. (2004) Gæðastjórnun, slátrun og vinnsla á eldisþorski (Quality management, slaughtering and processing of farmed cod). In: Björnsson, B. and Gunnarsson, V.I. (eds) *Þorskeldi á Íslandi (Cod Farming in Iceland)*. Marine Research Institute, Report 111, Reykyavik, Iceland, pp. 127–144.

[26] Ólafsson, Th. Hradfrystihúsid-Gunnvör Ltd., Ísafjördur, Iceland, 2007, personal communication.

[27] Pálsson, J.Ö. Thóroddur Ltd., Tálknafjördur, Iceland, 2007, personal communication.

[28] Gunnarsson, V.I. Marine Research Institute, Reykjavík, Iceland, 2007, personal communication.

[29] Fjalestad, K.T. and Mortensen, A. (2005) Avlsarbeid (Breeding). In: Otterå, H., Taranger, G.L. and Borthen, J. (eds) *Oppdrett av torsk (Farming of Cod)*. Norsk Fiskeoppdrett AS, Bergen, pp. 49–61.

[30] Hamre, K. (2006) Nutrition in cod (*Gadus morhua*) larvae and juveniles. *ICES Journal of Marine Science* 63, 267–274.

[31] Taranger, G.L., Aardal, L., Hansen, T. and Kjesbu, O.S. (2006) Continuous light delays sexual maturation and increases growth of Atlantic cod (*Gadus morhua* L.) in sea cages. *ICES Journal of Marine Science* 63, 365–375.

[32] Hamre, K. and Mangor-Jensen, A. (2006) A multivariate approach to optimization of macronutrient composition in weaning diets for cod (*Gadus morhua*). *Aquaculture Nutrition* 12, 15–24.

8. Technical sheet – Atlantic halibut (*Hippoglossus hippoglossus*)

GAMETES AND DEVELOPING EGGS	
Egg characteristics	Diameter 3.06–3.49 mm, pelagic[1,2]
Milt and sperm characteristics	Sperm concentrations: 2×10^{11} to 6×10^{11} spz/ml).[3] Sperm motile at stripping. Fertilization with frozen sperm efficient[2]
Egg yield (eggs/kg female)	Multiple spawner (up to 15 batches during spawning season), ≈ 40% of body mass,[2] 40,000 eggs/l[1,4] (*c.*8–12,000 eggs/kg female)
Incubation time	Hatch at ≈ 82 degree days[2] at 6°C
Optimal T°C	≈ 4–6°C[2]
Fertilization success	60–93%, depends on egg and sperm quality[5]
Overall incubation success	10–70%[2]
LARVAE–JUVENILE STAGE	
Larval size at hatching	≈ 6–7 mm[1,2]
Optimal T°C	4–6°C during yolk-sac stage (which lasts for 265 degree days),[5] 10–12°C at first-feeding[2] and 12–15°C during the first 6 months after metamorphosis[7]
Rearing units	Silo tanks (2–3 m^3) for large-scale incubation,[2] circular fibreglass tanks (1.5 × 1 m) for first-feeding[8]
Time until first-feeding	First-feeding at 230–265 degree days[2,9]
Live feed needed	Semi-intensive method (zooplankton and *Artemia*)[2,8,11] or intensive method (*Artemia*)[2,8,11]
Survival from hatching to weaning	10–70%,[8,10] highly variable, commercial producers work with estimates around 10%[9]
ON-GROWING STAGE	
Commercial size	3–5 kg[11,12]
Years to reach commercial size	2.5–3.5 years from post-weaning[11,12]
Rearing units	Sea cages with shelves,[12,13] fibreglass tanks.[12,14] shallow raceways[15]
Rearing density	30–100 kg/m^{-2} [12,16]
Susceptibility to disease	Moderate/low[17,18]
Optimal T°C	12–15°C (0–50 g),[7,19,20,21] 10–12°C (50–500 g), [19,21] 8–10°C (> 500 g)[13,21]
Diet	Extruded pellets 40–63% proteins,10–25% fat[22,23]

continued

REPRODUCTION	
Period	Ovarian growth is initiated during September–October.[6] Spawning commences during February–April[6,24]
Fertilization mode	Stripping and external fertilization[6,9,24,25]
Fertilization protocol	Eggs (in ovarian fluid) and milt mixed in seawater (10 l sea water, 1 ml milt and 1 l eggs) at 6°C; milt activated in seawater before mixing with eggs,[2,6,9,24,26] incubated for 10 min and then moved to large incubators. 'Dry' milt can be stored at 4°C for *c.*24 h[2,26]
Time to reach sexual maturity	3–5 years under culture conditions,[27,28] lower age at first maturity in males[27,28] 5–7 years in the wild[1]
Optimal T°C	Ovarian growth: 5–7°C,[2,6,24] spawning: ≈ 6°C[2,24]
Control of reproduction	Spawning time can be shifted by photoperiod manipulation[25]
COMMERCIALIZATION	
Product characteristics	Rich and tasty fillet;[1,9,14] long shelf life and firm texture, [29] no parasites or internal bones, 76.1% humidity, 0.5% fat, 22.6% proteins, 0.9% ash[29]
Flesh and/or egg yield	Fillet yield: 54–60%[30]
Price/kg of whole fish, gutted	Europe US$10.90–16.30 (depending on size, highest for fish > 5 kg)

RESEARCH AVENUES

– Improvement in larval rearing[2,6, 8,11,26]
– Optimization of feeding efficiency and feed composition[7,13,14,30]
– Genetic characterization and selection programmes[31,32]
– Optimization of rearing methods[7,11,12, 5,16,25,27,28,31]
– Fish health, development of vaccines[17,18]
– Product development[29,30]

[1] Haug, T. (1990) Biology of the Atlantic halibut, *Hippoglossus hippoglossus* (L., 1758). *Advances in Marine Biology* 26, 1–70.
[2] Mangor-Jensen, A., Harboe, T., Shields, R.J., Gara, B. and Naas, K.E. (1998) Atlantic halibut, *Hippoglossus hipppoglossus* L., larvae cultivation literature, including a bibliography. *Aquaculture Research* 29, 857–886.
[3] Tvedt, H.B., Benfeya, T.J., Martin-Robichaud, D.J. and Power, J. (2001) The relationship between sperm density, spermatocrit, sperm motility and fertilization success in Atlantic halibut, *Hippoglossus hippoglossus. Aquaculture* 194, 191–200.
[4] Haug, T. and Gulliksen, B. (1988) Fecundity and egg sizes in ovaries of female Atlantic halibut, *Hippoglossus hippoglossus* (L.). *Sarsia* 73, 259–261.
[5] Opstad, I., Bergh, Ø. and Skiftesvik, A.B. (1998) Large-scale rearing of Atlantic halibut, *Hippoglossus Hippoglossus* L., yolk sac larvae: effects of flow rate on growth, survival and accumulation of bacteria. *Aquaculture Research* 29, 893–898.

[6] Riple, G.L. (2000) Ovarian development and egg viability aspects in turbot (*Scophthalmus maximus*) and Atlantic halibut (*Hippoglossus hippoglossus*). Dr. Scient. Thesis, University of Bergen, ISBN 82-7744-079-0.
[7] Jonassen, T.M., Imsland, A.K., and Stefansson, S.O. (1999) The interaction of temperature and size on growth of juvenile Atlantic halibut. *Journal of Fish Biology* 54, 556–572.
[8] Harboe, I., Mangor-Jensen, A., Naas, K.E. and Næss, T. (1998) A tank design for first feeding of Atlantic halibut, *Hippoglossus hippoglossus* L., larvae. *Aquaculture Research* 29, 919–923.
[9] Shields, R.J., Gara, B. and Gillespie, M.J.S. (1999) A UK perspective on intensive hatchery rearing methods for Atlantic halibut (*Hippoglossus Hippoglossus* L.). *Aquaculture* 176, 15–25.
[10] Solbakken, J.S. and Pittman, K. (2004) Photoperiodic modulation of metamorphosis in Atlantic halibut (*Hippoglossus hippoglossus* L.). *Aquaculture* 232, 613–625.
[11] Olsen, Y., Evjemo, J.O. and Olsen, A. (1999) Status of the cultivation technology for production of Atlantic halibut (*Hippoglossus hipppoglossus*) juveniles in Norway/Europe. *Aquaculture* 176, 3–13.
[12] Brown, N. (2002). Flatfish farming systems in the Atlantic region. *Reviews in Fisheries Science* 10, 403–419.
[13] Karlsen, Ø. (2004) Ongrowing in tanks and sea cages. In: Mangor-Jensen, A. and Holm, J.C. (eds) *Håndbok i kveteoppdrett (Manual for Halibut Farming)*. Marine Research Institute, Bergen, Norway, pp. 145–150 (in Norwegian).
[14] Tuene, S. and Nortvedt, R. (1995) Feed intake, growth and feed conversion efficiency of Atlantic halibut, *Hippoglossus Hippoglossus* (L.). *Aquaculture Nutrition* 1, 27–35.
[15] Oiestad, V. (1999) Shallow raceways as a compact, resource-maximizing farming procedure for marine fish species. *Aquaculture Research* 30(11–12), 831–840.
[16] Bjornsson, B. (1994) Effects of stocking density on growth-rate of halibut (*Hippoglossus-hippoglossus* L.) reared in large circular tanks for 3 years. *Aquaculture* 123, 259–270.
[17] Bergh, Ø., Nilsen, F. and Samuelsen, O.B. (2001) Diseases, prophylaxis and treatment of the Atlantic halibut *Hippoglossus hippoglossus*: a review. *Diseases of Aquatic Organisms* 48, 57–74.
[18] Grotmol, S., Totland, G.K., Thorud, K. and Hjeltnes, B.K. (1997) Vacuolating encephalopathy and retinopathy associated with a nodavirus-like agent: a probable cause of mass mortality of cultured larval and juvenile Atlantic halibut *Hippoglossus hippoglossus. Diseases of Aquatic Organisms* 29, 85–97.
[19] Björnsson, B. and Tryggvadóttir, S.V. (1996) Effect of size on optimal temperature for growth and growth efficiency of immature Atlantic halibut (*Hippoglossus hippoglossus* L.) *Aquaculture* 142, 33–42.
[20] Jonassen, T.M., Imsland, A.K., Kadowaki, S., and Stefansson, S.O. (2000) Interaction of temperature and photoperiod on growth of Atlantic halibut, *Hippoglossus hippoglossus* L. *Aquaculture Research* 31, 219–227.
[21] Aune, A., Imsland, A.K. and Pittman, K. (1997) Growth of juvenile halibut (*Hippoglossus hippoglossus* L.) under a constant and switched temperature regime. *Aquaculture Research* 28, 931–939.
[22] Hemre, K., Øfsti, A., Næss, T., Nortvedt, R. and Holm, J.C. (2003) Macronutrient composition of formulated diets for Atlantic halibut (*Hippoglosssus hippoglossus* L.) juveniles. *Aquaculture* 227, 233–244.
[23] Helland, S.J. and Grisdale-Helland, B. (1998) Growth, feed utilization and body composition of juvenile Atlantic halibut (*Hippoglossus hippoglossus*) fed diets differing in the ratio between macronutrients. *Aquaculture* 166, 49–56.
[24] Norberg, B., Valkner, V., Huse, J., Karlsen, I. and Lerøy-Grung, G. (1991) Ovulatory rhythms and egg viability in the Atlantic halibut (*Hippoglossus hippoglossus*). *Aquaculture* 97, 365–371.
[25] Björnsson, B.Th., Halldórsson, Ó., Haux, C., Norberg, B. and Brown, C.L. (1998) Photoperiod control of sexual maturation of the Atlantic halibut (*Hippoglossus hipppoglossus*): plasma thyroid hormone and calcium levels. *Aquaculture* 166, 117–140.

[26] Mangor-Jensen, A. (2004) Eggfasen (The egg phase). In: Mangor-Jensen, A. and Holm, J.C. (eds) *Håndbok i kveteoppdrett (Manual for Halibut Farming).* Marine Research Institute, Bergen, Norway, pp. 22–29 (in Norwegian).
[27] Imsland, A.K. and Jonassen, T.M. (2003) Effects of exposure to extended photoperiods on long-term growth and age at first maturity in turbot and Atlantic halibut. *Aquaculture International* 11, 463–475.
[28] Imsland, A.K. and Jonassen, T.M. (2005) The relation between age at first maturity and growth of Atlantic halibut (*Hippoglossus hippoglossus* L.) reared under four different light regimes. *Aquaculture Research* 36, 1–7.
[29] Ruff, N., Fitzgerald, R.D., Cross, T.F. and Kerry, J.P. (2002) Comparative composition and shelf-life of fillets of wild and cultured turbot (*Scophthalmus maximus*) and Atlantic halibut (*Hippoglossus hippoglossus*). *Aquaculture International* 10, 241–256.
[30] Nortvedt, R. and Tuene, S. (1998) Body composition and sensory assessment of three weight groups of Atlantic halibut (*Hippoglossus hippoglossus*) fed three pellet sizes and three dietary fat levels. *Aquaculture* 161, 295–313.
[31] Imsland, A.K. and Jonassen, T.M. (2002) Regulation of growth in turbot (*Scophthalmus maximus* Rafinesque) and Atlantic halibut (*Hippoglossus hippoglossus* L.): aspects of environment × genotype interactions. *Reviews in Fish Biology and Fisheries* 11, 71–90.
[32] Jonassen, T.M., Imsland, A.K., FitzGerald, R., Stefánsson, M.Ö., Bonga, S.W., Ham, E. van, Nævdal, G. and Stefansson, S.O. (2000) Geographic variation in growth and food conversion efficiency of juvenile Atlantic halibut related to latitude. *Journal of Fish Biology* 56, 279–294.

9. Technical sheet – Channel catfish (*Ictalurus punctatus*)

GAMETES AND DEVELOPING EGGS	
Egg characteristics	Diameter 3.5–4.5 mm, demersal and adhesive when laid. Starting as yellowish and turning to brown and then reddish/orange during incubation[1]
Milt and sperm characteristics	Milt difficult to strip from males, generally male is sacrificed and testes are macerated in diluent for artificial fertilization[2]
Egg yield (eggs/kg female)	Approximately 6000[2]
Incubation time	135–175 degree days[3]
Optimal T°C	25–28°C[1]
Fertilization success	Highly variable
Overall incubation success	0–80%[4]
LARVAE–JUVENILE STAGE	
Larval size at hatching	6.4–11.8 mm[3]
Optimal T°C	25–28°C[5]
Rearing units	Troughs[1]
Time until first-feeding	3–5 days[1]
Survival from hatching to weaning	Highly variable 0–80%, typically estimate 40–70%[6]
ON-GROWING STAGE	
Commercial size	0.5–1.0 kg[6]
Years to reach commercial size	2 years from hatching[6]
Rearing units	Earthen ponds[6]
Rearing density	1200–10,000 kg/ha
Susceptibility to disease	Approximately 45% of inventory loss on catfish farms is due to disease[7]
Optimal T°C	25–30°C[8]
Diet	Floating diet, 28–35% protein, 6–10% fat[9]
REPRODUCTION	
Period	Spawning commence during April–May. Timing is influenced by prevailing temperature, then oocyte development is halted until cool temperatures are experienced[10]
Fertilization mode	External fertilization[10]
Fertilization protocol	Chosen males and females are placed together in spawning ponds together with spawning containers. Fertilized egg masses are collected from ponds[1]
Time to reach sexual maturity	~ 3 years, males and some females may mature at 2 years, but majority of fish mature at 3+[10]
Optimal T°C	25–27°C[1]
Control of reproduction	Temperature appears to be the primary environmental cue. Photoperiod may have a secondary role[10]

continued

COMMERCIALIZATION	
Product characteristics	Boneless, skinless, uniformly sized muscle product. Desirable flavour and texture profiles[11]
Flesh and/or egg yield	45% for fillets[11]
Price/kg of whole fish, gutted	US$2.80 received by processors[12]

RESEARCH AVENUES
– Measurement of genetic variation and estimation of correlations between traits[13]
– Genomic resources for mapping quantitative trait loci (QTLs)[14]
– Improving the economic performance of catfish feeds[15]
– Characterize interactions among fish size and feeding schedule on economic performance
– Managing ponds to reduce off-flavour episodes[16]
– For fish health, candidate genes such as the toll-like receptors are being tested for association with disease resistance[17]
– *In vitro* immune function assays are being developed

[1] Avery, J.L. and Steeby, J.A. (2004) Hatchery management. In: Tucker, C.S. and Hargreaves, J. A. (eds) *Biology and Culture of Channel Catfish.* Elsevier, Amsterdam, pp. 245–165.
[2] Kelly, A.M. (2004) Broodfish management. In: Tucker, C.S. and Hargreaves, J.A. (eds) *Biology and Culture of Channel Catfish.* Elsevier, Amsterdam, pp. 129–144.
[3] Makeeva, A.P. and Emel'yanova, N.G. (1993) Early development of channel catfish, *Ictalurus punctatus. Journal of Icthyology* 33, 87–103.
[4] Small, B.C., Wolters, W.R. and Bates, T.D. (2001) Factors affecting catfish egg hatching success. *National Warmwater Aquaculture Center News* 4, 10–11.
[5] Tucker, C.S. and Robinson, E.H. (1990) *Channel Catfish Farming Handbook.* Van Nostrand Reinhold, New York.
[6] Tucker, C.S., Avery, J.L. and Heikes, D. (2004) Culture methods. In: Tucker, C.S. and Hargreaves, J.A. (eds) *Biology and Culture of Channel Catfish.* Elsevier, Amsterdam, pp. 166–195.
[7] Hawke, J.P. and Khoo, L.H. (2004) Infectious diseases. In: Tucker, C.S. and Hargreaves, J.A. (eds) *Biology and Culture of Channel Catfish.* Elsevier, Amsterdam, pp. 405–496.
[8] Tucker, C.S. and Hargreaves, J.A. (2004) Pond water quality. In: Tucker, C.S. and Hargreaves, J.A. (eds) *Biology and Culture of Channel Catfish.* Elsevier, Amsterdam, pp. 215–278.
[9] Li, M.H., Robinson, E.H. and Manning, B.B. (2004) Nutrition. In: Tucker, C.S. and Hargreaves, J.A. (eds) *Biology and Culture of Channel Catfish.* Elsevier, Amsterdam, pp. 279–323.
[10] Silverstein, J.T. and Small, B.C. (2004) Reproductive physiology. In: Tucker, C.S. and Hargreaves, J.A. (eds) *Biology and Culture of Channel Catfish.* Elsevier, Amsterdam, pp. 69–94.
[11] Marshall, D.L. (2004) Processing. In: Tucker, C.S. and Hargreaves, J.A. (eds) *Biology and Culture of Channel Catfish.* Elsevier, Amsterdam, pp. 585–600.
[12] Engle, C.R. and Hanson, T.R. (2004) Marketing and economics. In: Tucker, C.S. and Hargreaves, J.A. (eds) *Biology and Culture of Channel Catfish.* Elsevier, Amsterdam, pp. 601–633.
[13] Bosworth, B.G., Wolters, W.R., Wise, D.J. and Klesius, P.H. (2004) Genetic effects for growth and resistance to *Edwardsiella ictaluri* estimated from diallel cross among USDA-102 strain channel catfish *Ictalurus punctatus*, blue catfish *Ictalurus furcatus*, and their F1 hybrid. *Journal of the World Aquaculture Society* 35, 420–426.
[14] Li, R.W. and Waldbieser, G.C. (2006) Production and utilization of a high-density oligonucleotide microarray in channel catfish, *Ictalurus punctatus. BMC genomics [electronic resource]* 7, 134.
[15] Li, M.H., Peterson, B.C., Janes, C.L. and Robinson, E.H. (2006) Comparison of diets containing various fish meal levels on growth performance, body composition, and insulin-like

growth factor-I of juvenile channel catfish *Ictalurus punctatus* of different strains. *Aquaculture* 253, 628–635.
[16] Zimba, P.V. and Gitelson, A. (2006) Remote estimation of chlorophyll concentration in hyper-eutrophic aquatic systems: model tuning and accuracy optimization. *Aquaculture* 256, 272–286.
[17] Baoprasertkul, P., Peatman, E., Somridhivej, B. and Liu, Z. (2006) Toll-like receptor 3 and TICAM genes in catfish: species-specific expression profiles following infection with *Edwardsiella ictaluri*. *Immunogenetics* 58, 817–830.

10. Technical sheet – Barramundi (*Lates calcarifer*)

GAMETES AND DEVELOPING EGGS	
Egg characteristics	Diameter 0.74–0.80 mm, buoyant[1]
Milt and sperm characteristics	N/A
Egg yield (eggs/kg female)	250,000[1]
Incubation time	12–17 h[2]
Optimal T°C	27–30°C[2]
Fertilization success	N/A
Overall incubation success	N/A
LARVAE–JUVENILE STAGE	
Larval size at hatching	2.6 mm[2]
Optimal T°C	Larvae, 28–30°C;[1] juveniles, 27–36°C[3]
Rearing units	Intensive culture, conical dark tanks up to 10 m^3; extensive culture, marine or brackish ponds[1,4]
Time until first-feeding	2–3 days[1]
Live feed needed	
Survival from hatching to weaning	Intensive culture, from hatching to 10 mm 15–40%; extensive culture, 40%[1]
ON-GROWING STAGE	
Commercial size	400 g–3 kg, plate size (400–500 g); banquet (~ 1 kg); fillet (~ 3 kg)
Years to reach commercial size	Plate size, 6–9 months; banquet, 8–12 months; fillet, 18–24 months from hatching
Rearing units	Cage culture, 2 × 2 × 2 m up to 16 × 16 × 8 m; recirculating systems; ponds < 2 m deep
Rearing density	15–25 kg/m^3; recirculating systems, 15 kg/m^3
Susceptibility to disease	High[1,8,9,10]
Optimal T°C	27–36°C (juveniles);[3] 27–30°C (broodstock)[1]
Diet	Extruded pellets, < 200 g juvenile 50–55% protein, DE 15–17 kJ/g; > 200 g fish, 45% protein, DE 17–20 kJ/g[6,7]

continued

REPRODUCTION	
Period	September–March[1]
Fertilization mode	External fertilization[1]
Fertilization protocol	Fertilized eggs are collected in 300 μm mesh net either inside (airlift collectors) or outside (overflow collectors) the spawning tank; for caged spawned fish, cages are lined with fine mesh 'hapa'[2]
Time to reach sexual maturity	Protandrous hermaphrodites, males 3–4 years; females 5–7 years[5]
Optimal T°C	28–29°C in Australia; 27–34°C in Malaysia and Thailand
Control of reproduction	Spawning can be manipulated by temperature and photoperiod to occur year round with the aid of hormonal injections at a dose of 3–5 μg/kg for a single spawn or 10–25 μg/kg for 2–4 consecutive spawns[11,12,13].
COMMERCIALIZATION	
Product characteristics	White firm flesh
Flesh and/or egg yield	45–50%
Price/kg of whole fish, gutted	US$7.80/kg
RESEARCH AVENUES	
– Genetic selection for disease resistance – Optimizing growth efficiency under suboptimal environmental conditions – Environmental impacts of cage culture – Fish health	

[1] Rimmer, M.A. (2003) Barramundi. In: Lucas, J.S. and Southgate, P.C. (eds) *Aquaculture: Farming Aquatic Animals and Plants.* Blackwell Publishing, Oxford, UK, pp. 364–381.
[2] Rimmer, M.A. (2006) Cultured aquatic species information programme – *Lates calcarifer.* Cultured Aquatic Species Fact Sheets, FAO, Rome (http://www.fao.org/figis/servlet/static?dom=culturespecies&xmml=Lates_calcarifer.xml, accessed 1 March 2007).
[3] Katersky, R.S. and Carter, C.G. (2005) Growth efficiency of juvenile barramundi, *Lates calcarifer*, at high temperatures. *Aquaculture* 250, 775–780.
[4] Kolkovski, S., Curnow, J. and King, J. (2004) Intensive rearing system for fish larvae research. I. Marine fish larval rearing system. *Aquaculture Engineering* 31, 295–308.
[5] Tucker, J.W. Jr, Russell, D.J. and Rimmer, M.A. (2002) Barramundi culture: a success story for aquaculture in Asia and Australia. *World Aquaculture* 33(3), 53–59.
[6] Boonyaratpalin, M. and Williams, K. (2002) Asian sea bass, *Lates calcarifer.* In: Webster, C.D. and Lim. C. (eds) *Nutrient Requirements and Feeding of Finfish for Aquaculture.* CAB International, Wallingford, UK, pp. 40–50.
[7] Glencross, B.D. (2006) Nutritional management of barramundi, *Lates calcarifer* – a review. *Aquaculture Nutrition* 12, 291–309.
[8] Azad, I.S., Jithendrana, K.P., Shekhara, M.S., Thirunavukkarasua, A.R. and Penab, L.D. de la (2006) Immunolocalisation of nervous necrosis virus indicates vertical transmission in hatchery produced Asian sea bass (*Lates calcarifer* Bloch) – a case study. *Aquaculture* 255, 39–47.

[9] Owens, L. (2003) Barramundi. In: Lucas, J.S. and Southgate, P.C. (eds) *Aquaculture: Farming Aquatic Animals and Plants*. Blackwell Publishing, Oxford, UK, pp. 199–236.
[10] Anderson, I.G. and Thomas, A. (1995) Recent advances in the bacterial disease of barramundi. *Proceedings of the Australian Barramundi Farming Workshop*, Queensland, Australia.
[11] Schipp, G. (2006) Barramundi farming in the Northern Territory. Darwin Aquaculture Centre, Department of Primary Industry, Fisheries and Mines (DPIFM), Northern Territory, Australia.
[12] Garrett, R.N. and O'Brien, J.J. (1994) All-year-round spawning of hatchery barramundi in Australia. *Austasia Aquaculture* 8, 40–42.
[13] Kungvankij, P. (1987) Induction of spawning of sea bass (*Lates calcarifer*) by hormone injection and environmental manipulation. In: Copland, J.W. and Grey, D.L. (eds) *Management of Wild and Cultured Sea Bass/Barramundi* (*Lates Calcarifer*). ACIAR Proceedings No. 20. Australian Centre for International Agricultural Research, Canberra, pp. 120–122.

11. Technical sheet – Haddock (*Melanogrammus aeglefinus*)

GAMETES AND DEVELOPING EGGS	
Egg characteristics	1.3–1.6 mm diam., transparent, spherical without oil globule, pelagic, buoyant[1,2]
Milt and sperm characteristics	1.84–13.15 × 10^9 spz/ml, correlated positively with spermatocrit, sperm motile for over an hour after exposure to seawater,[3] cryopreservation procedures exist[4]
Egg yield (eggs/kg female)	480,000–930,000 for 2.1–3.1 kg females[5]
Incubation time	20 days at 2°C and 9 days at 10°C,[6] 70–80 degree days[7]
Optimal T°C	5–8°C
Fertilization success	0–100% depending on egg and sperm quality; average 81%[8]
Overall incubation success	If egg cleavage is symmetrical > 97%.[9] Average survival to hatch 54%[8]
LARVAE–JUVENILE STAGE	
Larval size at hatching	3–4 mm
Optimal T°C	Starts at 6°C and then is moved to 12°C through the larval period[10]
Rearing units	Circular tanks
Time until first-feeding	When mouth opens, which depends on rearing temperature, 4–5 dph at 6°C[11]
Survival from hatching to weaning	Variable: up to 33% (S. Neil, personal communication), usually around 5%
ON-GROWING STAGE	
Commercial size	2–2.5 kg[7]
Years to reach commercial size	3 years posthatch or less[7]
Rearing units	6 × 6 × 3 m in May and transferred to 12 × 12 × 6 m cages July. Mesh is 210/80 standard for smolts[7]
Rearing density	40 kg/m^3 (S. Neil, personal communication)
Susceptibility to disease	Nodavirus;[12] sunburn; ISA
Optimal T°C	Unknown
Diet	Lipid levels should be low. Moist feed has provided better growth[13,14,15]

continued

REPRODUCTION	
Period	April to June[2]
Fertilization mode	External fertilization – tank spawning
Fertilization protocol	Haddock spawn in tanks, fertilized eggs are collected passively with egg collection scoops placed in tanks. Embryos are disinfected after removal from the tanks
Time to reach sexual maturity	3–5 years in the wild,[2] 2 years for males and 3 years for females held in captivity
Optimal T°C	Developmental stage dependent, colder water (< 10°C) early in development with greater range in temperature after weaning
Control of reproduction	Photothermal manipulation of reproduction
COMMERCIALIZATION	
Product characteristics	White flesh premium product
Flesh and/or egg yield	34–38%[7]
Price/kg of whole fish, gutted	US$3.50/kg whole fish US wildfish domestic landings 2005/06.[16] Retail fillet prices high around US$15/kg in North America
Other commercialization avenues	Sold as fresh and frozen fillets, fresh and frozen whole fish, salted, headless split and smoked fish (finan haddies). Frozen chowder is now marketed commercially in the USA
RESEARCH AVENUES	
– Improvements in larval survival through weaning – Nutritional requirements and feed development throughout life history – Vaccination development and trials – Suppression of early maturation	

[1] Fahay, M.P. (1983) Guide to the early stages of marine fishes occurring in the western North Atlantic Ocean, Cape Hatteras to the southern Scotian Shelf. *Journal of the Northwest Atlantic Fisheries Sciences* 4, 1–423.
[2] Scott, W.B. and Scott, M.G. (1988) Atlantic Fishes of Canada. *Canadian Bulletin of Fisheries and Aquatic Sciences* No. 219.
[3] Rideout, R.M., Trippel, E.A. and Litvak, M.K. (2004) Relationship between sperm density, spermatocrit, sperm motility and spawning date in wild and cultured haddock. *Journal of Fish Biology* 65, 319–332.
[4] Rideout, R.M., Trippel, E.A. and Litvak, M.K. (2004) The development of haddock and Atlantic cod sperm cryopreservation techniques and the effect of sperm age on cryopreservation success. *Journal of Fish Biology* 65, 299–311.
[5] Trippel, E.A., Doherty, C.M., Wade, J. and Harmon, P.R. (1998) Controlled breeding technology for Haddock (*Melanogrammus aeglefinus*) in mated pairs. *Bulletin of the Aquaculture Association of Canada* 98(3), 30–35.
[6] Martell, D.J., Kieffer, J.D. and Trippel, E.A. (2005) Effects of temperature during early life history on embryonic and larval development and growth in haddock. *Journal of Fish Biology* 66, 1558–1575.

[7] Frantsi, C., Lanteigne, C., Blanchard, B., Alderson, R., Lall, S., Johnson, S., Leadbeater, S., Martin-Robichaud, D. and Rose, P. (2002) Haddock culture in Atlantic Canada. *Bulletin of the Aquaculture Association of Canada* 102(1), 31–34.
[8] Warrington, S.J. (2003) Hatchery technology for haddock at the St. Andrews Biological Station. In: Aiken, D.E. (ed.) *Early Rearing of Haddock: State of the Art.* Aquaculture Association of Canada Special Publication 7, pp. 35–42.
[9] Rideout, R.M., Trippel, E.A. and Litvak, M.K. (2004) Predicting haddock embryo viability based on early cleavage patterns. *Aquaculture* 230, 215–228.
[10] Perez-Casanova, J.C., Murray, H.M., Gallant, J.W., Ross, N.W., Douglas, S.E. and Johnson, S.C. (2006) Development of the digestive capacity in larvae of haddock (*Melanogrammus aeglefinus*) and Atlantic cod (*Gadus morhua*). *Aquaculture* 251, 377–401.
[11] Downing, G. and Litvak, M.K. (2001) The effect of light intensity and spectrum on the incidence of first feeding by larval haddock. *Journal of Fish Biology* 59, 1566–1578.
[12] Bricknell, I.R., Bron, J.E. and Bowden, T.J. (2006) Diseases of gadoid fish in cultivation. *ICES Journal of Marine Science* 63, 253–266.
[13] Nanton, D.A., Lall, S.P. and McNiven, M.A. (2001) Effects of dietary lipid level on liver and muscle deposition in juvenile haddock, *Melanogrammus aeglefinus. Aquaculture Research* 32, 225–234.
[14] Kim, J.-D. and Lall, S.P. (2001) Effects of dietary protein level on growth and utilization of protein and energy by juvenile haddock (*Melanogrammus aeglefinus*). *Aquaculture* 195, 311–319.
[15] Kim, J.D., Lall, S.P. and Milley, J.E. (2001) Dietary protein requirements of juvenile haddock (*Melanogrammus Aeglefinus* L.). *Aquaculture Research* 32, 1–7.
[16] National Marine Fisheries Service (2006) *Fisheries of the United States 2006* (http://www.st.nmfs.gov/st1/fus/fus06/fus_2006.pdf, accessed 3 June 2007).

12. Technical sheet – Striped bass (*Morone saxatilis*)

GAMETES AND DEVELOPING EGGS	
Egg characteristics	Diameters of ripe eggs range from 0.9–1.6 mm[1,2] and are semi-buoyant, transparent, a single large oil droplet, a lightly granulated yolk mass and a wide perivitelline space[3]
Milt and sperm characteristics	Milt volume: 1.0–15 ml from 1–10 kg fish.[4] Sperm concentrations: 40–128 × 10^9 spz/ml.[5] Sperm motile at stripping, 25–50% motile after 24 h extended and refrigerated and < 15% motile after 48 h when extended and refrigerated.[6] Sperm can efficiently be cryopreserved[7]
Egg yield (eggs/kg female)	100,000 between the ages of 4–10[8]
Incubation time	48 h at 16°C, 27–37 degree days[9,10]
Optimal T°C	16–18°C[8]
Fertilization success	25–90%, depending on fresh versus cryopreserved sperm and egg quality[10,11]
Overall incubation success	0–90%[4]
LARVAE–JUVENILE STAGE	
Larval size at hatching	1.6–2.0 mm[3]
Optimal T°C	16–18°C[9]
Rearing units	Fertilized ponds,[12] aquaria or circular fibreglass tanks[13]
Time until first-feeding	5 days posthatch, (*Artemia* nauplii for intensive culture)[13]
Live feed needed	*Artemias* and rotifers
Survival from hatching to weaning	0–50% for ponds, 25–40% typical,[12] highly variable under intensive culture[13]
ON-GROWING STAGE	
Commercial size	0.45–1.36 kg[14]
Years to reach commercial size	1.5–2 years in ponds, 0.7–1 years in intensive culture tanks[14]
Rearing units	Earthen ponds 65.6%, tanks or raceways 25%, net pens 9.4%[15]
Rearing density	Pond density between 75,000–100,000 fry (phase I) per hectare;[16] on-growing fingerlings to market (phase II) between 25,000–60,000/ha;[17] finishing (phase III), if required to achieve market demands, between 5600–6600 kg/ha[16]
Susceptibility to diseases	Low to moderate (increased susceptibility with intensification of culture conditions and/or decreased environmental quality)[18]
Optimal T°C	Mid to upper 20s°C[19]
Diet	Floating or sinking diets, 40–55% protein,[20] 5–17% fat,[21] and 6–8 kcal energy/g protein[22]

continued

REPRODUCTION	
Period	Varies by latitude; between September–December, peak spawning usually coinciding with water temperatures of 16–17°C[23]
Fertilization mode	Iteroparous, group synchronous spawners, with external fertilization[23]
Fertilization protocol	Tank spawning[24] or *in vitro* fertilization:[25] strip eggs into dry pan, add sperm then add water and mix gently; allow to stand for 2–3 min and decant excess water
Time to reach sexual maturity	2–3 years under culture conditions (females 1 year later than males); in the wild, males mature in 3–5 years and females in 6–7 years[23]
Optimal T°C	Ovarian growth: 10–16°C; spawning 16–17°C[23]
Control of reproduction	Spawning time can be shifted by photoperiod and temperature manipulation[23]
COMMERCIALIZATION	
Product characteristics	Fine flavoured, premium quality food fish with the majority (75–80%) sold as iced, whole fish and the bulk of the remaining fish sold live[14]
Flesh and/or egg yield	When filleted (< 5% of total production), 29–50%[14]
Price/kg of whole fish	US$6.60/kg *live*/FOB farm; US$5.50/kg *fresh*/FOB farm (estimates from the Striped Bass Growers Association Annual Survey 2006)
RESEARCH AVENUES	
– Development of domestic broodstock and genetic improvement[15,24,25] – Development of intensive larviculture techniques[24,26] – Develop methodology to assess gamete quality prior to fertilization[24,25] – Develop methods for short-term storage of sperm and cryopreservation of eggs and embryos[25,27] – Optimize/develop complete diets for striped bass[15,24,28]	

[1] Scott, W.B. and Crossman, E.J. (1973) *Freshwater Fishes of Canada.* Bulletin of the Fisheries Research Board of Canada, Ottawa, 966 pp.

[2] Woods, L.C. III, Bennett, R.O. and Sullivan, C.V. (1992) Reproduction of a domestic striped bass brood stock. *The Progressive Fish-Culturist* 54(3), 184–188.

[3] Mansueti, R.J. (1958) Eggs, larvae and young of the striped bass, *Roccus saxatilis. Chesapeake Biological Laboratory Contribution* 112, Solomons, Maryland, USA, 12 pp.

[4] Woods, L.C. III. UM-Crane Aquaculture Facility, Maryland, 2007, personal communication.

[5] Vuthiphandchai, V. and Zohar, Y. (1999) Age-related sperm quality of captive striped bass *Morone saxatilis. Journal of the World Aquaculture Society* 30, 65–72.

[6] Jenkins-Keeran, K. and Woods, L.C. III (2002) An evaluation of extenders for the short-term storage of striped bass milt. *North American Journal of Aquaculture* 64, 348–256.

[7] He, S. and Woods, L.C. III (2004) Changes in motility, ultrastructure, and fertilization capacity of striped bass *Morone saxatilis* spermatozoa following cryopreservation. *Aquaculture* 236, 677–686.

[8] Lewis, R.M. and Bonner, R.R. Jr (1966) Fecundity of the striped bass, *Roccus saxatilis* (Walbaum). *Transactions of the American Fisheries Society* 95, 328–331.
[9] Rees, R.A. and Harrell, R.M. (1990) Artificial spawning and fry production of striped bass and hybrids. In: Harrell, R.M., Kerby, J.H. and Minton, R.V. (eds) *Culture and Propagation of Striped Bass and Its Hybrids.* Striped Bass Committee, Southern Division, American Fisheries Society, Bethesda, Maryland, pp. 43–77.
[10] Setzler-Hamilton, E.M., Boyton, W.R., Wood, K.V., Zion, H.H., Lubbers, L., Mountford, N.K., Frere, P., Tucker, L. and Mihursky, J.A. (1980) Synopsis of biological data on striped bass, *Morone saxatilis.* NOAA Technical Report, NMFC 433. Washington, DC.
[11] Woods, L.C. III and Sullivan, C.V. (1993) Reproduction of striped bass, *Morone saxatilis* (Walbaum), broodstock: monitoring maturation and hormonal induction of spawning. *Aquaculture and Fisheries Management* 24, 211–222.
[12] Brewer, D.L. and Rees, R.A. (1990) Pond culture of phase I striped bass fingerlings. In: Harrell, R.M., Kerby, J.H. and Minton, R.V. (eds) *Culture and Propagation of Striped Bass and Its Hybrids.* Striped Bass Committee, Southern Division, American Fisheries Society, Bethesda, Maryland, pp. 99–120.
[13] Nicholson, L.C., Woods, L.C. III and Woiwode, J.G. (1990) Intensive culture techniques for the striped bass and its hybrids. In: Harrell, R.M., Kerby, J.H. and Minton, R.V. (eds) *Culture and Propagation of Striped Bass and Its Hybrids.* Striped Bass Committee, Southern Division, American Fisheries Society, Bethesda, Maryland, pp. 141–157.
[14] Rawles, D.D., Fernandes, C.F., Flick G.J. and Ammerman, G.R. (1997) Processing and food safety. In: Harrell, R.M. (ed.) *Striped Bass and Other Morone Culture.* Elsevier Science B.V., Amsterdam, pp. 329–356.
[15] Harrell, R.M. and Webster, D.W. (1997) An overview of *Morone* culture. In: Harrell, R.M. (ed.) *Striped Bass and Other Morone Culture.* Elsevier Science B.V., Amsterdam, pp. 1–10.
[16] Harrell, R.M. (1997) *Morone* pond culture. In: Harrell, R.M. (ed.) *Striped Bass and Other Morone Culture.* Elsevier Science B.V., Amsterdam, pp. 75–97.
[17] Jenkins, W.E., Smith, T.I.J., Stokes, A.D. and Smiley, R.A. (1989) Effect of stocking density on production of advanced juvenile hybrid striped bass. *Proceedings of the Annual Conference Southeastern Association of Fish and Wildlife Agencies* 42, 56–65.
[18] Plumb, J.A. (1997) Infectious diseases of striped bass. In: Harrell, R.M. (ed.) *Striped Bass and Other Morone Culture.* Elsevier Science B.V., Amsterdam, pp. 271–313.
[19] Tomasso, J.A. (1997) Environmental requirements and non-infectious diseases. In: Harrell, R.M. (ed.) *Striped Bass and Other Morone Culture.* Elsevier Science B.V., Amsterdam, pp. 253–270.
[20] Gatlin, D.M. III (1997) Nutrition and feeding of striped bass and hybrid striped bass. In: Harrell, R.M. (ed.) *Striped Bass and Other Morone Culture.* Elsevier Science B.V., Amsterdam, pp. 235–251.
[21] Nematipour, G.R., Brown, M.L. and Gatlin, D.M. III (1992) Effects of dietary energy:protein ratio on growth characteristics and body composition of hybrid striped bass. *Aquaculture* 107, 359–368.
[22] Woods, L.C. III, Yust, D., McLeod, C. and Subramanyam, M. (1995) Effects of dietary protein:energy ratio on weight gain, body composition, serum glucose and triglyceride levels, and liver function of striped bass. *Water Science and Technology* 31, 195–203.
[23] Sullivan, C.V., Berlinsky, D.L. and Hodson, R.G. (1997) Reproduction. In: Harrell R.M. (ed.) *Striped Bass and Other Morone Culture.* Elsevier Science B.V., Amsterdam, pp. 11–73.
[24] Harrell, R.M., Kerby, J.H., Smith, T.I.J. and Stevens, R.E. (1990) Striped bass and striped bass hybrid culture: the next twenty-five years. In: Harrell, R.M., Kerby, J.H. and Minton, R.V. (eds) *Culture and Propagation of Striped Bass and Its Hybrids.* Striped Bass Committee, Southern Division, American Fisheries Society, Bethesda, Maryland, pp. 253–261.

[25] Woods, L.C. III (2005) Striped bass and hybrid striped bass culture. In: Kelly, A.M. and Silverstein, J. (eds) *Aquaculture in the 21st Century. American Fisheries Society Symposium 46*, Bethesda, Maryland, pp. 339–353.
[26] Kerby, J.H., Woods, L.C. III and Huish, M.T. (1983) Culture of the striped bass and its hybrids: a review of methods, advances and problems. In: Stickney, R.R. and Meyers, S.P. (eds) *Proceedings of the Warmwater Fish Culture Workshop.* Louisiana State University, Baton Rouge, Louisiana, pp. 23–54.
[27] He, S. and Woods, L.C. III (2003) The effects of osmolality, cryoprotectant and equilibration time on striped bass *Morone saxatilis* sperm motility. *Journal of the World Aquaculture Society* 34(3), 255–265.
[28] Small, B.C., Soares, J.H. Jr and Woods, L.C. III (2000) Optimization of feed formulation for mature female striped bass. *North American Journal of Aquaculture* 62(4), 290–293.

13. Technical sheet – Rainbow trout (*Oncorhynchus mykiss*)

GAMETES AND DEVELOPING EGGS	
Egg characteristics	3–6 mm[1,2,3]
Milt and sperm characteristics	Density: 3.2–6.5 × 10^9 cells/ml; optimal sperm/ova ratio: 2.5–5.5 × 10^6 spz/ova; can be cryopreserved[4]
Egg yield (eggs/kg female)	800–1000[2]
Incubation time	290–330 degree days[1,5]
Optimal T°C	6–8°C,[6] 8–10°C,[1] can be as high as 15°C[7]
Fertilization success	> 90%[8]
Overall incubation success	50–95%[1,5,9]
LARVAE–JUVENILE STAGE	
Larval size at hatching	12–20 mm,[1] ≈ 100 mg[10]
Optimal T°C	12–15°C[7]
Rearing units	Troughs at first feeding,[7] then raceways for grow-out (> 20 g)[6]
Time until first-feeding	10–12 days (at 15°C)[5]
Live feed needed	None, dry commercial feed administered after yolk-sac resorption[6]
Survival from hatching to weaning	90%[1]
ON-GROWING STAGE	
Commercial size	Pan size: 250–400 g; flesh: 3–5 kg.[1,6] In USA, 650–900 g[7]
Years to reach commercial size	Pan-size: 10–13 months(Rodriguez Souto and Fernandez Villanueva, 2003; 12 months to 800 g[7]
Rearing units	Indoor or outdoor rectangular, circular or cylindroconical tanks,[6] sea cages,[3] earthen ponds[5]
Rearing density	20–40 kg/m^3,[1,11] 9.6 kg/l/min in serial raceways[5]
Susceptibility to disease	High, unless vaccinated[7]
Optimal T°C	15°C[1,5,6,12]
Diet	44% proteins, 22% lipids[5]
REPRODUCTION	
Period	August–November
Fertilization mode	External fertilization[2]
Fertilization protocol	Stripping and mixing of the sexual products, dry method[6]
Time to reach sexual maturity	2 years[5]
Optimal T°C	10–13°C[6,13]
Control of reproduction	Thermal and photoperiodic manipulation[5,6]
COMMERCIALIZATION	
Product characteristics	Red-orange flesh,[2] tasty,[14] highly valued,[1] 19% proteins, 5–7% fat, 2% ash[5]
Flesh and/or egg yield	48%[15]
Price/kg of whole fish, gutted	US$4.95–5.85/kg[6]

continued

RESEARCH AVENUES
– Controlling or preventing common diseases by developing appropriate vaccines and vaccine delivery methods for mass inoculation so that every fish does not have to be injected individually – Developing new feed formulations using ingredients from sustainable sources to reduce dependency on marine protein and oil – Developing innovative and economical methods of reusing water or somehow raising more fish in the water supplies

[1] Rodríguez Villanueva, X.L. and Fernandez Souto, B. (2003) *European Fish Farming Guide.* Xunta de Galicia, Conselleria de Pesca E Asuntos Maritimos, Pontevedra, 74 pp.
[2] Scott, W.B and Scott, M.G. (1988) Atlantic fishes of Canada. *Canadian Bulletin of Fisheries and Aquatic Sciences* 219, 731 pp.
[3] Edwards, D.J. (1978) *Salmon and Trout Farming in Norway.* Fishing News Book Limited, Farnham, Surrey, UK, 195 pp.
[4] Lahnsteiner, F. (2000) Cryopreservation protocols for sperm of salmonid fishes. In: Tiersch, T.R. and Makik, P.M. (eds) *Cryopreservation in Aquatic Species. Advances in World Aquaculture,* Volume 7. World Aquaculture Society, Baton Rouge, Louisiana, pp. 91–100.
[5] Hardy, R.W., Fornshell, C.G. and Brannon, E.L. (2000) Rainbow trout culture. In: Stickney, R.R. (ed.) *Encyclopedia of Aquaculture.* John Wiley and Sons, Inc., New York, pp. 716–722.
[6] Dubé, P. and Mason, E. (1995) Trout culture in Atlantic Canada. In: Boghen, D. (ed.) *Cold-Water Aquaculture in Atlantic Canada.* The Canadian Institute for Research on Regional Development, Université de Moncton, NB, Canada, 672 pp.
[7] Liley, N.R., Tamkee, P., Tsai, R. and Hoysak, D.J. (2002) Fertilization dynamics in rainbow trout (*Oncorhynchus mykiss*): effect of male age, social experience, and sperm concentration and motility on in vitro fertilization. *Canadian Journal of Fisheries and Aquatic Sciences* 59, 144–152.
[8] Estay, F. and Diaz, N.F. (1994) Analysis of reproductive performance of rainbow trout in a hatchery in Chile. *Progressive Fish-Culturist* 56, 244–249.
[9] Troutlodge (2007) Early rearing of rainbow trout. Technical bulletin (www.troutlodge.com, accessed 23 March 2007).
[10] R. Hardy, University of Idaho, USA, 2007, personal communication.
[11] Alanara, A. and Brannas, E. (1996) Dominance in demand-feeding behavior in Arctic charr and rainbow trout: the effect of stocking density. *Journal of Fish Biology* 48, 242–254.
[12] Slinger, S.J. and Hilton, J.W. (1984) Feeding and nutrition. In: Castledine, A.J., Cho, C.Y., Fergusson, H.W., Hynes, J.D. and Slinger S.J. (eds) *Handbook of Fish Culture.* Ministry of Environment and Ministry of Natural Resources, Toronto, Ontario.
[13] Piper, R.G., McElwains, I.B., Orme, L.E., McCraven, J.P., Fowler, L.G. and Leonard, J.R. (1982) *Fish Hatchery Management.* US Department of the Interior, Field and Wildlife Service, Washington, DC.
[14] Scott, W.B. and Crossman, E.J. (1973) *Freshwater Fishes of Canada.* Supply and Services Canada, Ottawa.
[15] Straus, 1991. *The Seafood Handbook, Seafood Standards.* Seafood Business, Rockland, Massachusetts, USA, 144 pp.

14. Technical sheet – Tilapia (*Oreochromis* sp.)

GAMETES AND DEVELOPPING EGGS	
Egg characteristics	Ovoid or pear shaped, 2.8 – 4.3 mm ø, 1.9 – 3.5 mg[1], negatively buoyant[2]
Milt and sperm characteristics	20–25 μm, 10^8 to 10^{10} sperm/ml[3], can be cryoprotected[4]
Egg yield (eggs/kg female)	Several dozen to three thousand, 2–4 eggs/g of body weight[5]
Incubation time	2–6 days[5–6]
Optimal T °C	25–32 °C[6]
Fertilization success	68.6 %[7], 88.8%[8]
Overall incubation success	80 to 90%[6]
LARVAE–JUVENILE STAGE	
Larval size at hatching	5–6 mm[2]
Optimal T °C	25–32 °C[6]
Rearing units	Maternal mouth brooding, hapas, concrete or fibreglass tank, earthen ponds[2]
Time until first-feeding	8–10 days post hatch[5]
Live feed needed	No. Omnivorous: periphyton, phyto- and zooplankton, powdered feed[2]
Survival from hatching to weaning	93%[9], 80–90% common[10]
ON-GROWING STAGE	
Commercial size	200–250 g for developing country domestic markets[6–11], 750–1200 g for processing and international trade[12]
Years to reach commercial size	4 months (250 g), 12 months (1 kg)[10]
Rearing units	Ponds or lakes, cages, flow-through raceways, recirculated systems[13–14–15]
Rearing density	1000 kg/ha in extensive ponds, 3500–42000 kg/ha[13], over 100 kg/m^3 in cages and intensive recirculating systems[10]
Susceptibility to diseases	Low, more tolerant than other commonly cultured fish (viral, bacterial and parasitic problems mostly when stressed)[5]
Optimal T °C	25–30 °C[16–17–18]
Diet	Grain based pelleted diets, Naturally omnivores, feeding on algae, aquatic plants, small invertebrates, detrital material and associated bacterial films[5]. Ponds can be fertilized with chemical fertilizers, chicken litter. cow and swine manure[13]

continued

REPRODUCTION	
Period	Breeding at > 20 °C and spawning at < 22 °C. Temperate regions: spawning in april-may; in tropical regions spawning continue year-round[5]
Fertilization mode	Mouthbrooders, external fertilization[5]
Fertilization protocol	Natural spawning in ponds. Removal of the fertilized eggs from female's mouth with a wash bottle or female incubate the eggs in her mouth[6–19]
Time to reach sexual maturity	3–5 months at 11 to 20 cm[5]
Optimal T °C	20–35 °C[20-21-22-23]
Control of reproduction	All male populations, predators[6–24]
COMMERCIALIZATION	
Product characteristics	White flesh, fine textured, mild flavoured[26]; 78% water, 20% protein, 1.70% fat, 0.9% ash[25],
Flesh yield*	33%[27], 35–43%[6]
Price/kg of whole fish, gutted	US-Europe (US$2.30–3.00/kg)[26–28], Egypt (US$2.00/kg)[6]
Other commercialization avenues	Sashimi grade fillets, value added products, leather products from skins[12–26]
RESEARCH AVENUES	
– Complete vegetable diets for commercial growout farms – All male production without feeding hormone – Improved growth rates and disease resistance through selective breeding – Super intensive culture in recirculating systems – Best management practices – Traceability and food safety	

[1] Macintosh, D.J. and Little, D.C. (1995) Nile Tilapia (*Oreochromis niloticus*). In: Bromage, N.R. and Roberts, R.J. (eds) *Broodstock Management and Egg and Larval Quality*. Blackwell Science, Oxford, UK, pp. 277–320.

[2] Green, B.W. (2006) Fingerling production systems. In: Lim, C. and Webster, C.D. (eds) *Tilapia Biology, Culture and Nutrition*. Food Products Press, Haworth Press Inc., New-York, pp. 181–202.

[3] Chao, N.H., Chao, W.C., Liu K.C. and Liao, I.C. (1987) The properties of tilapia sperm and its cryopreservation. *Journal of Fish Biology* 30 (2), 107–118.

4 Godinho, H.P., Amorim, V.M.D.C., Peixoto, M.T.D. (2003) Cryopreservation of the semen of Nile tilapia *Oreochromis niloticus*, var. *chitralada*: cryoprotectants, spermatozoa activating solutions and cryogenic refrigerator. *Revista Brasileira de Zootecnia* 32(6) suppl.1, 1537–1543.

[5] Shelton, W.L. and Popma, T.J. (2006) Biology. In: Lim, C. and Webster, C.D. (eds) *Tilapia Biology, Culture and Nutrition*. Food Products Press, Haworth Press Inc., New-York, pp. 1–27.

[6] El-Sayed, A.-F.M. (2006) *Tilapia Culture*. CABI Publishing, Wallingford, UK.

[7] Myers, J.M. and Hershberger, W.K. (1991) Artificial spawning of tilapia eggs. *Journal of the World Aquaculture Society* 22(2): 77–82.

[8] Lee, J.C. (1979) Reproduction and hybridization of three cichlid fishes, *Tilapia aurea* (Steindachner), *T. hornorum* (Trewavas) and *T. nilotica* (L.) in aquaria and in plastic pools. Doctoral Dissertation. Auburn University, Auburn, Alabama, USA.

[9] Bodganova, L.S. (1970) The transition of *Tilapia mossambicus* Peters larvae to active feeding. *Journal of Ichthyology* 10(3), 241–248.
[10] Fitzsimmons, K. (2005) Tilapia culture. In: Kelly A.M. and Silverstein, J. (eds) *Aquaculture in the 21st Century.* American Fisheries Society, Symposium 46, Bethesda, Maryland. pp. 563–590.
[11] Engle, C.R. (2006) Marketing and economics. In: Lim, C. and Webster, C.D. (eds) *Tilapia Biology, Culture and Nutrition.* Food Products Press, Haworth Press Inc., New-York, pp. 619–645.
[12] Fitzsimmons, K. (2006a) Prospect and potential for global production. In: Lim, C. and Webster, C.D. (eds) *Tilapia Biology, Culture and Nutrition.* Food Products Press, Haworth Press Inc., New-York, pp. 51–72.
[13] Green, B.W. and Bauer Duke III, C. (2006) Pond production. In: Lim, C. and Webster, C.D. (eds) *Tilapia Biology, Culture and Nutrition.* Food Products Press, Haworth Press Inc., New-York, pp. 253–257.
[14] Soderberg, R.W. (2006) Culture in flowing water. In: Lim, C. and Webster, C.D. (eds) *Tilapia Biology, Culture and Nutrition.* Food Products Press, Haworth Press Inc., New-York, pp. 289–307.
[15] Schmittou, H.R. (2006) Cage culture. In: Lim, C. and Webster, C.D. (eds.) *Tilapia Biology, Culture and Nutrition.* Food Products Press, Haworth Press Inc., New-York, pp. 313–347.
[16] Balarin, J.D. and Haller, R.D. (1982) The intensive culture of tilapia in tanks, raceways and cages. In: Muir, J.F. and Roberts, R.J. (eds) *Recent Advances in Aquaculture.* Croom Helm, London and Canberra, and Westview Press, Boulder, Colorado, pp. 267–355.
[17] Chervinski, J. (1982) Environmental physiology of tilapias. In: Pullin, R.V.S and Lowe-McConnell, R.H. (eds) *The Biology and Culture of Tilapias.* ICKLARM Conference Proceedings No 7, ICLARM, Manila, Philippines, pp. 119–128.
[18] Philippart, J.-C.L. and Ruwet, J.-C.L. (1982) Ecology and distribution of tilapias. In: Pullin, R.V.S and Lowe-McConnell, R.H. (eds) *The Biology and Culture of Tilapias.* ICLARM Conference Proceedings No 7, Manila, Philippines, pp. 15–59.
[19] Watanabe, W.O., Fitzsimmons, K. and Yi, Y. (2006) Farming tilapia in saline waters. In: Lim, C. and Webster, C.D. (eds) *Tilapia Biology, Culture and Nutrition.* Food Products Press, Haworth Press Inc., New-York, pp. 347–427.
[20] Rothbar, S.M. (1979) Observation on the reproduction behaviour of *Tilapia zilii* and several *Sarotherodon* spp under aquarium conditions. *Bamidgeh* 31, 35–43.
[21] Popma, T.J. and Lovshin, L.L. (1996) Worldwide prospects for commercial production of tilapia. Research and Development series, No. 41, Department of fisheries and Allied Aquaculture, Auburn University, Auburn, Alabama.
[22] Green, B.W., Verrica, K.L. and Fitzpatrick, M.S. (1997) Fry and fingerling production. In: Egna, H.S. and Boyd, C.E. (eds) *Dynamics of Pond Aquaculture,* CRC Press, Boca Raton, Florida, pp. 215–243.
[23] Mironova, N.V. (1975) Energy expenditure on egg production in young *Tilapia mossambica* and the influence of maintenance conditions on their reproductive intensity. *Journal of Ichthyology* 17(4), 627–633.
[24] Lim, C. and Webster, C.D. (eds) (2006) *Tilapia Biology, Culture and Nutrition.* Food Products Press, Haworth Press Inc., New-York.
[25] U.S. Department of Agriculture (USDA) (2005) National Nutrient Database for Standard Reference, Release 18. Nutrient Data Laboratory Home Page. http://www.nal.usda.gov/fnic/foodcomp
[26] Fitzsimmons, K. (2006b) Harvest, handling and processing. In: Lim, C. and Webster, C.D. (eds) *Tilapia Biology, Culture and Nutrition.* Food Products Press, Haworth Press Inc., New-York, pp. 607–619.

[27] Snir, I.J. (2001) Value added tilapia products: deliberate policy or no other choice. In: Subasinghe, S. and Singh, T. (eds) *Tilapia: Production, Marketing and Technological Developments.* Proceedings of the Tilapia 2001 International Technical and Trade Conference on Tilapia. Infofish, Kuala Lumpur, Malaysia, pp. 84.
[28] Fitzsimmons, K. (2000) Future trends of tilapia aquaculture in the Americas. In: Costa-Pierce B.A. and Rakocy, J.E. (eds) *Tilapia Aquaculture in the Americas*, Vol. 2, The World Aquaculture Society, Baton Rouge, Louisiana, United States.

15. Technical sheet – Winter flounder (*Pleuronectes americanus*)

GAMETES AND DEVELOPING EGGS	
Egg characteristics	Diameter 0.7–0.8 mm, demersal eggs, adhesive[1,2,3]
Milt and sperm characteristics	Milt volume ≈ 20 ml/season,[4,5] sperm concentrations: ≈ 10×10^8 spz/ml,[4] sperm can be motile in testis, [4] cryopreservation efficient[5]
Egg yield (eggs/kg female)	300,000–500,000,[3,6] large females can produce over 3.5 million eggs[3]
Incubation time	≈ 2700 degree hours at 8°C[3,7]
Optimal T°C	Incubation 6–12°C[3,7,24]
Fertilization success	50–90%, depends on egg and sperm quality[3]
Overall incubation success	30–80%[1,3]
LARVAE–JUVENILE STAGE	
Larval size at hatching	≈ 2.4 mm[3,6]
Optimal T°C	14–16°C in the first 30 days[1,3,6]
Rearing units	Fibreglass or concrete tanks of various sizes[3,8]
Time until first-feeding	First-feeding at 3–5 days posthatch (dph)[9]
Live feed needed	Enriched rotifers from 5 dph, *Artemia* from 18–21 dph, formulated feed from *c.*26–33 dph,[3,8,9] microencapsulated feed also used during and after weaning[10]
Survival from hatching to weaning	40–80%,[3,8,9] rearing the larvae though to metamorphosis presents few problems as compared to other flatfish species[3]
ON-GROWING STAGE	
Commercial size	From 500 g (30 cm)[11]
Years to reach commercial size	In wild, 2.5 years to reach 500 g and 3.5 years to reach 1 kg,[3] 1–2 year in culture[3]
Rearing units	Shallow tanks of fibreglass,[3,12,13] shallow raceways,[10] sea cages[3,14,15,16]
Rearing density	> 200% bottom coverage,[17] i.e. ≈ 25 kg/m^2
Susceptibility to diseases	Moderate/low,[17,21] currently no vaccines available
Optimal T°C	12–17°C[3,22]
Diet	Extruded pellets,[3,8,10,23,26] 45–50% proteins, 10–15% fat[23]

continued

REPRODUCTION	
Period	Spawning during Jan–May[3,24] depending on latitude
Fertilization mode	Stripping easy, dry or wet fertilization used[1,3,9]
Fertilization protocol	Negative buoyant eggs, de-adhesion protocols used to prevent eggs from clumping[1,2,3]
Time to reach sexual maturity	2–3 years in the wild[24] (males earlier), 2 years in captivity[12]
Optimal T°C	Spawning: 4–6°C[6,24]
Control of reproduction	Spawning time is shifted by photoperiod (latitude) and temperature[6,24]
COMMERCIALIZATION	
Product characteristics	Rich and tasty fillet;[3,6,24] long shelf life and firm texture[25]
Flesh and/or egg yield	Fillet yield: ≈ 35–45% (for 1 kg fish)[26]
Price/kg of fillet	≈ US$10–12 (per kg fresh fillet)
Other commercialization avenues	Antifreeze proteins,[3,27] stock enhancement[12]
RESEARCH AVENUES	
– Optimization of diets, feeding efficiency and feed composition[3,9,13] – Optimization of rearing methods[1,3,10,17] – Fish health, development of vaccines[18,19,20,21] – Product development[3,12,15,16,26,27]	

[1] Ben Khemis, I., Audet, C., Fournier, R. and de la Noue, J. (2003) Early weaning of winter flounder (*Pseudopleuronectes americanus* Walbaum) larvae on a commercial microencapsulated diet. *Aquaculture Research* 34, 445–452.

[2] Laroche, W.A. (1981) Development of larval smooth flounder, *Liopsetta putnami,* with a redescription of development of winter flounder *Pseudopleuronectes americanus* (Family Pleuronectidae). *Fishery Bulletin* 78, 897–909.

[3] Litvak, M.K. (1999) The development of winter flounder (*Pleuronectes americanus*) for aquaculture in Atlantic Canada: current status and future prospects. *Aquaculture* 176, 55–64.

[4] Shangguan, B. and Crim, L.W. (1999) Seasonal variations in sperm production and sperm quality in male winter founder, *Pleuronectes americanus*: the effects of hypophysectomy, pituitary replacement therapy, and GnRH-A treatment. *Marine Biology* 134, 19–27.

[5] Harmin, S.A., Crim, L.W. and Wiegand, M.D. (1995) Plasma sex steroid profiles and the seasonal reproductive cycle in male and female winter flounder, *Pleuronectes americanus. Marine Biology* 121, 601–610.

[6] Scott, W.B. and Scott, M.G. (1988) Atlantic fishes of Canada. *Canadian Bulletin of Fisheries and Aquatic Sciences* 219, 731 pp.

[7] Buckley, L.J. (1982) Effects of temperature on growth and biochemical composition of larval winter flounder *Pseudopleuronectes americanus. Marine Ecology Progress Series* 8, 181–186.

[8] Lee, G.W.Y. and Litvak, M.K. (1996) Weaning of metamorphosed winter flounder (*Pleuronectes americanus*) reared in the laboratory, comparison of two commercial artificial diets on growth, survival and conversion efficiency. *Aquaculture* 144, 251–263.

[9] Mercier, L., Audet, C., de la Noue, J., Parent, B., Parrish, C.C. and Ross, N.W. (2004) First feeding of winter flounder (*Pseudopleuronectes americanus*) larvae: use of *Brachionus plicatilis* acclimated at low temperature as live prey. *Aquaculture* 229, 361–376.
[10] Montgolfier, B. de., Audet, C. and Lambert, Y. (2005) Growth of early juvenile winter flounder (*Pseudopleuronectes americanus* Walbaum). *Aquaculture Research* 36, 1595–1601.
[11] Hendrickson, L., Nitschke, P. and Terceiro, M. (2006) Winter flounder (*Pseudopleuronectes americanus*). In: Status of Fishery Resources off the Northeastern US, NEFSC – Resource Evaluation and Assessment Division (http://www.nefsc.noaa.gov/sos/spsyn/fldrs/winter/, accessed 1 July 2007).
[12] Fairchild, E.A., Rennnels, N., Howell, W.H. and Wells, R.E. (2007) Gonadal development and differentiation in cultured juvenile winter flounder, *Pseudopleuronectes americanus. Journal of the World Aquaculture Society* 38, 114–121.
[13] Fredette, M., Batt, J. and Castell, J. (2000) Feeding stimulant for juvenile winter flounders. *North American Journal of Aquaculture* 62, 157–160.
[14] Fairchild, E.A. Fleck, J. and Howell, H.W. (2005) Determining an optimal release site for juvenile winter flounder *Pseudopleuronectes americanus* (Walbaum) in the Great Bay Estuary, NH, USA. *Aquaculture Research* 36, 1374–1383.
[15] Litvak, M.K. (1994) Winter flounder. *Bulletin of the Aquaculture Association of Canada* 94, 4–8.
[16] Litvak, M.K. (1996) Development of winter flounder for aquaculture (*Pleuronectes americanus*), state of the art. *Bulletin of the Aquaculture Association of Canada* 96, 24–26.
[17] Fairchild, E.A. and Howell, W.H. (2001) Optimal stocking density for juvenile winter flounder, *Pseudopleuronectes americanus. Journal of the World Aquaculture Society* 32, 300–308.
[18] Levin, M.A., Wolke, R.E. and Cabelli, V.J. (1972) *Vibrio anguillarum* as a cause of disease in winter flounder (*Pseudopleuronectes americanus*). *Canadian Journal of Microbiology* 18, 1585–1592.
[19] Murchelano, R.A. (1975) The histopathology of in rot disease in winter flounder from the New York Bight. *Journal of Wildlife Diseases* 11, 263–268.
[20] Barker, D.E., Cone, D.K. and Burt, M.D.B. (2002) *Trichodina murmanica* (Ciliophora) and *Gyrodactylus pleuronecti* (Monogenea) parasitizing hatchery-reared winter flounder, *Pseudopleuronectes americanus* (Walbaum): effects on host growth and assessment of parasite interaction. *Journal of Fish Diseases* 25, 81–89.
[21] Olivier, G. (2002) Disease interactions between wild and cultured fish – perspectives from the America Northeast (Atlantic Provinces). *Bulletin of the European Association of Fish Pathologists* 22, 103–109.
[22] Frame, D. (1973) Conversion efficiency and survival of young winter flounder (*Pseudopleauronectes americanus*) under experimental conditions. *Transactions of the American Fisheries Society* 3, 614–617.
[23] Hebb, C.D., Castell, J.D., Anderson, D.M. and Batt, J. (2003) Growth and feed conversion of juvenile winter flounder (*Pleuronectes americanus*) in relation to different protein-to-lipid levels in isocaloric diets. *Aquaculture* 221, 439–449.
[24] Bigelow, H.B. and Shroeder, W.C. (1953) Fishes of the Gulf of Maine. *Fisheries Bulletin of the Fish and Wildlife Service* 53, Washington, DC, USA, 577 pp.
[25] Field, C.E., Pivarnik, L.F., Barnett, S.M. and Randjr, A.G. (1986) Utilization of glucose oxidase for extending the shelf-life of fish. *Journal of Food Science* 51, 66–70.
[26] Hebb, C.D. (1998) Nutritional studies on protein utilization by juvenile winter flounder (*Pleuronectes americanus*) and haddock (*Melanogrammus aeglefinus*). Msc thesis, Nova Scotia Agricultural College, Truro and Dalhousie University, Halifax, Nova Scotia, 158 pp.
[27] Fletcher, G.L. (1977) Circannual cycles of blood plasma freezing point and Na+ and Cl– concentrations in Newfoundland winter flounder (*Pseudopleuronectes americanus*), correlation with water temperature and photoperiod. *Canadian Journal of Zoology* 55, 789–795.

16. Technical sheet – Pollack (*Pollachius pollachius*)

GAMETES AND DEVELOPING EGGS	
Egg characteristics	Diameter: 1.1–1.2 mm, pelagic[1]
Milt and sperm characteristics	Milt volume: 0.4–25 ml, sperm concentration: 1–9 × 10^9 spz/ml, movement duration: 8–10 min,[2] no cryopreservation data available
Egg yield (eggs/kg female)	From 26,000 at 12°C to 600,000 at 8°C[3]
Incubation time	1700 degree hours[3]
Optimal T°C	8–10°C[3]
Fertilization success	From 8% at 12°C to 30% at 8°C[3]
Overall incubation success	From 10 to 40%[3,4]
LARVAE–JUVENILE STAGE	
Larval size at hatching	3.4–3.7 mm[2]
Optimal T°C	15°C[5] for larvae
Rearing units	150–250 l cylindroconical tanks[5]
Time until first-feeding	3–4 days at 14–16°C[4]
Live feed needed	*Artemias*
Survival from hatching to weaning	10–30%[6]
ON-GROWING STAGE	
Commercial size	30 cm–300 g[2]
Years to reach commercial size	1.5 years[2]
Rearing units	Sea cages, cylindrical 15 m^3 tanks[2,7]
Rearing density	20 kg/m^3
Susceptibility to disease	N/A
Optimal T°C	12–15°C for 150 g fish[7]
Diet	Protein 56%, fat 12%[7] (sinking)
REPRODUCTION	
Period	Vitellogenesis from November to January, spawn from February to May, depending on location[8,9]
Fertilization mode	External fertilization[4]
Fertilization protocol	No artificial fertilization technique available
Time to reach sexual maturity	0.7 kg (2 years) in males and 1.6 kg (3 years) in females under culture conditions[8]
Optimal T°C	Spawning: 8–10°C[3]
Control of reproduction	Spawning time can be shifted by photoperiod and temperature manipulation[6,8]
COMMERCIALIZATION	
Product characteristics	Rich and tasty fillets, low lipid content (0.6–0.8%)[2]
Flesh and/or egg yield	46%[2]
Price/kg of whole fish gutted	France: US$6/kg[10]
Other ommercialization avenues	None

continued

RESEARCH AVENUES
Incubation and weaning phases must be improved. The growing out phase is not mastered.

[1] Russell F.R.S. (1976) *The Eggs and Planktonic Stages of British Marine Fishes.* Academic Press, London, 576 pp.
[2] Suquet M. (2001) *Le lieu jaune* (Pollachius pollachius): *biologie, pêche, marché et potentiel aquacole.* Ressources de la Mer, Ifremer (ed), Paris, 35 pp.
[3] Suquet M., Normant Y., Gaignon J.L., Quéméner L. and Fauvel C. (2005) Effect of water temperature on individual reproductive activity of pollack (*Pollachius pollachius*). *Aquaculture* 243, 113–120.
[4] Suquet, M., Petton, B., Normant, Y., Dosdat, A. and Gaignon, J.L. (1996) First rearing attempts of pollack, *Pollachius pollachius. Aquatic Living Resources* 9, 103–106.
[5] Petton B. IFREMER, Plouzané, France, 2001, personal communication.
[6] Buchet, V., Suquet, M., Omnes, M.H., Petton, B., Normant, Y., Severe, A., Quemener, L., Barone, H., Fauvel, C. and Gaignon, J.L. (2002) Pollack (*Pollachius pollachius*): acquisition of biological data in a breeding purpose. In: *Seafarming Today and Tomorrow.* European Aquaculture Society, Special publication 32, Trieste, Italy, pp. 165–166.
[7] Person-Le Ruyet, J., Buchet, V., Vincent, B., Le Delliou, H. and Quemener, L. (2006). Effects of temperature on the growth of pollack (*Pollachius pollachius*) juveniles. *Aquaculture* 251, 340–345.
[8] Omnes, M.H., Severe, A., Barone, H., Suquet, M., Buchet, V., Le Roux, A., Gaignon, J.L., Fostier, A. and Fauvel C. (2002) Growth and reproductive performances of juveniles and sexually mature pollack *Pollachius pollachius* in different conditions. In: *Seafarming Today and Tomorrow.* European Aquaculture Society, Special publication 32, Trieste, Italy, pp. 394–395.
[9] Quero, J.C. and Vayne, J.J. (1997) *Les poissons de mer des pêches françaises.* Delachaux and Niestlé, Lausanne-Paris, 304 pp.
[10] Ofimer (2006) Ventes sous criées: bilan 2005/2004 (http://www.ofimer.fr/99_up99load/2_actudoc/965d1_01.pdf, accessed 9 October 2006).

17. Technical sheet – Atlantic salmon (*Salmo salar*)

GAMETES AND DEVELOPING EGGS	
Egg characteristics	~ 6 mm diameter; 4000– 6000 eggs/l
Milt and sperm characteristics	Sperm density: ~ 2–20 spz × 10^9/ml (range 0.19–55 spz × 10^9/ml); volume per ejaculate: ~ 3–8 ml/kg; osmolality: ~ 190–270 mOsm/kg[1,2,3,4,5]
Egg yield (eggs/kg female)	600–1500 eggs/kg
Incubation time	500–530 degree days from fertilization to hatch; 230 degree days to eyed stage
Optimal T°C	6°C
Fertilization success	Normally 75–100%
Overall incubation success	80–90% survival to hatch; survival to the eyed stage: ~ 60–75%[4,5,6]
LARVAE–JUVENILE STAGE	
Larval size at hatching	15–25 mm
Optimal T°C	Larvae, 6–8°C; juvenile, 6–10°C, higher T° can be used to increase growth
Rearing units	Eggs/fry: hatching silo, trough and basket system, trough and flow system, hexhatch; juveniles: 1–4 m diameter tank[7]
Time until first-feeding	290 degree days
Survival from hatching to weaning	97–98 %
ON-GROWING STAGE	
Salinity	Seawater
Seawater adaptation mode	Parr–smolt transformation, i.e. preadaptive development of seawater tolerance in freshwater[8]
Commercial size	3–7 kg
Years to reach commercial size	2–3 years from stripping (12–18 months from 50–100 g smolt)
Rearing units	Predominantly cage culture, circular or square, 5–20 m deep
Rearing density	< 25 kg/m^3. Norway: 5–15 kg/m^3; Australia (Tasmania) and Chile: 8–10 kg/ m^3
Susceptibility to disease	High
Optimal T°C	~ 15°C
Diet	1–2.5 kg fish: 19 g protein/mJ, 2.5–5.0 kg fish: 16–17 g protein/mJ[9,10]

continued

REPRODUCTION	
Period	Autumn and early winter in FW
Fertilization mode	External
Fertilization protocol	Dry method. Stripping of males and females, mixing of sexual products for 1 min, then add equal volumes of fresh water and leave for 1.5 min. Wash off excess milt quickly, then transfer eggs to incubator
Time to reach sexual maturity	1–3 years at sea
Optimal temperature	~ 10–14°C during vitellogenesis, 5–11°C during final oocyte maturation[11,12,13]
Control of reproduction	Temperature, photoperiod and drug administration[13,14,15]
COMMERCIALIZATION	
Product characteristics	Moist, light and tender texture and red flesh[16]
Flesh yield	From gutted whole fish: 73–77%
Price/kg of whole fish, gutted	NOK17.5 (2003) – NOK43.1 (2006)[17]
RESEARCH AVENUES	
– Intensive smolt production in recirculated aquaculture systems – Reduce general production losses during the sea phase – Tailoring of flesh quality to consumer demands – Improve production traits through selective breeding – New raw materials as feed ingredients – 'Ecological' production	

[1] Gjerde, B. (1984) Variation in semen production of farmed Atlantic salmon and rainbow trout. *Aquaculture* 40, 109–114.

[2] Piironen, J. (1985) Variations in the properties of milt from the Finnish landlocked salmon (*Salmo salar m. sebago* Girard) during a spawning season. *Aquaculture* 48, 337–350.

[3] Piironen, J. and Hyveärinen, H. (1983) Composition of the milt of some teleost fishes. *Journal of Fish Biology* 22, 351–361.

[4] Aas, G.H., Refstie, T. and Gjerde, B. (1991) Evaluation of milt quality of Atlantic salmon. *Aquaculture* 95, 125–132.

[5] Estay, F., Vergara, C. and Diaz, N. F. (1999) Reproductive performance of cultured Atlantic salmon *Salmo salar* L. 1758 in Chile. *Aquaculture Research* 30, 759–764.

[6] Gjedrem, T. (1993) *Fiskeoppdrett. Vekstnæring for distrikts-Norge.* Landbruksforlaget, Oslo (in Norwegian).

[7] Kindness, R.B. (2002) Broodstock management and early freshwater stages. In: Stead, S.M. and Laird, L. (eds) *Handbook of Salmon Farming.* Springer, London, pp. 37–64.

[8] McCormick, S.D. and Saunders, R.L. (1987) Preparatory physiological adaptations for marine life of salmonids: osmoregulation, growth, and metabolism. *American Fisheries Society Symposium* 1, 211–229.

[9] Einen, O. and Roem, A.J. (1997) Dietary protein/energy ratios for Atlantic salmon in relation to fish size: growth, feed utilization and slaughter quality. *Aquaculture Nutrition* 3, 115–126.

[10] Sinnott, R. (2002) Fish farming and feed companies. In: Stead, S.M. and Laird, L. (eds) *Handbook of Salmon Farming.* Springer Praxis Publishing, Chichester, UK, pp. 105–185.

[11] Taranger, G. L. and Hansen, T. (1993) Ovulation and egg survival following exposure of Atlantic salmon, *Salmo salar* L., broodstock to different water temperatures. *Aquaculture and Fisheries Management* 24, 151–156.
[12] King, H. R. and Pankhurst, N. W. (2000) Ovulation of Tasmanian Atlantic salmon maintained at elevated temperatures: implications of climate change for sustainable industry development. In: Norberg, B., Kjesbu, O.S., Taranger, G.L., Anderson, E. and Stefansson, S.O. (eds) *Proceedings of the Sixth International Symposium on Reproductive Physiology of Fish.* University of Bergen, Bergen, Norway, pp. 396–398.
[13] King, H.R. and Pankhurst, N.W. (2004) Effect of short-term temperature reduction on ovulation and LHRHa responsiveness in female Atlantic salmon (*Salmo salar*) maintained at elevated water temperatures. *Aquaculture* 238, 421–436.
[14] Taranger, G.L., Stefansson, S.O., Oppedal, F., Andersson, E., Hansen, T. and Norberg, B. (2000) Photoperiod and temperature affect spawning time in Atlantic salmon (*Salmo salar* L.). In: Norberg, B., Kjesbu, O.S., Taranger, G.L., Anderson, E. and Stefansson, S.O. (eds) *Proceedings of the Sixth International Symposium on Reproductive Physiology of Fish.* University of Bergen, Bergen, Norway, p. 345.
[15] Taranger, G.L., Vikingstad, E., Klenke, U., Mayer, I., Stefansson, S.O., Norberg, B., Hansen, T., Zohar, Y. and Andersson, E. (2003) Effects of photoperiod, temperature and GnRHa treatment on the reproductive physiology of Atlantic salmon (*Salmo salar* L.) broodstock. *Fish Physiology and Biochemistry* 28, 403–406.
[16] Farmer, L.J., McConnell, J.M. and Kilpatrick, D.J (2000) Sensory characteristics of farmed and wild salmon (*Salmo salar* L.). *Aquaculture* 187, 105–125.
[17] Statistics Norway (www.ssb.no/english, accessed 31 October 2007).

18. Technical sheet – Arctic charr (*Salvelinus alpinus L.*)

GAMETES AND DEVELOPING EGGS	
Egg characteristics	4–5 mm diameter[1,2]
Milt and sperm characteristics	Sperm concentration ~ 4–18 × 10^9/ml, osmolality; ~ 225–280 mosmol/l, glucose concentration; ~ 1–9 mmol/l[3]
Egg yield (eggs/kg female)	1400–3770,[4] 1500–2600[5]
Incubation time	400–500 degree days[6,7]
Optimal incubation temperature	4–7°C[8]
Fertilization success	Close to 100%[9]
Overall incubation survival	76–96% to hatch[6,10,11]
LARVAE–JUVENILE STAGE	
Larval size (length) at hatching	14.9 mm,[11] 20 mm[5]
Optimum temperature for growth	Start-feeding of larvae: 6–8°C,[12] larger juveniles 14°C[13] or 15°C[14]
Rearing units	Small circular tanks with centre-drain standpipe[5]
Time from hatching until first-feeding (swim-up)	At 6–8°C: 40–45 days[12]
Live feed needed	No[13,15]
Survival from hatching to weaning	70–90%[16,17,18]
ON-GROWING STAGE	
Water salinity applied in commercial culture	Mostly fresh water, in some degree brackish water and/or shorter periods in full seawater[19]
Seawater adaptation mode	Parr–smolt transformation, i.e. preadaptive development of seawater tolerance in fresh water[20,21,22] or acclimation to gradual increasing salinities[23,24]
Commercial size	Pan-size: 200–300 g[25] or 2–3 kg[5]
Years to reach commercial size	Pan size: 1–2 years,[26] 2–3 kg: 3 years[5]
Rearing units	Net pens in lakes or circular tanks with centre-drain standpipe,[5] open sea cages or closed or semi-closed sea cages supplied with fresh water[27]
Rearing density	40–200 kg/m^3 [5,7,28,29]
Susceptibility to disease	Hardy fish,[7] more tolerant than other salmonids[5,30]
Optimal temperature for growth	12–13°C,[31] 13.4°C[32]
Diet	54% protein, 20% lipids, 13% carbohydrate[18]

continued

REPRODUCTION	
Period	From mid-July to January, depending on the strain[5]
Fertilization mode	External
Fertilization protocol	Stripping of males and females, mixing of sexual products for 2–3 min, dry method[5]
Time to reach sexual maturity	2–5 years depending on strain[5]
Optimal temperature	4–5°C[2,25]
Control of reproduction	Reproduction cycle can be manipulated by photoperiod,[33,34] temperature[2,35] and drug administration[36]

COMMERCIALIZATION	
Product characteristics	Orange-red flesh,[1] flaky texture, rich and creamy taste, very versatile and refined product[37]
Flesh yield	56–65%[38,39,40]
Price/kg of whole fish, gutted	5.80–6.60 CAD,[6] ~ NOK50[41]
Other commercialization avenues	Production of juveniles for enhancement of wild populations and larger fish for put-and-take or catch-and-release sport fisheries

RESEARCH AVENUES
– Improving intensive production in recirculated aquaculture systems – Improving quality and survival of eggs and juveniles – Improving flesh quality – Improving production traits through selective breeding – Optimize production environment with respect to photoperiod, water temperature and salinity

[1] Scott, W.B. and Scott, M.G. (1988) Atlantic fishes of Canada. *Canadian Bulletin of Fisheries and Aquatic Sciences* 219, 731pp.

[2] Jobling, M., Johnsen, H.K., Pettersen, G.W and Henderson, R.J. (1995) Effect of temperature on reproductive development in Arctic charr, *Salvelinus alpinus* (L.). *Journal of Thermal Biology* 20, 157–165.

[3] Aste, C.B., Audet, C. and Noüe, J. de la (2002) Effects of temperature and salinity on the reproductive success of Arctic Charr, *Salvelinus alpinus* (L): egg composition, milt characteristics and fry survival. *Aquaculture Research* 33, 299–309.

[4] Groot, C. (1996) Salmonid life history. In: Pennell, W. and Barton, B.A. (eds) *Principles of Salmonid Culture*. Elsevier, Amsterdam, pp. 97–230.

[5] Johnston, G. (2002) *Arctic Charr Aquaculture*. Fishing News Books, Oxford, UK.

[6] Swift, D.R. (1965) Effect of temperature on mortality and rate of development on the eggs of Windmere char (*Salvelinus alpinus*). *Journal of the Fisheries Research Board of Canada* 22(4), 919–917.

[7] Delabbio, J. (1995) Arctic charr culture in Atlantic Canada. In: Boghen, D. (ed.) *Cold-Water Aquaculture in Atlantic Canada*. The Canadian Institute for Research on Regional Development, Université de Moncton, NB, Canada, 672 pp. 85–106.

[8] Guillard, J., Gillet, C. and Champigneulle, A. (1992) Principales caractéristiques de l'élevage de l'omble chevalier (*Salvelinus alpinus* L.) en eau douce. *Bulletin Français de la Pêche et de la Pisciculture* 325, 47–68.
[9] Morin, R. (1996) Guide sur l'élevage des salmonidés, Fascicule 3 Reproduction, Incubation et Alevinage. Gouvernement du Québec, Ministère de l'Agriculture, des Pëcheries et de l'Alimentation du Québec.
[10] Steiner, V. (1984) Experiments towards improving the culture of Arctic charr (*Salvelinus alpinus*). In: Johnson, L. and Burns, B. (eds) *Biology of the Arctic Charr.* University of Manitoba Press, Winnipeg, pp. 509–521.
[11] Baroudy, E. and Elliott, J.M. (1994) Racial differences in eggs and juveniles of Windmere charr, *Salvelinus alpinus. Journal of Fish Biology* 45, 407–415.
[12] Wallace, J.C. and Aasjord, D. (1984) The initial feeding of Arctic charr (*Salvelinus alpinus*) alevins at different temperatures and under different feeding regimes. *Aquaculture* 38, 19–33.
[13] Jobling, M., Jørgensen, E.H., Arnesen, A.M. and Ringø, E. (1993) Feeding, growth, and environmental requirements of Arctic charr: a review of aquaculture potential. *Aquaculture International* 1, 20–46.
[14] Brännäs, E. and Wiklund, B.-S. (1992) Low temperature growth potential of Arctic charr and rainbow trout. *Nordic Journal of Freshwater Research* 67, 77–81.
[15] Dick, T.A. and Yang, X. (2002) Arctic char, *Salvelinus alpinus.* In: Webster, C.D. and Lim, C.E. (eds) *Nutrient Requirements and Feeding of Finfish for Aquaculture.* CAB International, Wallingford, UK, pp. 203–218.
[16] Koskela, J., Pirhonen, J., Lavens, P., Sorgeloos, P., Jaspers, E. and Ollevier, F. (1991) Growth and survival of first-feeding Arctic charr, *Salvelinus alpinus*, fed a dry diet supplemented with artemia. *Fish and Crustacean Larviculture Symposium, European Aquaculture Society, special publication* 15, Ghent, Belgium, pp.164–166.
[17] Dumas, S., Blanc, J.M., Audet, C. and Noüe, J. de la (1992) The early development of hybrids between brook charr (*Salvelinus fontinalis*) and Arctic charr (*Salvelinus alpinus*). *Aquaculture* 108, 21–28.
[18] Tabachek, R.-A.L. (1993) A review of the feeding and nutrition of Arctic charr, *Salvelinus alpinus* (L.), and priorities for future research. *Bulletin of the Aquaculture Association of Canada* 93(1), 10–17.
[19] Arnesen, A.M., Tromso, Norway, 2007, personal communication.
[20] Arnesen, A.M., Halvorsen, M. and Nilssen, K.J. (1992) Development of hypoosmoregulatory capacity in Arctic charr (*Salvelinus alpinus*) reared under either continuous light or natural photoperiod. *Canadian Journal of Fisheries and Aquatic Sciences* 49, 229–237.
[21] Halvorsen, M., Arnesen, A.M. and Nilssen, K.J. (1993) Osmoregulatory ability of anadromous Arctic charr, *Salvelinus alpinus* (L.), migrating towards the sea. *Aquaculture and Fisheries Management* 24, 199–211.
[22] Jørgensen, E.H., Aas-Hansen, Ø., Moriyama, S., Iwata, M. and Strand, J.E.T. (2007) The parr–smolt transformation of Arctic charr is comparable to that of Atlantic salmon. *Aquaculture* 273, 227–234.
[23] McCormick, S.D. and Saunders, R.L. (1987) Preparatory physiological adaptations for marine life of salmonids: osmoregulation, growth, and metabolism. *American Fisheries Society Symposium* 1, 211–229.
[24] Duston, J., Astatkie, T. and Murray, S.B. (2007) Effect of salinity at constant 10°C on grow-out of anadromous Arctic charr from Labrador. *Aquaculture* 273(4), 679–686.
[25] Jobling, M., Tveiten, H. and Hatlen, B. (1998) Cultivation of Arctic charr: an update. *Aquaculture International* 6, 181–196.
[26] Morin, R. (2007) Ministère de l'Agriculture, des Pêcheries et de l'Alimentation, Direction de l'innovation et des technologies, personal communication.

[27] Sæther, B.S. and Siikavuopio, S. (2005) Cage culture of Arctic charr. In: Adam, C, Bean, C., Brännes, E., Fraser, D., Gillet, C., Kummer, H., Moore, M., Pursiainen, M., Sæther, B.s., Sigurgeirsson, O., Siikavuopio, S. and Thorarensen, H. (eds) Arctic charr. Guidelines and standardized protocols. *The charr network*, www.charrnet.org (accessed 22 Ocober 2007).
[28] Papst, M.H. and Hopky, G.E. (1984) Development of an Arctic charr (*Salvelinus alpius* L.) broodstock. *Aquaculture* 43, 323–332.
[29] Siikavuopio, S.I. and Jobling, M. (1995) The effects of stocking density on survival and growth of wild-caught Arctic charr. *Nordic Journal of Freshwater Research* 71, 419–423.
[30] Robbins, N.J., Meaney, B., Haedrich, R., Goddard, S., Pepper, V. and McKeever, T. (1990) Arctic charr culture technology in northern Europe: an opportunity for technology transfer to Newfoundland. Newfoundland Department of Fisheries Industry Support Service Report No 44, Newfoundland, Canada.
[31] Papst, M.H. and Hopsky, G.E. (1983) Growth of Arctic charr (*Salvelinus alpinus* L.) in a pilot commercial rearing system. *Canadian Technical Report in Fisheries and Aquatic Sciences*, 1182.
[32] Sutterlin, A.M. and Stevens, E.D. (1992) Thermal behaviour of rainbow trout and Arctic charr in cages moored in stratified water. *Aquaculture* 102, 65–75.
[33] Gillet, C. (1993) Egg production in arctic charr (*Salvelinus alpinus* L.) broodstock: effects of photoperiod on timing of ovulation and egg quality. *Canadian Journal of Zoology* 72, 334–338.
[34] Frantzen, M., Tveiten, H., Arnesen, A.M., Damsgård, B. and Johnsen, H. (2004) Photoperiod manipulation of reproduction in Arctic charr (*Salvelinus alpinus* L.): effects on sex steroids and timing of maturation. *Aquaculture* 240, 561–574.
[35] Eriksson, L. and Wicklud, B. (1989) Culturing of Arctir charr. Development of Agriculture, Swedish University of Agricultural Science, Sweden.
[36] Krieger, K. (1991) Arctic charr farming in Canada. In: *From Securing Eggs to the Grow-out and Delivery of Products to the Customer. Arctic Charr Workshop Proceedings*. Aquaculture Industry Development Report, British Columbia, Canada, pp. 3–7.
[37] Wray, T. (1998) Arctic char is here. *Seafood International*, March 1998, 75–76.
[38] Hatlen, B., Arnesen, A.M. and Jobling, M. (1996) Muscle carotenoid concentrations in sexually maturing and immature Arctic charr, *Salvelinus alpinus* (L.). *Aquaculture Nutrition* 2, 207–212.
[39] Hatlen, B., Arnesen, A.M., Jobling, M., Siikavuopio, S. and Bjerkeng, B. (1997) Carotenoid pigmentation in relation to feed intake, growth and social interactions in Arctic charr, *Salvelinus alpinus* (L.) from two anadromous strains. *Aquaculture Nutrition* 3, 189–199.
[40] Jorgensen, E.H., Johansen, S.J.S. and Jobling, M. (1997) Seasonal patterns of growth, lipid deposition and lipid depletion in anadromous Arctic charr. *Journal of Fish Biology* 51, 312–326.
[41] Siikavuopio, S., Tromso, Norway, 2007, personal communication.

19. Technical sheet – Brook charr (*Salvelinus fontinalis*)

GAMETES AND DEVELOPING EGGS	
Egg characteristics	3.3–5 mm[1,2]
Milt and sperm characteristics	Can be cryopreserved[3]
Egg yield (eggs/kg female)	600[4]
Incubation time	503 degree days[5]
Optimal T°C	6°C[6] and 8°C at eyed stage[7]
Fertilization success	Close to 100%[8]
Overall incubation success	70–96%[9,10]
LARVAE–JUVENILE STAGE	
Larval size at hatching	14.6 mm, 0.072–0.078 g[11]
Optimal T°C	7–13°C;[12] 13°C[13]
Rearing units	Raceways[12]
Time until first-feeding	24–35 days[14,15]
Live feed needed	Formulated feed can be administered at the resorption of vitelline reserves[12]
Survival from hatching to weaning	95%[15]
ON-GROWING STAGE	
Commercial size	200 g[16]
Years to reach commercial size	2[16]
Rearing units	Indoor or outdoor rectangular, circular or cylindroconical tanks[12]
Rearing density	25 kg/m^3 (Dubé and Mason, 1995); 30 kg/m^3 [17,18]
Susceptibility to disease	High, susceptible to many bacterial and viral diseases and parasites[2,19,20]
Optimal T°C	15°C;[12] 16°C[6]
Diet	35–50% protein[12]
REPRODUCTION	
Period	October–April[12]
Fertilization mode	External fertilization[1]
Fertilization protocol	Stripping and mixing of the sexual products, dry method[12]
Time to reach sexual maturity	Males: 1 year[10,21,22] (44% mature after 1 year). Females: 2 years
Optimal T°C	7–13°C;[23] 9°C[6]
Control of reproduction	Thermal and photoperiodic manipulation[12,24,25,26]
COMMERCIALIZATION	
Product characteristics	74.5% moisture, 18% proteins, 5% lipids, 1.2% ash, 0.16% salt[27]
Flesh and/or egg yield	48–50.5%[28,29]
Price/kg of whole fish, gutted	US$4.95–5.85/kg[12]
Other commercialization avenues	None

continued

RESEARCH AVENUES
– Advances in breeding programmes – Development of species specific diets – Research on saltwater cage aquaculture

[1] Scott, W.B and Scott, M.G. (1988) Atlantic fishes of Canada. *Canadian Bulletin of Fisheries and Aquatic Sciences* 219, 731 pp.

[2] Scott, W.B. and Crossman, E.J. (1973) Freshwater fishes of Canada. *Bulletin of Fisheries Research Board of Canada* 184, 966 pp.

[3] Lahnsteiner, F. (2000) Cryopreservation protocols for sperm of salmonid fishes. In: Tiersch, T.R. and Mazik, P.M. (eds) *Cryopreservation in Aquatic Species.* World Aquaculture Society, Baton Rouge, Louisiana, pp. 91–100.

[4] Coad, B.W. (1993) *Guide des poissons marins de pêche sportive de l'Atlantique canadien et de la Nouvelle-Angleterre.* Broquet, La Prairie, 400 pp.

[5] Dumas, S., Blanc, J.M., Audet, C. and Noüe, J. de la (1995) Variation in yolk absorption and early growth of brook charr *Salvelinus fontinalis* (Mitchill), arctic charr, *Salvelinus alpinus* (L.), and their hybrids. *Aquaculture Research* 26, 759–764.

[6] Hokanson, K.E.F., McCormick, J.H., Jones, B.R. and Tucker, J.H. (1973) Thermal requirements for maturation, spawning, and embryo survival of the brook trout, *Salvelinus fontinalis. Journal of the Fisheries Research Board of Canada* 30, 975–984.

[7] Marten, P.S. (1992) Effect of temperature variation on the incubation and development of brook trout eggs. *The Progressive Fish-Culturist* 54, 1–6.

[8] Morin, R. (1996) Guide sur l'élevage des salmonidés, Fascicule 3 Reproduction, Incubation et Alevinage. Gouvernement du Québec, Ministère de l'Agriculture, des Pëcheries et de l'Alimentation du Québec.

[9] Savaria, J.Y. (1998) Amorce d'un programme de sélection génétique chez deux souches d'omble de fontaine en fonction des critères de croissance et de l'âge à la maturation sexuelle. MSc thesis, Université du Québec à Rimouski, Rimouski, Canada, 97 pp.

[10] Chimits, P. (1963) Un hybride pour la salmoniculture en eau froide: l'omble saumon. *Bulletin Français de la Pêche et de la Pisciculture* 35, 147–153.

[11] Le François, N.R., Blier, P.U., Adambounou, L.T. and Lacroix, M. (1999) Exposures to low-level ionizing radiation: effects on biochemical and whole-body indices of growth in juvenile brook charr (*Salvelinus fontinalis*). *Journal of Experimental Zoology* 283, 315–325.

[12] Dubé, P. and Mason, E. (1995) Trout culture in Atlantic Canada. In: Boghen, D. (ed.) *Cold-Water Aquaculture in Atlantic Canada.* The Canadian Institute for Research on Regional Development, Université de Moncton, NB, Canada, 672 pp.

[13] Dwyer, W.P., Piper, R.G. and Smith, C.E. (1983) Brook trout growth efficiency as affected by temperature. *Progressive Fish-Culturist* 45, 161–163.

[14] Pomerleau, R. (1973) *L'omble de fontaine.* Ministère du Tourisme, de la Chasse et de la Pêche, Direction générale de la Chasse et de la Pêche, Service de la Faune, Québec, 8 pp.

[15] Dumas, S., Blanc, J.M., Audet, C. and Noüe, J. de la (1992) The early development of hybrids between brook charr (*Salvelinus fontinalis*) and Arctic charr (*Salvelinus alpinus*). *Aquaculture* 108, 21–28.

[16] Morin, R. (2007) Ministère de l'Agriculture, des Pêcheries et de l'Alimentation, Direction de l'innovation et des technologies, personal communication.

[17] Vijayan, M.M. and Leatherland, J.F. (1988) Effect of stocking density on the growth and stress-response in brook charr, *Salvelinus fontinalis. Aquaculture* 75, 159–170.

[18] Vijayan, M.M., Ballantyne, J.S. and Leatherland, J.F. (1990) High stocking density alters the energy metabolism of brook charr, *Salvelinus fontinalis. Aquaculture* 88, 371–381.

[19] Cipriano, R.C., Marchant, D., Jones, T.E. and Schachte, J.H. (2002) Practical application of disease resistance: a brook trout fishery selected for resistance to furunculosis. *Aquaculture* 206, 1–17.
[20] Perry, G.M.L., Tarte, P., Croisetière, S., Belhumuer, P. and Bernatchez, L. (2004) Genetic variance and covariance for 0+ brook charr (*Salvelinus fontinalis*) weight and survival time of furunculosis (*Aeromonas salmonicida*) exposure. *Aquaculture* 235, 263–271.
[21] Lafleur, P.E., Larrivée, M. and St-Gelais, S. (1986) La production de salmonidés en eau salée. In: Colloque sur l'aquiculture. Perspectives d'avenir en aquaculture, Saint-Hyacinthe, QC, Canada. 14, 15 March 1986. MAPAQ, pp. 143–156.
[22] McCormick, S.D. and Naiman, R.J. (1984) Some determinants of maturation in brook trout, *Salvelinus fontinalis. Aquaculture* 43, 269–278.
[23] Piper, R.G., McElwain, I.B., Orme, L.E., McCraren, J.P., Fowler, L.G., and Leonard, J.R. (1982) Fish hatchery management. US Department of the Interior, Fish and Wildlife Service, Washington, DC.
[24] Henderson, N.E. (1963) Influence of light and temperature on the reproductive cycle of the eastern brook trout, *Salvelinus fontinalis* (Mitchill). *Journal of the Fisheries Research Board of Canada* 20, 859–897.
[25] Carlson, A.R. and Hale, J.G. (1973) Early maturation of brook trout in the laboratory. *The Progressive Fish-Culturist* 35, 150–153.
[26] Holcombe, G.W., Pasha, M.S., Jensen, K.M., Tietge, J.E. and Ankley, G.T. (2000) Effects of photoperiod manipulation on brook trout reproductive development, fecundity, and circulating sex steroid concentrations. *North American Journal of Aquaculture* 62, 1–11.
[27] Coulombe, N., Carbonneau, M.-E. and Renaud, N. (2005) Évaluation sensorielle des Ombles de fontaine et des Ombles chevalier en bassin d'eau douce et d'eau salée. MAPAQ – Pêcheries, DIT – Rapport d'analyse 2005–01, 25 pp.
[28] Straus, K. (1991) *The Seafood Handbook, Seafood Standards.* Seafood Business, Rockland, Massachusetts, USA, 144 pp.
[29] Rasmussen, R.S. and Ostenfeld, T.H. (2000) Effect of growth rate on quality traits and feed utilization of rainbow trout (*Oncorhynchus mykiss*) and brook trout (*Salvelinus fontinalis*). *Aquaculture* 184, 327–337.

20. Technical sheet – Red drum (*Sciaenops ocellatus*)

GAMETES AND DEVELOPING EGGS	
Egg characteristics	Diameter 0.8–1.0 mm, pelagic, size varies with salinity, buoyant at spawning[1,2]
Milt and sperm characteristics	No data; can be preserved by cryopreservation[3]
Egg yield (eggs/kg female)	Multiple spawner: 1.7 million/kg female/year; individual spawns yield ~ 0.1 million eggs/kg[4,5]
Incubation time	18–29 h[2]
Optimal T°C	25–30°C[6]
Fertilization success	99%[5]
Overall incubation success	94–99%[5]
LARVAE–JUVENILE STAGE	
Larval size at hatching	2.2 mm yolk-sac stage; 25 μg dry weight[1,7]
Optimal T°C	25–30°C[6]
Rearing units	Conical tanks,[8] ponds[9]
Time until first-feeding	36–48 h[1]
Live feed needed	Enriched rotifers[1]
Survival from hatching to weaning	Up to 84%[8]
ON-GROWING STAGE	
Commercial size	1–2 kg in North America[10,11]
Years to reach commercial size	1–1.5, depending on temperature[10,11,12,13]
Rearing units	Earthen ponds, raceways, cages and net pens[10,11,12]
Rearing density	0.5–2.2 kg/m² (1–1.5 kg fish)[13,14,15,16]
Susceptibility to diseases	Low (parasites, mainly *Amyloodinium* and some bacteria such as *Vibrio*)[12,17,18]
Optimal T°C	24–30°C[10,12,13]
Diet	Floating or slow sinking diets, 35–45% protein, 7–11% lipid, 15 kJ/g diet energy[10,19]
REPRODUCTION	
Period	September–November in wild but year round using photothermal cycling[4]
Fertilization mode	External fertilization
Fertilization protocol	Natural tank spawning at 12 light:12 dark and 24–26°C[4,5]
Time to reach sexual maturity	3 years; 4–5 kg in wild but precocious spawning in the lab at 19.5 months; 2.9 kg[20,21]
Optimal T°C	24–26°C[4,5]
Control of reproduction	Photothermal conditioning or hormonally induced[21]

continued

COMMERCIALIZATION	
Product characteristics	Fillets are white, flaky, mild and no strong fish taste. Perfect for US market[11] Fillets are 77.5% water, 19.2% protein and 1.25% lipid[22]
Flesh and/or egg yield	34% skin-on fillets, 28% skinless fillet[11]
Price/kg of whole fish, gutted	US$4.32–7.41/kg whole fish, US$10.00/kg fillet (wholesale)[11,15,16]
Other commercialization avenues	Live fish for fee fishing, for golf course lakes, for stock enhancement

RESEARCH AVENUES

- Broodstock diet
- Egg quality
- Complete larval production without live prey
- Improve juvenile production in RAS, grading, diets, energy use
- Overwintering techniques in ponds
- Lowering feed cost
- Lower intraperitoneal fat on high energy diets
- Market research

[1] Holt, J., Johnson, A.G., Arnold, C.R., Fable, W.A. Jr and Williams, T.D. (1981) Description of eggs and larvae of laboratory reared red drum, *Sciaenops ocellata. Copeia* 1981(4), 751–756.
[2] Holt, G.J. (2005) Red drum aquaculture. In: Kelly, A.M. and Silverstein, J. (eds) *Aquaculture in the 21st Century. American Fisheries Society Symposium* 46, 457–463.
[3] Wayman, W.R., Thomas, R.G. and Tiersch, T.R. (1998) Refrigerated storage and cryopreservation of red drum sperm (*Sciaenops ocellatus*). *Aquaculture Research* 29, 267–273.
[4] Arnold, C.R. (1988) Controlled year-round spawning of red drum *Sciaenops ocellatus* in captivity. *Contributions in Marine Science* 30, 65–70.
[5] Arnold, C.R., Bailey, W.H., Williams, T.D., Johnson, A. and Lasswell, J.L. (1979) Laboratory spawning and larval rearing of red drum and southern flounder. *Proceedings of the 1977 Annual Conference Southeastern Association of Fish and Wildlife Agencies* 31, 437–440.
[6] Holt, J., Godbout, R. and Arnold, C.R. (1981) Effects of temperature and salinity on egg hatching and larval survival of red drum, *Sciaenops ocellata. Fisheries Bulletin* 79(3), 569–573.
[7] Lee, W.Y., Holt, G.J. and Arnold, C.R. (1984) Growth of red drum larvae in the laboratory. *Transactions of the American Fisheries Society* 113, 243–246.
[8] Holt, G.J. (1993) Feeding larval red drum on microparticulate diets in a closed recirculating water system. *Journal of the World Aquaculture Society* 24(2), 225–230.
[9] Colura, R.L., Hysmith, B.T. and Stevens, R.E. (1976) Fingerling production of striped bass (*Morone saxatalis*), spotted seatrout (*Cynoscion nebulosis*), and red drum (*Sciaenops ocellatus*) in saltwater ponds. *Proceedings of the World Mariculture Society Symposium* 7, 79–92.
[10] Gatlin, D. (2000) Red drum culture. In: Stickney, R.R. (ed.) *The Encyclopedia of Aquaculture.* John Wiley and Sons, New York, pp. 736–742.
[11] Miller, W.W. (1995) *Red Drum. Aquaculture Curriculum Guide.* National Council for Agricultural Education, USDA, Iowa, 117 pp.
[12] Matlock, G.C. (1990) Life history of red drum. In: Chamberlain, G.W., Miget, R.J. and Haby, M.G. (eds) *Red Drum Aquaculture.* Texas A and M University Sea Grant College Program, TAMU-SG-90-603, College Station, Texas, USA, pp. 1–21.

[13] Arnold, C.R., Holt, G.J. and Thomas, P. (1988) *Red Drum Aquaculture. Proceedings of a Symposium on the Culture of Red Drum and Other Warm Water Fishes.* University of Texas at Austin Marine Science Institute, *Contributions in Marine Science* 30 (Supplement), 197 pp.
[14] Hopkins, J.S. (1990) Semi-intensive grow-out in saltwater. In: Chamberlain, G.W., Miget, R.J. and Haby, M.G. (eds) *Red Drum Aquaculture.* Texas A and M Sea Grant College Program, College Station, Texas, pp. 170–177.
[15] Davis, J. T. (1991) *Red Drum: Production of Food Fish.* Publication No. 322, Southern Regional Aquaculture Center, Stoneville, Mississippi.
[16] Thacker, S.G., Griffin, W., Rahman, A. and Lambregts, J. (1991) Modeling the financial risk of intensive redfish *Sciaenops ocellatus* aquaculture in Texas. *Journal of the World Aquaculture Society* 22(3), 42A–43A.
[17] Johnson, S.K. (1990) Recognition and control of diseases common to grow-out aquaculture of red drum. In: Chamberlain, G.W., Miget, R.J. and Haby, M.G. (eds) *Red Drum Aquaculture.* Texas A and M University Sea Grant College Program, College Station, Texas, pp. 113–130.
[18] Hawke, J.P. (1991) An overview of the diseases of cultured hybrid striped bass and red drum. *Louisiana Aquaculture Conference 1991,* Louisiana Cooperative Extension Service and Louisiana Agricultural Experiment Station, LSU Agricultural Center, Baton Rouge, pp. 52–54.
[19] Gatlin, D.M. (2002) Red drum, *Sciaenops ocellatus.* In: Webster, C. and Lim, C. (eds) *Nutrient Requirements and Feeding of Finfish for Aquauculture.* CAB International, Wallingford, UK, pp. 147–158.
[20] Arnold, C.R. (1991) Precocious spawning of red drum. *Progressive Fish Culturist* 53, 50–51.
[21] Thomas, P., Arnold, C.R. and Holt, J. (1995) Red drum and other sciaenids. In: Bromage, N.R. and Roberts, R.J. (eds) *Broodstock Management and Egg and Larval Quality.* Blackwell Scientific Publishers, Berlin, pp. 118–137.
[22] Davis, D.A. Lazo, J.P. and Arnold, C.R. (1999) Response of juvenile red drum (*Sciaenops ocellatus*) to practical diets with medium chain triglycerides. *Fish Physiology and Biochemistry* 21, 235–248.

21. Technical sheet – Turbot (*Scophthalmus maximus*)

GAMETES AND DEVELOPING EGGS	
Egg characteristics	Diameter 0.98–1.20 mm, pelagic[1,2]
Milt and sperm characteristics	Milt volume: ≈ 1.6 ml. Sperm concentrations: 38 × 10^9 spz/ml).[3] Sperm motile at stripping. Fertilization with frozen sperm efficient[2]
Egg yield (eggs/kg female)	300,000–500,000[1,9]
Incubation time	2450 degree hours at 10°C to 1800 degree hours at 20°C, equivalent to 6–7 days at 13°C[1,3]
Optimal T°C	≈ 13°C[1,3]
Fertilization success	50–100%, depends on egg and sperm quality[1,3]
Overall incubation success	30–80%[1,3]
LARVAE–JUVENILE STAGE	
Larval size at hatching	≈ 3 mm, 0.1–0.2 mg[1,2,4,5]
Optimal T°C	18–19°C in the first 60 days,[1,4,6] 16–21°C in a semi-intensive system[6]
Rearing units	Fibreglass tanks of 2 × 2 × 0.5 m[1] or larger circular tanks[6]
Time until first-feeding	Standardized industrialized production is rotifers from 2–12 dph, *Artemia* 8–28 dph, formulated feed from *c.*28–35 dph[1,6,7]
Live feed needed	
Survival from hatching to weaning	20–80%,[1,5] highly variable. Commercial producers: estimates 20–30%[1,5,6]
ON-GROWING STAGE	
Commercial size	From 0.7 kg in China, > 1 kg in Europe[1]
Years to reach commercial size	1–2 years from hatching[1,8]
Rearing units	Shallow large tanks of concrete[2,7,12] or fibreglass,[2,7] shallow raceways[10,12]
Rearing density	100–300% bottom coverage,[19] i.e. 25–40 kg/m² (50–300 g fish);[11] 25–40 kg/m² (750 g fish);[11] < 70 kg/m² (0.8–3 kg fish)[11,13]
Susceptibility to diseases	Moderate/low[21,22,23,24]
Optimal T°C	22–22°C (0–50 g);[14,15] 19°C (50–100 g),[16,17] 16–17°C (> 100 g)[19]
Diet	Floating diet, 49–52% proteins, 12% fat[20]

continued

REPRODUCTION	
Period	Ovarian growth is initiated during February–May. Spawning May–August[25,26]
Fertilization mode	External fertilization[25,26]
Fertilization protocol	Eggs (in ovarian fluid) and milt are incubated for 5–10 min before transfer to seawater[25,26]
Time to reach sexual maturity	2–4 years under culture conditions[25] (lower age at first maturity in males)[27] 4–6 years in the wild[27,28]
Optimal T°C	Ovarian growth: 15°C; spawning: 12–13°C[3,9,25]
Control of reproduction	Spawning time can be shifted by photoperiod and T° manipulation[3,9,25,29]
COMMERCIALIZATION	
Product characteristics	Rich and tasty fillet,[1] long shelf life and firm texture,[30,31] no parasites or internal bones, 74.5–79.0% humidity, 0.7–30% fat, 17.4–19.2% proteins, 1.2–3.6% ash[30,31]
Flesh and/or egg yield	Fillet yield: ≈ 30% (650 g fish),[31] 45–55% (> 1.5 kg fish)[32]
Price/kg of whole fish, gutted	Europe US$10.90–19 (depending on size, highest for fish > 3 kg)
Other commercialization avenues	N/A
RESEARCH AVENUES	
– Optimization of feeding efficiency and feed composition[11,14,18] – Genetic characterization and selection programmes[11,14] – Optimization of rearing methods[11,14,19, 27, 29,31] – Fish health, development of vaccines[21,22,23,24] – Product development[11,31]	

[1] Person-Le Ruyet, J., Baudin-Laurencin, F., Devauchelle, N., Métailler, R., Nicolas, J.-L., Robin, J. and Guillaume, J. (1991) Culture of turbot (*Scophthalmus maximus*). In: McVey, J.P. (ed.) *CRC Handbook of Mariculture. Vol. 2. Finfish Aquaculture.* CRC Press, Boston, Massachusetts, pp. 21–41.

[2] Iglesias, J., Rodriguez-Ojea, G. and Peleteiro, J.B. (1991) Effect of light and temperature on the development of turbot eggs (*Scophthalmus maximus* L.). *International Council for the Exploration of the Sea.* CM 1991, F:51 (in mimeo).

[3] Leclercq, D. (1994) Turbot broodstock management: a key point to mid-term progress of the turbot industry. In: Lavens, P. and Remmerswaal, R.A.M. (eds) *Turbot Culture: Problems and Prospects.* European Aquaculture Society, Oostende, Belgium, pp. 3–13.

[4] Person-Le Ruyet, J. (1989) The hatchery rearing of turbot larvae (*Scopthalmus maximus,* L.). *Cuadernos da área de ciencias marinas* 3, 57–91.

[5] Minkoff, G. and Broadhurst, A.P. (1994) Intensive production of turbot, *Scophthalmus maximus*, fry. In: Lavens, P. and Remmerswaal, R.A.M. (eds) *Turbot Culture: Problems and pPospects.* European Aquaculture Society, Oostende, Belgium, pp. 14–31.

[6] Riaza, A. and Hall, J. (1993) Large scale production of turbot. In: Reinertsen, H., Dahle, L.A., Jorgensen, L. and Tvinnereim, K. (eds) *Fish Farming Technology*. A.A. Balkema, Rotterdam, The Netherlands, pp. 147–148.
[7] Production cycle of turbot, illustrated schematically (www.fao.org/figis).
[8] A.K. Imsland. Akvaplan-niva, Iceland Office, Kopavogur, Iceland, 2007, personal communication.
[9] Stoss, J.H. and Røer, J.E. (1993) Controlled reproduction in turbot (*Scopthalmus maximus*). In: Reinertsen, H., Dahle, L.A., Jørgensen, L. and Tvinnereim. K. (eds) *Fish Farming Technology*. A.A. Balkema, Rotterdam, The Netherlands, pp. 29–34.
[10] Oiestad, V. (1999) Shallow raceways as a compact, resource-maximizing farming procedure for marine fish species. *Aquaculture Research* 30(11–12), 831–840.
[11] Person-Le Ruyet, J. (2002) Turbot (*Scopthalmus maximus*) grow-out in Europe: practices, results and prospects. *Turkish Journal of Fisheries and Aquatic Sciences* 2, 29–39.
[12] Brown, N. (2002) Flatfish farming systems in the Atlantic region. *Reviews in Fisheries Science* 10, 403–419.
[13] Martinez-Tapia, C. and Fernández-Pato, C.A. (1991) Influence of stock density on turbot (*Scophthalmus maximus* L.) growth. *International Council for the Exploration of the Sea, C.M.* F:20, 7 pp.
[14] Imsland, A.K., Foss, A., Stefansson, S.O. and Nævdal, G. (2000) Hemoglobin genotypes of turbot (*Scophthalmus maximus*): variation in growth and optimal temperature for growth. *Fish Physiology and Biochemistry* 23, 75–61.
[15] Imsland, A.K., Foss, A., Gunnarsson, S., Berntssen, M., FitzGerald, R., Bonga, S.W., van Ham, E., Nævdal, G. and Stefansson, S.O. (2001) The interaction of temperature and salinity on growth and food conversion in juvenile turbot (*Scophthalmus maximus*). *Aquaculture* 198, 353–367.
[16] Burel, C., Person-Le Ruyet, J., Gaumet, F., Le Roux, A., Sévère, A. and Boeuf, F. (1996) Effects of temperature on growth and metabolism in juvenile turbot. *Journal of Fish Biology* 49, 678–692.
[17] Imsland, A.K., Sunde, L.M., Folkvord, A. and Stefansson, S.O. (1996) The interaction between temperature and size on growth of juvenile turbot. *Journal of Fish Biology* 49, 926–940.
[18] Imsland, A.K., Wergeland, T., Jonassen, T. and Stefansson, S.O. (2006a) Does malpigmentation improve growth in juvenile turbot (*Scophthalmus maximus*) and halibut (*Hippoglossus hippoglossus*). *Aquaculture Research* 37, 306–312.
[19] Irwin, S., O'Halloran, J. and FitzGerald, R.D. (1999) Stocking density, growth and growth variation in juvenile turbot, *Scophthalmus maximus* (Rafinesque). *Aquaculture* 178, 78–88.
[20] Guillaume, J., Coustans, M.F., Mæetailler, R., Person-Le Ruyet, J. and Robin, J. (1991) Flatfish, turbot, sole and plaice. In: Wilson, R.P. (ed.) *Handbook of Nutrient Requirements of Finfish*. CRC Press, Boston, Massachusetts, pp. 77–82.
[21] Mulcahy, M.F. (2002) Diseases of flatfish. *Bulletin of the European Association of Fish Pathologists* 22, 86–94.
[22] Hellberg, H., Koppang, E.O., Torud, B. and Bjerkas, I. (2002) Subclinical herpesvirus infection in farmed turbot *Scophthalmus maximus. Diseases of Aquatic Organisms* 49, 27–31.
[23] Vazquez Branas, M., Morales, J.C. and Estepa, A. (1994) A sandwich ELISA to detect VHSV and IPNV in turbot. *Aquaculture International* 2, 206–217.
[24] Novoa, B., Figueras, A., Puentes, C.F., Ledo, A. and Toranzo, A.E. (1993) Characterization of a birnavirus isolated from diseased turbot cultured in Spain. *Diseases of Aquatic Organisms* 15, 163–169.
[25] Devauchelle, N., Alexandre, J.C., Le-Corre, N. and Letty, Y. (1987) Spawning of turbot (*Scophthalmus maximus*) in captivity. *Aquaculture* 69, 159–184.
[26] McEvoy, L.A. (1989) Reproduction of turbot (*Scophthalmus maximus*) in captivity. Seminar on technology of turbot culture, O. Castro, Sada (La Coruna, Spain). *Seminario de Estudos Galegos* 3, 9–28.

[27] Imsland, A.K., Folkvord, A., Grung, G.L., Stefansson, S.O. and Taranger, G.L. (1997) Sexual dimorphism in growth and maturation of turbot, *Scophthalmus maximus* (Rafinesque 1810). *Aquaculture Research* 28, 101–114.

[28] Jónsson, G. (1992) *Íslenskir fiskar* (Icelandic fishes) Reykjavík: Fjölvaútgáfan, 568 pp. (in Icelandic).

[29] Imsland, A.K., Dragsnes, M. and Stefansson, S.O. (2003) Exposure to continuous light inhibits maturation in turbot (*Scophthalmus maximus*). *Aquaculture* 219, 911–919.

[30] Ruff, N., Fitzgerald, R.D., Cross, T.F. and Kerry, J.P. (2002) Comparative composition and shelf-life of fillets of wild and cultured turbot (*Scophthalmus maximus*) and Atlantic halibut (*Hippoglossus hippoglossus*). *Aquaculture International* 10, 241–256.

[31] Regost, C., Arzel, J., Cardinal, M., Robin, J., Laroche, M. and Kaushik, S.J. (2001) Dietary lipid level, hepatic lipogenesis and flesh quality in turbot (*Psetta maxima*). *Aquaculture* 193, 291–309.

[32] TURPRO. SME-Cooperative Research Contract 508070, Biological optimization and development of processing methods for turbot farming (http://www.fp6.eurocean.org/record.jsp?open=144&a=1135393930620., accessed 20 June 2007).

22. Technical sheet – Sole (*Solea solea* and *S. senegalensis*)

GAMETES AND DEVELOPING EGGS	
Egg characteristics	Diameter 1.0–1.60 mm, pelagic[1,2,3]
Milt and sperm characteristics	Milt volume very low; 5–80 μl,[4] sperm concentrations: 20–60 × 10^6 spz/ml.[4] Higher production and spermatozoa density in wild-captured broodstocks[4]
Egg yield (eggs/kg female)	140,000–200,000[5,6]
Incubation time	≈ 800 degree hours at 19°C[7] (*senegalensis*); ≈ 1400 degree hours at 12–13°C[5,8] (*solea*)
Optimal T°C	≈ 19°C[7] (*senegalensis*), 12–15°C[5,8] (*solea*)
Fertilization success	50–100%, depends on egg and sperm quality[1,3]
Overall incubation success	30–80%[5,8]
LARVAE–JUVENILE STAGE	
Larval size at hatching	≈ 2.2–2.9 mm[6,7] (*senegalensis*), 4–5 mm[7,8] (*solea*)
Optimal T°C	19–24°C in the first 60 days[6,9]
Rearing units	Fibreglass or concrete tanks of various sizes[6,8]
Time until first-feeding	First feeding at 3 days posthatch (dph)[6,8]
Live feed needed	Industrialized production is *Artemia* from 3–40 dph, formulated feed from *c.*28–35 dph[6,10]
Survival from hatching to weaning	40–80%,[6,8] rearing the larvae though to metamorphosis presents few problems as compared to other flatfish species
ON-GROWING STAGE	
Commercial size	From 125 g[8]
Years to reach commercial size	1–2 years from hatching[6,8]
Rearing units	Shallow large tanks of concrete or fibreglass,[6,8,11] shallow raceways, [12] earthen ponds[7]
Rearing density	Low, < 100% bottom coverage[6,7,13] < 10 kg/m^2 for fish < 100 g[14]
Susceptibility to disease	Moderate,[15,16,17,18] earlier problems due to black batch necrosis (BPN) in *S. solea*[8]
Optimal T°C	16–22°C[6,8,19,20] (*solea*); 22–27°C[6,7] (*senegalensis*)
Diet	Oligochaete worms,[8] extruded pellets,[6,8,21] 50–55% proteins, 12–16% fat[21]

continued

REPRODUCTION	
Period	Spawning commences during March–mid-May[5,6] (*solea*) and April–June (*senegalensis*)[7,22]
Fertilization mode	Natural spawning,[6,8,22] stripping appears not feasible[5,8]
Fertilization protocol	Eggs collected in the water column[2,3,5,8]
Time to reach sexual maturity	4–6 years in the wild,[24] 3–4 years in captivity[2,5,8]
Optimal T°C	Spawning: 8–12°C[5,6,23] (*solea*), 18–20°C[3,7] (*senegalensis*)
Control of reproduction	Spawning time can be shifted by photoperiod and temperature manipulation[5,22,23]
COMMERCIALIZATION	
Product characteristics	Rich and tasty fillet;[6,8,25] long shelf life and firm texture,[26] 75.5% humidity, 5.1% fat, 16.2% proteins, 2.9% ash[27]
Flesh and/or egg yield	Fillet yield: 30–35%[27,28]
Price/kg of whole fish, gutted	€10–16
Other commercialization avenues	N/A
RESEARCH AVENUES	
– Optimization of feeding efficiency and feed composition[5,8,11] – Development of formulated feed for broodstocks[5,8] – Optimization of rearing methods[5,13] – Fish health, development of vaccines, solving fin erosion problem[5,15,16,17,18] – Product development[5,8,25,27]	

[1] Houghton, R.G., Last, J.M. and Bromley, P.J. (1985) Fecundity and egg size of sole (*Solea solea* (L.)) spawning in captivity. *Journal du Conseil International pour l'Exploration de la Mer* 42, 162–165.

[2] Baynes, S.M., Howell, B.R. and Beard, T.W. (1993) A review of egg production by captive sole, *Solea solea* (L.). *Aquaculture and Fisheries Mangagement* 24, 171–180.

[3] Bedoui, R. (1995) Rearing of *Solea senegalensis* (Kaup, 1958) in Tunisia. *Cahiers Options Méditerranéenes* 16, 31–39.

[4] Cabrita, E., Soares, F. and Dinis, M.T. (2006) Characterization of Senegalese sole, *Solea senegalensis*, male broodstock in terms of sperm production and quality. *Aquaculture* 261, 967–975.

[5] Devauchelle, N., Alexandre, J.C., Le-Corre, N. and Letty, Y. (1987) Spawning of sole (*Solea solea*) in captivity. *Aquaculture* 66, 125–147.

[6] Imsland, A.K., Foss, A., Conseição, L.E.C., Dinis, M.T., Kamstra, A., Schram, E., Daanbar, D., Rema, P. and White, P. (2004) A review of the culture potential of *Solea solea* and *S. senegalensis*. *Reviews in Fish Biology and Fisheries* 13, 379–409.

[7] Dinis, M.T., Ribeiro, L., Soares, F. and Sarasquete, C. (1999) A review on the cultivation potential of *Solea senegalensis* in Spain and in Portugal. *Aquaculture* 176, 27–38.

[8] Howell, B.R. (1997) A re-appraisal of the potential of the sole, *Solea solea* (L.), for commercial cultivation. *Aquaculture* 155, 359–369.

[9] Fonds, M. (1979) Laboratory observations on the influence of temperature and salinity on development of the eggs and growth of the larvae of *Solea solea* (Pisces). *Marine Ecology Progress Series* 1, 91–99.
[10] Day, O.J., Howell, B.R., Aksnes, A. and Nygård, E. (1999) Recent advances in the weaning of sole, *Solea solea* (L.). *Abstracts of contributions presented at the International Conference Aquaculture Europe 1999.* European Aquaculture Society, Oostede, Belgium, Special Publication No. 27, pp. 40–41.
[11] Engrola, S., Conceição, L.E.C., Pereira, R. and Dinis, M.T. (2002) Growth and diet utilisation of sole, *Solea senegalensis* reared under different feeding regimes. *Book of Abstracts of the 10th International Symposium on Nutrition and Feeding in Fish*, 2–7 June 2002, Rhodes, Greece.
[12] Oiestad, V. (1999) Shallow raceways as a compact, resource-maximizing farming procedure for marine fish species. *Aquaculture Research* 30(11–12), 831–840.
[13] Howell, B.R (1998) The effect of stocking density on growth and size variation in cultured turbot, *Scophthalmus maximus*, and sole *Solea solea. ICES CM*, 1998/L:10.
[14] Schram, E., Van der Heul, J.W., Kamstra, A. and Verdegem, M.C.J. (2006) Stocking density-dependent growth of Dover sole (*Solea solea*). *Aquaculture* 252, 339–347.
[15] Baynes, S.M. and Howell, B.R. (1993) Observations on the growth, survival and disease resistance of juvenile common sole, *Solea solea* (L.), fed *Mytilus edulis* L. *Aquaculture and Fisheries Mangagement* 24, 95–100.
[16] Bernadet, J.F., Campbell, A.C. and Buswell, J.A. (1990) *Flexibacter maritimus* is the agent of 'black patch necrosis' in Dover sole in Scotland. *Diseases of Aquatic Orgamisms* 8, 233–237.
[17] Zorrilla, I., Balebona, M.C., Moriñigo, M.A., Sarasquete, C. and Borrego, J.J. (1999) Isolation and characterization of the causative agent of pasteurellosis, *Photobacterium damsela* ssp. *piscicida*, from sole, *Solea senegalensis* (Kaup). *Journal of Fish Diseases* 22, 167–172.
[18] Zorrilla, I., Arijo, S., Chabrillon, M., Diaz, P., Martinez-Manzanares, E., Balebona, M.C. and Moriñigo, M.A. (2003) *Vibrio* species isolated from diseased farmed sole, *Solea senegalensis* (Kaup), and evaluation of the potential virulence role of their extracellular products. *Journal of Fish Diseases* 26, 103–108.
[19] Irvin, D.N. (1973) The growth and survival of Dover sole, *Solea solea,* and some observations on the growth of the plaice *Pleuronectes platessa*, considered at various temperatures. PhD thesis, University of Liverpool, UK, 186 pp.
[20] Fonds, M. (1976) The influence of temperature and salinity on growth of young sole *Solea solea* L. *10th European Symposium on Marine Biology* 1, Ostend, Belgium, pp.109–125.
[21] Coutteau, P., Robles, R. and Spruyt, W. (2001) Ongrowing feed for Senegal sole (*Solea senegalensis* Kaup). *Abstracts of contribution presented at the International Conference Aquaculture Europe 2001.* European Aquaculture Society, Special Publication No. 29, Ostend, Belgium, pp. 58–59.
[22] Dinis, M.T., Soares, F., Ribeiro, L., Engrola, S., Cação, P., Aragão, C., Pousão-Ferreira, P. and Conceição, L.E.C. (2003) Broodstock management and larval rearing of Senegalese sole (*Solea senegalensis*). *Book of Abstracts of the World Aquaculture 03,* Salvador, Brazil, 19–23 May 2003.
[23] Lenzi, M. and Salvatori, R. (1989) Management of a module for sole eggs production. In: De Pauw, N., Jaspers, E. Ackefors, H. and Wilkins, N. (eds) *Aquaculture: A Biotechnology in*

Progress. Proceedings of the Aquaculture Europe 1987 Conference. European Aquaculture Society, Bredene, Belgium, pp. 549.
[24] Bromley, P.J. (2003) The use of market sampling to generate maturity ogives and to investigate growth, sexual dimorphism and reproductive strategy in central and south-western North Sea sole (*Solea solea* L.). *ICES Journal of Marine Science* 60, 52–65.
[25] Brown, N. (2002) Flatfish farming systems in the Atlantic region. *Reviews in Fisheries Science* 10, 403–419.
[26] Pranzetti, L., Martinoli, S., Piergiovanni, L. and Galli, A. (2001) Influence of active packaging on the shelf-life of minimally processed fish products in a modified atmosphere. *Packaging Technology and Science* 14, 267–274.
[27] Dias, J., Rueda-Jasso, R., Panserat, S., Conceiçao, L.E.C, Gomes, E.F. and Dinis, M.T. (2004) Effect of dietary carbohydrate-to-lipid ratios on growth, lipid deposition and metabolic hepatic enzymes in juvenile Senegalese sole (*Solea senegalensis*, Kaup). *Aquaculture Research* 35, 1122–1130.
[28] France fish (2000) (http://www.francefish.com/poissons/Sole.htm, accessed March 2007).

23. Technical sheet – Gilthead seabream (*Sparus auratus*)

GAMETES AND DEVELOPING EGGS	
Egg characteristics	Diameter 0.9–1 mm, pelagic. Spherical and transparent, single oil droplet with a diameter of 230–240 μm. Positive buoyancy at 35–37‰ salinity
Milt and sperm characteristics	5.5–17.5 × 10^9 spz/ml[1]
Egg yield (eggs/kg female)	≈ 800,000[2]
Incubation time	48–60 h after fertilization at 16–17°C
Optimal T°C	16–17°C
Fertilization success	90–95%
Overall incubation success	Hatching rate 70–80%. 350,000 2-days old larvae/kg female[2]
LARVAE–JUVENILE STAGE	
Larval size at hatching	Total length of around 3 mm
Optimal T°C	16–25°C
Rearing units	Round fibreglass tanks of 6–10 m³, semi-closed or flow-through system[3,4]
Time until first-feeding	3–5 days posthatching (total length of around 4 mm)
Live feed needed	First feeding on rotifers, followed by *Artemia* nauplii and/or microparticulates
Survival from hatching to weaning	30 ± 10% (end of weaning)
ON-GROWING STAGE	
Commercial size	Commonly ~ 400 g. Small size (200–300 g) and large size (1–1.5 kg) fish available on the markets
Years to reach commercial size	2–3 years
Rearing units	Sea cages, raceways or ponds
Rearing density	20–30 kg/m³ (cages) to 40–50 kg/m³ (raceways)
Susceptibility to disease	'Winter disease' associated with low water temperatures (< 10–11°C).[3] Some ectoparasites and endoparasites.[1,4] 'Myxobacteriosis' caused by *Tenacibaculum maritimum* and pasteurellosis (*Photobacterium damselae*)[5]
Optimal T°C	22–25°C
Diet	Pressed or extruded: protein (47–52%), fat (11–18%). Some formulas contain orange and yellow pigments (xanthoplylls).

continued

REPRODUCTION	
Period	December–March
Fertilization mode	External fertilization
Fertilization protocol	25 μl sperm/1 ml eggs (~ 1000 eggs)[1]
Time to reach sexual maturity	Protandrous hermaphrodite, functional male in the first 2 years, at sizes over 30 cm some become females. Males reach maturity at 1–2 years (100–300 g), females at around 3 years (> 600 g)
Optimal T°C	Gametogenesis: with decreasing temperatures from 20°C to 15–17°C; spawning: 15–17°C[1]
Control of reproduction	Spawning period can be shifted by photoperiod and temperature manipulation
COMMERCIALIZATION	
Product characteristics	White, firm and tasty flesh. Good shelf life, 18% proteins, total lipids 6–10%, 1.2–1.5% ash, 70–75% moisture[6]
Flesh and/or egg yield	Fillet yield: 40–48%
Price/kg of whole fish, gutted	Commonly traded as whole fresh fish. Italy, Spain and Greece (main consumers): US$5.40–8.15/kg with market fluctuations
Other commercialization avenues	Hybrids between *S. aurata* and other Sparids, such as *Diplodus* spp., *Pagrus major* or *Dentex dentex*
RESEARCH AVENUES	
– Fish nutrition and immunology – Fish welfare and product quality – Selective breeding, sterilization and hybridization (inter-specific)	

[1] Cabrita, E., Robles, V., Cuñado, S., Wallace, J.C., Sarasquete, C. and Herráez, M.P. (2005) Evaluation of gilthead sea bream, *Sparus aurata*, sperm quality after cryopreservation in 5 ml macrotubes. *Cryobiology* 50(3), 273–284.

[2] Moretti, A., Pedini Fernandez-Criado, M., Cittolin, G. and Guidastri, R. (1999) *Manual on Hatchery Production of Seabass and Gilthead Seabream, Vol. 1.* Food and Agriculture Organization of the United Nations, Rome, 205 pp.

[3] Bovo, G., Borghesan, F., Comuzzi, M., Ceschia, G. and Giorgetti, G. (1995) 'Winter disease' in reared sea-bream: preliminary observations. *Bolletino Della Societa Italiana Di Patologia Ittica* 17, 2–l 1.

[4] Papapanagiotou, E.P. and Trilles, J.P. (2001). Cymothoid parasite *Ceratothoa parallela* inflicts great losses on cultured gilthead sea bream *Sparus aurata* in Greece. *Diseases of Aquatic Organisms* 45, 237–239.

[5] Salati, F., Cubadda, C., Viale, I. and Kusuda, R. (2005). Immune response of sea bass *Dicentrarchus labrax* to *Tenacibaculum maritimum* antigens. *Fisheries Science* 71, 563567.

[6] Grigorakis, K., Alexis, M.N., Taylor, K.D.A. and Hole, M. (2002) Comparison of wild and cultured gilthead sea bream (*Sparus aurata*); composition, appearance and seasonal variations. *International Journal of Food Science and Technology* 37, 477–484.

24. Technical sheet – Tunas: Atlantic bluefin tuna (*Thunnus thynnus*), Southern bluefin tuna (*T. maccoyii*) and Pacific bluefin tuna (*T. orientalis*)*

*Text in **bold**, *underlined* and in *italic* refers respectively to *T. thynnus*, *T. orientalis* and *T. maccoyii* only

GAMETES AND DEVELOPING EGGS	
Egg characteristics	1 mm
Milt and sperm characteristics	N/A
Egg yield (eggs/kg female)	**93 oocytes/g**,[1] *57 oocytes/g*[1]
Incubation time	**48 degree days**
Optimal T°C	**> 24°C**
Fertilization success	<u>91.8–100%</u>[7]
Overall incubation success	<u>95%</u>
LARVAE–JUVENILE STAGE	
Larval size at hatching	3–4 mm
Optimal T°C	<u>25°C</u>
Rearing units	<u>20–30 m³ concrete or fibreglass reinforced plastic tanks</u>[7]
Time until first-feeding	<u>3 days</u>
Live feed needed	<u>Rotifers, *Artemias*, newly hatched stripped knifejaw *Oplegnathus fasciatus* larvae</u>
Survival from hatching to weaning	*Larval survival estimated at 52%/day*,[2] <u>1% in the first year</u>[7]
ON-GROWING STAGE	
Commercial size	80 kg, *<u>40 kg</u>*
Years to reach commercial size	1.5–2.5 years for15 kg fish to reach 80 kg market size, 6–10 months for 80 kg fish to fatten up, *4–8 months of on-growing from a 10–20 kg juvenile to double their initial weight*, <u>over 2 years of on-growing from 200–300 g and 6–9 months of on-growing from 15–50 g to a 30% size increase</u>[4,8]
Rearing units	*Floating ring cages (40 m in diameter)*
Rearing density	*2 kg/m³*, <u>5 kg/m³</u>
Susceptibility to disease	**Medium**,[6] *low*
Optimal T°C	N/A
Diet	**Baitfish (mackerel, herring, pilchard, sprat, anchovy)**, *Australian sardine (Sardinops neopilchardus)*[1]

continued

REPRODUCTION	
Period	**Multiple spawning grounds (Balearic Islands, Sicily and the Gulf of Mexico), June–July (Mediterranean), May (West Atlantic)** *September–October and February–March*[9]
Fertilization mode	External fertilization, batch spawning
Fertilization protocol	N/A
Time to reach sexual maturity	**4–5 years (females 110–120 cm) East Atlantic, 8 years West Atlantic**, 8–12 years, *8–14 years*
Optimal T°C	**> 24°C**
Control of reproduction	N/A
COMMERCIALIZATION	
Product characteristics	*Toro, outside high fat layer of pink meat; Akami, inner layer, bright red colour due to lower fat content*[5]
Flesh and/or egg yield	N/A
Price/kg of whole fish, gutted	
Other commercialization avenues	*Australia production is based on a quota of 5265 Mt*
RESEARCH AVENUES	
– ***Reproduction of captive broodstock***[3] – Development of commercial feeds – Larval rearing techniques – Fish health, immune response and disease resistance, parasites	

[1] Aguado-Gimenez, F. and Garcia-Garcia, B. (2005). Changes in some morphometric relationships in Atlantic bluefin tuna (*Thunnus thynnus thynnus* Linnaeus, 1758) as a result of fattening process. *Aquaculture* 249, 303–309.

[2] Davis, T.L.O., Lyne, V. and Jenkins, G.P. (1991) Advection, dispersion and mortality of a patch of southern bluefin tuna larvae *Thunnus maccoyii* in the East Indian Ocean. *Marine Ecology Progress Series* 73, 33–45.

[3] Elizur, A., Stehr, H., Stehr, M., Deichmann, M. and Stone, S. (2006) Clean seas tuna – a vision for southern bluefin tuna. In: *Australasian Aquaculture 2006, Innovation in Aquaculture*, Adelaide, 27–30 August 2006, Conference CD.

[4] Farwell, C.J. (2001) Tunas in captivity. In: Block, B.A. and Stevens, E.D. (eds) *Tuna: Physiology, Ecology and Evolution.* Academic Press, San Diego, California, pp. 391–412.

[5] Gibson, D.J.M. (1984) Southern bluefin tuna for the sashimi market. New Zealand Ministry of Agriculture and Fisheries, Wellington.

[6] Mladineo, I., Miletic, I. and Bocina, I. (2006) *Photobacterium damselae* subsp. *piscicida* outbreak in cage-reared Atlantic bluefin tuna *Thunnus thynnus. Journal of Aquatic Animal Health* 18, 51–54.

[7] Sawada, Y., Okada, T., Miyashita, S., Murata, O. and Kumai, H. (2005) Completion of the Pacific bluefin tuna *Thunnus orientalis* (Temminck et Schlegel) life cycle. *Aquaculture Research* 36, 413–421.

[8] Smart, A. and Sylvia, P.C. (2006) Current status of bluefin tuna farming in North America. In: *Australasian Aquaculture 2006, Innovation in Aquaculture*, Adelaide, 27–30 August 2006, Conference CD.

[9] Young, P.C. (2001) *Southern Bluefin Tuna: Its Propagation in Australia. 1999/396.* Fisheries Research and Development Corporation, Deakin, Australia.

Index

Note: Page references in *italic* refer to tables in the text; those in **bold** refer to figures